This book belongs to
Barbara J. Clement

second edition

cellular microbiology

second edition

cellular microbiology

Editors

Pascale Cossart
Institut Pasteur, Paris, France

Patrice Boquet
INSERM, Nice, France

Staffan Normark
Karolinska Institute, Stockholm, Sweden

Rino Rappuoli
Chiron Vaccines, Siena, Italy

ASM PRESS

WASHINGTON, D.C.

Address editorial correspondence to ASM Press, 1752 N St. NW, Washington, DC 20036-2904, USA

Send orders to ASM Press, P.O. Box 605, Herndon, VA 20172, USA
Phone: 800-546-2416; 703-661-1593
Fax: 703-661-1501
E-mail: books@asmusa.org
Online: www.asmpress.org

Library of Congress Cataloging-in-Publication Data

Cellular microbiology / editors, Pascale Cossart . . . [et al.].—2nd ed.
    p. cm.
  Includes bibliographical references and index.
  ISBN 1-55581-302-X
  1. Virulence (Microbiology) 2. Infection. 3. Host-bacteria relationships. 4. Pathology, Cellular. I. Cossart, Pascale.

QR175.C43 2004
571.9'36—dc22

2004054763

10 9 8 7 6 5 4 3 2 1

Cover and interior design: Susan Brown Schmidler

Cover photographs: (Top) Mammalian cells infected by *Listeria monocytogenes* (red) in the process of polymerizing actin (green) (courtesy of Helene Bierne and Pascale Cossart). (Bottom left) Mammalian cells infected by *Listeria monocytogenes* (red) in the process of polymerizing actin (green) (courtesy of Christine Kocks and Pascale Cossart). (Bottom right) Monolayer of MDCK cells stained with antibodies against the junction protein ZO-1 (red). The green and elongated cells are transfectants expressing a GFP (green fluorescent protein) fusion with the CagA protein of *Helicobacter pylori* (see chapter 16). (Courtesy of F. Bagnoli, M. Amieva, and A. Covacci.)

Contents

CHAPTER 8

Host Cell Membrane Structure and Dynamics 157

LYNDA M. PIERINI AND FREDERICK R. MAXFIELD

CHAPTER 18

Interaction of Pathogens with the Innate and Adaptive Immune System 425

EMIL R. UNANUE AND ENNIO DE GREGORIO

CHAPTER 23

Use of Simple Nonvertebrate Hosts To Model Mammalian Pathogenesis   543

COSTI D. SIFRI AND FREDERICK M. AUSUBEL

Contributors

KLAUS AKTORIES
Institut für Pharmakologie und Toxikologie der Albert-Ludwigs-Universität
Freiburg, Hermann-Herder-Strasse 5, D-79104 Freiburg, Germany

FREDERICK M. AUSUBEL
Department of Molecular Biology, Massachusetts General Hospital, Boston, MA
02114, and Department of Genetics, Harvard Medical School, Boston, MA 02115

DAVID N. BALDWIN
Departments of Biochemistry and of Microbiology & Immunology, Stanford
University Medical School, Stanford, CA 94035-5307

KENNETH BELL
Pathogen Sequencing Unit, Wellcome Trust Sanger Institute, Wellcome Trust
Genome Campus, Hinxton, Cambridge, CD10 1SA, United Kingdom

STEPHEN BENTLEY
Pathogen Sequencing Unit, Wellcome Trust Sanger Institute, Wellcome Trust
Genome Campus, Hinxton, Cambridge, CB10 1SA, United Kingdom

AVRI BEN-ZE'EV
Department of Molecular Cell Biology, Weizmann Institute of Science, Rehovot
76100, Israel

ALEXANDER D. BERSHADSKY
Department of Molecular Cell Biology, Weizmann Institute of Science, Rehovot
76100, Israel

PATRICE BOQUET
Unité INSERM 452, Faculté de Médecine, 28 Avenue de Valombrose, 06102 Nice,
France

MICHAEL CAPARON
Department of Molecular Microbiology, Washington University School of
Medicine, St. Louis, MO 63110-1093

ANA CERDEÑO-TARRAGA
Pathogen Sequencing Unit, Wellcome Trust Sanger Institute, Wellcome Trust
Genome Campus, Hinxton, Cambridge, CD10 1SA, United Kingdom

G. SINGH CHHATWAL
Department of Pathogenicity and Vaccine Research, Biozentrum der Technischen
Universität, Spielmannstrasse 7, D-38106 Braunschweig, Germany

SU L. CHIANG
Department of Microbiology and Molecular Genetics, Harvard Medical School,
200 Longwood Avenue, Boston, MA 02115

PETER J. CHRISTIE
Department of Microbiology and Molecular Genetics, The University of Texas
Health Science Center at Houston, 6431 Fannin, Houston, TX 77030

PASCALE COSSART
Unité des Interactions Bactéries-Cellules, Institut Pasteur, 28 rue du Dr. Roux,
Paris F-75015, France

ANTONELLO COVACCI
Chiron Vaccines, Via Fiorentina 1, 53100 Siena, Italy

LISA CROSSMAN
Pathogen Sequencing Unit, Wellcome Trust Sanger Institute, Wellcome Trust
Genome Campus, Hinxton, Cambridge, CD10 1SA, United Kingdom

CHANTAL DE CHASTELLIER
Centre d'Immunologie de Marseille-Luminy, Inserm-CNRS-Université de la
Méditerranée, 13288 Marseille Cedex 09, France

ENNIO DE GREGORIO
Chiron Vaccines, Via Fiorentina 1, 53100 Siena, Italy

MICHELA FELBERBAUM-CORTI
Department of Biochemistry, University of Geneva, 1211 Geneva 4, Switzerland

B. BRETT FINLAY
Biotechnology Laboratory and the Departments of Biochemistry & Molecular
Biology and Microbiology & Immunology, University of British Columbia,
Vancouver, British Columbia, Canada, V6T 1Z3

RALUCA FLUKIGER-GAGESCU
Department of Biochemistry, University of Geneva, 1211 Geneva 4, Switzerland

ÅKE FORSBERG
Department of Molecular Biology, Umeå University, SE-901 87 Umeå, Sweden,
and Department of Medical Countermeasures, Swedish Defence Research Agency,
FOI NBC-Defence, SE-901 82 Umeå, Sweden

MATTHEW S. FRANCIS
Department of Molecular Biology, Umeå University, SE-901 87 Umeå, Sweden

BENJAMIN GEIGER
Department of Molecular Cell Biology, Weizmann Institute of Science, Rehovot
76100, Israel

JEAN GRUENBERG
Department of Biochemistry, University of Geneva, 1211 Geneva 4, Switzerland

MATTHEW HOLDEN
Pathogen Sequencing Unit, Wellcome Trust Sanger Institute, Wellcome Trust
Genome Campus, Hinxton, Cambridge, CD10 1SA, United Kingdom

ILONA IDANPAAN-HEIKKILA
Skirball Institute, Department of Microbiology and Kaplan Cancer Center, New
York University School of Medicine, 540 First Avenue, New York, NY 10016

FRANK LAFONT
Department of Genetics and Microbiology, C.M.U., 1 rue Michel-Servet, CH-1211
Geneva 4, Switzerland

MARC LECUIT
Unité des Interactions Bactéries-Cellules, Institut Pasteur, 28 rue du Dr. Roux,
Paris F-75015, France

STEPHEN LORY
Department of Microbiology and Molecular Genetics, Harvard Medical School,
200 Longwood Avenue, Boston, MA 02115

MARK MARSH
Cell Biology Unit, MRC-Laboratory for Molecular Cell Biology and Department of
Biochemistry and Molecular Biology, University College London, Gower Street,
London, WC1E 6BT, United Kingdom

VEGA MASIGNANI
Chiron Vaccines, Via Fiorentina 1, 53100 Siena, Italy

FREDERICK R. MAXFIELD
Department of Biochemistry, Weill Medical College of Cornell University, 1300
York Avenue, New York, NY 10021

SANDRA J. MCCALLUM
Departments of Biochemistry and of Microbiology & Immunology, Stanford
University Medical School, Stanford, CA 94035-5307

TIMOTHY K. MCDANIEL
Illumina, Inc., 9390 Towne Centre Drive, Suite 200, San Diego, CA 92122

JEREMY E. MOSS
Skirball Institute, Department of Microbiology and Kaplan Cancer Center, New
York University School of Medicine, 540 First Avenue, New York, NY 10016

JULIAN PARKHILL
Pathogen Sequencing Unit, Wellcome Trust Sanger Institute, Wellcome Trust
Genome Campus, Hinxton, Cambridge, CD10 1SA, United Kingdom

DANA PHILPOTT
Unité INSERM 452, Faculté de Médecine, 28 Avenue de Valombrose, 06102 Nice,
France

LYNDA M. PIERINI
Department of Surgery, Weill Medical College of Cornell University, 1300 York
Avenue, New York, NY 10021

JAVIER PIZARRO-CERDÀ
Unité des Interactions Bactéries-Cellules, Institut Pasteur, 28 rue du Dr. Roux, Paris F-75015, France

MARIAGRAZIA PIZZA
Chiron Vaccines, Via Fiorentina 1, 53100 Siena, Italy

KLAUS T. PREISSNER
Institute for Biochemistry, Medical Faculty, Justus-Liebig-Universität, Friedrichstrasse 24, D-35392 Giessen, Germany

RINO RAPPUOLI
Chiron Vaccines, Via Fiorentina 1, 53100 Siena, Italy

JENNIFER R. ROBBINS
Department of Biology, Xavier University, Cincinnati, OH 45207

DAVID G. RUSSELL
Cornell University, College of Veterinary Medicine, C5171 Veterinary Medical Center, Ithaca, NY 14853

PHILIPPE J. SANSONETTI
Unité de Pathogénie Microbienne Moléculaire, INSERM U389, Institut Pasteur, 28 rue du Dr. Roux, 75724 Paris Cedex 15, France

KURT SCHESSER
Department of Microbiology and Immunology, University of Miami School of Medicine, Miami, FL 33136

MOHAMMED SEBAIHIA
Pathogen Sequencing Unit, Wellcome Trust Sanger Institute, Wellcome Trust Genome Campus, Hinxton, Cambridge, CD10 1SA, United Kingdom

COSTI D. SIFRI
Division of Infectious Diseases, Massachusetts General Hospital, Boston, MA 02114

FREDERICK S. SOUTHWICK
Division of Infectious Diseases, College of Medicine, University of Florida, Box 100277, ARB, Gainesville, FL 32611

JULIE A. THERIOT
Departments of Biochemistry and of Microbiology & Immunology, Stanford University Medical School, Stanford, CA 94035-5307

NICHOLAS THOMSON
Pathogen Sequencing Unit, Wellcome Trust Sanger Institute, Wellcome Trust Genome Campus, Hinxton, Cambridge, CD10 1SA, United Kingdom

GUY TRAN VAN NHIEU
Unité de Pathogénie Microbienne Moléculaire, DR2 Inserm, 28 rue du Dr. Roux, 75724 Paris Cedex 15, France

EMIL R. UNANUE
Department of Pathology and Center for Immunology, Washington University School of Medicine, St. Louis, MO 63110

RAPHAEL H. VALDIVIA
Department of Molecular Genetics and Microbiology, Duke University Medical Center, 273 Jones Building, Research Drive, Box 3580, Durham, NC 27710

HANS WOLF-WATZ
Department of Molecular Biology, Umeå University, SE-901 87 Umeå, Sweden

ELI ZAMIR
Department of Molecular Cell Biology, Weizmann Institute of Science, Rehovot 76100, Israel

ARTURO ZYCHLINSKY
Max Planck Institute for Infection Biology, Campus Charite Mitte, Schumannstrasse 21/22, Berlin 10117, Germany

Foreword

The term "cellular microbiology" was coined in 1996 by P. Cossart, P. Boquet, S. Normark, and R. Rappuoli (1) to describe an emerging scientific discipline that bridged the disciplines of cell biology and microbiology. The idea began to blossom at a scientific meeting held in the summer of 1991 in Arolla, Switzerland. The meeting brought together scientists who were exploring the use of cell biological tools to answer questions of bacterial pathogenesis and cell biologists who were discovering that the fundamental questions of cell biology could be productively addressed by using microbial pathogens and bacterial toxins as cellular probes. It was an exciting scientific meeting where there was a good deal of discussion and a search for a common language and common ideas. On a personal note, I experienced difficulty in addressing cell biologists some years earlier when Brett Finlay and I submitted our first paper to the *Journal of Cell Biology*. The topic of the paper was the entry of *Salmonella* into polarized epithelial cells. One of the reviewers made the comment that the paper was "difficult to understand because the particles added by the authors were alive!" The book *Cellular Microbiology* was a continuation of the mutual education process by cell biologists and microbiologists. This second edition is yet another milestone in the evolution of this discipline. There is now a successful journal with the same name and, indeed, other books on this same topic, including books devoted to the particular laboratory methods of cellular microbiology. As in the first edition, the chapters in this book contain information about fundamental cell biology and fundamental microbiology, as well as articles that show the marriage of the two—cellular microbiology. However, in this edition, the field has passed from its infancy to, well, at least puberty, with all of its attendant vigor, uncertainty, and growing pains.

Purified bacterial toxins and viruses were the first useful microbial reagents used by cell biologists. It was not necessary to know much in the way of microbiology to use them as biochemical probes of cell biological function. In particular, the dinucleotide-ribosylating enzymes secreted by the cholera-producing *Vibrio*, the diphtheria bacillus, and *Bordetella pertussis* (the agent of whooping cough) were used to explore the function of the large heterotrimeric G proteins. As described in this book, there is now a feast of toxin reagents, ranging from those that directly interact with the microfilament network, including the small GTP-binding proteins like Rac, Rho, and Cdc42, to those that induce apoptosis. Similarly, it was exciting to discover that the classical tetanus and botulinum neurotoxins acted identically to each other as proteases for the SNAP25, synaptobrevin, and syntaxin targets. This was a key factor in establishing the models for vesicles docking to membranes. At the same time, the medical microbiologist must examine the pathogenesis of infection to understand how one (tetanus) leads to muscle contraction and the other (botulism) leads to flaccid paralysis. For the cellular microbiologist, it is not sufficient to understand only the precise mode of action of a bacterial toxin in terms of the host effect; it is also crucial to understand the role of the toxins in the pathogenesis of infection. What do these lethal toxins do for the microbe? How did these specific eukaryotic poisons evolve during prokaryote evolution? The tendency of the classical cell biologist is to look at the products of toxin virulence genes as purified reagents. The classical microbiologist may view these proteins as protective antigens from the standpoint of vaccine development. The clinician views the toxins as the causative factors of disease. The cellular microbiologist may be interested in all of these facets but must put the toxin into the perspective of the pathogenesis of infection and the utility of the toxin for bacterial survival, persistence, and transmission.

The first time a cell biologist sees a video of a bacterium like *Salmonella* or *Shigella* enter an animal cell, with the attendant cataclysmic eruption of cytoskeletal elements, there is a sense of awe that so tiny an organism can precipitate such a rapid and dramatic host cell response. However, the rapid assembly and disassembly of cytoskeletal elements are the same as those that occur in such essential host cell functions as phagocytosis, cell division, and adhesion to the extracellular matrix and substratum. How does the bacterium take over the control of these basic, essential eukaryotic cellular functions? We now know that pathogens like *Salmonella, Shigella,* and *Yersinia* inject specific effector proteins directly into the host cell cytoplasm via a proteinaceous appendage assembled on the bacterial surface that looks like a hypodermic needle. The genes encoding this secretory apparatus and the effector proteins have led us to view the evolution of pathogenicity in an entirely new light. Moreover, in this second edition, one learns that a greater number of pathogenic organisms have been observed exploiting the host cell cytoskeleton, tight junctions, and other fundamental host functions to gain entry into a host cell or even to spread from cell to cell.

When the first edition was written, there was an appropriate emphasis on the fact that pathogens such as *Salmonella, Yersinia, Shigella,* and some enteropathogenic *Escherichia coli* strains possessed blocks of genes that distinguished them from their related commensal brethren. These blocks of genes, which are found as contiguous large DNA insertions in the bacterial chromosome, were called pathogenicity islands or were defined as an integral part of an extrachromosomal element like a plasmid or bacteriophage. In many cases, these blocks of genes encode a specialized secretory pathway that is configured by evolution to dispense specific effector proteins to the bacterial surface or to effect the transfer of specific effector proteins through a protein needle-like structure that acts as a conduit into the host cell membrane and cytoplasm. The effector proteins delivered in this way are extraordinarily keyed to interact with cytoskeletal elements, to enzymatically catalyze tyrosine phosphorylation or dephosphorylation of host proteins, or even to induce apoptosis. As this second edition documents, we are learning more and more about the properties of these specialized virulence proteins. We now understand that the genes found on chromosomal islands and plasmid genes associated with virulence have significant sequence and functional homology in bacteria ranging from pathogens of plants to dedicated pathogens of humans only. We can trace the evolution of these blocks. They appear to have been transmitted by horizontal gene transfer and possess a DNA base composition that overall is strikingly different from that of the bulk of the chromosomal genes. It is as if the island DNA once resided in a microbe that was at best distantly related to the pathogenic bacteria that plague us and in which it is now found. Moreover, while the first pathogenicity islands belonged mostly to type III secretory pathways, it is now clear that other secretory pathways, such as the type IV system used to transfer DNA between cells, also can be co-opted to deliver effector proteins to eukaryotic cells.

Thus, the study of the cellular microbiology of bacterial invasion has provided us with new ways to examine the fundamental aspects of the eukaryotic cytoskeleton. It has led as well to a deeper understanding of the evolution of pathogenicity and host-parasite relationships. As the bacterial effector proteins become purified and better studied, they will prove to be important tools for probing the nature of animal cells. We are just beginning to fully appreciate the influence of the effector molecules on the cell. Some interact with the tight junction, and others interact and manipulate host cell signaling, often "telling" the cell to stay calm while the microbe assaults the host. To the clinician, these studies provide new clues to how to prepare measures to thwart infectious agents of disease and, in some cases, explain paradoxical symptoms associated with particular infectious diseases.

Many pathogens breach the membrane barrier and intersect the intracellular trafficking network of the host cell. The bacteria initially are surrounded by a vesicular membrane much like any other ingested particulate. Some bacteria modify this compartment to prevent acidification, whereas others require the normal acidification to progress further with the infectious cycle. Some bacteria, like *Shigella, Listeria,* and the spotted-fever rickettsiae, dissolve the vesicular membrane to begin life in the host cell cytoplasm by using a fascinating interaction between synthesized bacterial proteins and host cell cytoskeletal components, as well as a highly developed mechanism for intracellular travel that in some cases undercuts the host cell's tight junctions. Other bacteria, like *Salmonella, Mycobacterium,* and *Legionella,* stay encased in host cell membranes but have developed ways to obtain the nutrients required to persist and replicate, as well as to modify the trafficking pattern of the host to avoid destruction by lysosomal fusion. It seems that all aspects of the biology of the eukaryotic cell have been fathomed by some microbe that has learned to change subtly and not so subtly to aid microbial domination. It is the job of the cellular microbiologist to discover these interactions and to put them into a broad biological perspective. This focus on the intricacies of the host-parasite interaction has given us a wealth of information. Since the first edition, there have been an increasing number of microbial systems used as experimental tools to explore normal cell function. In doing so, there is a greater appreciation of the intricacies of the host-microbe interactions that occur as two living creatures seek to survive one another. The rules of engagement are complicated, and there is often a fine line between a strained mutual survival and disaster for one or the other participant.

This volume, like the first edition, is written for the classical cell biologist and the classical microbiologist. Of course, the editors hope it will be useful and will appeal to students of all biological disciplines. The chapters are organized to provide both the fundamental "classical" background of each discipline and the cellular microbiology aspect. We are in an era of bacterial genomics. Since the first edition, the full DNA sequences of most important pathogenic microorganisms have been completed. In parallel, the full chromosomal DNA sequences of bacterial hosts, from worms to humans, also have been completed. It is of some note that many of the workhorses of genetics—the worm, the fruit fly, and the zebra fish—have become very useful surrogate hosts for a number of human and animal pathogens. Many investigators initially were concerned about whether these surrogate models were actually relevant to human disease. These smaller and less evolved hosts are not substitutes for a mammalian host, but it has been encouraging to learn that many of the same virulence factors that have proved important in human disease also play a role in infection of the surrogate models. In addition, novel virulence factors that are essential to infection of these surrogates often have been found to play a role in mammalian infection. Of course, while most of this volume focuses on the interactions between pathogenic microbes and their animal hosts, it also recognizes that the animal body is the home for countless other species. I suggested in the first edition that cellular microbiology would be a wellspring of information which would permit us to see more clearly the role of organismal interactions in shaping the evolution of both microbes and hosts, just as the DNA code provided us with the Rosetta stone that permitted all biologists to speak the same language. I think the years have proved this wishful thinking to be correct. Bacterial pathogenicity traces its roots back to the first time amoebas learned to prey on bacteria and vice versa. As my students like to say, "All life, including us, is part of the food chain." We are all prey. This book provides the continuing organized report of an exciting facet of biology. I hope the reader will find it as exciting as the practitioners who wrote the following chapters do.

STANLEY FALKOW
Stanford University
Stanford, California

Reference

1. **Cossart, P., P. Boquet, S. Normark, and R. Rappuoli.** 1996. Cellular microbiology emerging. *Science* **271:**315–316.

Preface

To better study the complex phenomena that occur in nature, scientists have artificially split their analysis into many disciplines, each addressing a limited number of issues. Reductionism has been and continues to be a necessary step to better handle the complexity of the phenomena studied and to cope with the limits of the technologies available. Oversimplification in biology, however, is often a dangerous route that analyzes phenomena under artificial conditions far from real life. Our generation of microbiologists did not escape this trend. We have learned to isolate and grow bacteria and viruses under laboratory conditions that these organisms would never encounter in vivo. Similarly, cell biologists have learned to grow cells in perfectly aseptic tissue cultures, forgetting that in real life these cells, when part of functional tissues and organs, may interact with a variety of microbes.

The name "cellular microbiology" was used for the first time in 1996 (1) to describe a new emerging discipline in which it was no longer enough to study pathogenic bacteria in artificial laboratory conditions such as agar plates and Erlenmeyer flasks, but they had to be studied in the real world, while interacting with the cells of their hosts.

The new name was immediately successful: pioneering scientific meetings entitled "cellular microbiology" (3, 4, 6) were organized; a new scientific journal (2, 5) was initiated; sections of scientific journals were named "cellular microbiology"; and cellular microbiology has also been the name given to many new research departments.

Two books on cellular microbiology were published in 2000. The first edition of this book has been very successful and has been selected as a text of choice in many courses. In today's world, however, even very innovative textbooks become rapidly obsolete because of technological progress. Since the first edition, the field of microbiology has gone through perhaps the most dramatic revolution since Pasteur. In less than 3 years most of the genomes of pathogenic bacteria became available. Microarray technology has allowed global analysis of gene expression, and proteomics has advanced biochemical studies to universal observation of the protein world. Hence the need for a new edition of the book, which incorporates genomic and proteomic studies of the microbial world. It is clear that genomics, proteomics, and postgenomics have not changed the concept of cellular microbiology; rather, they have made the field easier to approach and more easily accessible to all scientists. These new technologies powerfully allow the study of gene and protein expression of both the pathogen and the host within the same experiment. Clearly, cellular microbiology has entered the era of "systems biology" with a resurrection of bacterial and also cellular physiology. Last but not least, the in vivo imaging revolution allows the spatiotemporal study of microbes interacting with cells, not only in tissue-cultured cells but also in ex vivo explants and in animals. All of these concepts are now included in this new edition. Exploitation of the cytoskeleton plasticity was previously discussed in many chapters. The role of the plasma membrane structure, which is now increasingly documented, is also introduced. Finally, we have crossed the boundaries of the bacterial world and included new pathogens such as viruses and new hosts such as plant cells.

The second edition of this textbook is an attempt to provide to students and scientists the vision of a new science where microbiology and cell biology merge into a mature discipline.

PASCALE COSSART
PATRICE BOQUET
STAFFAN NORMARK
RINO RAPPUOLI

References

1. **Cossart, P., P. Boquet, S. Normark, and R. Rappuoli.** 1996. Cellular microbiology emerging. *Science* **271:**315–316.

2. **Falkow, S.** 1999. Cellular microbiology is launched. *Cell. Microbiol.* **1:**3–6.

3. **Niebuhr K., and S. Dramsi.** 1999. EMBO-EBNIC Workshop on Cellular Microbiology "Host cell-pathogen interactions in infectious disease." *Cell. Microbiol.* **1:**79–84.

4. **Reyrat, J. M., and J. Telford.** 1999. When microbes and cell meet. *Trends Microbiol.* **7:**187–188.

5. **Stephens, R. S., P. J. Sansonetti, and D. Sibley.** 1999. Cellular microbiology—a research agenda and an emerging discipline. *Cell. Microbiol.* **1:**1–2.

6. **Sweet, D.** 1999. Microbial fusion. *Trends Cell Biol.* **9:**239–240.

1

Microbial Pathogens: an Overview

PASCALE COSSART, JAVIER PIZARRO-CERDÀ, AND MARC LECUIT

This chapter—which does not claim to be exhaustive—is an introduction to the main human microbial pathogens, with a brief description of the clinical features of the disease and emphasis on the cell biology of the infectious process. For each pathogen, we have also tried to indicate if genetic tools are available. Thus, each section is the "identity card" of the major human pathogens, as seen by cellular microbiologists, and contains references to the chapters where these organisms are mentioned or discussed. Some systems are now paradigms in the field and are thus described in greater detail than those that are only starting to be analyzed. Figure 1.1 gives a schematic description of some of the most striking examples of molecular interactions between microbial and cellular components, and Figure 1.2 shows electron micrographs of the structures and modes of action of some of the pathogens discussed in this chapter. The status of genome-sequencing work for these pathogens is given in Table 1.1.

Bacteria

Strict Intracellular Bacteria

Strict intracellular pathogens cannot be cultivated in broth medium and can replicate only in vivo or in tissue-cultured cells in vitro.

Chlamydia

Chlamydia trachomatis is responsible for genital infections (serovars D to K) and ophthalmic infections (serovars A to C) in humans. It is a leading cause of sexually transmitted disease, causing infertility and extrauterine pregnancy (chronic salpingitis) and also infectious blindness in developing countries. *Chlamydia pneumoniae* causes a community-acquired pneumonia. There is increasing evidence for association of *C. pneumoniae* with atherosclerosis. *Chlamydia psittaci* is primarily an animal pathogen, and only in rare cases is it responsible for human respiratory tract infections, oropharyngitis, and atypical pneumonia.

Chlamydia has a biphasic developmental cycle with two morphologically different forms: the elementary body (EB), which is the infectious form

1

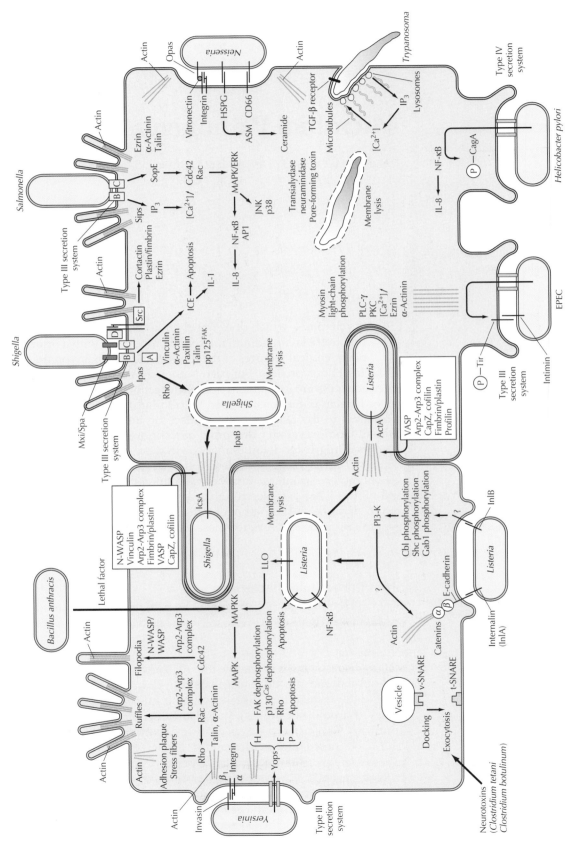

Figure 1.1   Schematic drawing of microbial factors and cellular targets.

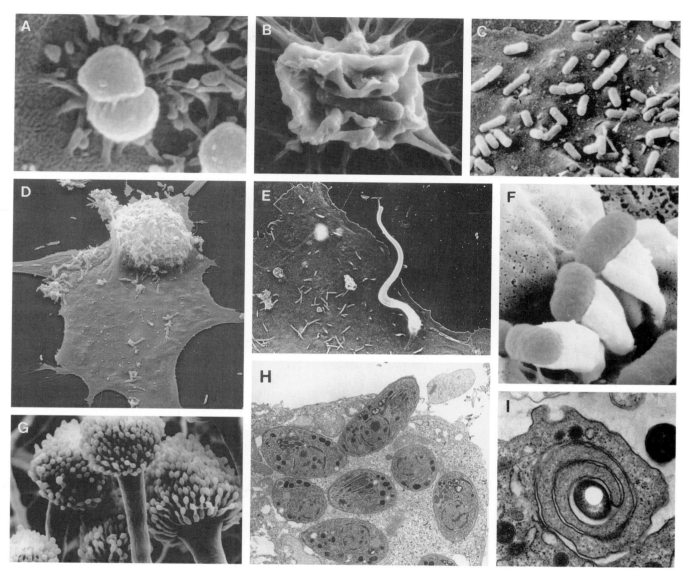

Figure 1.2 Examples of scanning and transmission electron micrographs. **(A)** *N. meningitidis* adhering to a cultured epithelial cell (courtesy of X. Nassif). **(B)** *S. flexneri* entering via macropinocytosis in a cultured epithelial cell by inducing membrane ruffling (courtesy of P. Sansonetti). **(C)** *L. monocytogenes* adherent to and invading a cultured cell (from our laboratory). **(D)** *B. henselae* adherent to and invading a cultured cell (courtesy of C. Dehio). **(E)** *T. cruzi* penetrating a cultured epithelial cell (courtesy of N. Andrews and E. Robbins). **(F)** EPEC on the top of pedestals induced on a cultured cell (courtesy of B. Finlay). **(G)** Conidial heads of *A. fumigatus* (courtesy of J. P. Latgé). **(H)** *T. gondii* tachyzoites free in the cytoplasm of a cultured macrophage, one escaping the cell (reproduced with permission from J. P. Dubey, D. S. Lindsay, and C. A. Speer, *Clin. Microbiol. Rev.* **11:**267–299, 1998). **(I)** *L. pneumophila* invading via coiling phagocytosis a cultured macrophage (reproduced with permission from M. Horvitz, *Cell* **36:**27–33, 1984).

and is metabolically inactive, and the reticulate body (RB), which results from the differentiation of the EB in the parasitophorous vacuole (also termed inclusion). The RB is a vegetative cell type that is able to divide up to the point where it changes to infectious bodies, which are liberated from the cells and are able to infect other cells.

Table 1.1 Genomes that have been sequenced[a]

Organisms	Length (bp) (date sequenced)
Bordetella bronchiseptica	5,338,400 (8/10/03)
Bordetella parapertussis	4,773,510 (8/10/03)
Bordetella pertussis	4,086,186 (8/10/03)
Borrelia burgdorferi	910,724[b] (12/17/97)
Brucella melitensis	
Chr 1	2,117,144 (8/1/02)
Chr 2	1,177,787 (8/1/02)
Brucella suis	
Chr 1	2,107,792 (10/1/02)
Chr 2	1,207,381 (10/1/02)
Campylobacter jejuni	1,641,481 (2/10/00)
Chlamydia pneumoniae	1,230,230 (3/10/99)
Chlamydia trachomatis	1,042,519 (5/20/98)
Coxiella burnetii	1,995,275 (4/29/03)
Enterococcus faecalis	3,000,000 (3/28/03)
Haemophilus influenzae	1,830,138 (7/25/95)
Helicobacter pylori J99	1,643,831 (1/12/99)
Leishmania major	
Chr 1	270,000 (7/29/98)
Chr 3	385,000 (7/25/03)
Leptospira interrogans	
Chr 1	4,332,241 (4/24/03)
Chr 2	358,943 (4/24/03)
Listeria monocytogenes	2,944,528 (10/26/01)
Listeria innocua	3,011,209 (10/26/01)
Mycobacterium bovis	4,345,492 (6/24/03)
Mycobacterium leprae	3,268,203 (2/22/01)
Mycobacterium tuberculosis	4,411,529 (6/11/98)
Mycobacterium avium	4,829,781 (1/30/04)
Mycoplasma genitalium	580,073 (10/30/95)
Mycoplasma pneumoniae	816,394 (11/15/96)
Neisseria meningitidis	2,300,000 (9/19/01)
Plasmodium falciparum	
Chr 1 to 14	23,000,000 (10/03/02)
Pseudomonas aeruginosa	5,900,000 (9/10/01)
Porphyromonas gingivalis	2,200,000 (12/07/01)
Rickettsia conorii	1,268,755 (9/14/01)
Rickettsia prowazekii	1,111,523 (11/12/98)
Salmonella enterica	
serovar Typhimurium	4,857,432 (10/25/01)
serovar Typhi	4,809,037 (10/25/01)
Staphylococcus aureus	2,800,000 (10/04/01)
Streptococcus agalactiae	2,211,485 (11/15/02)
Streptococcus mutans	2,200,000 (10/25/02)
Streptococcus pneumoniae	2,200,000 (10/03/01)
Streptococcus pyogenes	1,980,000 (9/19/01)
Treponema pallidum	1,138,006 (3/6/98)
Tropheryma whipplei	(2/24/03)
Ureaplasma urealyticum	750,000 (1/10/00)
Vibrio cholerae	2,500,000 (9/10/01)
Yersinia pestis	4,380,000 (10/15/01)

[a] For a complete list, please visit http://www.ncbi.nlm.nih.gov:80/genomes/static/eub_g.html. For work in progress, visit http://www.ncbi.nlm.nih.gov/genomes/InProgress.html.
[b] One chromosome, two circular plasmids, and nine linear plasmids.

The *Chlamydia* vacuole is unique in that it is largely devoid of known host protein markers, in particular, those present in vesicles of the endocytic pathway (only the phosphoserine-binding 14-3-3 protein has been associated with *C. trachomatis* inclusions). The lipid composition of the vacuole includes host sphingolipids and cholesterol acquired during the transit of exocytic vesicles from the Golgi apparatus to the plasma membrane.

The sequence of the genome of *C. trachomatis* and *C. pneumoniae* has permitted the identification of potential virulence genes. Several eukaryotic chromatin-associated domain proteins were recognized, suggesting a eukaryotic-like mechanism for chlamydial nucleoid condensation and decondensation. The presence of a number of genes encoding proteins normally found in eukaryotes, such as serine/threonine protein kinases or phosphatases, suggests a complex evolution for this obligate intracellular parasite.

Coxiella burnetii

The category B bioterrorism pathogen *Coxiella burnetii* is the etiological agent of Q (query) fever, an aerosol-borne disease. The reservoir is mainly farm animals, in which the disease manifests as abortion epidemics. In humans, *C. burnetii* is responsible for an acute flu-like illness complicated by endocarditis in chronic disease instances. A correlation has also been suggested between *C. burnetii* infections and onset of atherosclerosis, chronic-fatigue syndrome, and other cerebrovascular incidents.

C. burnetii displays antigenic variation similar to the smooth-rough variation in the family *Enterobacteriaceae*: the highly virulent smooth phase I organisms are found in infected hosts and insect vectors, while the less virulent rough phase II bacteria are obtained after multiple passages in laboratory animals but may also be present in infected hosts. Like *Chlamydia, C. burnetii* exhibits a developmental cycle consisting of two morphologically distinct cell types: a metabolically inactive extracellular "small-cell variant" and a metabolically active intracellular "large-cell variant."

Entry of virulent *C. burnetii* into host cells is mediated by interaction of unknown bacterial ligands with the $\alpha V \beta 5$ integrin. The parasite then intersects the autophagy pathway and occupies a mature phagosome, whose low pH is required to stimulate the growth and replication of the bacterium. The small GTPase Rab7 participates in the biogenesis of the parasitophorous vacuole. The lysosomal glycoproteins LAMP1 and LAMP2, the acid hydrolase cathepsin D, the tetraspan protein CD63, and the vacuolar H^+ ATPase are also present on the vacuolar membrane. The *Coxiella* vacuole has the capacity to fuse with fluid-phase markers containing vesicles or intracellular vacuoles containing *Leishmania amazonensis, Trypanosoma cruzi*, or *Mycobacterium avium*, indicating that it displays unusual fusogenicity.

C. burnetii expresses a type IV secretion system that functions similarly to components of the *Legionella pneumophila* Dot/Icm system (see below). The sequence of the *C. burnetii* genome reveals the important presence of pseudogenes and insertion sequences, suggesting that its obligate intracellular lifestyle may be a recent event.

Ehrlichia

Ehrlichia chaffeensis is one of the three members of a genus newly recognized as responsible for human diseases. Ehrlichiae are responsible for arthropod-borne disease characterized by headache, fever, and chills.

E. chaffeensis is an obligatory intracellular bacterium of monocytes or macrophages. It replicates in nonlysosomal intracellular slightly acidic vesicles called morulae, where different morphological forms of Ehrlichiae seem to coexist. Morulae in *E. chaffeensis*-infected cells are stained by antibodies against transferrin receptor, a marker for early endosomes, but are not labeled with lysosomal markers. Iron acquisition (hypothetically delivered by the transferring receptor) is essential for intracellular *E. chaffeensis* growth.

Mycobacterium leprae
Mycobacterium leprae is discussed in the section Mycobacteria, below.

Rickettsias
The genus *Rickettsia* contains two main subgroups, the typhus group, whose main member is *Rickettsiae prowazekii,* responsible for epidemic typhus (a hemorrhagic fever), and the spotted-fever group, which includes *R. rickettsii*, the agent of Rocky Mountain spotted fever, and *R. conorii*, the etiological agent of Mediterranean spotted fever. Both groups are transmitted to humans by arthropod vectors; fleas transmit typhus, whereas ticks transmit spotted-fever rickettsias.

These bacteria are able to lyse the phagocytic vacuole and to replicate in the cytosol. Actin polymerization and protein phosphorylation are required for entry into target cells. Early interaction of invading *R. conorii* with the cellular protein Ku70 has been detected, but its role in the internalization process remains to be determined. The role of phospholipase A_2 in the escape from the vacuole has been reported for *R. prowazekii,* but it remains unclear whether this activity is of rickettsial origin or if it is a host latent phospholipase activated on infection. Some rickettsias like *R. conorii* and *R. rickettsii* are able to polymerize actin and to move intra- and intercellularly, by a process very similar to that used by *Shigella* and *Listeria*.

Analysis of the factors involved in virulence has been hampered by the absence of genetic tools. However, the sequences of the *R. prowazekii* and the *R. conorii* genomes have been determined, highlighting the presence of several ATP/ADP translocases, probably of plant origin, that allow importation of ATP from the infected host cell. A single gene has its best match on eukaryotes, suggesting that rickettsias have no particular tendency to evolve by acquiring genes from their hosts. A single protein presents an overall organization similar to the protein ActA of *Listeria* (see below) and could be involved in actin polymerization.

Facultative Intracellular Bacteria (Entry and Multiplication in Phagocytic Cells)

Legionella pneumophila
Legionella pneumophila was relatively recently (1976) discovered as the etiological agent of Legionnaires' disease, a disease characterized by infiltrative pneumonia, particularly in immunocompromised patients. It belongs to a genus of aquatic bacteria consisting of over 40 species requiring a protozoan host (*Legionella* can also survive within biofilms in building water systems). It exhibits two morphologically distinct phenotypes: a "replicative form" (RF) found in its specialized intracellular compartment within target cells and a "mature intracellular form" (MIF) with cyst-like properties which could enable the bacteria to persist as a highly infectious particle in the environment between hosts.

L. pneumophila is capable of growing within human pulmonary alveolar macrophages, and uptake into these cells may occur by an intriguing coiling phagocytosis that does not require the activity of the phosphoinositide 3-kinase. After uptake, the *L. pneumophila*-containing phagosome can be observed associated successively with smooth vesicles, mitochondria, ribosomes, and finally the endoplasmic reticulum, where bacterial multiplication takes place. Fusion with lysosomes is normally prevented (despite the transient presence of rab7 in the *L. pneumophila*-containing phagosome), but it has been proposed that vacuoles of exponentially growing bacteria can acquire lysosomal characteristics.

Biogenesis of the *L. pneumophila* replication vacuole requires the activity of two virulence islands containing the *dot/icm* genes that encode a type IV secretion system (also involved in the induction of apoptosis in macrophages). The only recognized substrate of this transporter is the protein Ralf, which is an exchange factor for the ADP ribosylation factor 1; however, *ralf* null mutants infect macrophages as efficiently as wild-type bacteria. A Lsp type II secretion system seems to be required also for intracellular multiplication.

Mycobacteria

Mycobacterium tuberculosis and *Mycobacterium leprae* are the agents of human tuberculosis and leprosy, respectively. *Mycobacterium bovis* is responsible for bovine tuberculosis and was used early in the 20th century to generate the bacillus Calmette-Guérin (BCG) that has since been used worldwide for human vaccination. Inadequate treatment and immunodeficiency (AIDS patients) have contributed to the emergence of strains resistant to multiple antibiotics and to the consequent resurgence of tuberculosis.

M. leprae is the etiological agent of leprosy, a long-lasting infection still prevalent in some developing countries and characterized by cutaneous and nervous system lesions. Two forms of the disease are known: the tuberculoid and the lepromatous forms, which reflect the effects of active cellular immunity and anergy, respectively. *M. leprae* cannot be grown in vitro but can be grown only in mouse footpads and in the cold-blooded armadillo. The neural tropism of *M. leprae* is attributable to the specific binding of two bacterial ligands, the 21-kDa *M. leprae* laminin-binding protein (ML-LBP21) and the phenolic glycolipid-1 (PGL-1), to the globular domains of the laminin-2 α_2-chain present at the basal lamina of Schwann cells. Laminin-2 binds the peripheral protein α-dystroglycan, which is itself associated with the transmembrane protein β-dystroglycan and consequently could be able to induce *M. leprae* internalization by recruiting dystrophin/utrophin and actin at the site of bacterial entry.

M. tuberculosis causes chronic pulmonary infections and also focalized infections (tuberculoma), such as vertebral Pott's disease or central nervous system infections (meningitis and brain tuberculoma). *M. tuberculosis* can be grown in vitro and used to infect tissue-cultured phagocytic cells. Entry into macrophages can proceed through bacterial interaction with many phagocytic receptors, such as the complement receptor CR3, for which a cholesterol-rich environment seems to be required for appropriate mycobacterial internalization. Intracellular replication is achieved in a compartment that excludes the vacuolar ATPase but that remains connected with the early endosomal network and the biosynthetic pathway, as revealed by its association with rab5, the transferring receptor, calmodulin, and newly synthesized LAMP-1 or cathepsin D. Coronin I (also named

TACO) has been found stably associated with BCG-containing phago-somes, and it was proposed that this protein could prevent the fusogenicity of the *M. tuberculosis* compartment (an observation that is not correlated with the sustained fusion of the bacterial vacuoles with early endosomes). The divalent cation transporter NRAMP1 (for natural-resistance-associated macrophage protein 1) confers resistance to macrophages against the intra-cellular growth of *Mycobacterium,* but its molecular mechanism of antibac-terial activity is controversial.

The sequence of the *M. tuberculosis* genome reveals that approximately 30% of the genes code for lipids synthesis or lipid metabolism, suggesting that its thick waxy cell coat and the ability to catabolize fatty acids are im-portant adaptations to a parasitic lifestyle. The genome of *M. bovis* is >99.95% identical to that of *M. tuberculosis,* indicating that differential gene expression (rather than specific gene content) is related to the animal tropism of these species. The genome of *M. leprae* has also been sequenced, and it shows that extensive gene deletion and decay have eliminated many important metabolic activities, giving rise to a reduced genome of which half contains pseudogenes.

Nocardia

Nocardia species belong to the actinomycete group (high-G+C-content gram-positive bacteria). These soil organisms are responsible for two op-portunistic infections in humans: nocardiosis, caused essentially by *Nocar-dia asteroides* and more rarely by *N. brasiliensis* and *N. caviae,* and mycetoma, caused by *N. madurae* and *N. pelletieri.* Nocardiosis is characterized by a chronic pulmonary infection which can, via the lymph and the blood, dis-seminate to other tissues, in particular, to the brain. Mycetomas are severe cutaneous infections which can invade the underlying tissues including the bones. Nocardiosis is sensitive to antibiotic treatment. Surgery is usually re-quired for mycetomas. *N. asteroides* has been reported to reside in nonacid-ified vacuoles of macrophages.

Facultative Intracellular Bacteria (Entry and Multiplication in Nonphagocytic Cells)

Bartonella

The genus *Bartonella* contains 19 species, seven of which have been associ-ated with human disease. The human-specific species *B. bacilliformis* (re-sponsible for Carrion's disease, Oroya fever, and verruga peruana) and *B. quintana* (trench fever, bacillary angiomatosis, and endocarditis) and the zoonotic species *B. henselae* (cat scratch disease, bacillary angiomatosis and peliosis, endocarditis, bacteremia with fever, and neurorenitis) are respon-sible for most human infections. These three *Bartonella* species are also unique among all known bacterial pathogens in their ability to cause an-giogenic lesions that manifest as Kaposi's sarcoma-like lesions of the skin (*B. bacilliformis*), bacillary angiomatosis (*B. quintana* and *B. henselae*), or a cys-tic form in the liver and spleen (primarily *B. henselae*). These lesions com-prise proliferating endothelial cells, bacteria, and mixed infiltrates of macrophages and neutrophils. Bartonellae are found as aggregates sur-rounding and within endothelial cells.

Invasion of human umbilical vein endothelial cells (HUVEC) by *B. bacil-liformis* depends on the small GTPases Rho, Rac, and Cdc42. *B. henselae* seems to be internalized by an actin-depending process into HUVEC in groups of hundreds of bacteria in a structure called the invasome. Bacterial factors seem to be directly responsible for both the mitotic stimulation and apoptosis inhibition of endothelial cells, but their identity has not been established. A paracrine loop of endothelial cell proliferation by the vascular endothelial growth factor (VEGF) secreted by macrophages and neutrophils also contributes to the process of angiogenesis.

An intracellularly activated type IV secretion system, encoded by the *virB* locus, has been identified in *B. henselae,* but no function has been assigned yet to this apparatus. In the related species *B. tribocorum,* a VirB/VirD4 type IV secretion system has been shown to be essential for intraerythrocytic infection in rats.

Brucella

The genus *Brucella* contains eight species named *Brucella abortus, B. melitensis, B. suis, B. canis, B. ovis, B. neotomae, B. pinnipediae,* and *B. cetaceae.* All are pathogenic for mammals, but only the first four are known to be responsible for human disease, inducing a polymorphic illness characterized by both generalized and localized infections (septicemia and osteoarticular and neurological disorders). In livestock, extensive replication occurs in cells of the genitourinary tract, leading to abortions and sterility.

B. abortus entry in epithelial cells is inhibited by cytochalasin D, nocodazole, wortmannin, and genistein, suggesting a role for actin, microtubules, phosphatidylinositol 3-kinase, and protein tyrosine kinases in host cell invasion. Cellular expression of dominant negative forms of Rho, Rac, and Cdc42 diminishes the level of bacterial internalization; however, only Cdc42 is directly activated on *B. abortus* contact with host cells. In phagocytic and nonphagocytic cells, after a transient passage through early endosomal compartments characterized by the presence of the early endosomal markers EEA1 and the small GTPase Rab5, *B. abortus* is distributed in multimembrane autophagosome-like vacuoles associated with the lysosomal membrane proteins LAMP1 and LAMP2. Later during infection, *B. abortus* replicates in a compartment related to the endoplasmic reticulum of host cells. Normal Rab5 (but not Rab7) expression is required for this retrograde intracellular trafficking. The type IV secretion system encoded by the virB operon is also essential for the biogenesis of the *B. abortus*-containing vacuole.

The *B. abortus* two-component system BvrR/BvrS (necessary for cell invasion and intracellular survival) probably regulates cell envelope changes necessary to transit between extracellular and intracellular environments by controlling the expression of outer membrane proteins such as Omp25 and Omp3b, which have counterparts in members of the *Rhizobiaceae* family. The sequence of the *B. melitensis* and *B. suis* genomes indeed reveals fundamental similarities between these animal pathogens and plant symbionts, such as *Sinorhizobium meliloti* and *Agrobacterium tumefaciens,* suggesting that *Brucella* evolved from a soil/plant-associated ancestral bacteria.

Francisella tularensis

Francisella tularensis is the causative agent of tularemia, a disease affecting primarily wild animals. The bacterial true natural reservoir in the environment is not known, but the capability of *Francisella* to invade protozoa suggests that its lifestyle could be similar to that of *Legionella*. Transmission to humans occurs by tick bites or consumption of contaminated game leading to chronic ulceroglandular lesions, which can be complicated by dissemination to the bloodstream and endotoxemia. The related species

F. philomiragia is often associated with natural water systems and is generally considered to be pathogenic only in immunosuppressed individuals.

In laboratory animals, *F. tularensis* has the capacity to invade and grow in nonphagocytic cells, particularly hepatocytes. In vitro, infection of macrophage cell lines has been the most extensively analyzed mechanism. In these cells, *F. tularensis* resides in an acidic compartment, whose low pH facilitates the availability of the iron essential for *Francisella* growth. *F. tularensis* has the capacity to vary its lipopolysaccharide (LPS) when present in the vacuole of a macrophage. The normal LPS fails to stimulate the production of significant levels of nitric oxide, thereby allowing intracellular multiplication. The bacterium can spontaneously vary its LPS to induce macrophages to produce increased levels of NO, thereby suppressing intramacrophagic growth. This is the first report of a phase variation phenomenon which modulates intracellular growth and the innate immune response.

The level of expression of four proteins with apparent molecular masses of 20, 23, 55, and 70 kDa appears to be upregulated within macrophages, but the functions of these molecules are unknown. The 23-kDa protein is also induced on exposure of bacterial cells to oxidative stress and its mutation impairs intracellular growth of *F. tularensis* or fails to inhibit the Toll-like, receptor-induced secretion of tumor necrosis factor by macrophages. However, this molecule presents no homology to any known protein.

Listeria monocytogenes

The gram-positive bacterium *Listeria monocytogenes* is responsible for severe food-borne infections in humans and animals. Listeriosis is characterized by central nervous system and fetoplacental infections affecting primarily immunocompromised individuals, the elderly, neonates, and pregnant women. The genus *Listeria* contains five other species: the animal pathogen *L. ivanovii* and the nonpathogenic *L. innocua, L. seeligeri, L. welshimeri,* and *L. murrayi.*

In vitro, *L. monocytogenes* infects a wide variety of phagocytic and nonphagocytic cell types. Entry occurs by zipper-type phagocytosis, which requires an active actin cytoskeleton. It is mediated by at least two surface proteins, internalin (InlA) and InlB, which by themselves are sufficient to allow entry into mammalian cells. The receptor for internalin is the cell adhesion molecule E-cadherin (a proline at its position 16th is critical for InlA binding) which interacts with β- and α-catenins and with the myosin VIIa to promote bacterial internalization. InlB interacts with three cellular partners: glycosaminoglycans, the receptor for the globular head of the complement molecule C1q (gC1q-R), and c-Met. Activation of c-Met leads to recruitment to the entry site of three phosphorylated proteins (Gab1, Shc, and Cbl) and of the phosphoinositide 3-kinase. Downstream events of the InlB/c-Met interaction include a cycle of phosphorylation/dephosphorylation of cofilin by the LIM kinase-1, facilitating a dynamic control of actin polymerization during bacterial invasion. A signaling synergy between InlA and InlB is suspected to take place in raft-enriched microdomains at the plasma membrane of target cells.

After lysis of the internalization vacuole by the bacterial pore-forming toxin listeriolysin O (LLO) (and in some cells by the phosphoinositide-phospholipase C or the phosphatidylcholine phospholipase C), bacteria move intra- and intercellularly via an actin-based motility mediated by the bacterial surface protein ActA, which is sufficient to induce actin polymerization and movement (Figure 1.3). The ActA protein is a composite protein: the N-terminal part acts in concert with the Arp2/3 complex to stimulate actin polymerization in vitro, while the central proline-rich region and the C-terminal part are homologous to the actin-binding protein zyxin. ActA, as zyxin, binds VASP (vasodilator phosphoprotein), which would recruit profilin and actin to stimulate the actin polymerization process at the N-terminal part of the protein.

▶ *For Figure 1.3, see color insert.*

Intracellular movement leads to direct cell-to-cell spread. This event probably occurs in the liver, when bacteria move from macrophages to hepatocytes and induce apoptosis by a process which has not been deciphered at the molecular level. In contrast, in an endothelial cell line, apoptosis is mediated by LLO. During infection of HeLa cells, mitogen-activated protein kinases (MAPKs) are activated, with the MEK1/ERK2 pathway being required for entry while LLO stimulates both ERK1 and ERK2. Infection of cultured endothelial cells or macrophages activates NF-κB. Cell wall and virulence factors are involved in this process.

Most of the genes involved in virulence are clustered in a pathogenicity islet of 15 kb which is also present in *L. ivanovii* and *L. seeligeri* (however, in *L. seeligeri* the virulence genes are not expressed). Sequence analysis of the genomes of *L. monocytogenes* and *L. innocua* reveals a close relationship to *Bacillus subtilis,* suggesting a common origin for the three species, with multiple virulence gene acquisition and deletion events in the *Listeria* species.

Salmonella

The genus *Salmonella* contains several species. Among them, the species *S. enterica* has received the most attention; it contains the two widely studied serovars, *S. enterica* serovar Typhi and *S. enterica* serovar Typhimurium. Serovar Typhi is a human enteroinvasive bacterium that causes septicemia after dissemination to the liver and spleen. It is responsible for a variety of indirect systemic symptoms that comprise typhoid fever. Serovar Typhimurium is an enteric pathogen causing gastroenteritis in humans. In mice, it is the etiological agent of a systemic infection similar to typhoid fever and thus is used as a model to study the pathophysiology of typhoid fever.

Serovar Typhimurium encodes at least two type III secretion systems located in the pathogenicity islands named SPI-1 and SPI-2. The SPI-1-encoded system is required for entry into epithelial cells and secretes directly to the host cell cytosol at least 13 known proteins, several of which interact in different ways with the actin cytoskeleton. SopE and SopE2 function as guanosine exchange factors (GEFs) for Cdc42 and Rac while SopB (a phosphoinositide phosphatase) activates these GTPases by fluxing cellular phosphoinositides. Cdc42 and Rac activation leads to important cytoskeletal rearrangements that contribute to the formation of membrane ruffles that engulf the bacteria and enhance their internalization. The bacterial protein SptP acts as a GTPase-activating protein (GAP) for Cdc42 and Rac, and antagonizes the activity of SopE, SopE2, and SopB, leading to a recovery of the normal cell architecture. Two other SPI-1-secreted proteins, SipA and SipC, modulate entry by interacting directly with actin cytoskeleton; SipC nucleates actin and bundle actin filaments while SipA decreases the critical concentration required for actin polymerization, inhibits depolymerization, and enhances SipC activity.

Once inside epithelial cells, *Salmonella* remains in a membrane-bound compartment that undergoes a complex maturation process. The SPI-2 type III secretion system, which activity is favored by vacuolar acidification, translocates bacterial proteins across the *Salmonella*-containing compartment membrane into the host cell cytosol and is required for intracellular survival (several of its effectors contribute to modify the vacuolar biogenesis). Transient interactions with the early endosomal network are revealed by the presence of rab5, EEA1, and the transferring receptor in *Salmonella* vacuoles. Subsequently, rab7 controls the fusion of Lamp-1/Lamp-2/cd63-enriched vesicles with the bacterial-containing compartments, bypassing interactions with late endosomal network and lysosomes. Coincident with the onset of bacterial replication in epithelial cells is the appearance of tubular extensions from the *Salmonella* vacuoles that depend on the expression/translocation of the SPI-2-secreted protein sifA (for *Salmonella*-induced filaments), a protein that is required for maintaining the integrity of the vacuolar membrane. Though characterization of the vacuolar biogenesis in macrophages has been controversial, the current hypothesis is that maturation proceeds as in epithelial cells, with the most important difference being the absence of Sifs in phagocytic cells. Induction of apoptosis in macrophages, exclusion of the NADPH oxidase from the *Salmonella* vacuole, and recruitment of F-actin around the bacterial replication compartment depend on SPI-2-encoded effectors. Components of the PhoP-PhoQ regulon are also required for vacuolar biogenesis.

The genomes of serovars Typhi and Typhimurium have been sequenced and reveal an important rate of horizontal gene transfer: 11% of the gene complement of serovar Typhimurium is not found in serovar Typhi and 29% is missing from *Escherichia coli* K-12.

Shigella

The genus *Shigella* contains four species, *Shigella dysenteriae, S. flexneri, S. boydii,* and *S. sonnei,* which are responsible for bacillary dysentery, a bloody and purulent diarrhea reflecting the ability of this bacterium not only to invade and extensively damage the intestinal epithelium but also to cause local inflammation by triggering polymorphonuclear transmigration.

Both in vivo and in vitro studies have led to a detailed description of the successive steps of the infectious process, which probably starts at the level of the M cells, which underlie the Peyer's patches. Then bacteria reach the underlying lamina propria and invade either the enterocytes or the resident macrophages, where they induce apoptosis with concomitant release of interleukin-1, a proinflammatory cytokine which contributes to the attraction of neutrophils at the site of infection. In epithelial cells, bacteria enter by a process morphologically similar to that of *Salmonella* (membrane ruffles and intense rearrangements of the cytoskeleton). Then the bacteria are trapped in a membrane-bound vacuole, which they lyse to reach the cytosolic compartment, where they start moving. When the bacteria reach the plasma membrane, they induce the formation of protuberances that allow direct cell-to-cell spread of the bacteria. Most *Shigella* virulence genes are carried by the virulence plasmid, which on two divergent operons, encodes a type III secretion system responsible for the secretion of four invasion proteins, IpaA, IpaB, IpaC, and IpaD. In the bacterium, IpaB and IpaD prevent secretion. When secretion is induced on cell contact, IpaB and IpaC associate and might insert in the mammalian cell membrane. β_1-integrins seem to play a role in the early contact of IpaB, IpaC, and IpaD with the cell. IpaB also appears to interact with CD44, the acid hyaluronic receptor. This receptor interacts with ezrin after its activation by rho. This event would then allow IpaA translocation within the eukaryotic cell and its interaction with vinculin, which in turn would recruit α-actinin and somehow down-regulate the actin cytoskeletal rearrangements just after the internalization step. Entry also requires the small GTPases Cdc42 and Rac and the Src kinase, which would phosphorylate cortactin. The actin-bundling protein fimbrin (plastin) would play a role in organizing the actin filaments. Taken together these data suggest that the entry process mimicks the formation of a pseudo-adhesion plaque.

Campylobacter

Within the genus *Campylobacter* there are 15 bacterial species (including *C. coli, C. upsaliensis, C. laris,* and *C. fetus*) associated with enteritis, septic abortion, septic arthritis, and other diseases. *C. jejuni* is a leading cause of bacterial diarrhea ranging from mild abdominal symptoms to severe invasive enteritis. Extradigestive manifestations such as arthritis and endocarditis can occur. In addition, *C. jejuni* is now recognized as the most identifiable infection preceding Guillain-Barré syndrome, an acute postinfection immune-mediated disorder affecting the peripheral nervous system due to autoantibodies to GM_1 ganglioside, a membrane component of peripheral nerves (some *C. jejuni* strains have an LPS with a terminal tetrasaccharide similar to that in GM_1 ganglioside).

    C. jejuni adheres to and invades tissue-cultured cells. *C. jejuni* seems to have several proteins acting as adhesins, including CBF1 and the structural subunit of flagella, flagellin (FlaA). The fibronectin-binding protein CadF is required for binding of *C. jejuni* to epithelial cells, and this interaction enhances bacterial internalization. The protein Cjp29 is also required for invasion and is a putative effector of a type IV secretion system. Another factor implicated in invasion, the secreted protein CiaB, possesses similarities with type III secreted proteins. Full secretion of Cia protein and invasion depend also on intact flagellar synthesis genes. Epithelial cell invasion involves microtubules but not microfilaments (as shown by its insensitivity to cytochalasin D) and depends on host protein phosphorylation and an increase in free cellular calcium concentration (internalization is blocked by calcium chelators). Like several other enteropathogens (various *E. coli* strains and a few *Shigella* species), *C. jejuni* produces a cytolethal distending toxin, which induces actin assembly and arrest of cell division at G_2 in infected Chinese hamster ovary cells.

    The genome of *C. jejuni* reveals an almost complete lack of classical operon structure and repetitive DNA. *C. jejuni* can produce a capsular polysaccharide and has an extensive capacity for lipo-oligosaccharide phase variation and phase-variable gene expression.

Tropheryma whippelii

Tropheryma whippelii was recently identified as the etiological agent of Whipple's disease, a chronic digestive and systemic disorder characterized by fever, weight loss, abdominal pain, diarrhea, polyarthralgia, and lymphadenopathy. It is the first bacterial agent which has been indirectly identified by PCR and sequencing. It is a gram-positive actinomycete that is not closely related to any known genus. *T. whippelii* is found intracellularly in macrophages from duodenal biopsies of Whipple's disease patients, and in vitro it has been shown to invade and proliferate in HeLa cells within an intracellular acidic compartment that contains Lamp-1 and the vacuolar ATPase but that is devoid of cathepsin D.

Nonsporulating Extracellular Bacteria

Acinetobacter

The genus *Acinetobacter* includes seven named and nine unnamed genospecies. Acinetobacters can use a wide variety of substrates as their sole energy source; hence, they are generally ubiquitous and are found frequently in soil, water, dry environments, and hospitals. *A. baumannii* is responsible for human nosocomial infections, especially in patients with catheters in intensive care units. It is also responsible for respiratory and urinary tract infections and septicemia. *A. lwoffii* is a normal component of flora on the skin, oropharynx, and perineum in about 20 to 25% of healthy individuals, but it has been associated with chronic gastritis. Limited information is available concerning the virulence factors of acinetobacters and their association to host cells: *A. baumannii* has been reported to induce apoptosis of epithelial cells by a caspase-3-dependent mechanism and expresses adhesins of the mannose-resistant type, while certain *A. lwoffii* strains also express adhesins of the mannose-sensitive type.

Aeromonas

Aeromonas species (*A. hydrophila-punctata*, *A. hydrophila*, *A. sobria*, and *A. caviae*) are estuarine ubiquitous bacteria in fresh water and fish. Human infections are mostly opportunistic and are usually limited to the digestive tract, but they can also appear as septicemias, meningitidis, endocarditis, and osteomyelitis. *Aeromonas* produces many virulence factors, including adhesins, hemolysin, enterotoxins, and proteases. Aerolysin is a 48-kDa, channel-forming protein secreted as an immature molecule (proaerolysin) of 52 kDa that binds to the glucan core of GPI-anchored proteins in mammalian cell plasma membranes. After being processed by soluble or membrane-anchored proteases, aerolysin oligomerizes to a heptameric pore that allows entry of small ions into the cell and induces G-protein activation in granulocytes, apoptosis in T cells, and vacuolation of the endoplasmic reticulum in epithelial cells.

Bordetella

Bordetella bronchiseptica, *B. parapertussis*, and *B. pertussis* colonize the respiratory tract of mammals. *Bordetella pertussis* is strictly a human pathogen and is the etiological agent of whooping cough, which was a very common disease before the large immunization programs. Clinical manifestations are usually divided into three stages: the catarrhal and highly communicable stage, mimicking a common cold; the paroxysmal stage, marked by severe and repetitive paroxysmal cough, with excess mucus production, vomiting, and convulsions (neurological damage and bronchopneumonia are possible complications); and, finally, a recovery stage marked by a progressive diminution of the coughing episodes. The vaccine used for many years was whole killed bacteria, which caused numerous side effects, the most serious of which was irreversible brain damage. Concern about side effects has led to the development of a novel acellular vaccine.

Bordetellae have many different virulence factors which are under the control of a pivotal activator, the BvgA-BvgS two-component regulator system. Virulence factors include adhesins, of which filamentous hemagglu-tinin (FHA), pertussis toxin (Ptx, exclusive of *B. pertussis*), pertactin (PRN), tracheal colonization factor, and fimbriae mediate binding to ciliated cells and macrophages. Bordetellae produce several toxins. Ptx is an AB-type toxin acting on large G proteins. Ptx catalyzes the ADP-ribosylation of the α_1 subunit of G protein and prevents the G_i complex from interacting with receptors. The complex thus remains in a GTP-bound form, unable to in-hibit adenyl cyclase, thus leading to constitutive production of cyclic AMP (cAMP). Bordetellae produce adenylate cyclase (ACase), which enters cells and also increases the host cell cAMP level. ACase is an invasive cyclase that also increases the host cell cAMP level. It is active only in the presence of calmodulin. Ptx and ACase may contribute to symptoms of the disease, such as death of ciliated cells and mucin secretion. Another possible activ-ity is to lower the oxidative burst in polymorphonuclear neutrophils (PMNs) and macrophages, thus allowing bacterial survival. There is evi-dence that ACase induces apoptosis in macrophages. Other toxins are der-monecrotic toxin and tracheal cytotoxin, a peptidoglycan fragment that kills ciliated cells. Tracheal cytotoxin and LPS contribute to eliciting an inflam-matory response. Bordetellae can invade and survive within professional and nonprofessional phagocytes in vitro (and *B. pertussis* has been detected inside alveolar macrophages in children with AIDS), suggesting that the pathogenesis of the pertussis infection may also involve an intracellular stage.

The genomes of *B. bronchiseptica, B. parapertussis,* and *B. pertussis* have been sequenced. Adaptation to specific hosts (humans and sheep in the case of *B. parapertussis* and only humans in the case of *B. pertussis*) seems to be a consequence of gene loss and inactivation (and not gain of function) from a *B. bronchiseptica*-like ancestor. *B. bronchiseptica* genome contains a type IV pilus biosynthesis operon that is absent in the other two species. The three bordetellae encode a novel type III secretion system: *B. bronchiseptica* en-codes the full intact operon that is expressed; the operon in *B. parapertussis* contains two pseudogenes and is not expressed; in *B. pertussis,* the operon contains no clearly identifiable pseudogenes, but it does not seem to be ex-pressed. In the three bordetellae, the genes encoding Ptx are immediately followed by genes encoding a type IV secretion system involved in its ex-port; however, the subunit *ptxB* gene in *B. parapertussis* is represented by a pseudogene, and lack of expression in both *B. parapertussis* and *B. bron-chiseptica* is due to promoter inactivation mutations.

Borrelia

Borrelia burgdorferi sensu lato is a bacterial complex of spirochetes that comprise at least 10 genospecies. Three genospecies (*B. burgdorferi* sensu stricto, *B. garinii,* and *B. afzelii*) are commonly associated with Lyme dis-ease, a tick (*Ixodes scapularis*)-borne disease endemic in certain temperate-forest areas where the vector lives. It is characterized by three successive phases if untreated: (i) a cutaneous phase (spreading annular rash), in which the inoculation region becomes inflamed and erythematous (ery-thema migrans); (ii) an acute phase with lesions in the joints, heart, nervous and musculoskeletal systems, and skin; and (iii) a chronic phase with chronic arthritis of large joints, chronic fatigue, paralysis, and dementia.

In addition to its chromosome, *Borrelia* has two circular plasmids and nine linear plasmids, all of which have been sequenced. A large portion of its genome (more than 8%) is devoted to the production of lipoproteins that play different roles during the infection process: the lipoprotein Erp inhibits activation of the complement cascade; the lipoprotein P66 binds β3-chain integrins and mediates attachment of *Borrelia* to different mammalian host cells including platelets; the lipoproteins DbpA and DbpB bind decorin, a proteoglycan that decorates collagen fibers, and mediate the adherence of bacteria to collagen in skin and other tissues; the lipoprotein Bpk32 binds fibronectin and could also be important in the early phases of infection by promoting adhesion to the skin extracellular matrix; the lipoprotein Bgp is involved in binding to glycosaminoglycans; and finally, *B. burgdorferi* may invade endothelial cells, and OspA and OspB, the two major outer membrane lipoproteins, may play a role in this process.

Treponema

Spirochetes of the genus *Treponema* colonize a wide range of animal hosts and vary considerably in their pathogenic properties: *T. pallidum* subsp. *pallidum* is the etiological agent of syphilis while *T. phagedenis* is a normal commensal of the human urogenital tract, and *T. denticola* is associated with adult human periodontitis.

Syphilis is a sexually transmitted disease characterized by three clinical phases: (i) a phase characterized by a lesion (chancre); (ii) a phase in which bacteria penetrate the mucosa and enter the bloodstream, leading to fever, rash, and mucocutaneous lesions; and (iii) a phase in which bacteria can invade the heart, the musculoskeletal system, and the central nervous system. *T. pallidum* can attach to endothelial cells and a bacterial 47-kDa antigen stimulates the expression of intracellular cell adhesion molecule 1 (ICAM-1), in turn attracting polymorphonuclear neutrophils which may disrupt the endothelial barrier. These cells are responsible for the strong inflammation characterizing syphilis. *T. pallidum* has been shown to adhere to fibronectin- and laminin-coated surfaces, and the bacterial protein Tp0751 is a laminin-binding protein. The *T. pallidum* genome encodes a family of 12 proteins named Tpr that are homologous to the major surface protein (Msp) of *T. denticola* and at least three members (TprF, TprI, and TprK) are predicted to be associated with the outer membrane.

Periodontitis is a chronic inflammatory disease of the gums and gingival tissues that can lead to bone resorption and tooth loss. *T. denticola* stimulates both epithelial cell proliferation and apoptosis through the MAP kinases Erk1 and Erk2 (cell survival) and JNK and p38 (cell death). *T. denticola* expresses a chymotrypsin-like protease (CTLP) that has been implicated in cytotoxicity and degradation of fibronectin. By disrupting intercellular junctions, CTLP has also been shown to mediate migration of *T. denticola* through model basement membranes. The pore-forming major surface protein (Msp) induces cytopathic effects as well. *T. denticola* expresses an outer membrane lipid that is different from LPS, and it confers resistance to human β-defensins.

Corynebacterium

The genus *Corynebacterium* comprises over 50 species. Corynebacteria have been isolated from a wide range of environments (dairy products, soil, sewage, sediments, aquatic sources, and animals), but the vast majority of novel species have originated from human clinical sources. *C. diphtheriae* is the etiological agent of diphtheria, a disease with direct local effects of bacterial infection on the upper respiratory tract (formation of membranes and inflammation due to neutrophil attraction at the site of bacterial multiplication) and systemic effects (cardiac and neurological) due to the diffusible diphtheria toxin. Nontoxigenic *C. diphtheriae* strains have been increasingly recognized as a cause of infectious processes that may range from cutaneous lesions and pharyngitis to severe invasive disease.

Adhesion to host tissues could be mediated by exposed sugar residues, hydrophobins, or fimbriae. Two potential bacterial adhesins of 67 and 72 kDa have been reported to bind to human erythrocyte membranes. Actin-dependent invasion and intracellular survival of *C. diphtheriae* in Hep-2 cells have also been described. The diphtheria toxin, whose gene is iron regulated and carried on a phage, belongs to the AB toxin family which enter sensitive cells by receptor-mediated endocytosis. The enzymatically active A fragment subsequently enters the cytosol, where it inhibits protein synthesis by inactivating the elongation factor 2. The diphtheria toxin receptor is the heparin-binding epidermal growth factor (EGF)-like precursor (HB-EGF), which forms a complex with membrane protein DRAP27/CD9.

Escherichia coli

The species *E. coli* includes the nonpathogenic *E. coli* K-12, which has been widely used as a model in bacterial physiology and genetics. In humans, *E. coli* is the major constituent of the digestive commensal flora. Some *E. coli* strains have acquired specific virulence factors and are thus responsible for mild to severe digestive or urinary tract infections or for septicemia and meningitis. Depending on the symptoms that they are associated with, they are classified in four main categories: enteropathogenic *E. coli* (EPEC), enterohemorrhagic *E. coli* (EHEC), enteroinvasive *E. coli* (EIEC), and enterotoxinogenic *E. coli* (ETEC).

EPEC strains are a major cause of diarrhea in the developing world. EPEC infection is characterized by an intimate attachment to host epithelia. This attachment leads to the formation of attaching and effacing (A/E) lesions characterized by the degeneration of the epithelial brush border and the formation of actin-rich pedestals. The first step of these events is loose attachment via the bundle-forming pilus, resulting in localized adherence. Several proteins are then secreted by a type III secretion system: EspA polymerizes to form an extension to the needle complex which interacts with host cells; EspD is believed to be the main component of the translocation pore; EspB recruits α-catenin at the site of bacterial attachment; EspE (also known as Tir, for *t*ranslocated *i*ntimin *r*eceptor) becomes tyrosine phosphorylated and acts as a receptor for the outer membrane protein intimin. The Tir-intimin interaction results in intimate attachment and recruitment of a number of cytoskeletal elements such as α-actinin, ezrin, talin, plastin, vinculin, and villin which induce the formation of actin-rich pedestals on which bacteria reside. N-Wasp targets the Arp2/3 complex to Tir, which is required for pedestal elongation. Intimate attachment is accompanied by an increase in intracellular Ca^{2+} levels, activation of protein kinase C (PKC) and phospholipase C-γ, and phosphorylation of myosin light chain. This last event plays a key role in altering intestinal epithelium permeability.

The second category of *E. coli* that has received considerable attention is EHEC, which is responsible for large food-borne outbreaks of bloody diarrhea and hemolytic-uremic syndrome. The most notorious EHEC serotype is O157:H7. EHEC shares a number of virulence factors with EPEC that are responsible for the A/E lesions and the initial watery diarrhea associated with EHEC infections. Genes for Tir, intimin, the type III secretion system, and the EspA, EspB, and EspD secreted proteins are, as in EPEC, located on a 35-kb pathogenicity island called the locus of enterocyte effacement (LEE). In contrast to EPEC, Tir tyrosine phosphorylation is not necessary for pedestal formation by EHEC, but additional type III secreted factors are required. The most severe symptoms observed in EHEC infections are caused by the potent cytotoxin known as Shiga toxin or Vero toxin, which is phage encoded. This toxin consists of one A subunit and five identical B subunits. The B subunit binds to a glycolipid receptor, enters cells via clathrin-coated pits, and is transported to the endoplasmic reticulum. The A subunit inhibits host cell protein synthesis by an N-glycosidase activity which removes an adenine residue from the 28S rRNA. Infarction of the intestinal mucosa leads to bloody diarrhea.

Other virulence factors might be responsible for disease, including a 104-kDa protein with homology to a family of proteins called autotransporters, which includes immunoglobulin A1 (IgA1) protease of *Neisseria gonorrhoeae*, the vacuolating cytotoxin (VacA) of *Helicobacter pylori*, and pertactin of *B. pertussis*. The sequence of the genomes of *E. coli* K-12 and O157:H7 interestingly differs in many more loci than previously anticipated.

Haemophilus

The genus *Haemophilus* includes several gram-negative human pathogens, including *H. influenzae*, which is one of the leading causes of upper respiratory tract infections and otitis in children, sinusitis in children and adults, pneumonia in adults, and lower respiratory tract infections in adults with chronic obstructive pulmonary disease. *H. influenzae* is also responsible for meningitis. Encapsulated bacteria cause septicemias and meningitis, whereas nonencapsulated *H. influenzae* strains cause localized infections (e.g., middle ear infections). Peritrichous pili mediate attachment to respiratory mucus, to heparin-binding extracellular matrix proteins, and to oropharyngeal epithelial cells and erythrocytes; the erythrocyte receptor for one of the pili is the Anton receptor (An^{Wj}), a blood group antigen. The autotransporter proteins Hap, HMW1/HMW2, and Hia/Hsf constitute another group of adhesins: N-linked oligosaccharide chains with sialic acid in $\alpha 2$–3 configuration are recognized by the HMW1 molecule, but the receptors for the other autotransporters have not been identified yet. The outer membrane proteins P2 and P5 can interact with nasopharyngeal mucin; P5, in addition, can bind CEACAM1, a member of the carcinoembryonic antigen (CA) family of cell adhesion molecules. A minor role in adherence can also be attributed to the surface-associated lipoprotein OapA, responsible for colony opacity. A 46-kDa protein has also been implicated in adherence to phosphatidylethanolamine (PE) and gangliotriosyl-ceramide (Gg_3).

H. influenzae can invade epithelial cells in a process related to macropynocitosis that is dependent on actin polymerization and partially dependent on PI 3-kinase activity. Entry is associated with increases of cytosolic levels of Ca^{2+}. Interaction between *H. influenzae* lipooligosaccharide and the platelet-activating factor (PAF) receptor can lead also to bacterial internalization in host cells. A third pathway of invasion has been proposed that involves the host β-glucan receptors. Paracytosis has been observed, and the lipoprotein encoded by the open reading frame HI0638 could be necessary for this process.

The genome of *H. influenzae* was the first bacterial genome sequenced in 1995. This project was a hallmark in the genomics field because it highlighted that physical mapping was not necessary for completing the sequence of a bacterial genome. Shotgun cloning had an enormous repercussion in the way other genomes (including the human genome) were sequenced in the years to come.

Helicobacter pylori

H. pylori infection is the most common bacterial human infection. It is particularly prevalent in developing countries (90% compared with 30% in industrialized countries). Infection is acquired mainly during childhood, and, to date, there has been no conclusive evidence for the existence of an animal reservoir. Long-term infection is associated with chronic gastritis and peptic ulceration, and an epidemiological link has been established with mucosa-associated lymphoid tissue lymphoma and gastric adenocarcinoma. However, it remains unclear why infection with *H. pylori* is associated with such a wide diversity of pathologic findings extending from mild to very severe disease. Most likely, a combination of host, bacterial, and environmental factors determines the eventual clinical outcome.

   H. pylori possesses factors essential for colonization of the gastric mucosa, such as motility and urease activity. Less is known about how *H. pylori* is able to persist in the stomach and avoid elimination by the host immune response. It is likely that expression of variable and multiple adhesins with different binding specificities contributes to the persistence of the organism. Among the genetically characterized adhesins, the Lewis b binding adhesin (BabA) is now recognized as one of the members of a large family of 32 outer membrane proteins capable of phenotypic phase variation. Certain bacterial factors present only on a subset of strains are associated with gastric lesions: the protein VacA inserts itself into the plasma membrane of epithelial cells and forms an anion-selective channel which is capable of releasing bicarbonate, chloride, and urea from the cell cytosol. Eventually, the VacA toxin channel is internalized and changes the anion permeability of late endosomes, with the enhancement of the vacuolar ATPase proton-pumping activity (leading to endosomal swelling). Another bacterial factor that participates in the induction of proinflammatory responses is CagA, a protein encoded in the *cag* pathogenicity island (*cag* PAI). CagA is delivered into host cells through a type IV secretion system encoded in the *cag* PAI, associates with the plasma membrane, and undergoes tyrosine phosphorylation by the Src family of protein kinases. Phosphorylated CagA recruits SHP-2 and Grb2, inducing cell elongation; nonphosphorylated CagA can also associate with the protein ZO-1 and induce disruption of the epithelial apical-junctional complex.

Klebsiella pneumoniae

Klebsiella pneumoniae is the causative agent of nosocomial and community-acquired pneumonia. *Klebsiella* strains are usually resistant to a wide spectrum of antibiotic agents. Virulence is associated with a capsule that confers resistance to phagocytosis by PMNs and macrophages. It is also mediated by fimbrial adhesins, which mediate binding to collagen. Bacterial internalization has been detected in ileocecal cells, and it has been hypothesized that survival within epithelial cells may serve as a reservoir from which reinfection of the host can take place.

Leptospira

Leptospirosis is one of the most widespread zoonoses in the world and has emerged as an important public health problem in developing countries. Its severe disease form, Weil's syndrome, is an acute febrile illness associated with multiorgan system complications, including jaundice, renal failure, meningitis, and pulmonary hemorrhage, with a mortality rate that may exceed 15%. Leptospirosis is caused by spirochetes belonging to the genus *Leptospira,* which includes the saprophytic species *L. biflexa* and the pathogenic species *L. interrogans.* Pathogenic *Leptospira* species are highly motile and invasive organisms which rapidly disseminate to target organs after entering the host, usually through abrasions in the skin or mucous membranes.

Several bacterial factors that contribute to the pathogenesis of the disease induced by *L. interrogans* have been identified. Secretion of sphingomyelinase C (SphA) and pore-forming hemolysins (SphH) are probably associated with the hemolytic anemia observed in leptospirosis patients. The only putative leptospiral adhesin identified to date is a virulence-associated surface protein that binds fibronectin. Translocation through polarized MDCK monolayers without disruption of tight junctions and invasion of Vero cells has been described. Several surface-associated molecules that could mediate pathogen-host cell interactions are the integral membrane protein Ompl1, the lipoproteins LipL32 and LipL41, and the peripheral membrane protein P31$_{LipL45}$. Surface-exposed proteins LigA and LigB, belonging to the bacterial immunoglobulin superfamily, have also been associated with pathogenic *Leptospira* species.

Mycoplasmas

Mycoplasmas are eubacteria lacking a cell wall. They are the smallest known self-replicating organisms (length, 0.3 μm) with a genome size ranging from 580 kb for *Mycoplasma genitalum* to 1,350 kb for *M. mycoides.* The human pathogen *M. pneumoniae* is known to colonize ciliated lung epithelial cells and is responsible for an atypical pneumonia. The less well-characterized *M. genitalum* is also a human pathogen and is associated with nongonococcal urethritis and pelvic inflammatory disease.

M. pneumoniae and *M. genitalum* have a specialized tip structure implicated in adhesion to cells, gliding motility, and cell division. Detergent treatment of whole cells reveals that the *M. pneumoniae* tip possesses a cytoskeleton-like network of high molecular weight proteins that are thought to anchor adhesins at the bacterial surface. Three adhesins, P1 (170 kDa), P30 (32 kDa), and P116 (116 kDa), are involved in *M. pneumoniae* attachment to host cells. The surface antigen MgPa (150 kDa) is the major adhesin of *M. genitalum* and is homologous to P1. For other species, the adhesive properties are probably associated with lipoproteins.

Neisseriae

N. gonorrhoeae is the etiological agent of the most common sexually transmitted disease, gonorrhea, which is a highly contagious genital infection of males and females and occasionally becomes an invasive infection, leading, by hematogenous seeding, to localized infections such as septic arthritis. *N. meningitidis* is a commensal in 10 to 30% of healthy individuals, yet it is one of the most harmful pathogenic bacteria in susceptible humans. In young people, it causes purulent meningitis, with a very high mortality rate and a high incidence of sequelae. It also occasionally causes purpura fulminans, a lethal septicemic syndrome associated with septic shock.

One of the most striking features of *Neisseria* species is their capacity to vary their surface components by phase variation and/or antigenic variation. The modulation of the outer membrane lipooligosaccharide and pili is thought primarily to reflect an immune evasion mechanism. Variation of the opacity-associated (Opa) proteins facilitates the bacterial interaction with different host cell types. Two major classes of Opa proteins have been identified: Opa$_{HS}$ proteins, which mediate attachment and invasion of epithelial cells by recognition of host heparan sulfate proteoglycans (invasion can be enhanced by vitronectin), and Opa$_{CEA}$ proteins, involved in the attachment of bacteria to cells expressing members of the carinoembryonic antigen (CEA)-related cell adhesion molecule (CEACAM) family that harbor the CD66 epitope. CEACAM1 is the most broadly distributed and supports bacterial invasion and trafficking across epithelial layers, entry into endothelial cells, and engulfment by professional phagocytes. CEACAM3 is expressed exclusively by neutrophils and supports nonopsonic bacterial uptake dependent on PI 3-K and Rac activity. Type IV pili are also important for adhesion through binding of CD46, which is phosphorylated by c-Yes upon bacterial attachment, leading probably to cortical plaque formation. Type IV pili also trigger the release of Ca^{2+} from intracellular stores and exposure of Lamp1 at the cellular surface, which is cleaved by the bacterial IgA1 protease (modifying the host cell lysosomes and preventing phagolysosomal fusion). *N. meningitidis* is a human-specific pathogen. This species specificity is in part mediated by the species-specific interaction of bacterial pili with CD46. Recently, a transgenic mouse model expressing human CD46 has been reported as an animal model for studying *N. meningitidis* pathophysiology.

Pasteurella

The genus *Pasteurella* contains a nonhomogeneous group of gram-negative bacteria of which *P. multicocida* and *P. haemolytica* are the most important species. *P. multicocida* induces localized inflammation, pneumonia, bacteremia, and atrophic rhinitis in mammals (it can lead also to infections in humans). The *P. multicocida* toxin (PMT) is the main bacterial virulence factor. PMT binds to a ganglioside-type cell surface receptor and has to be internalized via receptor-mediated endocytosis to elicit its toxic effects, including mitogenic activation dependent on mitogen-activated protein kinase (MAPK) ERK and also heterotrimeric G-protein activation leading to Plc-β1 stimulation, mobilization of intracellular Ca^{2+}, and induction of protein kinase C activity. In addition, Rho is also activated by PMT, resulting in the formation of stress fibers, focal adhesions, and tyrosine phosphorylation of focal adhesion kinase and paxillin. Vimentin has also been shown to bind PMT.

Pseudomonas aeruginosa

Pseudomonas aeruginosa is a ubiquitous environmental gram-negative bacillus that behaves as an opportunistic pathogen when host defenses are compromised. *P. aeruginosa* is able to colonize multiple niches by using many environmental compounds as sources of energy. Nosocomial infections due to *P. aeruginosa* present a wide variety of clinical features and locations. *P. aeruginosa* is responsible for severe infections in burn patients and deleterious chronic lung infections in cystic fibrosis patients.

Bacteria stimulate cells through a c-Src/Ras/MEK1/MAPK/pp90^{rsk}-signaling pathway that leads to NF-κB (p65/p50) activation. Activated NF-κB binds to the 5'-flanking region of the *MUC2* gene and stimulates *MUC2* mucin transcription (which contributes to airway obstruction and death by mucus overproduction). A type III secretion system is used to deliver the bacterial molecules ExoS, ExoT, ExoU, and ExoY into host cells. ExoS and ExoT are highly related and have dual functions, acting as both a GAP and ADP-ribosyltransferase (ADPRT): ExoS inactivates Rac1 through its GAP activity and targets Ras and RalA through its ADPRT activity (down-regulating macrophage functions such as phagocytosis and adherence). ExoU acts as a membrane lytic and cytotoxic phospholipase, and ExoY has been identified as an adenylate cyclase. Interestingly, the activities of these enzymes depend on host cell factors: the 14-3-3 protein FAS for ExoS and ExoT and unidentified proteins for ExoU and ExoY.

Staphylococci

Staphylococci are very common in the environment and are highly resistant to hostile conditions, such as heat, desiccation, or salinity. They are often commensals on the skin or mucosa of humans and animals.

The most pathogenic species is *Staphylococcus aureus*, which is responsible for cutaneomucosal polymorphic infections (ranging from simple local erythemas to abscesses which can disseminate to secondary sites and cause generalized infections), toxic shock syndromes (mainly affecting young women using vaginal tampons), and food-borne toxic infections that occur very rapidly after ingestion of contaminated food. If *S. aureus* is also often the cause of nosocomial infections, resulting in the symptoms described above, nosocomial infections are also often due to coagulase-negative staphylococci such as *S. epidermidis* and, in rare cases, *S. haemolyticus*. *S. epidermidis* has the capacity to colonize plastic catheters. Urinary tract infections due to another coagulase-negative species, *S. saprophyticus*, appear to be the second most common urinary tract infections among young women in the United States.

The virulence of *S. aureus* is due mostly to the secretion of several toxins and enzymes. Some toxins, such as alpha, beta, gamma, and delta toxins, have a dermonecrotic effect; others, such as leucocidin, are cytotoxic for and destroy polymorphonuclear cells. Some toxins, such as exfoliatin and toxic shock syndrome toxin type 1 (TSST-1), have an exceptional capacity to rapidly diffuse into tissues and produce disease while the bacteria remain confined to the initial site of infection. The roles of the many extracellular enzymes secreted by *S. aureus* are not understood. The coagulase would be responsible for the formation of a clot. Those of other enzymes (hyaluronidase, deoxyribonuclease, lipase, esterase, and catalase) have not been clearly established.

In addition to secreted proteins, *S. aureus* has a number of surface proteins which may play a role in infections such as protein A, which binds the Fc fragment of Igs and abrogates the action of specific Igs, and several proteins which bind extracellular matrix proteins such as fibronectin (FnbA, FnbB), collagen (Cna), fibrinogen (ClfA, Fib), vitronectin (60-kDa protein), elastin (EbpS), and probably others.

There is a major concern about the increasing proportion of methicillin-resistant strains as well as the appearance of vancomycin-resistant isolates.

Some toxins, and in particular TSST-1, are superantigens. Superantigens link the antigen receptor of T cells with major histocompatibility complex (MHC) class II by interacting with some Vβ subunits of T cells and the MHC class II molecule of antigen-presenting cells. This leads to the activation (cytokine secretion and proliferation) or deletion of a large subpopulation of Vβ T cells in vivo and in vitro. The consequences in vivo are shock or immunosuppression, i.e., reduced antibody levels against some staphylococcal products and persistence of *Staphylococcus* cells in the host. Another explanation for this persistence derives from the recent discovery that staphylococci that were assumed to be exclusively extracellular can enter cultured epithelial cells. One report even describes escape from the internalization vacuole, probably due to the action of some of the toxins described above and induction of apoptosis.

Streptococci

Streptococci are divided into many subgroups. The most important clinically are group A streptococci (GAS), whose prototype is *Streptococcus pyogenes*, group B streptococci (GBS), whose prototype is *S. agalactiae*, and *S. pneumoniae*. Streptococci have long been considered extracellular pathogens, and their capacity to bind to extracellular matrix proteins or to produce a wide range of toxic molecules was considered the major features for their virulence. It has been recognized recently that the three groups described here have the capacity to invade tissue-cultured cells.

S. pyogenes. *S. pyogenes* is responsible for cutaneomucosal infections (the classic tonsilitis, scarlet fever, erysipelas), skin infections (impetigo), and occasionally endocarditis and postinfectious manifestations, such as immune complex glomerulonephritis and rheumatic fever. The initial encounter of GAS with the host is with the pharynx or the skin.

S. pyogenes produces a wide variety of surface-associated and secreted components which are important for their virulence, including the M protein, fibronectin-binding proteins, hyaluronic acid capsule, extracellular enzymes, and toxins which allow streptococci to colonize and/or cause invasive diseases. It has been suggested that expression of different proteins constitutes the basis for the cellular tropism of different *S. pyogenes* isolates; the fibronectin-binding proteins have been implicated in bacterial attachment to epithelial cells of the upper respiratory tract, whereas the M protein mediates binding to skin cells. Recent studies show that the fibronectin-binding protein (Sfb1 or protein F1) and also protein M mediate internalization. The surface protein SDH, a surface glyceraldehyde-3-phosphate dehydrogenase, can trigger signal transduction events. Other factors may be involved in internalization. Invasion of deeper tissues requires the expression of several toxins or enzymes; *S. pyogenes* produces well-known toxins: streptolysin O, a pore-forming toxin which binds to cholesterol, oligomerizes, and thus permeabilizes cells (hence, its use in cell biology), and the pyrogenic exotoxins SPEA and SPEB. These molecules, like TSST-1 produced by *S. aureus* and the staphylococcal enterotoxin SE, display superantigen activity.

S. agalactiae. *S. agalactiae* causes neonatal pneumonia, sepsis, and meningitis. This pathogen can translocate within brain microvascular endothelial cells, suggesting that direct invasion and injury of the blood-brain barrier allow bacteria to gain access to the central nervous system, leading to meningitis.

S. pneumoniae. *S. pneumoniae* is a major cause of meningitis and septicemia in children and of pneumonia and septicemia in adults. Pneumococci undergo spontaneous phase variation marked by switching from opaque to transparent colony morphotypes. Transparent strains are adapted to nasopharyngeal adherence and display less capsule, more surface choline, and more of the adhesin CbpA than do opaque strains. Opaque strains show improved survival in the bloodstream and bear more capsule, less choline, and more of the protective antigen PspA. Pneumococci were the first bacteria shown to have surface proteins attached to the bacteria via choline residues present in the cell wall. The presence of choline in the cell wall has also been described for *Haemophilus, Pseudomonas, Neisseria,* and *Mycoplasma* species. Choline-binding proteins include several important virulence factors: PspA, a well-described antigen; LytA, the autolytic enzyme; and CbpA, a major adhesin. The Cbp proteins contain 3 to 10 repeats which bind choline. CbpA participates in binding of pneumococci to activated human cells and in nasopharyngeal colonization.

The major pneumococcal toxin pneumolysin acts as a protective antigen and also activates complement.

CbpA is absolutely required for pneumococcal traffic across the blood-brain barrier; this entry involves cytokine activation of the cerebral endothelial cells, which then passively let the pneumococci follow the PAF receptor-recycling pathway, which leads them into the cell. How pneumococci drive the endocytosed vesicle across the cell to release bacteria into the cerebrospinal fluid remains to be understood.

The genome sequence has been completely determined.

Vibrio cholerae

The genus *Vibrio* contains a large number of closely related bacterial species, including the etiological agent of cholera, *V. cholerae.* Cholera is a serious epidemic disease that is acquired by drinking water contaminated with human feces or by eating food washed in contaminated water. The disease is characterized by an aqueous diarrhea essentially due to cholera toxin (CT), a potent AB-type ADP-ribosylating toxin containing one A enzymatic subunit and five identical B subunits. The toxin is secreted when *V. cholerae* is on the intestinal mucosa and it binds to a host cell ganglioside GM_1, which is a sialic acid-containing oligosaccharide covalently attached to a ceramide lipid found on many cell types. The A1 subunit of CT ADP-ribosylates the α-subunit of a G_s protein leading to constitutive activation of adenylate cyclase. The high levels of cyclic AMP alter the activities of sodium and chloride transporters and are responsible for the aqueous diarrhea.

In addition to CT, *V. cholerae* virulence requires the expression of the toxin-coregulated pili (TCP). The membrane-associated response regulator ToxR regulates both CT and TCP. The CT gene is encoded by the filamentous bacteriophage CTXɸ, which uses TCP as receptor and infects *V. cholerae* cells more efficiently within gastrointestinal tracts of mice than in broth media. TCP and the gene regulator *toxT* are found in a 39.5-kb *Vibrio* pathogenicity island (VPI), absent in nontoxigenic strains of *V. cholerae.*

Yersiniae

The genus *Yersinia* contains many species, but only three of them are pathogenic for rodents or humans: *Y. pestis, Y. pseudotuberculosis,* and *Y. enterocolitica. Y. pestis,* the etiological agent of bubonic and respiratory plague, is transmitted by fleas, in contrast to *Y. pseudotuberculosis* and *Y. enterocolitica,* which are present in contaminated food products. *Y. pseudotuberculosis* causes adenitis and septicemias while *Y. enterocolitica* is responsible for a broad range of gastrointestinal syndromes in humans. Despite differences of infection routes, the three species have a common tropism for lymphoid tissues.

Enteropathogenic *Yersinia* (*Y. pseudotuberculosis* and *Y. enterocolitica*), after initial gastrointestinal colonization, translocates across the ileum into deeper host tissues. In animal infection models, bacteria are internalized by M cells, which are intercalated into the epithelium overlaying ileal lymphoid follicles called Peyer's patches. Efficient translocation across Peyer's patches requires interaction between the 986-amino-acid bacterial outer membrane protein invasin (the first bacterial invasion factor ever described) and its receptor β_1-integrin. Clustering of β_1-integrin receptors will lead to recruitment/phosphorylation of the focal adhesion kinase (FAK) and the proline-rich tyrosine kinase 2 (Pyk2). Downstream effects of Pyk2 include activation of Src and subsequent phosphorylation of p130^{Cas}, which complexes with Crk, leading to Rac1 activation. Actin, the Arp2/3 complex, WASP family members, and microtubules are also necessary for invasion.

Later during infection, bacteria remain extracellular. A type III secretion system is involved in the impairment of phagocytosis by disturbing the dynamics of the cytoskeleton through the injection of *Yersinia* outer membrane proteins (Yops) into the host cell cytosol. YopH is a protein tyrosine kinase phosphatase that dephosphorylates Fak and p130^{Cas} in focal adhesion complexes, in addition to the Fyn-binding protein (Fyb) and the SKAP55 homologue (SKAP-HOM) in adhesion-signaling complexes. YopE exerts a GAP activity on Rho, Rac, and CDC42, and YopT cleaves these same GTPases close to their carboxyl terminus, releasing them from their membrane anchor. YpkA/YopO is activated by actin binding and binds to RhoA and Rac1, but its mode of action is unclear. YopP/YopJ downregulates the inflammatory response of macrophages, epithelial cells, and endothelial cells by blocking the MAPK and the nuclear factor-κB (NF-κB) pathways. Finally, YopM is a leucine-rich repeat protein that targets the cell nucleus, but its function and partners are unknown.

Sporulating Extracellular Bacteria

Bacillus anthracis

Anthrax is a disease resulting from infection by spores of the gram-positive bacteria *Bacillus anthracis*. Three different clinical manifestations are recognized: cutaneous anthrax, the most common form of the disease, results from infection through a cut or abrasion of the skin; digestive anthrax involves digestion of contaminated food; and pulmonary anthrax is the most severe form of the disease. Following contact with an animal host, environmental stimuli that are not fully understood induce the conversion of the spore to the vegetative actively dividing stage of the bacteria.

The plasmid pX02 encodes a poly-D-glutamic acid capsule that surrounds the bacterial vegetative form and prevents phagocytosis by host immune cells. The plasmid pX01 encodes the anthrax toxin, a tripartite molecule consisting of the PA (protective antigen), EF (edema factor), and LF (lethal factor) polypeptides that associate either as PA plus LF (lethal toxin or LeTx) or as PA plus EF (edema toxin or EdTx). PA binds to a type I membrane protein (named anthrax toxin receptor or ATR) with an extracellular von Willebrand factor A domain and mediates delivery of the enzymatic components EF and LF to the cytosol. EF is an adenylate cyclase that raises the levels of cyclic AMP, causing disruption of water homeostasis and swelling or edema (and also inhibits phagocytosis). LF is a zinc-dependent protease that cleaves MAPK and causes lysis of macrophages.

Clostridia

Clostridia are anaerobic bacteria that are widespread in the environment and belong to the normal flora of the digestive tract. They can enter organisms either by the oral route or through the skin via wounds. Clostridia produce various toxins and hydrolytic enzymes that are responsible for the local lesions and general symptoms of infection. Most of the clostridial toxins damage cell membranes or disrupt the cytoskeleton. Two *Clostridium* groups (*C. tetani* and *C. botulinum*) produce potent neurotoxins.

C. botulinum is a ubiquitous bacterium that contaminates food products, in particular, preserved foods. It produces the botulinum neurotoxin, which is responsible for flaccid paralysis and death. Botulinum neurotoxins are a group of closely related protein toxins (seven different serotypes) which show absolute tropism for the neuromuscular junction, where they bind to unidentified receptors. They enter into the cytoplasm of motoneurons and cleave their intracellular targets, which are VAMP/synaptobrevin, a protein of synaptic vesicles, or syntaxin for serotype C or SNAP25 for serotypes C, A, and E. This cleavage results in the blockade of transmitter release. *C. botulinum* also produces the C3 toxin, which has been widely used in cell biology due to its capacity to inhibit Rho function. This species can now be transformed.

C. difficile is responsible for a highly contagious nosocomial antibiotic-associated diarrhea called pseudomembranous colitis. It produces many toxins, encoded by chromosomal genes, including two large cytotoxins, ToxA and ToxB, which both act on small G proteins and are, like C3, very useful tools in cell biology. It also produces one enterotoxin. *C. difficile* has never been transformed.

C. perfringens is responsible for food poisoning and wound contaminations leading to clostridial myonecrosis (gas gangrene), cellulitis, intra-abdominal sepsis, and postabortion and postpartum infections (puerperal fever). It produces the largest number of potential virulence factors of any bacterium, including alpha, beta, epsilon, and iota toxins. However, all strains do not produce all these toxins. The best-characterized toxin is alpha toxin, which is a phospholipase C, and the pore-forming toxin perfringolysin.

C. tetani, a soil bacterium, is the etiological agent of tetanus, a fatal neurological disease that is due to tetanus toxin. This neurotoxin is responsible for spastic paralysis and death. Its receptor has not been identified. The toxin binds to neuronal cells and enters the cytosol, where it specifically cleaves vesicle-associated membrane protein/synaptobrevin at a single peptide bond. This selective proteolysis prevents the assembly of the neuroexocytosis apparatus and consequently the release of neurotransmitter. *C. tetani* has recently been transformed.

Parasites

Entamoeba histolytica

The genus *Entamoeba* includes two related species, the commensal *E. dispar* and the pathogenic *E. histolytica,* the latter being the agent of amebic dysentery and visceral amebiasis, the second leading cause of death due to parasitic disease. It is transmitted to humans primarily by ingestion of contaminated water and food. *E. histolytica* is an anaerobic ameba, and the trophozoites lack mitochondria. Cyst germination and cell growth occur on mucosal cells. Continuous growth leads to ulceration of the intestinal mucosa, causing diarrhea and severe intestinal cramps. The diarrhea is then replaced by a condition referred to as dysentery, characterized by intestinal bloody and mucoid exudates. If the condition is not treated, trophozoites of *E. histolytica* can migrate to the liver, lungs, bones, and brain, where large abscesses may appear.

   E. histolytica trophozoites adhere to colonic epithelial cells through the galactose/N-acetylgalactosamine-specific lectin, a complex of a disulfide-linked 170-kDa and 35/31-kDa subunits, and an associated 150-kDa protein. Host cell killing can take place through cytolysis by amebaphores, a family of at least three isoforms of a 77-amino-acid polypeptide, structurally related to granulisins and NK-lysins produced by mammalian T cells, that can make pores in lipid bilayers. *E. histolytica* trophozoites can also induce programmed cell death through Fas- and tumor necrosis factor alpha-independent pathways. Cysteine proteinases secreted by trophozoites digest extracellular matrix proteins and facilitate *E. histolytica* invasion into and within the submucosal tissues. Trophozoites entering portal circulation are protected from the complement cascade by the amebic Gal/GalNAc lectin, which has a region with antigenic cross-reactivity with CD59, a membrane inhibitor of the complement component C5b-9.

Leishmania Species

Leishmania spp. are protozoa that exhibit two morphological stages: a promastigote found in sand fly vectors and an intracellular amastigote found in vertebrate hosts. There are approximately 21 species of *Leishmania* transmitted by about 30 phlebotomine sand fly vectors, but only six species cause the three main forms of the disease in humans. *L. tropica* and *L. major* cause dermal leishmaniasis, consisting of cutaneous ulcers usually localized to the initial site of the *Phlebotomus* sand fly bite: there the amastigotes replicate in the reticuloendothelial system and the lymphoid cells of the skin. *L. donovani* and *L. infantum* are the cause of visceral leishmaniasis, a disease in which the protozoa multiply in the liver, spleen, lymph nodes, and intestine. Mucocutaneous leishmaniasis is caused by *L. braziliensis* and *L. mexicana* and is spread by the *Lutzomya* sand fly: the primary lesion occurs at the initial site of the bite on the skin, but the infection also involves the mucosal system of the nasal and buccal cavity, causing degeneration of the cartilaginous and soft tissues.

   Metacyclic promastigotes exploit opsonization by complement to gain entry into macrophages through CR1 and CR3 receptors. *L. major* and *L. mexicana* have also been shown to use immunoglobulin opsonization as a means of entering host cells through the Fc receptor. *L. major* can attach through a lectin-like receptor that recognizes the LPG. *L. amazonensis* can bind to heparin sulfate and fibronectin receptor (as the surface protein gp63 appears to mimic fibronectin). Phagocytosed promastigotes reside in a parasitophorous vacuole (PV) that initially exhibits reduced interactions with endosomal compartments possibly through the release of protozoan lipophosphoglycan (LPG) that reduce the fusogenic properties of the phagosomal membrane. Later during infection the PV undergoes fusion with late endosomal compartments to become a phagolysosome, where the LPG may act as a degradation barrier protecting the amastigotes from hydrolytic enzymes. *L. donovani* has been shown to impair tyrosine phosphorylation of JAK1, JAK2, and STAT1 in response to interferon gamma (IFN-γ), possibly involving the activation of the cellular protein tyrosine phosphatase SHP-1 (*L. donovani* promastigotes also express an endogenous phosphatese to impair cellular phosphorylation). *Leishmania* also decreases macrophage protein kinase C (PKC) activity to enhance its survival.

Plasmodium

Plasmodium belongs to the Apicomplexa group of protozoa that comprises unicellular obligate intracellular parasites. *Plasmodium falciparum* (and, to a much lesser extent, *P. vivax*) is the agent of human malaria. Insight into the disease pathogenesis has also been provided by study of the rodent malaria caused by *P. berghei* and *P. yoelii,* and the rhesus malaria caused by *P. knowlesi.* Anopheles mosquitoes inject *Plasmodium* sporozoites into the subcutaneous tissue (and less frequently directly into the bloodstream) and from there the sporozoites travel to the liver. Within the hepatocyte each sporozoite develops into tens of thousands of merozoites, which can invade red blood cells (RBC) on release from the liver. Encompassed in a parasitophorous vacuolar membrane (PVM) within the RBC, the parasites develop from ring stages to trophozoites, and finally to schizonts: rupture of schizonts releases up to 24 merozoites into the bloodstream, which initiate a new round of schizogony. Disease, which begins only once the asexual parasite multiplies in RBC, is characterized initially by fever, and it can evolve to severe malaria distinguished by metabolic acidosis, severe anemia, respiratory distress, and cerebral malaria. A small portion of the asexual parasites converts to gametocytes that are essential for transmitting the infection to other individuals through female anopheline mosquitoes, but cause no disease.

Cellular infection involves the sequential exocytosis of three different apical secretory organelles (rhoptries, micronemes, and dense granules) that define the Apicomplexa phylum. Many proteins involved in parasite recognition of host cells and penetration are secreted from rhoptries and micronemes. Invasion of hepatocytes implicates the thrombospondin domains on the circumsporozoite protein and on the thrombospondin-related adhesive protein (TRAP), which bind specifically to heparin sulfate proteoglycans. On the surface of hepatocytes, the tetraspanin CD81 protein is also necessary for sporozoite infectivity. *P. vivax* invades only Duffy blood group-positive erythrocytes through expression of a single protozoan receptor, the Duffy-binding protein (DBP). *P. falciparum* possesses several homologues of DBP, including the erythrocyte-binding antigen-175 (EBA-175) that binds glycophorin A, EBA-140 (or BAEBL) that binds glycophorin C, and the *P. falciparum* erythrocyte membrane protein 1 (PfEMP1) that binds several receptors including CD36, chondroitin sulfate A (CSA), vascular cell adhesion molecule (VCAM), and the intercellular adhesion molecule 1 (ICAM-1). PfEMP1 is encoded by the large and diverse *var* gene family, which is involved in clonal antigenic variation and has a central role in *P. falciparum* pathogenesis since PfEMP1 is expressed at the infected erythrocyte surface and allows the homing of infected erythrocytes in several tissues, including the placenta and the brain. A *P. falciparum* paralogue of DBP is the MAEBL protein, of which the receptor is unknown. A second family of adhesion molecules is the reticulocyte-binding-like (RBL) protein group, of which multiple variants are present in the different *Plasmodium* species.

Gliding motility and host cell invasion depend on the parasite's acto-myosin system located beneath the plasma membrane of invasive stages. TRAP acts as a transmembrane link connecting bound extracellular ligands to intracellular actin; the actin/TRAP complex is redistributed from the anterior to the posterior end of the parasite by the class XIV unconventional Myosin A, leading to its forward movement on a substrate or to penetration of a host cell.

Inside erythrocytes, which are metabolically inert cells, *Plasmodium* proteins secreted from dense granules induce the formation of a tubovesicular network that extends from the PVM into the cytoplasm and to the plasma membrane of RBC and is involved in the trafficking of nutrients and waste products. The family of early-transcribed membrane proteins (ETRAMPs) seems to define functionally different domains in the PVM that dynamically change during the morphological changes associated with the parasite development.

Toxoplasma gondii

Toxoplasma gondii is another obligate intracellular parasite of the Apicomplexa group. Clinical toxoplasmosis is most frequent in newborns whose mothers contracted the infection during pregnancy; this condition results in abortions or congenital malformations. Postnatally acquired toxoplasmosis is usually asymptomatic but results in lifelong persistence of slowly replicating parasites. In immunocompromised patients (AIDS patients or transplant recipients), reactivation of latent parasites leads to intracerebral toxoplasmic abscesses and sometimes diffuse encephalitis and generalized infection.

T. gondii can invade a broad range of host cells, including phagocytic and nonphagocytic cells.

Invasion is an active process driven by the parasite, which usually takes about 5 to 10 s; the host cell is a passive spectator (no membrane ruffling, no participation of host actin, and no tyrosine phosphorylation of host proteins). Extracellular parasites display gliding motility on the surface of host cells. This twisting motion depends on the parasite cytoskeleton (actin and myosin) and is necessary for cell invasion. Entry is initiated by reorientation of the parasite to create contact between its apical end and the host cell plasma membrane. Passage in the host cell involves the formation of a region of very close apposition of the two plasmalemmas located at the site of entry, called the moving junction. The moving junction has the morphological characteristics of a tight junction, although it has not been characterized at the molecular level. When formed, the PV does not fuse with lysosomal compartments and is rapidly modified by the parasite secretory proteins in a compartment suitable for intracellular multiplication. Note that when parasites are primed by the immune response and phagocytosed (opsonized), they are killed; only active invasion leads to parasite development. The PV formed by live parasites has nonselective pores that allow the diffusion of molecules of up to 1.2 kDa. The vacuole is surrounded by a layer of endoplasmic reticulum and mitochondria, which may be used for the metabolic requirements of the parasite.

Trypanosoma cruzi

Trypanosoma cruzi is the etiological agent of Chagas' disease in humans. The vector for *T. cruzi* is an insect of the *Reduviidae* family. Transfusion can also lead to contamination in areas of endemic infection (South America). In the digestive lumen of these insects, trypomastigotes differentiate in epimastigotes, which replicate extensively by binary fission. When liberated in insect feces, newly differentiated trypomastigotes actively penetrate the vertebrate mucosa or cutaneous wounds.

T. cruzi is intracellular in vertebrates but extracellular in insects. It enters cells by a mechanism characterized by a smooth diving into the cell. This process is coupled to the recruitment of lysosomes at the site of entry. While inhibitors of microtubules such as nocodazole inhibit entry, the actin cytoskeleton does not seem to be involved in the entry process. Disruption of the actin cytoskeleton by cytochalasin D even increases the rate of entry. Availability of intracellular Ca^{2+} is a prerequisite for internalization. Ca^{2+} is probably released from intracellular stores upon inositol triphosphate production, suggesting the action of a phospholipase C. Ca^{2+} transients are thought to be required for rearrangements of actin microfilaments, which might facilitate lysosome access to the plasma membrane. The Ca^{2+}-signaling activity expressed by tissue culture trypomastigotes is generated through the action of a cytosolic enzyme, *T. cruzi* oligopeptidase B, which probably acts as a processing enzyme that generates an active Ca^{2+} agonist. The role of this peptidase has recently been confirmed by using oligopeptidase B gene knockout mutants. Signaling through transforming growth factor β receptors is also required for *T. cruzi* trypomastigote invasion.

T. cruzi does not stay in the parasitophorous vacuole. It is released in the cytosol less than 2 h through the action of a 65-kDa pore-forming toxin active at pH 5 and of a transialidase/neuraminidase that is thought to disorganize the vacuolar membrane by desialylation of the vacuole constituents and to facilitate the action of the toxin. Once released into the cytosol, the trypomastigotes are ready to differentiate in amastigotes and to replicate. After lysis of the first infected cells, the parasites migrate into deeper tissues and/or circulate in the blood before invading other cells. Recent advances in genetic manipulation (targeted gene replacement) should help identify the molecular mechanisms underlying trypanosome infections.

Yeasts and Molds

Aspergillus fumigatus

Aspergillus fumigatus is one of the most common of the airborne saprophytic fungi. Humans and animals constantly inhale numerous conidia of this fungus. The conidia are normally eliminated in immunocompetent hosts by innate immune mechanisms, and aspergilloma and allergic bronchopulmonary aspergillosis, both uncommon clinical syndromes, are the only infections observed in such hosts. With the increasing number of immunosuppressed patients, there has been a dramatic increase in severe and fatal invasive aspergillosis, now the most common mold infection worldwide. The lung is the site of infection by *Aspergillus* spp., and since alveolar macrophages are the major resident cells of the lung alveoli, along with neutrophils, which are actively recruited during inflammation, they are the two main cells involved in the phagocytosis of *A. fumigatus*. Several lines of evidence suggest that nonoxidative mechanisms are essential for the killing of conidia by the alveolar macrophages of the immunocompetent host. The second line of phagocytic cells, the neutrophils, plays a role in containing the conidia that resist intracellular killing and germinate. Contact between neutrophils and hyphae triggers a respiratory burst, secretion of reactive oxygen intermediates, and degranulation. Killing of hyphae required oxidants, but oxidant release by PMNs could not mediate hyphal killing without concomitant fungal damage by granule constituents. The cell biology events involved in the phagocytosis of *A. fumigatus* conidia and hyphae and the down-regulation of the immune system by immunosuppressive drugs are poorly understood. Several strategies are available to produce single or multiple mutants of *A. fumigatus*, which is a haploid organism. The classical method involves the disruption of the gene of interest by the insertion of an antibiotic resistance gene. A *PYRG* blaster, very similar to the *URA* blaster developed for *Candida albicans*, has been developed for *A. fumigatus*. Screening randomly STM-REMI-generated mutants for loss of virulence has failed in *A. fumigatus*.

Efforts to sequence the genome of *A. fumigatus* are under way.

Candida Species

Candida is the most frequent cause of human mycosis. Several species of *Candida* cause human infections, and the major etiological agent of candidiasis is *C. albicans*. *C. albicans* is a commensal of the gastrointestinal and genitourinary tracts. It causes infections in immunocompromised patients or in patients with local predisposing factors. A wide range of infection types have been reported. The most common diseases are superficial oropharyngeal and vulvovaginal candidiasis. The most severe forms of the disease are hematogenously disseminated candidiasis, which occurs in immunocompromised patients. Natural resistance to infection is believed to rely on polymorphonuclear cells and on mononuclear phagocytes. The alternative complement pathway plays an important role in enhancing the effect of phagocytic cells via opsonization of the organism. PMNs can ingest *C. albicans*. Once in the phagosome, *Candida* yeasts are killed in the immunocompetent host. Typical morphological features of *Candida* include budding yeasts and pseudohyphae with branching and production of oval blastospores. With the exception of *C. glabrata*, all *Candida* species produce hyphae and pseudohyphae in tissue. Most of the efforts aimed at understanding the virulence of this pathogen at the molecular level have been centered on the identification of genes regulating the dimorphism of *C. albicans*. *C. albicans* is naturally diploid. A *URA* blaster method has been developed, allowing successive rounds of gene disruption.

The *C. albicans* genome sequence will soon be available.

Cryptococcus neoformans

Infections caused by the encapsulated basidiomycetous yeast *Cryptococcus neoformans* are initiated by inhalation of the yeast into the lungs and show a remarkable propensity to spread to the brain and meninges (causing cryptococcal meningitis). Immunosuppression, especially AIDS, is the leading predisposing factor for cryptococcosis. Known virulence factors are the polysaccharide capsule and the phenol oxidase responsible for melanin formation from phenolic substrates. Human PMNs and macrophages are able to ingest and kill yeasts in vitro. Antibody- and complement-mediated opsonization is essential during phagocytosis. Evidence that encapsulated and acapsular yeasts may differently trigger certain macrophage functions also provides an important model to study the cell biology of receptor signaling at the macrophage cell surface. Gene inactivation strategies have been developed. They vary for the different serotypes, and different DNA delivery systems are used. In all cases, the rate of homologous integration is very low (1 in 1,000 or less). Therefore, strategies for selecting these rare events have been developed.

Histoplasma capsulatum

Histoplasmosis is caused by *Histoplasma capsulatum,* which is found in temperate and tropical climates. *H. capsulatum* grows as a saprophyte and is acquired by inhalation of airborne conidia. Most infections are mild or subclinical, except in immunosuppressed individuals, especially AIDS patients, who are at high risk for disseminated histoplasmosis. *H. capsulatum* is typically dimorphic; the pathogenic phase is the yeast form, which is obtained at 37°C. Although biochemical and molecular differences between the yeast and mycelial forms have been observed, the mechanism of dimorphism, as for *Candida,* remains poorly understood. The primary host cell is the macrophage, in which *H. capsulatum* proliferates in a phagolysosome. The yeast form is able to modulate the pH of this intracellular compartment, avoiding damage by the degradative lysosomal enzyme, thus allowing intramacrophage survival and eventually leading to macrophage death. A calcium-binding protein and a cell wall α1-3-glucan contribute to intracellular survival. Homologous transformation is a rare event in *H. capsulatum,* and transforming DNA mostly integrates randomly.

Pneumocystis carinii

Pneumocystis carinii is a frequent cause of pneumonia in immunocompromised patients, especially AIDS patients. Historically, this extracellular pathogen was thought to be a protozoan, but recent molecular data suggested that it is a fungus. The reservoir and the mode of acquisition of the infection remain undetermined, although molecular analysis has shown the presence of *P. carinii* in the air. In infected lungs, *P. carinii* is intimately associated with alveolar cells, suggesting that it may use cell-based nutrients or stimuli to proliferate. This attachment involves the major surface glycoprotein and fibronectin. One of the striking features of *P. carinii* is its ability to generate surface variation via programmed gene rearrangements. The genes involved in this process encode various isoforms of the major surface glycoprotein. Data from many studies indicate that infection is host species specific. One of the major obstacles in *Pneumocystis* research is the lack of a method of sustained in vitro culture of this organism. The absence of axenic growth has limited the extent of molecular studies.

A sequencing project of the entire *P. carinii* genome is in progress.

Selected Readings

Alberts, B., D. Bray, J. Lewis, M. Raff, K. Roberts, and J. D. Watson. 1994. *Molecular Biology of the Cell,* 3rd ed. Garland, New York, N.Y.

Cossart, P., P. Boquet, S. Normark, and R. Rappuoli. 1996. Cellular microbiology emerging. *Science* **271:**315–316.

Cossart, P., and J. Miller (ed.). 1999. Host-microbe interactions: bacteria. *Curr. Opin. Microbiol.* **2:**1–106.

Finlay, B., and P. Cossart. 1997. Exploitation of mammalian host cell functions by bacterial pathogens. *Science* **276:**718–725.

Finlay, B., and S. Falkow. 1997. Common themes in microbial pathogenicity revisited. *Microbiol. Mol. Biol. Rev.* **61:**136–169.

Finlay, B., and P. Sansonetti (ed.). 1998. Host-microbe interactions: bacteria. *Curr. Opin. Microbiol.* **1:**1–129.

Henderson, B., M. F. Wilson, R. McNab, and A. J. Lax. 1999. *Cellular Microbiology: Bacteria-Host Interactions in Health and Disease.* John Wiley & Sons, Inc., New York, N.Y.

Mandell, G. L., J. E. Bennett, and R. Dolin. 2000. *Principles and Practice of Infectious Diseases,* 4th ed. Churchill Livingstone, New York, N.Y.

Miller, V. L., J. B. Kaper, D. A. Portnoy, and R. R. Isberg (ed.). 1994. *Molecular Genetics of Bacterial Pathogenesis.* ASM Press, Washington, D.C.

Mims, C. 2000. *Mims' Pathogenesis of Infectious Disease,* 4th ed. Academic Press, Inc., New York, N.Y.

Salyers, A. A., and D. D. Whitt. 1994. *Bacterial Pathogenesis: a Molecular Approach.* ASM Press, Washington, D.C.

Sussman, M. (ed). 2001. *Molecular Medical Microbiology.* Academic Press, Inc., New York, N.Y.

Wilson, M., R. McNab, and B. Henderson. 2002. *Bacterial Disease Mechanisms: an Introduction to Cellular Microbiology.* Cambridge University Press, Cambridge, United Kingdom.

Hacker, J., and J. Heesemann (ed). 2002. *Molecular Infection Biology: Interactions between Microorganisms and Cells.* John Wiley & Sons, Inc., New York, N.Y.

2

Bacterial Human Pathogen Genomes: an Overview

Stephen Bentley, Mohammed Sebaihia, Nicholas Thomson, Matthew Holden, Lisa Crossman, Kenneth Bell, Ana Cerdeño-Tarraga, and Julian Parkhill

By the end of 2002 the public DNA databases contained 87 complete bacterial genomes, the majority of which belong to human pathogens (Box 2.1). The aim of this chapter is to give an overview of what has been learned from analysis of bacterial human pathogen genomes. With bacterial genomes being completed and published at an ever-increasing rate, it would be impossible to make this chapter truly current, but we feel the "end of 2002" cutoff leaves us with a good representative group which can be feasibly covered in this format. Clearly this represents a large amount of information, so for the most part we have restricted discussion to the primary genome publications as referenced in Table 2.1. Much of the basic genome data are also summarized in Table 2.1.

The chapter ends with concluding remarks looking at the major findings from bacterial pathogen genomics to date, the impact on treatment and disease control, and finally a discussion of possible directions for future research.

Note: Throughout this chapter protein-coding sequence is referred to as CDS.

Alphaproteobacteria

Rickettsia prowazekii

As the causal agent of epidemic typhus, *Rickettsia prowazekii* was responsible for millions of deaths in the wake of the First and Second World Wars. The *R. prowazekii* genome sequence brought new insights into its pathogenicity and its likely relationship with mitochondria.

Its unusually small 1.1-Mb genome is the result of reductive evolution, a process which appears to be ongoing in *Rickettsias*. The *R. prowazekii* genome displays evidence of small- and large-scale deletion of DNA, as well as many mutations resulting in frameshifting or premature termination of erstwhile protein-coding genes. Consequently, 24% of the genome is noncoding. Such gene loss has also meant that *R. prowazekii* is unable to synthesize amino acids or nucleosides, both of which must be scavenged from the host. Loss of genes for energy metabolism has left *R. prowazekii*

35

BOX 2.1

Gathering pace of bacterial pathogen genome sequencing

1995
Haemophilus influenzae KW20
Mycoplasma genitalium G-37
1996
Mycoplasma pneumoniae M129
1997
Helicobacter pylori 26695
Borrelia burgdorferi B3
1998
Mycobacterium tuberculosis H37Rv
Treponema pallidum
Chlamydia trachomatis DUW
Rickettsia prowazekii Madrid E
1999
Helicobacter pylori J99
Chlamydophila pneumoniae CWL029

2000
Campylobacter jejuni
Neisseria meningitidis MC58
Chlamydia pneumoniae AR39
Chlamydia trachomatis MoPn
Neisseria meningitidis Z2491
Chlamydia pneumoniae J138
Vibrio cholerae
Pseudomonas aeruginosa PAO1
Ureaplasma urealyticum
2001
Escherichia coli O157:H7 EDL933
Mycobacterium leprae
Escherichia coli O157:H7 Sakai
Pasteurella multocida PM70
Streptococcus pyogenes SF370
Staphylococcus aureus N315 (MRSA)
Staphylococcus aureus Mu50 (VRSA)
Mycoplasma pulmonis UAB CTIP
Streptococcus pneumoniae TIGR4

Rickettsia conorii Malish 7
Mycobacterium tuberculosis CDC1551
Yersinia pestis CO-92
Streptococcus pneumoniae R6
Salmonella serovar Typhi CT18
Salmonella serovar Typhimurium LT2
Listeria monocytogenes EGD
2002
Clostridium perfringens
Streptococcus pyogenes MGAS8232
Fusobacterium nucleatum
Staphylococcus aureus MW2
Streptococcus pyogenes MGAS315
Yersinia pestis KIM5
Streptococcus agalactiae 2603V/R
Streptococcus agalactiae NEM316
Shigella flexneri 2a 301
Streptococcus mutans
Mycoplasma penetrans HF-2
Escherichia coli UPEC-CFT073

with a repertoire of genes for ATP production similar to that seen in mitochondria. Phylogenetic analysis of genes encoding ribosomal proteins and for the subunits of NADH dehydrogenase indicates that *R. prowazekii* is more closely related to some mitochondria than it is to other alphaproteobacteria.

In terms of typical virulence genes, *R. prowazekii* seems sparsely equipped. It encodes several VirB-like protein homologues which in *Agrobacterium tumefaciens* are involved in the transfer of T-DNA into the host cell, while in *Helicobacter pylori* they are involved in the export of a factor which induces interleukin secretion in gastric epithelial cells. The genome also contains homologues of capsule biosynthesis genes implicated in virulence in *Staphylococcus aureus*.

Rickettsia conorii

Rickettsia conorii, the causal agent of Mediterranean spotted fever, is also an obligate intracellular parasite and is thought to have diverged from *R. prowazekii* between 40 and 80 million years ago. The two genomes are almost entirely colinear with some small rearrangements near the replication terminus. This synteny facilitates the identification of pseudogenes. Several "split genes" were uncovered where an intact copy in one species was split into several open reading frames (ORFs) in the other. In addition, 229 pseudogenes were found in *R. prowazekii* that contained partial similarity to intact CDSs in *R. conorii*. Many of the split genes retain codon properties associated with functional genes, and many of those in *R. conorii* were shown to have transcripts associated with them, indicating recent decay.

Brucella melitensis

The 3.3-Mb genome of *Brucella melitensis* comprises two circular replicons of 2.1 and 1.2 Mb. The presence on both of typical α-proteobacterial replication

Table 2.1 Bacterial animal pathogen genomes completed and published by the end of 2002

Taxonomic group/strain	Size (bp)	Geometry[a]	% G+C	CDSs[b]	% coding[c]	Accession no.[d]	Primary publication
Alphaproteobacteria							
Rickettsia prowazekii Madrid E	1,111,523	C	30.9	834	75.4	AJ235269	Andersson *Nature* **396:**133–140 (1998)
Rickettsia conorii Malish 7	1,268,755	C	32.7	1,374	87.4	AE006914	Ogata *Science* **290:**347–350 (2000)
Brucella melitensis 16M	3,294,931	2 × C	58.3	3,198	86.8	AE008917-8	DelVecchio *PNAS* **99:**443–448 (2002)
Brucella suis 1330	3,315,173	2 × C	58.2	3,264	84.3	AE014291-2	Paulsen *PNAS* **99:**13148–13153 (2002)
Betaproteobacteria							
Neisseria meningitidis serogroup B MC58	2,272,351	C	53.1	2,025	77.7	AE002098	Tettelin *Science* **287:**1809–1815 (2000)
N. meningitidis serogroup A Z2491	2,184,406	C	53.3	2,121	80.7	AL157959	Parkhill *Nature* **404:**502–506 (2000)
Gammaproteobacteria							
Haemophilus influenzae Rd	1,830,138	C	38.8	1,709	85.6	L42023	Fleischmann *Science* **269:**496–512 (1995)
Xylella fastidiosa 9a5c	2,679,306	C	53.8	2,766	83.4	AE003849	Simpson *Nature* **406:**151–157 (2000)
Vibrio cholerae	4,033,464	2 × C	48.1	3,828	86.7	AE003852-3	Heidelberg *Nature* **406:**477–483 (2000)
Pseudomonas aeruginosa PAO1	6,264,403	C	67.1	5,566	89.3	AE004091	Stover *Nature* **406:**959–964 (2000)
Escherichia coli O157:H7 EDL933	5,228,970	C	51.5	5,349	87.5	AE005174	Perna *Nature* **409:**529–533 (2001)
E. coli O157:H7 RMID 0509952	5,498,450	C	51.6	5,361	88.0	BA000007	Hayashi *DNA Res.* **8:**11–22 (2001)
E. coli CFT073	5,231,428	C	51.4	5,379	90.6	AE014075	Welch *PNAS* **99:**17020–17024 (2002)
Pasteurella multocida PM70	2,257,487	C	41.0	2,014	89.0	AE004439	May *PNAS* **98:**3460–3465 (2001)
Yersinia pestis CO-92	4,653,728	C	48.9	4,008	80.2	AL590842	Parkhill *Nature* **413:**523–527 (2001)
Y. pestis KIM5	4,600,755	C	48.9	4,090	83.3	AE009952	Deng *J. Bacteriol.* **184:**4601–4611 (2002)
Salmonella serovar Typhi CT18	4,809,037	C	53.3	4,600	83.1	AL513382	Parkhill *Nature* **413:**848–852 (2001)
Salmonella serovar Typhimurium LT2	4,857,432	C	53.3	4,452	86.8	AE006468	McClelland *Nature* **413:**852–856 (2001)
Shigella flexneri 2a 301	4,607,203	C	51.9	4,434	81.0	AE005674	Jin *Nucleic Acids Res.* **30:**4432–4441 (2002)
Epsilonproteobacteria							
Helicobacter pylori 26695	1,667,867	C	39.6	1,566	89.6	AE000511	Tomb *Nature* **388:**539–547 (1997)
H. pylori J99	1,643,831	C	39.9	1,491	90.4	AE001439	Alm *Nature* **397:**176–180 (1999)
Campylobacter jejuni	1,641,481	C	39.9	1,654	90.4	AL111168	Parkhill *Nature* **403:**665–668 (2000)
Firmicutes							
Mycoplasma genitalium G37	580,074	C	31.7	480	90.7	L43967	Fraser *Science* **270:**397–403 (1995)
Mycoplasma pneumoniae M129	816,394	C	31.6	688	90.7	U00089	Himmelreich *Nucleic Acids Res.* **24:**4420–4449 (1996)
Mycoplasma penetrans HF 2	1,358,633	C	26.5	1,037	88.7	BA000026	Sasaki *Nucleic Acids Res.* **30:**5293–5300 (2002)

(*continued*)

Table 2.1 Bacterial animal pathogen genomes completed and published by the end of 2002 (continued)

Taxonomic group/strain	Size (bp)	Geometry[a]	% G + C	CDSs[b]	% coding[c]	Accession no.[d]	Primary publication
Mycoplasma pulmonis UAB CTIP	963,879	C	27.3	782	90.6	AL445566	Chambaud *Nucleic Acids Res.* **29:**2145–2153 (2001)
Ureaplasma urealyticum	751,719	C	25.7	611	91.1	AF222894	Glass *Nature* **407:**757–762 (2000)
Streptococcus pyogenes SF370	1,852,441	C	39.2	1,696	83.7	AE004092	Ferretti *PNAS* **98:**4658–4663 (2001)
S. pyogenes MGAS8232	1,895,017	C	39.2	1,845	85.2	AE009949	Smoot *PNAS* **99:**4668–4673 (2002)
S. pyogenes MGAS315	1,900,521	C	39.2	1,865	85.7	AE014074	Beres *PNAS* **99:**10078–10083 (2002)
Staphylococcus aureus N315 (MRSA)	2,814,816	C	33.6	2,593	83.5	BA000018	Kuroda *Lancet* **357:**1225–1240 (2001)
S. aureus Mu50 (VRSA)	2,878,040	C	32.9	2,714	83.8	BA000017	Kuroda *Lancet* **357:**1225–1240 (2001)
S. aureus MW2	2,820,462	C	33.5	2,632	83.6	BA000033	Baba *Lancet* **359:**1819–1827 (2002)
Streptococcus pneumoniae TIGR4	2,160,837	C	40.6	2,094	82.7	AE005672	Tettelin *Science* **293:**498–506 (2001)
S. pneumoniae R6	2,038,615	C	40.5	2,043	87.0	AE007317	Hoskins *J. Bacteriol.* **183:**5709–5717 (2001)
Listeria monocytogenes EGD	2,944,528	C	38.4	2,855	89.0	AL591824	Glaser *Science* **294:**849–852 (2001)
Clostridium perfringens	3,031,430	C	29.5	2,660	83.4	BA000016	Shimizu *PNAS* **99:**996–1001 (2002)
Streptococcus agalactiae 2603V/R	2,160,267	C	36.1	2,124	86.5	AE009948	Tettelin *PNAS* **99:**12391–12396 (2002)
S. agalactiae NEM316	2,211,485	C	36.1	2,134	87.7	AL732656	Glaser *Mol. Microbiol.* **45:**1499–1513 (2002)
Streptococcus mutans UA159	2,030,921	C	37.5	1,960	85.9	AE014133	Ajdic *PNAS* **99:**14434–14439 (2002)
Spirochetes							
Borrelia burgdorferi B31	910,724	L	28.8	851	93.6	AE000783	Fraser *Nature* **390:**580–586 (1997)
Treponema pallidum	1,138,011	C	52.6	1,031	92.7	AE000520	Fraser *Science* **281:**375–388 (1998)
Actinobacteria							
Mycobacterium tuberculosis H37Rv	4,411,529	C	65.9	3,919	90.6	AL123456	Cole *Nature* **393:**537–544 (1998)
M. tuberculosis CDC1551	4,403,836	C	65.8	4,187	90.6	AE000516	Fleischmann *J. Bacteriol.* **184:**5479–5490 (2002)
Mycobacterium leprae	3,268,203	C	60.1	2,720	49.6	AL450380	Cole *Nature* **409:**1007–1011 (2001)
Chlamydiae							
Chlamydia trachomatis DUW	1,042,519	C	41.7	897	90.0	AE001273	Stephens *Science* **282:**754–759 (1998)
Chlamydophila pneumoniae CWL029	1,230,230	C	41.3	1,052	88.4	AE001363	Kalman *Nat. Genet.* **21:**385–389 (1999)
Chlamydia pneumoniae AR39	1,229,853	C	41.3	1,110	89.0	AE002161	Read *Nucleic Acids Res.* **28:**1397–1406 (2000)
C. trachomatis MoPn	1,072,950	C	40.7	904	90.0	AE002160	Read *Nucleic Acids Res.* **28:**1397–1406 (2000)
C. pneumoniae J138	1,226,565	C	41.4	1,069	89.8	BA000008	Shirai *Nucleic Acids Res.* **28:**2311–2314 (2000)
Fusobacteria							
Fusobacterium nucleatum	2,174,500	C	27.4	2,077	89.1	AE009951	Kapatral *J. Bacteriol.* **184:**2005–2018 (2002)

[a] C, circular; L, linear.
[b] Number of CDSs according to annotation in the public database entry.
[c] Based on annotation.
[d] This number can be used to retrieve the genome sequence and annotation from the public databases (National Center for Biotechnology Information [NCBI], European Molecular Biology Laboratory [EMBL], and DDBJ).

origins, rRNA operons, and genes for housekeeping functions indicates that both are legitimate chromosomes. Comparison with related genomes suggests that the two chromosomes arose from a single ancestral chromosome by recombination between rRNA operons.

Typical virulence-related genes identified on the chromosome include those encoding adhesins, invasins, and hemolysins, as well as likely type III, IV, and V secretion system proteins. Notably, the type III secretion system appears to be derived from a now defunct flagella regulon. The major economic impact of brucellosis is due to abortion in pregnant livestock. Metabolic reconstruction using the genome reveals an erythriol catabolism pathway. This polyol is a key sugar produced by the pregnant host and may represent a specific environmental adaptation.

Brucella suis

The genome of *Brucella suis* 1 strain 1330 is highly similar to that of *B. melitensis* (98 to 100% identity at the nucleotide level) and largely syntenic. The differences between the two genomes result from either single-nucleotide polymorphisms (SNPs) (7307) or one of the following categories: (i) genes that shared <95% identity (encoding mostly hypothetical and surface-exposed proteins); (ii) species-specific genes; (iii) genes intact in one genome but carrying frameshift mutations in the other; and (iv) differently predicted hypothetical genes. Large regions of difference are due to integrated bacteriophages. Almost half of the bacteriophage-encoded hypothetical proteins are predicted to be surface exposed and hence likely to contribute to the differences in host specificity and pathogenicity. Comparison with the genomes of other α-proteobacteria suggests that *B. suis* chromosome 1 (Chr1) is related to the main circular chromosomes of *A. tumefaciens*, *Sinorhizobium meliloti*, and *Mesorhizobium loti* while chromosome 2 (Chr2) resembles *S. meliloti* megaplasmids and the *A. tumefaciens* linear chromosome.

Betaproteobacteria

Neisseria meningitidis

Neisseria meningitidis is one of the primary causative agents of bacterial meningitis and is usually found as a commensal, living harmlessly in the throat of a proportion of the human population. Broadly speaking, serogroup A strains are responsible for pandemic and epidemic meningitis in the developing world while serogroup B strains are associated with sporadic meningitis in the developed world.

In the genomes of *N. meningitidis* serogroup A strain Z2491 and serogroup B strain MC58, tens of genes are predicted to be phase variable due to replication-mediated changes in the length of short homo- or heteropolymeric repeat sequences. Many of these encode proteins expressed on the surface of the cell or are involved in the synthesis of surface structures, such as polysaccharides. Also, a novel form of variation has been postulated involving the C terminus of surface proteins in which alternate 3' sequences could be added to expressed genes by recombination between local repeats embedded in alternate C-terminal "cassettes." A third form of variation involves a family of dispersed short repeats termed NIMEs, which occur in discrete arrays associated with genes encoding surface proteins. It is suggested that these arrays might act to target recombination of DNA taken up from the environment to specific genes, thus enhancing the rate of change of those genes in a population.

Gammaproteobacteria

Haemophilus influenzae

Haemophilus influenzae is a common commensal of the human respiratory tract. On the basis of their capsular polysaccharides, *H. influenzae* isolates are classified into either six serotypes (a to f) or nontypeable strains. The nontypeable strains are associated with noninvasive respiratory tract infections such as otitis media, whereas the encapsulated strains, particularly those expressing the type B capsule, are responsible for more serious infections, including meningitis, pneumonia, and septicemia.

The 1.83-Mb chromosome of the nonpathogenic *H. influenzae* Rd strain was the first bacterial genome to be sequenced and encodes an estimated 1,743 CDSs. At this early stage in the genomic era, these CDSs were compared with relatively small protein databases. On the basis of sequence similarity, 1,007 CDSs were assigned functions, 347 were similar to proteins of undefined function, and 389 had no database match.

Three genes of the tricarboxylic acid (TCA) cycle are missing (citrate synthase, isocitrate dehydrogenase, and aconitase). This explains the requirement for high levels of glutamate in culture media. Eight fimbrial genes, previously identified in the pathogenic serotype B strains as being involved in adhesion, are deleted from the genome of the nonpathogenic Rd strain.

Aside from its obvious implications on the studies of various aspects of *H. influenzae* biology and pathogenicity, the publication of the *H. influenzae* genome had enormous implications for genomic research because it heralded the success of whole-genome shotgun sequencing.

Vibrio cholerae

A free-living environmental microorganism in interepidemic periods, *Vibrio cholerae* is one of the most important opportunistic human pathogens. It causes the severe and highly lethal diarrheal disease, cholera. The disease has been at epidemic level in southern Asia for the past 1,000 years and has caused seven pandemics since 1817. The complete sequence of the two replicons of *V. cholerae* El Tor N16961 brought new insights to the origins of the genome, survival of the bacterium in the environment, and the pathogenic process.

The larger of the two chromosomes (Chr1, 3.0 Mb) contains most of the genes essential for growth and pathogenicity while the smaller chromosome (Chr2, 1.1 Mb) contains the majority of the genes for DNA repair and damage response and a high proportion (59%) of CDSs with no significant homologues in the databases. Three lines of evidence lead to the proposal that Chr2 originated from a plasmid "captured" by an ancestral *Vibrio* strain. First, Chr2 contains an integron island, typically found on plasmids, which contains a gene capture system and host addiction genes. The gene capture system seems to have facilitated the acquisition of several CDSs for potential virulence and drug resistance proteins. Second, the Chr1-encoded ParA, essential for partition and distribution of replicons to daughter cells during cell division, is most related to other chromosomally encoded ParAs, while Chr2-encoded ParA is most related to those encoded by plasmids, bacteriophages, and megaplasmids. Third, all the rRNA and tRNA genes are on Chr1, with a few redundant copies on Chr2.

The complete genome has brought further clarification regarding *V. cholerae* virulence determinants. Some *V. cholerae* replicons are thought to contain multiple loci for cholera toxin, but the El Tor N16961 genome contains a single copy on Chr1. Additionally, the genome sequence revealed a gene cluster similar to *rtxABCD* as well as several hemolysins, proteases, and lipases. *V. cholerae* El Tor N16961 encodes three type IV pilus systems: TCP, MSHA, and *pilA*. The TCP cluster produces a pilus that acts as a receptor for entry of the temperate filamentous phage CTXΦ. This phage contains the cholera toxin locus. The TCP locus itself appears to lie within a now defunct prophage. The MSHA locus comprises 14 genes and produces a pilus which is involved in biofilm formation but is not essential for pathogenesis. This gene cluster does not show evidence of recent lateral acquisition. The *pilA* locus does appear to have been recently acquired and besides encoding a pilus, it includes the only copy of *pilD*. PilD is a prepilin peptidase required for the correct processing of all three pili. This is a curious situation because it implies that TCP and MSHA rely on a recently acquired *pilD*.

Pseudomonas aeruginosa

On publication, the *Pseudomonas aeruginosa* PAO1 genome represented the largest bacterial genome sequence, consisting of a single large circular chromosome of 6.3 Mb. *P. aeruginosa* can be isolated from a diverse range of ecological niches, including the rhizosphere, and is an important opportunistic human pathogen. It possesses intrinsic resistance to many antibiotics and disinfectants and is a major cause of mortality among cystic fibrosis patients. The genome reveals a large variety of regulatory proteins and membrane transport systems, endowing the organism with greater environmental adaptability and widening the spectrum of stimuli detected and substrates utilized from the immediate environment. Many of these transport systems actively transport drugs across the membrane and include members of the major facilitator superfamily (MFS) and the resistance-nodulation-cell division family (RND).

In addition to its ability to survive in diverse ecological niches, *P. aeruginosa* is also a potent pathogen that is adept at persisting in the host. The PAO1 genome contains examples of type I, type II, and type III secretion systems that are involved in the export of virulence factors, including toxins, lipases, and proteases. The arsenal of virulence factors is also enhanced by invasion and colonization factors, such as adhesins similar to the filamentous hemagglutinin of *Bordetella pertussis,* and the genes responsible for the production of extracellular polysaccharide (EPS), alginate. The survival of this organism is also augmented by its ability to form biofilms, both in the environment and in the host. The production of these complex structures is promoted by EPS, which coats the cells, providing protection from outside stresses including desiccation and antibiotics.

Escherichia coli

The first pathogenic *Escherichia coli* genome to be sequenced was that of the enterohemorrhagic (EHEC) strain O157:H7. First isolated in 1982, it is recognized as being the major cause of diarrhea, hemorrhagic colitis, and, sporadically, hemolytic uremic syndrome. The genomes of isolates from two separate O157:H7 outbreaks have been published, as well as one from the uropathogenic *E. coli* (UPEC) attributed with causing neonatal meningitis/sepsis and the majority of urinary tract infections. All three genomes are essentially colinear, displaying conservation in both sequence and gene order. Conserved genes display >95% sequence identity and have been termed the core genes or backbone sequence.

In addition to the backbone sequence, both EHEC and UPEC carry an additional ~1.3 to 1.4 Mb of unique sequence. When comparing the non-pathogenic *E. coli* K-12 with EHEC O157:H7 strains, unique regions are described as K-islands (or K-loops) and O-islands (or O-loops), referring to regions unique to *E. coli* K-12 or O157:H7, respectively. With this notation *E. coli* O157:H7 was found to possess 296 O-islands (>19 kb), compared with 325 K-loops identified in *E. coli* K-12. Approximately half of the genes carried on the O-islands are phage related (48.2%). Of the remainder, 33% cannot be ascribed a function and 15% are related to virulence. Among the larger O-loops four are virulence related, encoding fimbrial proteins unique to O157:H7. A further five fimbrial clusters were also identified which were partially conserved in *E. coli* K-12, leaving open the possibility that they could also facilitate binding to strain-specific target sites. Along with the fimbrae, there are 14 adhesins/invasins, including the previously characterized intimin, carried on the EHEC locus of enterocyte effacement (LEE) pathogenicity island (PAI), and the Iha adhesin. In addition to the type III secretion system known to be located on LEE, a novel type III secretion system was also discovered. Interestingly, this novel type III secretion system is more similar to the *Salmonella enterica* subspecies *enterica* serovar Typhimurium (*Salmonella* serovar Typhimurium) secretion system carried on another PAI, SPI-1 (see below). In addition to the previously characterized phage-encoded Shiga-like toxins and enterohemolysin, two other toxins were identified. One of these toxin genes was predicted to encode a large RTX-family protein (5,292 residues) which was carried on an S-loop, alongside genes that may facilitate its secretion and activation.

Similar notation is used to describe the UPEC-specific islands (UI). Genes encoded within UIs included 12 distinct fimbrial systems, such as the two *pap* operons, which are known to be uropathogen specific. Several other fimbrial systems are present which are also present in K-12 and EHEC. However, even these ubiquitous fimbrial systems display a high level of sequence variation, suggesting that they may interact with different target sites. UIs also carry seven autotransporters, a novel RTX-family toxin, and five *fimE* and *fimB* recombinase systems, all of which are associated with aspects of host interaction or, in the case of the recombinases, phase switching (rapid random phenotypic variation). In total the O-islands of EHEC account for 25.3% (1,393 Mb) of the total genome. The analogous figure for UIs is 24.9% (1.303 Mb). Some 1,827 genes present in UPEC are absent from *E. coli* K-12. The equivalent figure for EHEC is 1,387. K-12 has 585 genes not present in the other two. Interestingly, only a small percentage (11%) of the 1,827 genes unique to UPEC are also found in EHEC. Analysis of the nucleotide composition and codon usage of the O-loops and UIs reveals an atypical G+C content (47 to 48%; this figure excludes phage-related genes for EHEC), with respect to the backbone sequence (50.05%). In addition, there was also a preponderance of rare codons within coding sequences in these regions. These observations are consistent with the hypothesis that pathogenic *E. coli* genotypes have evolved from a much smaller non-pathogenic relative by the acquisition of foreign DNA. This laterally acquired DNA has been attributed with conferring on the different genotypes the ability to colonize alternative niches within the host and the ability to cause a range of different disease outcomes.

Pasteurella multocida

Pasteurella multocida is part of the normal upper respiratory tract florae of a number of animals. It is a frequent cause of opportunistic infections in domestic livestock and usually infects humans via the scratch or bite of a cat or dog. The strain chosen for genome sequencing, Pm70, is an avian isolate chosen as being representative of the major clonal population isolated from diseased birds.

The organism has a single circular chromosome of approximately 2.26 Mb and is most closely related to *H. influenzae.* The molecular basis for virulence in *P. multocida* is not well understood, but some candidate virulence factors have been identified. Proteins encoded by two large CDSs, *pfhB1* and *pfhB2,* share similarities with FhaB, a filamentous hemagglutinin from *B. pertussis.* FhaB is involved in adherence to host cells and is a major component of the acellular vaccines used to protect against whooping cough. Approximately 10% of the CDSs are unique to *P. multocida,* while 71 and 69% have homologues in *H. influenzae* and *E. coli,* respectively.

In excess of 2.5% of the genome encodes genes involved with iron acquisition. This is consistent with the central role of iron in pathogenicity. One 9-kb region shares a significant degree of similarity with the high-pathogenicity island of *Yersinia pestis* involved in the high-affinity uptake of iron. This region is not present in *E. coli* or *H. influenzae,* suggesting that it has been either deleted from these organisms or acquired by *P. multocida* via horizontal gene transfer.

Yersinia pestis

Yersinia pestis has been responsible for three pandemics, one of which (known as the Black Death) wiped out one-third of the population of Europe. This pathogen is thought to have evolved very recently (1,500 to 20,000 years ago) from the relatively benign enteropathogen *Y. pseudotuberculosis.* This speciation event has seen not only the development of a complex lifestyle involving an insect vector but also a shift from being a pathogen of the gut to being a blood-borne pathogen. The complete genome (chromosome and two plasmids) of *Y. pestis* pathovar *Orientalis,* the strain responsible for the most recent pandemic, exposes many of the characteristic features of a bacterium undergoing rapid evolutionary change, including the presence of many pseudogenes, insertion sequences (3.7% of the genome), and gross chromosomal rearrangements. Gene acquisition has also had a major impact on the development of this pathogen; *Y. pestis* has acquired two plasmids essential for pathogenesis and multiple chromosomal loci exhibiting many of the characteristics of foreign DNA. These regions include the *hms* locus required to block the insect gut, forcing it to regurgitate its contents on biting a new host and thereby disseminating this bacterium. The *Y. pestis* genome also carries multiple insect toxins and genes similar to insect viral enhancins proposed to be important for escaping the confines of the midgut and colonization of the insect host. Other novel features of the genome consistent with this genus of pathogens include an additional chromosomally encoded type III secretion system (important for the delivery of protein effectors into host cells) and an array of fimbrae and adhesins (potentially important for evasion of the immune system or interactions with multiple hosts).

Yersinia pestis biovar *Mediaevalis* strain KIM is a highly characterized representative of strains dating back to the Black Death. Comparison of KIM and CO92 shows that up to 95% of the sequence is almost identical. The genomes of CO92 and KIM are also very similar in size, with CO92 being only about 50 kb larger than KIM. The majority of the differences between these two strains can be attributed to the fact that CO92 possesses an expanded population of IS elements and lacks an rRNA operon. The most significant finding is centered on genomic rearrangements in the KIM sequence. *Y. pestis* has the capacity to tolerate large-scale rearrangements within its genome. Most of the genome rearrangements detectable in KIM have occurred following symmetrical recombination around the origin of replication. This type of recombination is thought to have occurred so frequently that almost 50% of the KIM backbone genes have switched sides relative to the origin at some point during the evolution of this strain. In addition, three further regions were identified that have undergone multiple inversions with respect to CO92. The rearrangements in this instance are not centered around the origin and are thought to have been driven by recombination between homologous IS elements.

CO92 and KIM lack an identifiable *dnaC* gene. In *E. coli*, DnaC is responsible for guiding the DnaB helicase onto the DNA replisome. Although dispensable in some systems, it is postulated that the low generation time of *Y. pestis*, 1.25 h even when grown under ideal growth conditions, might negate the need to initiate multiple replication forks and, therefore, the requirement for DnaC. Other intriguing findings from the genome include two large clusters of flagellar genes, which are retained by both KIM and CO92. This is not particularly noteworthy until you consider that *Y. pestis* has never been seen to be motile. Remarkably, a mutation in a single gene within one of the clusters, the regulator *flhD*, is thought to account for the lack of motility in *Y. pestis*.

Salmonella enterica Subspecies *enterica*

The complete genome sequences of *Salmonella enterica* serovar Typhi (*Salmonella* serovar Typhi) strain CT18, the causative agent of typhoid fever, and *Salmonella enterica* serovar Typhimurium LT2 (*Salmonella* serovar Typhimurium), the causative agent of human gastroenteritis, were reported in back-to-back publications. The former infects only humans, whereas the latter has a broader host range and is used as a mouse model for typhoid fever. As expected from their classification as serovars of the same species, the genomes of *Salmonella* serovar Typhi and *Salmonella* serovar Typhimurium share a high degree of similarity. However, there are enough differences between them to account for their peculiarities in lifestyle and pathogenicity. The most notable difference is the presence of 204 pseudogenes in *Salmonella* serovar Typhi compared with only 39 in *Salmonella* serovar Typhimurium. The high number of pseudogenes in *Salmonella* serovar Typhi could be attributed to its restriction to humans as a host and might also suggest that this organism has gone through a recent evolutionary bottleneck. Comparison of the two genomes with other salmonellae, as well as other related enterobacteria, allowed the identification of a wide range of genetic insertions and deletions. Many of the large insertions encode virulence determinants and are termed *Salmonella* pathogenicity islands (SPIs). In addition to the previously described SPIs (SPI 1–5), several potential new candidates were identified. Some SPIs are shared between all or a few *Salmonella* serovars, but a few are unique to particular serovars. In addition to SPIs, both genomes contain several unique smaller blocks of genes that carry determinants for toxins, bacteriocin immunity, iron acquisition, and fimbriae. Furthermore, a set of previously unknown genes was found only in subspecies of *Salmonella* that infect warm-blooded animals. Thus, these unique genes(s) and the SPIs could determine the more esoteric aspects of *Salmonella* infections. In addition to the chromosomes, the sequences of the plasmids harbored by *Salmonella* serovar Typhi (pHCM1 and pHCM2) and *Salmonella* serovar Typhimurium (pSLT) were also revealed. The larger plasmid of *Salmonella* serovar Typhi, pHCM1, is thought to have acquired a significant number of genes, many of which confer multiple drug or heavy metal resistance. Conversely, pHCM2 is essentially cryptic, with no obvious role in *Salmonella* pathogenesis. Interestingly, a comparison of pHCM2 with the *Y. pestis* major virulence plasmid pMT1 showed that they shared >56% of their DNA backbone; however, genes known to be important for *Y. pestis* virulence are missing from pHCM2, and its role remains equivocal.

Shigella flexneri

It has been suggested on many occasions that *Shigella* spp. can simply be thought of as *E. coli* with a plasmid. Indeed, many previous studies have focused on the plasmids alone, and for good reason, given that many of the primary virulence determinants are located here. The *S. flexneri* serotype 2a strain 301 chromosome is 4.6 Mb in size and shares 3.9 Mb of common backbone sequence with *E. coli* K-12. However, like other enteric bacteria it is apparent from the comparison with *E. coli* K-12 that the *S. flexneri* genome has also undergone a significant number of translocation, inversion, and insertion/deletion events. The regions unique to *S. flexneri* from this comparison are termed *Shigella*-specific islands (SIs). There are 64 SIs greater than 1 kb in size and 13 that are greater than 5 kb. Many of the SIs share the characteristics of pathogenicity islands, encoding potential virulence determinants, bacteriophage, and IS element-related genes. The most notable of these SIs carry genes related to the type III secreted effector protein IpaH. Members of this family of proteins have been shown to modulate host gene expression. Notably, the genome contains 247 intact and 64 partial IS elements, more than reported for any other enteric bacteria at the time. Nearly half of these IS elements are of the IS1 family. Moreover, all the larger regions that had undergone translocation or inversion events were bordered by IS elements, indicating that the major driving force behind apparent rearrangements within the *S. flexneri* genome is recombination between related IS elements. Another feature of the genome that is associated with genetic flux is the large number of pseudogenes. As *S. flexneri* has evolved to colonize its fairly unique niche, it seems that selection has led to the disruption of genes either that are of no consequence or that would reduce the ability to colonize its current niche. Based on the genome comparisons, it is suggested that *S. flexneri* probably diverged from *E. coli* K-12 after *E. coli* O157:H7, further supporting the argument for reclassification of the genus.

Epsilonproteobacteria

Helicobacter pylori

Up to 50% of the human population is a carrier of *H. pylori*. Strains J99 and 26695 were isolated from patients with duodenal ulcer and gastritis, respectively, although the 26695 strain has been subject to laboratory manipulation. Both have a circular chromosome of about 1.6 Mb, with the strain 26695 chromosome 24 kb larger than that of J99. Strain-specific genes include restriction-modification genes and cell envelope components. A major feature of both strains is that several components of the recombination and repair pathways are absent. Additionally, tracts of dinucleotide and homopolymeric repeats are found both within and preceding several CDSs. It is believed that *H. pylori* probably uses recombination and slipped-strand mismatching at repeat sequences as a mechanism for antigenic variation and adaptive evolution to evade the host immune response. Repeat lengths can vary between strains, suggesting that this may lead to differential gene expression. An apparent lack of regulatory networks may be due to the fact that the organism is restricted to the gut environment.

A key feature of *H. pylori* is its ability to survive in a low-pH environment using a switched polarity positive-inside membrane potential. The severity of disease is partly attributed to the "cag" pathogenicity island, which encodes a set of factors for the secretion of cytotoxins that act on the host gut epithelium. At least five adhesins are expected to participate in attachment to gastric epithelial cells, two of which are outer membrane proteins. Cell surface lipopolysaccharide may be involved in molecular mimicry since the O antigen is known to mimic human blood group antigens. This may provide an additional means of evading the host immune response.

Campylobacter jejuni

Campylobacter jejuni is a microaerophilic flagellated bacterium and is the leading global cause of gastroenteritis. In addition, prior infection with *C. jejuni* has been implicated as a factor in Guillain-Barré neuromuscular paralysis. The chromosome is about 1.6 Mb with a low G+C content of 30.6% and is predicted to encode 1,654 proteins. Although the genome contains relatively few repeats, short variable tracts of nucleotides are found associated with genes encoding surface structure components. Rapid variation of these sequences may be rendered possible by the lack of DNA repair genes since homologues of *E. coli* mismatch repair and the SOS response pathways are absent. Swiftly changing surface structure components could provide a means to allow the organism to evade a host immune response. The high levels of variation in the *C. jejuni* genome mean that a definitive genomic sequence is not possible. This has prompted the nomenclature of *C. jejuni* as a "quasi-species." Despite the high levels of genomic variation, there are almost no IS elements or bacteriophage sequences. A further distinct feature of the chromosome is that few operons or gene clusters are present. Since operons and gene clusters can be regulated at the transcriptional level with a single regulatory protein in bacteria, the *C. jejuni* genome would seem to be unusual in this respect. In terms of regulatory networks, this organism possesses several two-component regulatory proteins and a homologue of fumarate/nitrate reduction, known as the aerobic/anaerobic switch regulator in *E. coli*. This relatively complicated regulatory network may be required for persistence of *C. jejuni* in the environment.

Firmicutes

Mycoplasma spp.

Mycoplasmas are the smallest and simplest self-replicating microorganisms known. They are unique in lacking a cell wall and are thought to have evolved from low G+C gram-positive bacteria by a process of reductive evolution. They are host-restricted pathogens that show tight host specificity and are considered to be the best examples of microorganisms with a minimal gene set. Published genomes include *Mycoplasma genitalium, M. pneumoniae, Ureaplasma urealyticum, M. pulmonis,* and *M. penetrans.*

M. genitalium exists in parasitic association with ciliated epithelial cells of human genital and respiratory tracts. At 0.6 Mb (encoding 470 proteins), it is the smallest sequenced bacterial genome and is therefore a key model for exploring the contentious issue of "what constitutes a minimal cell?."

M. pneumoniae is a human pathogen which causes atypical pneumonia, usually in children and young adults. The genes shared between *M. genitalium* and *M. pneumoniae* are arranged in six contiguous segments. Each segment is flanked by highly similar repetitive sequences. Although the order of orthologous genes within these segments is well conserved, the order of the segments themselves is not. These gross genomic rearrangements are most likely caused by homologous recombination between the repetitive sequences.

U. urealyticum is a common commensal of the urogenital tract of humans. This microorganism is also recognized as an opportunistic pathogen during pregnancy that can cause premature birth or spontaneous abortion. It generates almost all of its ATP via urea hydrolysis, and a large fraction of *U. urealyticum*-specific genes are dedicated to this process. Compared with other mycoplasmas, the genome of *U. urealyticum* contains a relatively large number of iron transporters.

M. pulmonis is the causative agent of respiratory mycoplasmosis in rats and mice and is therefore a good model for studying mycoplasmal respiratory infections, especially those caused by its close relative, the human pathogen *M. pneumoniae*. Interestingly, several of the genes previously suggested to be essential for a minimal cell are missing in the *M. pulmonis* genome.

M. penetrans is a newly identified species within the *Mycoplasma* genus. It infects the human urogenital and respiratory tracts and is frequently found in patients suffering from HIV-1 infections. As its name indicates, *M. penetrans* can penetrate human cells and persist within them for long periods. At 1.4 Mb, the genome of *M. penetrans* is the largest of the five sequenced *Mycoplasma* genomes. Comparison with the other sequenced members of this genus showed that there was little or no gene order conservation.

The relatively large size of the *M. penetrans* genome is attributed to the presence of a number of paralogous gene families. The largest of these is the p35 family, which encodes surface-exposed immunogenic and phase-variable lipoproteins. They are widely used for serological diagnostics and have been implicated in evasion of the host immune response. The p35-family of lipoproteins might have a role in extending the antigenic variability of this pathogen, thereby allowing for a more persistent infection. Unlike the other mycoplasmas, *M. penetrans* has a two-component regulatory system.

Streptococcus pyogenes

Alternatively known as group A streptococci (GAS), *Streptococcus pyogenes* is a strict human pathogen that causes a wide range of diseases from pharyngitis to necrotizing fasciitis. The wide range of diseases is reflected in an array of virulence factors including (i) M protein, fibronectin-binding protein (protein F), and lipoteichoic acid for adherence; (ii) hyaluronic acid capsule as an immunological disguise and an inhibitor of phagocytosis; (iii) M protein to inhibit phagocytosis invasins, such as streptokinase, streptodornase (DNase B), hyaluronidase, and streptolysins; and (iv) exotoxins, such as pyrogenic (erythrogenic) toxin, which causes the rash of scarlet fever and systemic toxic shock syndrome.

GAS strains are classified according to serological differences in the surface-exposed M protein, and >130 M protein serotypes have been identified. The genomes of three GAS strains, M1, M18, and M3, were published before the end of 2002. In addition to their serological differences, these three strains differ in the severity of the infection they elicit. M1 causes pharyngitis and invasive infections, M18 causes acute rheumatic fever, and M3 causes a severe invasive infection with high rates of morbidity and mortality. Publication of the genome of the nonpathogenic close relative, *Lactococcus lactis* subspecies *lactis* (LAB), has also been valuable for comparative studies.

Like GAS, LAB lacks a complete TCA cycle, consistent with the notion that these bacteria are exclusively fermentative. Although natural competence has not been demonstrated for either LAB or GAS, multiple genes showing high similarity to competence-associated genes have been identified in both genetic backgrounds. It is estimated that bacteriophage and transposon genes account for ~10% of their total genomes, suggesting that this is also a significant source of laterally acquired DNA. Three of the four identified prophage in M1 carry virulence-associated genes; although it is notably those of the 40 putative virulence-associated genes identified in GAS, none lie in pathogenicity islands of the type observed in *S. aureus*.

Comparison shows the M18 genome to be largely colinear with that of M1 with a ~1.7-Mb backbone punctuated with discrete regions of difference. The M18 genome contains 178 unique CDSs, compared with 112 in the serotype M1 strain. There are 24 regions of difference between the two chromosomes that are larger than 2 kb. Of these, one-third is the result of the direct insertion or variation of phage sequences. Another class of mobile genetic elements that also varies between the two strains and could also affect gene regulation and therefore phenotype is the IS elements.

At 1.9 Mb, the serogroup M3 genome is slightly larger, though still largely colinear with M1 and M18. The major differences among the three genomes are found in the number, composition, and different integration sites of prophages.

In total, the three GAS strains carry 15 prophages: four in M1, five in M18, and six in M3. Phylogenetic analysis grouped 12 of the 15 prophages into five separate lineages. In addition to the previously known virulence factors, some of the prophages carried by M3 were found to encode additional, novel virulence factors. These include the streptococcal superantigen, Ssa; a homologue of a streptococcal pyrogenic exotoxin, SpeK; and a phospholipase A2 similar to a toxin found in snake venom, Sla. Interestingly, the genes encoding SpeK and Sla were missing from M3 isolates collected between 1920 and 1984, but were present in 50 of 53 isolates collected after 1987. It is suggested that the severity of the disease caused by the M3 strain could result from the presence of these prophage-encoded virulence factors, showing the importance of bacteriophages in the evolution and emergence of novel and more virulent bacterial isolates.

Staphylococcus aureus

S. aureus causes a variety of suppurative (pus-forming) infections and toxinoses in humans and is a major cause of hospital-acquired (nosocomial) infection of surgical wounds and infections associated with indwelling medical devices. It also causes food poisoning by releasing enterotoxins into food and toxic shock syndrome by release of pyrogenic exotoxins into the bloodstream. The first genome publication for *S. aureus* described two complete chromosomes, one from a methicillin-resistant (MRSA) strain (N315) and the other from an MRSA strain which also carries vancomycin resistance (Mu50). Both of these isolates were from hospital-acquired infection. Genome highlights include three new classes of pathogenicity islands for toxic-shock-syndrome toxin, exotoxin, and endotoxin, respectively, and several candidates for new virulence factors. Lateral acquisition of genes from many other different species and the extreme diversity of superantigens seem crucial to the virulence of these strains and correlate with the ability of *S. aureus* to adapt to environmental pressures, such as antibiotics and the human immune system.

The genome of the community-acquired MRSA strain (MW2) shows only seven major regions of differences from the two hospital-acquired strains. These take the form of unique islands or allelic forms of islands common to all three strains. Like many community-acquired MRSA strains, MW2 is resistant to β-lactam antibiotics but is susceptible to many other antibiotic classes. Accordingly, the only drug-resistance locus on the chromosome is a novel SCC mec element in the genomic island nearest the origin of replication. Of the other genomic islands identified, two are prophages and four are similar to previously identified pathogenicity islands or contain virulence determinants. Notably, two of the islands were shown to excise from the chromosomes spontaneously and form extrachromosomal elements with the capacity to replicate. This observation highlights a possible route for the transfer of these elements between strains. Another feature of community-acquired strains that distinguishes them from their hospital-acquired counterparts is their greater virulence. Reflecting this, MW2 contains 19 additional virulence determinants (including 18 toxins), all but two of which are located in four of the seven genomic islands. One of the genomic islands also contains a putative lantibiotic biosynthetic cluster that might contribute to the success of the MW2 strain by modulating the growth of competing microorganisms, thus promoting colonization.

Streptococcus pneumoniae

Streptococcus pneumoniae is the most common cause of acute respiratory tract infection and is thought to account for 3 million child deaths per annum. The genome encodes an array of exoenzymes not only important for liberating substrates for its large number of transport and catabolic pathways, but also for facilitating host invasion via tissue damage. The primary determinant for immune evasion is capsular polysaccharide (CPS). The CPS biosynthetic pathway has not been defined for the first sequenced strain, *S. pneumoniae* TIGR4, but a good candidate for the biosynthetic genes cluster is present in the genome. Although naturally competent, *S. pneumoniae* appears to resist the uptake of DNA from closely related bacteria by the production of polymorphic restriction enzymes. An invertible region was detected in the genome shotgun data which is thought to direct the production of restriction enzymes with altered specificities. The TIGR4 genome carries a large number of IS elements (representing 5% of the genome), but they appear to disrupt only three CDSs. This suggests that *S. pneumoniae* is under strong selective pressure to retain its current complement of genes.

The R6 strain was the second *S. pneumoniae* genome to be completed. It is a derivative of the type 2 capsule strain that was used by Avery and colleagues to identify DNA as the transforming principle and, consequently, the genetic material. Because R6 is an avirulent laboratory strain, comparative genomics should be useful in identifying some of the virulence determinants that make *S. pneumoniae* such a significant pathogen. The R6 genome is approximately 122 kb smaller than that of TIGR4 and encodes 193 fewer proteins. One of the most obvious features of the strain that accounts for the lack of pathogenicity is the absence of a CPS biosynthetic gene cluster. Although R6 has been propagated in the benign conditions of the laboratory and is considered nonpathogenic, the genome sequence contains known virulence determinants, such as hyaluronidase, immunoglobulin A1 protease, and pneumolysin. The natural genetic competence of *S. pneumoniae* has resulted in an eclectic genome. Horizontal gene transfer is proposed as the likely source of several genes in the R6 chromosome that have gram-negative, but not gram-positive, homologues. The likely source of these is thought to be gram-negative bacteria occupying the same niche in the human respiratory system.

Three percent of the *S. pneumoniae* chromosome is composed of three classes of repeat elements. Two of the repeat classes, BOX and RUP, are thought to form stable secondary structures, although their exact function has yet to be confirmed. The repeats are predominantly intergenic, indicating a possible role in transcriptional termination or transcript stability. About 5% of the *N. meningitidis* genome is composed of intergenic DNA repeats. Like *S. pneumoniae*, *N. meningitidis* is naturally competent, but its repeats have been postulated to be involved in antigenic variation through horizontal gene exchange.

Listeria monocytogenes

Although *Listeria monocytogenes* and *Listeria innocua* occupy the same ecological niches, the former is an intracellular pathogen that causes a serious food-borne disease (listeriosis) and the latter is nonpathogenic. Both genomes were sequenced in parallel, facilitating useful comparison. Both have chromosomes of about 3 Mb, but *L. innocua* also harbors an 81.9-kb plasmid that encodes heavy metal resistance. The chromosomes are colinear with detectable synteny extending to *Bacillus subtilis* and *S. aureus* chromosomes. The mosaic-like genome might be shaped in part by DNA uptake because it contains the components of a DNA-competence system. This is in clear contrast to the previously held view that *Listeria* are not naturally competent. The *Listeria* genomes contain a wide variety of components that modulate environmental and host interactions, reflecting the variable environmental conditions of its niche.

One category of proteins that are abundant in both genomes is surface and secreted proteins. For example, *L. monocytogenes* contains 41 LPXTG surface-anchored proteins. Absent from the *L. innocua* genome are internalins associated with cell invasion. Another group of abundant surface proteins is the carbohydrate-transporting, phosphoenolpyruvate-dependent phosphotransferase system (PTS) proteins. Significantly, *L. monocytogenes* contains eight more than *L. innocua*, highlighting the role of carbohydrate utilization in pathogenesis. As one would expect for organisms adapting to an array of environmental conditions, both *Listeria* genomes contain an abundance of transcriptional regulators.

Clostridium perfringens

The anaerobic, spore-forming, gram-positive *Clostridium perfringens* type A is the major cause of gas gangrene and is also known as a food-poisoning bacterium because of its heat tolerance and toxin production. The genome consists of a circular 3.0-Mb chromosome encoding 2,660 proteins and a single plasmid of 54 kb. Among the previously unidentified virulence factors are five hyaluronidases that break down hyaluronic acid and chondroitin sulfate and facilitate spreading in the flesh. Other notable components of its arsenal are five putative hemolysins, four *Bacillus cereus* exotoxin-like proteins, an α-clostripain, and a surface protein with similarity to *S. aureus* collagen-binding protein. The last of these is proposed to mediate bacteria-host interactions, thus promoting colonization in vivo. Metabolic reconstruction from the genome reveals that this strict anaerobe does not contain TCA cycle or respiratory chain proteins but does contain components of a fermentation pathway. Interestingly, the conversion of pyruvate to acetyl-CoA leads to the end products of hydrogen and carbon dioxide, both of which contribute to the anaerobic environment and promote growth and survival. Complementing its fermentative nature, there is also a plethora of enzymes involved in the metabolism and transport of sugars. By contrast, the capacity of the organism to synthesize amino acids is rather limited, highlighting the difficulty that this organism has in growing in an amino acid-limited environment.

A facet of the biology of clostridia that promotes survival is their ability to adapt to environmental conditions and to form spores. Previously, the genome sequence of *Clostridium acetobutylicum* revealed that the number and diversity of sporulation and germination components are less than those found in the archetypal *B. subtilis*. A similar complement is also found in *C. perfringens*, enhancing the notion that members of the genus *Clostridium* have a distinct sporulation mechanism.

Streptococcus agalactiae

Streptococcus agalactiae, a group B streptococcus, normally resides as a commensal within the gut and is the leading cause of septicemia, pneumonia, and meningitis in newborns, following vertical transmission from the mother. There are eight serotypes of *S. agalactiae* (Ia, Ib, Ia/c, and II to VI), five of which (Ia, Ib, II, III, and V) are responsible for the majority of human invasive disease. The genome sequences of strains from two serotypes, III (strain NEM316) and V (strain 2603 V/R), were determined by two separate groups. Serotype III principally causes neonatal infections, whereas serotype V is an emerging pathogen that causes invasive infections among nonpregnant adults. On the basis of gene similarities and conservation of gene order, it was found that *S. agalactiae* is more closely related to *S. pyogenes* than to *S. pneumoniae*. Of the genes unique to *S. agalactiae*, which include many virulence genes, the majority were located within regions that bear the characteristics of mobile genetic elements; some of these were possibly acquired laterally from other bacteria residing within the gut. Genome analysis also revealed large-scale gene duplication in *S. agalactiae*, which might confer functional diversity on this organism.

Streptococcus mutans

Streptococcus mutans is the principal agent of dental caries. Analysis of the genome of strain UA159 revealed that *S. mutans* is capable of metabolizing and transporting a remarkable range of carbohydrates. In addition, the genome carries all the genes necessary for the synthesis of all required amino acids. The end products of carbohydrate fermentation, such as lactic acid, contribute to the acidification of the local niche (oral cavity) and consequently lead to dental caries. To tolerate this acidification, *S. mutans* relies on a proton-pumping ATPase, the components of which can be found in the genome. Attachment to the tooth is facilitated by the array of encoded glucan-producing and glucan-binding proteins. Other identified potential virulence factors include several surface proteins (adhesins) and extracellular enzymes (proteases).

Spirochetes

Borrelia burgdorferi

Borrelia burgdorferi is a tick-borne parasite and the causal agent of Lyme's disease. It is thought to infect through the skin, producing varied symptoms including rashes, fever, neurological complications, and arthritis. The genome is unusual, comprising a 910-kb linear chromosome with centrally located replication origin and another 533 kb of DNA distributed across 17 linear and circular plasmids. In line with its lifestyle and small genome size, *B. burgdorferi* has a marked lack of genes for biosynthetic pathways and is likely to rely entirely on the host for supply of amino acids, fatty acids, enzyme cofactors, and nucleosides. There is also a sparsity of obvious virulence genes but a notable wealth of chemotaxis genes, some of which are duplicated across different plasmids. It has been suggested that they may be important in the pathogenic process with possible differential expression of duplicated genes in the various tissues. Other genes are also found on more than one plasmid. It is suggested that such sources of sequence identity between plasmids may be a basis for recombination that may lead to antigenic variation. Evidence for recombination comes from plasmid pseudogenes, which bring the coding capacity of the plasmids down to only 71% compared with 93% for the chromosome.

Treponema pallidum

The causal agent of venereal syphilis, *Treponema pallidum*, was the second spirochete to have its genome sequenced to completion. Like *B. burgdorferi*, *T. pallidum* is an obligate human parasite. Its recalcitrance to culture has made development of diagnostics and vaccines difficult. Unlike *B. burgdorferi*, the *T. pallidum* genome has a single replicon, a 1.1-Mb circular chromosome predicted to encode 1,041 proteins; 46% have orthologues in *B. burgdorferi*. More than 70 of these orthologues appear to be unique to spirochetes, but none occur in the *B. burgdorferi* plasmids. The fact that over half the *T. pallidum*-predicted CDSs do not have an orthologue in *B. burgdorferi* gives a feeling for how distantly related the two bacteria are and the fact that 80% of these are of unknown biological function indicates that there is still much to learn about *Treponema* biology.

Similarities in complement of genes for DNA repair, transport, and energy metabolism, along with a lack of genes for biosynthetic pathways, suggest that *B. burgdorferi* and *T. pallidum* may have undergone convergent reductive evolution. Although both genomes prioritize broad-spectrum transporters, they are often composed of different transport protein families.

The microaerophilic nature of *T. pallidum* correlates with a metabolic requirement for oxygen coupled with a lack of enzymes for protection against its destructive properties. The lack of proteins in the *T. pallidum* outer membrane causes problems for immune detection and vaccine development. It is of great interest, then, that the genome sequence revealed a large family of proteins (TprA–L) which may function as porins or adhesions.

Actinobacteria

Mycobacterium tuberculosis

Mycobacterium tuberculosis primarily infects the lungs and is transmitted via water droplets in the breath. It causes a chronic illness by growing slowly within the host with occasional periods of dormancy. The bacterial cells seem to combat the immune system with their thick, waxy cell envelope and their ability to persist within macrophages.

The 3,924 protein products predicted for the *M. tuberculosis* strain H37Rv genome include extensive sets of genes for aerobic and anaerobic respiration, allowing for survival in both the aerobic lung environment and microaerophilic intracellular locations. The genome encodes a remarkable 250 enzymes for lipid metabolism, five times more than in *E. coli,* correlating with the range of lipophilic compounds made by *M. tuberculosis* as well as the breakdown of host material. Many polyketide synthase and siderophore genes have also been identified in the genome and may have roles in pathogenesis.

Interest in vaccine targets has been focused on the identification of 90 lipoproteins and two novel protein families named PE and PPE after their conserved N-terminal motifs. PE and PPE proteins have conserved N-terminal regions of about 110 and 180 residues, respectively, which may be involved in targeting the proteins to the cell surface; these are often followed by a variable-length, repetitive C-terminal portion. It has been suggested that surface-expressed PE and PPE family proteins may interfere with immune processing of antigens and/or their repetitive nature may allow for antigenic variation by recombination between genes or by strand slippage during replication.

Only three single-gene virulence determinants were known prior to the elucidation of the chromosome sequence. One of them, *mce,* was found to be present at four different loci. Consistent with a role in host cell invasion, the *mce* gene is flanked by genes for integral membrane and secreted proteins at each locus. Also identified were a homologue of smpB (necessary for *Salmonella* serovar Typhimurium intracellular survival) and a range of lipases, esterases, and proteases.

M. tuberculosis strain H37Rv is a model strain which has been passaged many times in the laboratory. Strain CDC1551 is a recent clinical isolate. Comparative genomic analysis of H37Rv and CDC1551 focused on looking for large-scale polymorphisms (LSPs; inserted sequences, >10 bp) and single-nucleotide polymorphisms (SNPs). Within the H37Rv sequence, 37 LSPs are unique with respect to CDC1551 and, conversely, 49 within the CDC1551 genome. Of the LSPs found to be coding, some carried genes such as adenylate cyclases and paralogues of *moaA* and *moaB* (involved in molybdenum cofactor biosynthesis). However, almost half of the LSPs were found to encode PPE and PE family proteins. To determine the stability of these LSPs within the larger population, 17 LSP-specific probes were used to challenge 169 clinical isolates, which were grouped by restriction fragment length polymorphism analysis. Isolates falling within the same group were thought to share a common ancestor and possibly be linked by a recent transmission or outbreak event. All the clinical isolates tested lacked at least one of the 17 LSPs, showing a significantly higher level of variability within the wider population than previously anticipated. It was also noted that isolates within the same cluster shared the same LSP fingerprint, whereas comparison of isolates between clusters showed clear differences. This suggests that the loss of these LSPs had occurred independently and multiple times. In addition, as these polymorphisms were not observed to occur within the relatively short timescale of an outbreak event, they are likely to have been (and continue to be) occurring over a long timescale. Like the LSPs, SNPs were also indicators of a level of variation higher (by threefold) than previously reported.

Of the 1,075 SNPs detected within the combined 8.8 Mb of *M. tuberculosis* sequence, there was an unusual ratio of nonsynonymous versus synonymous substitutions. Possible contributing factors include reduced selective pressure on the laboratory strain, low recombination rates, and recent passage through an evolutionary bottleneck.

Mycobacterium leprae

The genome analysis of *Mycobacterium leprae*, the leprosy bacillus, provides a fascinating insight into reductive evolution. The *M. leprae* genome is predicted to contain only 1,604 active coding sequences in a genome of 3.3 Mb. Its analysis was greatly facilitated by comparison with the highly related, but larger, *M. tuberculosis* genome (4.4 Mb). Had *M. leprae* exhibited a comparable coding capacity to *M. tuberculosis* it would have been expected to encode ~3,000 genes, implying that, following its divergence from its closest relative, *M. leprae* could have lost >2,000 genes. In its current form, analysis of the *M. leprae* genome revealed the presence of 1,116 pseudogenes or gene remnants, representing 41% of the total gene set. Although the *M. leprae* genome sequence is unparalleled in the extent of observable genome decay, it is not the first example of its type. Andersson et al. showed that 25% of the 1.1-Mb genome of *R. prowazekii*, the causative agent of typhus, was noncoding. Like *M. leprae*, *R. prowazekii* is an obligate intracellular parasite; however, the effect of genome decay is different in the two organisms. It would be expected that such intracellular parasites would dispense with anabolic pathways, relying on the host for the majority of nutrients. In *R. prowazekii* this is indeed the case; it lacks, for example, all determinants for nucleoside biosynthesis and many for amino acid biosynthesis. *M. leprae*, by contrast, has followed another route: it retains an almost complete set of anabolic pathways, making very nearly all of its own nucleotides and amino acids. What it has lost are mainly catabolic pathways and genes involved in energy metabolism, indicating that it survives on a very restricted diet of nutrients and energy sources, effectively starving in a nutrient-rich environment.

Chlamydiae

Chlamydia and *Chlamydophila* spp.

The *Chlamydiaceae* is a phylogenetically distinct bacterial lineage, encompassing two genera and several important human and animal pathogens. However, their obligate intracellular nature and a lack of genetic tools have obscured their biology. These factors, coupled with their small genome size, have made them attractive targets for genome sequencing: by the end of 2000, the sequences of five different chlamydial genomes had been published.

Chlamydia trachomatis is an important sexually transmitted pathogen, which can also cause the ocular disease trachoma. *Chlamydophila pneumoniae* (formerly *Chlamydia pneumoniae*) causes acute respiratory tract infections and is also implicated in chronic conditions such as adult-onset asthma and arthrosclerosis.

The different chlamydial species share much in the biology of their infection and developmental processes. Their infectious developmental form, called the elementary body (EB), is regarded as being metabolically inactive. After invasion of the host cells, *Chlamydia* spp. grow within an intracellular vacuole (called an inclusion) and differentiate into a metabolically active form known as the reticulate body (RB). Following multiple rounds of cell division, RBs differentiate into EBs, which are released to infect more host cells.

Thus, key questions are: how do EBs avoid immunosurveillance and gain entry to host cells, what defines the differentiation of EBs into RBs and back again, and what is the basis for different tissue tropisms and disease symptoms?

The two strains of *C. trachomatis* and three of *C. pneumoniae* that had been sequenced by the end of 2002 share many general genome features. As would be expected for obligate intracellular pathogens, the genomes are streamlined. They range from 1.04 to 1.27 Mb in size (with 40 to 41% G+C content), and genes for steps in pathways such as de novo amino acid, nucleotide, and cofactor synthesis are often absent, suggesting a general reliance on host-derived intermediates. In addition, central metabolic pathways such as glycolysis and TCA pathways are not wholly complete, but alternative routes can be hypothesized from genes that are present.

Peptidoglycan was thought to be absent from the chlamydial cell envelope so the presence of all necessary genes for peptidoglycan synthesis in the genome is surprising and suggests that it may be produced in small amounts or at a particular stage in the developmental cycle. There is also a type III secretion system along with candidate effectors, which may aid entry into the cell.

There is no evidence of recent horizontal gene transfer from other bacteria in any of the chlamydial genomes, although there is some evidence of acquisition from the eukaryotic host. This is consistent with their lifestyle as obligate intracellular parasites with very limited contact with other bacteria. The genome sequences also reveal a large family of polymorphic outer membrane protein genes, previously identified only in the sheep pathogen *Chlamydophila abortus*. Retention of so many paralogues (some of which are pseudogenes) in such small genomes is striking.

The *C. trachomatis* genomes are very similar to one another, but there is greater divergence than between the *C. pneumoniae* genomes. The degree of divergence was perhaps less than expected since one is a human serovar D isolate (a sexual tract pathogen), whereas the other is a mouse pneumonitis strain (MoPn). The degree of similarity is encouraging for the use of the mouse strain as an infection model. Most differences are concentrated in a 50-Kb plasticity zone near the terminus. The most interesting differences are that the MoPn genome has lost three putative tryptophan synthetase genes with respect to serovar D, whereas a putative toxin gene in the MoPn genome has apparently decayed into a pseudogene in serovar D. It is thought that tryptophan synthesis aids intracellular survival since tryptophan levels are depleted due to the action of interferon-γ from the immune response. Serovar D may be more dependent on synthesis of its own tryptophan since its mode of transmission (sexual) requires it to be more persistent than the respiratory MoPn strain.

The *C. pneumoniae* genomes are bigger, having more than 200 genes not present in *C. trachomatis*, whereas *C. trachomatis* has about 70 unique genes. This may reflect *C. pneumoniae*'s greater invasiveness and broader tissue host range. Also the expansion of the *pmp* gene family may be relevant here: *C. pneumoniae* has 21 *pmp* genes as compared with 9 in *C. trachomatis*, accounting for 22% of the increased coding capacity of the former. *C. trachomatis*, does, however, have a number of HKD family genes which are suggested to have a role in interference with the host cells' normal processes. *C. pneumoniae* also lacks the tryptophan synthesis genes.

The percentage identity for interspecies orthologues is about 65%, supporting the reclassification of *C. pneumoniae* as *Chlamydophila*. Despite the divergence at sequence level there is a high degree of synteny, with most rearrangements being movement or inversion of blocks of genes most commonly around the terminus. There are no insertion sequences or other repetitive elements to facilitate more extensive rearrangements.

Fusobacteria

Fusobacterium nucleatum

Fusobacterium nucleatum is an opportunistic human pathogen that is found naturally as part of the complex oral microflora. It has an unusual extended cell morphology and high surface adherence properties that allow it to stick to various host surfaces such as tooth enamel. By binding specifically to *Fusobacterium*, periodontal disease-causing bacteria gain a foothold on the tooth surface and can go on to cause decay. Analysis of the genome sequence highlights contradictory indications as to whether this is a gram-positive or gram-negative bacterium. Cluster analysis of the predicted genes and 16S rRNA suggests that *F. nucleatum* is most closely related to gram-positive bacteria. However, *F. nucleatum* possesses the membrane structure and organization of a gram-negative bacterium and has been classified as such. Known to secrete few proteins, the genome sequence shows that there are no type II, III, or IV secretion systems, although there is a functional Sec-dependent export pathway, another typical gram-positive feature. The main energy source for *F. nucleatum* appears to come from the fermentation of glutamate. This ultimately leads to the production of butyrate, which contributes to halitosis, and might also contribute to pathogenicity by fueling other members of the oral microbiota. Although amino acids are the preferred substrate for growth, *F. nucleatum* lacks the genes to synthesize many of them directly and, as a probable consequence of this, many of the predicted genes (6%) are thought to be involved in acid import. Lack of a lysine biosynthetic pathway leaves *F. nucleatum* unable to make *meso*-diaminopimelate, an essential component of peptidoglycan. In its place, *F. nucleatum* substitutes a very unusual amino acid, *meso*-lanthionine, which has been put forward as a possible target for chemotherapeutic intervention. The genome sequence has been used to piece together the route of synthesis of *meso*-lanthionine and has revealed all but one of the enzymes involved.

Concluding Remarks

It can be seen from the summaries presented that bacterial pathogen genomes have provided researchers with an unprecedented wealth of fundamental information on the functional (and nonfunctional) content of pathogen genomes. In terms of the benefits to basic science, the payoff has been huge. Genome sequencing has fundamentally altered the rate and range of projects that can be undertaken at the bench.

Genome sequencing has allowed novel insights into fundamental questions of bacterial metabolism and evolution. Perhaps one of the most surprising findings has been that bacteria contain pseudogenes and are not always the dense, streamlined systems expected. Even more surprising is the fact that these nonfunctional genes were found not just in obligate parasites, but also in organisms capable of survival outside the host. Another common and unexpected theme is the enormous differences in gene content revealed by comparisons between apparently closely related organisms, such as different strains of *E. coli* and *S. enterica*.

As with any new technology, these benefits will take longer to feed through to applied or clinical research. However, the effect is beginning to be felt at this level. Genomic sequencing has already been successfully applied to vaccine discovery and molecular forensics. Stepping briefly outside the prokaryotic world, genomic sequence data have allowed the development of new antimalarial drugs. It can be confidently expected that over the next decade or so, pathogen genome sequences will have an impact on clinical research and practice at every level, from diagnostics to vaccines to drugs.

The future for bacterial genomics itself is more difficult to predict. In the early stages, it was expected that it would be cost-effective only to sequence one representative of each of the major human pathogens. However, sequencing technologies continue to improve and get cheaper, year-on-year, continually changing the cost-benefit equation. In many cases, complete sequences are now being generated for five or more strains of a single species, and many more obscure pathogens are being tackled. In addition, sequencing is expanding to commensal and probiotic organisms, in addition to many animal pathogens, plant pathogens, and environmental species. Comparative data from projects like these continue to feed back into our understanding of human pathogens. As sequencing continues to get cheaper and easier in the future, we can only speculate as to where the field will be in another 7 years.

Selected Readings

Ajdic, D., W. M. McShan, R. E. McLaughlin, G. Savic, J. Chang, M. B. Carson, C. Primeaux, R. Tian, S. Kenton, H. Jia, S. Lin, Y. Qian, S. Li, H. Zhu, F. Najar, H. Lai, J. White, B. A. Roe, and J. J. Ferretti. 2002. Genome sequence of *Streptococcus mutans* UA159, a cariogenic dental pathogen. *Proc. Natl. Acad. Sci. USA* **99:**14434–14439.

Alm, R. A., L. S. Ling, D. T. Moir, B. L. King, E. D. Brown, P. C. Doig, D. R. Smith, B. Noonan, B. C. Guild, B. L. deJonge, G. Carmel, P. J. Tummino, A. Caruso, M. Uria-Nickelsen, D. M. Mills, C. Ives, R. Gibson, D. Merberg, S. D. Mills, Q. Jiang, D. E. Taylor, G. F. Vovis, and T. J. Trust. 1999. Genomic-sequence comparison of two unrelated isolates of the human gastric pathogen *Helicobacter pylori. Nature* **397:** 176–180.

Andersson, S. G., A. Zomorodipour, J. O. Andersson, T. Sicheritz-Ponten, U. C. Alsmark, R. M. Podowski, A. K. Naslund, A. S. Eriksson, H. H. Winkler, and C. G. Kurland. 1998. The genome sequence of *Rickettsia prowazekii* and the origin of mitochondria. *Nature* **396:**133–140.

Baba, T., F. Takeuchi, M. Kuroda, H. Yuzawa, K. Aoki, A. Oguchi, Y. Nagai, N. Iwama, K. Asano, T. Naimi, H. Kuroda, L. Cui, K. Yamamoto, and K. Hiramatsu. 2002. Genome and virulence determinants of high virulence community-acquired MRSA. *Lancet* **359:**1819–1827.

Beres, S. B., G. L. Sylva, K. D. Barbian, B. Lei, J. S. Hoff, N. D. Mammarella, M. Y. Liu, J. C. Smoot, S. F. Porcella, L. D. Parkins, D. S. Campbell, T. M. Smith, J. K. McCormick, D. Y. Leung, P. M. Schlievert, and J. M. Musser. 2002. Genome sequence of a serotype M3 strain of group A *Streptococcus:* phage-encoded toxins, the high-virulence phenotype, and clone emergence. *Proc. Natl. Acad. Sci. USA* **99:** 10078–10083.

Chambaud, I., R. Heilig, S. Ferris, V. Barbe, D. Samson, F. Galisson, I. Moszer, K. Dybvig, H. Wroblewski, A. Viari, E. P. Rocha, and A. Blanchard. 2001. The complete genome sequence of the murine respiratory pathogen *Mycoplasma pulmonis. Nucleic Acids Res.* **29:**2145–2153.

Cole, S. T., R. Brosch, J. Parkhill, T. Garnier, C. Churcher, D. Harris, S. V. Gordon, K. Eiglmeier, S. Gas, C. E. Barry III, F. Tekaia, K. Badcock, D. Basham, D. Brown, T. Chillingworth, R. Connor, R. Davies, K. Devlin, T. Feltwell, S. Gentles, N. Hamlin, S. Holroyd, T. Hornsby, K. Jagels, A. Krogh, J. McLean, S. Moule, L. Murphy, K. Oliver, J. Osborne, M. A. Quail, M.-A. Rajandream, J. Rogers, S. Rutter, K. Seeger, J. Skelton, R. Squares, S. Squares, J. E. Sulston, K. Taylor, S. Whitehead, and B. G. Barrell. 1998. Deciphering the biology of *Mycobacterium tuberculosis* from the complete genome sequence. *Nature* **393**:537–544.

Cole, S. T., K. Eiglmeier, J. Parkhill, K. D. James, N. R. Thomson, P. R. Wheeler, N. Honore, T. Garnier, C. Churcher, D. Harris, K. Mungall, D. Basham, D. Brown, T. Chillingworth, R. Connor, R. M. Davies, K. Devlin, S. Duthoy, T. Feltwell, A. Fraser, N. Hamlin, S. Holroyd, T. Hornsby, K. Jagels, C. Lacroix, J. Maclean, S. Moule, L. Murphy, K. Oliver, M. A. Quail, M.-A. Rajandream, K. M. Rutherford, S. Rutter, K. Seeger, S. Simon, M. Simmonds, J. Skelton, R. Squares, S. Squares, K. Stevens, K. Taylor, S. Whitehead, J. R. Woodward, and B. G. Barrell. 2001. Massive gene decay in the leprosy bacillus. *Nature* **409**:1007–1011.

DelVecchio, V. G., V. Kapatral, R. J. Redkar, G. Patra, C. Mujer, T. Los, N. Ivanova, I. Anderson, A. Bhattacharyya, A. Lykidis, G. Reznik, L. Jablonski, N. Larsen, M. D'Souza, A. Bernal, M. Mazur, E. Goltsman, E. Selkov, P. H. Elzer, S. Hagius, D. O'Callaghan, J. J. Letesson, R. Haselkorn, N. Kyrpides, and R. Overbeek. 2002. The genome sequence of the facultative intracellular pathogen *Brucella melitensis*. *Proc. Natl. Acad. Sci. USA* **99**:443–448.

Deng, W., V. Burland, G. Plunkett III, A. Boutin, G. F. Mayhew, P. Liss, N. T. Perna, D. J. Rose, B. Mau, S. Zhou, D. C. Schwartz, J. D. Fetherston, L. E. Lindler, R. R. Brubaker, G. V. Plano, S. C. Straley, K. A. McDonough, M. L. Nilles, J. S. Matson, F. R. Blattner, and R. D. Perry. 2002. Genome sequence of *Yersinia pestis* KIM. *J. Bacteriol.* **184**:4601–4611.

Ferretti, J. J., W. M. McShan, D. Ajdic, D. J. Savic, G. Savic, K. Lyon, C. Primeaux, S. Sezate, A. N. Suvorov, S. Kenton, H. S. Lai, S. P. Lin, Y. Qian, H. G. Jia, F. Z. Najar, Q. Ren, H. Zhu, L. Song, J. White, X. Yuan, S. W. Clifton, B. A. Roe, and R. McLaughlin. 2001. Complete genome sequence of an M1 strain of *Streptococcus pyogenes*. *Proc. Natl. Acad. Sci. USA* **98**:4658–4663.

Fleischmann, R. D., M. D. Adams, O. White, R. A. Clayton, E. F. Kirkness, A. R. Kerlavage, C. J. Bult, J. F. Tomb, B. A. Dougherty, J. M. Merrick, et al. 1995. Whole-genome random sequencing and assembly of *Haemophilus influenzae* Rd. *Science* **269**:496–512.

Fleischmann, R. D., D. Alland, J. A. Eisen, L. Carpenter, O. White, J. Peterson, R. DeBoy, R. Dodson, M. Gwinn, D. Haft, E. Hickey, J. F. Kolonay, W. C. Nelson, L. A. Umayam, M. Ermolaeva, S. L. Salzberg, A. Delcher, T. Utterback, J. Weidman, H. Khouri, J. Gill, A. Mikula, W. Bishai, W. R. Jacobs Jr., J. C. Venter, and C. M. Fraser. 2002. Whole-genome comparison of *Mycobacterium tuberculosis* clinical and laboratory strains. *J. Bacteriol.* **184**:5479–5490.

Fraser, C. M., S. Casjens, W. M. Huang, G. G. Sutton, R. Clayton, R. Lathigra, O. White, K. A. Ketchum, R. Dodson, E. K. Hickey, M. Gwinn, B. Dougherty, J.-F. Tomb, R. D. Fleischmann, D. Richardson, J. Peterson, A. R. Kerlavage, J. Quackenbush, S. Salzberg, M. Hanson, R. van Vugt, N. Palmer, M. D. Adams, J. Gocayne, J. Weidman, T. Utterback, L. Watthey, L. McDonald, P. Artiach, C. Bowman, S. Garland, C. Fujii, M. D. Cotton, K. Horst, K. Roberts, B. Hatch, H. O. Smith, and J. C. Venter. 1997. Genomic sequence of a Lyme disease spirochaete, *Borrelia burgdorferi*. *Nature* **390**:580–586.

Fraser, C. M., J. D. Gocayne, O. White, M. D. Adams, R. A. Clayton, R. D. Fleischmann, C. J. Bult, A. R. Kerlavage, G. Sutton, J. M. Kelley, et al. 1995. The minimal gene complement of *Mycoplasma genitalium*. *Science* **270**:397–403.

Fraser, C. M., S. J. Norris, G. M. Weinstock, O. White, G. G. Sutton, R. Dodson, M. Gwinn, E. K. Hickey, R. Clayton, K. A. Ketchum, E. Sodergren, J. M. Hardham, M. P. McLeod, S. Salzberg, J. Peterson, H. Khalak, D. Richardson, J. K. Howell, M. Chidambaram, T. Utterback, L. McDonald, P. Artiach, C. Bowman, M. D. Cotton, C. Fujii, S. Garland, B. Hatch, K. Horst, K. Roberts, M. Sandusky, J. Weidman, H.

O. Smith, and J. C. Venter. 1998. Complete genome sequence of *Treponema pallidum*, the syphilis spirochete. *Science* **281**:375–388.

Glaser, P., L. Frangeul, C. Buchrieser, C. Rusniok, A. Amend, F. Baquero, P. Berche, H. Bloecker, P. Brandt, T. Chakraborty, A. Charbit, F. Chetouani, E. Couve, A. de Daruvar, P. Dehoux, E. Domann, G. Dominguez-Bernal, E. Duchaud, L. Durant, O. Dussurget, K. D. Entian, H. Fsihi, F. Garcia-del Portillo, P. Garrido, L. Gautier, W. Goebel, N. Gomez-Lopez, T. Hain, J. Hauf, D. Jackson, L. M. Jones, U. Kaerst, J. Kreft, M. Kuhn, F. Kunst, G. Kurapkat, E. Madueno, A. Maitournam, J. M. Vicente, E. Ng, H. Nedjari, G. Nordsiek, S. Novella, B. de Pablos, J. C. Perez-Diaz, R. Purcell, B. Remmel, M. Rose, T. Schlueter, N. Simoes, A. Tierrez, J. A. Vazquez-Boland, H. Voss, J. Wehland, and P. Cossart. 2001. Comparative genomics of *Listeria* species. *Science* **294**:849–852.

Glaser, P., C. Rusniok, C. Buchrieser, F. Chevalier, L. Frangeul, T. Msadek, M. Zouine, E. Couve, L. Lalioui, C. Poyart, P. Trieu-Cuot, and F. Kunst. 2002. Genome sequence of *Streptococcus agalactiae*, a pathogen causing invasive neonatal disease. *Mol. Microbiol.* **45**:1499–1513.

Glass, J. I., E. J. Lefkowitz, J. S. Glass, C. R. Heiner, E. Y. Chen, and G. H. Cassell. 2000. The complete sequence of the mucosal pathogen *Ureaplasma urealyticum*. *Nature* **407**:757–762.

Hayashi, T., K. Makino, M. Ohnishi, K. Kurokawa, K. Ishii, K. Yokoyama, C. G. Han, E. Ohtsubo, K. Nakayama, T. Murata, M. Tanaka, T. Tobe, T. Iida, H. Takami, T. Honda, C. Sasakawa, N. Ogasawara, T. Yasunaga, S. Kuhara, T. Shiba, M. Hattori, and H. Shinagawa. 2001. Complete genome sequence of enterohemorrhagic *Escherichia coli* O157:H7 and genomic comparison with a laboratory strain K-12. *DNA Res.* **8**:11–22.

Heidelberg, J. F., J. A. Eisen, W. C. Nelson, R. A. Clayton, M. L. Gwinn, R. J. Dodson, D. H. Haft, E. K. Hickey, J. D. Peterson, L. Umayam, S. R. Gill, K. E. Nelson, T. D. Read, T. Tettelin, D. Richardson, M. D. Ermolaeva, J. Vamathevan, S. Bass, H. Qin, I. Dragoi, P. Sellers, L. McDonald, T. Utterback, R. D. Fleishmann, W. C. Nierman, and O. White. 2000. DNA sequence of both chromosomes of the cholera pathogen *Vibrio cholerae*. *Nature* **406**:477–483.

Himmelreich, R., H. Hilbert, H. Plagens, E. Pirkl, B. C. Li, and R. Herrmann. 1996. Complete sequence analysis of the genome of the bacterium *Mycoplasma pneumoniae*. *Nucleic Acids Res.* **24**:4420–4449.

Hoskins, J., W. E. Alborn Jr., J. Arnold, L. C. Blaszczak, S. Burgett, B. S. DeHoff, S. T. Estrem, L. Fritz, D. J. Fu, W. Fuller, C. Geringer, R. Gilmour, J. S. Glass, H. Khoja, A. R. Kraft, R. E. Lagace, D. J. LeBlanc, L. N. Lee, E. J. Lefkowitz, J. Lu, P. Matsushima, S. M. McAhren, M. McHenney, K. McLeaster, C. W. Mundy, T. I. Nicas, F. H. Norris, M. O'Gara, R. B. Peery, G. T. Robertson, P. Rockey, P. M. Sun, M. E. Winkler, Y. Yang, M. Young-Bellido, G. Zhao, C. A. Zook, R. H. Baltz, S. R. Jaskunas, P. R. Rosteck Jr., P. L. Skatrud, and J. I. Glass. 2001. Genome of the bacterium *Streptococcus pneumoniae* strain R6. *J. Bacteriol.* **183**:5709–5717.

Jin, Q., Z. Yuan, J. Xu, Y. Wang, Y. Shen, W. Lu, J. Wang, H. Liu, J. Yang, F. Yang, X. Zhang, J. Zhang, G. Yang, H. Wu, D. Qu, J. Dong, L. Sun, Y. Xue, A. Zhao, Y. Gao, J. Zhu, B. Kan, K. Ding, S. Chen, H. Cheng, Z. Yao, B. He, R. Chen, D. Ma, B. Qiang, Y. Wen, Y. Hou, and J. Yu. 2002. Genome sequence of *Shigella flexneri* 2a: insights into pathogenicity through comparison with genomes of *Escherichia coli* K12 and O157. *Nucleic Acids Res.* **30**:4432–4441.

Kalman, S., W. Mitchell, R. Marathe, C. Lammel, J. Fan, R. W. Hyman, L. Olinger, J. Grimwood, R. W. Davis, and R. S. Stephens. 1999. Comparative genomes of *Chlamydia pneumoniae* and *C. trachomatis*. *Nat. Genet.* **21**:385–389.

Kapatral, V., I. Anderson, N. Ivanova, G. Reznik, T. Los, A. Lykidis, A. Bhattacharyya, A. Bartman, W. Gardner, G. Grechkin, L. Zhu, O. Vasieva, L. Chu, Y. Kogan, O. Chaga, E. Goltsman, A. Bernal, N. Larsen, M. D'Souza, T. Walunas, G. Pusch, R. Haselkorn, M. Fonstein, N. Kyrpides, and R. Overbeek. 2002. Genome sequence and analysis of the oral bacterium *Fusobacterium nucleatum* strain ATCC 25586. *J. Bacteriol.* **184**:2005–2018.

Kuroda, M., T. Ohta, I. Uchiyama, T. Baba, H. Yuzawa, I. Kobayashi, L. Cui, A. Oguchi, K. Aoki, Y. Nagai, J. Lian, T. Ito, M. Kanamori, H. Matsumaru, A. Maruyama, H. Murakami, A. Hosoyama, Y. Mizutani-Ui, N. K. Takahashi, T. Sawano, R. Inoue, C. Kaito, K. Sekimizu, H. Hirakawa, S. Kuhara, S. Goto, J. Yabuzaki, M. Kanehisa, A. Yamashita, K. Oshima, K. Furuya, C. Yoshino, T. Shiba, M. Hattori, N. Ogasawara, H. Hayashi, and K. Hiramatsu. 2001. Whole genome sequencing of meticillin-resistant *Staphylococcus aureus*. *Lancet* **357**:1225–1240.

May, B. J., Q. Zhang, L. L. Li, M. L. Paustian, T. S. Whittam, and V. Kapur. 2001. Complete genomic sequence of *Pasteurella multocida*, Pm70. *Proc. Natl. Acad. Sci. USA* **98**:3460–3465.

McClelland, M., K. E. Sanderson, J. Spieth, S. W. Clifton, P. Latreille, L. Courtney, S. Porwollik, J. Ali, M. Dante, F. Du, S. Hou, D. Layman, S. Leonard, C. Nguyen, K. Scott, A. Holmes, N. Grewal, E. Mulvaney, E. Ryan, H. Sun, L. Florea, W. Miller, T. Stoneking, M. Nhan, R. Waterston, and R. K. Wilson. 2001. Complete genome sequence of *Salmonella enterica* serovar Typhimurium LT2. *Nature* **413**:852–856.

Ogata, H., S. Audic, V. Barbe, F. Artiguenave, P. E. Fournier, D. Raoult, and J. M. Claverie. 2000. Selfish DNA in protein-coding genes of *Rickettsia*. *Science* **290**:347–350.

Parkhill, J., M. Achtman, K. D. James, S. D. Bentley, C. Churcher, S. R. Klee, G. Morelli, D. Basham, D. Brown, T. Chillingworth, R. M. Davies, P. Davis, K. Devlin, T. Feltwell, N. Hamlin, S. Holroyd, K. Jagels, S. Leather, S. Moule, K. Mungall, M. A. Quail, M.-A. Rajandream, K. M. Rutherford, M. Simmonds, J. Skelton, S. Whitehead, B. G. Spratt, and B. G. Barrell. 2000. Complete DNA sequence of a serogroup A strain of *Neisseria meningitidis* Z2491. *Nature* **404**:502–506.

Parkhill, J., G. Dougan, K. D. James, N. R. Thomson, D. Pickard, J. Wain, C. Churcher, K. L. Mungall, S. D. Bentley, M. T. Holden, M. Sebaihia, S. Baker, D. Basham, K. Brooks, T. Chillingworth, P. Connerton, A. Cronin, P. Davis, R. M. Davies, L. Dowd, N. White, J. Farrar, T. Feltwell, N. Hamlin, A. Haque, T. T. Hien, S. Holroyd, K. Jagels, A. Krogh, T. S. Larsen, S. Leather, S. Moule, P. O'Gaora, C. Parry, M. Quail, K. Rutherford, M. Simmonds, J. Skelton, K. Stevens, S. Whitehead, and B. G. Barrell. 2001. Complete genome sequence of a multiple drug resistant *Salmonella enterica* serovar Typhi CT18. *Nature* **413**:848–852.

Parkhill, J., B. W. Wren, K. Mungall, J. M. Ketley, C. Churcher, D. Basham, T. Chillingworth, R. M. Davies, T. Feltwell, S. Holroyd, K. Jagels, A. V. Karlyshev, S. Moule, M. J. Pallen, C. W. Penn, M. A. Quail, M.-A. Rajandream, K. M. Rutherford, A. H. van Vliet, S. Whitehead, and B. G. Barrell. 2000. The genome sequence of the food-borne pathogen *Campylobacter jejuni* reveals hypervariable sequences. *Nature* **403**:665–668.

Parkhill, J., B. W. Wren, N. R. Thomson, R. W. Titball, M. T. Holden, M. B. Prentice, M. Sebaihia, J. D. James, C. Churcher, K. L. Mungall, S. Baker, D. Basham, S. D. Bentley, K. Brooks, A. M. Cerdeno-Tarraga, T. Chillingworth, A. Cronin, R. M. Davies, P. Davis, G. Dougan, T. Feltwell, N. Hamlin, S. Holroyd, K. Jagels, A. V. Karlyshev, S. Leather, S. Moule, P. C. Oyston, M. Quail, K. Rutherford, M. Simmonds, J. Skelton, K. Stevens, S. Whitehead, and B. G. Barrell. 2001. Genome sequence of *Yersinia pestis*, the causative agent of plague. *Nature* **413**:523–527.

Paulsen, I. T., R. Seshadri, K. E. Nelson, J. A. Eisen, J. F. Heidelberg, T. D. Read, R. J. Dodson, L. Umayam, L. M. Brinkac, M. J. Beanan, S. C. Daugherty, R. T. DeBoy, A. S. Durkin, J. F. Kolonay, R. Madupu, W. C. Nelson, B. Ayodeji, M. Kraul, J. Shetty, J. Malek, S. E. Van Aken, S. Riedmuller, H. Tettelin, S. R. Gill, O. White, S. L. Salzberg, D. L. Hoover, L. E. Lindler, S. M. Halling, S. M. Boyle, and C. M. Fraser. 2002. The *Brucella suis* genome reveals fundamental similarities between animal and plant pathogens and symbionts. *Proc. Natl. Acad. Sci. USA* **99**:13148–13153.

Perna, N. T., G. Plunkett III, V. Burland, B. Mau, J. D. Glasner, D. J. Rose, G. F. Mayhew, P. S. Evans, J. Gregor, H. A. Kirkpatrick, G. Posfai, J. Hackett, S. Klink, A. Boutin, Y. Shao, L. Miller, E. J. Grotbeck, N. W. Davis, A. Lim, E. T. Dimalanta, K. D. Potamousis, J. Apodaca, T. S. Anantharaman, J. Lin, G. Yen, D. D. Schwartz, R. A Welch, and F. R. Blattner. 2001. Genome sequence of enterohaemorrhagic *Escherichia coli* O157:H7. *Nature* **409**:529–533.

Read, T. D., R. C. Brunham, C. Shen, S. R. Gill, J. F. Heidelberg, O. White, E. K. Hickey, J. Peterson, T. Utterback, K. Berry, S. Bass, K. Linher, J. Weidman, H. Khouri, B. Craven, C. Bowman, R. Dodson, M. Gwinn, W. Nelson, R. DeBoy, J. Kolonay, G. McClarty, S. L. Salzberg, J. Eisen, and C. M. Fraser. 2000. Genome sequences of *Chlamydia trachomatis* MoPn and *Chlamydia pneumoniae* AR39. *Nucleic Acids Res.* **28**:1397–1406.

Sasaki, Y., J. Ishikawa, A. Yamashita, K. Oshima, T. Kenri, K. Furuya, C. Yoshino, A. Horino, T. Shiba, T. Sasaki, and M. Hattori. 2002. The complete genomic sequence of *Mycoplasma penetrans*, an intracellular bacterial pathogen in humans. *Nucleic Acids Res.* **30**:5293–5300.

Shimizu, T., K. Ohtani, H. Hirakawa, K. Ohshima, A. Yamashita, T. Shiba, N. Ogasawara, M. Hattori, S. Kuhara, and H. Hayashi. 2002. Complete genome sequence of *Clostridium perfringens*, an anaerobic flesh-eater. *Proc. Natl. Acad. Sci. USA* **99**:996–1001.

Shirai, M., H. Hirakawa, M. Kimoto, M. Tabuchi, F. Kishi, K. Ouchi, T. Shiba, K. Ishii, M. Hattori, S. Kuhara, and T. Nakazawa. 2000. Comparison of whole genome sequences of *Chlamydia pneumoniae* J138 from Japan and CWL029 from USA. *Nucleic Acids Res.* **28**:2311–2314.

Simpson, A. J., F. C. Reinach, P. Arruda, F. A. Abreu, M. Acencio, R. Alvarenga, L. M. Alves, J. E. Araya, G. S. Baia, C. S. Baptista, M. H. Barros, E. D. Bonaccorsi, S. Bordin, J. M. Bove, M. R. Briones, M. R. Bueno, A. A. Camargo, L. E. Camargo, D. M. Carraro, H. Carrer, N. B. Colauto, C. Colombo, F. F. Costa, M. C. Costa, C. M. Costa-Neto, L. L. Coutinho, M. Cristofani, E. Dias-Neto, C. Docena, H. El-Dorry, A. P. Facincani, A. J. Ferreira, V. C. Ferreira, J. A. Ferro, J. S. Fraga, S. C. Franca, M. C. Franco, M. Frohme, L. R. Furlan, M. Garnier, G. H. Goldman, M. H. Goldman, S. L. Gomes, A. Gruber, P. L. Ho, J. D. Hoheisel, M. L. Junqueira, E. L. Kemper, J. P. Kitajima, J. E. Krieger, E. E. Kuramae, F. Laigret, M. R. Lambais, L. C. Leite, E. G. Lemos, M. V. Lemos, S. A. Lopes, C. R. Lopes, J. A. Machado, M. A. Machado, A. M. Madeira, H. M. Madeira, C. L. Marino, M. V. Marques, E. A. Martins, E. M. Martins, A. Y. Matsukuma, C. F. Menck, E. C. Miracca, C. Y. Miyaki, C. B. Monteriro-Vitorello, D. H. Moon, M. A. Nagai, A. L. Nascimento, L. E. Netto, A. Nhani Jr., F. G. Nobrega, L. R. Nunes, M. A. Oliveira, M. C. de Oliveira, R. C. de Oliveira, D. A. Palmieri, A. Paris, B. R. Peixoto, G. A. Pereira, H. A. Pereira Jr., J. B. Pesquero, R. B. Quaggio, P. G. Roberto, V. Rodrigues, A. J. de M Rosa, V. E. de Rosa Jr., R. G. de Sa, R. V. Santelli, H. E. Sawasaki, A. C. da Silva, A. M. da Silva, F. R. da Silva, W. A. da Silva Jr., J. F. da Silveira, M. L. Silvestri, W. J. Siqueira, A. A. de Souza, A. P. de Souza, M. F. Terenzi, D. Truffi, S. M. Tsai, M. H. Tsuhako, H. Vallada, M. A. Van Sluys, S. Verjovski-Almeida, A. L. Vettore, M. A. Zago, M. Zatz, J. Meidanis, and J. C. Setubal. 2000. The genome sequence of the plant pathogen *Xylella fastidiosa*. The *Xylella fastidiosa* Consortium of the Organization for Nucleotide Sequencing and Analysis. *Nature* **40**:151–157.

Smoot, J. C., K. D. Barbian, J. J. Van Gompel, L. M. Smoot, M. S. Chaussee, G. L. Sylva, D. E. Sturdevant, S. M. Ricklefs, S. F. Porcella, L. D. Parkins, S. B. Beres, D. S. Campbell, T. M. Smith, Q. Zhang, V. Kapur, J. A. Daly, L. G. Veasy, and J. M. Musser. 2002. Genome sequence and comparative microarray analysis of serotype M18 group A *Streptococcus* strains associated with acute rheumatic fever outbreaks. *Proc. Natl. Acad. Sci. USA* **99**:4668–4673.

Stephens, R. S., S. Kalman, C. Lammel, J. Fan, R. Marathe, L. Aravind, W. Mitchell, L Olinger, R. L. Tatusov, Q. Zhao, E. V. Koonin, and R. W. Davis. 1998. Genome sequence of an obligate intracellular pathogen of humans: *Chlamydia trachomatis*. *Science* **282**:754–759.

Stover, C. K., X. Q. Pham, A. L. Erwin, S. D. Mizoguchi, P. Warrener, M. J. Hickey, F. S. Brinkman, W. O. Hufnagle, D. J. Kowalik, M. Lagrou, R. L. Garber, L. Goltry, E. Tolentino, S. Westbrock-Wadman, Y. Yuan, L. L. Brody, S. N. Coulter, K. R. Folger, A. Kas, K. Larbig, R. Lim, K. Smith, D. Spencer, G. K. Wong, Z. Wu, I. T. Paulsen, J. Reizer, M. H. Saier, R. E. Hancock, S. Lory, and M. V. Olson. 2000. Complete genome sequence of *Pseudomonas aeruginosa* PA01, an opportunistic pathogen. *Nature* **406**:959–964.

Tettelin, H., V. Masignani, M. J. Cieslewicz, J. A. Eisen, S. Peterson, M. R. Wessels, I. T. Paulsen, K. E. Nelson, I. Margarit, T. D. Read, L. C. Madoff, A. M. Wolf,

M. J. Beanan, L. M. Brinkac, S. C. Daugherty, R. T. DeBoy, A. S. Durkin, J. F. Kolonay, R. Madupu, M. R. Lewis, D. Radune, N. B. Fedorova, D. Scanlan, H. Khouri, S. Mulligan, H. A. Carty, R. T. Cline, S. E. Van Aken, J. Gill, M. Scarselli, M. Mora, E. T. Iacobini, C. Brettoni, G. Galli, M. Mariani, F. Vegni, D. Maione, D. Rinaudo, R. Rappuoli, J. L. Telford, D. L. Kasper, G. Grandi, and C. M. Fraser. 2002. Complete genome sequence and comparative genomic analysis of an emerging human pathogen, serotype V *Streptococcus agalactiae. Proc. Natl. Acad. Sci. USA* **99:**12391–12396.

Tettelin, H., K. E. Nelson, I. T. Paulsen, J. A. Eisen, T. D. Read, S. Peterson, J. Heidelberg, R. T. DeBoy, D. H. Haft, R. J. Dodson, A. S. Durkin, M. Gwinn, J. F. Kolonay, W. C. Nelson, J. D. Peterson, L. A. Umayam, O. White, S. L. Salzberg, M. R. Lewis, D. Radune, E. Holtzapple, H. Khouri, A. M. Wolf, T. R. Utterback, C. L. Hansen, L. A. McDonald, T. V. Feldblyum, S. Angiuoli, T. Dickinson, E. K. Hickey, I. E. Holt, B. J. Loftus, F. Yang, H. O. Smith, J. C. Venter, B. A. Dougherty, D. A. Morrison, S. K. Hollingshead, and C. M. Fraser. 2001. Complete genome sequence of a virulent isolate of *Streptococcus pneumoniae. Science* **293:**498–506.

Tomb, J.-F., O. White, A. R. Kerlavage, R. A. Clayton, G. G. Sutton, R. D. Fleischmann, K. A. Ketchum, H. P. Klenk, S. Gill, B. A. Dougherty, K. Nelson, J. Quackenbush, L. Zhou, E. F. Kirkness, S. Peterson, B. Loftus, D. Richardson, R. Dodson, H. G. Khalak, A. Glodek, K. McKenney, L. M. Fitzegerald, N. Lee, M. D. Adams, E. K. Hickey, D. E. Berg, J. D. Gocayne, T. R. Utterback, J. D. Peterson, J. M. Kelley, M. D. Cotton, J. M. Weidman, C. Fujii, C. Bowman, L. Watthey, E. Wallin, W. S. Hayes, M. Borodovsky, P. D. Karp, H. O. Smith, C. M. Fraser, and J. C. Venter. 1997. The complete genome sequence of the gastric pathogen *Helicobacter pylori. Nature* **388:**539–547.

Welch, R. A., V. Burland, G. Plunkett III, P. Redford, P. Roesch, D. Rasko, E. L. Buckles, S. R. Liou, A. Boutin, J. Hackett, D. Stroud, G. F. Mayhew, D. J. Rose, S. Zhou, D. C. Schwartz, N. T. Perna, H. L. Mobley, M. S. Donnenberg, and F. R. Blattner. 2002. Extensive mosaic structure revealed by the complete genome sequence of uropathogenic *Escherichia coli. Proc. Natl. Acad. Sci. USA* **99:**17020–17024.

3

Cell Biology: an Overview

Dana Philpott and Patrice Boquet

Mechanisms controlling basic cellular functions, such as cell division, motility, adherence, differentiation, cell death, and the detection of potential cell dangers, are extremely highly conserved throughout the animal kingdom. Many of these mechanisms are parasitized by microbes when they colonize or invade host cells. This chapter aims to introduce briefly some of these basic cell biology mechanisms.

Microbes first encounter the outside of the cell. Many of the proteins on the outside of the cell are used for communication or for cell adhesion. Extracellular messengers bind to cell surface receptors; this induces modifications in these membrane receptor proteins, including dimerization or oligomerization and autophosphorylation. The signals are transduced by nucleotide binding or reversible phosphorylation, and signaling networks are assembled by the interaction of specific modular protein domains. Cascades of kinases are the effector molecules of cell signaling systems in the cytosol and the nucleus.

Once inside the cell, microbes are often caught up in the membrane trafficking system, which is used primarily to move proteins throughout the cell. Intracellular vesicles are formed by budding from a donor membrane, and targeting and fusion to an acceptor compartment are mediated by the assembly and disassembly of protein complexes. Special motor proteins move these vesicles, their cargo, mRNAs, and regulatory proteins associated into large molecular complexes. Movement is along intracellular tracks made by the cytoskeleton. The cytoskeleton is made of building blocks, such as actin or tubulin, which polymerize into supramolecular structures (either dynamic or stable) that are also involved in motility, maintenance of cell shape, and resistance to external stresses. The cytoskeleton is also important in the adhesion of cells to their substratum or to neighboring cells, since this involves activation and clustering of transmembrane proteins and their reversible attachment to the cytoskeleton. Loss of this attachment is one of the many reasons why cells self-destruct. Cell suicide is carried out primarily by proteases; other proteases are activated when cells need to terminate the life span of regulatory proteins or are undergoing starvation conditions. Any decision on cell suicide is largely dependent on the status of the cell cycle clock, which is driven by the periodic activation of cyclin-dependent kinases, while associated checkpoint mechanisms preserve genome integrity.

General Cell Organization

▶ *For Figure 3.1, see color insert.*

Cell organization is outlined in Figure 3.1. Cells are enclosed by a membrane (shown under a magnifying glass in the figure) (1) made up of a lipid bilayer into which channels for the transport of anions, cations, and water are embedded. Certain saturated glycolipids of the membrane, together with cholesterol, may be organized into discrete, highly ordered lipid microdomains with different properties named "lipid rafts." The cytoplasm contains a nucleus (2) surrounded by a nuclear envelope which is punched by holes named "nucleopores" that allow signaling molecules, containing nuclear localization signals, to get into the nucleus and to control gene transcription (3). Signals from outside the cell bind to membrane receptors (4), triggering phosphorylation of kinase cascades (5), which phosphorylate cytoplasmic proteins or, upon entry into the nucleus (6), phosphorylate nuclear factors regulating DNA transcription (7) and cell division factors (see Signal Transduction and Cell Regulation and Cell Cycle and Organelle Inheritance, below). Microtubules (8) are used as tracks for vesicles carrying cargo or during mitosis to pull apart chromosomes (9) (see The Cytoskeleton, below) or mitochondria (10), which provide energy to the cell or become a central executioner for cell death. Cells are attached to each other (11) or to their substratum (12) by adhesion molecules that are linked to the action cytoskeleton and which can engage, via their cytoplasmic tails, signaling cascades (13) (see Cell Adhesion and Morphogenesis, below). The nuclear envelope is continuous with the endoplasmic reticulum (ER) (14), into which secreted proteins are translocated during their synthesis (see Organelle Biogenesis, below) and which also serves as a tank of calcium ions, which can be released in the cytosol for signaling purposes (15). Secreted proteins are glycosylated and transported via the Golgi apparatus to the cell membrane into vesicles (16) (see Membrane Traffic, below). Cells take up molecules from the external medium by different endocytic pathways (17). These molecules are either degraded in lysosomes (L in Figure 3.1), recycled back to the plasma membrane, or routed to other compartments such as the Golgi apparatus. Cytosolic or membrane-associated proteins can be modified by the small molecule ubiquitin (18) and proteolyzed by the proteasome (P in Figure 3.1). Ubiquitinylation of proteins may also serve as a tag to deliver molecules to specific cell compartments or to induce the binding of signaling complexes (19). Actin filaments, which can be branched or not (20), form wide membrane extensions called lamellipodia (21) or long protrusions called filopodia, both of which are involved in cell locomotion (see The Cytoskeleton, below). All of these cellular processes can be studied by a variety of techniques, including the application of the activator and inhibitor chemicals, some of which are listed in Table 3.1. Many of these processes can also be visualized in real time by video microscopy of cells transfected with DNA encoding fusion proteins with the autofluorescent protein such as the green fluorescent protein (GFP). DNA encoding proteins fused to autofluorescent proteins are available through contacts with different laboratories and from biological companies such as Clonetech *(http://www.bdbiosciences.com/clontech/archive/JUL03UPD/Living ColorHcRed.shtml).*

Table 3.1 Main inhibitors and activators used in cell biology

Cellular target	Inhibitor or activator
Targets of inhibitors	
Tyrosine kinases	Genistein/tyrphostin/herbimycin
Ser/Thr kinases	Staurosporine
Protein kinase A	8-Bromo-cyclic AMP
Protein kinase C	pan PKC:GFX
Myosin kinase	Butanedione monoxime
PI3-K	LY294002
EGF-R (ERBB1-4)	PD158780
ERBB2	TAK165
Phosphatases (general)	Okadaic acid/vanadate
Calcineurin	Cyclosporin/FK506
Phospholipase C	Neomycin sulfate
Cyclo-oxygenase	Indomethacin
ATP synthesis	Apyrase
Sodium/potassium membrane ATPases	Ouabain
Chloride channels	NPPB
Vesicular ATPases	Bafilomycin A1/concanamycin A
Sodium/proton antiporter	Amiloride
Calcium ions	BAPTA
Calmodulin	Trifluoperazine
ER calcium entry pumps	Thapsigargin
IP_3 calcium release	D-myo-Inositol-1,4,5-P_3
Farnesylation and geranylation	Lovastatin
Sterol and fatty acid biosynthesis	Cerulenin
Actin polymerization	Cytochalasin D/ latrunculin A or B
Tubulin polymerization	Nocodazole
MAPK (MEK)	PD 98059/$Bacillus\ anthracis$ lethal factor
p38 Kinase	SB203580
Myosin light-chain kinase	ML-7/ butanedione monoxime
NF-κB (proteasome inhibition)	Lactacystin; MG-132
Cysteine proteases	E-64
pan-Caspase	Z-VAD
Serine/threonine proteases	Dichlorocoumarin
Protein synthesis inhibitor	Cycloheximide
Zinc-dependent proteases	Phosphoramidon
Glycoprotein processing	Castanospermine
Lysosomal alpha mannosidases	Swainsonine
Heterotrimeric G proteins G_o/G_i	Pertussis toxin
Rho A, B, C	$Clostridium\ botulinum$ exoenzyme C3
Rho + Rac + Cdc42	$C.\ difficile$ toxin B
Ras+Ral+ Rap	$C.\ sordellii$ lethal toxin
GTPase farnesylation/gernylation	Lovastatin/sinvastin
v-SNAREs/cellubrevin	Tetanus toxin/botulinum toxin B
t-SNAREs	Botulinum toxin C
SNAP25	Botulinum toxin A
Caveolae endocytic pathway	Filipin/ cyclodextrin

(continued)

Table 3.1 Main inhibitors and activators used in cell biology *(continued)*

Cellular target	Inhibitor or activator
Targets of inhibitors *(continued)*	
Coated pit-coated vesicle endocytic pathway	Chlorpromazine
COP1 binding to ARF-GTP	Ilimaquinone
ARF1 GEF (Golgi transport)	Brefeldin A
Cell cycle inhibitor S phase	Hydroxyurea/afidicholine
Mitotic arrest	Nocodazole
Blocking cholesterol in late endosomal membranes	U18666A
Blocking the activation of N-WASP	Wiskostatin
Macropinocytosis	Amiloride
Targets of activators	
Actin polymerization	Jasplakinolide
Actin filament stabilization	Phalloidin
Protein kinase C	Phorbol esters
Protein kinase A	Dibutyryl cyclic AMP
Tyrosine kinases	D-*erythro*-Sphingosine
Adenylate kinase	Forskolin
Phospholipase A_2	Melittin
GTP-binding protein (general)	GTPγS
Heterotrimeric G proteins (general)	Mastoparan
Heterotrimeric αG_s proteins	Cholera toxin
Rho + Rac + Cdc42	*Escherichia coli* CNF1 toxin
MT stabilization	Taxol/ epothilone
Calcium cytosolic entry	Ionomycin/A23187

Structure of the Cell Membrane

Cell membranes are made of a double leaflet of lipids (the lipid bilayers). All eukaryotic membranes contain glycerophospholipids (phosphatidylserine, -ethanolamine, and -inositol), cholesterol, and sphingolipids. Cholesterol and sphingolipids are in minor amounts in internal membranes and in high amounts in the plasma membrane (40 mol% of cholesterol in the plasma membrane, 15 mol% of sphingolipids). Most of the sphingolipids are in the outer leaflet of the plasma membrane. An asymmetry of lipids is constantly achieved by active transport in living cells between the two leaflets, but cholesterol seems to have the same concentration in the outer or inner leaflet. Cell membranes are fluid structures at physiological temperatures due to the *cis* double bond present in glycerophospholipids, which prevent the close packing of the lipidic acyl chains. The presence of the rigid molecule cholesterol in the membrane increases the close packing of phospholipid acyl chains and, therefore, the thickness and the impermeability to small ions of a membrane without changing notably its fluidity. Also the sphingolipids, which have no *cis* double bond, form ordered lipid structures. The possible relative abundance of cholesterol or/and sphingolipids in certain domains of the plasma membrane may increase locally the ordered structure of the membrane: a process that has been successfully named "lipid rafts." Major residents of lipid rafts are proteins attached to the cell surface by covalent linkage to glycosylphosphatidylinositol (GPI) lipid anchors. It has been pushed forward that GTP-binding proteins and nonreceptor tyrosine, kinases which are anchored to the inner membrane leaflet due to their covalent modification with multiple acyl chains, may selectively partition in lipid rafts. This may allow lipid rafts to be "signaling platforms" coupling events from outside the cell to signaling pathways inside the cell. The classical technique to examine if a protein is included in lipid rafts is based on the observation that upon solubilization of membrane by the nonionic detergent Trition X-100 at 4°C, only lipid-raft-associated proteins remain bound to lipids and therefore "float" at the surface of a density gradient. Cholesterol-sequestering agents, such as nystatin or beta-methylcyclodextrin, a drug that removes cholesterol from membranes, are used to disrupt lipid rafts in vivo. The actin cytoskeleton associated with the cytoplasmic tail of transmembrane proteins may limit the surface of lipid rafts, a process named "pickets and fence."

Signal Transduction and Cell Regulation

Detection and Initial Transduction

Molecular messengers that alter cell behavior must be detected, and their message must be transmitted and translated. Messengers are detected by membrane receptors, and a primary means of transmission is by phosphorylation or dephosphorylation of target proteins by specific kinases or phosphatases. Phosphorylation or dephosphorylation allows activation or deactivation of proteins involved in the regulation of either structural molecules or DNA transcription. Coordination of recognition motifs between different pathways is achieved to create a signaling network. Deactivation occurs by GTP hydrolysis, dephosphorylation, or endocytosis followed by degradation of membrane receptors.

Membrane receptors can be divided into three main classes according to their response to ligand binding: the ligand-gated ion channels open a selective pore (these channels are, however, usually restricted to the nervous system, and so they are not discussed here further); the seven transmembrane receptors are linked to heterotrimeric GTP-binding proteins, and various receptors are linked to enzyme cascades. Many of the receptors in the last group have enzymes as part of their cytoplasmic extension, such as the large group of tyrosine receptors (TKRs). The TKRs dimerize upon ligand binding, and they link to other kinases and small GTP-binding proteins of the $p21^{ras}$ superfamily. Other receptors have no intrinsic enzymatic activity but can recruit cytosolic kinases. The next sections first address the activation of the receptors and the proteins directly associated with them and then discuss the methods by which these signals are amplified and transmitted throughout the cell. Some of the same transmission mechanisms are shared by many different receptors, even those of different types.

The members of the family of seven transmembrane (or serpentine) receptors, such as the receptor for the growth factor thrombin, are linked by their cytoplasmic tails to heterotrimeric GTP-binding proteins (also designated as large GTPases due to their molecular mass of 45 kDa). A large GTPase is a complex of three proteins—α, β, and γ—in which α is the GTP-binding protein. When the α subunit is linked to GDP, the $\alpha\beta\gamma$ complex is formed and is associated with the cytoplasmic tail of the serpentine receptor. A serpentine receptor activated by its ligand will provoke, through a transmembrane modification in its shape, the removal of GDP bound on α. The α subunit, now free of GDP, binds GTP, since there is a large excess of this nucleotide in the cell. Binding of GTP to the α subunit separates α from βγ. Activated α subunit (free to diffuse in the cytosol) and the βγ complex (associated with the membrane) then leave the serpentine receptor to bind and regulate different proteins (or protein complexes) implicated in signaling pathways. Once the α subunit has activated its downstream target protein, GTP is hydrolyzed into GDP, α-GDP reassociates with βγ, and the complex returns to the serpentine receptor.

TKRs, such as the receptor for epidermal growth factor, have a single transmembrane domain. Their cytoplasmic tails have tyrosine kinase activity. Ligand binding induces TKR dimerization; the cytoplasmic tails of the TKRs then cross-phosphorylate each other on specific tyrosine residues. The phosphorylated tyrosines are bound by adapter proteins (such as Grb2) using a specific motif called SH2 (*src* homology 2) on the adapter protein. The adapter proteins have other binding motifs such as the SH3 domain, which binds to polyproline residues. This links the receptors to activators (called guanine exchange factors (GEFs) of small GTPases of the p21^{ras} superfamily. SH2 domains also direct the binding of other enzymatic activities, such as phosphoinositide 3-kinase (PI3-K), the Src tyrosine kinase, or phospholipase C-γ (PLC-γ).

Other types of receptors that span the membrane only once turn on different signaling pathways. For example, receptors that bind the immune system signaling molecules, including interleukins and interferons, are usually linked to the JAK (Janus kinase)/STAT (signal transducers and activators of transcription) pathways. As with TKRs, initial activation is by dimerization. The cytosolic kinase JAK associates with the cytosolic portion of the activated membrane receptor and phosphorylates the transcription factor STAT. Phosphorylated STAT proteins leave the receptor, dimerize, and migrate into the nucleus, where they bind DNA and activate the transcription of certain genes.

Secondary Signal Transduction

This section addresses the events occurring after ligand binding and receptor activation. These events include signaling by GTPases, kinases, phosphoinositide kinases, and calcium.

As mentioned above, G proteins have varied targets, depending on the identity of the particular G protein. Some G-protein α subunits attach to the enzyme adenyl cyclase, either increasing or decreasing its production of the signaling molecule cyclic AMP. An increase in the amount of cyclic AMP turns on protein kinase A (PKA), which has many targets. Other α subunits attach to and turn on phospholipase C-β (PLC-β), which cleaves phosphatidylinositol-4,5-bisphosphate [PtdIns(4,5)P$_2$ or PIP$_2$] into inositol-1,4,5-triphosphate (InsP$_3$) and diacyl-glycerol (DAG). Both molecules eventually turn on kinases. InsP$_3$ triggers the release of calcium, which binds to a protein called calmodulin. The calcium-calmodulin then binds and activates a kinase called calcium/calmodulin-dependent protein kinase. DAG directly turns on certain isoforms of the Ser/Thr-specific protein kinase C (PKC), which sets in motion a cascade of protein kinase activation, thus turning on the mitogen-activated protein kinases (MAPK).

Responses to the Ras superfamily of GTPases are also diverse, depending on which of the five branches—Ras, Rho, Rab, Ran, and ADP-ribosylating factors (ARFs)—is activated. (ARF was first discovered as a factor that was indispensable for cholera toxin to exert its ADP-ribosyltransferase activity.) Members of each of these branches control a different set of cellular mechanisms. Ras controls cell division and cell differentiation, Rho controls actin cytoskeleton organization, Rab controls intracellular vesicular traffic, Ran controls the transport of proteins through nuclear pores, and ARF1–4 controls the formation or budding of intracellular vesicles, whereas ARF6 is involved in the fusion of endosomes to the plasma membrane.

These small GTPases (named for their low molecular mass of approximately 21 kDa) are timer molecules. They are defined as "on" when bound to GTP and "off" when the GTP is hydrolyzed into GDP. Small GTPases are activated by stimulation of a GEF, which promotes the exchange of the bound GDP for GTP as described for large G proteins. Small GTPases are deactivated by GTP hydrolysis. GTP affects the shape of two GTPase domains called switch 1 and switch 2. In the GTP-bound form of the small G protein, the switch 1 region (a polypeptide of 10 residues) moves so that it binds to a protein or protein complex termed the effector, which often has Ser/Thr kinase activity. This activates the kinase, which turns on other kinases such as the members of the MAPK family (Figure 3.2). The switch 2 domain contains amino acids crucial for GTP hydrolysis. These residues are not, however, properly oriented for active GTP hydrolysis until a GTPase-

Figure 3.2 Important signaling pathways: the MAPK and NF-κB pathways. The MAPK pathways are kinase cascades that end in activation of a terminal kinase (either ERK, P38, or SAPK/JNK). These terminal kinases phosphorylate and activate proteins that control the transcription of SRF/AP1, ATF2, C-jun-dependent genes. Stimuli like LPS or TNFα activate the Nemo complex that phosphorylates the NF-κB inhibitor IκB. The phosphorylated IκB is polyubiquitinated and degraded by the proteasome. Removal of IκB from NF-κB uncovers a nuclear localization signal on NF-κB. This factor then enters into the nucleus and plays the role of a transcription factor. The ultimate response of the cell activation of these pathways is indicated at the bottom of the figure.

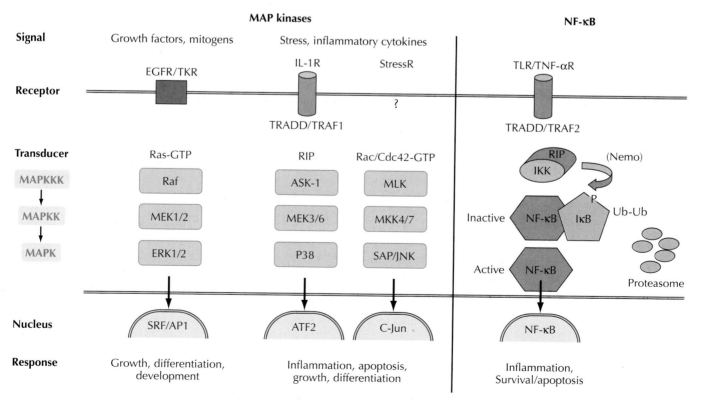

activating protein binds. This protein introduces an arginine residue called the arginine finger into the GTP-bound form of the small G protein, allowing fast hydrolysis of GTP.

The Rho subfamily of GTPases encompasses Cdc42, Rac, and Rho. These proteins induce the formation of various actin-rich structures, including protrusions called filopodia (by Cdc42-GTP), a dense actin mesh on the cell periphery that makes up lamellipidia (by Rac-GTP), and stress fibers and focal adhesion complexes (by Rho-GTP). The mechanism by which Cdc42 induces actin polymerization at the plasma membrane requires its binding to a protein named N-WASP (neuronal Wiskott-Aldrich syndrome protein). This protein, activated by Cdc42-GTP, harbors a recognition motif at its amino terminus, which binds polyproline stretches contained within proteins (such as zyxin or vinculin) that associate with receptors such as integrins involved in the attachment of cells to the extracellular matrix. N-WASP also contains specific motifs that recruit the actin-nucleating Arp2/3 complex and/or profilin, an accelerator of F-actin polymerization. The Rac-dependent nucleation and polymerization of actin is mediated by the protein Scar/WAVE, which like Cdc42 in its active conformation recruits Arp2/3. At a difference with Cdc42, there is no direct interaction between ScarWAVE and Rac-GTP. The Rac-GTP molecule indirectly causes activation of Scar/WAVE by inducing the dissociation of ScarWAVE with an inhibitory molecular complex. Rho-GTP activates a Ser/Thr kinase (Rho kinase [Rok]). Rok phosphorylates and inactivates myosin light-chain phosphatase (MLCP), so that the activity of the opposing myosin light-chain kinase (MLCK) predominates. Phosphorylation of myosin by MLCK leads to activation of this motor protein so that it binds to actin filaments and causes them to contract.

The autophosphorylation of a TKR, followed by adapter and then Ras-GEF binding, leads to the activation of Ras. Ras in turn stimulates the first kinase of the MAPK cascade, Raf (Figure 3.2). The next kinases in this cascade, the MAPKKs (MEKs), are dual Ser/Thr and Tyr kinases of narrow specificity. MEKs activate MAPKs by phosphorylation (Figure 3.2). (The particular MAPK in the pathway downstream of Raf is also called an ERK for extracellular signal-regulated protein kinase.) MAPKs are Ser/Thr kinases with broader specificity than MEKs. Scaffolding proteins such as JIP-1 (see Box 3.1) or MP1 bind a specific sequence of MAPKs and may channel their activities to certain sites. MAPKs phosphorylate cytosolic proteins such as microtubule-binding proteins and transcription factors required for cell division or differentiation. Parallel and very similar to the MAPK cascade of kinases, the JNK/SAPK and p38 kinase pathways are controlled at the level of the cell membrane by receptors implicated in the detection of stresses, such as osmotic shock, or inflammatory cytokines, such as interleukin-1 (IL-1). Apparently, the small GTP-binding proteins that turn on the c-Jun pathway are Rac and Cdc42 (Figure 3.2).

The TKRs can also bind PLC-γ. As with the serpentine receptors and PLC-β, this produces DAG (activating PKC) and InsP$_3$ (releasing calcium). Other kinases of note include AKT/PKB and tyrosine kinases such as p125^{fak} (focal adhesion kinase), p60^{src}, or p130^{cas}, which are usually associated with focal contacts and are important in cross talk with TKRs or serpentine receptors and the activation of kinase cascades such as the MAPKs.

Receptors for different cytokines, including tumor necrosis factor α (TNFα) and IL-1, as well as the signaling receptor for lipopolysaccharide (LPS, a Toll-like receptor; see Toll and Toll-Like Receptors), trigger what is

BOX 3.1

Signaling specificity through molecular scaffolds

The MAP kinases, which include ERK, JNK, and p38, are a family of related proteins involved in signal transduction. These proteins integrate and relay extracellular stimuli to control various cellular responses to the environment. ERKs are activated mainly by mitogenic stimuli, whereas JNK and p38 relay information in response to cytokines and cell stress. Each of these pathways is composed of a cascade of kinases where successive phosphorylation events regulate the activity of each kinase in the pathway. The MAPK cascades demonstrate a high degree of homology both in the general organization of the modules and also at the level of the primary sequence of these proteins. For example, ERK1 has a 60% similarity at the level of primary sequence with both JNK and p38. How then does the cell control the activity of these kinases and keep the different pathways distinct to maintain the specificity of the response? One way that the cell prevents potential cross talk between these different enzyme cascades is to compartmentalize the signaling through proteins called "scaffolds" that can

regulate these multienzyme complexes. First described in yeast, scaffold proteins were shown to be composed of multiple interaction domains capable of selectively binding different components of a particular signaling module. The function of scaffolds is twofold. These proteins insulate the module from activation by irrelevant stimuli and also favor rapid and non-amplified transmission of the signal.

In mammalian cells, JIP-1 is one example of a scaffold protein that regulates the activation of the JNK signaling cascade. JIP-1 selectively interacts with multiple components of the JNK signaling pathway through distinct binding domains. There are distinct binding domains present on JIP-1 for the selective interaction of mixed-lineage protein kinase (MLK), MAP kinase, kinase 7 (MKK7), and JNK. Recruitment and binding of these factors to JIP-1 provide a platform for the successive activation of these kinases, thereby facilitating the linear transduction of the signal and the ultimate phosphorylation of the target of this pathway, c-jun. Once c-jun is phosphorylated, it can associate with transcription factor partner proteins like c-fos, bind DNA, and promote gene expression.

Scaffold proteins like JIP-1 are not only involved in signal propagation, however; these proteins have also been shown to be required for the regulation of signaling pathways and inactivating the different kinases. JIP-1 also recruits phosphatases that can inactivate MKK7 and JNK to inhibit the phosphorylation of c-jun.

The next important question remaining about JIP-1 and other scaffold proteins is how do these proteins regulate signaling pathways in a spatial/temporal fashion? As mentioned above, the scaffold JIP-1 binds to multiple members of the JNK signaling module, proteins involved in both the activation and inactivation of the cascade. In terms of timing, perhaps these proteins bind sequentially to maintain the linearity of the response? Another question that remains unanswered is where are these interactions taking place within the cell? A recent study showed that JIP-1 is moved along microtubules by the molecular motor protein, kinesin. Is this the site of JIP-1 regulation of the JNK signaling module or is JIP-1 using microtubules to shuttle to a particular subcellular location?

known as an NF-κB response (Figure 3.2), which is the initiation of transcription of various genes whose protein products are involved in the development of inflammation. The induction of these genes is regulated by the transcription factor, NF-κB. NF-κB is held in an inactive state in the cytosol through its binding to inhibitory proteins, called IκBs. These inhibitory proteins bind and mask a nuclear localization signal present within the NF-κB molecule. For induction of the NF-κB pathway by TNFα, for example, ligand binding induces the formation of what is likely a trimeric receptor complex. The cytoplasmic domain of the TNF receptor can then interact with an adaptor molecule called TRADD, which in turn recruits TRAF2 and the serine/threonine kinase, Rip, to the cytoplasmic portion of the receptor complex. Rip is then responsible for the activation of the IKK complex, a high molecular weight complex composed of two kinases, IKKα and IKKβ, and a regulatory subunit, IKKγ or NEMO. The kinases of the IKK complex phosphorylate IκB on serine residues, and this targets IκB for ubiquitination and subsequent degradation of this molecule by the proteosome. NF-κB is thus liberated and it can then translocate to the nucleus, bind to

DNA, and mediate the expression of a variety of genes, many of which are involved in the development of inflammation (Figure 3.2).

Proteins regulated by phosphorylation at Ser/Thr or Tyr residues require dephosphorylation to return to their active or inactive initial states. One example of such regulation is control of the cell cycle kinase Cdc2 by Cdc25 and other phosphatases. Activities of Ser/Thr phosphatases and Tyr phosphatases are regulated by a combination of targeting and regulatory subunits.

As noted briefly above, ions such as calcium are important intermediates in signaling cascades. Cells tightly regulate their intracellular level of calcium via calcium-binding proteins and extrusion mechanisms. This allows calcium to be used in localized cellular regulation such as muscle contraction, actin filament organization, membrane hyperpolarization, and regulated exocytosis, but it also means that calcium released outside the cell or from intracellular stores can be used as a rapidly mobilizable messenger. To maintain a low cytosolic calcium concentration, cytosolic calcium is actively pumped into the ER. Calcium is sequestered in the ER by binding to specialized proteins (such as calsequestrin or calreticulin) or into mitochondria, which specifically pump calcium ions in high concentrations (for instance, in proximity to the ER or the plasma membrane). Several mechanisms may introduce small bursts of calcium into the cytosol, either from the ER or across the plasma membrane. In nonexcitable cells, the InsP$_3$ pathway is the major system used to induce a burst of calcium in the cytosol. Membrane receptors coupled to either heterotrimeric GTPases or TKRs activate PLC-β (heterotrimeric G protein) or PLC-γ (TKRs) that convert PIP$_2$ into InsP$_3$ and DAG. InsP$_3$ binds the ryanodine receptor on the ER membrane, triggering the release of calcium. Receptors that induce InsP$_3$ formation cause a slow release of calcium into the cytosol. Conversely, stimulation of certain membrane receptors induces a fast release of calcium into the cytosol through the activation of plasma membrane Ca^{2+} channels (CRAC). These channels are activated when the ER calcium store is depleted, thus rapidly replenishing the calcium in the cell. Excitable cells also contain voltage-dependent calcium channels that enable them to increase their calcium concentration very rapidly. One important signaling pathway involving calcium during T-lymphocyte stimulation by antigen-presenting cells (APC) is the NFAT pathway. T-cell receptors activate a PLC-γ which produces InsP$_3$, releasing calcium from the ER. Depletion of ER calcium activates CRAC which, by increasing abundantly the cytosolic concentration of calcium, activates the calcium-activated phosphatase calcineurin. Calcineurin by dephosphorylating the transcription factor NFAT allows its translocation into the nucleus where it stimulates the transcription of important immunological factors such as IL-2. Immunoblocking agents like cyclosporin A or FK506 inhibit calcineurin activation by blocking its phosphatase activity and thereby NFAT activation.

The second group of small molecule signals is the family of 3,4,5- and 4-phosphorylated inositol lipids (PIP$_3$ and PIP$_2$). Two kinases, termed 3-kinase and 5-kinase (the numbers refer to the position of phosphorylation on the inositol ring), dominate this regulation by producing PIP$_3$ and PIP$_2$, respectively. The PI3-K 85-kDa regulatory subunit binds to activated TKRs (such as the epidermal growth factor receptor) on phosphorylated tyrosines and activates the 110-kDa catalytic subunit. Its product, PIP$_3$, is involved in the activation of kinases, which are pivotal in certain signaling cascades. For instance, the insulin receptor (insulin being a major hormone controlling glucose and lipid metabolisms) activates the PI3-K p110 via p85. The PIP3 produced stimulates a PIP3-dependent kinase (PDK) which turn on the AKT/PKB kinase. AKT induces glycogen, fatty acids, and protein synthesis. The protein PTEN is a phosphatase, which dephosphorylates PIP3. PTEN is a tumor suppressor because it blocks, among other regulatory proteins, the activation of AKT, resulting in maintenance of apoptotic mechanisms pivotal to inhibit cell transformation. In contrast to PI3-K, PI5-K phosphorylates phosphatidylinositol-4-phosphate to produce PIP_2. PIP_2 modulates the activity of many actin-binding proteins such as profilin, gelsolin, vinculin, CapZ, and cofilin, thereby regulating the assembly and disassembly of the actin cytoskeleton. It is also involved in the binding of regulatory or structural proteins with membranes.

Toll and Toll-Like Receptors

Detection and Initial Signal Transduction

The innate immune system, which is present in all multicellular organisms, is the first line of defense against microbial infection. This system is capable of immediately initiating a defense response when it encounters a signature of microbial infection. In contrast to the adaptive immune system, which is present only in vertebrates, the innate system relies on a defined number of germ line-encoded receptors, termed pattern recognition receptors or PRRs, that are capable of recognizing these microbial signatures. Because of this lack of diversity in the PRRs, the microbial factors that are recognized are structural patterns common to many microbes. For example, one of these pathogen-associated molecular patterns (PAMPs) recognized by PRRs is LPS, which is a structural component of the outer membrane of gram-negative bacteria.

The study of innate immunity in mammals intensified after the discovery of Toll and the role of this protein in innate immune defense in *Drosophila melanogaster*. In 1996, it was demonstrated that flies deficient in Toll were highly susceptible to fungal and then later to gram-positive bacterial infections. The first discovery of Toll-related proteins in mammals in 1998 was quickly followed by the demonstration that the mammalian Toll-like receptor 4 (TLR4) was the signaling receptor in a protein complex responsible for the recognition of LPS and cellular responses leading from this event. In humans, 10 TLRs exist and research is now focused on the identification of which microbial ligand or PAMP is recognized by each TLR or combination of TLRs (Table 3.2).

Toll of *Drosophila* is not classified as a bona fide PRR since it does not directly bind a microbial PAMP. Instead, Toll binds an endogenous ligand called spaetzle that is generated by a proteolytic cascade, similar in concept to the complement cascade in mammals. For fungal recognition, the PAMP/PRR combination is not known, whereas for gram-positive bacterial recognition, the upstream PRR is a peptidoglycan recognition protein called PGRP-S. Recent findings have shown that the PAMP recognized by PGRP-S is indeed peptidoglycan. PGRP-S appears to recognize a specific peptidoglycan motif associated with gram-positive bacteria.

Drosophila also possess a sensing system for gram-negative bacteria that is also dependent on peptidoglycan recognition. Although the signaling receptor implicated in this sensing system has yet to be identified, flies deficient in a cytoplasmic signaling protein called Imd (for immune deficiency) are highly sensitive to gram-negative bacterial infections. Imd is related to Rip in the TNFα signaling cascade described earlier in this chapter, again

Table 3.2   TLRs and their ligands

TLR family member	Ligands (origin)
TLR1 (with TLR2)	Triacyl lipopeptides (bacteria, mycobacteria)
TLR2	Lipoproteins / lipopeptides (various bacteria)
	Peptidoglycan (gram-positive bacteria)
	Lipoteichoic acid (gram-positive bacteria)
	Lipoarabinomannan (mycobacteria)
	Zymosan (fungi)
	Atypical LPS (*Leptospira interrogans* and *Porphyromonas gingivalis*)
	HSP70 (host)
TLR3	Double-stranded RNA (viruses)
TLR4	LPS (gram-negative bacteria)
	Taxol (plants)
	Viral proteins
	HSP60 (bacteria)
	HSP60 / HSP70 (host)
	β-Defensin (host)
TLR5	Flagellin (bacteria)
TLR6 (with TLR2)	Diacyl lipopeptides (mycoplasma)
TLR7	Imidazoquinoline (synthetic compound)
TLR8	Imidazoquinoline (synthetic compound)
TLR9	CpG DNA (bacteria)
TLR10	?

highlighting the similarities in signal transduction systems in mammals and flies. As with Toll, upstream of Imd is a peptidoglycan recognition protein, PGRP-LC. This PRR is specific for a peptidoglycan motif characteristic of gram-negative bacteria. Stimulation of this pathway then leads to the production of antimicrobial peptides, the majority of which are gram-negative bacterial specific.

More recently, an intracellular pathogen-sensing system was uncovered in mammals. This system relies on cytosolic proteins called Nods, which are structurally similar to a protein family in plants, called R proteins that are involved in disease resistance against pathogen infection. Nod1 detects peptidoglycan motifs from gram-negative bacteria while Nod2 senses the minimal component of peptidoglycan, muramyl dipeptide, which makes it a general sensor of bacteria. Mutations in the gene encoding Nod2 were recently shown to be associated with the inflammatory bowel disease, Crohn's disease, suggesting that defects in sensing bacterial peptidoglycan in the cytoplasmic compartment contribute to the etiology of this disease.

Secondary Signal Transduction: NF-κB Signaling

Cell signaling downstream of Toll, TLRs, and the IL-1 receptor in mammals is remarkably similar owing to their conserved cytoplasmic domain, called the Toll/IL-1 receptor or TIR domain. Upon ligand binding to the extracellular portion of these receptors, the TIR domains recruit TIR-containing adapter molecules, like MyD88 and dMyD88, in mammals and flies, respectively. This recruitment process sets up a cascade of further recruitment of adapter molecules and activation of kinases that ultimately leads to the degradation of the inhibitory molecules that keep the Rel/NF-κB family of transcription factors in an inactive state in the cytoplasm. In mammals, the inhibitory protein, IκB, is degraded, releasing NF-κB that can then translocate to the nucleus and drive the expression of genes, many of which are involved in the proinflammatory response. Cactus is the inhibitory protein in *Drosophila* that is degraded, releasing the transcription factor Dif/dorsal, which drives the expression of antimicrobial peptides.

Nod proteins also activate NF-κB but through a pathway distinct from that of TLRs. Once activated, Nod proteins oligomerize and recruit a serine/threonine kinase called Rip2. Related to Rip in the TNFα pathway, Rip2 can directly interact with the γ subunit of the IKK complex, leading to activation of the kinase molecules, IKKα and IKKβ, and the subsequent release of active NF-κB.

Membrane Traffic

Cells contain many distinct membrane-bound compartments or organelles. Each organelle has specialized functions and therefore needs a unique combination of lipids and proteins. Vesicles transport proteins and lipids between compartments. Donor compartments, such as the ER and Golgi complex, have specific mechanisms that allow selective packaging of proteins (or cargo) into vesicles. Vesicles can then recognize the membranes of acceptor compartments. Cargo molecules transported to the new compartment can either stay there, be transported to another organelle, or be recycled to the donor compartment.

Proteins that leave the donor compartment ER associate with vesicle proteins. The p23/24 family of transmembrane proteins are probably the major molecules that, using their luminal extensions, bind cargo molecules in the ER and the *cis* part of the Golgi. The cytoplasmic tail of p23/24 also allows a direct or indirect interaction with the coat protein complex COP1, which is involved in the formation or budding of the vesicle. Budding from the Golgi apparatus involves three sets of proteins: ARF1, COP1, and p23/24. The GTP-bound form of ARF1 is favored by ARNO, a GEF for ARF1, and associates with the vesicle donor membrane through a myristic acid modification of the amino terminus. This allows binding of the COP1 coat proteins to both ARF1 and the cytosolic extension of p23/24. Binding of COP1 induces budding. Binding of COP1 also seems to activate the hydrolysis of GTP into GDP on ARF. After GTP hydrolysis, ARF1-GDP is then released from the vesicle, inducing the loss of the coat protein complex COP1.

Vesicles containing secreted proteins migrate to the acceptor compartment, guided by the proteins GM130 and p115 (see Cell Cycle and Organelle Inheritance, below). A family of highly conserved proteins called SNAP receptors (SNAREs) perform targeting and fusion of vesicles to the acceptor compartment. Vesicles display vesicle-associated SNAREs (v-SNAREs) on their surface. Each v-SNARE can interact with two t-SNAREs (target for v-SNARE) localized on the acceptor compartment. This tight complex, stable even after heating at 90°C or treatment with ionic detergents such as sodium dodecyl sulfate, is formed by coiled-coiled associations and glutamine-arginine interactions. The binding of a v-SNARE to t-SNAREs is used not only for recognition and docking of vesicles to membranes but also for fusion of the lipid bilayers. Upon vesicle fusion, the v-SNARE–t-SNAREs complex is dissociated by *N*-ethylmaleimide-sensitive factor (NSF) in the presence of an adapter protein (soluble NSF attachment protein [SNAP]) and ATP. NSF may interfere with the glutamine-arginine ionic interactions between v-SNARE and t-SNAREs and thus prime the dissociation of the complex.

Rab GTPases are also important in vesicle docking and fusion. More than 30 different Rabs have been isolated to date. Rab function appears to involve shuttling between membranes (a location favored by the GTP-bound form) and the cytosol (in the GDP-bound form). One important function of Rabs might be to control the final docking and fusion of vesicles. Ypt1, a yeast protein that binds Rab-GTP, activates the v-SNARE–t-SNARE association by removing Sec1p, a yeast protein bound to the t-SNARE. This allows the SNARE proteins to associate. Rab5 seems to control early endosome fusion by allowing the binding of a specific protein called EEA1 (early endosome-associated protein). Rab5-GTP associates with early endosomes and binds the amino terminus of EEA1. The carboxy terminus of EEA1 contains a specific domain that binds to the membrane lipid phosphatidylinositol-3-phosphate, which is generated by PI3-K. Thus, EEA1 can cross-link early endosomes so that they are ready for fusion. Other roles for Rab proteins have been described, including control of organelle locomotion on microtubules (by a complex of Rab6 and Rabkinesin) and control of exocytosis (by a complex of Rab3 and Rabphilin). Rab GTPases seem to be associated with specific discrete domains of the vesicular surface. This might help to recruit molecular complexes at localized points of the vesicular surface for outward or inward vesiculations or membrane-fusion events, for instance.

Endocytosis is the process by which extracellular ligands are captured by cells and routed to different cytosolic compartments. Endocytosis is mediated by different mechanisms. The first discovered involves the formation of a cytoplasmic clathrin coat forming pits at the surface of the cell and vesicles. Clathrin-associated adapter proteins incorporate cargo proteins into clathrin-coated pits. Receptors on which ligands to be endocytosed are bound (i.e., transferin receptor) contain special sequences on their cytoplasmic tail such as YXXf (where X is any amino acid and f is a bulky hydrophobic amino acid). This sequence recruits the clathrin adapter protein AP2, allowing the assembly of the clathrin coat. Beside small linear motifs of internalization in the cytoplasmic tails of certain receptors, monoubiquitination of proteins is also used not only as a signal of cargo recruitment but also as a tag for targeting membrane receptors to multivesicular bodies (MBVs). MBVs are endocytic compartments that mature from early endosomes and by a process of intraluminal vesiculation (probably by selective incorporation or clustering into the lipidic membrane of a conical-shape lipid lysobisphosphatidic acid [LBPA]) produce numerous internal small vacuoles. These intraluminal vesicles are used to sort proteins along the degradative pathway.

Upon the discovery of the clathrin pathway of endocytosis, other systems of endocytosis have been described, such as noncoated pit pathways, the caveosome pathway, and the macropinocytic pathway. How these new endocytic pathways interact with early endosomes, MBVs, late endosomes, or the Golgi apparatus is still under debate.

Endocytosis is a key event during cell signaling. Not only can activated receptors be classically down-regulated by their routing to the degradative pathway (lysosomes) but they also can induce different specific signaling pathways when they are localized either in the plasma membrane or in internal compartments. Endocytosis is also involved in establishing cell polarity or the spreading of certain signaling molecules (morphogens) which are secreted by a restricted group of cells.

Organelle Biogenesis

Although vesicles transport many proteins throughout the cell, other proteins arrive at distinct locations directly. These targeting systems allow the ER, mitochondria, and peroxisomes to accumulate their own particular combinations of proteins and lipids so that they can carry out their specialized functions. Each organelle selects its own set of proteins by using specific receptors. Proteins synthesized on ribosomes are targeted to the ER, mitochondria, or peroxisomes by specific sets of amino acids called signal sequences. The proteins are then transferred through membranes by a translocating apparatus so that they end up in the lumen of the organelle or spanning the membrane. The main mechanism to deliver proteins into the ER is via the signal recognition peptide and its receptor to deliver nascent proteins from the ribosome (a cotranslational mechanism). The nascent polypeptide chain is translocated into the ER lumen through the hydrophilic channel Sec61p. Misfolded proteins in the ER can be retrotransported by the Sec61p channel and upon ubiquitination degraded by the proteasome. Very small peptides can be transferred from the cytosol to the ER lumen through the TAP channel. Transfer of proteins into mitochondria or peroxisomes uses posttranslational mechanisms. For mitochondria, this involves unfolding the protein followed by recognition of a mitochondrial outer membrane receptor (Tom 70 for proteins without a signal sequence and Tom 20 for those with a signal sequence). Tom receptors associated with a TOM protein complex, which forms a transmembrane pore through which the protein is transferred. Depending on a second signal in the sequence of the translocating protein, the protein can either stay in the mitochondrial intermembrane space or be transferred to the mitochondrial matrix by another translocating complex called TIM (TIM22 for proteins incorporated in the mitochondrial membrane and TIM 23 for proteins translocated into the matrix).

Compared with our knowledge of the delivery of proteins to organelles, we know relatively little about how lipids traffic between the different membranes. One fundamental point is that organelles share a set of identical lipids, but the relative abundance of these lipids is unique to each organelle. Lipids are synthesized at different sites in the cell. The ER produces glycerolipids and cholesterol, and the Golgi apparatus produces sphingolipids. Lipids travel between organelles by the secretory and endocytic pathways via vesicle-mediated transport or via lipid-carrier proteins present in the cytosol (protein lipid transporter). Direct contact between membranes, such as those of the ER and mitochondria, may also allow the transfer of lipids.

The Cytoskeleton

The cell cytoplasm is structured by a cytoskeleton encompassing three types of elements: actin filaments (microfilaments), intermediate filaments (IFs), and microtubules (MTs). These are classified according to their diameters of 6 nm (actin), 10 nm (IFs), or 23 nm (MTs). Actin filaments provide the driving force for cells to move and divide. IFs form an intracellular scaffold which allows cells to resist external stresses. MTs segregate chromosomes during mitosis and serve as tracks along which motor proteins transport vesicles loaded with cargo (Figure 3.3).

There are more than 50 different IF genes in humans, but all IFs have the same basic structure—a conserved head and tail with a long α-helical cen-

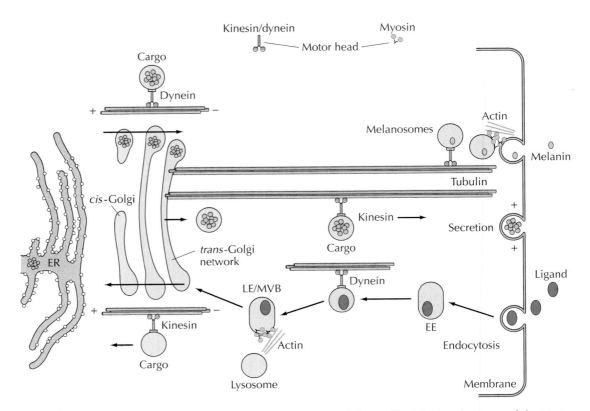

Figure 3.3 Motor proteins and intracellular traffic. The involvement of the motor proteins kinesin, dynein, and myosin in cellular motility of intracellular vesicles is shown.

ter. The α-helices of two monomers associate and intertwine to form a parallel coiled-coil rod. Association of two dimers produces a tetrameric protofibril, and association of several protofibrils forms a filament 10 nm in diameter. IFs include the keratins (the major structural protein of hair and epidermal cells), the nuclear lamins (which link chromatin to the nuclear membrane), vimentin (a cytosolic scaffolding molecule found in fibroblasts), and neurofilaments (which form the backbone of neuronal axons and determine the axonal diameter). The main role of IFs is to provide protection against mechanical stresses. For that purpose, IFs are anchored to the cell membrane and to the rest of the cytoskeleton, forming a dense mesh. IFs attach to cell-cell adhesion points called desmosomes via the protein desmoplakin and form a network with actin filaments, MTs, and myosin by binding to the cross-linker protein plectin.

Actin filaments are composed of two twisted linear actin protofilaments made of polymerized actin monomers. An actin filament is a dynamic structure with a functional polarity. The barbed or plus end of the filament is the fast-growing end, where there is rapid association of actin monomers associated with ATP. The pointed or minus end is the depolymerizing end, where ADP-associated actin monomers dissociate. Association and dissociation of actin monomers are controlled by regulatory actin-binding molecules. At the barbed ends, profilin and thymosin regulate polymerization by controlling actin monomers bound to ATP, whereas capping proteins by binding to the pointed ends terminates the elongation. ADF/cofilin is a molecule that severs and depolymerizes actin-ADP-containing filaments promoting the dissociation of ADP-actin from the filament.

When extracellular signals activate, for instance, certain tyrosine kinase receptors, Rac and Cdc42 GTPases, are stimulated together with production of PIP_2. This activates proteins of the WASP/Scar/WAVE family which bind Arp2/3 and actin monomer on a preexisting actin filament. Arp2/3 nucleates a new actin filament which is branched with an angle of 45° onto the preexisting one. This new filament will push the membrane forward and thus will drive the formation of a membrane protrusion. Nucleation and polymerization of unbranched actin filaments can also be initiated at the level of the cytoplasmic face of the plasma membrane specifically by Rho-GTP via activation of proteins containing the so-called "formin" domains. The mechanochemical properties of myosins—the actin motor—are also needed for motility. Myosin motors form a large family of proteins. The type II myosin proteins have two actin-binding motor heads, which bind ATP and are linked via an elongated coiled-coiled domain. At the hinge region between the motor head and the elongated domain lie the myosin light chains, which activate myosin on actin filaments toward the plus end and are involved in actin contractility. Myosins read the actin filament in a linear fashion and bind sites separated by 36 nm (a distance dictated by the architecture of the two twisted linear actin protofilaments). This distance is too great for the myosin head to span, so after binding and hydrolysis of ATP, the motor detaches and jumps from one binding site to the other. The fact that myosin motors detach from the actin filament implies that many myosin motors (usually several hundred) must work together to ensure that movement along the actin filament is efficient.

MTs are produced by polymerization of dimers of α- and β-tubulin. The amino acid sequences of the α- and β-tubulin monomers are 50% homologous to each other, but it is β-tubulin that binds to GTP and hydrolyzes it to yield GDP. Dimers of α- and β-tubulin are nucleated (by assembly of two or three dimers) from centrosomes that contain a special type of tubulin (γ-tubulin), which stimulates this step. Thirteen protofilaments associate laterally to form a hollow cylindrical MT 25 nm in diameter.

MTs are dynamic structures that grow or shrink rapidly. The faster-growing plus end exposes β-tubulin, whereas the slower-growing minus end exposes α-tubulin. MT dynamics are thought to be accounted for by two phenomena. MTs are said to treadmill when there is a constant incorporation of GTP-bound tubulin dimers at one end of the filament and a balanced loss of GDP-bound dimers at the opposite end. Dynamic instability, however, describes the abrupt transitions between extended states of MT growth and shrinking at one end of the MT. The transition from shrinking to growing is called rescue, and the transition from growing to shrinking is called catastrophe.

Kinesins and dyneins are motor proteins that drive the motion of vesicles, organelles, and perhaps regulatory complexes along MTs (Figure 3.3). Kinesins move toward either the plus or minus ends of MTs, whereas dyneins move toward the minus ends. The structural organization of dyneins and kinesins is similar to that of type II myosin; there are two motor heads linked by an elongated neck domain through coiled-coil interactions. Kinesin movement has been extensively analyzed, although the exact sequence of events is not yet known. Kinesins bind only to the β-tubulin subunits of MTs. Each β-subunit is separated by 8 nm, which corresponds to the spacing between the two heads of a kinesin. The kinesin is thought to move hand-over-hand. In one theory, one head is thought to remain bound as the second head hydrolyzes its bound ATP, lets go of its β-tubulin subunit, and then swings to the next free β-tubulin. Another theory postulates that ATP binding to the rearward head induces a conformational change, producing a modification in the rearward neck structure that displaces the forward head from its tubulin subunit. The rearward head then binds to this β-tubulin subunit, while the forward head attaches to the next β-tubulin subunit.

Grafting experiments with the next and motor regions of different kinesins show that the directionality of kinesins is determined not by their heads but rather by the adjacent neck regions. It must be stressed that kinesins (and dyneins), unlike myosin motors, never detach from MTs during their movement cycle, since one head is always associated with a β-tubulin subunit. This is why a single kinesin is capable of moving a vesicle or other cargo for considerable distances. For the mobilization of certain organelles, such as lysosomes or melanosomes, kinesins and myosins cooperate. Long-distance journeys of these organelles are due to a kinesin, whereas the final short distance that remains to be done to deliver the organelle to its acceptor membrane relies on myosin (like myosin V A). Mutations in kinesins or myosins may lead to severe human pathologies, such as an immune deficiency, skin depigmentation, or sensory deficits.

MTs are stabilized by specific proteins called MT-associated proteins, which bind MTs in a nucleotide-insensitive manner. These proteins are regulated by phosphorylation, which decreases their MT-stabilizing activities. MTs can also be destabilized by specific factors. Indeed, certain kinesins can induce dramatic destabilization of MTs.

Cell Adhesion and Morphogenesis

Multiprotein adhesion complexes link cells to the extracellular matrix (ECM) or to one another. The adhesion complexes usually contain transmembrane receptors, signal transduction proteins, linker proteins, and cytoskeletal proteins.

Cells are attached to the ECM through focal adhesions. These structures coordinate cell adhesion and cell motility. Highly motile cells do not contain easily observable focal adhesions, probably because these structures are transient. Focal adhesions are prominent in adherent stationary cells. The assembly of focal adhesions is regulated both by binding to the ECM and by intracellular signaling. Binding to the ECM is mediated mostly by the integrin class of transmembrane receptors. Integrins form a large family of heterodimeric transmembrane proteins with different α and β subunits. The affinity of integrins for the ECM can be modulated; this process of activation is the main mechanism for regulating the binding of cells to their substratum. Binding affinity is increased by the binding of proteins (such as the plaque proteins talin and paxillin or tyrosine kinases such as $p60^{src}$, $p125^{fak}$, or $p130^{cas}$) to the integrin cytoplasmic tail. This binding induces a conformational change, which is propagated across the membrane to cause modification of the extracellular ligand-binding site.

Intracellular signaling networks regulate focal adhesions mainly through the Rho GTPase. Rho-GTP induces the binding of plaque proteins to actin filaments, the subsequent clustering of integrins in focal adhesions by an actin- and myosin-dependent process, and the bridging of nonintegrin transmembrane receptors and actin filaments by ezrin, radixin, or moesin (ERM). Rho-GTP appears to act through two mechanisms. It activates PI5-K, which increases the concentration of PIP_2. The PIP_2 binds to a protein called vinculin, unfolding it and thereby unmasking its actin- and integrin-binding domains. Integrin clustering, however, is induced by actin filament contraction that follows Rho-GTP activation of Rok. Rho-GTP activation of ERM proteins may occur either by PIP_2 production and binding or by Rok phosphorylation.

Cell-cell adhesion is mediated at adherens junctions by cadherins, which are transmembrane calcium-dependent homophilic adhesion receptors. The E-cadherins, expressed in epithelial tissue, have been the most extensively studied and are required for epithelial cells to remain tightly associated. To be functional, cadherins must form complexes with both the cytoplasmic catenin proteins and the actin cytoskeleton. There are two types of catenins: α and β. The α-catenin is required for cadherin-cadherin interaction but also links cadherins to actin microfilaments. The β-catenin links α-catenin to the cytoplasmic tail of cadherin. Phosphorylation of β-catenins, in response to growth factors, induces a loss of adhesion.

The β-catenin is also used during embryonic development by the Wnt signaling pathway. Wnt, a secreted glycoprotein, binds to the Frizzled receptor. In the absence of Wnt, β-catenin is down-regulated by a degradation complex (containing the kinase-GSK3 which phosphorylates the β-catenin, leading to its ubiquitination and proteasome degradation). In the presence of Wnt, the protein Discheveled is activated and removes the GSK3 from the degradative complex, thereby blocking β-catenin degradation. The β-catenin is then translocated into the nucleus and acts as a DNA transcription factor.

Desmosomes are intercellular junctions in epithelia and cardiac muscle that are linked to IFs. The adhesion receptors of desmosomes—the desmogleins and desmocollins—are members of the cadherin superfamily. These receptors bind to IFs via specific proteins (desmoplakin and plakoglobin). Desmosomes and IFs operate together to maintain cell integrity during extracellular mechanical stresses.

Occludins and claudins of tight junctions are the most important elements for the formation of a permeability barrier in tissues such as epithelia and endothelia. Tight junctions are used to regulate the permeability of the paracellular space between adjacent cells and to divide the cell membrane into two distinct biochemical and functional areas. The transmembrane protein complex occludin-claudins contributes to the barrier function of tight junctions and interacts with two cytosolic proteins, ZO-1 and ZO-2. Several cytoskeletal proteins (such as cingulin) link actin filaments with ZO-1 and ZO-2.

The Rho family of GTP-binding proteins is involved in the regulation of both tight and adherens junctions. Rac may affect tight junctions by controlling the formation of actin filaments that are linked to the ZO-1/ZO-2/occludin/claudin complex. Rac and Cdc42 may regulate adherens junctions by modulating the adhesive properties of E-cadherins. Rho-GTP may control the formation of focal adhesion points on the basal face of epithelial cells. Rab GTPases are probably another important regulator of adherens junctions. Rabs, by regulating membrane traffic, may control the flux of membrane-associated receptors and effector molecules that affect cadherin-based adhesion.

Adhesion mechanisms are associated with morphogenesis. Indeed, similar molecular components are required for adhesion and for dynamic tissue changes such as the mesenchymal-epithelial transition or blastocyst formation. During the mesenchymal-epithelial transition, mesenchymal cells that are loosely attached progressively form a monolayer made of adherent cells with both tight and adherens junctions in a process called compaction. Cadherins are the major proteins that induce compaction. Adhesion mechanisms also operate in cell rearrangements, such as the formation of elongated tubular cells during blood vessel formation.

Apoptosis

Cells can self-destruct via an intrinsic program of cell death. Apoptosis or programmed cell death is characterized by blebbing of the plasma membrane, condensation of the cytoplasm and nucleus, and cellular fragmentation into apoptotic bodies. At the molecular level, apoptosis is characterized by degradation of chromatin, first into large fragments and subsequently into pieces of less than 200 bases. Apoptosis minimizes the leakage of cytoplasmic proteins, such as proteases that could damage adjacent cells, since apoptotic cells are immediately engulfed by macrophages. This feature distinguishes apoptosis from necrosis. Cell necrosis, which usually results from physical trauma, causes cells to swell and lyse, releasing cytosolic proteins that may stimulate an inflammatory process.

Apoptosis is divided into three phases. Detection of the apoptotic signal by sensors is followed by conversion of the signal so that it can trigger the execution phase of apoptosis. Finally comes the execution phase itself.

There are two major apoptotic pathways in mammalian cells, the death-receptor pathway and the mitochondrial pathway. The death-receptor pathway is triggered by members of the death-receptor superfamily, which includes the Fas receptor (also known as CD95), tumor necrosis factor receptor I (TNFR1), and TNF-related apoptosis-inducing ligand receptor (TRAIL-R). Binding of the ligand induces receptor clustering and the formation of the death-inducing receptor complex through the recruitment of various adapter molecules to the cytoplasmic domain of the receptor. Through a process known as "induced proximity" these events ultimately result in the activation of the execution molecules, the caspases. Caspases are a family of cysteine proteases that are synthesized as inert proenzymes. Upon activation, these molecules are cleaved to release the active enzymes. Caspases bring about the visible changes within the cell that characterize apoptotic cell death by activating various pathways leading to cell destruction. For instance, caspases lead to the activation of the nuclease responsible for the degradation of DNA into the fragments characteristic of apoptosis, as well as the activation of enzymes that cleave nuclear lamins resulting in nuclear shrinkage and budding.

Cells can also undergo apoptosis through what is known as the mitochondrial pathway. This pathway is activated by various cues of cell damage, including DNA damage. Signals of cell damage converge on the activation of a group of proapoptotic proteins, the Bcl-2 family members. These proteins somehow mediate the release of cytochrome c from the mitochondrial membrane, possibly by forming a channel within the membrane. Apart from its more innocuous role as an electron carrier in mitochondrial oxidative phosphorylation, cytochrome c is a central player in the induction of apoptosis. Once released into the cytosol, cytochrome c binds to Apaf-1, which then recruits and activates caspase-9, forming the "apoptosome" and leading to the activation of the executioner caspases and subsequent cell destruction.

Macrophages efficiently recognize, engulf, and destroy apoptotic cells, thereby avoiding the onset of inflammatory processes and resulting in a "silent" death of the cell. Apoptotic cells are recognized by macrophages since cells dying from apoptosis display a modified lipid composition in their cytoplasmic membranes. Phosphatidylserine, which is usually a lipid characteristic of the inner leaflet of the cytoplasmic membrane, is exposed to the surface where it is recognized by macrophages leading to cell engulfment and destruction. Detection of surface phosphatidylserine is also used as a diagnostic tool in research to identify apoptotic cells.

Cell Cycle and Organelle Inheritance

The cell cycle is an orderly progression of duplication (e.g., DNA synthesis during S phase) and division (e.g., chromosome segregation during M phase or mitosis), separated by two gap phases: G_1 before S phase and G_2 after S phase. When a cell prepares to divide, it must coordinate the duplication and division steps. To ensure good quality control, the transitions between phases must be governed by a decision-making process that assesses such things as the quality of the duplicated DNA and the successful reorganization of the cellular machinery. During G_1, the cell must make sure that it has reached a sufficient size and has sufficient nutrients to complete a cell cycle. Cells that do no meet the criteria may decide to stop division and stay quiescent or differentiate. In G_2, the cell must dissolve its nuclear membrane, duplicate its centrosome, and assemble an array of MTs into a spindle that will pull apart or segregate the duplicated chromosomes. Intracellular membrane-bound compartments such as the ER and the Golgi apparatus must also be duplicated. All these events are triggered by Ser/Thr kinases called cyclin-dependent kinases (CDKs). The periodic activation and deactivation of CDKs ensure the progression of the cell division cycle. CDKs are activated by association with regulatory subunits called cyclins and by phosphorylation and dephosphorylation of the CDKs. Cyclins are ubiquitinated and proteolyzed by the proteasome to deactivate CDKs.

The G_1/S cell cycle checkpoint initiates the cell cycle. Two CDKs (CDK4/6-cyclin D and CDK2-cyclin E) and the transcription complex Rb and E2F control this checkpoint. Binding of Rb to E2F blocks the transcription of factors implicated in the cell progression. CDK4/6 and CDK2 by phosphorylation of Rb dissociate Rb from E2F, allowing the transcription to start. Many signaling cascades (such as MAPK) exert on this checkpoint positive or negative controls by modulating the activity of either CDK4/6 or CDK2.

In G_2, a CDK called Cdc2 associates with cyclin B but accumulates in an inactive state due to the phosphorylation of a tyrosine residue located in the Cdc2 catalytic site. Cdc2-cyclin B is dephosphorylated, and thus activated, by the phosphatase Cdc25; it then induces mitosis by phosphorylating various proteins including those involved in chromatin condensation. Cdc2 is deactivated by proteolysis of cyclin B by the proteasome machinery.

Checkpoint pathways check the quality of DNA duplication and segregation and stop the progression of the cell cycle if there is a problem. DNA damages are recognized by the complex ATM-ATR. Between G_1 and S, ATM-ATR stimulates the p53 protein (a transcriptional activator named the guardian of the genome). The p53 protein activates the p21 protein, which is an inhibitor of Cdk2-cyclin complexes. Between G_2 and M, ATM-ATR inhibits Cdc25.

The total number of division cycles that a cell can undergo is limited, often to about 40 or 50. The maximum number of divisions may be partially controlled by the length of the telomeres, which are the ends of chromosomes. DNA duplication fails to reproduce the very end of the telomere (by the enzyme telomerase) in each cell cycle, leading to a progressive shortening. When telomeric DNA becomes too short, the cell becomes quiescent or apoptosis is triggered and the cell is eliminated.

During G_2, each organelle must double in size so that later it can be correctly divided between the daughter cells. The strategy used for organelle inheritance is based on having multiple copies of organelles (such as mitochondria, lysosomes, or chloroplasts) randomly dispersed in the cytosol. During cytokinesis, these organelles are separated equally among daughter cells. The system of nuclear and ER membranes and the Golgi apparatus are normally single-copy organelles, but they also use this strategy by fragmenting before division and re-forming after division. Fragmentation is driven by Cdc2-dependent phosphorylation of lamins (which link chromatin to the inner face of the nuclear envelope) and of GM130, a vesicle-docking protein on the acceptor Golgi compartment. Phosphorylation of GM130 prevents normal vesicle fusion from taking place, even as the vesicle formation involved in Golgi transport continues. The result is vesiculation of the Golgi.

BOX 3.2

RNAi—a reverse genetic tool to study gene function

Various tools are available to study the role of a gene or protein in a particular cellular response. Chemical inhibitors of enzymes or kinases can be applied; however, depending on the inhibitor, there can be nonspecific effects on other signaling pathways and responses, thus making it difficult to form any conclusions. Dominant-negative molecules can also be genetically constructed and overexpressed in cells where these modified proteins can interfere with the signaling cascades of a particular response. Indeed, this approach is often more informative and more flexible than using chemical inhibitors; however, it also is a method that suffers from nonspecificity. Dominant-negative molecules may titrate signaling proteins that are involved in a distinct cellular response, thus affecting the interpretation of the results. Another more direct method of examining gene function is to create mice deficient in the gene of interest and examining cellular responses in the context of the absence of this gene. This technique, while be-

ing straightforward and often highly informative, is limited in its application because of the difficulty in the technique itself and the time and expense of generating the knockout animals. Recently, another tool for studying gene function has become available to scientists—RNA-mediated interference (RNAi) technology. This technology was first used to study gene function in plants, but it now can be applied to mammalian cells. RNAi, also known as posttranscriptional gene silencing, is the inhibition of the expression of specific genes by double-stranded RNA. It is an evolutionary conserved mechanism likely having an essential role in mediating responses to exogenous RNA (like that from viruses) and in stabilizing the genome by sequestering repetitive sequences (such as transposons). RNAi is triggered by short interfering RNA (siRNA) molecules. These siRNAs can be generated in three ways: (i) long double-stranded RNA molecules are processed inside the cell to form siRNA by an enzyme called Dicer; (ii) chemically synthesized siRNAs can be transfected into cells; and (iii) siRNA sequences can be incorporated into ex-

pression plasmids or retroviral vectors and delivered into cells to generate the siRNA molecules. These last two techniques are useful for the study of mammalian gene function since the first method of using long double-stranded RNA molecules to generate siRNA can trigger an antiviral response in mammalian cells. Once the siRNA molecules are generated, they are incorporated into a complex known as the RNA-induced silencing complex (RISC) that binds to the target mRNA, which is then degraded by a nuclease.

Although not completely suppressing the expression of a particular gene, siRNA results in the "knockdown" of gene expression, an outcome that can be beneficial especially when the complete absence of a particular gene results in death of the cell or animal. Moreover, because of the high degree of sequence specificity to knockdown expression of a particular gene, siRNA has important medical implications. Conceptually, siRNA may be used to treat diseases that are caused by dominantly acting mutant alleles, such as in certain cancers, as well as combating viral infections.

After mitosis, phosphorylation events performed by Cdc2 are reversed and vesicles reassociate around chromatin. Golgi-derived vesicles self-associate into stacks, which are actively transported by motor proteins along MTs to the pericentriolar area of the cell. The stacks then fuse with each other to re-form the conventional Golgi structure.

Many cellular regulatory processes may now be more thoroughly investigated using RNAi, a reverse genetic tool to study gene function (see Box 3.2).

Conclusion

To conclude this chapter, it is worth quoting Georges Palade, a cell biology pioneer: "Although cells appear to have a large variety of mechanisms to fulfill their tasks, when they have found an efficient system they use it repetitively for different purposes." For example, nucleotide binding, reversible phosphorylation, proteolysis, and protein oligomerization are repeatedly used in many pathways. In unraveling how microorganisms interact with cells or how virulence factors damage the eukaryotic host, we should remember this fact.

Selected Readings

Bigay, J., P. Gounon, S. Robineau, and B. Antony. 2003. Lipid packing sensed by ArfGAP1 couples COPI coat disassembly to membrane bilayer curvature. *Nature* **426:**563–566.

> With artificial liposomes having a similar composition to that of Golgi membranes, the authors have shown that the COPI-type coat (which mediates vesicle trafficking between the endoplasmic reticulum and the Golgi apparatus) by provoking a membrane curvature-stimulated coat disassembly. Coat assembly is due to Arf1GTP and disassembly by hydrolysis of Arf1-GTP into Arf-GDP by ArfGAP1. By using liposomes of different diameters they observed that ArfGAP1 activity was greatest when liposomes had a small diameter. The results hint that membrane curvature serves as a sensor for controlling the timing of Arf1-GTP hydrolysis by ArfGAP1.

Bonifacino, J. S., and B. S. Glick. 2004. The mechanisms of vesicle budding and fusion. *Cell* **116:**153–166.

Brummelkamp, T. R., R. Bernards, and R. Agami. 2002. A system for stable expression of short interfering RNAs in mammalian cells. *Science* **296:**550–553.

Burridge, K., and K. Wennerberg. 2004. Rho and Rac take center stage. *Cell* **116:**167–179.

> A complete review with the last developments concerning the role of the small GTPases of the Rho family.

Carlier, M. F., C. le Clainche, S. Wiesner, and D. Pantaloni. 2003. Actin-based motility: from molecules to movement. *BioEssays* **4:**336–345.

> An important review focusing on the biochemical properties of the proteins involved in actin-based motility showing how they are used to generate force production based on in vitro reconstituted systems.

Chamaillard, M., S. E. Girardin, J. Viala, and D. J. Philpott. 2003. Nods, Nalps and Naip: intracellular regulators of bacterial-induced inflammation. *Cell. Microbiol.* **5:**581–592.

Dykxhoorn, D. M., C. D. Novina, and P. A. Sharp. 2003. Killing the messenger: short RNAs that silence gene expression. *Nat. Rev. Mol. Cell. Biol.* **4:**457–467.

Elbashir, S. M., J. Harborth, W. Lendeckel, A. Yalcim, K. Weber, and T. Tuschl. 2001. Duplexes of 21-nucleotide RNAs mediate RNA interference in cultured mammalian cells. *Nature* **411:**494–498.

> The authors for the first time show that 21-nucleotide siRNA duplexes specifically suppress expression of endogenous and heterogenous genes in mammalian cell lines, providing a new tool for studying gene function.

Gonzalez-Gaitàn, M., and H. Stenmark. 2003. Endocytosis and signaling: a relationship under development. *Cell* **115**:513–521.

Heemels, M.-T. (ed.). Nature insight: apoptosis 2000. *Nature* **407**:707–810 (multiple articles on apoptosis).

Johnson, E. 2002. Ubiquitin branches out. *Nat. Cell Biol.* **4**:E295–E298.

Kadonaga, J. T. 2004. Regulation of RNA polymerase II transcription by sequence specific DNA binding factors. *Cell* **116**:247–257.

Munro, S. 2003. Lipid rafts: elusive or illusive? *Cell* **115**:377–388.

A provocative but very comprehensive review of one of the most important emerging concepts of cell biology concerning the structure of cell lipidic membranes.

Murk, J. L. M., B. Humbel, U. Ziese, G. Posthuma, J. M. Griffith, J. W. Slot, B. Koster, A. Verkleij, H. J. Geuze, and M. J. Kleijmeer. 2003. Endosomal compartmentalization in three dimensions: implication for membrane fusion. *Proc. Natl. Acad. Sci. USA* **100**:13332–13337.

By electron tomography of cryoimmobilized cells, the authors demonstrate that in multivesicular endosomes the inner membranes are free vesicles. Hence, protein transport from inner to outer membranes cannot occur laterally in the plane of the membranes but requires fusion between the two membrane domains. This implies the existence of a machinery that mediates fusion between the exoplasmic leaflets of the membranes involved.

Murray, A.W. 2004. Recycling the cell cycle: cyclins revisited. *Cell* **116**:221–234.

Pawson, T. 2004. Specificity in signal transduction: from phophotyrosine-SH2 domain interactions to complex cellular systems. *Cell* **116**:191–203.

Pawson, T., and P. Nash. 2003. Assembly of cell regulatory systems through protein interaction domains. *Science* **300**:445–452.

An interesting review on how proteins involved in regulatory processes associate with one another and with phospholipids, small molecules, or nucleic acids.

Pollard, T. D., and G. Borisy. 2003. Cellular motility driven by assembly and disassembly of actin filaments. *Cell* **112**:453–465.

Ridley, A. J., M. A. Schwartz, K. Burridge, R. A. Firtel, M. H. Ginsberg, G. Borisy, J. T. Parsons, and A. R. Horwitz. 2003. Cell migration: integrating signals from front to back. *Science* **302**:1704–1709.

Takeda, K., T. Kaisho, and S. Akira. 2003. Toll-like receptors. *Annu. Rev. Immunol.* **21**:335–376.

Zerial, M., and H. McBride. 2001. Rab proteins as membrane organizers. *Nat. Rev. Mol. Cell Biol.* **2**:107–117.

Videos and Signaling Pathways Available through Websites

Videos of the main cell biology processes can be viewed at the *J. Cell Biol.* website: http://www.jcb.org/misc/annotatedvideo.shtml.

Major cell signaling pathways are depicted in the *Cancer Genome Anatomy Project* website: http://cgap.nci.nih.gov/Pathways.

4

Extracellular Matrix and Host Cell Surfaces: Potential Sites of Pathogen Interaction

KLAUS T. PREISSNER AND G. SINGH CHHATWAL

Tissue cells produce and release a variety of macromolecules, which form a complex structural network within the extracellular space. This extracellular matrix (ECM) not only serves as a structural support for resident cells but also provides a support for infiltrating pathogenic bacteria to colonize, particularly in the setting of injury or trauma. Similarly, integral cell surface or cell-associated adhesive components of the host are often recognized by pathogenic bacteria in a tissue- or cell-specific manner, and pathogens come into contact with host tissue fluids that also contain a variety of adhesive components. The initiation of infection requires bacterial adherence to nonphagocytic cells or the ECM, and many invasive microorganisms enter host cells after binding to specific surface structures (Box 4.1). Microbial binding may lead to structural and/or functional alterations of host proteins and to activation of cellular mechanisms that influence tissue and cell invasion by pathogens. Not only do particular mammalian receptors facilitate the entry of pathogens into host cells, but also microbes may escape certain antibiotics. Interactions with soluble or immobilized host components can mask the microbial surface and thereby interfere with antigen presentation and provide an overall immune evasion strategy.

Bacteria express surface-associated adhesion molecules, generally termed adhesins, that recognize the eukaryotic cell surface, the ECM protein, or carbohydrate structures. Nonadherent prokaryotes are rapidly cleared by the local nonspecific host defense mechanisms (peristalsis, ciliary movement, and fluid flow) or by turnover of epithelial cells and the mucus layer. Many of these adhesins, especially in gram-positive bacteria, also facilitate their invasion in eukaryotic cells. The identification and characterization of host cell and ECM adhesion components and complementary bacterial adhesins and invasins are thus central to understanding the molecular aspects of pathogenesis and, given the emergence of multidrug-resistant bacteria, will lead to alternative targets for antimicrobial therapy.

How do bacteria get into touch with the host tissue?

A number of specific bacterial interactions with host components determine the initial contact of pathogens and the subsequent infection cycle in the host organism. The connective tissue matrix and host cell surfaces provide a support for infiltrating pathogenic bacteria to colonize and invade, particularly under conditions of injury or trauma. Adhesion molecules found in the ECM, such as collagens, fibronectin, and other matrix proteins, or adhesion receptors, such as integrins on host cells, can interact directly with bacteria through their surface adhesins and facilitate the entry of pathogens into tissues. As a consequence, pathogens may also activate cellular mechanisms which influence their invasion. In contrast, nonadherent microbes are cleared by local nonspecific host defense mechanisms.

Structural and Functional Components of the Extracellular Matrix

The major structural components of the eukaryotic ECM are collagens that form different types of interstitial or basement membrane networks, including fibril-forming collagens (types I, II, and III) and the two-dimensional collagen type IV network (Table 4.1). For further organization of the ECM, additional collagenous and noncollagenous glycoproteins, proteoglycans, hyaluronan, and many other components, such as growth factors and proteases, become associated with interstitial or basement membrane ECM, giving rise to their specialized structure and function at different locations in the body. The main portion of interstitial ECM, which determines the specific character of each tissue or organ, is produced and deposited by different embedded connective tissue cells. Consequently, the tropism of pathogenic microorganisms is determined largely by the local composition of this complex network of host-derived factors.

More recently, research in molecular cell biology has indicated that in a variety of cellular systems, cellular shape, orientation, differentiation, and metabolism are intimately linked to and determined by the ECM. Of particular interest are cellular receptors for ECM components, such as integrins and proteoglycans that serve to make cell-to-cell or cell-to-ECM contacts, thereby providing a physical link between the cellular cytoskeleton (inte-

Table 4.1 Structural and functional components of the ECM

Interstitial ECM
    Fibril-forming collagens (I, II, III)
    FACIT (e.g., collagen IX)
    Fibronectin fibrillar network
    Elastin/fibrillin microfibrils
    Hyaluronan
Basement membranes
    Collagen (IV) network
    Laminin network
    Nidogen
Interstitial ECM and basement membranes
    Proteoglycans
    Adhesive and counter-adhesive glycoproteins
    Growth factors
    Proteases
    Transglutaminase

rior) and the ECM (exterior). The time-dependent modification and rebuilding of ECM at different locations in the body is essential for inflammation and wound healing. At these sites, the provisional ECM may provide bacterial entry and colonization, whereas intact tissues or epithelia, which are covered by a variety of adhesive and mucoidal components, are mostly protected against bacterial infection. Furthermore, pathologically disturbed ECM can often be the basis for the initiation and progression of various organ or tissue defects, resulting in the increased susceptibility to microbial pathogens. Conditions such as atherosclerosis, liver cirrhosis, glomerulonephritis and glomerulosclerosis, scleroderma, lung fibrosis, and micro- and macrovascular complications of diabetes are associated with dysfunctional ECM and may provide exposed sites for bacterial adherence. Besides stimulating inflammation, pathogens may directly cause alterations of the ECM, such as in rheumatic diseases.

Collagens

More than 25 different collagen gene products have been defined on a molecular basis, and most (>90%) of the body's collagens are of the fibril-forming types I to III. These collagens are produced and assembled during normal wound-healing processes but also contribute to the formation of granulation tissue or various types of fibrosis. Processing and posttranslational modifications of collagens are essential to achieve their triple-helical conformation, which is needed, for example, for mechanical stability in various fibrillar networks. Tissue transglutaminase-dependent cross-linking of ECM components with each other or with the eukaryotic cell surface adds to the tight linkage of epithelial cells to their underlying basement membrane. Some bacteria, such as *Staphylococcus aureus,* are covalently connected to host connective tissue fibronectin by an analogous mechanism. There are three different mechanisms by which cells (and bacteria) can interact with the various forms of collagen: (i) recognition of both triple-helical and denatured collagen (as in wound-healing areas or at sites of ECM degradation) in a conformation-independent fashion, (ii) a triple-helix-dependent binding mechanism, and (iii) a type of recognition that requires fibrillar structures. In contrast to adhesive glycoproteins, such as fibronectin or laminin, which have defined cell-binding domains (see below), collagens provide multiple interaction sites for cells (and bacteria) along the triple-helical structures within the fibrils or networks. Due to the tight association of other adhesion proteins, such as fibronectin, vitronectin, von Willebrand factor, laminin, nidogen, and proteoglycans with collagens to form supramolecular aggregates of variable structure and composition, the interaction with host cells or bacteria is not determined solely by the collagen component.

S. aureus is the most frequent cause of bacterial arthritis and osteomyelitis. Infection is initiated by hematogenous spread or by direct inoculation after trauma or surgery, and collagenous proteins present a common target for pathogenic bacteria that recognize the typical triple-helical structural motifs. The expression of a specific collagen-binding adhesin on S. aureus was shown to be necessary and sufficient for adhesion to cartilage. Moreover, matrix-associated bone sialoprotein provides another target for these microorganisms in inflamed areas of connective tissue. Contacts with host components often are initiated by different gram-negative bacteria utilizing their pili. As an example, fimbriae of uropathogenic *Escherichia coli* and the type III fimbriae of *Klebsiella pneumoniae* bind specifically to collagen type IV or V, respectively. Other bacterial adhesins, such as polymeric

YadA, which confers pathogenic functions of *Yersinia enterocolitica* associated with joint diseases, recognize multiple ECM components, including fibrillar collagens, at sites distinct from the tripeptidic Gly-X-Y repeats.

Group A streptococci interact with and adhere to collagen fibers directly or through recruitment mediated by fibronectin (Figure 4.1). The interaction of streptococci with collagen leads to bacterial aggregation, adherence, colonization, and resistance to phagocytosis and may therefore represent a novel streptococcal colonization and immune invasion mechanism. In infectious endocarditis, which is characterized by the formation of septic masses of platelets on the surfaces of heart valves and which is most commonly caused by streptococci, collagen-like "platelet aggregation-associated protein" of *Streptococcus sanguis,* as well as direct interactions with host ECM collagens, enhances platelet accumulation and subsequent bacterial colonization.

Laminins

As a major constituent of basement membranes, laminins are the first ECM protein to be produced during embryogenesis. Through its specific interactions with type IV collagen, proteoglycans, and other ECM components, as well as with several cell types, laminin fulfills a central structural and functional role within basement membranes. This 900-kDa glycoprotein has a cruciform structure and is composed of three different chains, α, β, and γ, linked to each other by disulfide bridges. Due to variations in chain composition, there are at least 10 different isoforms of laminin, which are produced by a large variety of cell types and whose biosynthesis appears to be tightly linked to the mesenchymal-epithelial-cell transition during tissue differentiation. Specific interaction sites exist for self-association and for binding of other basement membrane components or cell surface receptors, which promote disparate biological activities, including cell attachment, chemotaxis, neurite outgrowth, and enhancement of angiogenesis.

Exposure of laminin to pathogenic bacteria is most frequently seen at damaged or inflamed tissues. For example, following epithelial-cell denudation in the wounded human respiratory tract, *Pseudomonas aeruginosa* cells can attach to exposed basement membrane laminin via nonpilus adhesins. This pathogen is responsible for the most common infection observed in patients with cystic fibrosis, and further damage of the respiratory epithelium is mediated by bacterial as well as host-derived proteinases. Administration of protease inhibitors may dampen the fatal outcome of this genetic disorder. Moreover, both lipopolysaccharide and a protein adhesin of *Helicobacter pylori* mediate a specific interaction of this gastroduodenal pathogen with laminin, thereby interfering with the binding between gastric mucosal cells and laminin receptor, with possible loss of mucosal integrity. While *Streptococcus* strains (such as *S. viridans*) associated with endocarditis frequently express a laminin-binding adhesin(s), this bacterial receptor has a lower expression frequency in oral streptococcal isolates.

Elastin and Fibrillins

The elasticity of lung, skin, and other tissues, particularly blood vessels, is achieved by the recoiling property of elastic fibers embedded into the ECM, forming up to 50% of its dry mass. The main component (90%) of elastic fibers is elastin, which is a highly cross-linked polymer of the nonproteolytically modified, hydrophobic precursor tropoelastin. The other components of elastic fibers are 10- to 12-nm-thick microfibrils, which are composed of different

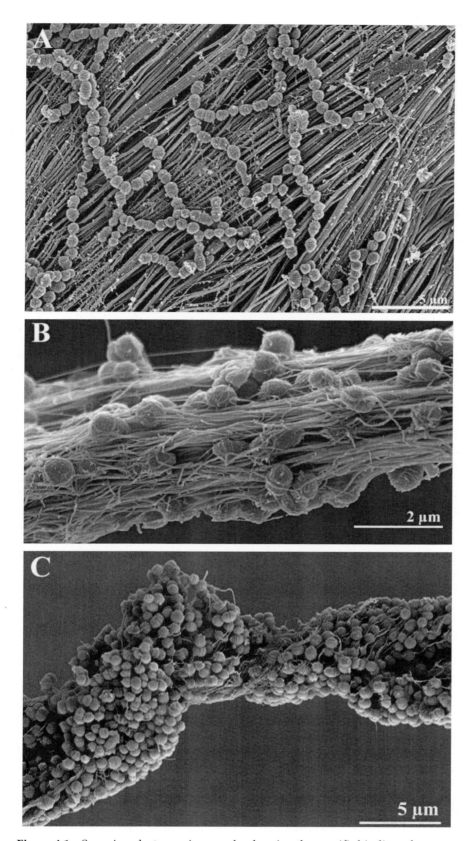

Figure 4.1 Scanning electron micrographs showing the specific binding of strepto-cocci to the ECM protein fibronectin in association with collagen **(A)**, indicating that fibronectin-mediated collagen recruitment leads to matrix deposition on and be-tween streptococcal cells and to their adherence on fibrillar collagen. The interaction of streptococci to fibrillar collagen alone leads to bacterial aggregation adherence and colonization **(B, C)**. Photograph: M. Rohde (Braunschweig, Germany).

fibrillins and other associated glycoproteins. A number of mutations within the fibrillin genes are known, and the clinical phenotype of Marfan syndrome is characterized by malformations of the skeletal muscle, cardiovascular, and ocular systems and the occurrence of bacterial endocarditis. As with many other ECM components, *S. aureus* expresses an elastin-binding protein that contributes to the colonization of elastin-containing organs, such as the lungs, skin, and blood vessels, by this bacterial species.

Proteoglycans and Hyaluronan

Proteoglycans represent the most abundant, heterogeneous, and functionally most versatile nonfibrillar component of ECM, and their properties are determined by the variable length and structures of the glycosaminoglycan side chains, which are covalently attached to a specific core polypeptide. Besides the most prominent secreted extracellular heparan proteoglycan, perlecan, which is found in all basal membranes of the body, decorin and biglycan are two smaller ECM proteoglycans with leucine-rich repeats. The specific binding of *Borrelia burgdorferi* to decorin via two adhesins may increase the adherence of the spirochete to collagen fibers in skin and other tissues.

Hyaluronan is a high-molecular-weight nonsulfated polysaccharide with a high-hydration capacity that is found in the ECM of most animal tissues. It binds tightly to the chondroitin sulfate proteoglycan aggrecan or other proteoglycan members and several link proteins. These aggregates determine the viscoelastic properties of cartilage or other tissues. Cell surface hyaluronan receptors, or "hyaladherins," such as the family of CD44 isoforms on a variety of normal and transformed cell types, contribute to ECM assembly and turnover or are crucial for cellular invasion. Recognition and binding of hyaluronan by *Treponema denticola* mediate the adherence of spirochetes to epithelial cells in periodontal tissue. Because gram-positive bacteria, such as group A streptococci, contain capsular hyaluronan, their binding to proteoglycans or CD44 may strengthen their interactions with host tissue. Likewise, capsular polysaccharides of *E. coli* K5 are structurally related to heparan sulfate and may contribute to the interaction of this bacterium with heparin-binding host proteins to gain access to ECM or cell surface sites. Conversely, hyaluronate- or heparin-degrading lyases of other microorganisms contribute to tissue destruction and may allow the penetration of pathogens.

Adhesive Glycoproteins

Upon vessel wall injury, particularly at sites of wound healing, initial adhesion of platelets to the exposed subendothelium and subsequent platelet aggregation are dependent on adhesive glycoproteins present in the subendothelial cell matrix as well as those stored inside platelets and secreted during this initial phase of hemostatic plug formation. In addition to their strong attachment-promoting activity, residing predominantly in the Arg-Gly-Asp (RGD)-containing epitope and being the predominant recognition site for integrins, these multifunctional proteins are of major importance in the initial adherence phase of pathogens. The circulating forms of these adhesive proteins may differ from those in the subendothelium and α-granules of platelets, due to alternative splicing (fibronectin), differences in the state of polymerization (von Willebrand factor), different conformational forms (vitronectin), or the transition into a self-aggregating molecule (fibrinogen-fibrin).

Fibronectin

Fibronectin is a ubiquitous adhesive matrix protein which is essential for the adhesion of almost all types of cells. The 30-kDa amino-terminal fragment contains the major acceptor site for factor XIIIa-mediated cross-linking and also bears the binding sites for heparin, fibrin, and bacteria, including *S. aureus*. Fibronectin serves as the prototype of adhesion proteins that bind specifically to microorganisms. It binds to over 16 bacterial species, and the interaction with gram-positive bacteria has been extensively characterized (Table 4.2). In particular, most *S. aureus* isolates, as well as various streptococcal strains, specifically bind and adhere to fibronectin, and the interaction is mediated by several different bacterial adhesins with highly homologous recognition motifs for the adhesion protein, also allowing host cell entry (Figure 4.2). In addition to the amino-terminal fragment of fibronectin, cooperative binding sites for staphylococci and streptococci are located within the fibrin-binding carboxy terminus. A streptococcal fibronectin-binding protein I (SFBI) uses a unique mechanism to mediate adherence and invasion. This protein has two fibronectin-binding domains, a repeat region (essential for adherence) which binds exclusively to the 30-kDa amino-terminal fragment of fibronectin and a spacer region (required for invasion) which binds to the 45-kDa collagen-binding fragment of fibronectin, and cooperative binding is observed. This fibronectin-binding protein therefore represents the highly evolved prokaryotic molecule that exploits the host fibronectin not only for extracellular targeting but also for a subsequent activation that leads to efficient cellular invasion (Figure 4.3). Although most characterized interactions with gram-negative bacteria involve recognition of lectins and carbohydrate structures in the host tissue, several types of fimbriae of enterobacteria exhibit specific interactions with fibronectin, laminin, or other adhesion proteins. Certain strains of *E. coli* provide surface-associated "curlin" subunits that mediate fibronectin-binding during wound colonization. In addition to fibronectin-mediated adherence of bacteria at sites of blood clots or damaged tissue, pathogens are able to colonize artificial devices, such as intraocular lenses, prosthetic cardiac valves, vascular grafts, prosthetic joints, and intravascular catheters, via biomaterial-adsorbed matrix proteins, particularly fibronectin and fibrinogen-fibrin. Synthetic peptides representing the fibronectin-binding domain of *S. aureus* adhesins, as well as antibodies against the fibrinogen-binding domain of *S. aureus* "clumping factor" (see below), have been successful in blocking bacterial colonization on implanted foreign materials.

Vitronectin

The multifunctional adhesive glycoprotein vitronectin is synthesized predominantly by liver cells. It is found as a single-chain polypeptide with a molecular mass of 78 kDa in the circulation and becomes associated as a multimeric, heparin-binding form with different ECM sites, particularly during tissue or vascular remodeling. Histochemical studies suggest that vitronectin is deposited in a fibrillar pattern in loose connective tissue, in association with dermal elastic fibers in skin and with renal tissue and the vascular wall at sites of arteriosclerotic lesions. Vitronectin functions as an inhibitor of cytolytic reactions of the terminal complement pathway and serves as a major regulator in cellular interactions related to migration and invasion.

Vitronectin has equivalent effects on cellular adhesion and bacterial binding to those demonstrated for fibronectin, yet specific interactions with

Table 4.2 Specific interactions of bacterial species with host ECM components

Microorganism	Collagen types I and II	Collagen type IV	Elastin	Fibronectin	Vitronectin	Fibrinogen-fibrin	Laminin	Thrombospondin	Bone sialoprotein II	Glycosamino-glycans
Aeromonas hydrophila	×									
Candida albicans		×		×	×	×				
Escherichia coli	×			×	×	×				
E. coli fimbriae		×								
Helicobacter pylori				×	×		×			×
Leishmania spp.				×						×
Mycobacterium tuberculosis										
Neisseria gonorrhoeae					×					
Plasmodium falciparum				×				×		×
Pneumocystis carinii				×						
Porphyromonas gingivalis					×					
Staphylococcus aureus	×		×	×	×	×	×	×	×	×
Staphylococcus (coagulase negative)			×	×	×				×	×
Streptococcal groups C and G				×	×	×				
Streptococcus dysgalactiae					×			×		
Streptococcus equi					×	×				
Streptococcus sanguis				×						
Streptococcus pyogenes		×		×	×	×	×			×
Streptococcus pneumoniae					×					
Treponema denticola				×			×			
Treponema pallidum				×	×					
Trypanosoma cruzi										×
Vibrio cholerae			×							
Yersinia spp.				×						

bacteria are likely to occur in damaged or altered tissues where the protein is deposited. Specific interactions of vitronectin with various strains of staphylococci and group A and G streptococci, as well as with gram-negative bacteria, have been described (Table 4.2). The adhesin(s) of *S. aureus* responsible for vitronectin interaction may also bind to heparan sulfate. The adherence of streptococci to the luminal side of cultured endothelial cells or to epithelial cells is mediated by fibronectin-independent vitronectin-specific interactions. Hemopexin-type repeats in vitronectin, as well as in hemopexin itself, have been identified as primary binding sites for group A streptococci. For internalization, gonococci expressing the OpaA adhesin require interaction with heparan sulfate proteoglycans on the host cell surface followed by entry in a vitronectin-dependent fashion.

Fibrinogen-Fibrin

The fibrin precursor of fibrinogen is a major 350-kDa plasma glycoprotein, composed of two identical sets of three polypeptide chains, and serves as predominant macromolecular substrate for thrombin in the blood-clotting cascade. Together with vitronectin and other adhesion proteins, the fibrin clot constitutes the majority of the initial provisional ECM network for sealing a wound site, thereby protecting this area against infiltration by opportunistic, infectious microorganisms. Cell surface receptors for fibrinogen that belong to the family of integrins have been identified on mammalian cells, of which the platelet $\alpha_{IIb}\beta_3$-integrin (glycoprotein IIb/IIIa complex) is principally required for platelet aggregation. $\alpha_M\beta_2$-Integrin (Mac-1, complement receptor 3) on phagocytes may also recognize the ligand in situations of wound healing and defense in which phagocytotic clearance of fibrin(ogen)-associated clot or (bacterial) cell fragments is required. Due to its dimeric structure, fibrinogen may serve as a bridging component between surface receptors on different cells or other extracellular sites once they become exposed. Likewise, *S. aureus* adherence to endothelial cells is mediated predominantly by fibrinogen as a bridging molecule leading to acute endovascular infections. Although *S. aureus* can induce platelet aggregation via a fibrinogen-dependent mechanism, this process is independent of the aforementioned principal $\alpha_{IIb}\beta_3$-integrin-binding interactions with fibrinogen.

Different bacterial fibrinogen-binding proteins mediate pathogen colonization in wounds or catheters; the proteins expressed by *S. aureus* are the best characterized. In particular, "clumping factor" serves to recognize the carboxy-terminal portion of the fibrinogen α-chain in a manner analogous to $\alpha_{IIb}\beta_3$-integrin binding. Moreover, the homology between metal ion-dependent adhesion sites of integrin subunit α_{IIb} or α_M or clumping factor and an integrin-like protein from *Candida albicans* indicates common mechanisms of fibrinogen binding in mammalian cells, lower eukaryotes, and prokaryotes. Staphylocoagulase serves as an additional fibrinogen-binding factor, which is not involved in bacterial clumping but, due to prothrombin binding and conversion, serves to promote fibrin formation or bacterial attachment onto fibrinogen-coated surfaces. Fibrinogen binding to streptococci of groups A, C, and G leads to inhibition of complement fixation and subsequent phagocytosis, indicating an important role for this interaction. Bacterial colonization is reduced in a mouse mastitis model by vaccination with *S. aureus* fibrinogen-binding proteins, providing a new concept for antimicrobial therapy.

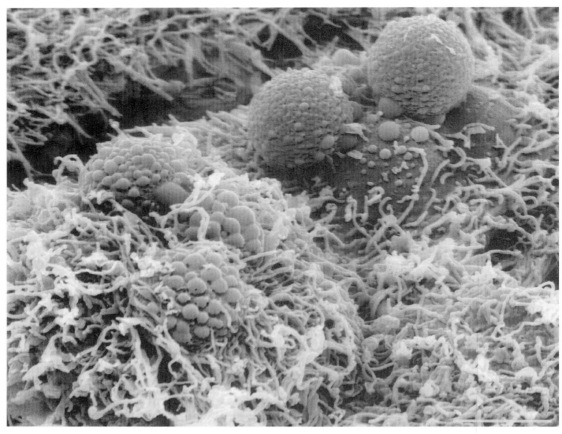

Figure 4.2   Scanning electron micrographs demonstrating the interaction between latex beads, coated with fibronectin-binding streptococcal adhesin SFBI, and HEp-2 cells. Different stages of adhesion, engulfment, and internalization are apparent. Bar, 3 μm. Photograph: M. Rohde (Braunschweig, Germany).

In addition to staphylococcal or streptococcal surface proteins that interact with ECM components, proteins released by these bacterial species can directly influence fibrin formation or dissolution. While staphylocoagulase binds prothrombin and mediates its conversion to thrombin, staphylokinase and streptokinase interact stoichiometrically with plasminogen, resulting in plasmin formation, whereby the former fibrinolytic agent acts in a fibrin-specific manner. These strategies apparently allow effective fixation and subsequent penetration of bacteria into wound areas.

Thrombospondin and Other "Matricellular" Proteins

Members of the thrombospondin family of extracellular proteins are found in different tissues. Together with structurally unrelated members of the tenascin protein family as well as with osteonectin (SPARC, BM40) or osteopontin, they associate with collagen fibrils or basement membranes, promoting divergent cellular functions including counteradhesive activities. While thrombospondin interactions with bacteria have been studied to some extent, bacterial binding to the other matricellular proteins remains to be analyzed. In particular, *S. aureus* adherence to activated platelets, to blood clots, or to ECM in pyogenic infections is mediated by heparin-dissociable interactions with thrombospondin. Importantly, thrombospondin binds *Plasmodium falciparum*-parasitized erythrocytes and, together with its cell surface receptor, CD36, mediates their adherence to endothelial and

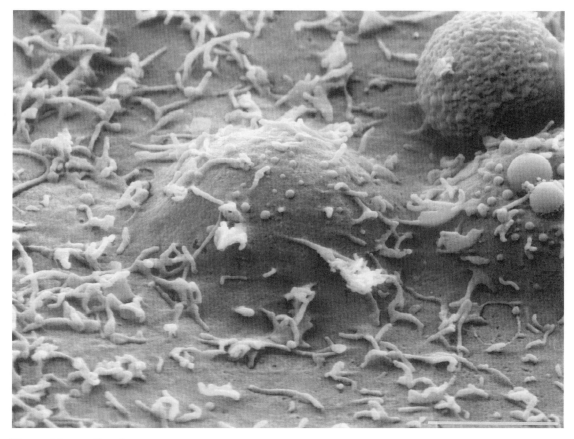

Figure 4.2 *continued*

other cells; it has thus has been implicated as an adhesive mediator of malaria infection.

ECM Degradation

A prerequisite for bacteria to invade normal tissues or wound sites is the degradation of matrix proteins, accomplished by bacterial collagenases or elastases, as produced by various clostridial strains. Host-derived plasmin appears to play a prominent role in pathogen-mediated tissue destruction as well. In situ, urokinase-type and tissue-type plasminogen activators, which are produced and secreted by a variety of eukaryotic cells, particularly under inflammatory conditions, are responsible for plasmin formation by limited proteolysis. A number of bacteria, such as *Yersinia pestis,* produce plasminogen activators that activate host-derived plasminogen. The spirochete *B. burgdorferi,* the causative agent of Lyme disease (characterized by inflammatory manifestations in the skin, joints, heart, and central and peripheral nerve systems), induces monocytes to secrete urokinase-type plasminogen activator. Moreover, due to the availability of plasmin(ogen) receptors on the surface of this microorganism, accelerated plasmin formation and protection against inactivation by host inhibitors is achieved, resulting in efficient pathogen invasion. Likewise, other pathogens, such as *E. coli,* group A, C, and G streptococci, and *Neisseria gonorrhoeae,* express surface receptors for plasmin(ogen) that facilitate pericellular proteolysis and invasion. The structural similarity of neutrophil-derived polypeptide "defensins" to plasminogen kringle motifs suggests that their antimicrobial activity can be related to

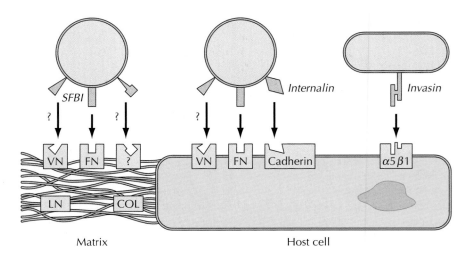

Figure 4.3   Schematic representation of bacterial interactions with host cells and ECM mediated by adhesion molecules. COL, collagens; FN, fibronectin; LN, laminin; VN, vitronectin; SFBI, fibronectin-binding streptococcal adhesin.

interference with plasmin formation, thereby preventing the spread of infection. Moreover, some bacterial glycolytic enzymes, such as α-enolase or glyceraldehyde-3-phosphate dehydrogenase (which exhibit highly conserved structures as compared with their eukaryotic counterparts), are expressed on the outer surface of streptococci and staphylococci, respectively, and serve as multifunctional binding proteins for, e.g., host plasmin(ogen) or ECM components. Besides certain signaling interactions with host cells, glyceraldehyde-3-phosphate dehydrogenase also serves as a transferrin receptor. As an acute-phase reactant in host defense, the circulating broad-spectrum proteinase inhibitor α_2-macroglobulin can eliminate complexed proteinases via receptor-mediated endocytosis and may do so with microbial enzymes as well. Interestingly, various strains of streptococci exhibit specific binding to α_2-macroglobulin and may thereby gain access to host tissues, possibly via the α_2-macroglobulin receptor.

Cell Surfaces and Bacterial Interactions

The bacterial tropism to colonize the restricted range of hosts, tissues, and cell types is dependent, among other factors, on the availability of specific host cell receptors for a given bacterial adhesin(s). Although not all bacterium-host cell interactions result in cellular entry, adherence to epithelial mucous membranes, particularly of the respiratory, gastrointestinal, and urogenital systems, or entry via the eye, ear, or wound sites is a prerequisite for the infectious process and is considered an important virulence factor. The predominant parts of the accessible host cell surface glycocalyx are membrane-anchored glycoconjugates, including glycoproteins, proteoglycans, and glycolipids. In addition, secreted and epithelial surface-associated mucin glycoproteins are thought to act in a lubricative and protective manner by shielding the gastrointestinal, respiratory, and urogenital tracts against physical damage, dehydration, and bacterial infection. Despite the common oligosaccharide structures in the different groups of glycoconjugates, their high compositional diversity allows variability among species, among cell types, and as a function of tissue differentiation. Conversely, the pathogens that colonize the mucosal surfaces are also rich in carbohydrates, and lectin-like activities are often characteristic of pilus-containing gram-negative and eukaryotic microorganisms. Mannose is frequently present in N-linked saccharides of glycoproteins and is recognized by a large number of bacteria known to bind mannose. However, pathogens may also express glycolipid-specific adhesins, indicating that a combination of recognition specificities confers bacterial tropism. Cell surface glycoconjugates thus function as nonselective coreceptors, and pathogen internalization is only achieved together with the primary bacterial receptor.

Lectins, Proteoglycans, and Mucins

Two major families of animal lectins, the calcium-dependent C-type lectins and the S-type lectins, have been classified based on their carbohydrate-binding structures and functional properties. The C-type lectins are either soluble extracellular proteins or integral membrane proteins and include receptors involved in the uptake of plasma glycoproteins in the liver, such as the asialoglycoprotein receptor; adhesion receptors, such as selectins involved in leukocyte homing; and collectins, such as the serum mannan-binding protein involved in complement activation and phagocytosis. High-affinity binding is conferred by multivalent interactions between multimeric C-type lectins and complementary complex saccharides on the bacterial surface. The mannan-binding proteins present on granulocytes and macrophages and in serum are the best characterized, with the latter being able to activate complement in an antibody-independent manner after encountering cognate carbohydrates on microbial surfaces. Complex mannose structures present on bacteria, yeasts, and host-derived oligosaccharides of viruses play a role in pathogenesis by binding to mannan-binding proteins. Evidence for the biological relevance and host defense property of these interactions is based on the fact that individuals with a genetic deficiency in the mannan-binding protein have increased susceptibility to certain infectious diseases. Additional members of the C-type lectins (also designated the collectin family) include conglutinin and the lung-associated proteins, surfactant proteins A and D. Attachment of *Pneumocystis carinii* or *S. aureus* to specialized epithelium or alveolar macrophages, respectively, is mediated by these collectins through interactions between their C-type lectin domains and bacterial carbohydrates. Because the collagen-type helices of these collectins are recognized by the collectin receptor (also designated C1q receptor), which is present on the lung epithelium and on many other (phagocyte) cell types, they serve a bridging function in bacterium-host cell adherence.

The S-type lectins include several soluble proteins that are characterized by their affinity for lactose and other β-galactosides. They are composed of either two identical or two different carbohydrate-binding subunits, which serve as bridges between bacterial and host cell carbohydrates at various mucosal epithelial surfaces in a calcium-independent manner. Due to the widespread appearance of β-galactosides as structural determinants of bacterial carbohydrates, S-type lectins found in the intestines, lungs, and kidneys, as well as on macrophages, are able to recognize various pathogens.

By analogy to ECM proteoglycans, membrane-associated cell surface proteoglycans of the integral membrane type (syndecan family), as well as the glycosylphosphatidylinositol-anchored type (glypican family), directly interact with matrix proteins, proteinase inhibitors, and their target enzymes, growth factors, or lipoproteins. Consequently, cellular proteoglycans are thought to contribute to cell adhesion, cell growth, and the control of proteolytic and lipolytic pathways, particularly in the vasculature, by influencing the local concentration, stability, conformation, activity, and clearance of various protein ligands.

In particular, cell-associated heparan sulfate proteoglycans are used by several microbes, such as *Plasmodium, Leishmania,* and *Trypanosoma* species, to enter host tissue. Bacterial binding is significantly reduced by secreted heparinases that may either dissociate cell-adherent microorganisms or weaken cell-ECM interactions and decrease the stability of the ECM. All these processes may increase bacterial penetration into host tissues. Moreover, soluble heparan sulfate (or heparin) effectively interferes with *Trypanosoma cruzii* invasion mediated by the adhesin penetrin, indicating that successful bacterial invasion depends on the interplay between soluble and immobilized host factors. *Chlamydia trachomatis* expresses a heparan sulfate-like glycan that links the bacterium to host cell heparin-binding proteins, thereby using a trimolecular complex for adherence and invasion. Finally, the epithelial-cell mucosal barriers in the body provide different high-molecular-weight mucin glycoproteins (containing 50 to 80% carbohydrate) which exhibit considerable genetic polymorphism among individuals. In addition to their protective properties, particular mucins, such as episialin (MUC1), serve cell-adhesive functions in normal and tumor tissues or mediate selective binding of, e.g., *Haemophilus influenzae* with airway epithelium.

Glycolipids

Several hundred eukaryotic cell surface glycolipid structures are known, and they belong to different core saccharide series such as the lacto-, the ganglio-, the globo-, or the muco-series. These core structures often contain terminal antigenic determinants, including the blood group ABO(H) and Lewis antigens or sialic acid and sulfate groups. Despite the high diversity of saccharide structures linked to the membrane-anchored lipid tail, each glycolipid has only one saccharide, as opposed to glycoproteins and proteoglycans, which often contain several different oligosaccharides linked to the same polypeptide. Because blood group antigens are also expressed on cells comprising the mucosal surfaces of the intestine and other organs, increased bacterial adherence and also risk of infectious diseases may be associated with the appearance of these glycolipid components, as exemplified by the binding of *H. pylori* to the mucous cell Lewis[b] blood group antigen. In addition, most common urinary tract infections with *E. coli* involve galactose-specific adherence. The phagocytosis (termed lectinophagocytosis) of bacteria by animal cells such as peritoneal macrophages and polymorphonuclear leukocytes is mediated by bacterial surface lectins behaving as glycolipid receptors. The specificity of these interactions is undefined.

In particular, lactosyl-ceramide is present in the colon epithelium but not in the small intestinal epithelium and serves to establish the adhesion of gram-negative and gram-positive bacteria to their specific host cell-binding factors. In contrast to the terminally placed recognition sequence for bacteria, glycolipid "isoreceptors" contain an internal recognition structure that may affect the overall binding of microorganisms with regard to their affinity and specificity. Conversely, some gram-negative mucosal pathogens express glycolipids that are identical in their chemical and antigenic properties to carbohydrates of glycosphingolipids present in host cells or tissues. The availability of capsular polysaccharides of the lactoneo-series type with similar antigenicity to host glycolipids suggests that these structures are involved in the mechanism of survival or invasion of bacteria in human tissues.

Cell Surface Adhesion Receptors

Physiological processes in the host that are related to cell activation, proliferation, differentiation, or cellular motility require cell-cell or cell-ECM contacts and are differentially mediated by adhesion receptors, including selectins, cadherins, and immunoglobulin family members, as well as ubiquitously expressed heterodimeric integrins. Integrins provide a physical linkage between the extracellular environment and the intracellular cytoskeleton; they are intimately involved in signal transduction pathways and thereby directly control the principal cellular processes and cell survival. Integrins appear to be a common target for pathogens and for subsequent manipulation of the host cell machinery to gain microbial entry. The general strategy of microbes to engage these essential host adhesion molecules during the course of pathogenesis may result in the loss of important cellular functions.

Integrins of the β_1 subclass are present on epithelial, mesenchymal cells and on circulating blood cells, β_2-integrins are expressed exclusively on leukocytes and are up-regulated during inflammatory processes and infectious diseases, and β_3-integrins along the vasculature and on platelets are the major target receptors for interaction with pathogenic microorganisms (Table 4.3). Three different strategies may be used by bacteria to infiltrate the host system and thereby promote intracellular survival. (i) Microorganisms express an RGD-containing or RGD-like peptide that serves as a recognition site for integrins (ligand mimicry), such as the *E. coli* outer membrane protein intimin, which interacts with β_1-integrins, or *B. burgdorferi* proteins, which bind to activated integrins on platelets. Some microorganisms, such as *Mycobacterium avium* and *Candida* strains, express proteins whose structure resembles that of integrin (receptor mimicry). Alternatively, complex mimicry of host cellular recognition molecules is found in bacterial surface structures (such as the filamentous hemagglutinin of *Bordetella pertussis*) that may interact with glycoconjugates, heparin, and β_2-integrin or its ligands. (ii) The mi-

Table 4.3 Microbial interactions with host adhesion receptors

Microorganism and ligand	Integrins ($\alpha\beta$) and other receptors
Yersinia invasin	$\alpha_3\beta_1$, $\alpha_4\beta_1$, $\alpha_5\beta_1$, $\alpha_6\beta_1$, $\alpha_v\beta_1$
Histoplasma capsulatum	$\alpha_L\beta_2$, $\alpha_M\beta_2$, $\alpha_x\beta_2$
Leishmania LPG, LPS	$\alpha_M\beta_2$, $\alpha_x\beta_2$
Borrelia burgdorferi	$\alpha_{IIb}\beta_3$
Neisseria meningitidis Opc protein	NCAM
Neisseria gonorrhoeae	CD66, proteoglycans
Plasmodium falciparum	ICAM-1, VCAM-1, E-selectin, CD36
Staphylococcus aureus Eap	ICAM-1
Bordetella pertussis	P-selectin, E-selectin, $\alpha_M\beta_2$
Shigella flexneri	$\alpha_5\beta_1$
Listeria monocytogenes	E-cadherin
Adenovirus penton base	$\alpha_v\beta_3$, $\alpha_v\beta_5$
HIV-1[a] Tat protein	$\alpha_5\beta_1$, $\alpha_v\beta_3$

[a] HIV-1, human immunodeficiency virus type 1.

croorganism binds to a non-RGD recognition site on the integrin (ancillary recognition). (iii) The bacterium binds to a natural integrin ligand, leading to bacterial entry (masking), as exemplified by *N. meningitidis*. Bacterial mimicry of the natural ligands for leukocyte integrins also takes into account the fact that endotoxin, coagulation factor X, and complement C3bi serve as RGD-independent ligands for β_2-integrins. Once the bacterium is attached to the host cell, the subsequent fate of the pathogen is determined by the activation of the receptor, the ligand used by the bacterium, the signal transduction pathway from the integrin to the cytoskeleton, or other host cell responses such as degranulation. As a prerequisite for entry of *Yersinia* species into host cells, high-affinity binding of the bacterial adhesin invasin to various β_1-integrins allows effective competition with low-affinity binding of host ligands, such as fibronectin, and subsequent invasion can interfere with signal transduction via accessory proteins (Figure 4.3). Interestingly, the F1 adhesin of *Streptococcus pyogenes* that, together with fibronectin and integrins, allows entry of bacteria into host cells can also inhibit fibronectin matrix assembly and thereby disturbs the host's defense machinery.

The β_2-integrins on circulating blood cells are crucial for the arrest of leukocytes at the endothelium and their subsequent recruitment into tissues toward an inflammatory or infectious stimulus. In the rare condition, leukocyte adhesion deficiency, patients suffer from severe bacterial infections, affecting the oral, respiratory, and urogenital mucosa as well as the intestine and skin. Microorganisms that must survive within tissue macrophages are often internalized via β_2-integrins. Pathogens do so by either synthesizing a ligand(s) that is recognized by β_2-integrins or binding to integrin ligands such as C3bi (e.g., *Legionella* and *Leishmania* species), which subsequently induce phagocytic uptake without activating the macrophage antimicrobial cytotoxic machinery. Several lines of evidence have documented that integrin-mediated internalization of microorganisms involves the host cell cytoskeleton, differing from surface adhesion of microbes on phagocytic or nonphagocytic cells.

Additional cell surface adhesion molecules that are recognized by microbes include endothelial counterreceptors of integrins, such as ICAM-1, VCAM-1, or E-selectin; CD36; and the hyaluronan receptor CD44 (Table 4.3). The pathogens may either directly bind to these cytoskeleton-associated host receptors or interact with host-derived secreted polysaccharides or adhesion proteins, which subsequently interact with the above-mentioned receptors and thereby indirectly mediate the binding of the microorganism or support internalization. Of particular interest is a secretable extracellular adherence protein (Eap) from *S. aureus* that can rebind to the bacterial surface and thereby mediates multiple interactions with complex host ECM structures. Due to its specific binding to endothelial cell-expressed ICAM-1, not only does Eap attach *S. aureus* to the vessel wall, but isolated Eap inhibits the recruitment of host leukocytes to, for example, the site of wound infection or inflammation or interferes with T-cell-mediated responses. This dual role of Eap may help to understand the pathogenetic causes in nonhealing wounds and allows the generation of a new type of anti-inflammatory substance based on a bacterial product. The hyaluronan receptor CD44 mediates bacterial phagocytosis in human neutrophils by a cytoskeleton-dependent mechanism, a process that was augmented by the CD44 ligand hyaluronan. Calcium-dependent adhesion molecules, such as cadherins involved in homotypic intercellular junction interactions, are utilized by *Listeria* species for invasion. E-cadherin on epithelial cells was

Novel antimicrobial strategies to circumvent the emergence of antibiotic resistance

The emergence of pathogens resistant to conventional antimicrobial therapy, as well as harmful side reactions due to cross-reactivities between bacterial surface proteins and host factors, requires novel strategies for the prevention and treatment of infectious diseases. Based on the identification of adhesion and invasion mechanisms of pathogens and the relevant components involved in these processes at different host niches, novel "antiadhesion therapies" can be designed that include substances such as soluble bacterial adhesins or host receptor analogues, antibodies against bacterial adhesins, and low-molecular-weight antagonists. Since each microorganism can use multiple mechanisms of adhesion to initiate infection, effective antiadhesion drugs should contain cocktails of different inhibitors. These strategies are, however, complicated by the fact that not all bacterial adhesin genes are present in every strain and that environmental factors can affect the expression of these genes. Thus, the knowledge of strategies used by microorganisms for their tissue and cellular entry may be converted into beneficial therapeutic regimens.

shown to be targeted by surface-located internalin of *Listeria monocytogenes* as a prerequisite for cellular entry.

Conclusion

The emergence of pathogens resistant to conventional antimicrobial substances and the possible harmful side effects of cross-reactivities between, for example, streptococcal M proteins and host factors or the large number of distinct serotypes require novel strategies for the prevention and treatment of infectious diseases. Based on the identification of adhesion and invasion mechanisms of pathogens and the relevant components involved in these processes at different host niches, novel therapeutic interventions can be designed. These are summarized as "antiadhesion therapies" for microbial diseases and include substances such as soluble bacterial adhesins or host receptor analogues, antibodies against bacterial adhesins, and low-molecular-weight adhesion/invasion antagonists (Figure 4.3). Among these, a new class of virulence factors of gram-positive bacteria include anchorless adhesins and invasins that are secreted by bacteria by a yet unknown mechanism and reassociate with the microbial surface. These ECM-binding bacterial proteins or adhesins can be used as candidates in an antiadhesin vaccine, and proof-of-principle approaches have already been successful in different animal models of infection or sepsis. These strategies are, however, complicated by the fact that not all adhesin genes are present in every strain and that environmental factors can affect the expression of these genes. Because each microorganism can use multiple mechanisms of adhesion to initiate infection, effective antiadhesion drugs may contain cocktails of different inhibitors (Box 4.2).

In several vascular diseases, vessel injury promotes the development of wounds and the predisposition toward bacterial infections, as observed in diabetes-associated progressive nonenzymatic glycation of proteins and lipids. Certain invasive bacteria may utilize these modified tissues for binding, as was shown for *Pseudomonas* soil strains. Moreover, the presence of bacteria in degenerated vessels suggests an association with atherosclerosis, and, indeed, dental infections pose a serious risk for acute myocardial infarction. Although causal relationships are not sufficiently clarified in these cases, antibacterial interventions may also be of benefit for the associated vascular complications.

Selected Readings

Chavakis, T., M. Hussain, S. M. Kanse, G. Peters, R. G. Bretzel, J.-I. Flock, M. Herrmann, and K. T. Preissner. 2002. *Staphylococcus aureus* extracellular adherence protein (Eap) serves as anti-inflammatory factor by inhibiting the recruitment of host leukocytes. *Nat. Med.* **8:**687–693.

In this "trick and treat" paper, Eap was found to inhibit leukocyte-endothelial cell interaction in addition to its adherence function for bacteria. Eap thereby not only protects bacteria from host defense mechanisms but can act, independent of bacteria, as an efficient anti-inflammatory factor for new therapeutic strategies.

Cheung, A. L., M. Krishnan, E. A. Jaffe, and V. A. Fischetti. 1991. Fibrinogen acts as a bridging molecule in the adherence of *Staphylococcus aureus* to cultured human endothelial cells. *J. Clin. Investig.* **87:**2236–2245.

This is another example in which the symmetric, divalent fibrinogen molecule is able to cross-link bacteria with host cells and thereby promote cell-cell contact. Similarly, fibrinogen serves to bridge platelets (in platelet aggregation) as well as leukocytes and endothelial cells.

Chhatwal, G. S. 2002. Anchorless adhesins and invasins of Gram-positive bacteria: a new class of virulence factors. *Trends Microbiol.* **10:**205–208.

Chhatwal, G. S., K. T. Preissner, G. Müller-Berghaus, and H. Blobel. 1987. Specific binding of the human S protein (vitronectin) to streptococci, *Staphylococcus aureus,* and *Escherichia coli. Infect. Immun.* **55:**1878–1883.

Coleman, J. L., J. A. Gebbia, J. Piesman, J. L. Degen, T. H. Bugge, and J. L. Benach. 1997. Plasminogen is required for efficient dissemination of *B. burgdorferi* in ticks and for enhancement of spirochetemia in mice. *Cell* **89:**1111–1119.

In plasminogen knockout mice the dissemination and invasion of bacteria were significantly reduced, making a strong case for the contribution of this fibrinolytic factor in the early stages of microbe-host tissue interactions.

Flock, J. I. 1999. Extracellular-matrix-binding proteins as targets for the prevention of *Staphylococcus aureus* infections. *Mol. Med. Today* **5:**532–537.

Hynes, R. O. 1992. Integrins: versatility, modulation, and signaling in cell adhesion. *Cell* **69:**11–25.

Karlsson, K. A. 1995. Microbial recognition of target-cell glycoconjugates. *Curr. Opin. Struct. Biol.* **5:**622–635.

Kuusela, P., T. Vartio, M. Vuento, and E. B. Myhre. 1984. Binding sites for streptococci and staphylococci in fibronectin. *Infect. Immun.* **45:**433–436.

Mengaud, J., H. Ohayon, P. Gounon, R.-M. Mege, and P. Cossart. 1996. E-cadherin is the receptor for internalin, a surface protein required for entry of *L. monocytogenes* into epithelial cells. *Cell* **84:**923–932.

A classical case of bacteria-host cell contact in which a bacterial surface protein is specifically recognized by a host adhesion receptor, with consequences for subsequent cellular functions. The knowledge of the atomic structure of the participating contact sites may give sufficient information for the design of new drugs that are able to block bacterial invasion.

Ofek, I., I. Kahane, and N. Sharon. 1996. Toward anti-adhesion therapy for microbial diseases. *Trends Microbiol.* **4:**297–299.

Pancholi, V., and V.A. Fischetti. 1998. a-Enolase, a novel strong plasmin(ogen) binding protein on the surface of pathogenic streptococci. *J. Biol. Chem.* **273:**14503–14515.

Patti, J. M., and M. Höök. 1994. Microbial adhesins recognizing extracellular matrix macromolecules. *Curr. Opin. Cell Biol.* **6:**752–758.

Relman, D., E. Tuomanen, S. Falkow, D. T. Golenbock, K. Saukkonen, and S. D. Wright. 1990. Recognition of a bacterial adhesin by an integrin: macrophage CR3 (αMβ2, CD11b/CD18) binds filamentous hemagglutinin of *Bordetella pertussis. Cell* **61:**1375–1382.

Roberts, D. D., J. A. Sherwood, S. L. Spitalnik, L. J. Panton, R. J. Howard, V. M. Dixit, W. A. Frazier, L. H. Miller, and V. Ginsburg. 1985. Thrombospondin binds falciparum malaria parasitized erythrocytes and may mediate cytoadherence. *Nature* **318:**64–66.

Virji, M. 1996. Microbial utilization of human signalling molecules. *Microbiology* **142:**3319–3336.

5

Bacterial Adherence to Cell Surfaces and Extracellular Matrix

B. Brett Finlay and Michael Caparon

An essential step in the successful colonization and production of disease by microbial pathogens is their ability to adhere to host cell surfaces and the underlying extracellular matrix. Although different pathogens can adhere to nearly any site in the body, each pathogen usually colonizes one or a few particular sites in the host. The ability to adhere to specific host molecules enables a pathogen to target itself to a particular tissue within the body, thereby giving the pathogen tissue tropism. In the past few years, it has become apparent that many pathogens selectively adhere to cell adhesion receptors, which then triggers signaling pathways in the host cell. Such adherence can also be the first step in penetrating further into the body by initiating invasion through activation of these cell signaling pathways (see chapter 6). If the pathogen also produces toxins that cause tissue damage (see chapter 13), destruction of host cells will expose the extracellular matrix, to which many pathogens can bind. Finally, most pathogens possess many molecules that mediate adherence to host cells and the extracellular matrix (adhesins). Often these adhesins are synergistic in their function, thereby enhancing adherence, while others appear to be functionally redundant. Alternatively, they can be differentially regulated such that specific adhesins are expressed under different environmental conditions, enabling a pathogen to express different adhesins as it progresses through different sites within its host. Not all adhesins are essential virulence factors, and the specific role of a particular adhesin in disease has been surprisingly difficult to define due to the complication of multiple adherence factors in most pathogens. Many adhesins are found in both nonpathogenic and virulent strains. However, in most cases adhesins seem to play some role in disease, often by increasing the virulence of a pathogen. If more than one adhesin is deleted from a pathogen, the effect on virulence is usually more pronounced than deletion of any single adhesin.

The choice of host cell substrate that a pathogen can adhere to is large. The mammalian cell surface contains many proteins, glycoproteins, glycolipids, and other carbohydrates that could potentially serve as a receptor for an adhesin. Additionally, the extracellular matrix provides a rich source of glycoproteins for adhesins to bind to and even initiate signaling, and im-

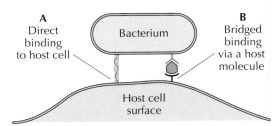

Figure 5.1 Bacterial adherence to host cell surfaces. Bacteria can adhere to host cell surfaces either directly by using a surface-anchored adhesin or indirectly by binding a soluble host component that then serves as a bridge between the bacterial adhesin and the natural host receptor for that molecule.

planted devices remain a major target for bacterial adherence. Although many bacterial adhesins have been identified, only a few of their cognate receptors have been identified. These receptors encompass a wide array of cell surface and matrix molecules, although they generally include a carbohydrate moiety exposed on the host molecule.

Bacterial adhesins are usually proteins, located either at the tip of a scaffold-like structure on the bacterial surface (called a pilus or fimbriae), or anchored in the bacterial membrane but surface exposed (nonfimbrial or afimbrial adhesins). It is becoming apparent that the molecular machinery needed to build pili, or that needed to transport an afimbrial adhesin to the bacterial surface, is often conserved. This machinery is often complex in gram-negative organisms due to the difficulties of transporting molecules across the two membranes. Despite this conservation, the receptor specificity of the adhesin is dictated by a portion of the adhesin exposed at the pilus tip or on the bacterial surface. The concepts of bacterial adhesion are illustrated below by examining a few well-characterized examples of adhesins, both fimbrial and nonfimbrial in nature. Alternatively, many pathogens are capable of binding soluble host molecules such as matrix proteins, complement, etc. These molecules can then serve as a bridge between the bacterium and host cell surface when these molecules bind to their natural receptor on host cell surfaces (Figure 5.1).

Gram-Negative Fimbriae

P Pili: a Model Fimbria

One of the best-studied examples of pili and its associated assembly machinery is P pili (pyelonephritis-associated pili), which are encoded by *pap* genes. *Escherichia coli* strains that express P pili are associated with pyelonephritis, which arises from urinary tract colonization and subsequent infection of the kidney. It is thought that P pili are essential adhesins in this disease process.

The *pap* operon is a useful paradigm since it contains many conserved features that are found among various pilus operons, and illustrates the complexity of assembling an adhesin on the bacterial surface (Figure 5.2). Two molecules guide newly synthesized pilus components to the bacterial surface. PapD, a conserved chaperone molecule, has an immunoglobulin-like fold, which is necessary to transport several pilus subunits from the cytoplasmic membrane to the outer membrane. In addition, it is necessary to prime the pilus subunit for assembly, which drives fiber formation. PapC serves as an outer membrane usher and accepts molecules from PapD. The major subunit of the pilus rod is PapA, which is anchored in the outer mem-

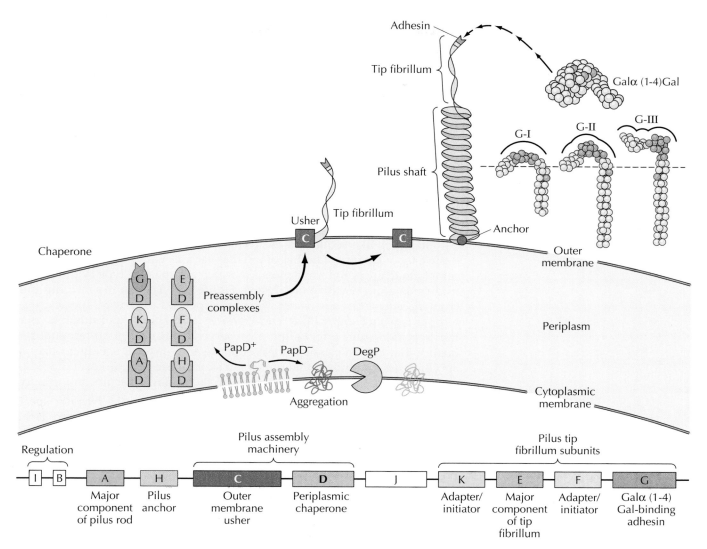

Figure 5.2 Schematic of P pilus operon and biogenesis. (Adapted from *Cell* **73:**887–901, 1993.)

brane by PapH. At the distal end of the pilus rod is the tip fibrillum (a thinner filamentous structure), composed of PapE, and the actual molecule that mediates adherence (i.e., tip adhesin), PapG. Two other proteins, PapF and PapK, are involved in tip fibrillum synthesis.

P pili bind to the α-D-galactopyranosyl-(1-4)-β-D-galactopyranoside moiety present in a globoseries of glycolipids which are found on host cells lining the upper urinary tract. However, there are three adhesin variants of PapG: G-I, G-II, and G-III, which recognize three different but related Galα-(1–4)-Gal receptors. It is thought that different hosts and tissue may contain differences in distribution of these receptors, and differential expression of the PapG adhesins at the pilus tip could enhance tissue and host specificity for adherence.

Although the host receptor varies for different bacterial pili, the general concepts provided by the P pilus operon and its assembly mechanism are conserved in many other pilus systems and components are often interchangeable. For example, the PapD chaperone also can modulate the assembly of type 1 pili, a very common pilus type, which mediate binding to mannose-containing molecules on the host cell surface. There is a large family of such periplasmic PapD-like chaperones that are necessary for assembly of several pili, including K88, K99, and *Haemophilus influenzae* pili. Additionally, homologous chaperones are needed for several afimbrial adhesins, including filamentous hemagglutinin (FHA) and the pH 6 antigen from *Yersinia pestis*. The gene organization among such pilus operons also is usually conserved. Thus, type I and P pili have very similar operons and functionally analogous sequences that can be aligned. Yet, they bind to quite different carbohydrates on the cell surface, a common theme for many adhesins, allowing pathogens to conserve organelle mechanisms yet vary receptor specificity.

Type IV Pili

Although *pap*-like sequences are common throughout gram-negative adhesins, there are other families of pili that use alternative biogenesis and assembly machinery to form a pilus. One such group is the type IV pili found in diverse gram-negative organisms (Figure 5.3). This family includes pili from *Pseudomonas aeruginosa*, *Neisseria* species, *Moraxella* species, enteropathogenic *E. coli* (EPEC), and *Vibrio cholerae*. Type IV pili subunits contain specific features, including a conserved, unusual amino-terminal sequence that lacks a classic leader sequence and, instead, usually utilizes a specific leader peptidase that removes a short, basic peptide sequence. Several possess methylated amino termini on their pilin molecules and usually contain pairs of cysteines that are involved in intrachain, disulfide bond formation near their carboxy termini. The pilus assembly genes are also members of the type II secretion system, a general secretion system used to transport molecules across the gram-negative envelope, which is needed to assemble this complex organelle on the bacterial surface. It has been proposed that the pilin molecules located at the tip have different sequences exposed than those that are packed into repeating structures within a pilus, and it is these exposed regions that may function as the adhesins. However, analogous to the P pilus tip adhesin, a separate tip protein (PilC for *Neisseria gonorrhoeae*) may function as a tip adhesin for these pili. A membrane glycoprotein (CD46) has been proposed as the receptor for *N. gonorrhoeae* pili. Alterations in the pilus subunit can also affect adherence levels. For example, although *P. aeruginosa* strains usually express only one pilus subunit, this subunit can vary significantly between strains, which affects their adhesive capacities. At least for one strain of *P. aeruginosa*, the pilus mediates binding to the disaccharide β-GalNAc(1–4)βGal in asialo-GM$_1$ gangliosides on host cell surfaces. Additionally, the pilin subunit of *N. gonorrhoeae* that is expressed is always changing due to genetic switching of the expressed pilin gene, which leads to antigenic variation. Different pilin sequences may lead to tissue tropism for adherence and different invasion capacities into epithelial cells, in addition to providing antigenic variation on the bacterial surface. The crystallographic structure of the *N. gonorrhoeae* and *P. aeruginosa* pilin is providing additional clues about how this molecule functions in adherence.

Curli

Diarrheagenic and other *E. coli* strains produce thin aggregative fimbriae termed curli, a descriptive name based on their curved appearance (Figure 5.3). In addition, curli are homologous to thin aggregative fimbriae pro-

Afimbrial adhesion:

Bald bacterial surface, afimbrial adhesins embedded in surface.

Pap fimbriae:

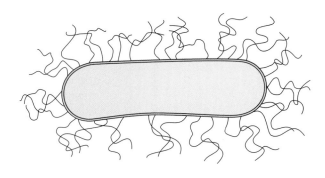

Very hairy surface, thin filaments protruding from surface.

Type IV bundle-forming pilus:

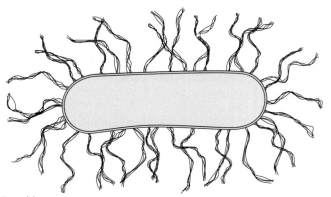

Ropelike structures made of many individual "threads" intertwined; then ropes are tangled.

Curli:

Coiled surface structure intertwined. Not ropelike like type IV. Curved/curled.

Figure 5.3   Schematic of various gram-negative adhesins.

duced by *Salmonella enteritidis* and appear to share a common ancestry. These fimbrial structures are assembled by a mechanism that is surprisingly different than other fimbrial assembly systems and that does not employ the complex assembly machinery typical of most fimbriae. Instead, they use a self-assembly mechanism which involves the production of a bacterial cell surface-bound nucleator, followed by polymerization of curlin (the structural repeating subunit of curli, which is secreted out of the bacteria) condensing on the nucleator. Curlin can be supplied by the nucleator-presenting cell or by adjacent bacteria, driven by a diffusion gradient. Once assembled, curli are quite stable structures and difficult to dissociate. Curli have been reported to bind to several host proteins, including fibronectin, laminin, plasminogen, and tissue-type plasminogen activator (t-PA).

Pathogens and Their Nonfimbrial Adhesins

Escherichia coli Afimbrial Adhesins

E. coli contains many adhesins and can cause several diseases, including meningitis, diarrhea, and urinary tract infections. There is a family of adhesins that are associated with *E. coli* and that cause the latter two diseases. Members of this family include F1845 and Dr, which are fimbrial adhesins, and two nonfimbrial adhesins (afimbrial adhesins Afa-I and Afa-III) (Figure 5.3). All four of these adhesins use the Dr^a blood group antigen present on decay-accelerating factor (DAF) on erythrocytes and other cell types as their receptor, although they appear to recognize different epitopes of the Dr antigen. The Dr adhesin, but not the other three, also binds type IV collagen.

Much like other fimbrial adhesins, expression and production of these adhesins require five or six gene products. These include a periplasmic chaperone, an outer membrane anchor protein, one or two transcriptional regulators, and the adhesin. At least for Afa-III, there appear to be two adhesins encoded within this operon (AfaD and AfaE). One perplexing question has been why some adhesins in this family form fimbriae (F1845 and Dr), yet others form nonfimbrial adhesins on the bacterial surface (Afa-I and Afa-III). It appears that the sequence of the adhesin molecule dictates whether or not it will be assembled into a fimbriae, and genetically switching the genes encoding these adhesins switches the adhesin type. It is possible that the *E. coli* afimbrial adhesins have evolved from the related fimbrial adhesins but have been altered such that the properties needed to polymerize a pilus are missing, yet the adhesin domain remains anchored on the bacterial surface. It is now possible to test the role of assembling an adhesin in a fimbrial structure versus a nonfimbrial adhesin in bacterial disease.

In addition to the variety of "classical" adhesins employed by various pathogenic *E. coli*, EPEC and enterohemorrhagic *E. coli* (EHEC) (two diarrheagenic strains) employ a very clever adhesion strategy. These pathogens utilize a type III secretion system to insert a bacterial molecule (Tir) directly into host cell membranes. An outer membrane protein, intimin, then binds Tir, thereby locking the bacterium onto the host cell surface. Surprisingly, the regions of Tir exposed to the mammalian cytoplasm then bind actin cytoskeleton-linking components, thereby driving significant actin accumulation beneath the adherent bacteria. This results in the formation of a "pedestal" beneath the raised organisms, although the purpose of such cytoskeletal recruitment in disease has not been established.

Helicobacter pylori

Helicobacter pylori is a human-specific pathogen that adheres to epithelial cells from the gastrointestinal tract but not to cells from the nervous system or urogenital tract. Additionally, it shows preferential adherence to the gastric mucosa rather than the colon, which correlates with the site of *H. pylori* infections (stomach). This organism also preferentially adheres to surface mucous cells of the gastric mucosa and does not bind to host cells that are found deeper within the mucosal layer. *H. pylori* binds to the Lewis[b] blood group antigen that is expressed on cells in the stomach epithelium, presumably explaining the tissue specificity of this adherence. Interestingly, *H. pylori* preferentially infects people that are Lewis[b] positive, but other individuals can also be infected. *H. pylori* also has several other adhesins which mediate adherence to other host receptors, although the details of these adhesins and their receptors have not been fully characterized. The bacterial molecule responsible for mediating Lewis[b] binding is an outer membrane protein that belongs to a large family of related *H. pylori* outer membrane proteins recently identified using the entire genome sequence. Other members of this family also appear to function as adhesins and porins. Thus, this family of adhesins requires no specialized assembly machinery and is anchored directly in the outer membrane with the adhesin domain exposed to the extracellular surface. Interaction of *H. pylori* with gastric mucous cells results in intimate attachment to the host cell surface, and it has been reported that such attachment results in localized actin polymerization and host cell signaling beneath the adherent organisms, much like that seen with enteropathogenic *E. coli* (chapter 6).

Bordetella pertussis

Bordetella pertussis, the causative agent of whooping cough, is a respiratory mucosal pathogen that possesses several potential adherence factors that exemplify the complexity of bacterial adherence to host cell surfaces. Potential *B. pertussis* adherence factors include at least four fimbrial genes and several nonfimbrial adhesins, including FHA, pertactin, pertussis toxin, BrkA, and BipA. FHA is a large (220-kDa) secreted molecule that has several domains that are homologous to other bacterial adherence molecules or to eukaryotic sequences that mediate cell-cell adhesion. FHA is homologous to two high-molecular-weight, nonpilus adhesins from *H. influenzae*, another pathogen that adheres to respiratory surfaces. FHA also contains an RGD tripeptide sequence that is a characteristic eukaryotic recognition motif that binds to host cell surface integrins. This RGD sequence appears to be involved in FHA binding to the leukocyte integrin CR3, which mediates bacterial uptake into macrophages without triggering an oxidative burst. It has been proposed that, by possessing this RGD sequence, FHA mimics host molecules (at a molecular level). Additionally, this RGD sequence induces enhanced *B. pertussis* binding to monocytes by activating a host signal transduction complex that normally up-regulates the CR3-binding activity. (Most bacterial molecules that mediate adherence to integrins, such as *Yersinia* invasin, do not possess RGD sequences, although their receptor affinity is often greater than the native host ligand.) Both FHA and one of the binding subunits of pertussis toxin (S2) share homologous sequences that mediate binding to lactosylceramides, which suggests that these two proteins might use similar motifs to bind to host molecules. Pertussis toxin also mimics eukaryotic adhesive molecules. For example, the S2 and S3 subunits have several features in common with eukaryotic selectins. By mimicking host molecules, these bacterial adhesins can effect responses in the host that enhance "desired" (from the bacterial viewpoint) interactions with host cells. These examples highlight the ability of bacteria to mimic host molecules and to use this mimicry to enhance their pathogenesis. At least three other *B. pertussis* molecules are involved in adherence to host cells. Both pertactin and BrkA, which are 29% identical, contain RGD motifs that may be involved in adherence. Additionally, BrkA is also involved in serum resistance. BipA is homologous to *Yersinia* invasin and intimin of pathogenic *E. coli*, although its receptor has not been identified.

Neisseria Species

N. gonorrhoeae and *N. meningitidis* are two mucosal pathogens that have developed sophisticated and overlapping mechanisms to adhere to host cell surfaces. As discussed above, these bacteria produce a type IV pilus which mediates adherence to cell surfaces. In addition to pili, they also produce a family of outer membrane proteins called opacity proteins (Opa), which also mediate adherence and invasion by activating host cell signaling. *Neisseria* species can express up to 12 different Opa proteins, with each being expressed independently of the others. This variable expression is thought to contribute to its antigenic variation and possibly altered adherence characteristics. A family of Opa proteins (Opa$_{CEA}$) binds to members of the carcinoembryonic (CEA)-related cell adhesion molecule (CEACAM) gene family. CEACAMs are surface glycoproteins involved in cell-cell interactions. Members of this eukaryotic transmembrane glycoprotein family show high levels of similarity and belong to the immunoglobulin superfamily. Although most Opa proteins bind to the CEA antigens, their affinity for individual members of this family vary, and these differences may contribute to differential adherence and invasion of these pathogens in different tissues. It appears that *Neisseria* species use the downstream signaling capacity of the CEACAM family to trigger various signals in host cells, which can mediate invasion. The Opa$_{HS}$ family also mediates adhesion and invasion but by binding to heparan sulfate proteoglycans. These proteins normally bind to a variety of extracellular matrix proteins and are involved in cell adhesion events. Lipopolysaccharide (lipooligosaccharide, LOS) also contributes to *Neisseria* adherence and invasion into host cells, possibly by interacting with the asialoglycoprotein receptor (ASGP-R). LOS can be sialyted, which inhibits bacterial invasion. Thus, this pathogen expresses several different adhesins which are involved in binding to host cell surfaces.

Yersinia Species

Pathogenic *Yersinia* species contain several adhesins which mediate adherence to the host cell surface or matrix. Perhaps the best-studied adhesin, which can also mediate invasion under certain conditions, is invasin. This outer membrane protein adheres tightly to members of the β_1-integrin family, including the fibronectin receptor (see chapter 6 for details). In addition to invasin, *Yersinia enterocolitica* also expresses another adhesion, which mediates low levels of invasion called Ail (attachment invasion locus). The host receptor for this outer membrane protein has not been identified. Finally, YadA of *Y. enterocolitica* mediates binding to cellular fibronectin, but not to plasma fibronectin, which may provide a possible mechanism for the organism to adhere to tissue rather than to bind to circulating molecules. YadA also mediates adherence to various collagens and laminins, as well as mucus. The loss of YadA decreases *Y. enterocolitica* virulence in mice by 100-fold, which suggests that adherence to these molecules may potentiate disease (see below). However, it is not needed for virulence in *Y. pseudotuberculosis* and actually hinders virulence in *Y. pestis*. Yersiniae also express a flexible fimbria whose assembly genes are homologous to Pap pili and other related fimbria. In *Y. pestis*, this organelle is produced at pH 6, and is thus called the pH 6 antigen (a homologue [Myf] is also found in *Y. enterocolitica*). Recent reports indicate that it binds apolipoprotein B-containing lipoproteins. Studies with the pH 6 antigen indicate that it is needed for full virulence of *Y. pestis* in mice, although its exact role in disease has not been defined.

Haemophilus influenzae

Nontypeable (i.e., lacking type b capsular polysaccharide) *H. influenzae* is a common respiratory pathogen that causes ear infections, sinusitis, and conjunctivitis. It has developed a complement of several adhesins that mediate colonization of the upper respiratory tract. For example, it has a pilus that is similar to P pili that mediates adherence to fibronectin and heparin-binding growth-associated molecule. It also has a family of several adhesins that belong to the autotransporter group. Autotransporters are synthesized as preproteins, encoding a signal sequence (which mediates passage across the inner membrane), a translocator domain (which forms a pore in the outer membrane), and a passenger domain (which passes through the pore onto the bacterial surface). The Hap autotransporter can cleave itself from the bacterial surface but also mediates adherence to selected extracellular matrix proteins, including fibronectin, collagen, and laminin. The Hia autotransporter also mediates adherence to a variety of respiratory cells, but its receptor has not been defined. Another two high-molecular-weight adhesins (HMW1 and HMW2) also belong to the autotransporter family and mediate bacterial adherence. Although the two proteins are homologous, they appear to have different cell-binding activities and receptors. In addition to the autotransporter family of adhesins, this pathogen also has two outer membrane proteins (P2 and P5), as well as LOS, that mediate adherence, demonstrating the complexity of adhesins employed by a mucosal pathogen that interacts with respiratory surfaces.

Mycobacterium Species

Mycobacterium species also bind fibronectin using three related bacterial molecules (called the BCG85 complex). At least two other fibronectin-binding molecules also have been described for *Mycobacterium* species. One of these has been described for *M. avium* and *M. intracellulare* as a 120-kDa protein recognized by antibodies against β_1-integrins, which indicates that these bacteria may express integrin-like molecules that may mediate matrix adherence. A fibronectin attachment protein (FAP) from *M. avium* has been characterized and shown to be highly conserved in *M. leprae, M. tuberculosis,* and other mycobacteria.

M. leprae has a specific neural tissue tropism to Schwann cells of the peripheral nerve. It has been established that this tropism is due to bacterial binding of a host matrix molecule (laminin), which serves as a bridge between the bacterium and a host receptor, β_4-integrin. Although the bacterial component was not identified, it was shown that *M. leprae* binds specifically to the G domain of the soluble laminin α_2-chain. Laminin in turn binds to its cell surface receptors, including $\alpha_6\beta_4$-integrin. Not only does this provide a good model for determining tissue tropism mediated by bacterial adherence, it also provides an excellent example of a pathogen binding a soluble matrix protein which serves as a bridge to its host cell receptor, thereby mediating bacterial adherence.

Staphylococcus and *Streptococcus*

As described for *M. leprae,* a common theme in the adherence of many gram-positive cocci, including many species of staphylococci and streptococci, involves bacterial recognition of host proteins that, in turn, associate with the surface of target host cell or tissue. The host proteins that are most often bound by these bacteria include components of the extracellular matrix. Binding to fibronectin, collagen, fibrinogen, vitronectin, laminin, bone sialoprotein, elastin, and thrombospondin have all been reported for various staphylococcal and streptococcal species (see Table 5.1). Numerous adhesins that recognize a specific extracellular matrix component and several that can recognize multiple components have been characterized at the molecular level. In some cases, the bacterium may not discriminate between recognition of soluble, immobilized, or tissue-specific forms of the extracellular matrix component and will bind to all forms equally. However, in other cases, a specific form of the matrix component is exclusively bound. This latter instance almost always involves preferential recognition of tissue

Table 5.1 Selected examples of the interactions between some gram-positive bacteria and various components of the extracellular matrix[a]

ECM[b] molecule	Microorganism	Adhesin
Fibronectin	*Staphylococcus aureus*	FnbA, FnbB
	Coagulase-negative staphylococci	
	Group A streptococci	PrtF1/Sfb1, PrtF2, GAPDH, ZOP, LTA, SOF/SfbII, FNB54, 28-kDa antigen, M3 protein
	Group B streptococci[c]	
	Group C streptococci[c]	FnBA, FnBB
	Group G streptococci[c]	FnB, GfbA
	Streptococcus pneumoniae[c]	
	Streptococcus gordonii[c]	CshA
	Enterococcus faecalis	
Collagen	*S. aureus*	Cna
	Group A streptococci	57-kDa protein
	Streptococcus mutans	
Fibrinogen	*S. aureus*	ClfA, Fib
	Group A streptococci	M proteins, FNB54
	Streptococcus parasanguis[c]	FimA
Vitronectin	*S. aureus*	60-kDa protein
	Group A streptococci	
Laminin	*S. aureus*	
	Group A streptococci	
	S. gordonii	145-kDa protein
Bone sialoprotein	*S. aureus*	
Elastin	*S. aureus*	EbpS
Thrombospondin	*S. aureus*	
Multiple-binding activity	*S. aureus*	Map
	Group A streptococci	GAPDH
	Group C streptococci	FIA

[a] A more comprehensive list can be found in Langermann et al. (1997).
[b] ECM, extracellular matrix.
[c] Binding is done preferentially to an immobilized form of the extracellular matrix component.

or immobilized forms of the component over soluble forms. As described above for YadA of *Y. enterocolitica*, this level of discrimination may be useful in situations where the adhesin encounters both soluble and immobilized forms of the extracellular matrix component. Preferential recognition of the immobilized form would prevent competition for binding with the soluble form to allow the microorganism to most efficiently adhere to the target substrate.

Binding to extracellular matrix components may contribute to the tropism of streptococcal and staphylococcal infections, particularly when the bacteria recognize extracellular matrix components that are distributed only among defined compartments within the host (e.g., laminin in basement membrane or bone sialoprotein in bone). In other cases, the extracellular matrix components that are recognized are broadly distributed. An example of this latter class is binding to fibronectin. This extracellular matrix glycoprotein is present in most tissues and fluids of the host, is found in many secretions, and is a prominent component of a wound. Fibronectin can be encountered as a soluble form, can be found as an immobilized form bound to receptors on various cell types, or can be found incorporated into many different types of extracellular matrices. Given this broad distribution, it is not surprising that many different species of streptococci, staphylococci, and enterococci have evolved the ability to interact with fibronectin.

Perhaps the best-characterized fibronectin-binding proteins of gram-positive cocci are members of a large family that includes protein F1, Sfb, protein F2, and serum opacity factor (SOF) produced by the group A streptococcus; FnbA and FnbB of *Staphylococcus aureus*; FnBA and FnBB of *Streptococcus dysgalactiae* (a group C streptococcus); FnB of *Streptococcus equisimilas* (a group G streptococcus); and GfbA of human-associated group G streptococcal isolates. In general, these proteins are of the class that does not discriminate between binding to soluble and immobilized forms of fibronectin, binding is highly specific for fibronectin, and they bind to fibronectin in an essentially irreversible manner. Furthermore, they bind fibronectin at very high affinity and apparent K_d values in the 1.0 nM range are not uncommon. When the corresponding genes are cloned and expressed in *E. coli*, the proteins retain the ability to bind fibronectin (although they are not expressed on the *E. coli* cell surface), and when purified from *E. coli*, the proteins can competitively inhibit the binding of fibronectin not only to the homologous strain but also to other gram-positive cocci which express a member of this family. Finally, the binding of fibronectin by the members of this adhesin family involves protein-protein interaction rather than recognition of a carbohydrate moiety of the fibronectin glycoprotein.

Comparison of the DNA sequences of the genes which encode these adhesins has revealed that the deduced structures of the corresponding proteins share a similar domain architecture (the structure of protein F1 is shown in Figure 5.4). Sequences at their carboxy termini are characteristic of proteins from gram-positive cocci that are sorted and displayed on the cell surface by a pathway first described for the M protein of the group A streptococcus and protein A of *S. aureus*. These include a proline- and lysine-rich domain (W in the Figure 5.4) thought to be important for interaction with the gram-positive cell wall, a domain rich in hydrophobic amino acids (M in the figure) that may be involved in interaction with the membrane and a short positively charged carboxy terminus. A characteristic highly conserved pentapeptide "LPXTG" motif is located between the cell wall interaction domain and the membrane interaction domain. Studies with protein A have shown that this sequence is recognized by the sorting pathway following secretion of the nascent polypeptide across the membrane, with the result that the protein is cleaved following the threonine residue of the LPXTG motif. The threonine then becomes covalently coupled by a pentaglycine bridge to a lysine residue of the cell wall peptidoglycan. The end product of this sorting pathway is a cell surface protein that is covalently tethered to the cell wall by its carboxy terminus with its amino terminus exposed to the environment.

The amino termini of the proteins in the protein F family contain a signal sequence characteristic of gram-positive bacteria. This is followed by a large domain that contains unique sequences not shared between the different members of this family that can constitute up to 80% of the total sequence of the protein (U in the figure). This region can actually be quite divergent between different alleles of the same family member. For the most part, the function of this region is unknown. A notable exception is found

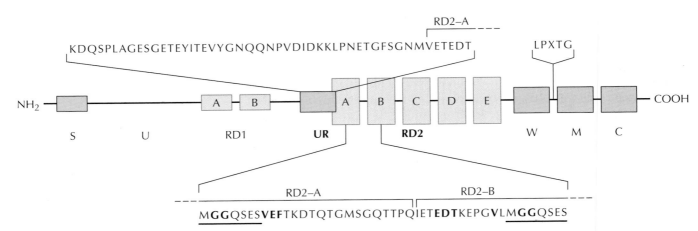

Figure 5.4 Domain architecture of protein F. Protein F is a member of a large family of related surface proteins found among the gram-positive cocci that bind to fibronectin. The signature feature of this family is a repetitive domain consisting of 32 to 44 amino acids that is repeated two to six times in tandem. This domain consists of 37 amino acids in protein F and is known as RD2. Sequences highly conserved in the repetitive region from different relatives of protein F are highlighted in bold type. This repetitive domain binds to the amino-terminal 29-kDa fibrin-binding domain of fibronectin, and the minimal functional binding domain of RD2 includes 44 amino acid residues derived from the carboxy-terminal segment of one repeat and the amino-terminal segment of the adjacent repeat, as is shown underneath RD2. This binding unit begins and terminates with amino acid motif MGGQSES (underlined). Other common features include a signal sequence (S), a long amino-terminal nonrepetitive region of sequence that is unique to different family members (U), and a second repetitive region of unknown function (called RD1 in protein F). The proteins also share features common to many surface proteins of gram-positive cocci, including a proline- and lysine-rich domain that may interact with the cell wall (W), a hydrophobic domain that may interact with the membrane (M), a short tail of charged amino acids (C), and an LPXTG motif (LPATG in protein F) that serves as a processing site, in which the protein becomes cleaved after the threonine and is cross-linked to a lysine residue of the peptidoglycan cell wall. Some members of this family contain an additional domain for binding to fibronectin which can be located in various regions amino terminal to the repetitive domain. In protein F, this domain is known as UR, and it is composed of 43 amino acids immediately N terminal to RD2, as is shown above UR. The ability of UR to bind to fibronectin also has an absolute requirement for the first six amino acids from the first repeat unit of RD2. UR binds to a region of fibronectin that includes the amino-terminal 29-kDa domain and the adjacent 40-kDa collagen-binding domains. The various domains are not shown to scale.

in SOF, where this region contains the catalytic domains that are involved in cleaving apolipoprotein AI in serum, which then initiates a cascade of events that renders the serum opalescent. The unique amino-terminal domain is often followed by another domain of unknown function that frequently consists of a sequence motif that is repeated several times in tandem (RD1 in the figure).

The signature feature of the protein F family is a repetitive domain located just amino-terminal of the cell wall attachment domains (RD2 in the figure). This signature domain consists of a 32- to 44-amino-acid motif that can be repeated up to six times in tandem. Considerable variation in the number of repeats exists between different members of the family and even can vary extensively between alleles of the same gene. This level of variation in not uncommon in repetitive regions of other surface proteins from gram-positive cocci. For example, the M proteins of the group A streptococci exhibit a similar variation. A large body of evidence, much of which initially came from the analysis of FnbA of *S. aureus,* has strongly implicated this repetitive domain in binding to fibronectin. However, further study has suggested that the mechanisms by which this domain interacts with fibronectin may not be identical for all family members. While a single 37-amino-acid repeat unit of Sfb of *S. pyogenes* was an effective inhibitor of fibronectin binding to the native protein, a single repeat unit of the very closely related protein F1 was not. Further analysis of protein F1 revealed that the minimal functional binding unit consists of a 44-amino-acid region derived from the carboxy-terminal end of one repeat and the amino-terminal region of the adjacent repeat (see Figure 5.4). This binding element begins and ends with a direct repeat of the sequence motif MGGQSES.

Additional studies on protein F1 have revealed that the binding of fibronectin to this protein was much more complicated than originally thought and may involve an additional domain of the adhesin. This was based on the observation that while the full-length protein could completely inhibit the binding of fibronectin to streptococcal cells, the repeat region was capable of only partial inhibition, even when tested at very high concentrations. A second domain was subsequently found, which could also partially inhibit binding, and when a mixture of the two domains was tested, inhibition similar to that obtained with the full-length protein F1 was obtained. This second domain, called UR (see Figure 5.4), contains the 43 amino acids located immediately N terminal to the repetitive domain. In addition, an additional six amino acids derived from the first repeat of the repetitive domain are absolutely required for UR's binding activity. Other than these six residues, there is no obvious homology between UR and the repetitive domain. Protein F2 was also found to use a two-domain mechanism to bind fibronectin. However, while its second binding domain is also located amino-terminal to the repetitive domain, the two domains are separated by 100 amino acids. Also, there is no obvious homology between the second domain of protein F2 and UR.

Investigations of the domains of fibronectin that are recognized by these proteins have strongly implicated the very amino terminus of the protein which is contained on a 29-kDa protease-cleavage fragment that also includes binding sites for fibrin and heparin. The isolated 29-kDa domain will bind to all staphylococci and streptococci that express adhesins of the protein F family, and considerable evidence suggests that this domain of fibronectin recognizes the repetitive domain of the microbial adhesins. In contrast, this 29-kDa region of fibronectin does not bind to the UR domain of protein F1. Instead, UR interacts with a region of fibronectin that must include both the 29-kDa domain and the adjacent 40-kDa collagen-binding domain on the same polypeptide fragment. Thus, not only does protein F1 contain two domains that can bind independently to fibronectin, these two domains recognize distinct sites in the fibronectin molecule.

The principal function of fibronectin and many other extracellular matrix proteins is to serve as a substrate for the adherence of eukaryotic cells. It is interesting that the gram-positive cocci have exploited the host's own cell adhesion molecules to mediate their adherence to the host (Table 5.1). This strategy has an additional advantage that makes use of the fact that most extracellular matrix components themselves have a broad range of different ligands with which they can interact. By binding an extracellular matrix protein, which can, in turn, bind to large numbers of different cells, structures, and tissues, the bacterium can gain an extensive adhesive potential. For example, by binding to fibrinogen, *S. aureus* gains the capacity to bind to endothelial cells, and by binding to soluble fibronectin, *S. pyogenes* can bind much more efficiently to a collagen-containing structure like the dermis.

Role of Adherence in Disease

Adhesins play an important role in disease and represent the interface between the pathogen and the host cell. In many cases the precise role that individual adhesins play in the pathogenesis of specific diseases has been established. For example, inactivation of the gene which encodes Cna, a collagen-binding adhesin of *S. aureus*, results in a mutant with a considerably diminished capacity to cause septic arthritis in an animal model. Similarly, construction of a mutant of *E. coli*, which can express a P pilus that is intact except for the loss of the PapG tip adhesin, produces a mutant which can infect the bladder but loses the ability to infect the kidney. Also, studies which have successfully used adhesins like PapG as vaccines have shown that novel antimicrobial therapies can be based on understanding the precise role that adhesins play in the pathogenic process.

However, in many instances, the precise role that adhesins contribute to disease has been less clear. One reason may be that due to the importance of adherence, many pathogens express multiple different adhesins, each of which may contribute unique or overlapping specificities of host cell recognition. This level of complexity is illustrated in infection by the invasive pathogen *Salmonella enterica* serovar Typhimurium. While adherence plays a key role in establishing infection by *Salmonella* serovar Typhimurium, this pathogen can encode up to 13 distinct fimbrial operons, each of which may participate in adherence. In addition, the *Salmonella* serovar Typhimurium invasion loci (*inv*) can also mediate adherence, which occurs as a prerequisite to invasion into nonphagocytic cells. The lpf fimbriae provide tissue specificity by mediating adhesion to Peyer's patches in mice, the site of bacterial entry in the intestine. Although strains containing *lpf* mutations are attenuated, a much stronger attenuation and lack of colonization occur if strains containing a double mutation in *lpfC* and *invA* are used in the oral infections. However, if this strain is delivered by the intraperitoneal route, it is fully virulent, indicating that these adhesins and invasins are involved in intestinal colonization and not the later systemic events. Attenuation and lack of colonization are also seen following oral challenge with strains containing a *pefC* mutation. Thus, it seems that several adhesins are involved in intestinal colonization, rather than being mediated by only a single bacterial adhesin.

Y. enterocolitica and *Y. pseudotuberculosis* provide similar examples of the complexity of bacterial adherence and its involvement in disease. In *Y. enterocolitica*, invasin is needed for efficient penetration of the intestinal epithelium. However, YadA is also needed for persistence in Peyer's patches and for the bacterium to cause disease in mice infected either orally or intraperitoneally. Ail's role in intestinal colonization is not as pronounced. In *Y. pseudotuberculosis*, invasin mutants are unable to efficiently penetrate murine Peyer's patches and instead colonize on the luminal surface of the intestinal epithelium. Mutants in YadA or the pH 6 antigen also show decreased binding to the luminal surface, indicating that these two molecules also participate in intestinal colonization.

Additionally, the recently realized concept that adhesins often trigger signaling in host cells, has not been incorporated into disease mechanisms. Why adherent organisms would want to trigger specific signal transduction pathways, and even resulting cytoskeletal rearrangements, is not understood (although they are often used for invasion). Perhaps it is the price a pathogen pays for adhering to an extracellular matrix protein or membrane protein, or alternatively the pathogen has realized the potential of activating a signaling pathway that it can then exploit. These are critical questions that need to be addressed in the future to understand adherence and disease better.

The complexity of the adherence process, as illustrated by the examples of *Yersinia* and *Salmonella,* has frustrated most attempts to base novel antimicrobial therapies on intervening in this critical step of host cell-pathogen interaction. In addition to the complex issue of multiple adhesins, individual adhesins themselves often have evolved strategies to evade host immune responses. Examples of these are antigenic variation of the major immunodominant domains of the adhesins, as occurs in the type IV pili of the neisseriae, and the localization of a single copy of the adhesin subunit as a minor component at the very tip of a large heteropolymeric fiber, as occurs with P pili and type 1 pili. However, by unraveling the molecular details of adhesin structure, organization, and assembly, progress is being made. Perhaps the best example of this has come from studies on type 1 pili. Through knowledge of the structure of the adhesive organelle, the identification of the adhesive subunit of the organelle (see above), and through advances in expression and purification of the unstable adhesive subunit, it has been possible to test the efficacy of a vaccine composed solely of the adhesive subunit. Initial results from trials in animal models of infection of the bladder have demonstrated that this approach can be successful. Thus, continued study of other adhesins and this critical component of pathogen-host cell interaction holds similar promise for augmenting our arsenal of antimicrobial agents.

Selected Readings

Baumler, A. J., R. M. Tsolis, P. J. Valentine, T. A. Ficht, and F. Heffron. 1997. Synergistic effect of mutations in *invA* and *lpfC* on the ability of *Salmonella typhimurium* to cause murine typhoid. *Infect. Immun.* **65:**2254–2259.

An interesting manuscript that begins to probe the complexity of adhesions on disease. Several more studies have shown that *Salmonella* has many adhesions and their roles are complex.

Evans, D. J., Jr., and D. G. Evans. 2000. *Helicobacter pylori* adhesins: review and perspectives. *Helicobacter* **5:**183–195.

Finlay, B. B., and S. Falkow. 1997. Common themes in microbial pathogenicity. II. *Mol. Biol. Microbiol. Rev.* **61:**136–169.

A good overview of mechanisms of bacterial pathogenicity.

Hanski, E., and M. G. Caparon. 1992. Protein F, a fibronectin-binding protein, is an adhesin of the group A streptococcus, *Streptococcus pyogenes*. *Proc. Natl. Acad. Sci. USA* **89:**6172–6176.

Identification and characterization of protein F and its binding substrate.

Hauck, C. R. 2002. Cell adhesion receptors: signaling capacity and exploitation by bacterial pathogens. *Med. Microbiol. Immunol. (Berlin)* **191:**55–62.

Hauck, C. R., and T. F. Meyer. 2003. 'Small' talk: Opa proteins as mediators of Neisseria-host-cell communication. *Curr. Opin. Microbiol.* **6:**43–49.

Hultgren, S. J., S. Abraham, M. Caparon, P. Falk, J. St. Geme, and S. Normark. 1993. Pilus and nonpilus bacterial adhesins: assembly and function in cell recognition. *Cell* **73:**887–901.

A strong overview of bacterial adhesion mechanisms.

Kenny, B., R. DeVinney, M. Stein, D. J. Reinscheid, E. A. Frey, and B. B. Finlay. 1997. Enteropathogenic *E. coli* (EPEC) transfers its receptor for intimate adherence into mammalian cells. *Cell* **91:**511–520.

Langermann, S., S. Palaszynski, M. Barnhart, G. Auguste, J. S. Pinkner, J. H. Burlein, P. Barren, S. Koenig, S. Leath, C. H. Jones, and S. J. Hultgren. 1997. Prevention of mucosal *Escherichia coli* infection by FimH-adhesin-based systemic vaccination. *Science* **276:**607–611.

> Proof of concept that adhesins can be blocked by vaccination.

Menozzi, F. D., K. Pethe, P. Bifani, F. Soncin, M. J. Brennan, and C. Locht. 2002. Enhanced bacterial virulence through exploitation of host glycosaminoglycans. *Mol. Microbiol.* **43:**1379–1386.

Nougayrede, J. P., P. J. Fernandes, and M. S. Donnenberg. 2003. Adhesion of enteropathogenic *Escherichia coli* to host cells. *Cell Microbiol.* **5:**359–372.

Ozeri, V., A. Tovi, I. Burstein, S. Natanson-Yaron, M. G. Caparon, K. M. Yamada, S. K. Akiyama, I. Vlodavsky, and E. Hanski. 1996. A two-domain mechanism for group A streptococcal adherence through protein F to the extracellular matrix. *EMBO J.* **15:**898–998.

Patti, J. M., B. L. Allen, M. J. McGavin, and M. Höök. 1994. MSCRAMM-mediated adherence of microorganisms to host tissues. *Annu. Rev. Microbiol.* **48:**585–617.

Patti, J. M, T. Bremell, D. Krajewska-Pietrasik, A. Abdelnour, A. Tarkowski, C. Ryden, and M. Höök. 1994. The *Staphylococcus aureus* collagen adhesin is a virulence determinant in experimental septic arthritis. *Infect. Immun.* **62:**152–161.

Pepe, J. C., M. R. Wachtel, E. Wagar, and V. L. Miller. 1995. Pathogenesis of defined invasion mutants of *Yersinia enterocolitica* in a BALB/c mouse model of infection. *Infect. Immun.* **63:**4837–4848.

> An interesting report of the complexity of adhesins and invasins and their contribution to disease.

Rambukkana, A., J. L. Salzer, P. D. Yurchenco, and E. I. Tuomanen. 1997. Neural targeting of *Mycobacterium leprae* mediated by the G domain of the laminin-alpha2 chain. *Cell* **88:**811–821.

St. Geme, J. W., III. 2002. Molecular and cellular determinants of non-typeable *Haemophilus influenzae* adherence and invasion. *Cell Microbiol.* **4:**191–200.

6

Molecular Basis for Cell Adhesion and Adhesion-Mediated Signaling

Benjamin Geiger, Avri Ben-Ze'ev, Eli Zamir, and Alexander D. Bershadsky

This chapter addresses the complex molecular interrelationships between cell adhesion and the transduction of transmembrane signals that affect cell adhesion and fate. It is shown here that adhesion sites such as focal contacts and cell-cell adherens junctions contain multimolecular protein complexes that participate both in the physical assembly of adhesion sites and the associated cytoskeleton and in the transduction of long-range growth, differentiation, and survival signals. The network of molecular interactions of the different adhesions, their involvement in the interaction with the cytoskeleton, and their particular role in adhesion-mediated signaling are discussed.

Structural and Functional Diversity of Cell Adhesions

Adhesive interactions of cells exert major short- and long-term effects on cell shape and fate and thus play a central role in the assembly of multicellular organisms. Cell adhesion in metazoan organisms is a complex and molecularly diversified process involving a multitude of molecular systems. This diversity is expressed at many levels. There are many distinct types of extracellular surfaces with which cells interact in vivo. These include networks of extracellular matrix (ECM) molecules as well as the membranes of adjacent cells. Within each group of adhesions, there is further molecular complexity and diversity. There are numerous distinct ECM molecules, which interact with cells via families of specific adhesion receptors. Among the important ECM adhesion receptors are syndecans, dystroglycan complex, hyaluronan receptors, and a large family of integrins, each one specific for certain matrix components. Cell-cell adhesion is also mediated by a multitude of transmembrane receptor molecules including immunoglobulin superfamily cell adhesion molecules (CAMs), selectins, and cadherins. Some of these receptors are Ca^{2+} dependent, some Ca^{2+} independent, some involved in homophilic interactions, where the interacting molecules present on the two partner cells are identical, while others are involved in heterophilic interactions. Moreover, the long-term effects of adhesive interactions on cell activity and fate may differ greatly from one adhesion site to the other. For example, some adhesions promote cell growth while others may suppress it; adhesion to different matrices can have distinct effects on cell motility, differentiation, or survival. The objective of this chapter is to discuss the structure-signaling relationships in two types of cytoskeleton-associated adhesions, namely, integrin-mediated cell-matrix adhesions, such as "focal adhesions" (also known as "focal contacts"), and cadherin-mediated cell-cell adhesions known as "adherens junctions." Recent studies have shed much light on the molecular properties of these adhesive sites, providing an insight not only into their structure but also into their signaling activities.

These adhesion structures differ in the nature of the adhesion surface and adhesion receptors involved in their formation. Focal contacts and related matrix adhesions are widely distributed in tissues, involving different ECM molecules and receptors of the heterodimeric integrin family (Figure 6.1). Based on their molecular composition and subcellular localization, integrin-mediated adhesions were recently divided into several distinct groups. These include "classical" focal adhesions, located mainly at the cell periphery and associated with rigid ECM; fibrillar adhesions, formed with pliable fibronectin fibers; small focal complexes, formed under the leading

Figure 6.1 Scheme depicting the structure of a typical integrin molecule. Integrins are composed of two noncovalently associated subunits designated α and β. Sixteen different α variants and eight different β variants have been identified in mammals to date. The integrin heterodimer has a globular head containing a ligand-binding pocket formed by both subunits. Two extended stalks correspond to the C-terminal parts of the α and β subunits. Some α subunits are cleaved posttranslationally to give heavy (extracellular) and light (transmembrane) chains linked by S—S bonding. The majority of integrin ligands contain the Arg-Gly-Asp (RGD) motif as the minimal sequence necessary for the integrin binding. Divalent cations, especially calcium, play an important role in the regulation of ligand binding, and the extracellular domain of the α subunit contains multiple repeats of EF-hand Ca^{2+}-binding sites. The adhesion plaque proteins talin, α-actinin, paxillin, and FAK interact with the short cytoplasmic domain of the β subunit.

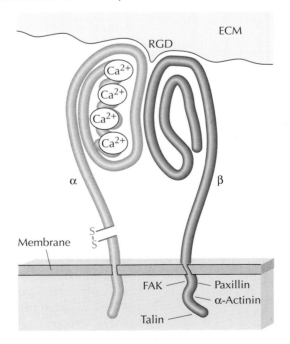

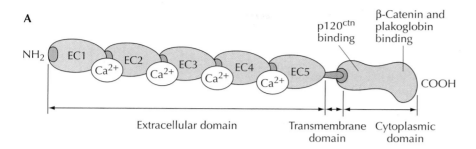

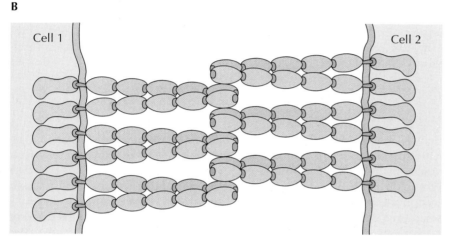

Figure 6.3   Scheme depicting classical cadherins involved in the formation of cell-cell adherens junctions. **(A)** A classical cadherin molecule contains five homologous extracellular domains, denoted EC1 through EC5. Ca^{2+} ions bind to the regions between the EC domains and probably contribute to the rigid elongated shape of the molecule. The cytoplasmic domain of the classical cadherin is highly conserved and contains a binding site for the armadillo family proteins β-catenin and plakoglobin, while another family member, p120^{ctn} protein, binds to the juxtamembrane region of the cadherin tail. **(B)** Cadherin molecules exist as parallel dimers. It is thought that cadherin dimers on one cell make homophilic contacts with the dimers on a neighboring cell, forming a zipper-like structure. The EC1 domain participates both in dimer formation and in homophilic adhesive interactions. According to one model, adhesive specificity depends on this domain of cadherin. According to another model, EC1 through EC3 domains are involved in homophilic adhesion, as shown in Figure 6.9.

lamellae of migrating cells; and podosomes, present in osteoclasts and macrophages (Figure 6.2). In adherens junctions, on the other hand, cell-cell adhesion is mediated by members of the homophilic, Ca^{2+}-dependent cadherin family (Figure 6.3). The integrin- and cadherin-dependent adhesions are both associated with the actin cytoskeleton via a submembrane network or "plaque," comprising the "anchor proteins." Examples of these relationships are provided in Figure 6.4, demonstrating that in cultured cells actin-containing stress fibers terminate at vinculin- and paxillin-containing focal adhesions. Double immunofluorescence labeling for vinculin and $\alpha_v\beta_3$-integrin indicated that the two proteins were colocalized in matrix adhesion sites, yet careful examination of such images reveals vinculin-rich structures which are devoid of integrin, corresponding to cell-cell adhesions. In cell-cell adherens junctions, the membrane receptors are various members of the cadherin family that are associated with the submembrane plaque via

▶ *For Figure 6.2, see color insert.*

▶ *For Figure 6.4, see color insert.*

▶ *For Figure 6.5, see color insert.*

▶ *For Figure 6.6, see color insert.*

β-catenin and the homologous protein plakoglobin (Figures 6.3 and 6.5). As pointed out above, some of the plaque components are shared by focal adhesions and adherens junctions, while others are unique to each type of adhesion site.

An interesting aspect of the structure and function of the submembrane plaque, which physically links the adhesion receptors of the integrin or cadherin families to the actin cytoskeleton, is its association with various signaling molecules, either constitutively or transiently. While it is still common to distinguish between "structural" and "signaling" molecules at adhesion sites, it becomes increasingly difficult to justify such a distinction. Several molecules, for example, focal adhesion kinase (FAK) and β-catenin, may be directly involved both in the mechanical interactions present at the adhesion sites and in the generation and transduction of long-range adhesion-mediated signals.

Transmembrane Interaction of the Extracellular Matrix with the Actin Cytoskeleton

Electron and immunofluorescence microscopy examination (Figure 6.4) reveals an abundance of actin microfilaments at the cytoplasmic side of focal adhesions. Elucidation of the molecular structure of these sites is based on two major types of data. The first includes immunocytochemical (usually immunofluorescence) localization of the various proteins, and the second is based on direct biochemical analysis of the molecular interactions between the various plaque components and between them and the cytoskeleton. Such approaches provide the basis for "interaction maps" similar to the one shown in Figure 6.6, in which a network of interactions, indirectly linking the actin cytoskeleton to the ECM, is presented. In this scheme, we may distinguish several molecular domains of the adhesion sites.

The external surface to which the membrane is attached at focal contacts and related matrix adhesions may contain several types of ECM molecules, such as fibronectin, vitronectin, and collagens. This diversity is quite intriguing, since it is still not known how matrix adhesions formed with molecularly different surfaces direct the assembly of molecularly and functionally different adhesion sites. Recent data strongly suggest that not only the chemical nature of the matrix, but also its mechanical characteristics and "dimensionality" (prevailing two-dimensional versus three-dimensional architecture) characteristics affect the composition, organization, and dynamics of the adhesion molecular complexes. Some mechanisms of these effects are discussed below.

The transmembrane domain of matrix adhesions consists of adhesion receptors, mainly different members of the integrin superfamily. As may be expected from the fact that these receptors can interact with different matrix molecules, this domain is also quite heterogeneous with respect to the integrin composition. Such differences may exist between different cells, each expressing a different set of integrins, or may even be detected between individual adhesion sites within single cells. For example, classical focal adhesions display relatively high levels of $\alpha_v\beta_3$ integrin (Figure 6.4), while fibrillar adhesions (Figure 6.2), associated with fibronectin fibrils, are enriched with $\alpha_5\beta_1$ integrin.

Another domain of focal adhesions is the submembrane plaque, which harbors a multitude of proteins, some of which directly bind to the cytoplasmic faces of different integrins (i.e., talin, α-actinin, filamin, tensin, and probably FAK). Other proteins of the plaque, such as vinculin or zyxin, can bind to these components either directly or indirectly. Interestingly, several of the "anchor proteins" which reside in the submembrane plaque (named after their capacity to anchor actin filaments in the membrane) have multiple binding sites and may interact alternatively or simultaneously with several proteins. For example, one of the major components of the plaque, vinculin, was shown to interact with talin, paxillin, tensin, α-actinin, vasodilator-stimulated phosphoprotein (VASP), actin, and other proteins. Zyxin, another plaque component, interacts with α-actinin and VASP. VASP, in addition, interacts with profilin, vinculin, and actin (Figure 6.6). Various actin-binding proteins, like profilin, fimbrin, α-actinin, and others, are enriched in adhesion plaques. Moreover, the transient interaction of vinculin with the Arp2/3 complex, a key component nucleating

Figure 6.7 Model depicting role of conformation changes in integrin, talin, and vinculin at the early stages of formation of integrin-mediated matrix adhesions. In the low-affinity state, the C-terminal portions of integrin α and β subunits interact with each other, which prevents interaction of the integrin β subunit with the talin head. In nonactive talin, the head interacts with the C-terminal portion of the rod, which masks the integrin-binding site in the head. The vinculin molecule consists of a globular head domain with the binding sites for talin, C-terminal tail domain with the binding sites for actin, and junctional domain with the binding site for the Arp2/3 complex. In closed conformation (left), the vinculin tail is attached to the head in such a way that the binding sites to talin, Arp2/3, and actin are masked. The signaling molecule PIP_2 binds vinculin and talin and induces a transition from the closed conformation to an extended one, rendering the binding sites accessible (middle). Integrin transition from a low-affinity to high-affinity state is accompanied by moving the α and β subunit cytoplasmic tails away from one another, allowing interaction of β subunit with talin. Conversely, talin activation may in turn promote integrin activation. Vinculin in the opened conformation can link actin nucleator Arp2/3 to talin (right), promoting the assembly of adhesion complexes in association with Arp2/3-induced actin polymerization and branching (right).

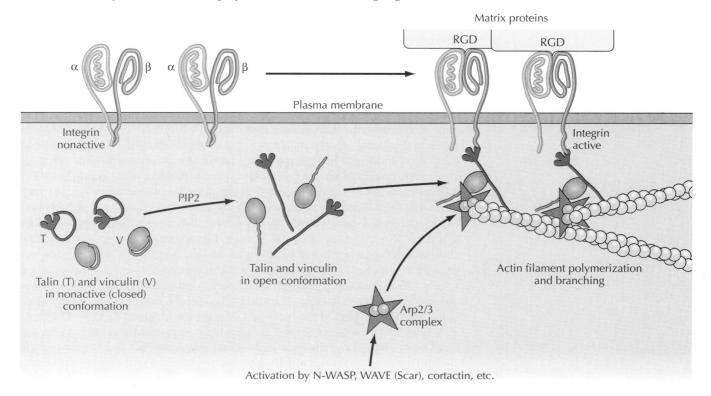

actin polymerization in lamellipodia, was registered (Figure 6.7), suggesting that a cross talk exists between formation of integrin-mediated matrix adhesions and actin polymerization at the cell leading edge. The presence of such diverse binding proteins suggests that the submembrane plaque is indeed a tightly packed three-dimensional meshwork, which may be highly diversified in its detailed molecular composition.

As pointed out above, some of the plaque proteins are signaling molecules, including serine/threonine- and tyrosine-specific protein kinases, their substrates, and a variety of adapter proteins. The involvement of these proteins in adhesion-dependent signaling is discussed below, but it is noteworthy that such molecules are also involved in the physical cross-linking of different components of the adhesion sites. For example, FAK is a multidomain protein that can interact with several focal adhesion proteins, including paxillin, c-Src, p130^{cas}, and Grb2 (Figure 6.6). While some of these interactions are apparently constitutive, others (for example, those mediated by Src homology 2 [SH2] domains on one protein and phosphotyrosine groups on a target protein) are regulated by specific phosphorylation and dephosphorylation events. In FAK- or Src-null cells the levels of tyrosine phosphorylation in focal adhesions are reduced, and consequently, the dynamics of focal adhesion assembly and disassembly are altered.

A potential mechanism for regulating the assembly of the submembrane plaque involves conformational changes in different plaque proteins. A most striking example is provided by recent structural studies of integrin itself, showing that this molecule can exist in two distinct conformations, one with low affinity to ECM ligands ("inactive") and another with high affinity ("active"). The inactive conformation differs from the active one by major changes in the extracellular portion of the α-β dimer, as well as by a "clasp" between the C-terminal portions of the two chains (Figure 6.7). Binding of the extracellular RGD ligand to the extracellular "head" of the α-β integrin dimer induces gross conformational changes that affect the transmembrane parts of the molecule, inducing separation of the α and β cytoplasmic tails, and exposing them for interaction with different intracellular partners such as components of the actin cytoskeleton. The transition from inactive to active conformation can be triggered not only by binding of a ligand to the extracellular part of the integrin dimer, but also by binding of an intracellular partner to the cytoplasmic portion of the molecule. Thus, the binding of the talin molecule to the cytoplasmic portion of the β subunit converts the whole integrin molecule into an active conformation (Figure 6.7). It is interesting that talin binding to integrin, in turn, requires a bulk conformational reorganization of talin. The integrin-binding site is located at the head (N-terminal) domain of talin and is masked in the inactive conformation by interaction with the C-terminal portion of the talin molecule. This "autoinhibition" can be reversed upon binding of phosphatidylinositol 4,5-bisphosphate (PIP$_2$) to talin. Another example of such a mechanism is displayed by vinculin, which apparently is folded "head-to-tail," forming a "closed" conformation in which its binding sites to several proteins (talin, α-actinin, actin, Arp2/3 complex) are hidden, while in the extended conformation these sites are exposed and available for binding (Figure 6.7). In this case also the transition from the folded to the extended state can be induced by PIP$_2$. This possibility is attractive since it may constitute an important mechanism for the cross talk between signaling processes that stimulate PIP$_2$ formation and the assembly of cell adhesions. PIP$_2$ can be produced locally upon signaling, since, as was recently shown, an enzyme responsible for its formation, phosphatidylinositol-4-phosphate 5-kinase, can be recruited to focal adhesions. The gross conformational changes that lead to alterations in binding and functional properties were described also for several other focal adhesion-related signaling components, such as ERM proteins (ezrin-radixin-moesin), diaphanous related formins (DRFs), Src family kinases, etc. The assembly of cell adhesions may be modulated also by regulated alterations in the cytoplasmic levels of various junction-associated proteins. Such modulation may be achieved either by controlled expression or by degradation. Thus, it was shown that changes in cell adhesion and cytoskeletal organization have a profound effect on the synthesis of such proteins as vinculin, α-actinin, and actin. The mechanisms underlying this process are regulated at both the transcriptional and posttranscriptional levels. A process which can have a marked effect on the levels of focal adhesion proteins is proteolytic degradation. The levels of FAK or talin in some systems are regulated by degradation. This degradation is carried out by a calcium-activated neutral protease, calpain. Calpain was shown to localize at adhesion plaques and can specifically cleave components, including talin, integrin, and FAK. Inhibition of calpain by pharmacological inhibitors, or through overexpression of its endogenous inhibitor, calpastatin, results in an inhibition of the natural disassembly of focal adhesions and, consequently, interferes with cell locomotion. In conclusion, it appears that molecular interactions in focal adhesions might be regulated by several "switches," affecting protein phosphorylation, conformation, and stability.

Signaling Components in Matrix Adhesions

The notion that adhesion to the ECM can trigger signaling events has long been recognized, since it was amply demonstrated that the spreading of cells on appropriate matrices is essential for cell growth, differentiation, macromolecular metabolism, and survival. However, an insight into the molecular basis for such adhesion-mediated signaling was obtained only recently, with the discovery that focal adhesions contain a multitude of signaling molecules, including kinases, phosphatases, kinase substrates, and a variety of adapter proteins. Signaling at the focal adhesions affects, first of all, the assembly and dynamics of these molecular complexes themselves, as well as the associated cytoskeletal structures. Moreover, cell adhesion to the ECM was shown to trigger a cascade of signaling events that together convert changes in integrins and integrin-associated complexes into specific changes in gene expression and in the stability of target proteins.

The major signaling molecules detected in focal adhesions are depicted in Figure 6.6. A pivotal component is the tyrosine kinase FAK, whose activation is an early event in adhesion-mediated signaling. FAK was reported to bind to the cytoplasmic domain of β-integrin and to other proteins localized at focal adhesions, including paxillin, talin, and p130^{cas}. Other protein-tyrosine kinases that are similar in their molecular structure to FAK were recently discovered; however, their role in cell adhesion is still unknown. The main function of FAK is suggested to be the transduction into the cell of signals generated at focal adhesions. FAK binds Grb2 and most probably participates in the integrin-dependent activation of the Ras/mitogen-activated protein kinase (MAPK) signaling pathway. Interestingly, it was found that an activated (membrane-tethered) form of FAK can rescue adhesion-deprived epithelial cells from apoptosis and, when overexpressed, can even transform MDCK cells.

Protein kinases of the Src family constitute another group of kinases, associated (at least transiently) with focal adhesions. pp60$^{v\text{-}src}$ was the first protein tyrosine kinase localized in the residual focal adhesions of Rous sarcoma virus-transformed cells. The association of v-*src* with focal adhesions can induce hyperphosphorylation of resident proteins and, consequently, disrupt the organization of these adhesion sites. The normal, proto-oncogenic counterpart of v-*src*, pp60$^{c\text{-}src}$, was later identified in association with adhesion plaques of normal fibroblasts. The association of c-Src with focal adhesions is apparently not constitutive, since its phosphorylation on tyrosine 527 by C-terminal Src kinase (CSK), another focal adhesion-associated tyrosine kinase, causes c-Src folding into a "closed" configuration and dissociation from the plaque. It was also reported that c-Src is activated in an integrin-dependent manner after cell adhesion and that fibroblasts derived from c-Src$^{-/-}$ mice exhibit a reduced rate of spreading on fibronectin. In addition to c-Src, other members of the Src family such as Fyn can be associated with focal adhesions. Interestingly, c-Src-null cells can form focal adhesions, yet these adhesions display an aberrant turnover and fail to evolve into fibrillar adhesions. Kinases of the FAK and Src families can interact with each other. In the focal adhesions, clustered FAK undergoes autophosphorylation at Tyr 397, enabling its binding to pp60^{src} or other SH2 domain-containing proteins. Src in turn can phosphorylate FAK, and this phosphorylation increases the kinase activity of FAK and enhances Grb2 binding and MAPK activation. In addition, Src and FAK may participate in the feedback regulation of the small Rho GTPase activity. Tyrosine phosphatases that can reverse the action of the kinases mentioned above can also be associated with focal adhesions. One member of this family associated with focal adhesions is LAR (leukocyte common antigen-related molecule), a transmembrane tyrosine phosphatase. Transmembrane receptor protein tyrosine phosphatase alpha (RPTP-α) associates with $\alpha_v\beta_3$ integrins and may participate in the activation of Src family kinases, abolishing CSK-mediated inhibitory phosphorylation. In addition, nonreceptor tyrosine phosphatases (e.g., Shp-2) were also shown to be associated with focal adhesions. These kinases and phosphatases regulate the phosphorylation levels of tyrosine on a number of focal adhesion components. A variety of tyrosine-phosphorylated proteins in the cell are localized at focal adhesions and cell-cell contact sites, as demonstrated by the staining of cells with fluorescent antiphosphotyrosine antibody, either without treatment or following a short inhibition of tyrosine dephosphorylation by vanadate. Upon adhesion of cells to the ECM, there is a group of proteins whose level of phosphorylation on tyrosine increases conspicuously. These include FAK, p130^{cas}, tensin, paxillin, tyrosine kinases of the Src family, etc. Paxillin was also shown to undergo strong serine phosphorylation following cell adhesion to the substrate. Several serine-threonine kinases, in addition to tyrosine kinases and phosphatases, are associated with focal adhesions (e.g., protein kinase C [PKC] that colocalizes with talin and vinculin). Yeast two-hybrid screen revealed a special serine/threonine kinase that binds to the cytoplasmic domain of β-integrin. Overexpression of this integrin-linked kinase (ILK) was shown to induce anchorage-independent cell proliferation. Other kinases that appear to play an important role in focal adhesion assembly and reorganization are p21-activated kinases (PAKs). These kinases are the effectors of small GTPases Rac and Cdc42 (see below), and, in particular, activated PAK1 accumulates at nascent focal adhesions (focal complexes) at the lamellipodia of fibroblasts.

Finally, phosphoinositides, PIP_2 and phosphatidylinositol 3,4,5-trisphosphate (PIP_3), as well as the enzymes involved in their metabolism, phosphatidylinositol 4-phosphate 5-kinase (PIP5-kinase) and phosphatidylinositol 3-kinase (PI3-kinase), respectively, are also associated with focal adhesions. Phospholipase C-γ, which degrades PIP_2, was also localized in these sites. SH2 domain of the regulatory p85 subunit of PI3-kinase was suggested to interact with phosphotyrosine residues on certain focal adhesion proteins, e.g., FAK.

Specialized sequence motifs mediate the molecular interactions of these signaling molecules with each other or with other components of the submembrane plaque (Figure 6.6). These include Src homology domain 2 (SH2), present on such proteins as Src family kinases, PI3-kinase, tensin, and SHP-2, which mediates their binding to specific tyrosine-phosphorylated sites on partner proteins. SH3 domains (present on Src family kinases, $p130^{cas}$, vinexin, etc.) interact with proline-rich motifs (present in FAK, paxillin, vinculin, $p130^{cas}$, etc.), affecting the formation of the multimolecular junctional complexes of the submembrane plaque. Additional interactions, such as the binding of zyxin to the cysteine-rich protein (CRP) and to paxillin, depend on protein–protein interactions mediated by the LIM zinc finger motif. It is noteworthy that the presence of these and additional binding motifs on multiple proteins suggests that the molecular organization of focal adhesions might be quite variable, due to competition between different proteins carrying similar binding motifs.

Molecular Interactions in Cell-Cell Adherens Junctions

The adhesion receptors in adherens junctions are cadherins. These transmembrane molecules belong to a multigene superfamily that consists of several subfamilies of adhesion receptors. Adherens junction-forming "classical cadherins" have a highly conserved cytoplasmic domain, which mediates the association of these cadherins with the actin cytoskeleton. Other cadherin subfamilies include desmosomal cadherins associated with intermediate filaments, "Flamingo"-type cadherins having a seven-pass transmembrane region with high homology to G-protein-coupled receptors, T-cadherins lacking a cytoplasmic tail, and a novel subfamily of neural cadherins, cytoplasmic portions of which are associated with the Src family kinase Fyn.

The immediate partners of the classic cadherins in the submembrane junctional plaque of adherens junctions are catenins, in particular β- and γ-catenin (plakoglobin), which bind to the same domain in the cadherin cytoplasmic tail, and $p120^{ctn}$ catenin that associates with the juxtamembrane region of the cadherin tail (Figures 6.3, 6.8, and 6.9). Additional components of the junctional plaque include α-catenin, vinculin, α-actinin, and other structural and signaling proteins (Figure 6.8). Similar to focal adhesion molecules, many of the junctional plaque components of adherens junctions are capable of interacting with multiple partners (Figure 6.8); for example, α-catenin, a vinculin homologue, can interact with β-catenin, plakoglobin, vinculin, α-actinin, ZO-1, and actin. Recent studies revealed new players in the cadherin-based molecular complexes: transmembrane adhesion molecules of the immunoglobulin superfamily called nectins and their cytoplasmic partner, afadin, were shown to associate with the cadherin-catenin complex via α-catenin. α-Catenin may associate with the actin cytoskeleton both by direct binding and via its actin-binding partners, such as α-actinin, vinculin, and ZO-1; afadin can also bind actin filaments. Some of

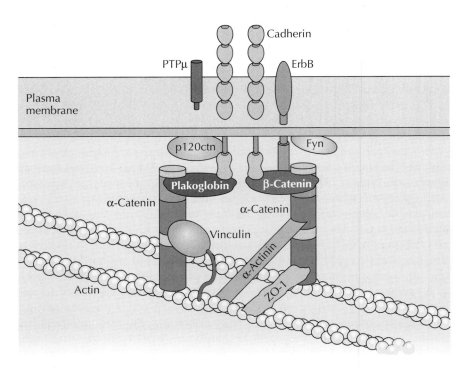

Figure 6.8 Protein interactions at adherens junctions (see text).

the molecular interactions present in these adhesion sites appear to be constitutive and depend mostly on the availability of the partner proteins. Other interactions, mainly those depending on phosphorylation events, can be subject to modulation by external stimulation or internal control mechanisms. For example, the affinity of β-catenin to cadherin can be reduced by β-catenin phosphorylation at tyrosine residue 654, which is a site of phosphorylation by growth factor receptor tyrosine kinases. In the coming pages, these and other signal transduction elements responsible for cadherin adhesion regulation are discussed.

Signaling at Cell-Cell Adherens Junctions

Adherens junctions contain a battery of plaque proteins, some of which are the same ones found in focal adhesions (e.g., vinculin, α-actinin, tensin, zyxin) while others (α- and β-catenin, plakoglobin, and p120^{ctn}) are apparently specific for cell-cell adhesions. It is noteworthy that adherens junctions are also associated with signaling molecules, including transmembrane receptor tyrosine kinases, such as the fibroblast growth factor (FGF) receptors, the hepatocyte growth factor-scatter factor (HGF/SF) receptor cMet, and the ErbB-family receptors, ErbB-1 (epidermal growth factor receptor) and ErbB-2. There are also receptor tyrosine phosphatases (e.g., PTPμ and DEP-1), as well as nonreceptor tyrosine kinases of the Src family, such as Fyn (Figure 6.8). In a few cases, the mechanisms of association of these signaling molecules with adherens junctions have been determined. The FGF receptor specifically interacts with the extracellular C4 domain of N-cadherin, while the localization of ErbB-family receptors depends on the interactions of their cytoplasmic tails with β-catenin and, perhaps, with a special scaffold protein erbin; PTPμ and DEP-1 bind to p120^{ctn}.

Formation of cadherin-mediated adhesions induces signaling events of several types: first, rapid signals that do not depend on transcription, some of which may trigger feedback loops that regulate the assembly of adherens junctions themselves; second, sustained signals that lead to alterations of the transcriptional profile and which determine stable phenotypic changes in contacting cells.

The rapid signals are often generated due to interactions of cadherin molecules with growth factor receptors. For example, formation of cadherin-mediated junctions in some cell types induces ErbB-1-dependent MAP kinase activation. In other cells, N-cadherin-mediated contacts activate an FGFR-mediated signaling cascade, most probably owing to prevention of internalization of the FGF receptors or direct activation of the receptor. Another general early response is the recruitment of PI3-kinase to newly formed cadherin complexes, leading to PI3-kinase activation. The molecular details of this process are not known; however, the p85 regulatory subunit of PI3-kinase can interact with receptor tyrosine kinases associated with cadherins or directly with β-catenin. PI3-kinase activation is an important step in the signaling loop that induces the assembly of cadherin-mediated cell-cell contacts; its inhibition hinders formation of these contacts.

Formation of cadherin-mediated cell-cell contacts has an immediate effect on the dynamics and organization of both the actin cytoskeleton and microtubules. The mechanisms of these regulatory events are complex and not yet understood in detail. In some cases, the transient recruitment of certain regulatory components into assembling adherens junctions can explain such effects. For example, cadherin was recently shown to associate with the actin-nucleating Arp2/3 complex whose recruitment to nascent cadherin adhesions may result in the activation of local actin polymerization during adherens junction assembly.

Changes in the activities of small GTPases of the Rho family induced by cadherin-mediated contacts are perhaps among the most important feedback signals affecting the process of assembly of these contacts. These changes are discussed in the next section.

Another mode of signaling from adherens junctions involves the junctional molecules β-catenin and plakoglobin, which belong to a highly conserved family of proteins possessing a central domain consisting of multiple "arm repeats" (named after the *Drosophila* homologue of β-catenin, armadillo). In adherens junctions, β-catenin plays an essential role in interconnecting the cadherin cytoplasmic domains, via α-catenin, to the actin cytoskeleton. In addition, β-catenin and armadillo play a major role in signal transduction by the Wg/Wnt pathway to regulate morphogenesis. When expressed in excess, or upon release from the membrane, β-catenin can translocate into the nucleus (Figure 6.5), where, together with specific transcription factors of the LEF/TCF family, it activates the transcription of specific target genes (Figure 6.9). Uncontrolled regulation of target gene expression by β-catenin is a major cause of its oncogenic action in various tumors. These genes include positive regulators of cell proliferation including Cyclin D1 and C-MYC, but also genes involved in the modulation of cell motility and invasion/metastasis, including the ECM components fibronectin and laminin α-2, various metalloproteinases, and cell adhesion receptors including Nr-CAM, CD44, and U-PAR. The regulation of catenin-mediated signaling may occur at several different levels. Best studied is the regulation of β-catenin levels by its controlled degradation by the ubiquitin-proteasome system via the Wg/Wnt pathway (Figure 6.9). Interestingly, cadherins may

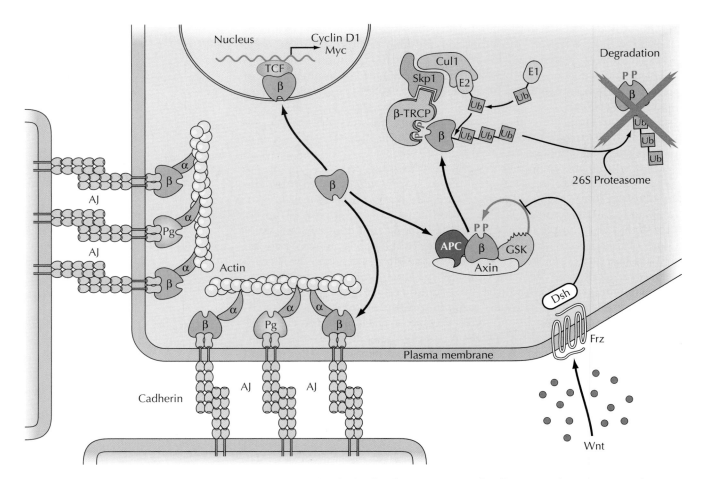

Figure 6.9   The dual role of β-catenin in cell adhesion and transcriptional activation. β-Catenin (β) and plakoglobin (Pg) bind to cadherin adhesion receptors, and via α-catenin (α) they associate with the actin cytoskeleton to form AJs. When the Wnt signaling pathway is inactive, free β-catenin is degraded by a complex including glycogen synthase kinase (GSK), adenomatous polyposis coli (APC), and Axin, which phosphorylate β-catenin (PP). This protein complex recruits β-TrCP, which, together with Skp1, Cul1, and the E1 and E2 ubiquitination components, mediates the ubiquitination of β-catenin (Ub) and directs it to degradation by the 26S proteasome. The binding of Wnt to Frizzled (Frz) receptors activates Wnt signaling, and disheveled (Dsh) inhibits β-catenin phosphorylation by GSK. This results in β-catenin accumulation in the nucleus, where it complexes with T-cell factor (TCF) and transactivates target genes such as Cyclin D1 and Myc (adapted from Conacci-

play an important role in controlling the fate of β-catenin. On the one hand, they can sequester the molecule by binding it to the plasma membrane (sequestering it away from the nucleus), thus potentially suppressing β-catenin-driven transactivation. On the other hand, the loss of cadherin expression can lead to nuclear localization of β-catenin and the activation of β-catenin target genes. This behavior is displayed during colon cancer cell metastasis at the leading edge of invasive cells that have lost E-cadherin expression but show nuclear β-catenin, while cells at the central and more differentiated part of the tumor express both E-cadherin and β-catenin at cell-cell junctions.

Cross Talk between Adhesion-Dependent Signaling and the Rho Family of GTPases

Small G proteins of the Rho family are major cellular regulators of the cytoskeleton. The Rho family is represented in humans by at least 20 proteins classified into seven subfamilies. Rho, Rac, and Cdc42h are the best-studied subfamilies, in which RhoA, Rac1, and Cdc42 are the most-studied members. It has become increasingly clear that many signals that affect functions of the actin cytoskeleton (and, most probably, microtubules too) are operating via activation or inhibition of certain members of the Rho family. Therefore, all adhesion-related events that depend on cytoskeletal activity are controlled by Rho GTPases. Moreover, recent studies indicate that besides cytoskeletal targets, Rho GTPases may control some components of the junctional structures themselves.

Initial information on the role of Rho family GTPases in adhesion emerged from experiments with constitutively active and dominant negative mutants of these proteins and with bacterial toxins that specifically inhibit or activate Rho GTPase functions (see chapter 14). It was shown that Rac1, responsible for the formation of branching networks of actin filaments in the lamellipodium, is necessary for the formation of focal complexes, the precursors of mature focal adhesions. RhoA activity is required for maturation of focal complexes into focal adhesions and the formation of actin- and myosin-rich stress fibers. Rac1, Cdc42, and RhoA are all necessary for the assembly of the cadherin-mediated cell-cell junctions, since dominant negative mutants of Rac1 and Cdc42 and botulinum C3 toxin that inhibits RhoA prevent this process.

Since formation of both cell-cell and cell-matrix adhesions requires Rho family GTPases, the maintenance of threshold levels of such activities is indispensable in the assembly of these adhesion structures. Swiss 3T3 fibroblasts, for example, cannot assemble mature focal adhesions in serum-free medium even after attachment to the ECM, since they require stimulation of RhoA activity by the serum factor lysophosphatidic acid (LPA). In the majority of cell types, however, the signals activating Rho GTPases are generated by the adhesive structures themselves, which provide a feedback that allows assembly and turnover of these structures.

Adhesion of cells to fibronectin induces the activation of Cdc42 and Rac1 and promotes its coupling with its target PAK. RhoA is also activated by adhesion to fibronectin, but the time course of this activation differs from that of Rac. While Rac is activated first after cell attachment, inducing the formation of lamellipodia and focal complexes, the activity of RhoA increases only later and induces the maturation of focal adhesions. Formation of cadherin-mediated cell-cell contacts also triggers Rac activation and, at least in some cells, the concomitant increase of Cdc42 activity, while the level of active Rho initially decreases. Both the temporal and the spatial distributions of Rho GTPase activities are strictly controlled during contact formation. Active (GTP-bound) forms of Rac and Cdc42 are localized at the tips of lamellipodia, where the formation of new focal complexes occurs, but mature focal adhesions do not contain small GTPases. Both Rac and Cdc42 are enriched in cadherin-mediated cell-cell junctions, while RhoA is not specifically associated with either cell-matrix or cell-cell adhesions.

The mechanisms involved in integrin- and cadherin-dependent regulation of Rho family G proteins are incompletely understood, but they are activated by a variety of guanine nucleotide exchange factors (GEFs) that catalyze the dissociation of GDP from inactive G proteins. Cytosolic GTP can then bind to G proteins and induce a conformational change, allowing their interaction with downstream effectors. GEFs that might be involved in Rho family protein activation include Tiam-1, Vav-2, and Dock180. In particular, integrin-mediated activation of Rac1 may occur when a FAK-Src-p130Cas-Dock180 signaling complex is assembled, while activation of Rac by vascular endothelial (VE)-cadherin apparently involves Tiam-1. One possible scenario for adhesion-dependent Rac/Cdc42 activation involves the recruitment of PI3-kinase to adhesion plaques that induces the accumulation of PIP_3, forming a docking site for the pleckstrin homology (PH) domains of putative exchange factors.

p120^{ctn} plays a special role in the cadherin-mediated activation of Rac. It was recently shown that binding of this protein to the cytoplasmic cadherin tail is required for Rac activation upon cadherin engagement. Blocking this pathway by either a mutation in the cadherin tail or the silencing of the p120^{ctn} gene prevents cadherin-mediated adhesion and actin reorganization. It is interesting that p120^{ctn} can activate Rac and Cdc42 not only at cell-cell junctions, but also at other locations, in a cadherin-independent manner. This activation may induce the formation of cytoplasmic protrusions and, consequently, augment cell motility. Thus, p120^{ctn} may control the motile behavior of cells; its junctional localization stabilizes cell-cell junctions and suppresses motility, while its localization to the free cell edges activates protrusion formation and cell migration.

Activation of Rac1 at early stages of adhesion is often accompanied by a transient decrease in RhoA activity. The mechanism underlying this decrease is incompletely understood, but it involves p190RhoGAP. This protein stimulates the GTPase activity of RhoA and triggers its transition from an active, GTP-bound to an inactive, GDP-bound form. p190RhoGAP is activated by Src family kinases, which, in turn, can be activated by receptor (cadherin or integrin) engagement.

Why is the regulation of Rho family GTPases so important for the proper formation of adhesion structures? To answer this question we should examine the activities of the major targets of these molecular switches and their diverse effects on adhesion events. A first group of such targets includes components of adhesion plaques. As indicated above, several types of focal adhesion and adherens junction proteins are activated by PIP_2 by a transition from a "closed" to an "open" conformation (Figure 6.7). Synthesis of PIP_2 is catalyzed by PIP5-kinase, a direct target of Rac1. Moreover, some targets of Rac1 and Cdc42 control the interactions between specific plaque proteins. For example, the Rac1/Cdc42 target IQGAP1 binds β-catenin and prevents its interaction with α-catenin. GTP-bound active forms of either Rac or Cdc42 can release IQGAP1 from β-catenin and thus trigger β-catenin–α-catenin interactions.

Another way by which Rho GTPases affect adhesion structures is by regulating the cytoskeleton. In particular, the formation of focal complexes and cadherin-mediated junctions depends on Rac1-driven actin polymerization. Rac1 activates a Scar/Wave protein, which in turn activates an Arp2/3 complex that nucleates the formation of the branching actin network in lamellipodia. In addition, Rac1, as mentioned above, promotes the production of PIP_2 following PIP5-kinase activation, and PIP_2 removes a capping protein from the actin filament plus ends, which also promotes polymerization. Finally, Rac1 activates PAK1, which activates LIM kinase, which, in turn, phosphorylates and inactivates the actin-depolymerizing factor ADF/cofilin.

Rho, which is indispensable for the formation of both focal adhesions and cell-cell adherens junctions, operates exclusively on the level of cytoskeletal regulation. Two (among many) Rho targets, the formin family protein Diaphanous (Dia1) and Rho-associated kinase (known also as Rho kinase, ROCK or ROK), are responsible together for Rho-mediated activation of focal adhesion assembly. Dia1 is a potent nucleator of actin filament polymerization, promoting linear, nonbranching filament growth. This effect of Dia1 is mediated by its direct interactions with the actin filament plus ends and with profilin. In addition, Dia1 is involved in the regulation of microtubule dynamics. Rho kinase appears to be a most potent regulator of myosin II-driven cell contractility. Activation of nonmuscle myosin II requires phosphorylation of its regulatory light chain by myosin light chain kinase (MLCK); the reverse process, dephosphorylation, is performed by myosin light chain phosphatase (MLCP). The major effect of Rho kinase on myosin II is the inactivation of MLCP by its phosphorylation. In some cases Rho kinase, similarly to MLCK, may also directly phosphorylate myosin light chain.

Thus, RhoA, via Rho kinase activation, increases the phosphorylation of myosin II light chain and stimulates cell contractility, while via Dia1 it affects actin polymerization and microtubule dynamics. Rac1 and Cdc42, in addition to their effect on actin polymerization, can also modulate microtubule dynamics and cell contractility, most probably via PAK family kinases. The way in which cell contractility and the microtubule system are related to the regulation of the assembly of adhesion structures is discussed in the following sections.

Tension-Dependent Regulation of Adhesion Structure Assembly

A most striking example of the involvement of cell contractility in the regulation of adhesion events is provided by focal adhesion formation. In a simplified way, this process could be regarded as a sequence of self-assembly events, whereby the different components are recruited, and they recruit other partners, to the integrin-containing adhesion site. Recent studies, however, suggest that adhesion to the ECM, per se, is insufficient to induce focal adhesion formation and that Rho activity is necessary for this process. It was shown that an increase in cell contractility, induced by Rho activation, is critical for adhesion-dependent signaling and that inhibition of myosin II activity blocks the formation of focal adhesions and stress fibers.

The importance of tension for triggering adhesion-dependent signal transduction is supported by recent findings where external forces were directly applied to cell-ECM adhesion sites by a microneedle, by stretching an elastic substrate, or by laser trapping of cell surface-attached beads covered with adhesion ligands. These mechanical manipulations of adhesion sites promote the recruitment of new components and growth of adhesion plaques and affect the strength of adhesion, the organization of the cytoskeleton, and the progress of downstream signaling. The plating of cells on soft flexible substrates suppresses tension development and inhibits the formation and tyrosine phosphorylation of focal adhesions. Collectively, these studies suggest that development of tension in the adhesion plaque serves as a checkpoint that regulates the growth of this structure. Thus, focal adhesions may function as "mechanosensors," informing the cell about the mechanical characteristics of the microenvironment and possibly about locally applied external forces. The role of mechanical forces in adherens junction formation is less clear, but there is evidence that these structures also have mechanosensory function.

Complex Cytoskeletal Cross Talk

Recent studies indicated that actomyosin contractility is greatly affected by the cross talk between different cytoskeletal networks. In particular, disruption of microtubules triggers myosin II-driven contractility. The mechanism at the basis of this effect might depend on both the mechanical characteristics of the microtubule network (a "tensegrity" model) and the ability of microtubules to deliver (or capture) regulatory molecules that affect contractility (a "signaling" model). Recent studies demonstrated an association of microtubules with the Rho exchange factor (GEF-H1), whose release upon microtubule disruption increases Rho activity and cell contractility. All in all, microtubules can efficiently restrict cell contractility.

This feature of microtubules may provide a mechanism for their effect on focal adhesions. Since myosin II-driven tension triggers the growth of focal adhesion and the associated actomyosin bundle, which, in turn, increases tension and promotes further growth of focal adhesion, negative regulators are also necessary to break this loop and prevent an indefinite growth of focal adhesions. Microtubules are fit ideally for this role as negative physiological regulators. This model suggests that microtubules are attracted to growing focal adhesions and reduce myosin II-driven contractility in their proximity. Such local relaxation could then restrict focal adhesion growth or even induce their disassembly.

Definitive molecular mechanisms responsible for microtubule directing to focal adhesions are not clear, but the Rho effector, Dia1, might be involved in this process based on its effects on microtubule dynamics. The yeast homologue of Dia1 (Bni1p) is also involved in the targeting of microtubule ends to cortical structures. Thus, Rho might coordinate the entire process of focal adhesion maturation. While via Rho kinase, it induces myosin II-driven tension that activates the focal adhesion mechanosensor, triggering an assembly process, via Dia1, it might promote actin polymerization supporting focal adhesion assembly. Furthermore, changes in microtubule dynamics, facilitating their growth in the direction of focal adhesions regulated by Dia1, may provide a means for the termination of focal adhesion growth.

Conclusions and Link to Microorganism Invasion

We outlined here the major players and principal mechanisms involved in the formation of focal adhesions and adherens junctions. We highlighted the fact that besides the transmembrane integrin and cadherin receptors, numerous structural and signaling components of the submembrane plaques and the cytoskeleton, as well as GTPases of the Rho family, are indispensable for proper organization of these processes. Apparently, this machinery with its complex structural features and sophisticated regulation is successfully exploited by bacteria and other microorganisms in the course of their invasion into cells. Since this subject is discussed in detail in other chapters of this volume, we mention here only a few examples to illustrate the general principles. *Listeria monocytogenes* uses a surface protein named internalin A (InlA) to adhere to and then enter cells. The receptor for this bacterial protein was shown to be E-cadherin. The process of internalization depends on the cytoplasmic part of the E-cadherin molecule and can be blocked by inhibitors of tyrosine phosphorylation and by actin-depolymerizing drugs. To make this process even more similar to cadherin-mediated adhesion, contact with bacteria was shown to stimulate both MAP kinase and PI3-kinase activities. Stimulation of PI3-kinase was shown to depend on another bacterial protein, internalin B (InlB), that interacts with the adherens junction-associated receptor tyrosine kinase, c-Met. After internalization, *Listeria* travels in the cytoplasm using its protein ActA that activates the Arp2/3 complex-mediated actin polymerization. Various microbial pathogens bind to host cell integrin receptors either directly or via integrin ligands. Blocking integrin function by antibodies specific to $\alpha_v\beta_5$- or $\alpha_v\beta_3$-integrin resulted in the abrogation of such bacterial internalization. Bacterial invasion, similar to formation of adhesion-like structures, often depends on the activity of Rho family GTPases. Some pathogenic *Escherichia coli* strains produce a protein toxin, named cytotoxic necrotizing factor 1 (CNF1), which permanently activates proteins belonging to the Rho family. In epithelial cells, the consequence of this activation is the rearrangement of the actin cytoskeleton and the promotion of an intense and generalized ruffling activity. The process of *Shigella* invasion is Rho dependent. A Rho-specific inhibitor abolishes *Shigella*-induced membrane folding and impairs the entry of these bacteria into cells. Bacteria use diverse methods to activate the Rho GTPases or to bypass the requirements in these proteins affecting directly their targets. In particular, *Shigella* delivers into cells IpaC protein that activates Cdc42 and Rac. Often bacterial proteins interact directly with proteins involved in adhesion plaque formation. It was shown that IpaA, another *Shigella* protein, rapidly associates with vinculin during bacterial invasion. An IpaA mutant defective for cell entry differs from wild-type *Shigella* in its ability to recruit vinculin. IpaA-vinculin interaction was suggested to initiate the formation of focal adhesion-like structures required for efficient invasion. Finally, even major cytoskeletal structures, like actin filaments and microtubules, could be the targets of bacteria during internalization pathway. For example, a recent study of the *Shigella* VirA effector protein, which is delivered via a type III secretion system, suggests that MT destabilization plays an important role in *Shigella* infection.

Thus, understanding of the principles underlying the regulation of cell adhesion can greatly contribute to deciphering the complex interrelationships between invading microorganisms and their host cells.

Selected Readings

Bershadsky, A. D., N. Q. Balaban, and B. Geiger. 2003. Adhesion-dependent cell mechanosensitivity. *Annu. Rev. Cell. Dev. Biol.* **19:**677–695.

Burridge, K., and K. Wennerberg. 2004. Rho and Rac take center stage. *Cell* **116:**167–179.

An excellent review, in which Rho proteins' cross talk with adhesion molecules is discussed. Additionally, the strategies of various bacterial pathogens to manipulate Rho protein activity are summarized.

Calderwood, D. A. 2004. Integrin activation. *J. Cell. Sci.* **117:**657–666.

In this review structural aspects of integrin activation and signaling are discussed.

Carlier, M. F., C. L. Clainche, S. Wiesner, and D. Pantaloni. 2003. Actin-based motility: from molecules to movement. *BioEssays* **25:**336–345.

Conacci-Sorrell, M., J. Zhurinsky, and A. Ben-Ze'ev. 2002. The cadherin-catenin adhesion system in signaling and cancer. *J. Clin. Invest.* **109:**987–991.

Cossart, P., J. Pizarro-Cerda, and M. Lecuit. 2003. Invasion of mammalian cells by Listeria monocytogenes: functional mimicry to subvert cellular functions. *Trends Cell. Biol.* **13**:23–31.

DeMali, K. A., and K. Burridge. 2003. Coupling membrane protrusion and cell adhesion. *J. Cell. Sci.* **116**:2389–2397.

  Possible association between actin-nucleating Arp2/3 complex and the components of cell–cell and cell–matrix adhesions is discussed.

Etienne-Manneville, S., and A. Hall. 2002. Rho GTPases in cell biology. *Nature* **420**:629–635.

  An archetypal review on the function of small Rho family GTPases in signal transduction.

Geiger, B., and A. Bershadsky. 2002. Exploring the neighborhood: adhesion-coupled cell mechanosensors. *Cell* **110**:139–142.

Geiger, B., A. Bershadsky, R. Pankov, and K. M. Yamada. 2001. Transmembrane extracellular matrix—cytoskeleton crosstalk. *Nat. Rev. Mol. Cell. Biol.* **2**:793–805.

Gooding, J. M., K. L. Yap, and M. Ikura. 2004. The cadherin-catenin complex as a focal point of cell adhesion and signalling: new insights from three-dimensional structures. *Bioessays* **26**:497–511.

Gruenheid, S., and B. B. Finlay. 2003. Microbial pathogenesis and cytoskeletal function. *Nature* **422**:775–781.

Hynes, R. O. 2002. Integrins: bidirectional, allosteric signaling machines. *Cell* **110**:673–687.

  A comprehensive review on the mechanisms of integrin signaling.

Perez-Moreno, M., C. Jamora, and E. Fuchs. 2003. Sticky business. Orchestrating cellular signals at adherens junctions. *Cell* **112**:535–548.

  An excellent review describing the molecular architecture of adherens junctions and signal transduction pathways associated with these structures.

Pollard, T. D., and G. G. Borisy. 2003. Cellular motility driven by assembly and disassembly of actin filaments. *Cell* **112**:453–465.

Small, J. V., B. Geiger, I. Kaverina, and A. Bershadsky. 2002. How do microtubules guide migrating cells? *Nat. Rev. Mol. Cell. Biol.* **3**:957–964.

Yap, A. S., and E. M. Kovacs. 2003. Direct cadherin-activated cell signaling: a view from the plasma membrane. *J. Cell. Biol.* **160**:11–16.

Yoshida, S., and C. Sasakawa. 2003. Exploiting host microtubule dynamics: a new aspect of bacterial invasion. *Trends Microbiol.* **11**:139–143.

Zamir, E., and B. Geiger. 2001. Molecular complexity and dynamics of cell-matrix adhesions. *J. Cell. Sci.* **114**:3583–3590.

7

Bacterial Signaling to Host Cells through Adhesion Molecules and Lipid Rafts

GUY TRAN VAN NHIEU, PHILIPPE J. SANSONETTI, AND FRANK LAFONT

The development of an infectious disease depends on the capacity of the causative microorganism to multiply and to colonize specific host tissues. Pathogenic microorganisms do so by virtue of specific virulence factors interacting with host cell components. If the pathogen grows extracellularly on the apical surface of an epithelium, a prerequisite for the infection will be the pathogen's ability to adhere to epithelial cells. This property allows the microbe to colonize host tissues by resisting mechanical clearing mechanisms or by conferring a selective advantage to the pathogen over the endogenous flora. For invasive and intracellular pathogens, entry into normally nonphagocytic cells can be used as a means to cross an epithelial layer and to penetrate deeper within host tissues where further interactions are established. The establishment of the disease will then depend on the microorganism's ability to control the host defense response. This chapter will focus on a few examples of the involvement of cell adhesion molecules during infection, in interactions involved in processes such as adhesion or invasion, or in controlling the process of host inflammation (Box 7.1).

Adhesion Molecules as Pathogen Receptors

Adhesion molecules are cell surface receptors that establish cell-to-cell interactions, or interactions between cells and the extracellular matrix. They are involved in a wide variety of processes such as embryogenesis, cell growth, and differentiation. Adhesion molecules are classically divided in five main groups: the integrins, the cadherins, the immunoglobulin (Ig) superfamily, the selectins, and the proteoglycans (Figure 7.1).

Integrins are well-studied adhesion molecules that are ubiquitously distributed and involved in processes such as cell adhesion to the extracellular matrix and cell-cell interaction; in vivo studies have shown their implication in various functions such as in cell migration, wound healing, and embryogenesis. Integrins consist of heterodimers bearing an α and a β chain. α Chains can associate with a given β chain, and a specific combination between an α and a β subunit confers ligand specificity. For instance,

Cell Adhesion Molecules and Bacterial Pathogens

Adhesion molecules participate in various fundamental processes such as cell adhesion, cell migration, or cell differentiation. They consist of families of receptors that allow the cell to sense and respond to its environment. Engagement of these receptors can result in signaling that leads to short-term cell responses such as cytoskeletal reorganization, or to long-term responses, such as de novo gene transcription and cytokine induction. Cell adhesion molecules are used by pathogenic microorganisms to attach to or to invade normally nonphagocytic cells, and these early interactions determine the physiopathology associated with these pathogens. During these very initial phases of host-pathogen interaction, diversion of host cell processes can occur, for example, by induction of a phagocytic process by an invasive organism, but may also translate in signaling via cell adhesion molecules to neighboring tissue in attempts by the pathogen to control the inflammatory response. In many instances, the cell responses involved do not appear to have physiological equivalents and the study of these interactions is likely to shed new light on the function of these receptors.

$\alpha_5\beta_1$ is a fibronectin receptor, whereas $\alpha_6\beta_1$ preferentially binds to laminin. This specificity, however, is relative since a given integrin heterodimer can often bind to several ligands and can show overlapping function with other integrins. To date, 18 α chains and 8 β chains have been characterized that associate to form 24 heterodimers. This number is likely to expand as more

Figure 7.1 Cell adhesion molecules: determinants of cell adhesion and cell–cell interactions. β_1 integrins bind the extracellular matrix (ECM) via their extracellular domain and associate with the actin cytoskeleton via the cytoplasmic tail of the β_1 subunit. Focal complexes are small β_1-integrin-containing adhesive structures that are formed upon activation of the small GTPase Rac. Focal adhesions, which are larger structures requiring Rho-dependent actomyosin contraction and connect to actin stress fibers, involve integrin clustering, tyrosylphosphorylation, and the recruitment of Src substrates. Cadherins establish homotypic interactions at cell-cell junctions and also associate with the cytoskeleton via their cytoplasmic tail. The formation of cadherin-based junctions depends on Rho GTPases. Members of the immunoglobulin (Ig) superfamily bind to a counter receptor, such as the LFA-1 ($\alpha_L\beta_2$) integrin or VCAM-1, on the surface of neutrophils and monocytes. Selectins bind to carbohydrate residues on the surface of leukocytes.

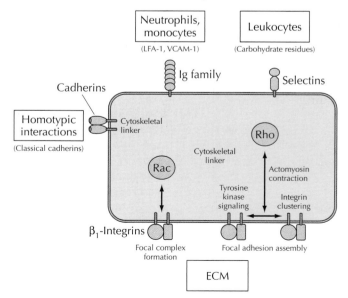

α and β chains have been identified, but whose function remains to be established. The original classification of integrins, based on the nature of the β chain, also reflects some extent of functional requirements. β_1-integrins are usually receptors for the extracellular matrix, β_2-integrins are expressed on the surface of leukocytes and participate in cell-cell interactions during inflammatory processes, whereas β_3-integrins are expressed in platelets and participate in blood clotting.

Different β_1-integrin-containing structures are involved in cell adhesive processes. Cell spreading involves the formation of small clusters called "focal complexes," usually located at the cell periphery, that are considered to be the precursor of "focal adhesions," which are larger and more complex structures that anchor the cytoskeleton. The assembly of focal adhesions is a complex process involving integrin clustering, the recruitment of actin-binding proteins, some of which, such as talin and α-actinin, can directly associate with the cytoplasmic tail of the integrin β_1 subunit, as well as tyrosine phosphorylation of focal adhesion substrates. It has been proposed that upon activation, the FAK tyrosine kinase associates with the cytoplasmic domain of β_1-integrins and is tyrosyl phosphorylated. This initial response allows the recruitment and the activation of Src kinases and the subsequent phosphorylation of Src substrates. The small GTPase Rho controls integrin clustering and focal adhesion formation through actomyosin contraction, whereas focal complexes are induced by Rac activation.

The cadherins are involved in cell-cell interactions at the level of intercellular junctions where they establish calcium-dependent homotypic interactions with cadherins of the adjacent cell. The classical (P, E, and N) cadherins consist of homodimers present at the adherens junctions of polarized cells. These molecules can interact with the cytoskeleton via binding of their cytoplasmic tail to catenins. The formation of cadherin junctions appears to be negatively regulated by tyrosine kinase signaling, whereas the small G proteins Rho and Rac are required for adherens junction formation. In some instances, the formation of intercellular junctions antagonizes focal adhesions assembly and cell adhesion to the extracellular matrix, perhaps by regulating Src kinase activity and by intersecting pathways involved in growth factor receptor signaling.

The immunoglobulin superfamily of adhesion molecules is mostly involved in cell-to-cell interactions and consists of receptors with extracellular domains sharing homology with immunoglobulins. They are involved in homotypic interactions such as nerve-cell adhesion molecule (NCAM)-NCAM binding in nerve cells, but also establish heterotypic interactions, in particular, during inflammatory processes. For example, the interaction of intercellular cell adhesion molecule 1 (ICAM-1) with the leukocyte factor antigen-1 (LFA-1) integrin ($\alpha_L\beta_2$) on the surface of lymphocytes is critical for the antigen-dependent activation of T cells. Interactions between ICAM-1 and Mac1 ($\alpha_M\beta_2$) or vascular cell adhesion molecule-1 (VCAM-1) ($\alpha_4\beta_1$) interactions on the surface of endothelial cells and leukocytes participate in the attachment and extravasation of polymorphonuclear cells through blood vessels during inflammation. Besides their role in inflammation, members of the immunoglobulin superfamily can also regulate growth factor responses. For example, association of ICAM-1 with the fibroblast growth factor (FGF)-receptor modulates this growth factor receptor's tyrosine kinase activity.

Finally, the selectins are a family of lectin receptors that interact with carbohydrate residues on the surface of endothelial cells and leukocytes. As opposed to the families of adhesion molecules described above and as inferred from mice knockout studies, the function of selectins appears to be limited to the vascular system. The main role of selectins appears to be the mediation of the attachment and the rolling of leukocytes on endothelial tissues during inflammatory processes, although they may also participate in lymphocyte homing in Peyer's patches.

As shown in Table 7.1, interactions between pathogens and the different families of adhesion molecules have been reported. Although this list is far from exhaustive, we have tentatively regrouped these examples in three main categories of interactions promoting (i) bacterial adhesion/and or low levels of internalization in host cells, (ii) efficient bacterial internalization by host cells, and (iii) bacterial interaction with phagocytic cells' adhesion molecules. In general, interactions leading to adhesion or low levels of internalization often consist of lectin-carbohydrate associations or indirect association of the bacteria to β_1-integrins via components of the extracellular matrix. On the other hand, integrin or cadherin receptors can also allow uptake of bacterial pathogens after direct interaction with a bacterial surface ligand. This latter situation may reflect a requirement for deeper subversion of the adhesion molecules function as none of these receptors have been implicated in phagocytic processes. It also suggests a role for the cell cytoskeleton during bacterial uptake as both integrins and cadherins associate

Table 7.1   Cell adhesion molecules as receptors for bacterial pathogens

Pathogen	Ligand	Counterligand/receptor
Adhesion/low levels of invasion		
Direct interaction		
Gram-negative bacteria	(Pilus, fimbrial) adhesins	Carbohydrate residues (ECM, integrins, Ig superfamily)
B. pertussis	FHA	CR3
	FimD	RGD binding integrins
	PT	P- and E-selectins
N. meningitidis	Capsule/polysialic acid	NCAM
N. gonorrhoeae/N. meningitidis	OpaA	HSPG, Vn, Fn
	Opa proteins	CDα66
B. burgdorferi		$\alpha_{IIb}\beta_3$
Bridging mechanism		
Staphylococcus	LTA, FnBP	Fn
Streptococcus	LTA, M protein, FnBP	Fn
Yersinia	YadA	Fn, Cn
N. gonorrhoeae	Opc	HSPG, Vn, Fn
Mycobacterium		Fn, Ln
Invasion		
Listeria	Internalin	E-cadherin
	InlB	gClqR, Met receptor
Shigella	IpaB, IpaC	CD44, $\alpha_5\beta_1$
Yersinia	Invasin	$\alpha_{3-6}\,v\beta_1$
Uptake by phagocytic cells		
Legionella	Momp	C3b
Mycobacterium	?	C3b
N. gonorrhoeae	Opa proteins	CD66

with actin filaments through their cytoplasmic part. This role, however, is not exclusive, as cell invasion by some bacteria, such as *Chlamydia* or *Campylobacter* species, is insensitive to cytochalasin, an inhibitor of F-actin, and do not require the actin cytoskeleton.

Receptors' Accessibility and Bacterial Fate

For most bacterial pathogens, the initial stage of the infectious process starts with interaction with a host epithelium. Pathogens that multiply extracellularly on the surface of an epithelium have to establish the interactions required for binding to and colonizing the epithelium (Figure 7.2, a). It may therefore not be by chance that such interactions very often consist of association between bacterial lectins and carbohydrate moieties of host cell surface glycoproteins. Such carbohydrate moieties are readily accessible to pathogens on the apical surface of the epithelium; they may be largely represented on glycoproteins expressed on the cell apical surface, or may constitute part of the mucin layer coating the epithelium.

Invasive enteropathogens, on the other hand, are confronted with the problem of invading or crossing a polarized epithelium that is tightly sealed by different junctional structures. These cell junctions are involved in maintaining the polarization between the apical and the basolateral surfaces, which are distinct in terms of their integral membrane component contents. Bacterial engagement of the proper receptor may therefore be a limiting fac-

Figure 7.2 Interactions of invading pathogens with an intestinal epithelium. The open arrows show potential interactions between the pathogen and cell adhesion molecules on the surface of various cell types during the invasion process. Bacteria that adhere to the apical surface of the epithelium often do so by establishing interactions with carbohydrate moieties or by a "bridging" mechanism (a). Invasive pathogens may enter via specialized cells, such as M cells present in the intestinal epithelium (b), and have elicited various ways to survive professional phagocytes (c). This encounter may result in the release of IL-1β and IL-18; and, in combination with IL-8 released by enterocytes, these cytokines contribute to recruit PMNs at the site of infection (d). Invasive pathogens that bind to cell receptors that are basolaterally distributed (open box) may then enter cells via the basolateral side, or via the apical side after receptor redistribution during a trauma or PMN transmigration (e).

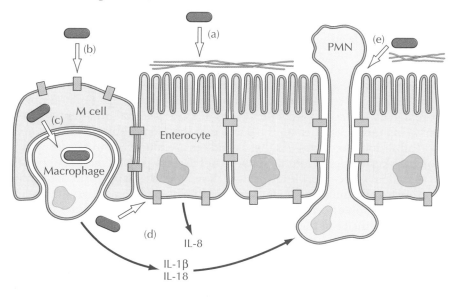

tor in the case of receptors such as cadherins or integrins that are preferentially expressed basolaterally. In these latter cases, bacterial internalization would require receptor recruitment to the apical cell surface. Such receptor redistribution from the basolateral surface to surfaces exposed apically has been reported for integrins following a mechanical trauma or during leukocyte transmigration across the epithelium (Figure 7.2, e). It is therefore conceivable that some factors such as local inflammation might paradoxically favor pathogen infection. Also, interaction of some pathogens with the apical surface of enterocytes may lead to the release of interleukin 8 (IL-8), which acts as a chemoattractant for polymorphonuclear cells (PMNs) and could thus favor bacterial accessibility to basolateral receptors (Figure 7.2, d). In the case of a healthy epithelium, however, bacterial invasion may be inefficient for pathogens that utilize receptors expressed on the basolateral side and invasion occurs preferentially at the levels of specific structures of the epithelium. A mucosal epithelium acts as a barrier against microorganisms but also needs to allow exchange with the lumen. In the case of the intestinal epithelium, for instance, sampling of lumenal contents takes place at the level of M cells, which are specialized absorptive cells with a poorly organized brush border, overlaying solitary lymph nodes, as well as lymphoid follicles or Peyer's patches. These cells play a role in the presentation of antigens sampled from the lumen to the immune system. The preferential invasion of enteropathogens such as *Yersinia*, *Listeria*, or *Shigella* via M cells, rather than by neighboring enterocytes, argues for the accessibility of receptors on the M cell's apical surface (Figure 7.2, b). Following the crossing of the epithelial layer via M cells, invasion can occur at the basal side of enterocytes. Pathogens need to devise schemes to survive the encounter with host defense cells and, in particular, against macrophages. In the case of *Shigella*, this encounter results in proinflammatory cytokine production, which attracts monocytes and polymorphonuclear cells to the site of infection (Figure 7.2, d). This influx of inflammatory cells results in destabilization of the epithelium and favors further invasion of the enterocytes from the lumen of the colon.

Pathogen Adhesion Determining Host Tissue Tropism

Direct Interaction

A widespread feature of pathogens is their ability to directly bind cell surface receptors via adhesins. These adhesins are exposed on the bacterial surface, possibly as integral component or exposed at the tip of pili or fimbriae, which are filamentous multimeric structures on the surface of gram-negative bacteria. The vast majority of adhesins are lectins and recognize sugar moieties of cell adhesion glycoproteins. For example, type I fimbriae expressed by various pathogenic strains of *Escherichia coli* preferentially recognize mannose residues on the surface of the extracellular matrix as well as on adhesion molecules. As these residues are found on numerous glycoproteins, these types of interactions are not usually considered to be specific for a particular receptor. Interestingly, however, they appear to determine the tropism of a microorganism for a specific host tissue, due to the preferential expression of certain carbohydrate moieties in some tissues and differential specificity of the bacterial adhesins.

Carbohydrate-lectin interactions can also mediate microbial attachment via cell surface proteoglycans. Proteoglycans, which are ubiquitously found on the surface of many cell types, are membrane glycoproteins with two major types of glycosaminoglycan chains, heparan or chondroitin sulfate, that have been implicated in cell-cell interactions as well as adhesion to the extracellular matrix. Based on sugar inhibition studies, various bacterial species have been shown to adhere to cells via proteoglycans. These include the spirochete *Borrelia burgdorferi,* the causative agent of Lyme disease, or *Helicobacter pylori,* which causes peptic ulcers. The OpaA protein on the surface of *Neisseria gonorrhoeae* has also been reported to bind heparan sulfate proteoglycans. It is possible that interactions of proteoglycans with bacterial ligands participate in the initial stages of adhesion and that other types of interactions are involved either in stabilization of the bacterial adhesion processes or in invasion. For example, OpaA was also shown to bind to fibronectin and vitronectin and, by these means, to indirectly associate with β_1-integrins (see "bridging" mechanism below). Also, other *Neisseria* Opa proteins bind to CD66 carcinoembryonic antigens, which are members of the Ig superfamily, and this interaction mediates bacterial internalization into host cells. Similarly, Opc proteins in *N. gonorrhoeae* bind to heparan sulfate proteoglycans and to vitronectin. These interactions, however, are unlikely to occur in capsulated *N. meningitidis* during initial adhesion to cells, as this step was shown to depend on a type IV pilus.

Adherence via ECM Proteins: Bridging Mechanism

Several pathogens have the ability to bind to proteins of the extracellular matrix. Often, bacteria have devised several means with which to achieve this. This is the case for gram-positive bacteria such as *Staphylococcus* or *Streptococcus* species, which can bind to fibronectin or collagen via specific fibronectin-binding proteins expressed at the bacterial surface as well as via their lipotechoic acid. This is also the case for some gram-negative bacteria such as *Pseudomonas* or *Yersinia* species. In the case of *Yersinia*, the YadA protein has been reported to bind fibronectin and collagen. This interaction appears to be important for bacterial attachment to insoluble forms of extracellular matrix proteins that can be found on the surface of an epithelium. Besides allowing the attachment of the pathogen to the matrix, binding to matrix proteins can also promote cell adhesion by a bridging mechanism in which the pathogen interacts with the extracellular matrix protein and, in turn, associates with receptors for extracellular matrix protein such as integrins. It is not clear if this type of bridging mechanism can result in internalization. It has been reported that fibronectin-coated particles can be ingested by cultured epithelial cells presumably via integrin receptors, but in this case uptake does not appear to be very efficient. As mentioned for *Neisseria*, this type of "bridging" mechanism may result in bacterial invasion, when associated with other types of bacterial-cell molecular interactions.

Signaling via Adhesion Molecules and Bacterial Internalization

Some pathogens have the ability to enter normally nonphagocytic cells. This property allows them to multiply within a local niche or to breach an epithelium to gain access to deeper tissues.

Invasin-Mediated Uptake of *Yersinia*

The invasin-mediated uptake of *Yersinia* is a well-characterized system of bacterial-induced phagocytosis by epithelial cells. Enteropathogenic *Yersinia* species such as *Y. enterocolitica* or *Y. pseudotuberculosis* are responsible for enteric diseases after ingestion of contaminated foodstuff by the host. *Yersinia* has the ability to cross the epithelium at the level of M cells of the terminal ileum to reach the lamina propria where it can multiply. *Yersinia* multiplies mostly extracellularly during these later stages of infection, but the invasin protein is critical for efficient bacterial internalization by M cells and crossing of the epithelium during the initial steps of the infectious process. Invasin is a bacterial surface protein that binds at least five members of the β_1 integrin family that is sufficient to promote bacterial internalization by epithelial cells. These receptors include the fibronectin receptor $\alpha_5\beta_1$, but as opposed to other bacteria that bridge the same receptor after associating with fibronectin, binding of invasin to β_1 integrin is followed by the bacterial internalization. A combination of biochemical and mutational analysis has allowed the localization of the receptor-binding domain to the C-terminal third of the protein. The crystal structure indicates that this region consists of two immunoglobulin-like domains. Interestingly, these domains are reminiscent of the RGD-peptide-containing domain and the synergistic domain of fibronectin that promote integrin binding. Although invasin does not contain an RGD sequence, two critical aspartate residues (Asp^{811} and Asp^{911}), each located on one domain, are exposed within protruding loops that may mimic the corresponding aspartate residues on the fibronectin synergistic and the RGD-containing domains. As opposed to fibronectin, the two invasin domains are predicted to form a "superdomain" with little flexibility. This lack of flexibility may explain in part the much higher affinity of invasin for the $\alpha_5\beta_1$-integrin than of fibronectin. This higher affinity, combined with the ability of invasin to multimerize, thus to cluster integrins upon receptor engagement, is critical for determining bacterial uptake in a zipper-like process. This process involves tyrosine kinase signaling and actin polymerization. Studies on the implication of the focal adhesion kinase (FAK) as well as mutagenesis of the β_1-cytoplasmic moiety suggest that signals transmitted via integrins during invasin-mediated bacterial internalization differ from signals leading to focal adhesion assembly. For example, inhibition of the small GTPase Rho, which regulates focal adhesion assembly, has little effect on invasin-mediated bacterial uptake. Conversely, the small GTPase Rac that induces the polymerization of actin and the formation of small focal complexes is critical for bacterial internalization.

InlA-Mediated Uptake of *Listeria*

Listeria monocytogenes, an enteroinvasive gram-positive bacterium responsible for meningitis and fetal systemic infection, provides another example of a single bacterial surface-ligand promoting entry in nonphagocytic cells. Like many invasive pathogens, *L. monocytogenes* has elicited several pathways to promote its uptake by epithelial cells that involve a family of related bacterial proteins, the internalins. The InlA protein, or internalin, mediates internalization of the bacterium by binding to the E-cadherin. As is observed for *Yersinia* invasin binding to integrins, *L. monocytogenes* internalization occurs in a "zippering" process with the formation of a tight phagosome surrounding the internalized bacterium (Figure 7.3A). The mechanism of InlA-mediated invasion requires the association of the cadherin cytoplasmic domain with catenins and association with the cell cytoskeleton. Remarkably, a single amino acid substitution at position 16 is sufficient to explain the differential binding of InlA to human E-cadherin, whereas no binding is observed for the murine E-cadherin. This specificity has allowed the development of a human E-cadherin transgenic mouse model to study in vivo listeriosis. For both *Yersinia* invasin and *Listeria* in-

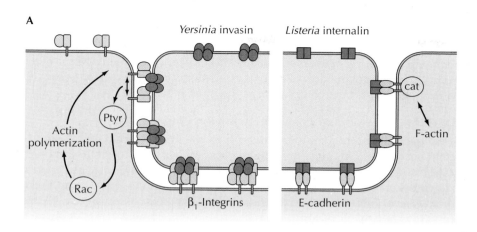

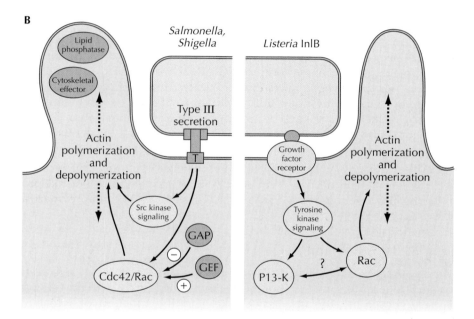

Figure 7.3 Modes of bacterial induced phagocytosis. **(A)** *Yersinia* invasin or *Listeria* internalin-mediated internalization results in bacteria surrounded by a tight phagosome. (Left) *Y. pseudotuberculosis,* high-affinity binding of invasin to integrins and integrin clustering are critical to drive the phagocytic process. The invasin-mediated bacterial internalization involves tyrosine kinase signaling, activation of the small GTPase Rac, and actin polymerization. (Right) Internalin mediates the internalization of *Listeria* by binding to E-cadherin. Bacterial uptake requires interaction between the cadherin cytoplasmic domain and catenins (cat). **(B)** (Left) Type III secretory apparatus allows the targeting of bacterial effectors in the cell cytosol by means of a protein complex, the "translocon," that inserts in the host cell membranes. A combination of bacterial type III effectors promotes bacterial internalization by regulating cytoskeletal reorganization at various levels. The IpaC protein, a component of the *Shigella* translocon (T), activates Src kinase signaling and actin polymerization dependent on the Cdc42 and Rac GTPases. The *Salmonella* SopE and SptP proteins are a GEF and a GAP, respectively, for the GTPases Cdc42 and Rac. Cytoskeletal effectors correspond to bacterial type III effectors that associate with and regulate the function of cytoskeletal proteins, the *Salmonella* SipA protein that binds to and stabilizes F-actin, and the *Shigella* IpaA protein that binds to the focal adhesion protein vinculin and induces actin depolymerization. Lipid phosphatase corresponds to the *Salmonella* SopB or the *Shigella* IpgD protein that hydrolyzes PIP_2. (Right) The *Listeria* InlB protein binds to and activates the Met receptor. InlB binding to Met leads to tyrosine kinase signaling, recruitment of adapter proteins and of the PI-3 kinase, activation of the Rac GTPase, and actin cytoskeleton reorganization that drives the phagocytic process.

ternalin, the entry process appears to be driven by incremental interactions between bacterial surface ligands and cell adhesion molecules. These interactions result in receptor recruitment and clustering at the bacterium-cell interface and transduction of signals controlling the cytoskeletal reorganization required for the completion of the uptake process.

Bacterial Invasion and Growth Factor Receptor Signaling

Some bacterial pathogens also trigger tyrosine kinase signaling by growth factor receptors. This type of signaling could be critical for actin polymerization required for the formation of membrane extensions or can modulate cytoskeletal reorganization during bacterial uptake. For example, entry of *N. meningitidis* into endothelial cells has been associated with the activation of Erb2, a receptor from the epidermal growth factor receptors' family, required for the downstream activation of a Src kinase, and the tyrosyl phosphorylation of the cytoskeletal protein cortactin. Although this signaling pathway is not critical for actin polymerization during *Neisseria* invasion, it is required for efficient bacterial internalization. Also, the *Listeria* InlB protein has been shown to induce bacterial entry by binding to the complement component receptor gC1qR, as well as Met, a tyrosine kinase receptor for the hepatocyte growth factor. The signaling cascade that is triggered during binding of InlB is similar to Met activation by hepatocyte growth factor (HGF) and results in tyrosine phosphorylation, the recruitment of the adapter molecules (Gab1, Shc, Cbl), and the activation of the phosphatidylinositol-3 kinase (PI-3K) and of the Rac GTPase, critical for actin polymerization. As HGF-dependent tyrosine kinase signaling was found to negatively regulate cadherin-based junctions, it is possible that InlB-mediated signaling cooperates with InlA-mediated entry by enhancing cadherin availability.

Type III Secretion and Bacterial Invasion

Some pathogens have elicited ways to bypass classical ways of activation via surface receptors and can potentially interfere directly with the cell machinery regulating the actin dynamics and the function of adhesion molecules. This is the case for gram-negative bacterial enteropathogens that use a specialized type III secretory apparatus, allowing the access of bacterial effectors to this cell machinery. Critical for this event, a "translocon," composed of type III secreted proteins that oligomerize and insert in the host cell membranes, allows the subsequent targeting of other bacterial effectors in the cell cytosol. For example, these bacterial effectors can act as a guanosine exchange factor (GEF) to activate, or conversely as a GTPase-activating protein (GAP) to down-regulate Rho family GTPases that regulate cytoskeletal reorganization. The fact that cell surface receptors are dispensable for this type of bacterial signaling is not totally true because interactions between components of the *Shigella* translocon and the $\alpha_5\beta_1$-integrin or the CD44 hyaluronic acid receptor were reported. As will be discussed further, this type of interaction may favor the insertion/oligomerization of the translocon within host cell membranes or signaling through lipid microdomains.

In the case of *Salmonella*, responsible for gastroenteritis, and *Shigella*, the causative agent of bacillary dysentery, this type of device allows bacterial internalization by epithelial cells involving an important reorganization of the actin cytoskeleton and the formation of large membrane extensions at the site of bacterium-cell contact. Upon cell contact, *Salmonella* type III secretion allows the translocation of a variety of bacterial effectors that reorganize the actin cytoskeleton to promote bacterial entry. Critical for this event, the "translocon," composed of the SipB and SipC proteins, allows the subsequent targeting of other bacterial effectors in the cell cytosol. The translocated SopE protein acts as an exchange factor for the Cdc42 and Rac GTPases, and activation of these GTPases leads to actin polymerization required for the formation of membrane extensions surrounding the bacterium. Other translocated *Salmonella* proteins further regulate these extensions; the SipA protein binds to and stabilizes actin filaments, thus favoring protrusion formation. The SopB protein, which bears an inositol-polyphosphatase and also appears to depress the levels of polyinositol (4,5) bisphosphate (PIP_2), is involved in membrane fusion events probably required for completion of the phagocytic process; the SptP protein is a dual enzyme that carries a tyrosylphosphatase activity as well as a GAP activity toward the Cdc42 and Rac GTPases. These different effectors act in concert to allow bacterial uptake. Interestingly, although *Shigella* invasion induces processes that resemble *Salmonella*-induced extensions, no bacterial GEFs or GAP proteins have been identified that regulate the activity of small Rho GTPases. Instead, it appears that activation of the Cdc42 and Rac GTPases and initial actin polymerization is due to the *Shigella* IpaC protein, a component of the "translocon" secreted by the type III apparatus. Also, the Src tyrosine kinase was shown to regulate actin polymerization as well as the down-regulation of *Shigella* entry structures. As for *Salmonella*, *Shigella* invasion results from the concerted action of several bacterial type III effectors; the IpaA protein interacts with vinculin, allows the formation of an adhesion structure at the intimate bacterium-cell contact site, and is also required for actin depolymerization required for efficient bacterial uptake. The IpgD protein, the *Shigella* homologue of the *Salmonella* SopB protein, is a PIP_2 phosphatase and destabilizes cortical actin cytoskeleton, thus favoring actin dynamics at the site of bacterial contact. The VirA protein interferes with microfilament formation by binding to tubulin oligomers and activates Rac-dependent actin polymerization, presumably through the described antagonistic relationship between regulators of the actin and microtubule cytoskeleton.

Type III secretion may also be used to prevent phagocytosis by professional phagocytes. For instance, the *Yersinia* Yop proteins act synergistically to interfere at different levels of the phagocytic process. YopH is a tyrosine phosphatase that targets focal adhesion substrates such as p120CAS and p125FAK. YopE acts as a GAP that negatively regulates the Rac GTPase. YopO/YpkA is a serine/threonine kinase that phosphorylates actin. YopT and also YopO/YpkA interfere with Rho function. *Yersinia* antiphagocytosis may thus result from the targeting of β_1-integrins by the invasin and YadA protein, combined with the inhibition of downstream signals by the type III secreted Yop effectors.

Enteropathogenic *E. coli* (EPEC) also provides another model in which signaling involves type III secreted bacterial effectors and interaction between a surface bacterial ligand and a receptor. Remarkably, this receptor corresponds to a self-designed bacterial receptor. EPEC strains, which are commonly responsible for diarrhea in children, have the ability to adhere to intestinal epithelial cells. Adhesion of EPEC to cultured cells is characterized by effacing lesions, corresponding to the disruption of microvilli and followed by the formation of a pedestal-like structure at the site of bacterial interaction with the host cell. Initial adherence of EPEC to cells is mediated by BFP, a type IV pili. This first step is followed by an intimate adherence of the microorganism to the cell surface. Intimate adhesion requires the type III secreted protein Tir that inserts in the cell membrane and acts as a receptor for intimin, a bacterial adhesin. Tir is then tyrosyl phosphorylated and this, in turn, allows the recruitment of the adapter protein Nck, which may activate actin polymerization by allowing recruitment of Rho GTPases and N-WASp. Interestingly, the Tir N terminus was reported to interact with vinculin and α-actinin, suggesting that this protein integrates determinants required for cytoskeletal anchoring as well as signals leading to actin polymerization that lead to pedestal formation.

Lipid Raft Microdomain Involvement in Pathogen-Host Cell Interactions

The finding of lipid microdomains, called rafts, sheds new light on the mechanisms used by pathogens to interact with host cells. Raft dynamics may allow the clustering of surface receptors helping weakly adherent bacteria to get internalized. Activation of signaling complexes within ligand-induced clusters of rafts can participate in mediating pathogen entry. Raft-dependent endocytic pathways may allow pathogens to find an intracellular niche where they can persist in escaping degradation.

Structure-Function of Lipid Rafts

Lipid rafts have changed our view of the classical membrane fluid mosaic model into a more complex system. Raft microdomains are described as dispersed liquid-ordered phase microdomains that diffuse laterally within the two-dimensional liquid-disordered phase membrane. Rafts are dynamic assemblies into which specific lipids, i.e., cholesterol, glycosphingolipids, and sphingomyelin, are enriched. These assemblies are fluid but more tightly packed than the surrounding bilayer. A subset of proteins specifically partitions with long-time residency in these rafts, e.g., lipid-anchored proteins associated either with the outer leaflet (glycosyl-phosphatidyl-anchored proteins [GPI-AP]; cholesterol-linked and palmitate-anchored proteins [Hedgehog]) or with the inner leaflet (doubly acylated proteins [Src-like kinase, Gα subunits of heterotrimeric G proteins, endothelial nitric oxide synthase]). The presence of cholesterol and sphingolipids with high melting temperature due to long and saturated fatty acid chains gives rafts the remarkable biophysical property to be resistant to solubilization by Triton X-100 in the cold. Besides the biochemical analysis of rafts through the characterization of detergent-resistant membranes (DRMs), rafts or clustered rafts at the cell surface have been visualized after the clustering of raft-associated proteins (or lipids) using antibodies followed by immunofluorescence labeling.

Rafts can be functionally considered as clustering devices that act as platforms implicated in sorting mechanisms and in compartmentalizing the membranes. This clustering allows the integration of external signals to modulate signaling cascades. They have been described not only at the plasma membrane, but also in the biosynthetic pathway (Golgi and vesicular carriers en route to the plasma membrane) and in the endocytic-phagocytic pathways (early and late endosomes, caveolae, and phagosomes). Interestingly, numerous signaling molecules have been shown to associate with rafts and DRMs. These include lipid signaling molecules, G-protein-coupled receptors, and adenylate cyclase, components of the tyrosine kinase-mitogen-activated protein kinase pathway, regulators of the intracellular calcium homeostasis, and molecular components linking the plasma membrane to the cytoskeleton.

Lipid Rafts and Bacterial Signaling

An emerging and expanding body of literature implicates rafts in bacterial invasion. In most instances, such demonstrations were performed by visualizing raft components at bacterial entry sites and using treatments that induce raft disorganization. For example, many adhering bacteria show some levels of internalization that are inhibited by raft-disrupting drugs such as filipin and methyl-β-cyclodextryl (MeCD). This is the case for uropathogenic strains of *E. coli* expressing the Afa/Dr adhesin, or type I fimbriae, which interact with GPI-APs, such as CD55, CD66e, or CD48 (Table 7.2). For *Campylobacter jejuni* invasion, the sensitivity to both MeCD and kinase (phosphatidylinositol-3-kinase and Akt protein kinase) inhibitors suggests the implication of signaling machineries localized within caveolae/clustered rafts activated during internalization. Cholesterol dependence has been equally reported for invasion of *Mycobacterium tuberculosis*, *M. bovis* BCG, and *M. kasasii* and of *Brucella suis* and *B. abortus*, although the implication of rafts may depend on the cell type (Table 7.2).

Cellular invasion by *Salmonella* or *Shigella* that involve type III secretion has also been reported to implicate rafts. It is possible that raft lipid composition favors the insertion or the oligomerization into host cell membranes of effectors that are targeted by the type III secretory apparatus, reminiscent of what has been proposed for some bacterial toxins. Alternatively, receptors that partition in rafts may be targeted by bacterial type III effectors. For example, *Shigella* invasin IpaB associates with DRMs upon binding to CD44, a receptor reported to link the cell surface to the actin cytoskeleton via the ERM family of proteins. Possibly, as for toxins, *Shigella* effectors can induce raft-mediated clustering establishing lipidic/proteinaceous platforms. These platforms could reinforce the weak adhesion of the bacterium by activation of signaling cascades required for the cytoskeleton-driven plasma membrane reshaping observed during the engulfment of the bacterium.

Rafts are important for bacterial internalization processes, but there are some indications that they also contribute to bacterial intracellular survival and persistence within phagosomes. It thus has been proposed that cell surface rafts, caveolae, and eventually a recruitment of intracellular caveolae-

Table 7.2 Lipid rafts and bacterial pathogens

Bacteria	Components mobilized within rafts at entry site	Cholesterol dependence, for binding/entry
FimH-expressing *E. coli*	CD48[a], caveolin, GM1	Yes
Afa/Dr DAEC	Apical side, CD55[a], CD66e basolateral side, $\alpha_5\beta_1$	Yes
C. trachomatis	Caveolin, GM1	Yes
C. jejuni		Yes
M. tuberculosis, M. bovis BCG	CR3[a]	Yes
M. kansasii	CR3[a], GPI-AP (CD 55, CD66b, CD16b, CD14)	Yes
Brucella spp.	GM1, GPI-AP (CD48)	Yes
Salmonella enterica serovar Typhimurium	GPI-AP (CD55)	Cell type specificity
Shigella flexneri	CD44, GPI-AP, and sphingolipid-dependent entry	Yes

[a] Identified cell surface receptor.

derived vesicles participate in the formation of the bacteria-encapsulating intracellular compartment in the case of FimH expressing *E. coli*. Similarly, internalization of *Chlamydia trachomatis* has been reported to be raft dependent and bacterial phagosomes contain caveolin. These phagosomes do not fuse with lysosomes and redirect the trafficking to the Golgi where chlamydial inclusions are observed. At this location, chlamydiae intercept host sphingolipids in transit from the Golgi to the plasma membrane. It is possible that rafts participate in sorting mechanisms during fusion with endocytic organelles.

Route of Internalization and Intracellular Survival within the Macrophage

The first line of defense that a pathogen has to face during host tissue invasion usually consists of polymorphonuclear neutrophils or resident macrophages. Once phagocytosed by these cells, killing of invading microorganisms occurs after fusion of the bacterial phagosome with lysosomes containing hydrolytic enzymes and toxic products, and also by the generation of toxic oxygen radicals during the respiratory burst. Intracellular microbial pathogens need therefore to elicit ways to survive the encounter with these cells and some do that so successfully that they even grow within macrophages. This is the case for *Mycobacterium* or *Legionella* species, for example, which avoid lysosome fusion and multiply intracellularly within "replicative phagosomes." It is unclear to what extent the pathway of uptake influences trafficking of the bacterial vacuole, but it is clearly established that it can modulate the generation of the oxidative burst. An invasion mechanism via specific adhesion molecules on the surface of professional phagocytes may represent a means for the pathogen to avoid being killed by oxygen radicals. For example, *N. gonorrhoeae* internalization by PMNs does not appear to be accompanied by the strong oxidative response that is usually associated with phagocytosis of opsonized particles. This particular feature may result from the specific Opa-mediated internalization via CD66 antigens. Similarly, many other intracellular pathogens such as *Legionella* and *Mycobacterium* species have the ability to bind the complement component C3 and are internalized by monocytes via the complement receptor CR3 ($\alpha_M\beta_2$). This may also be a means to avoid the generation of a strong oxidative response, which appears to require both activation of CR3 and Fc receptor ligation during opsonized phagocytosis.

Pathogen-Induced Up-Regulation of Adhesion Molecules

Bordetella pertussis, the causative agent of whooping caugh, binds to ciliated respiratory cells and leukocytes. This property is a prerequisite for *B. pertussis* infection, because it allows for colonization and destruction of the lung epithelium by the microorganism. Different bacterial factors are involved in *B. pertussis* adhesion to cells. These factors include the FimD fimbrial protein, which binds to β_1-integrins, the pertussis toxin which binds carbohydrate residues on cell surface glycoproteins, and the filamentous hemagglutinin (FHA) protein which binds the CR3 integrin. These different factors appear to act in a cooperative fashion to promote cell attachment. For instance, binding of the pertussis toxin to leukocytes results in up-regulation of CR3, thus reinforcing bacterial attachment via FHA-CR3 interactions (Figure 7.4). The situation is in fact more complicated since, in addi-

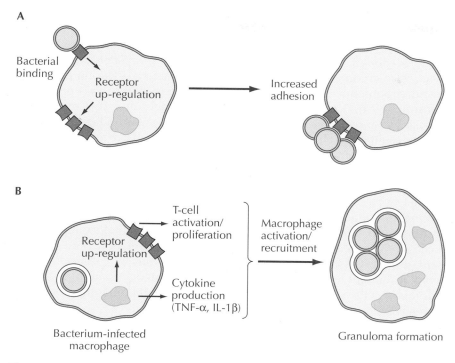

Figure 7.4 Two examples of up-regulation of adhesion molecules by bacterial pathogens. **(A)** Up-regulation of the CR3 molecule as a means to enhance attachment of *B. pertussis* to leukocytes. Initial interaction of the bacterium (closed circle) with the cell leads to up-regulation of CR3 (filled box) and strengthens bacterial adhesion via FHA–CR3 interactions. **(B)** Up-regulation of ICAM-1 (filled box) during *M. tuberculosis* infection leads to the recruitment of T cells and may indirectly participate in granuloma formation.

tion to CR3, FHA directly interacts with the integrin-associated proteins LRI/IAP, a complex that has been implicated in activation of neutrophil adhesive functions. This direct interaction results in cross-linking of the LRI/IAP complex and up-regulates CR3 binding to FHA. A similar mechanism of up-regulation of CR3 following cross-linking of the $\alpha_5\beta_1$-integrin by the *B. pertussis* FimD protein may also contribute to bacterial adhesion to monocytes. In the case of *Mycobacterium avium*, cross-linking of the $\alpha_V\beta_3$-integrin via bacterial ligands induces up-regulation of CR3 and not only results in enhanced bacterial attachment, but also enhanced uptake via the CR3 pathway.

Pathogens can also manipulate cell adhesion molecule expression in the course of the infection and influence cell-to-cell interactions. Such changes are observed during the host inflammatory response to the invading organism. Up-regulation of adhesion molecules, such as selectins or members of the immunoglobulin family, often occurs in endothelial tissues as a result of local inflammation induced at the site of infection. Modulation of expression of adhesion molecules at the cell surface can be induced by various bacterial components, such as surface proteins, bacterial lipopolysaccharide, or secreted toxins, and also via the release of cytokines by inflammatory cells. In some instances, the host inflammatory response and the changes in cell adhesive properties lead to the recruitment of cells that participate in the persistence of the infection. For example, *M. tuberculosis* is able to survive and to replicate within macrophages. During *M. tuberculosis* infection, granulomas may form, which consist of large multinucleated cells that probably correspond to the fusion of macrophages and T lymphocytes. The formation of these granulomas is a host defense response to clear out invading microorganisms, but in cases where bacterial clearing is not successful, granulomas can become a niche where *M. tuberculosis* persists or multiplies. Production of several cytokines, including IL-1 and tumor necrosis factor α (TNF-α) by inflammatory cells, as well as T-cell activation in response to antigen presented by accessory cells, has been shown to be essential for granuloma formation (Figure 7.4B). Also, ICAM-1 is up-regulated in granulomas either indirectly, in response to cytokine production, or directly by stimulation by bacterial components. As the association of ICAM-1 with LFA-1 (the $\alpha_L\beta_2$-integrin) on the surface of lymphocytes determines T-cell activation, the up-regulation of ICAM-1 may contribute to recruitment of T cells leading to the sustained inflammatory reaction induced by *Mycobacterium* infection and granuloma formation. Similarly, up-regulation of ICAM-1 expression on endothelial cells by *Chlamydiae pneumoniae* is thought to favor interactions between leukocyte and blood vessel endothelial cells which could account for local vascular inflammatory alterations linked to *C. pneumoniae* infection.

Future Challenges

Presented in this chapter are a few examples of bacterial manipulation of cell adhesion molecules that illustrate the vast diversity of strategies used by pathogens during the infectious process. Bacterial adhesion to cell adhesion molecules is often the result of interactions between lectins and sugar moieties, a bridging mechanism by which the pathogen binds a protein from the extracellular matrix. Bacterial invasion, on the other hand, appears to result from high-affinity interaction with specific receptors linking the cytoskeleton or from a combination of an adhesion mechanism with specific invasion determinants. The function of bacterial interactions with adhesion molecules is not limited to adherence and invasion, but can also play a role in the pathogen's ability to control and subvert the host defense system. The precise relevance of such interactions, interactions that are most often characterized in in vitro systems, may prove challenging. Such issues depend, in general, on the availability of animal models that faithfully reproduce the disease. A promising area of research along with the molecular characterization of pathogen–host interactions, is the use of transgenic animals expressing heterologous adhesion molecules that are specific receptors for a given pathogen in animals that are innately insensitive to the pathogen. This type of study should allow a better understanding of the role of a specific interaction in the development of infection.

Selected Readings

Aderem, A., and D. Underhill. 1999. Mechanisms of phagocytosis in macrophages. *Annu. Rev. Immunol.* **17:**593–623.

Angst, B., C. Marcozzi, and A. Magee. 2001. The cadherin superfamily: diversity in form and function. *J. Cell Sci.* **114:**629–641.

Brakebusch, C., and R. Fassler. 2003. The integrin-actin connection, an eternal love affair. *EMBO J.* **22:**2324–2333.

Chen, T., F. Grunert, A. Medina-Marino, and G. Ec. 1997. Several carcinoembryonic antigens (CD66) serve as receptors for gonococcal opacity proteins. *J. Exp. Med.* **185:**1557–1564.

Connell, H., M. Hedlund, W. Agace, and C. Svanborg. 1997. Bacterial attachment to uro-epithelial cells: mechanisms and consequences. *Adv. Dent. Res.* **11:**50–58.

Cornelis, G. 2002. The *Yersinia* Ysc-Yop 'type III' weaponry. *Nat. Rev. Mol. Cell Biol.* **3:**742–752.

Cossart, P., and H. Bierne. 2001. The use of host cell machinery in the pathogenesis of *Listeria monocytogenes. Curr. Opin. Immunol.* **13:**96–103.

Crossin, K., and L. Krushel. 2000. Cellular signaling by neural cell adhesion molecules of the immunoglobulin superfamily. *Dev. Dyn.* **218:**260–279.

Danen, E., and A. Sonnenberg. 2003. Integrins in regulation of tissue development and function. *J. Pathol.* **201:**632–641.

Duncan, M., J. Shin, and S. Abraham. 2002. Microbial entry through caveolae: variations on a theme. *Cell Microbiol.* **4:**783–791.

Ehlers, M., and M. Daffe. 1998. Interactions between *Mycobacterium tuberculosis* and host cells: are mycobacterial sugars the key? *Trends Microbiol.* **6:**328–335.

El Tahir, Y., and M. Skurnik. 2001. YadA, the multifaceted *Yersinia* adhesin. *Int. J. Med. Microbiol.* **291:**209–218.

Galan, J. 2001. *Salmonella* interactions with host cells: type III secretion at work. *Annu. Rev. Cell Dev. Biol.* **17:**53–86.

Geiger, B., A. Bershadsky, R. Pankov, and K. M. Yamada. 2001. Transmembrane extracellular matrix-cytoskeleton crosstalk. *Nat. Rev. Mol. Cell. Biol.* **2:**793–805.

Hauck, C., and T. Meyer. 2003. 'Small' talk: Opa proteins as mediators of *Neisseria*-host-cell communication. *Curr. Opin. Microbiol.* **6:**43–49.

Hoffmann, I., E. Eugene, X. Nassif, P. Couraud, and S. Bourdoulous. 2001. Activation of ErbB2 receptor tyrosine kinase supports invasion of endothelial cells by *Neisseria meningitidis. J. Cell Biol.* **155:**133–145.

Hueck, C. J. 1998. Type III secretion systems in bacterial pathogens of animals and plants. *Microbiol. Mol. Biol. Rev.* **62:**379–433.

Isberg, R., and P. Barnes. 2001. Subversion of integrins by enteropathogenic *Yersinia. J. Cell Sci.* **114:**21–28.

Kaukoranta-Tolvanen, S., T. Ronni, M. Leinonen, P. Saikku, and K. Laitinen. 1996. Expression of adhesion molecules on endothelial cells stimulated by *Chlamydia pneumoniae. Microb. Pathog.* **21:**407–411.

Kerr, J. 1999. Cell adhesion molecules in the pathogenesis of and host defence against microbial infection. *Mol. Pathol.* **52:**220–230.

Knodler, L., J. Celli, and B. Finlay. 2001. Pathogenic trickery: deception of host cell processes. *Nat. Rev. Mol. Cell Biol.* **2:**578–588.

Lopez Ramirez, G., W. Rom, C. Ciotoli, A. Talbot, F. Martiniuk, B. Cronstein, and J. Reibman. 1994. *Mycobacterium tuberculosis* alters expression of adhesion molecules on monocytic cells. *Infect. Immun.* **62:**2515–2520.

Merz, A., and M. So. 2000. Interactions of pathogenic neisseriae with epithelial cell membranes. *Annu. Rev. Cell. Dev. Biol.* **16:**423–457.

Parkos, C., S. Colgan, and J. Madara. 1994. Interactions of neutrophils with epithelial cells: lessons from the intestine. *J. Am. Soc. Nephrol.* **5:**138–152.

Parsons, J. 2003. Focal adhesion kinase: the first ten years. *J. Cell Sci.* **116:**1409–1416.

Patti, J., B. Allen, M. McGavin, and M. Hook. 1994. MSCRAMM-mediated adherence of microorganisms to host tissues. *Annu. Rev. Microbiol.* **48:**585–617.

Rostand, K., and J. Esko. 1997. Microbial adherence to and invasion through proteoglycans. *Infect. Immun.* **65:**1–8.

Sansonetti, P. 2002. Host-pathogen interactions: the seduction of molecular cross talk. *Gut* **5**(Suppl. 3):III2–III8.

Schilling, J., M. Mulvey, and S. Hultgren. 2001. Structure and function of *Escherichia coli* type 1 pili: new insight into the pathogenesis of urinary tract infections. *J. Infect Dis.* **183**(Suppl. 1):S36–S40.

Shen, Y., M. Park, and K. Ireton. 2000. InIB-dependent internalization of *Listeria* is mediated by the Met receptor tyrosine kinase. *Cell* **103:**501–510.

Shuman, H., and M. Horwitz. 1996. *Legionella pneumophila* invasion of mononuclear phagocytes. *Curr. Top Microbiol. Immunol.* **209:**99–112.

Stuart, E., W. Webley, and L. Norkin. 2003. Lipid rafts, caveolae, caveolin-1, and entry by *Chlamydiae* into host cells. *Exp. Cell Res.* **287:**67–78.

Tran Van Nhieu, G., R. Bourdet-Sicard, G. Duménil, A. Blocker, and P. J. Sansonetti. 2000. Bacterial signals and cell responses during *Shigella* entry into epithelial cells. *Cell. Microbiol.* **2:**187–193.

van der Goot, F. G., and T. Harder. 2001. Raft membrane domains: from a liquid-ordered membrane phase to a site of pathogen attack. *Sem. Immunol.* **13:**89–97.

Vazquez-Torres, A., and F. Fang. 2000. Cellular routes of invasion by enteropathogens. *Curr. Opin. Microbiol.* **3:**54–59.

Vleminckx, K., and R. Kemler. 1999. Cadherins and tissue formation: integrating adhesion and signaling. *BioEssays* **21:**211–220.

Wadstrom, T., and A. Ljungh. 1999. Glycosaminoglycan-binding microbial proteins in tissue adhesion and invasion: key events in microbial pathogenicity. *J. Med. Microbiol.* **48:**223–233.

Westerlund, B., and T. Korhonen. 1993. Bacterial proteins binding to the mammalian extracellular matrix. *Mol. Microbiol.* **9:**687–694.

Wheelock, M., and K. Johnson. 2003. Cadherins as modulators of cellular phenotype. *Annu Rev. Cell Dev. Biol.* **19:**207–235.

Xiong, J., T. Stehle, S. Goodman, and M. Arnaout. 2003. New insights into the structural basis of integrin activation. *Blood* **102:**1155–1159.

Host Cell Membrane Structure and Dynamics

LYNDA M. PIERINI AND FREDERICK R. MAXFIELD

One of the main functions of cell membranes is the physical separation of one compartment from another. By sequestering a volume of material from the surrounding milieu, membranes create specialized compartments whose contents can be tightly regulated. For example, the plasma membrane, which defines the cell boundary, delineates intracellular and extracellular environments, while intracellular membranes, such as those that make up the endoplasmic reticulum, the Golgi apparatus, the nucleus, mitochondria, endosomes, and intracellular transport vesicles, enclose and form functional compartments within cells. In addition to this three-dimensional compartmentalization of cell volumes, membrane bilayers can mediate two-dimensional compartmentalization. That is, proteins and lipids can be organized laterally in the plane of the membrane and/or across the membrane. This chapter will focus on the nature and function of these two-dimensional membrane domains and highlight several examples of microbial exploitation of host cell membrane organization.

Host Cell Plasma Membrane Organization

Plasma Membrane Lipids

The plasma membrane of mammalian cells consists of a lipid bilayer with associated proteins. The lipid bilayer is made up of a mixture of many different lipids whose individual properties contribute in a complex way to those of the membrane as a whole. There are roughly three categories of membrane lipids: glycerolipids (also known as phospholipids), sphingolipids, and sterols. Because chemical structure largely determines lipid behavior, the chemical structures of these classes of lipids are reviewed in the following sections.

Glycerolipids

The generic structure for a glycerolipid is a glycerol backbone with two nonpolar hydrocarbon chains (from fatty acids) esterified to the 1' and 2' positions, and a phosphorylated alcohol group esterified in the 3' position (Figure 8.1). Because their polar headgroups invariably contain a phosphate

Generic structure of a glycerolipid

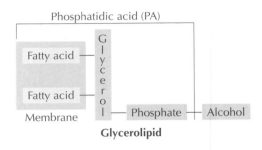

Phosphatidic acid (PA)

| Fatty acid | Glycerol |
| Fatty acid | |

Membrane

Phosphate — Alcohol

Glycerolipid

Structures of typical glycerolipid components

$$H_3C-(CH_2)_{14}-\overset{\overset{\displaystyle O}{\|}}{C}-OH$$

Palmitic acid

$$H_3C-(CH_2)_7-\underset{\underset{\displaystyle H}{|}}{C}=\underset{\underset{\displaystyle H}{|}}{C}-(CH_2)_7-\overset{\overset{\displaystyle O}{\|}}{C}-OH$$

Oleic acid

Fatty acids

$$H_2C\overset{1'}{-}OH$$
$$HC\overset{2'}{-}OH$$
$$H_2C\overset{3'}{-}OH$$

Gylcerol

$$HO-CH_2-CH_2-\overset{+}{N}(CH_3)_3$$

Choline

A typical phosphatidyl choline (PC) (1-palmitoyl-2-oleoyl-phosphatidyl choline)

PA

Nonpolar

Ester linkage

Glycerol

$$H_3C-(CH_2)_{14}-\overset{\overset{\displaystyle O}{\|}}{C}-O-CH_2$$

$$H_3C-(CH_2)_7-C=C-(CH_2)_7-\overset{\overset{\displaystyle }{|}}{\underset{\underset{\displaystyle O}{\|}}{C}}-O-\overset{|}{C}-H$$

Alcohol

$$H_2C-O-\overset{\overset{\displaystyle O}{\|}}{\underset{\underset{\displaystyle O^-}{|}}{P}}-O-CH_2-CH_2-\overset{+}{N}\overset{\overset{\displaystyle CH_3}{|}}{\underset{\underset{\displaystyle CH_3}{|}}{}}CH_3$$

Fatty acids

Phosphate

A typical phosphatidyl ethanolamine (PE)

$$\begin{array}{cccccccc} CH_2 & H_2C & H_2C & H_2C & H_2C & H_2C & H_2C & H_2C \\ CH_3 & CH_2 & CH_2 & CH_2 & CH_2 & CH_2 & CH_2 & CH_2 \end{array} \quad CH_2-\overset{\overset{\displaystyle O}{\|}}{C}-O-H_2C$$

$$\begin{array}{c} CH_3 \; H_2C \; H_2C \; H_2C \\ CH_2 \; CH_2 \; CH_2 \; CH_2 \end{array} HC=HC \begin{array}{c} H_2C \; H_2C \; H_2C \\ CH_2 \; CH_2 \; CH_2 \end{array} CH_2-\overset{}{\underset{\underset{\displaystyle O}{\|}}{C}}-O-CH$$

$$CH_2-O-\overset{\overset{\displaystyle O}{\|}}{\underset{\underset{\displaystyle O^-}{|}}{P}}-O-CH_2-CH_2-{}^+NH_3$$

cis double bond

Nonpolar | Polar

Figure 8.1 Structures of glycerolipids.

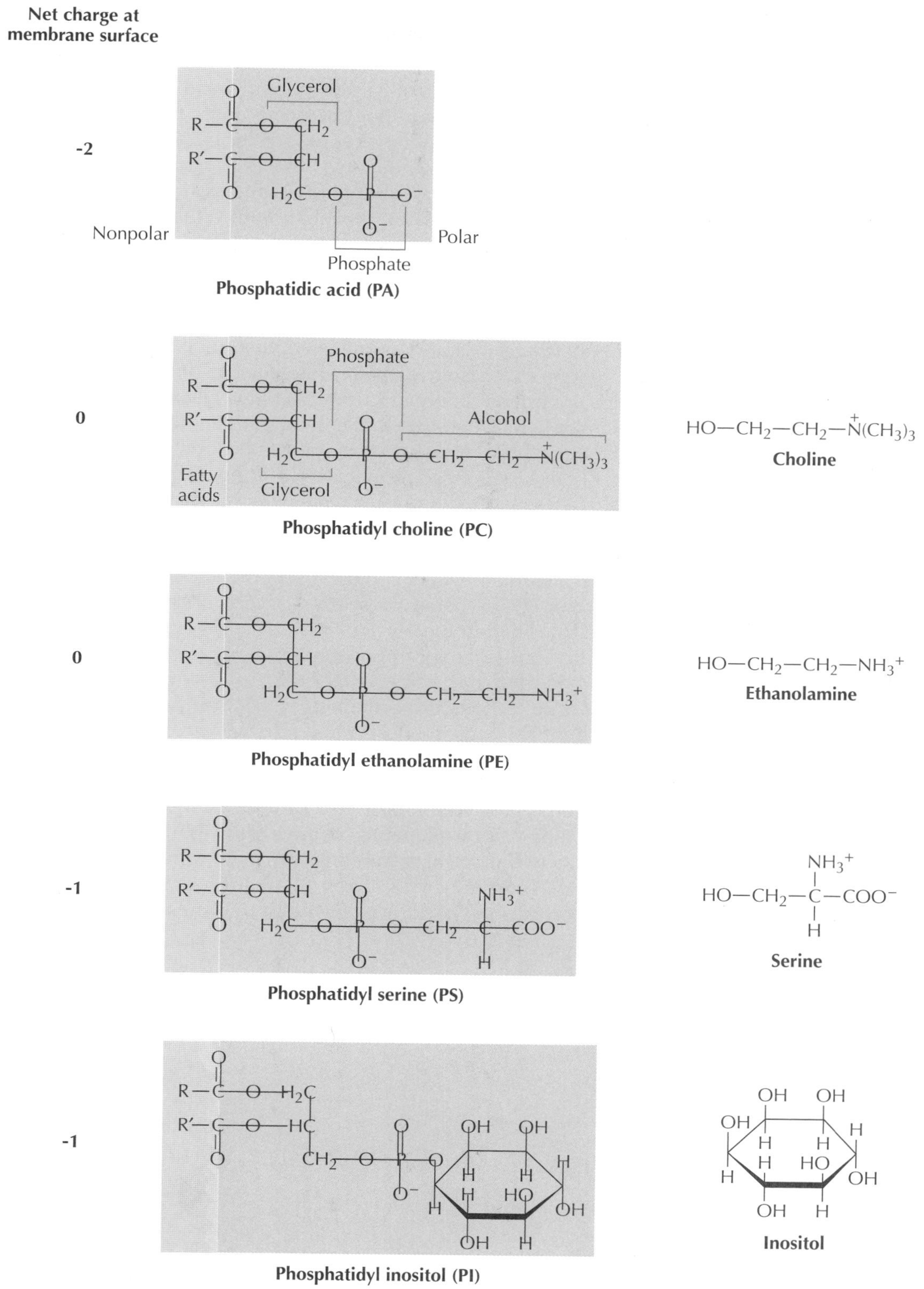

Net charge at membrane surface

-2

Glycerol

Nonpolar   Polar

Phosphate

Phosphatidic acid (PA)

0

Phosphate

Alcohol

$HO-CH_2-CH_2-\overset{+}{N}(CH_3)_3$
Choline

Fatty acids   Glycerol

Phosphatidyl choline (PC)

0

$HO-CH_2-CH_2-NH_3^+$
Ethanolamine

Phosphatidyl ethanolamine (PE)

-1

$HO-CH_2-\overset{NH_3^+}{\underset{H}{C}}-COO^-$
Serine

Phosphatidyl serine (PS)

-1

Inositol

Phosphatidyl inositol (PI)

Figure 8.1 *continued*

group, glycerolipids are also known as phospholipids. Specific classes of phospholipids are derived by esterification of an alcohol or a sugar to the phosphate. For example, in the absence of any modification of the phosphate, the phospholipid is known as phosphatidic acid (PA). PA itself is present in only small amounts in biological membranes, but it is a precursor for many of the major membrane glycerolipids. Esterification of the alcohols choline, ethanolamine, or serine or the sugar, inositol, to the phosphate headgroup of PA yields phosphatidylcholine (PC), phosphatidylethanolamine (PE), phosphatidylserine (PS), and phosphatidylinositol (PI), respectively (Figure 8.1). These polar headgroups dictate the charge carried by the whole phospholipid. At physiological pH values (pH ~7.4 for lipids at the plasma membrane), PC and PE are zwitterions. That is, they carry both a positive and a negative charge, and so they are electrically neutral. PS, on the other hand, has one positive charge and two negative charges, resulting in an overall negative charge. PI is also negatively charged. The polar headgroups of phospholipids are oriented outward from the bilayer, so the net charges of phospholipids are held at the surface of membranes. In this manner, the surface charge of the membrane is governed by its phospholipid composition. Importantly, PI can carry additional negative charges after phosphorylation by specific enzymes, and this represents one way in which the characteristics of the plasma membrane can be rapidly altered in response to a stimulus.

The classes of phospholipids, categorized according to their headgroups, contain a further level of complexity in the varied compositions of their hydrocarbon chains. The hydrocarbon chains arise from fatty acids that have been esterified to the glycerol backbone (Figure 8.1). Apart from the fact that fatty acid chains of phospholipids usually contain an even number of carbon atoms, the chains are structurally diverse, ranging in length from 12 to 26 carbons and containing none to six double bonds. Furthermore, the double bonds in unsaturated fatty acids can exist in two isomers, *cis* or *trans*, which can produce significant structural differences in otherwise identical fatty acids. An unsaturated fatty acid with a *trans*-double bond would behave similarly to a saturated fatty acid of the same length because their carbon chains are relatively straight. However, unsaturated fatty acids in biological membranes typically contain double bonds in the *cis*-isomer (Figure 8.2). This forces a kink in the hydrocarbon chain, which in turn perturbs the packing of the lipids and thus membrane struc-

Figure 8.2   Structures of a saturated and an unsaturated C_{18} fatty acid.

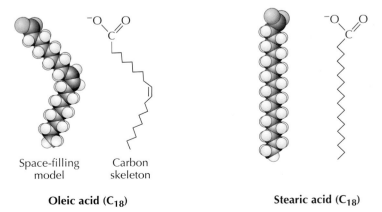

Space-filling model Carbon skeleton

Oleic acid (C_{18})   **Stearic acid (C_{18})**

ture. In this way, the fatty acid composition of an individual phospholipid can affect the average (or bulk) properties of the entire membrane.

The customary nomenclature for fatty acids consists of two numbers separated by a colon, with the first number representing the length of the fatty acid in carbon atoms and the second number indicating the number of carbon-carbon double bonds. Fatty acid configurations that occur frequently in biological systems have been given common names, such as myristic acid (14:0), palmitic acid (16:0), and oleic acid (18:1). In general, phospholipids contain tails of differing length, with one tail containing at least one *cis*-double bond and the other containing none, and there are distinct distributions of fatty acids among the different phospholipids. PCs are mainly composed of short (16 to 18 carbons) chains, with the unsaturated tail containing only one or two double bonds. In contrast, PE, PS, and PI often contain long polyunsaturated tails (e.g., C20:4). Although the distribution of fatty acids is peculiar to both the class of phospholipid and the membrane type, the wide variety of available structures leads to thousands of unique phospholipids in membranes.

Sphingolipids

Sphingolipids are a distinct class of membrane lipids that are built from a sphingosine backbone (Figure 8.3), as opposed to the glycerol backbone of phospholipids. Adding a fatty acid to the amino acid serine forms sphingosine. Sphingosine thus contains one of the two fatty acid tails of a sphingolipid; the other tail arises when the amino group at the polar end of sphingosine is esterified to a second fatty acid. Although the fatty acids of sphingolipids can be quite varied, they tend to be longer (24 carbons) with fewer double bonds (none or only one) than the fatty acids found in phospholipids. Also, the hydrocarbon tail contributed by the sphingosine backbone behaves similarly to a saturated fatty acid. These general features of the fatty acid tails of sphingolipids are important when considering how this class of lipids contributes to the bulk properties of the membrane as discussed later. Derivatives of sphingolipids include ceramides, sphingomyelins, glycosphingolipids, and gangliosides, which are produced by substitution of the amino and alcohol moieties of sphingosine. Ceramides are the simplest sphingolipids, being just sphingosines esterified to contain a second acyl chain (Figure 8.3). Sphingomyelins (SMs) are abundant in membranes and are generated by modification of the terminal hydroxyl group of ceramides with phosphocholine. SM is thus a sphingolipid and a phospholipid. Cerebrosides are analogues of sphingomyelin, with non-charged sugar groups, such as galactose and glucose, in place of the phosphocholine group of SM. Gangliosides are a subset of cerebrosides that arise when multiple sugars are attached to the cerebroside and at least one of the sugars ends with sialic acid. Lipids whose headgroup consists of one or more sugar residues, such as cerebrosides, gangliosides, and PI, are also known as glycolipids. In contrast to bacterial and plant glycolipids, which are formed from glycerol-based lipids, almost all glycolipids (PI being an exception) in animal cells are glycosphingolipids because they derive from ceramide.

Sterols

Cholesterol, the major sterol found in mammalian cell membranes, is an essential lipid, for it is required for normal cell function. Cholesterol comprises approximately 30 mol% of plasma membrane lipids, but it can con-

I
**Structure of
sphingosine**

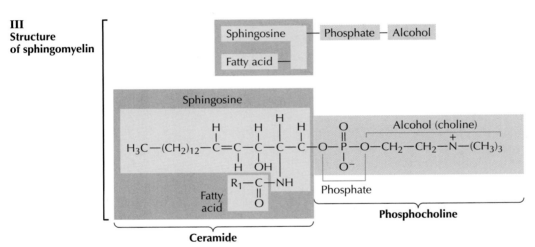

II
**Structure
of ceramide**

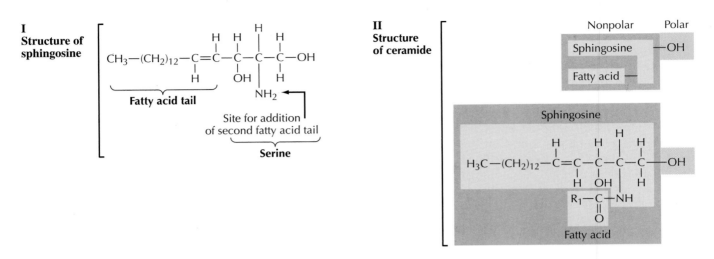

III
**Structure
of sphingomyelin**

IV
**Structure of a
cerebroside
(galactocerebroside)**

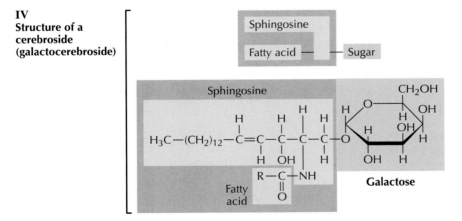

Figure 8.3 Structures of sphingolipids.

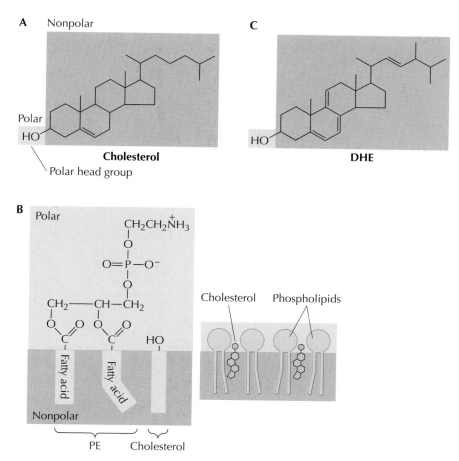

Figure 8.4 **(A)** Structure of cholesterol. **(B)** Position of cholesterol in a membrane. **(C)** Structure of the fluorescent sterol, dehydroergosterol (see text for details).

stitute a considerably higher fraction in some membranes, such as the plasma membrane of erythrocytes. The structure of cholesterol is very different from that of other membrane lipids (Figure 8.4). The body of cholesterol consists of a series of fused rings, which make the molecule quite rigid, and has important consequences for membrane properties. At one end of this rigid, planar ring system is a hydroxyl group, while at the other end is a hydrocarbon tail, so cholesterol, like other membrane lipids, has both hydrophilic and hydrophobic poles. This amphipathic nature of cholesterol affects its positioning within the lipid bilayer. When the hydroxyl group sits just next to the phospholipid ester carbonyl (Figure 8.4), the rigid body of cholesterol is situated alongside the fatty acid tails of neighboring phospholipids and can thus influence the properties of these tails. As for ceramides, the polar moiety of cholesterol (i.e., the hydroxyl group) is much smaller than the polar headgroups of other lipids, and so movement between layers (also known as leaflets) of the membrane bilayer can occur readily. This type of transbilayer movement is referred to as "flip-flop."

The major membrane lipids, along with a summary of their general structural characteristics, are given in Table 8.1.

Table 8.1   Major membrane lipids

Abbreviation	Full name	Class	Backbone	Polar headgroup	Typical acyl tail characteristics	
					Sn1	*Sn2*
PC	Phosphatidylcholine	Phospholipid Glycerolipid	Glycerol	Phosphorylated alcohol	16–18 carbons, unsaturated	16–18 carbons, 1–2 double bonds
PE	Phosphatidylethanolamine	Phospholipid Glycerolipid	Glycerol	Phosphorylated alcohol	16–18 carbons, unsaturated	~20 carbons, polyunsaturated
PS	Phosphatidylserine	Phospholipid Glycerolipid	Glycerol	Phosphorylated alcohol	18 carbons, unsaturated	~20 carbons, polyunsaturated
PI	Phosphatidylinositol	Phospholipid Glycerolipid Glycolipid	Glycerol	Phosphorylated sugar	18 carbons, unsaturated (e.g., C18:0)	~20 carbons, polyunsaturated (e.g., C20:4)
Cer	Ceramide	Sphingolipid	Sphingosine	Hydroxyl group	Sphingosine base (C18:1)	>24 carbons, 0–1 double bond
SM	Sphingomyelin	Sphingolipid Phospholipid	Sphingosine	Phosphorylated alcohol	Sphingosine base (C18:1)	>24 carbons, 0–1 double bond
	Ganglioside, e.g., GM1, Gb3	Sphingolipid Cerebroside Glycolipid	Sphingosine	Multiple sugars containing sialic acid	Sphingosine base (C18:1)	>24 carbons, 0–1 double bond
	Cholesterol	Sterol	Rigid ring structure	Hydroxyl group	NA[a]	NA

[a] NA, not applicable.

Plasma Membrane Structure

A common feature of all membrane lipids is their amphipathicity. Every lipid has both hydrophilic and hydrophobic character, and it is this bipolar nature coupled with the hydrophobic effect that drives bilayer formation in an aqueous environment. Lipids will spontaneously orient themselves to maximize interactions of their polar headgroups with water while minimizing exposure of their hydrophobic tails. Even under simplified artificial conditions, many lipids form bilayers, and these bilayers close on themselves to form sealed compartments. In this way, the edges of the bilayer sheet are eliminated and the hydrophobic tails of the lipids are completely protected from water. It is firmly established that membranes in biological systems have proteins associated with a lipid bilayer structure (Figure 8.5). However, an exact description of this structure is still elusive. For example, it is highly likely that the individual monolayers are interdigitated, with the acyl chains of some inner leaflet lipids intercalated with those of outer leaflet lipids, but the extent of interdigitation is uncertain. Likewise, uncertainty exists in the descriptions of lateral organization of lipids in the membrane. Do certain classes of lipids and/or proteins preferentially associate in the membrane to form separated, two-dimensional domains? If so, what drives the formation of these domains? What are their physical properties and biological functions? To begin answering these sorts of questions, it is first necessary to discuss the lipid and protein composition of the membrane.

Lipid Composition

Much of the information about the lipid composition of mammalian cell membranes comes from studies on the plasma membranes of human erythrocytes (red blood cells). The reason for this is that erythrocytes lend themselves to membrane analysis; they are easily isolated in large quanti-

Figure 8.5 Diagram of the bilayer lipid structure of the plasma membrane. Note the asymmetric distribution of lipids within each leaflet.

Extracellular

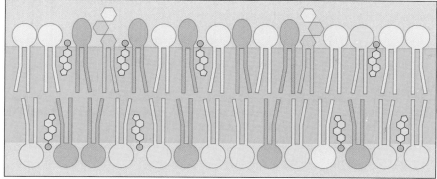

Intracellular

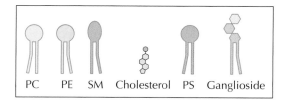

ties as a pure population, and they lack a nucleus or other internal organelles, so the plasma membrane is their only membrane. Analysis of erythrocyte and other mammalian cell plasma membranes has shown that plasma membranes typically contain ~15 to 20% cholesterol by weight. This is equivalent to approximately 30 mol% and represents a significant fraction of the total membrane lipids. The four main classes of phospholipids—phosphatidylcholine (PC), phosphatidylethanolamine (PE), phosphatidylserine (PS), and sphingomyelin (SM)—constitute ~50 to 60% of the total lipids (by weight), with SM comprising almost 20% on its own. Glycolipids, such as cerebrosides and gangliosides, on the other hand, make up only ~3 to 8% of the lipids of the plasma membranes of most cells, although they comprise a significantly higher fraction of the plasma membrane lipids of neurons. In fact, the lipid compositions of eukaryotic cell membranes generally differ depending on the cell type and the organelle. For instance, PI usually makes up just a few percent of the lipids in most cell types, but in brain tissue it can comprise up to 10% of the lipids. As another example, the cholesterol content of the intracellular membranes of the endoplasmic reticulum and the Golgi apparatus is lower than that of the plasma membrane, while the level of PC is higher. The functional significance of these compositional differences between membranes of various organelles is largely unknown.

All biological membranes exhibit transbilayer asymmetry regarding both lipid and protein composition. The best-studied case for lipid asymmetry is again the erythrocyte. PS is nearly exclusively (~98%) in the inner leaflet of the bilayer, while SM (~96%) and gangliosides (100%) are constrained to the outer membrane leaflet. The distributions of PE and PC are less extreme, with ~75 to 80% of each restricted to the inner and outer monolayers, respectively. In contrast, cholesterol flip-flops rapidly, and its transbilayer distribution is not certain. The membranes of intracellular organelles also manifest asymmetry of phospholipids and glycolipids, with their orientation preserved so that luminal- and cytosolic-facing lipids correspond to those in the plasma membrane with extracellular and intracellular orientations, respectively. This asymmetry is maintained by the action of specific transport proteins, known as lipid translocases or flippases. Table 8.2 summarizes the compositional makeup of mammalian cell membranes and the bilayer distributions of each lipid.

Table 8.2 Composition of cellular membranes

Lipid	Weight % of membrane[a]	Inner/outer leaflet distribution	
		Inner	Outer
PC	~25	~20–25	~75–80
PE	~10–15	~75–80	~20–25
PS	<5	~98	~2
SM	~10	~4	~96
Cerebroside, ganglioside	~3–8		~100
Cholesterol	~15–20	?[b]	?[b]
Other lipids	~20		

[a] Data from Keenan and Morré (1970).
[b] The bilayer distribution of cholesterol is unclear.

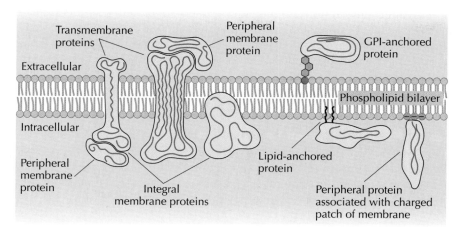

Figure 8.6 Depiction of the various ways in which proteins can associate with the plasma membrane.

Protein Composition

Lipids contribute just ~50% of the mass of the plasma membrane. The other half of membrane mass is from proteins. Because proteins are generally much larger than lipids, a membrane that is 50% protein by mass will have approximately 50 lipids per protein. It is easy to see how lipids can contribute significantly to the bulk properties of the membrane even though they comprise just half of the mass. Even so, discussion of membrane structure must necessarily include a description of its protein constituents. Membrane proteins can be categorized as integral, anchored, or peripheral based on the nature of their interaction with the membrane bilayer. Integral membrane proteins are those that can be released only by disruption of the bilayer. These proteins are either partially or completely embedded within the membrane (Figure 8.6). Many cell surface receptors fall into the latter group. They are amphipathic molecules that have at least one hydrophobic polypeptide stretch that traverses the membrane and hydrophilic polypeptides that extend into the aqueous environments on both sides of the membrane. By spanning and extending through the membrane, these transmembrane proteins are able to transduce signals between the extracellular and the intracellular milieus.

Other surface receptors are anchored just to the outer leaflet of the membrane via a glycolipid, specifically glycosylphosphatidylinositol (GPI), to which they are covalently coupled. GPI-anchored proteins, like all glycolipids, are found exclusively on the extracellular side of the plasma membrane. It is less obvious how these proteins, which penetrate only the outer leaflet of the membrane, transmit signals to the cell interior. There are many intracellular proteins that are analogous to the GPI-anchored proteins in that they are anchored to just one leaflet of the membrane. These cytosolic proteins have fatty acid modifications, such as myristic acid (C14:0) and/or palmitic acid (C16:0), or isoprenyl modifications, such as farnesyl (C_{15}) or geranylgeranyl (C_{20}), which insert into the inner leaflet of the plasma membrane.

Peripheral membrane proteins have a more tenuous association with the bilayer. The connection of these proteins to the membrane can be disrupted under fairly gentle conditions; simply changing the ionic strength of the surrounding solution can cause dissociation of certain peripheral membrane proteins. Peripheral association with the membrane can be mediated by either protein-protein or protein-lipid interactions. For example, some peripheral proteins are noncovalently associated with integral membrane proteins, and they do not necessarily come into direct contact with the lipid bilayer. Other peripheral proteins associate directly with the membrane by adsorbing to patches of charged lipids at the membrane surface.

Lateral Plasma Membrane Organization

Based in part on the experiments of Frye and Edidin (1970), Singer and Nicolson (1972) proposed a fluid mosaic model for plasma membrane organization. This model of membrane structure emphasized the fluid nature of the membrane and, consequently, the diffusional mobility of proteins and lipids within the lateral plane of the membrane. Furthermore, it was postulated that the fluid nature of the membrane prohibited long-range ordering of its lipids. However, studies (mostly on synthetic bilayers) showed that lipids could segregate laterally to form heterogeneous regions with varying degrees of fluidity within the membrane. Based on these early studies, alternative models were rapidly developed, and more recent studies have led to their refinement; most current models of membrane structure allow for an additional level of lateral organization imparted by the properties of the membrane constituents. It is now widely accepted that the plasma membrane is compartmentalized into two-dimensional domains that range in scale from tens of nanometers to tens of micrometers (Figure 8.7). Organization of the plasma membrane into these various domains could have important biological implications since the rate and extent of interactions between signaling molecules at the membrane can be regulated. For instance, compartmentalization of membrane components may serve to concentrate molecules into a small region of the membrane, thereby promoting their interaction. Alternatively, by excluding certain species from particular membrane compartments, interactions can be inhibited.

Varied forces, including protein-protein, lipid-protein, and lipid-lipid interactions, govern the formation and stabilization of lateral membrane domains. Figure 8.8 depicts the different ways in which the plasma membrane can be segregated into distinct domains. In brief, a meshwork of cytoskeletal proteins closely apposed to the membrane may act as a "membrane-skeleton fence" to transiently confine transmembrane proteins to subsections of the plasma membrane, or transmembrane proteins may interact directly with the cortical cytoskeleton to create a protein fence that partially corrals lipids and other proteins within a defined area. Charged lipid patches in the membrane may act as a docking site for certain proteins,

Figure 8.7 Schematic representation of microdomains in the plasma membrane.

Caveolin

Caveolae
(50 nm)

Transient
confinement zone
(100–300 nm)

GPI
anchor

Fluid lipid

Nanodomain
(20–30 nm)

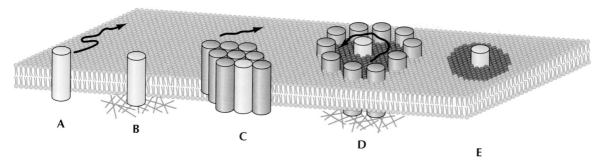

Figure 8.8 Representation of the plasma membrane depicting different mechanisms for compartmentalizing membrane constituents. **(A)** In the absence of any compartmentalization, proteins and lipids can diffuse freely within the plane of the membrane. **(B)** Some transmembrane proteins are prevented from moving because they interact either directly or indirectly with the underlying cytoskeleton. **(C)** Proteins can associate with other proteins in the membrane to form large complexes whose diffusion is restricted compared with monomeric proteins. **(D)** Proteins that are anchored to the cytoskeleton can coral in other proteins and lipids. **(E)** Membrane lipids can associate into domains that serve to trap or exclude other membrane constituents.

or classes of lipids may coalesce to form a domain that preferentially recruits or excludes other membrane constituents. It is likely that each type of compartmentalization contributes to the organization of the membrane into a complex patchwork of domains of varying size, stability, and function. It is also likely that domains in cellular membranes are regulated by some combination of several of the forces described. It is beyond the scope of this chapter to discuss each mechanism of membrane segregation in detail. Instead, this chapter will focus mainly on submicroscopic lipid heterogeneities, or lipid microdomains, in the membrane and how lipid-lipid interactions can control their formation.

Lipid Phases and Membrane Order

Because the plasma membrane is a complex mixture of many types of lipids and proteins, defining the physical state or phase of the membrane is difficult. What is clear is that lipids can move relatively freely in some but not all directions. For instance, the hydrophobic effect imposes a high free-energy barrier against lipids popping out of the membrane, while bulky polar headgroups limit their flip-flop between leaflets. In a seminal experiment, Frye and Edidin (1970) showed that membrane lipids are able to translate laterally, suggesting that the membrane behaves as a two-dimensional fluid. A true three-dimensional fluid or liquid state would allow movement equally in all directions. Because membrane lipids are largely constrained to move within the plane of the bilayer, they are said to exist in a liquid-crystalline (L_c) or liquid-disordered (L_d) phase. This phase has characteristics of both a liquid and a solid phase, for it has fewer degrees of freedom than a liquid state but more than the solid state. In the L_d phase, the acyl chains of the lipids are fluid and disordered, and the lipid can diffuse laterally in the membrane, but the lipids cannot move normal to the membrane. The solid state of isolated lipids, known as the gel phase, exists when the temperature is lowered below a critical level (the phase transition temperature). In the gel phase, lateral diffusion is slowed by several orders of magnitude and intramolecular motions (such as rotations, acyl chain movements, etc.) are largely abrogated. In other words, the lipids become stiff and immobile. Importantly, the gel phase is thought not to exist in biological membranes at physiological temperatures.

The physical properties of a membrane then depend in part on the nature of the acyl tails of its lipid constituents. One level of membrane ordering is instilled by the phase preference of the lipids. The phase state of isolated lipids depends on the temperature, and the temperature at which the transition between phases occurs (i.e., the phase transition temperature) depends on the nature of the lipids' acyl chains. For instance, the transition temperature between the gel and L_d phases increases with increasing chain length within a class of phospholipids, because van der Waals interactions between long hydrocarbon chains stabilize the tightly packed gel phase. By analogous reasoning, unsaturation (especially *cis*-double bonds) in the acyl chains destabilizes the gel state. A *cis*-double bond introduces a kink into the acyl chain that inhibits hydrocarbon chain packing of the gel phase and, consequently, lowers the transition temperature.

The effects of cholesterol on the physical properties of lipids are varied and sometimes opposing. In binary mixtures of cholesterol and phospholipids (e.g., PCs), cholesterol has divergent effects at different temperatures. The fatty acid chains of PCs in the L_d phase become less mobile (i.e., more ordered) when cholesterol is added because of van der Waals interactions with the rigid rings of cholesterol. In contrast, the fatty acid chains of PCs in the gel phase become less ordered upon addition of cholesterol, because tight tail packing is disrupted by insertion of cholesterol. Cholesterol then acts to attenuate the differences between the phases adopted by PCs at different temperatures. Furthermore, in model membranes containing mixtures of lipids, cholesterol addition at certain temperatures induces domain separation, with the formation of regions enriched in cholesterol relative to neighboring membrane patches. This could have important implications for biological membranes. If biological membranes were made of phospholipids alone, they would adopt the L_d phase because of the low transition temperatures of the phospholipids' short unsaturated acyl chains. But the plasma membrane, for example, has a high content of SM, glycolipids, and cholesterol. SM has long saturated acyl chains and a correspondingly high transition temperature, while cholesterol has complex effects on phase behavior and is postulated to interact directly with the headgroup of SM. The mixing of cholesterol with lipids that have disparate transition temperatures and unequal propensities to bind cholesterol raises the possibility that phase separation may occur in biological membranes.

In model membranes, consisting of binary or ternary mixtures of lipids and cholesterol, a distinct phase has been characterized that has properties between the gel and L_d phases. This phase is known as the liquid-ordered (L_o) phase. The acyl chains of lipids in the L_o phase are more ordered than those in the L_d phase, but the lipid molecules retain their lateral mobility, unlike gel phase lipids, which are immobile. The L_o phase can coexist with both L_d and gel phases. Its formation is temperature and cholesterol dependent, and there is indirect evidence that it can exist in biological membranes. The proposal that some biological membranes, or parts thereof, resemble the L_o phase provides a structural basis for understanding lateral lipid domains within these membranes.

Physical Properties of Lateral Lipid Domains

Lateral membrane organizations are known by an assortment of names, including (but not limited to) caveolae, rafts, detergent-resistant membrane domains (DRMs), and lipid domains. Caveolae are an important type of membrane domain, but they are clearly a specialized subset. For this reason, the name "caveolae" should be reserved for those domains that exhibit a particular set of well-defined physical properties, which are described in subsequent sections. The term "rafts" is widely used to describe any small, cholesterol-rich domain that contains a high fraction of order-preferring lipids and proteins. The term was coined to describe a model in which specialized lipid domains are small in both size and number. However, it is not clear that all lipid domains are small or that they represent just a minor fraction of the plasma membrane, so the image evoked by the term "raft" may not be the most suitable one for describing membrane domain organization in general. Further, because "raft" has become a generic name for any cholesterol-rich lipid microdomain, it leads to the idea that there are only two types of domains ("rafts" and "nonrafts"). This is clearly an oversimplification since many types of lipid domains with varying properties have already been described (see below). The name "detergent-resistant membranes" or DRM refers to membrane fractions that are resistant to solubilization by cold nonionic detergents. Therefore, DRM is useful only as an operational definition and is not appropriate to describe membrane organizations in whole cells. To circumvent the ambiguities inherent in these common names, we will refer to any lateral organization of lipids as a lipid or membrane domain, and we will distinguish between subsets of domains by specifying physical properties. For example, fluid membrane domains are collections of mainly disordered (L_d phase)-preferring lipids, whereas ordered membrane domains are enriched in order (L_o phase)-preferring lipids. Both fluid and ordered membrane domains can be further categorized based on their other physical properties, such as size, composition, and sensitivity to particular detergents (see below).

As with the nomenclature in the field, a consensus on the physical properties of lipid domains is absent. The explanation for this is manifold. First, biological membranes consist of hundreds of chemically distinct lipids and myriad proteins, which contribute in a complex way to the physical properties of the whole membrane. Second, cellular membranes are dynamic structures whose composition (and thus organization) is constantly in flux owing to vesicular traffic and lipid metabolism. Third, most lipid domains cannot be directly observed, making physical measurements technically challenging. Fourth, there appear to be many types of specialized lipid domains with distinct properties, so attempts to elaborate one set of properties to describe all domains will likely be futile. The diversity of putative microdomains found in cell membranes is depicted in Figure 8.7, and their physical properties are reviewed in the ensuing sections.

Size

Based on their size, lipid domains can be broadly categorized as microdomains or macrodomains. Microdomains are thought to have diameters that range from tens to hundreds of nanometers, whereas macrodomains may span several micrometers. Because microdomains are below the resolution of light microscopy, indirect techniques, with varying spatial and temporal resolution limits, have been used to try to define their size. Consequently, the dimensions of microdomains are still unknown. Caveolae represent a special type of microdomain (see Morphology), and their size has been directly measured using electron microscopy. Caveolae are 50 to 70 nm in diameter, while estimates for the diameters of other microdomains range from 30 to 300 nm, depending on the method used to measure them.

Macrodomains, on the other hand, are well above the resolution limits of light microscopy and so they can be readily visualized. Macrodomains are found in polarized cell types, whose function requires one end of the cell to adopt one functional role, and the other end to adopt a distinct functional role. Examples of polarized cells include epithelial cells, hepatocytes, neurons, and migrating cells (e.g., neutrophils), to name a few. In the first two examples, tight junctions prevent mixing of membrane components and physically define two distinct membrane macrodomains: apical and basolateral domains for epithelial cells, and the analogous bile canalicular and sinusoidal domains for hepatocytes. For migrating cells, macrodomains are less well defined because there are no long-lived stable structures, like tight junctions, to act as physical barriers for lipid mixing. As a result, the boundaries between domains in migrating cells are less apparent, but they can be detected and measured under the appropriate conditions. The macrodomains in each of these cell types can extend several micrometers or even tens of micrometers (Figure 8.9).

Morphology

As mentioned above, most types of membrane microdomains do not have morphological characteristics that readily distinguish them from surrounding lipids in the plasma membrane. The exception is caveolae, which represent a subset of lipid domains with a well-defined morphology. It is this unique attribute that has made their characterization far simpler than that of other types of lipid microdomains. Caveolae ("small caves" in Latin) are membrane invaginations that are easily visualized by electron microscopy. These structures are distinguished from other types of membrane invaginations, such as clathrin-coated pits, by their diminutive size and their distinctive coats. Caveolae appear striated in electron micrographs because their surfaces are studded with an integral membrane protein called cave-

Figure 8.9 Examples of cells with plasma membrane macrodomains. **(A)** The plasma membranes of epithelial cells are divided into apical and basolateral domains, which are kept segregated by protein barriers called tight junctions. **(B)** The plasma membranes of hepatocytes have macrodomains analogous to the apical and basolateral domains in epithelial cells; these domains are the canicular membrane and the sinusoidal/lateral membranes. **(C)** In migrating cells, like neutrophils, the cell front (leading lamella) represents one type of macrodomain, while the cell rear (the cell body and the uropod) represents a different type of macrodomain.

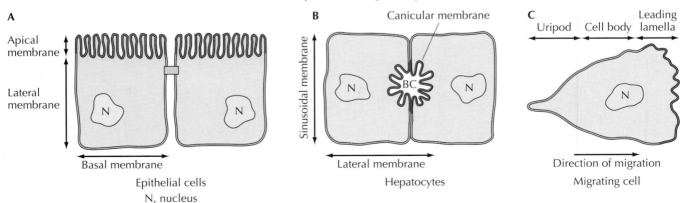

olin, and they are generally found in clusters, resembling grapes on a vine. No other type of microdomain can be so readily distinguished morphologically. On the other hand, apical macrodomains of polarized epithelial cells, or the bile canalicular compartments of hepatic cells, are characterized by abundant microvilli and a dense underlying F-actin meshwork that can be visualized by light and electron microscopy. Alternatively, fluorescence microscopy and probes for tight-junction components can delineate the macrodomains that comprise apical and basolateral plasma membrane compartments.

Composition

Membrane domains can be broadly categorized into two types based on their composition: those that are enriched in order-preferring lipids (e.g., L_o domains) and those that contain mainly disorder-preferring lipids (fluid or L_c or L_d domains). However, this is not to suggest that there are only two kinds of domains, for there could be numerous types of L_o domains which have distinct compositions and probably have diverse functions. Although it would be erroneous to consider all L_o domains as one, it is still possible to make some generalizations about L_o and L_d domains. L_d microdomains contain mainly phospholipids with either unsaturated or short acyl chains, which prefer a more fluid environment. In contrast, L_o microdomains are rich in glycosphingolipids and sphingomeylin, which contain long and saturated acyl chains that can pack into more ordered structures. L_o microdomains may also be enriched in cholesterol compared with L_d microdomains; however, the degree of cholesterol enrichment is unclear.

In addition to lipids, proteins that associate with the plasma membrane via lipid modifications can sort into microdomains based on the nature of their membrane association. For example, cross-linked GPI-anchored proteins associate with L_o microdomains, as do palmitylated and myristylated cytosolic proteins (e.g., Src family kinases). On the other hand, inner leaflet-associated proteins that are isoprenylated (e.g., GTP-binding proteins) appear to be relatively excluded from L_o microdomains, and for this reason are found in L_d microdomains.

As for lipid microdomains, macrodomains exhibit differences in composition. It has long been known that epithelial cells contain a different repertoire of proteins and lipids on their apical surfaces compared with their basolateral ones. In particular, glycolipids and cholesterol are more abundant in the apical membrane, and GPI-anchored proteins are found exclusively there in some polarized epithelia. Sphingomyelin distributes evenly between both apical and basolateral macrodomains, and the transmembrane protein CD44 localizes basolaterally. In migrating cells, differences in lipid properties between leading and trailing regions of the cells have been detected (see below), but there has been no detailed analysis of the lipid composition of membranes at each pole. Certain transmembrane proteins, such as CD44 and CD43, localize to the rear of migrating neutrophils, while the isoprenylated GTP-binding protein, Rac, is targeted almost exclusively to protruding membrane regions at the front of the cell. In migrating T cells, different gangliosides (GM3 and GM1) segregate to the leading and trailing portions of the cells, presumably in association with distinct L_o microdomains that coalesce to form distinct macrodomains. The lipid macrodomains may be important functionally. For instance, the apical lipids may provide an enhanced permeability barrier, and the lipid organization in migrating cells may help to localize functions such as lamellar protrusion.

Detergent Sensitivity

The compositional differences of membrane domains result in differential sensitivity to solubility by cold detergents. Treatment of unpolarized cells with nonionic detergents in the cold (4°C) reveals that a large fraction of the lipids in most of the plasma membrane are resistant to extraction. The cell surface takes on an appearance similar to Swiss cheese, with nearly three-fourths of the membrane area retained as an intact sheet that is punctuated by extracted regions. When the detergent-resistant membrane fractions are isolated and analyzed, they are found to exhibit characteristics of the L_o phase, and model membranes in the L_o phase are detergent resistant, so it is believed that the L_o phase confers detergent resistance to lipids. In this context, the extracted regions can be interpreted as those regions of the plasma membrane that correlate with L_d lipid domains, while the unextracted regions are membranes in the L_o phase. This interpretation comes with several caveats. First, it is likely that cold temperatures enhance the formation of ordered domains. Second, detergent extraction seems to cause coalescence of microdomains into larger macrodomains (visualized as the holes in the membrane for L_d domains or sheets of membrane for L_o domains). Third, interactions with the underlying cytoskeleton affect detergent resistance at 4°C in ways that are only partially understood. Some L_o domain components, such as CD44, are indirectly anchored to the underlying F-actin meshwork, and so their resistance to extraction by cold detergents is due in part to association with L_o domains and in part to their linkage to the cytoskeleton. It should also be noted that detergent resistance of individual molecules depends on their surrounding lipids. GPI-anchored proteins or glycosphingolipids that are normally detergent resistant will be solubilized if they are in a membrane that is made up of disordered lipids.

Resistance to extraction by cold nonionic detergents is one of the most widely used criteria for determining whether a membrane component is localized to L_o domains, and Triton X-100 is the detergent typically used. Notably, detergent-resistant membrane domains that are soluble in Triton X-100, but resistant to extraction by other nonionic detergents have been identified, indicating that it may be possible to subcategorize L_o domains based on their resistance to different detergents.

Dynamics

Macrodomains in epithelial cells and hepatocytes are long-lived structures that persist for much of the lifetime of the cells, whereas macrodomains in migrating cells are necessarily less stable. In the case of neutrophils, resting cells are roughly round in shape with no apparent polarization in morphology or membrane domains. Within seconds of exposure to a chemoattractant (e.g., bacterial peptides), neutrophils adopt a highly polarized phenotype with leading and trailing regions of the cell that comprise distinct macrodomains. Once polarized, migrating neutrophils continually reorganize their membranes as they reorient in response to environmental cues. The macrodomains in migrating neutrophils are stable in that they persist for several minutes as the cells translocate, but they are also dynamic enough to allow rapid formation and dissolution in response to extracellular signals.

The difficulty in detecting microdomains in living, unperturbed cells has led to the conclusion that these structures are highly dynamic.

Inner/Outer Leaflet Coupling

There is little information on how the organization of outer leaflet membrane components affects lipids and proteins associated with the inner leaflet of the plasma membrane. However, domains of many different shapes and sizes can be found in giant unilamellar vesicles composed of various lipid mixtures, and in all cases, the domains span the inner and outer leaflets of the membrane. This suggests that tight coupling exists between leaflets. In fact, the functional coupling of inner and outer membrane leaflets has been well documented for the immune receptors, FcεRI and the T-cell receptor. Some evidence that microdomains extend over both leaflets of the plasma membrane comes from copatching experiments in whole cells. Aggregation of outer-leaflet-anchored (e.g., GPI-anchored proteins) or transmembrane (e.g., CD44, FcεRI) L_o domain components causes coaggregation of inner membrane-associated signaling molecules (e.g., Lyn, Lck, Fyn, annexin II). In general, aggregation of outer-leaflet L_o domain markers causes coaggregation of inner leaflet molecules that associate with the membrane by palmitoyl and myristoyl acylations, but not molecules associated via isoprenyl groups. Because palmitoyl and myristoyl modifications, but not isoprenyl ones, are thought to target proteins to inner-leaflet L_o domains, this again supports the idea that the inner-leaflet membrane is organized into microdomains that are linked by some means to outer-leaflet domains. The mechanism of inner- and outer-leaflet membrane coupling is unknown, but possibilities include that there is interdigitation of lipid tails from each leaflet or that cholesterol forms transbilayer dimers.

Methods for Studying Membrane Organization

Model Membranes

Some of the strongest support for the hypothesis that phase-separated domains can exist in biological membranes comes from studies of model membranes. Phase separation of lipids has been best described in binary mixtures of an order-preferring phospholipid and cholesterol, which separate into cholesterol-rich and cholesterol-poor domains. Findings in two-component systems may not apply to multicomponent biomembranes. Cell membranes contain a high fraction of sphingolipids in addition to phospholipids and cholesterol, and the phases in these membranes are thought to be sphingolipid-rich L_o domains coexisting with phospholipid-rich L_d domains. Cholesterol is thought to facilitate the phase separation, as opposed to forming its own phase. To more closely mimic biological membranes, model membranes have been made with phospholipids, sphingolipids, and cholesterol. Supported planar monolayers and bilayers of lipid mixtures provide an attractive model system for studying lipid phase behavior. Such a system has been used to show that micrometer-sized domains are observable in supported membranes made from either a mixture of phospholipids, sphingolipids, and cholesterol, or renal brush-border membranes. Measurements of motional parameters and evaluation of detergent sensitivity can help to define the phase states of observed domains in model membranes (see below), and they provide important references for measurements made in cells.

The major concern with the approach above is that the behavior of lipids in supported membrane models may be influenced by the support. For instance, detergent extraction of supported lipid layers leaves L_o domains completely intact, with neither their size nor shape affected. It is likely that the support stabilizes L_o domains and prevents coalescence. To avoid this external influence, giant unilamellar vesicles (GUVs) can be used as an alternate model system. GUVs are very large (>50 μm) liposomes in which sizeable (>1 μm) domains form under appropriate conditions. Microscopically observable domains form in GUVs composed of PC-SM-cholesterol, a mixture of order- and fluid-preferring phospholipids with cholesterol, or in liposomes reconstituted from renal brush-border membranes. In addition to this persuasive evidence that phase separation can occur in biological membranes, studies on model membranes have suggested potential mechanisms of domain regulation. By varying the lipid composition of GUVs, it was found that formation of micrometer-sized domains was exquisitely sensitive to cholesterol content, and, at certain lipid compositions, critical point phase behavior was observed. At a critical point, coexisting phases cannot be distinguished because the phases become too similar. Importantly, as a system approaches a critical point, submicroscopic transient transitions between coexisting phases can occur. This critical-point phenomenon may replicate the state of biological membranes, with the membrane poised to undergo rapid reorganization in response to slight perturbations. Further studies in model membranes will undoubtedly continue to provide insights into domain organization and dynamics in living cells.

Fluorescent Probes for Studying Membrane Organization and Dynamics

Studying lipid organization and trafficking in living cells presents challenges that are not encountered in comparable protein studies because techniques for specifically tagging or altering endogenous lipids are problematic. Gangliosides and GPI-linked proteins have been studied using fluorescently conjugated specific antibodies as probes. Specific antibodies exist for some other lipid species, but lipids are small molecules relative to antibodies, so antibody binding would likely perturb the properties of the lipid. In some cases, the antibody epitope is accessible only after cellular fixation or subcellular fractionation, which eliminates the prospect of following lipid flow in living cells and may cause lipid redistribution. Alternatively, radiolabeled lipid precursors have been used to track lipid trafficking in cells. This technique also has limitations; data interpretation can be complicated since whole classes of lipids can be derived from a single labeled precursor. Furthermore, determining the subcellular distribution of lipids depends on isolation of pure populations of organelles, which is often difficult, and determination of lateral separation within a purified organelle cannot be done. As an alternative to radiolabeling or antibody labeling, many studies of lipid organization and dynamics in living cells have used fluorescently tagged lipid analogues and fluorescence microscopy.

Fluorescent probes can be attached to lipid analogues that mimic a naturally occurring counterpart, and these fluorescent analogues are then added exogenously to cell membranes. The relatively simple structures of lipid analogues limit the positions at which fluorescent probes can be attached. Basically, there are just two regions for probe attachment: along one of the hydrophobic tails or on the hydrophilic headgroup. Both the position at which a probe is placed and the nature of the probe itself can affect the properties of the lipid analogue and dictate the applications for its use. Here we will survey just a handful of commonly used probes and their applications (Table 8.3).

Table 8.3 Fluorescent probes for studying membrane organization and dynamics

Application	Probe
Label *trans*-Golgi	C_6-NBD-ceramide
	C_5-BODIPY-ceramide
Endocytic trafficking	C_6-NBD-SM
	C_5-BODIPY-SM
	C_6-NBD-PC
	C_5-BODIPY-PC
	C_6-NBD-PE
	Rhodamine-PE
	DiI-C_{16}
	FAST-DiI
	GPI-anchored proteins
	Natural ligands (e.g., folate, LDL)
Cholesterol distribution	DHE
	Filipin
Membrane domain organization	DiI-C_{16}
	FAST-DiI
	GPI-anchored proteins
	Natural ligands (e.g., folate)
	Fluorescently labeled cholera toxin

Fluorescent ceramide analogues are selective stains for the Golgi apparatus (specifically the *trans*-Golgi and the *trans*-Golgi network) in living and fixed cells. Because these probes reach their destination even if they are added after cellular fixation, they apparently traffic to the Golgi apparatus via nonvesicular transport. The most commonly used derivatives are C_6-NBD-ceramide and C_5-BODIPY-ceramide, which have their fluorescent moieties at the end of a short acyl chain extending from the *sn-2* position of the ceramide backbone. The NBD and BODIPY fluorophores have spectroscopic properties similar to fluorescein ("blue" excitation, "green" emission); however, BODIPY exhibits a concentration-dependent change in spectroscopic properties that can complicate its imaging while giving it certain unique advantages. At high concentrations, BODIPY emission shifts from 515 nm (green) to 620 nm (red) due to excimer (excited state dimer) formation. This can be detrimental or beneficial, depending on the experiment. The possibility of excimer formation can lead to difficulties in interpreting imaging data from double-label experiments, whereas in single fluorophore experiments, the shift in emission can be used in conjunction with ratio imaging to quantitatively measure the distribution of the BODIPY-labeled lipid.

C_6-NBD and C_5-BODIPY derivatives of SM and PC are used to study vesicle-mediated transport pathways. These probes are readily delivered to and removed from the exoplasmic leaflet of the plasma membranes of cells, allowing for detailed analyses of their endocytosis and exocytosis kinetics. Also, the exact intracellular routes traveled by these probes can be delineated using pulse-chase protocols and double-label experiments. For example, the plasma membrane of cells can be labeled with C_6-NBD-SM, and then the labeled cells incubated for a set time to allow internalization of labeled membrane (this is the "pulse" part of an experiment). The labeled cells are then incubated in the presence of fatty acid-free bovine serum albumin (BSA) to remove or "back exchange" the lipid probe from the plasma membrane, but not from intracellular membranes. Further incubation of these cells in the continued presence of BSA constitutes the "chase." By varying the pulse-and-chase periods and labeling the cells with both the lipid probe and a well-characterized marker of the endocytic pathway (e.g., transferrin; see below), intracellular itineraries and dynamics of the lipids can be discerned. These techniques along with C_6-NBD-SM have been used to monitor membrane trafficking after endocytosis. As for the ceramide derivatives, the NBD and BODIPY fluorophores are linked to the ends of short acyl chains of the SM and PC lipid analogues (Figure 8.10). The NBD moiety is less lipophilic than BODIPY, so it loops back toward the headgroup region, whereas BODIPY remains embedded in the bilayer (Figure 8.10). The perturbations that NBD causes in acyl chain alignment are likely to cause NBD probes to traffic differently from the endogenous lipids that they are meant to mimic. This is one important reason why it is useful to verify results with an assortment of related lipid probes whenever possible.

A different class of lipid probes comprises derivatives of PE. With fluorophores covalently attached to the amine of the headgroup, rather than at the end of one of the acyl chains, the acyl chain properties of these lipids are unaffected by the probe, although headgroup interactions could be affected (Figure 8.10). Additionally, tagging the headgroup of these lipids can change their overall shape, and thus their packing preferences. For example, PE normally has an inverted cone shape, because the cross section of the volume taken up by its tails is large compared to that of its small headgroup (Box 8.1). The cross-sectional area of the headgroup is increased by addition of a fluorophore such as rhodamine, and the lipid's shape is altered. Because lipid shape may affect sorting (Box 8.1), the trafficking of rhodamine-labeled PE may differ from that of unmodified PE. With this caveat in mind, PE derivatives are still useful probes for monitoring lipid trafficking, especially when preserving the properties of the acyl chains is important. Rhodamine- and NBD-labeled PE (N-Rh-PE, N-NBD-PE) have been used to study lipid trafficking in a number of cell types and their fates are distinct from that of the short-chain PC and SM analogues discussed above.

Figure 8.10　Schematic representation of the outer leaflet of the plasma membrane depicting how fluorescent lipid probes insert into the membrane. **(A)** NBD-C_6-HPC; **(B)** BODIPY-C_{12}-SM; **(C)** rhodamine-PE; **(D)** DilC$_{16}$.

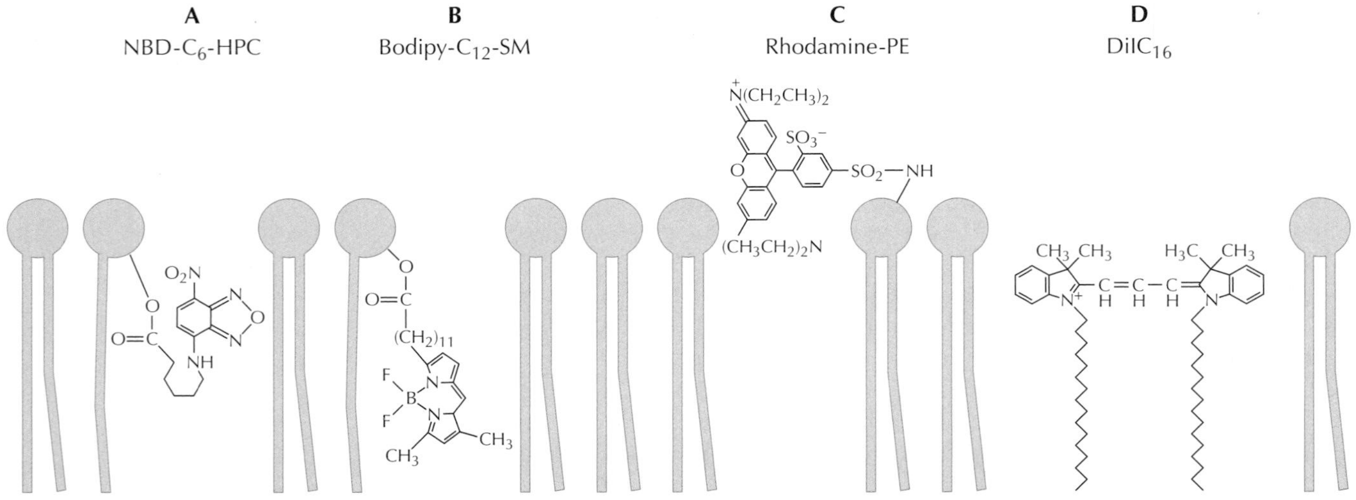

BOX 8.1

How to form a lipid domain

(A) Lipid domains may form as a consequence of both lipid and organelle geometries. Based on the ratio of the cross-sectional diameters of lipid headgroups versus their acyl chains, lipids can be categorized into three basic shapes: cylinder, cone, or inverted cone. Lipids of each of these shapes have a different propensity for membranes of differing curvature. For example, lipids that are cone shaped prefer convex membranes, while those that are shaped like inverted cones favor concave membranes. According to these preferences, lipids of a specific shape may gather in regions of an organelle that manifest a partic-

ular geometry. For example, sorting endosomes have a tubulovesicular morphology with highly curved regions of membrane (arrows) at the necks of tubular extensions and along the rims of invaginations. These highly curved membrane regions may accumulate lipids of a particular shape (gold-shaded membrane patches) or they may act as physical barriers for lipids of the wrong shape, allowing only lipids that can withstand the highly curved membrane to pass freely. This latter scenario would result in different lipid compositions for the tubular, vesicular, and invaginated portions of the organelle. **(B)** Phase separation of lipids provides another means for lipid domain for-

mation. For example, lipids with long, straight, and saturated acyl chains may coalesce and pack closely together to form a more ordered domain than lipids with unsaturated, crooked acyl chains. (See text for more information on phase separation.) **(C)** Finally, chainlength mismatch is another factor that may drive lipid domain formation. Lipids contain acyl chains of varying lengths, and membrane order can be perturbed when a lipid has significantly longer or shorter acyl chains than the surrounding lipids. Accordingly, lipids with acyl chains of similar lengths may coalesce to avoid intermingling of short- and long-chained lipids.

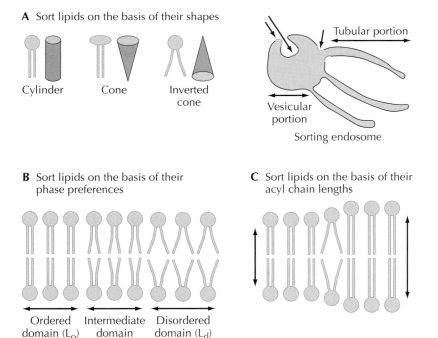

A Sort lipids on the basis of their shapes

Cylinder   Cone   Inverted cone

Tubular portion

Vesicular portion

Sorting endosome

B Sort lipids on the basis of their phase preferences

Ordered domain (L_o)   Intermediate domain   Disordered domain (L_d)

C Sort lipids on the basis of their acyl chain lengths

Ceramides, SM, PC, and PE are naturally occurring lipids, which consequently are subject to metabolic turnover. For example, labeled ceramides can be metabolized to labeled SM derivatives, and, conversely, fluorescent SM probes can be catabolized to fluorescent ceramides in vivo. In mammalian cells, ceramides are second messengers that are involved in a number of signaling cascades, so generation of ceramides may induce an unwanted signaling event. Even in the absence of signal transduction, production of a second species of fluorescently labeled lipid will obscure attempts to monitor trafficking of the original probe. To avoid the complicating contributions of metabolic turnover, nonnatural lipid analogues, which do not have naturally occurring counterparts, can be used. DiI lipid analogues, for example, are amphiphilic lipid mimetics composed of an indocarbocyanine headgroup and two hydrophobic alkyl chains. As with fluorescent PC, SM, and PE probes, the DiI analogues insert into the membrane and traffic as an integral part of the bilayer (Figure 8.10). The properties (e.g., length and degree of unsaturation) of the fatty acyl tails of each of these classes of lipid probes can be varied, and this, in turn, may affect their trafficking patterns.

Cholesterol is an important component of most membranes of mammalian cells, and disruption of its homeostasis or distribution in intracellular membranes leads to debilitating disease in humans. Visualizing the subcellular distribution of cholesterol and tracking its transport within cells are thus very important. Filipin has long been used to accomplish the former task. Filipin, a member of the same family of weak antibiotics as amphotericin B and nystatin, is a macrolide polyene extracted from *Streptomyces filipinensis*. The amphipathic nature of filipin allows ready insertion into membranes, where it is thought to bind directly to cholesterol. Filipin is fluorescent, making it a useful probe for detecting the presence and determining the distribution of cholesterol. Although filipin does not form large pores across the membrane bilayer, it does cause leakage of cellular components, probably due to microscopic defects in the highly perturbed lipid bilayer. For this reason its use is restricted to analyses of cholesterol distribution in dead (i.e., fixed) cells.

For dynamic analyses of cholesterol distribution in living cells, it is necessary to have a fluorescent analogue that faithfully mimics cholesterol, as opposed to a fluorescent probe, like filipin, that binds to cholesterol and potentially perturbs its distribution. Dehydroergosterol (DHE) is a naturally occurring fluorescent sterol found in sponge and fungal cells. The structure of DHE differs only slightly from that of cholesterol; it has three additional double bonds and an extra methyl group (Figure 8.4). Two of the additional double bonds are located within the steroid rings of DHE, and they form a conjugated double-bond system with a double bond that is also found in cholesterol; this conjugated double-bond system makes DHE fluorescent. Despite these structural differences, which make some changes in the sterol structure and thus affect its behavior in the membrane bilayer, DHE closely mimics the function and distribution of native cholesterol in both model and cellular membranes. Replacement of up to 85% of the endogenous sterol of some cultured fibroblasts with DHE has no significant effect on growth properties, membrane composition, or activities of membrane enzymes. However, it is not clear whether DHE can be esterified and metabolized in mammalian cells as efficiently as endogenous cholesterol; this may affect the overall cholesterol homeostatic mechanisms and alter the steady-state distribution of DHE. Despite this potential complication, DHE has proven to be a powerful tool for studying sterol distribution and dynamics in living cells.

Detergent Sensitivity

The popularity of this test for L_o domain association is largely due to the simplicity of the technique. Cells are bathed in ice-cold detergent-containing buffer and then the membrane preparations are separated into soluble and insoluble fractions based on density differences: insoluble cytoskeletally associated membrane fractions sink to the bottom of a sucrose density gradient, insoluble L_o domain-associated membrane constituents float to the low-density portions of the same gradient, and soluble membrane components remain at intermediate densities. Fractions from the gradient are then biochemically analyzed (e.g., via Western blotting) for the protein of interest.

Although Triton X-100 has been the most widely used detergent for testing for L_o domain association of membrane components, other detergents including Brij 96, Lubrol WX, and CHAPS have been used as well. Membrane components can exhibit differential sensitivity to these various detergents, and this differential sensitivity has been used to identify subclasses of L_o domains. For example, in Madin-Darby canine kidney (MDCK) cells (a polarized epithelial cell line), the transmembrane protein prominin is soluble in Triton X-100, but insoluble in Lubrol WX, while in the same cells, the GPI-linked alkaline phosphatase, PLAP, is insoluble in Triton X-100. Lubrol-insoluble prominin is located largely on microvilli, while PLAP is segregated to planar regions of the plasma membrane. Similarly, in neuronal cells, the GPI-anchored proteins Thy-1 and prion protein segregate to different regions on the cell surface and associate with membrane domains that are differentially sensitive to extraction by Triton X-100 and Brij 96. Note that in both cases spatial segregation of proteins in vivo correlates with differential detergent sensitivity in vitro.

Fluorescence Microscopy

Regional differences in plasma membrane properties can be uncovered by several methods. For instance, membrane components that are resistant to cold detergent (e.g., 0.5% Triton X-100) can be visualized by labeling the membranes before extraction with fluorescently conjugated antibodies against L_o domain markers, such as the transmembrane receptor, CD44, or the ganglioside, GM1. In the case of transmembrane L_o domain markers, care must be taken in the interpretation of the extraction data because retention in the membrane may be a consequence of association with the cytoskeleton instead of, or in addition to, L_o domain association. Differential extraction of the transmembrane protein at high (37°C) and low (4°C) temperatures can help to sort out the contributions from cytoskeletal linkages and lipid domains, respectively.

To avoid contributions from the cytoskeleton, fluorescent lipid analogues can be used as markers of lipid domains. The most widely used pair of domain markers is DiI-C_{16} and *FAST*-DiI. These DiI derivatives have the same indocarbocyanine fluorophore linked to two hydrocarbon tails that insert in the membrane bilayer like the fatty acids of lipids. DiI-C_{16} contains two 16-carbon-long saturated alkyl chains, while *FAST*-DiI has two 18-carbon-long doubly unsaturated chains. The double *cis*-unsaturations in the acyl tails of *FAST*-DiI introduce significant kinks into the tails and effectively shorten them. Based on these features, DiI-C_{16} is thought to have a preference for more ordered domains (L_o), while *FAST*-DiI prefers more disordered domains (L_d). DiI-C_{16} and *FAST*-DiI both fluoresce in the red, so they cannot be used to label L_o and L_d domains in the same cell. Instead, the green fluorescing dye, NBD, linked to a short acyl chain, can be used to label disordered domains in cells labeled with either DiI-C_{16} or *FAST*-DiI. Each of these lipid analogues labels the plasma membranes of living cells relatively uniformly, with no readily apparent segregation of the probes. When fluorescent lipid-labeled cells are extracted with cold Triton X-100,

FAST-DiI and NBD-C$_6$-sphingomyelin are completely removed from the cell surface, as would be expected for lipids in the L$_d$ phase. In contrast, a significant fraction of DiI-C$_{16}$ is retained in the detergent-resistant sheets of membrane that remain, suggesting that it is a good marker for L$_o$ domains.

Another method for revealing domain organization on the cell surface is to examine the localization of domain components that have been cross-linked with antibodies or toxins. Extensive cross-linking of plasma membrane components tends to cause their redistribution into large patches on the cell surface. These patches of membrane proteins or lipids are large enough to be visualized by fluorescence microscopy, allowing for easy analysis of the patching behavior of different membrane elements. When one component of L$_o$ domains is cross-linked, a second independently aggregated L$_o$ domain component sometimes redistributes into patches that are coincident with those formed by the first cross-linked component. For example, in T-hybridoma cells, cross-linking of the ganglioside GM1 with choleratoxin B induces cell surface patches that colocalize with patches formed by antibody cross-linked Thy-1 (a GPI-anchored protein). On the other hand, cross-linking of a GPI-anchored protein in fibroblasts did not lead to detectable patching of other GPI-anchored proteins or DiI-C$_{16}$. The differences in these behaviors are not well understood. Patching of outer membrane leaflet L$_o$ domain constituents can also influence inner membrane leaflet elements; cross-linking of the GPI-anchored PLAP copatches fyn, a cytosolic tyrosine kinase associated with the plasma membrane via double acylations. In addition to the lipid-lipid interactions revealed by copatching of lipids or lipids and lipid-anchored proteins, lipid–protein interactions can be uncovered by copatching experiments involving transmembrane proteins. Under some conditions, cross-linked IgE receptor (FcεRI) copatches with DiI-C$_{16}$ and aggregated ganglioside, GD1b; FcεRI is a transmembrane protein that localizes to L$_o$ domains, as does the fluorescent lipid analogue DiI-C$_{16}$ and the ganglioside GD1b. Importantly, patches of L$_o$ domain components do not colocalize with patches of L$_d$ domain components; aggregation of FcεRI does not induce copatching of *FAST*-DiI, and cross-linked PLAP does not colocalize with cross-linked transferrin receptor. So two patched membrane components *may* coalesce if they prefer the same lipid domain, but they segregate if they prefer different lipid environments.

Biophysical Characterization

The detection and characterization of lipid domains in unperturbed cells have proven to be a challenging problem for scientists. As mentioned in previous sections, attempts to directly visualize and make measurement on lipid domains in native cell membranes have been futile, presumably because naturally occurring lipid domains are small and transient. Measuring the physical characteristics (e.g., size or phase state) of lipid domains requires methods with resolution limits smaller than that of optical microscopy. These methods are necessarily indirect and the results they yield can have multiple interpretations. Despite this caveat, there are several examples of biophysical measurements of cell membranes that have contributed to our understanding of plasma membrane organization. These examples are discussed below, and the specific techniques that have been utilized are summarized in Table 8.4.

Single-particle tracking (SPT) measurements demonstrated the existence of membrane domains, termed "transient confinement zones" (TCZs), within the plasma membrane. SPT is a powerful technique that allows de-

Table 8.4   Fluorescence techniques for studying membrane properties

Technique	Spatial resolution	Temporal resolution
SPT	<10 nm	1 ms–10 s
FRAP	0.2–5.0 μm	>10 s
FCS	0.2–5.0 μm	1 ms–1 s
FRET	1–10 nm	0.1–10 s
Fluorescence microscopy imaging	>300 nm	0.1–10 s
Electron microscopy	<1 nm	Fixed time point

tection of the movements of individual proteins on the surface of living cells. In this technique, the protein of interest is specifically tagged with a fluorophore or gold particle, and then the motions of the fluorescent probe or gold particle are monitored over time via microscopy. The positions of the centroids of the fluorescent probe or gold particle (and thus the protein) can be determined with a precision of several nanometers, and these positions can be plotted as a function of time to yield a picture of the protein's movements across the cell surface. Analyses of protein movements provide diffusion rates and paths within the lipid bilayer. The diffusion rate and path traveled by a protein depend on both its size and environment, and so these data convey information about the organization of the plasma membrane (Figures 8.7 and 8.8). SPT was used to determine the diffusional behavior of various gold particle-tagged lipids and proteins on the plasma membranes of living cells. In these studies, the movements of particles attached to ganglioside GM1 or GPI-linked Thy-1 were found to be restricted to 200- to 300-nm-wide areas of the membrane for 10 to 15% of the time; the regions in which the proteins were trapped were called TCZs. The mechanism of confinement in TCZs remains unresolved because there are a number of other reasons, besides lipid phase separation, why membrane constituents might diffuse slowly and become trapped (Figure 8.8).

TCZs can recapture a protein tens of seconds after that protein has escaped, which means that TCZs are fairly long-lived. If TCZs are relatively stable structures with diameters in the range of 200 to 300 nm, shouldn't they be observable with high-resolution optical microscopy techniques? Not necessarily. To image TCZs by microscopy would require a probe (lipid or protein) that is significantly enriched in TCZs compared with the surrounding membrane. Even if ganglioside GM1 or GPI-anchored Thy-1, for example, satisfied this criterion, TCZs might still be concealed because they are likely to be just one type of many coexisting domains on the cell surface (Figure 8.7). Unless there are specific markers for each type of lipid domain, direct visualization of any one type of microdomain will remain elusive.

Fluorescence recovery after photobleaching (FRAP) is another method for determining the diffusion characteristics of a protein in the plasma membrane (Figure 8.11). FRAP takes advantage of a property of fluorescent probes that is normally considered a detriment: susceptibility to photobleaching. Photobleaching refers to the loss of fluorescence caused by photochemical reactions when the molecule is in the excited state. In FRAP, a small region on the surface of a cell is illuminated with high-intensity light, causing the fluorescent molecules in this region to become irreversibly photobleached. Subsequently, low-intensity illumination and fluorescence microscopy are used to monitor the movement of the fluorescent molecules that were originally outside of the photobleach region and, thus, un-

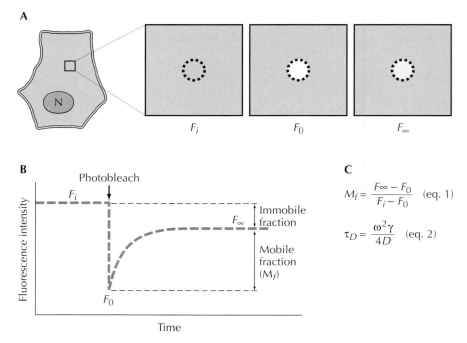

Figure 8.11 Principles of fluorescence recovery after photobleaching (FRAP). FRAP is used to study the diffusion properties of lipids and proteins in the plasma membranes of living cells. The diffusion rate of a protein depends on both the protein's size and environment, so physical characteristics of the plasma membrane can be inferred from the diffusion rates of its resident proteins. **(A)** In a FRAP experiment, first the plasma membrane is labeled with a fluorescent probe, and the initial fluorescence intensity (F_i) in a small region of the plasma membrane (dotted circle) is measured. Next, fluorescent probes within the small region of the membrane (dotted circle) are irreversibly photobleached (i.e., rendered nonfluorescent) by illumination with a strong laser; this drastically reduces the fluorescence intensity in that region (F_0). The fluorescence intensity within the photobleached region is then monitored over time. As unbleached fluorescent probes from the surrounding membrane diffuse into the photobleached region, the fluorescence intensity in that region recovers (F_∞). **(B)** These data can be graphed to display the change in fluorescence intensity within the measurement region over time, thereby providing the rate of fluorescence recovery. **(C)** Both the mobile fraction (M_f) and the diffusion constant (D) for the fluorescently labeled membrane constituent can be derived from the data by using equations 1 and 2, where τ_D is the diffusion time, ω is the radius of the focused laser beam, and γ is a correction factor.

bleached. With time, the unbleached fluorescent molecules will move into the photobleached region, causing the photobleached region to "recover" its fluorescence. FRAP measures both the rate and percentage of fluorescence recovery into the photobleached region, and these measurements, respectively, yield diffusion constant, D, and the mobile fraction, M_f, for the labeled molecules. These kinetic parameters provide similar information regarding plasma membrane organization as the measurements from SPT, but, importantly, FRAP measures the properties of whole populations of molecules, in contrast to SPT, which provides information on individual molecules.

The utility of FRAP for analyzing lateral heterogeneities in the plasma membrane is somewhat limited because the size of the photobleach region is typically several times larger than the smallest lipid domains. If proteins

BOX 8.2

What size is your domain?

The technique of fluorescence recovery after photobleaching (FRAP) involves monitoring the recovery of fluorescence into a photobleached region of the cell surface (see text and Figure 8.11 for more details). FRAP measures the average diffusion constant (D) for a population of fluorescently tagged molecules in the membrane and the fraction of molecules that are able to move in the lateral plane of the membrane (also known as the mobile fraction, M_f). The size of the membrane domains that are labeled with a particular probe can be estimated by varying the size of the region that is photobleached in the FRAP experiment. For example, the values of M_f will decrease with increasing size of the photobleached area if the lipid probe labels discrete domains in the membrane that cannot exchange lipids with neighboring membrane.

Furthermore, how the values of D vary with photobleach region size gives information about membrane domain size. If the photobleached regions are larger than the domains, then D will not vary with repeated measurement at different locations, whereas if the photobleached regions are smaller than the domains, then D will vary significantly because the small photobleach regions will sample different areas of the plasma membrane.

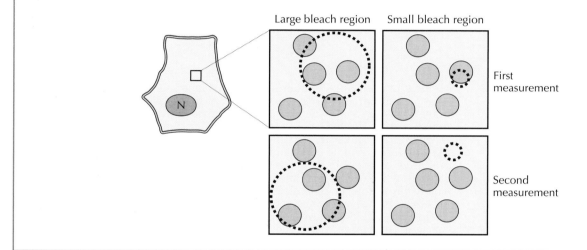

or lipids exhibit heterogeneous diffusion rates as they traverse dissimilar microdomains, these differences in diffusion will be averaged and will remain undetected when the measurement region is large compared with the microdomains. This limitation can be partially overcome by varying the size of the photobleach region, with the idea that the existence of membrane domains can be inferred from the dependence of M_f and D on the size of the diffusion area (Box 8.2).

A third technique for measuring diffusion of proteins in living cells is fluorescence correlation spectroscopy (FCS). In an FCS experiment, a small region of the plasma membrane is monitored for fluctuations in fluorescence intensity (Figure 8.12). The fluctuations in fluorescence intensity are caused by the diffusion of labeled molecules in and out of the measurement region. The fluctuations can be analyzed and fit to an autocorrelation function, which is used to extract statistical properties from the signal. The autocorrelation function of fluorescence intensity fluctuations provides two pieces of information about the labeled molecules: concentration and the diffusion constant. The molecule's in vivo concentration can be calculated from the amplitude of the autocorrelation function, for the amplitude is inversely proportional to the number of fluorescent molecules in the mea-

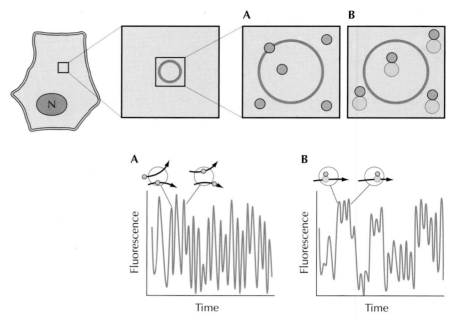

Figure 8.12 Principles of fluorescence correlation spectroscopy (FCS). FCS can be used to study protein-protein interactions in living cells. First, the plasma membrane is labeled sparsely with a fluorescent probe. It is important that the membrane is only sparsely labeled because FCS relies on detecting spontaneous fluctuations in the fluorescence intensity signal within a small detection area. Monomeric proteins **(A)** will diffuse rapidly in and out of the measurement region (gold circle) and this will produce spikes in the fluorescence intensity measurements (see graph A). In contrast, when one protein interacts with another **(B),** the complex diffuses more slowly through the measurement region and produces a more sustained rise in fluorescence intensity, causing the signal to fluctuate more slowly (see graph B). These fluctuations in fluorescence intensity over time are analyzed with an autocorrelation function. This type of analysis yields information about the concentration of the fluorescently labeled molecule and its diffusion rate. Although this figure depicts FCS measurements on membrane proteins, FCS can also be used to monitor interactions between cytosolic proteins. Also, dual-color FCS, in which two molecular species are labeled with distinct fluorophores, provides a sensitive means to detect molecular interactions. In brief, if the two molecules move independently, the fluctuations in their fluorescence will also be independent and will not correlate with each other. However, if some of the molecules move together in the same complex, then a fraction of their fluorescence fluctuations will correlate with each other.

surement area. The diffusion constant can be determined from the correlation time, which reflects the characteristic time for labeled molecules to move into and out of the measurement area.

The utility of FCS for studying membrane phase behavior has been established by studies on model membrane systems. FCS was used to characterize the phases of lipids in GUVs formed from various mixtures of phospholipids and cholesterol. The translational diffusion coefficients measured by FCS correlated systematically with the nature of the phase being probed in the GUVs, and, significantly, when there were coexisting phases, different diffusion coefficients could be detected. The use of FCS to investigate biological membranes is just beginning, and so there have been few studies to date. Future investigations with FCS should prove to be important for elucidating membrane organization.

A final technique for probing plasma membrane organization is fluorescence resonance energy transfer (FRET) (Figure 8.13). FRET requires two fluorophores with distinct but overlapping excitation and emission spectra; the emission spectrum of the donor fluorophore must overlap with the excitation spectrum of the acceptor fluorophore so that energy transfer can occur. To detect protein-protein interactions by FRET, one protein is labeled with the donor fluorophore and the other is labeled with the acceptor fluorophore. When these proteins are far apart, no energy transfer occurs be-

Figure 8.13   Principles of fluorescence resonance energy transfer (FRET). FRET is used to detect protein-protein interactions either at the membrane or in the cell cytosol. FRET requires two fluorescent probes with different but overlapping spectral properties. **(A)** The fluorescence emission of one probe, called the donor, must overlap with the absorbance of the other probe, called the acceptor. When this happens, excitation of the donor can lead to emission of the acceptor because energy is transferred from the donor to the acceptor. **(B)** The efficiency of energy transfer (E) depends on the distance (r) between the donor and acceptor pair and a characteristic radius (R_0) for that pair. For a donor-acceptor pair with an R_0 of ~5 nm, the efficiency of energy transfer falls to near zero by the time the distance between the pair reaches 10 nm. So when the donor and acceptor are far apart (>10 nm), excitation of the donor leads to fluorescence of the donor, but not the acceptor **(C)**. On the other hand, as the donor and acceptor move closer together (<10 nm), excitation of the donor causes increased emission of the acceptor and decreased emission of the donor **(C).** This means that FRET can be detected by measuring an increase in acceptor emission, a decrease in donor emission, or both.

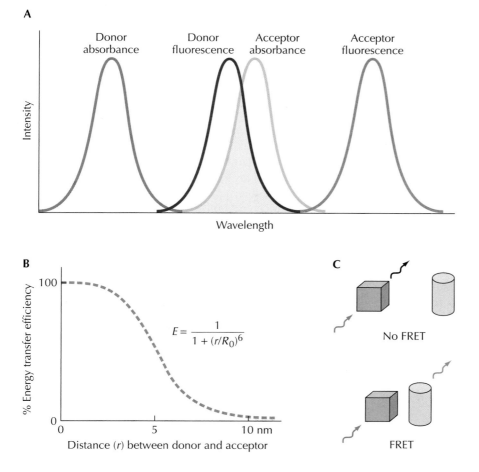

$$E = \frac{1}{1 + (r/R_0)^6}$$

cause energy transfer falls off steeply as the distance between the donor and acceptor pair increases. However, when these proteins move into close proximity (<10 nm) and bring their donor/acceptor pair of fluorophores near to each other, energy transfer occurs efficiently. This means that excitation of the donor fluorophore leads to an increase in the emission of the acceptor fluorophore and a concomitant decrease in emission of the donor fluorophore. By monitoring the sample for an increase in acceptor fluorescence intensity and/or a decrease in donor fluorescence intensity over time, the dynamic interactions of proteins can be detected in living cells. FRET measurements can thus be used to address several different biological questions, such as are molecules that appear colocalized on the level of fluorescence microscopy (~300-nm resolution) also colocalized on the molecular level (~10-nm resolution)? Do these molecular interactions change as a function of time or cellular stimulation? A variation of the former question is particularly relevant for studies on membrane domain organization. Are certain membrane components (e.g., GPI-anchored proteins or gangliosides) randomly distributed at the cell surface or are they clustered in microdomains? This question can be answered by determining the dependence of FRET on donor and acceptor surface density (Box 8.3).

Cholesterol Modulation: Acute and Metabolic Methods

Cholesterol maintains the integrity of lipid domains in the plasma membrane, and so modulation of the cholesterol content of the plasma membrane is a simple means of perturbing lipid domain organization. Cholesterol levels in the plasma membrane can be lowered acutely by treating cells with cholesterol-binding agents, such as filipin and methyl-β-cyclodextrin (MβCD). Filipin effectively lowers plasma membrane cholesterol levels by binding selectively to cholesterol and sequestering it into complexes in the plasma membrane (see Fluorescent Probes for Studying Membrane Organization and Dynamics for more information about filipin). Cholesterol sequestration by filipin induces structural disorder in the plasma membrane, which has a variety of consequences, including disruption of lipid domain organization, dispersal of caveolae, and, in some cases, cellular permeabilization. Because filipin can lead to cellular permeabilization, experimental results should be interpreted with caution.

MβCD sequesters cholesterol by a different mechanism than filipin. MβCD is a membrane-impermeable, water-soluble cyclic oligosaccharide with a hydrophobic cavity, which can bind hydrophobic compounds, such as cholesterol, thereby enhancing their solubility in aqueous solutions. MβCD binds cholesterol with a higher affinity than it binds other lipids, making MβCD a relatively specific acceptor of cholesterol. This property of MβCD makes it a useful tool for removing cholesterol from cellular membranes. When incubated with cells for short times, MβCD extracts cholesterol from the cells' plasma membranes. With longer incubation times, MβCD extracts cholesterol from both the plasma membrane and intracellular organelles that participate in trafficking itineraries through the plasma membrane. MβCD can also be used to raise the content of cholesterol in cell membranes. When MβCD is preloaded with cholesterol, it acts as a cholesterol donor and transfers its cholesterol to cell membranes.

Another way to alter membrane cholesterol levels is to inhibit cholesterol biosynthesis. Cholesterol biosynthesis begins with the formation of 3-hydroxy-3-methylglutaryl coenzyme A (HMG-CoA) from acetyl coenzyme A and acetoacetyl coenzyme A. HMG-CoA is then reduced to mevalonate,

BOX 8.3

Is your favorite protein clustered in plasma membrane microdomains?

FRET measurements can be used to determine whether proteins are clustered in domains or randomly distributed at the cell surface. First, the protein of interest must be labeled with donor and acceptor fluorophores, and then a series of experiments are required to determine whether FRET between the donor- and acceptor-

labeled proteins depends on the protein's density. For a protein that is randomly distributed on the cell surface, FRET will increase systematically as the surface density of the protein increases. However, if the protein is concentrated within microdomains on the cell surface, FRET will be independent of the protein's surface density. Changes in FRET can be either measured directly or monitored indirectly as changes in fluorescence anisotropy (i.e., a loss in polarization of the net emission) during energy transfer (see

Kenworthy et al. [2000] and Varma and Mayor [1998]). This type of approach has been used to study proteins, such as GPI-anchored proteins, that are predicted to reside in microdomains. Different studies using this methodology have reached different conclusions about the clustering of GPI-anchored proteins, possibly because of disparities in the experimental conditions (e.g., differences in cell types) or because only a small fraction of GPI-anchored proteins are constitutively clustered within microdomains.

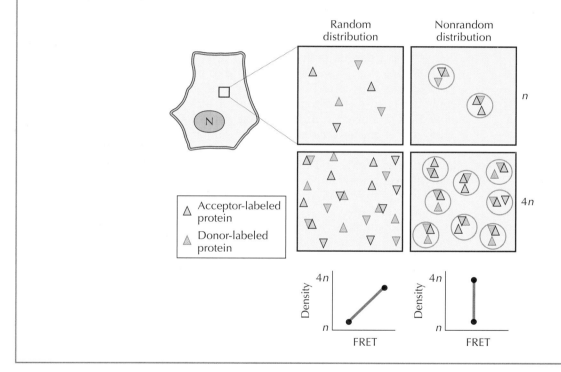

a precursor of isoprenoids and their derivatives, as well as cholesterol. The reduction of HMG-CoA to mevalonate is an irreversible reaction that is catalyzed by the enzyme, HMG-CoA reductase; this reaction represents the committed step in the cholesterol biosynthetic pathway. Consequently, cellular cholesterol content can be reduced by competitive inhibition of HMG-CoA reductase; a class of drugs known as HMG-CoA reductase inhibitors or statins can accomplish this very effectively. All statins contain an HMG-like moiety that may be in either an inactive (lactone) or active (hydroxy acid) form. Statins with inactive HMG-like moieties are enzymatically hydrolyzed to their active forms in vivo. Once active, the statins occupy the HMG-binding site of HMG-CoA reductase and part of the binding surface for CoA. Because statins inhibit the formation of mevalonate, isoprenoid as well as cholesterol synthesis is inhibited. Isoprenoids and their derivatives regulate a number of cellular processes, including cell proliferation, migration, and apoptosis. For this reason, statins have pleiotropic effects, and it would be an oversimplification to attribute the effects of statin treatment to disruption of domain organization alone. It is also an oversimplification to interpret the effects of membrane cholesterol modulation as an indication of dependence on L_o domains because cholesterol is a component of both L_o and L_d domains. Rather than equating cholesterol dependence with L_o domain dependence, it is more accurate to infer that membrane organization in general is important.

Lipid Organization on Intracellular Membranes

The Biosynthetic Pathway

The apical domain of polarized epithelial cells contains a high content of glycosphingolipids compared with basolateral membranes or plasma membranes of nonpolarized cells, and in many epithelial cells GPI-anchored proteins are found mainly within the apical domain. Similarly, in neurons, many GPI-anchored proteins are specifically targeted to the axonal membrane. Correct sorting of proteins and lipids is essential for generating and maintaining these compositionally and functionally distinct compartments in polarized cells. The mechanism of sorting of many transmembrane proteins has been well characterized and is thought to principally depend on specific sequence motifs within the proteins' cytoplasmic tails. GPI-anchored proteins and membrane lipids must rely on a different mechanism to sort to correct intracellular compartments and plasma membrane domains. Interaction of GPI-linked proteins with lipids can provide this alternative sorting mechanism, possibly involving the packaging of these proteins within lipid domains in the membranes of compartments along the biosynthetic route.

The biosynthetic route of secreted proteins, transmembrane proteins, and lipids that make up the plasma membranes of mammalian cells starts in the endoplasmic reticulum (ER), a complex network of membrane that extends throughout the cytoplasm and sequesters an internal volume (lumen) from the cytosol. Various types of protein modification, including glycosylations and GPI linkage, occur within the lumen of the ER, and nearly all membrane lipids (e.g., PC, PE, PS, and PI) are produced in the cytosolic leaflet of the ER. PC is then selectively transferred across the bilayer by a putative flippase to the inner leaflet, in this way establishing the asymmetry of lipids across the membrane bilayer. Cholesterol and ceramide are also produced in the ER membrane, but ceramide is not modified further into glycosphingolipids and sphingomyelin until it is transferred to the Golgi apparatus. The Golgi apparatus is completely segregated from the ER, so transport vesicles mediate traffic of proteins and lipids between these compartments. The Golgi apparatus consists of a stack of membrane-bound compartments, typically located near the nucleus. Each end of the stack has a different function and morphology. Lipids and proteins enter the *cis*-end of the Golgi complex in vesicles that have budded from the ER, undergo various modifications as they traverse the Golgi stacks (or cisternae), and then exit in vesicles that form from the tubular extensions that make up the *trans*-Golgi network (TGN). The vesicles that exit from the TGN have a variety of destinations, as indicated in Figure 8.14.

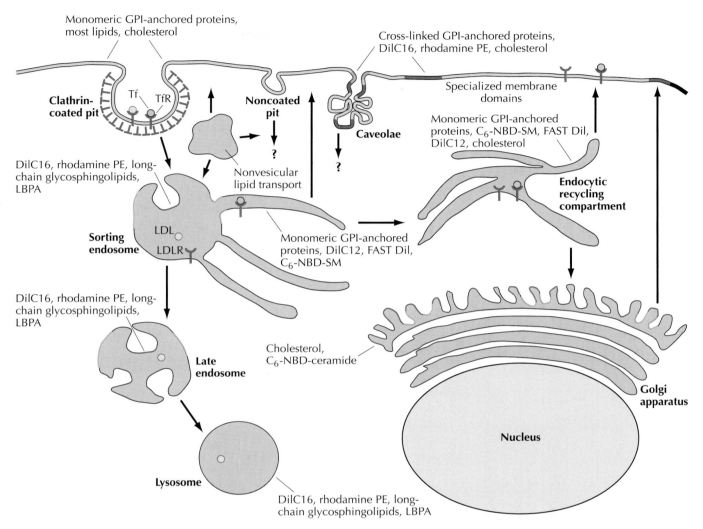

Figure 8.14 Intracellular trafficking pathways in a nonpolarized mammalian cell. Prior to internalization, many surface receptors, such as the receptors for transferrin (TfR) and LDL (LDLR), are recruited to clathrin-coated pits, while other membrane proteins may be recruited to separate specialized membrane structures, such as caveolae (see Figure 8.15). Once internalized via clathrin-coated pits, receptors with different destinations are separated from each other in sorting endosomes. Those molecules that are destined for lysosomal degradation are transported to late endosomes and lysosomes, while all other molecules are delivered either directly to the cell surface or first to the endocytic recycling compartment and then to the cell surface. This sorting process can be mediated either by specific signal sequences in the cytoplasmic tails of receptors or by the morphology of the sorting endosome (see Mukherjee et al. [1997] for review). Newly synthesized proteins and lipids are trafficked along the biosynthetic route. After synthesis in the ER (not depicted), molecules are delivered to the Golgi apparatus, where posttranslational modifications occur. When lipids and proteins reach the *trans*-Golgi network (TGN), they are sorted and packed into vesicles for further transport to various destinations, including the plasma membrane, regulated secretory granules (not shown), or endosomal/lysosomal compartments. Colored areas of membrane denote putative membrane domains, and the "?" symbols indicate uncharacterized steps.

The Endocytic Pathway

Once newly synthesized lipids and proteins reach the plasma membrane, they are not held static. The plasma membrane is a dynamic structure that is continually internalizing, recycling, and replenishing its components. Plasma membrane constituents are removed from the cell surface by incorporation into membrane invaginations that pinch off and form intracellular vesicles. Surface invaginations can have protein coats, such as clathrin or caveolin, that help to selectively recruit certain membrane components, or they can be uncoated, taking up bulk membrane and fluid-phase molecules (Figure 8.14). The itineraries of two prototypical transmembrane receptors, the low-density lipoprotein (LDL) receptor and the transferrin receptor (TfR), have been well characterized and are used to illustrate the canonical endocytic pathways in mammalian cells. The soluble ligands, LDL and Tf, bind to their receptors, LDLR and TfR, on the cell surface and then enter cells via clathrin-coated pits. Once internalized and pinched off from the plasma membrane, the vesicles rapidly become uncoated and fuse with early sorting endosomes. The slightly acidic (pH ~6) environment of the sorting endosome causes LDL to dissociate from its receptor, whereas Tf stays tightly bound to TfR throughout its intracellular passage. Sorting endosomes have a tubulovesicular morphology, which provides a geometric basis for sorting endosomal contents. Thin tubules radiate from the vesicular bodies of sorting endosomes so that the bulk of the volume is in the vesicular portion of the organelle, and most of the surface area is in the tubular portions. In this way, membrane components (e.g., LDLR) can be partially sorted from the fluid contents (e.g., LDL) of the organelle. Contents of sorting endosomes mainly have two possible fates: recycling back out to the cell surface or delivery to degradative compartments (i.e., late endosomes and lysosomes). Transport vesicles formed from the tubules of the sorting endosome are delivered either to a long-lived endocytic recycling compartment (ERC) or directly to the plasma membrane. The ERC, a highly tubular compartment often with a perinuclear localization, is the major site from which receptors (e.g., TfR and LDLR) are recycled back to the plasma membrane. On the other hand, proteins that are destined for degradation (e.g., LDL) remain in the vesicular bodies of the sorting endosomes, which transform into late endosomes. Delivery from late endosomes to lysosomes is less well understood, but there appears to be extensive trafficking among late endosomal compartments (including lysosomes). It is noteworthy that the endocytic pathway intersects the biosynthetic pathway at several points. See chapters 3 and 9 for an overview and detailed description of endocytic trafficking in eukaryotic cells.

Lipid Sorting and Membrane Traffic

Membranes along the biosynthetic and endocytic pathways have different lipid compositions, caused in part by differential localization of lipid biosynthetic and catabolic enzymes and in part by the preferential transport (possibly by incorporation into lipid domains) of specific lipids from their site of synthesis to other destinations. It appears that there are membrane inhomogeneities in the ER, but their nature must differ from L_o microdomains in the plasma membrane because the ER contains only low levels of cholesterol and no SM or glycosphingolipids. It has been suggested that ceramide-based microdomains might exist in ER membranes. The formation of L_o domains within Golgi membranes, especially the *trans*-Golgi cisternae, is supported by the observation that depletion of sphingolipids or cholesterol leads to missorting of GPI-anchored proteins to basolateral domains. Because the packaging of lipids and proteins into destination-specific vesicles occurs upon exit from the TGN, the missorting of GPI-anchored proteins after domain disruption suggests that GPI-anchored proteins are recruited to L_o microdomains within the TGN in mammalian cells. L_o microdomains and their associated GPI-anchored proteins are then packaged into vesicles specifically destined to the apical domain of polarized epithelial cells.

Each site of vesicle or tubule formation represents a potential site of lipid and protein sorting, and thus the formation of compositionally distinct membranes. Transport between endocytic compartments along the recycling and degradative pathways, as well as transport along the biosynthetic pathway (from the ER to the *cis*-end of the Golgi apparatus, between Golgi cisternae, and from the TGN to destination membranes), yields an opportunity to reorganize lipids and proteins. This sorting of membrane components can occur as a result of protein-protein, protein-lipid, or lipid-lipid interactions. For GPI-anchored proteins and lipids, lipid-lipid and lipid-protein interactions must necessarily predominate. Much of what is known about lipid sorting in cells is derived from studies utilizing fluorescent lipid analogues as tracers within the membrane. In every case where it has been examined carefully, there is evidence for lipid sorting in the formation of vesicles and tubules that bud from a donor organelle. For example, when a short-chain BODIPY-SM is incorporated into the plasma membrane at a low concentration, no excimers are detected in the plasma membrane, but excimers are detected in endosomes within seconds of formation, indicating that the BODIPY-SM became more concentrated as the vesicles formed. Lipid sorting during the formation of phagosomes was observed using two fluorescent lipid analogues with identical headgroups, DiI-C_{16} and *FAST*-DiI. DiI-C_{16} has long and saturated tails that prefer to reside in more ordered lipid domains, whereas *FAST*-DiI has unsaturated tails that preferentially partition into disordered regions. During the phagocytosis of beads, rat basophilic leukemia cells specifically excluded DiI-C_{16}, but not *FAST*-DiI, from the forming phagosomes. These same lipids have been used to demonstrate lipid sorting in endosomes; following endocytosis DiI-C_{16} is delivered efficiently to late endosomes, but *FAST*-DiI is delivered to the endocytic recycling compartment. From the endocytic recycling compartment of CHO fibroblasts, C_6-NBD-SM is returned to the plasma membrane with a $t_{1/2}$ of about 10 min, compared with a $t_{1/2}$ of 30 min for export of GPI-anchored proteins. When cellular cholesterol levels were reduced, both the C_6-NBD-SM and the GPI-anchored proteins returned to the plasma membrane with a $t_{1/2}$ of about 10 min.

Differences in partitioning into ordered versus disordered domains can partially explain the sorting of lipids in vesicle trafficking, but other factors must also be involved. Both GPI-anchored proteins and DiI-C_{16} have a preference for ordered domains, but DiI-C_{16} is sorted to late endosomes and GPI-anchored proteins go to the endocytic recycling compartment. Headgroup interactions are likely to play a role in lipid sorting, and these have been proposed to play a role in sorting of GPI-anchored proteins. Another factor that is likely to be important is the shape of lipids (e.g., cone versus inverted cone) since the process of vesicle budding imposes high curvature on membranes. An unusual lipid, lysobisphosphatidic acid (LBPA), is enriched in internal membranes of multivesicular bodies and late endosomes and may form another type of lipid domain within those membranes. The unusual shape of this lipid may play a key role in its enrichment in the highly curved internal membranes of these organelles.

In addition to trafficking among cellular organelles by incorporation into transport vesicles, lipids and cholesterol can also be transported by nonvesicular mechanisms, mainly in association with diffusible cytosolic proteins. Desorption of lipids with two long chains is energetically unfavorable, but cholesterol or lipids with a short acyl chain can desorb from membranes at an appreciable rate. Cholesterol and lipid carriers could consist of a large number of cytosolic proteins, each with low affinity and specificity for cholesterol. Alternatively, the carriers could be specialized cholesterol or lipid transport proteins. A family of high-affinity lipid and sterol carriers has been identified, of which one of the prototypes is the steroidogenic acute regulatory protein (StAR/StarD1), and the lipid- or sterol-binding domain in these proteins is called the StAR-related lipid transfer (START) domain. Nonvesicular transport of lipids is especially important for organelles such as mitochondria that do not engage in vesicular transport processes.

Membrane Domains, Toxin Trafficking, and Microbial Pathogenesis

There is growing support for the idea that L_o lipid domains in the plasma membranes of host cells are hijacked by disease-causing agents, including toxins, bacteria, and viruses. Before discussing specific examples, several caveats are warranted. Much of the evidence to date has been indirect, based largely on the effects of various cholesterol-modulating strategies on pathogen infectivity, and the resulting data have been interpreted in the context of simplified models of plasma membrane organization. This has led to the wholesale conclusion that cholesterol and/or L_o domains are required for the virulence of many types of infectious agents. Plasma membrane domains of host cells undoubtedly are involved in microbial pathogenesis, but the precise nature and role of these domains remain to be elucidated.

Figure 8.15 shows the various types of endocytosis that can be misappropriated by pathogens to enter cells, and it indicates which processes are susceptible to cholesterol depletion. Virtually every known mechanism of internalization is at least partially affected when plasma membrane cholesterol is altered. For example, cholesterol depletion inhibits, albeit incompletely, uptake mediated by clathrin-coated pits, possibly because of its effects on pit structure. Instead of forming invaginations, clathrin-coated pits become flattened on the underside of the plasma membranes of cells depleted of cholesterol. The processes of macropinocytosis and phagocytosis, which are clathrin independent, are also impeded when membrane cholesterol is lowered; the membrane ruffling that precedes macropinocytosis does not occur because actin reorganization is prevented, and a similar explanation may underlie the inhibition of phagocytosis. Finally, caveolar structure is abolished on removal of membrane cholesterol and presumably so is its function. So membrane cholesterol appears to be a general requirement for cellular internalization processes rather than a specific requirement for pathogen uptake. With this in mind, the next sections review some examples of pathogens that are thought to take advantage of host cell membrane organization to gain cellular entry or exit (Table 8.5).

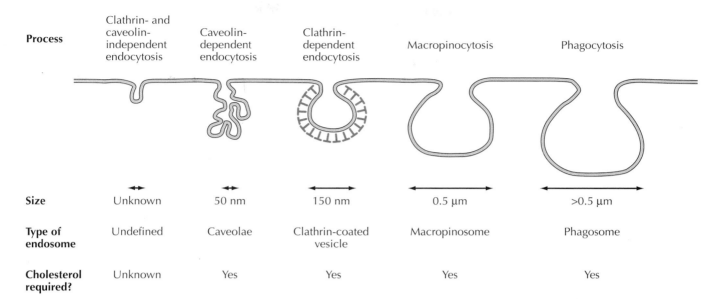

Process	Clathrin- and caveolin-independent endocytosis	Caveolin-dependent endocytosis	Clathrin-dependent endocytosis	Macropinocytosis	Phagocytosis
Size	Unknown	50 nm	150 nm	0.5 μm	>0.5 μm
Type of endosome	Undefined	Caveolae	Clathrin-coated vesicle	Macropinosome	Phagosome
Cholesterol required?	Unknown	Yes	Yes	Yes	Yes

Figure 8.15 Mechanisms of endocytosis. Endocytosis is the process by which cells sample their environment and turn over their membrane proteins and lipids. A variety of mechanisms have evolved to handle specific tasks, and each of these mechanisms has distinct molecular requirements. The best-studied endocytic pathway is clathrin-mediated endocytosis, which cells use to take up extracellular ligands that bind to specific cell surface receptors. Clathrin-coated surface invaginations ("pits") recruit cell surface receptors (and their bound ligands) and then pinch off to form vesicles. Uptake of larger particles (>0.5 μm), such as bacteria or dead host cells, is achieved through phagocytosis, an actin-dependent, clathrin-independent process. Other clathrin-independent processes include pinocytosis, caveolar uptake, and macropinocytosis. These processes are less well understood than clathrin-mediated endocytosis and phagocytosis. Macropinocytosis involves the formation of large actin-rich membrane ruffles that capture sizeable volumes of extracellular fluid as they fold back over the cell surface. The membrane ruffles that precede macropinosome formation are analogous to those that make up leading lamellae of migrating cells or that precede some forms of phagocytosis, and so it is likely that the molecular machinery for each of these processes is similar. Pinocytosis is a constitutive process by which cells take up small volumes of extracellular fluid, including soluble molecules in that fluid, and plasma membrane. Little is understood about pinocytosis and related processes. Finally, caveolar uptake utilizes membrane invaginations that contain the integral membrane protein, caveolin. Notably, the extent to which caveolae actually pinch off from the cell surface and become internalized is controversial. However, caveolar association of the protein dynamin, which is involved in vesicle budding, supports the idea that caveolae are internalized, although it may occur infrequently.

Toxin Uptake

Many bacterial toxins exert their action in the cytosols of host cells, but first they enter host cells by endocytosis and then they are translocated to the cytosol. An example of a well-studied bacterial toxin is cholera toxin (CT). CT is a pentavalent toxin that binds to surface-expressed ganglioside GM1 and causes its aggregation. When cholera toxin is bound to GM1, it can be visualized in caveolae in several cell types, and this has led to speculation that caveolae mediate its internalization. This may be the case for some cells, like endothelial cells, in which caveolae are abundant, but cells that do not contain caveolae can also internalize CT, indicating that alternate mechanisms for toxin entry exist. In some cells, complexes of CT and GM1 reside in non-

Table 8.5 Pathogens/toxins that exploit host cell membrane organization[a]

Pathogen/toxin	Host membrane component	Putative membrane domain-dependent step
Toxin		
Cholera toxin	GM1	Entry, intracellular transport
Shiga toxin	Gb3	Entry, intracellular transport
LPS	CD14 (GPI-linked)	Signal transduction
Tetanus toxin	GPI-anchored proteins	Entry, intracellular transport
Anthrax toxin	Anthrax toxin receptor	Entry
Aerolysin	GPI-anchored proteins	Pore formation
Streptolysin O	Cholesterol	Pore formation
Lysenin	Sphingomyelin	Pore formation
Vibrio cholerae cytolysin	Cholesterol, sphingomyelin	Pore formation
Bacteria		
Mycobacteria	Varied	Entry
FimH-*E. coli*	CD48	Entry
Shigella	CD44	Entry
Virus		
HIV	Varied	Entry/budding
Influenza	?	Budding
SV40	MHC class I	Entry
Marburg	Folate receptor (GPI-linked)	Entry/budding
Ebola	Folate receptor (GPI-linked)	Entry/budding
Semiliki Forest virus	Cholesterol, sphingolipids	Entry

[a] The list of pathogens and toxins in this table is not comprehensive.

caveolar, low-density, detergent-insoluble membranes, and disruption of membrane organization by addition of filipin inhibits toxin internalization in these cells. In contrast, even though CT is in detergent-insoluble membrane domains at the surface of hippocampal neurons, cholesterol-reducing treatments do not affect neuronal uptake of CT, but they do disrupt intracellular trafficking of the toxin. Many toxins are routed from endosomes through the Golgi apparatus to the ER, and from there they are released to the cytosol. Cholesterol depletion inhibits endosome to Golgi transport in neurons, suggesting that CT may utilize association with lateral membrane organizations to affect sorting along the correct intracellular pathway. Evidently, CT has multiple mechanisms for entering cells, including microdomain-independent means, but in the end membrane domain organization is likely to be important for CT to induce toxicity.

Anthrax toxin, produced by *Bacillus anthracis,* is a second example of a bacterial toxin that may require specialized organizations in host cell plasma membranes for virulence. Anthrax toxin is composed of three subunits; two subunits, edema factor and lethal factor, are responsible for the toxin's virulence, and the third subunit, PA83, is an 83-kDa protein that facilitates entry into cells. PA83 binds to a surface receptor on host cells known as anthrax toxin receptor (ATR). After binding to ATR, PA83 is proteolytically cleaved by furin-family proteases to a 63-kDa form of the subunit, termed PA63. PA63-ATR then oligomerizes and creates heptameric rings that are now able to recruit edema factor and lethal factor. ATR, carrying fully assembled anthrax toxin, is then internalized into endosomes, where the low pH environment permits channel formation and escape of edema factor and lethal factor to the cells' cytosol.

Lipid microdomains in the host cell's plasma membrane regulate the internalization step of anthrax intoxication. Heptameric PA63-ATR associates with low-density, detergent-resistant membrane domains and undergoes rapid clathrin-dependent internalization, whereas the monomeric PA83-ATR is not associated with low-density, detergent-resistant membrane domains and its internalization is slow. Evidence indicates that it is the oligomerization of ATR that promotes its association with lipid microdomains, and this in turn triggers a signal for efficient clathrin-dependent internalization.

A final example of a toxin that utilizes microdomains for cellular entry is VacA from *Helicobacter pylori*. VacA is synthesized as a 140-kDa protoxin, which is then processed to the 90-kDa mature form and released into the environment. VacA causes cytosolic degenerative vacuolation of many, but not all, cultured cell lines. VacA is internalized via an unidentified GPI-linked protein, and its vacuolating activity can be abolished if cells are first treated with the cholesterol-sequestering drug, nystatin. The dependence of VacA on surface-expressed GPI-linked proteins and plasma membrane cholesterol suggests that this toxin requires specialized lipid microdomains in the host cell plasma membrane to gain cellular entry. VacA utilizes an entry mechanism that is distinct from those used by either CT or anthrax toxin; VacA uptake is prevented when F-actin is depolymerized. Together, these examples illustrate that recruitment to detergent-resistant lipid domains per se does not trigger a particular endocytic process, but perhaps recruitment to specific subsets of these microdomains governs the precise entry mechanism that is utilized.

Bacterial Invasion

Attempts to define the role of lipid domains in bacterial invasion have been complicated by the requirement of host cell cytoskeletal rearrangements for bacterial entry mechanisms. An intimate connection between lipid domains and the underlying F-actin network is just beginning to be established, and until the subtleties of their interplay are more fully appreciated, it will be difficult to understand how lipid domains regulate bacterial infectivity. Three examples of bacteria that may require host cell membrane domains for infection are *Shigella*, FimH-expressing *Escherichia coli*, and *Mycobacterium* spp. In the case of *Shigella*, the invasin IpaB binds to CD44, which is a host cell receptor for hyaluronic acid. CD44 localizes to detergent-insoluble domains in host cell membranes and copurifies on sucrose gradients with IpaB from *Shigella*-infected cells. Cholesterol depletion of host cell membranes with MβCD inhibits binding and internalization of *Shigella*, but not of an *E. coli* strain expressing the *Yersinia* invasin. Similarly, *E. coli* that express the surface adhesin, FimH, bind to a membrane domain-associated protein, the GPI-linked CD48 on mast cells. Colocalization and copurification studies indicate that CD48-bound bacteria target to caveolin-containing lipid domains, and disruption of mast cell membrane organization with MβCD hinders entry of FimH-expressing *E. coli*, but not latex beads or FimH-negative *E. coli*. Finally, the uptake of *Mycobacterium* spp. by mouse macrophages and human neutrophils is blocked by MβCD treatment. Importantly, cholesterol modulation did not affect the phagocytosis of zymosan, serum-opsonized zymosan, or serum-opsonized *Mycobacterium* spp. by neutrophils, nor did it affect the internalization of a variety of other bacteria by macrophages. Although these data are suggestive of a role of membrane domains in bacterial entry, the dependence of pathogen uptake on distinct receptors or different phagocytic mechanisms may be the basis for the differential sensitivity to membrane cholesterol levels. The exact role of microdomains in bacterial invasion remains unclear.

Viral Entry and Budding

Simian virus 40 (SV40), a nonenveloped DNA virus, enters cells specifically by caveolar uptake. SV40 internalization can be completely blocked by expression of dominant-negative caveolin mutants, but it is unaffected by inhibition of clathrin-dependent endocytosis. Morphological, biochemical, and virological methods have been used to define the steps involved in caveolar uptake of SV40. SV40 first binds to class I major histocompatability complex (MHC-1) molecules, which are GPI linked and targeted to lipid domains. SV40–MHC-I complexes then diffuse laterally in the membrane until they are trapped by caveolae. Virus particles remain associated with caveolae for many minutes, until virus-laden caveolae eventually pinch off from the plasma membrane and move as caveolin-coated endocytic vesicles into the host cells' cytoplasm. Caveolae without associated SV40 remain at the plasma membranes of the host cells, suggesting that SV40 association initiates important internalization-triggering signaling events. Interestingly, MHC-I is not internalized along with SV40, implying that another receptor acts to bind SV40 tightly to the caveolar membrane and possibly initialize internalization signals. Regardless, there is compelling evidence that SV40 uses caveolae as entry portals for cellular infection.

There is also mounting evidence that enveloped viruses preferentially assemble and bud from specialized lipid domains at the plasma membranes of their host cells. Two examples of such viruses are influenza A virus and human immunodeficiency virus type 1 (HIV-1). Influenza A virus assembly involves the formation of an envelope membrane around the viral capsid. Analysis of the lipid composition of influenza virions indicates that the composition of the envelope membrane resembles the composition of host cell L_o microdomains; that is, envelope membranes are enriched in cholesterol and sphingolipids. Targeting to L_o microdomains is driven by hemagglutinin (HA) and neuraminidase (NA), the two major viral glycoproteins found in the envelope membrane. When influenza-infected cells are extracted with cold detergent, HA and NA cofractionate with detergent-resistant low-density membrane domains; while on whole cells, HA co-patches at the cell surface with cross-linked L_o domain components. HA and NA have intrinsic affinity for detergent-resistant microdomains, for they each exhibit detergent resistance when expressed in the absence of other viral proteins. These findings suggest that selective association of HA and NA with L_o lipid microdomains in host cell membranes mediates the assembly and budding of influenza virions.

As for influenza A virus, there is evidence that HIV-1 selects specific lipid domains for assembly and budding. HIV-1 primarily assembles and buds from the plasma membrane, but it can also assemble on intracellular membranes, possibly utilizing specialized domains on these internal membranes as well. Viral assembly begins on the cytoplasmic face of the plasma membrane, where the major core protein, Gag, is anchored. Oligomerization of Gag drives the assembly process, and Gag can form virus-like enveloped particles even in the absence of other viral components. The other major envelope component is the glycoprotein complex, Env. Env consists of two noncovalently associated subunits, gp120 and gp41. gp120 is an external protein that is noncovalently associated with gp41. gp41 is a transmembrane protein, whose cytoplasmic tail can interact directly with Gag.

The first indication that HIV-1 does not bud from random locations within host cell membranes came from analyses of the composition of HIV-1 envelope membranes. HIV-1 envelope membranes have a higher cholesterol/phospholipid ratio than that found in host cell plasma membranes. Because HIV-1 lacks the machinery to metabolize its own lipids, the lipids in the envelope membrane necessarily originate from the host cell, and the compositional differences between the envelope membrane and the plasma membrane indicate that a selection and/or sorting process is taking place during budding. The cholesterol-rich envelope membrane may derive from cholesterol-rich microdomains; however, HIV-1 envelope membranes have only low levels of SM compared with isolated L_o microdomains. So HIV-1 may select specialized SM-poor subsets of microdomains or it may sort SM out of the envelope membrane during budding.

In addition to the composition of the envelope membrane, the properties of its two major proteins, Gag and Env, provide further evidence that assembly and budding preferentially occur within L_o microdomains in host cell membranes. Both viral proteins have acyl modifications that are predicted to target to L_o microdomains; Gag is myristoylated and Env is doubly palmitoylated. Cold detergent extraction of HIV-infected cells shows that a fraction of Env is resistant to extraction, and this Env cofractionates with detergent-resistant, low-density membrane domains. In whole cells, Env colocalizes with a number of L_o domain components, including GPI-linked Thy-1, ganglioside GM1, and the lipid analogue $DiIC_{16}$, and several of these L_o domain components are incorporated into the envelope membranes of HIV-1 virions. In contrast, the L_d domain component CD45 and Env mutants, which cannot be acylated, do not integrate into HIV-1 virions.

Gag also has intrinsic affinity for lipid microdomains, and this affinity depends on Gag's myristylation and oligomerization. Interestingly, oligomerized Gag-containing membrane fractions float to a higher density on sucrose gradients than traditional low-density membrane domains; and, under certain conditions, Gag is relatively soluble in cold Triton X-100, but insoluble in cold Brij 98. So oligomeric Gag complexes may localize to a specialized subset of L_o domains, or their high protein density may affect both the density and the detergent sensitivity. A step-by-step mechanism for HIV-1 assembly has yet to be fully worked out, and the exact role that L_o microdomains play in virion formation is unknown, but it is clear that a high cholesterol content in the envelope membrane is critical for maintaining viral structure and infectivity. Selective budding from L_o microdomains provides a plausible mechanism by which HIV-1 can achieve the appropriate cholesterol/phospholipid ratio in its envelope membrane.

Conclusion

In conclusion, the term "membrane domain" is a generic one that can refer to any nonrandom organization of membrane components. Membrane domains can be lipid or protein based, large (>1 μm) or small (<100 nm), long-lived or transient, or detergent soluble or insoluble. Lipid microdomains are a subset of membrane domains that can themselves be subcategorized into varying types, such as liquid ordered versus disordered, or resistant to extraction by Triton X-100 versus Lubrol. Such heterogeneity in physical properties and composition is likely to be matched by heterogeneity in function. As additional techniques develop for studying microdomains in unperturbed cells, the functional significance of membrane domain diversity should become apparent. In the meantime, it is clear that the plasma membrane of mammalian cells is far from a sea of randomly distributed lipids that simply acts as a support for embedded proteins. Instead, membranes from intracellular organelles to transport vesicles to the plasma membrane have a high degree of two-dimensional organization that is likely to be important for regulating cellular functions. Interestingly, membrane domains probably exist in the membranes of nonmammalian cells as well. Yeast membranes contain glycosphingolipids and the sterol ergosterol, which are capable of forming liquid ordered domains, and detergent-resistant membranes have already been isolated from both yeast and parasite membranes. Hopefully, the study of lateral membrane organizations in nonmammalian cells will provide insights into the function and regulation of similar domains in mammalian cells, and vice versa.

Selected Readings

Fluid Mosaic Model

Frye, L. D., and M. Edidin. 1970. The rapid intermixing of cell surface antigens after formation of mouse-human heterokaryons. *J. Cell Sci.* **7:**319–335.

> This work provided important information about membrane structure by demonstrating that plasma membrane lipids can mix in two dimensions.

Singer, S. J., and G. L. Nicholson. 1972. The fluid mosaic model of the structure of cell membranes. *Science.* **175:**720–731.

> This seminal paper in cell biology proposed a model for cell membranes that is still the basis for current models.

Lateral Membrane Organization

Brown, D. A., and J. K. Rose. 1992. Sorting of GPI-anchored proteins to glycolipid-enriched membrane subdomains during transport to the apical cell surface. *Cell* **68:**533–544.

Brown, D. A., and E. London. 2000. Structure and function of sphingolipid- and cholesterol-rich membrane rafts. *J. Biol. Chem.* **275:**17221–17224.

Hao, M., S. Mukherjee, and F. R. Maxfield. 2001. Cholesterol depletion induces large scale domain segregation in living cell membranes. *Proc. Natl. Acad. Sci. USA* **98:**13072–13077.

> This characterization of the effects of cholesterol depletion on membrane organization in living cells provides crucial insights into lipid domain structure.

Holowka, D., and B. Baird. 2001. Fc(epsilon)RI as a paradigm for a lipid raft-dependent receptor in hematopoietic cells. *Semin. Immunol.* **13:**99–105.

Keenan, T. W., and D. J. Morré. 1970. Phospholipid class and fatty acid composition of Golgi apparatus isolated from rat liver and comparison with other cell fractions. *Biochemistry* **9:**19–25.

Maxfield, F. R. 2002. Plasma membrane microdomains. *Curr. Opin. Cell Biol.* **14:**483–487.

Rodgers, W., and J. K. Rose. 1996. Exclusion of CD45 inhibits activity of p56lck associated with glycolipid-enriched membrane domains. *J. Cell Biol.* **135:**1515–1523.

Seveau, S., R. J. Eddy, F. R. Maxfield, and L. M. Pierini. 2001. Cytoskeleton-dependent membrane domain segregation during neutrophil polarization. *Mol. Biol. Cell* **12:**3550–3562.

Sprong, H., P. van der Sluijs, and G. van Meer. 2001. How proteins move lipids and lipids move proteins. *Nat. Rev. Mol. Cell Biol.* **2**:504–513.

Methods for Studying Membrane Organization

Axelrod, D., D. E. Koppel, J. Schlessinger, E. Elson, and W. W. Webb. 1976. Mobility measurement by analysis of fluorescence photobleaching recovery kinetics. *Biophys. J.* **16**:1055–1069.

Dietrich, C., B. Yang, T. Fujiwara, A. Kusumi, and K. Jacobson. 2002. Relationship of lipid rafts to transient confinement zones detected by single particle tracking. *Biophys. J.* **82**:274–284.

Feigenson, G. W., and J. T. Buboltz. 2001. Ternary phase diagram of dipalmitoyl-PC/dilauroyl-PC/cholesterol: nanoscopic domain formation driven by cholesterol. *Biophys. J.* **80**:2775–2788.

> Using advanced imaging techniques, this paper examines the complex phase behavior of model three-component membranes.

Kenworthy, A. K., N. Petranova, and M. Edidin. 2000. High-resolution FRET microscopy of cholera toxin B-subunit and GPI-anchored proteins in cell plasma membranes. *Mol. Biol. Cell* **11**:1645–1655.

London, E., and D. A. Brown. 2000. Insolubility of lipids in Triton X-100: physical origin and relationship to sphingolipid/cholesterol membrane domains (rafts). *Biochim. Biophys. Acta* **1508**:182–195.

Mukherjee, S., X. Zha, I. Tabas, and F. R. Maxfield. 1998. Cholesterol distribution in living cells: fluorescence imaging using dehydroergosterol as a fluorescent cholesterol analog. *Biophys. J.* **75**:1915–1925.

Sheets, E. D., G. M. Lee, R. Simson, and K. Jacobson. 1997. Transient confinement of a glycosylphosphatidylinositol-anchored protein in the plasma membrane. *Biochemistry* **36**:12449–12458.

Varma, R., and S. Mayor. 1998. GPI-anchored proteins are organized in submicron domains at the cell surface. *Nature* **394**:798–801.

> FRET microscopy was used to investigate the cell surface distribution of GPI-anchored proteins. This paper shows that at least in some cases GPI-anchored proteins are clustered in domains at the plasma membrane.

Yechiel, E., and M. Edidin. 1987. Micrometer-scale domains in fibroblast plasma membranes. *J. Cell Biol.* **105**:755–760.

> An excellent example of the use of FRAP to study membrane organization.

Endocytosis

Chen, C. S., O. C. Martin, and R. E. Pagano. 1997. Changes in the spectral properties of a plasma membrane lipid analog. *Biophys. J.* **72**:37–50.

Mukherjee, S., R. N. Ghosh, and F. R. Maxfield. 1997. Endocytosis. *Physiol. Rev.* **77**:759–803.

> A comprehensive review of endocytosis in mammalian cells.

Mukherjee, S., T. T. Soe, and F. R. Maxfield. 1999. Endocytic sorting of lipid analogues differing solely in the chemistry of their hydrophobic tails. *J. Cell Biol.* **144**:1271–1284.

Pierini, L., D. Holowka, and B. Baird. 1996. Fc epsilon RI-mediated association of 6-micron beads with RBL-2H3 mast cells results in exclusion of signaling proteins from the forming phagosome and abrogation of normal downstream signaling. *J. Cell Biol.* **134**:1427–1439.

Sharma, P., S. Sabharanjak, and S. Mayor. 2002. Endocytosis of lipid rafts: an identity crisis. *Semin. Cell Dev. Biol.* **13**:205–214.

Membrane Domains and Pathogens

Abrami, L., S. Liu, P. Cosson, S. H. Leppla, and F. G. van der Goot. 2003. Anthrax toxin triggers endocytosis of its receptor via a lipid raft-mediated clathrin-dependent process. *J. Cell Biol.* **160**:321–328.

Duncan, M. J., J. S. Shin, and S. N. Abraham. 2002. Microbial entry through caveolae: variations on a theme. *Cell Microbiol.* **4:**783–791.

Pelkmans, L., and A. Helenius. 2002, Endocytosis via caveolae. *Traffic* **3:**311–320.

Rosenberger, C. M., J. H. Brumell, and B. B. Finlay. 2000. Microbial pathogenesis: lipid rafts as pathogen portals. *Curr. Biol.* **10:**R823–R825.

Sandvig, K., and B. van Deurs. 2002. Transport of protein toxins into cells: pathways used by ricin, cholera toxin and Shiga toxin. *FEBS Lett.* **529:**49–53.

9

Membrane Traffic in the Endocytic Pathway of Eukaryotic Cells

MICHELA FELBERBAUM-CORTI, RALUCA FLUKIGER-GAGESCU, AND JEAN GRUENBERG

Membrane Traffic in Eukaryotic Cells

Eukaryotic cells need to be in constant communication with their environment in order to perform most of their functions, such as the transmission of neuronal, metabolic, and proliferative signals and the uptake of nutrients or to protect the organism from microbial invasion, to name only a few. During a process called endocytosis, cell surface receptors and their ligands, as well as particles or solutes present in the extracellular space, can be taken up by vesicles that form at the plasma membrane, sorted to early endosomes, and then targeted to various intracellular destinations (Figure 9.1). Lysosomes are a common final destination for endocytosed macromolecules, where digestive enzymes degrade them. The resulting metabolites are then released into the cytoplasm where they can be recycled by incorporation into newly synthesized macromolecules.

Cellular homeostasis is maintained by balancing degradation and cell division with biosynthesis (Figure 9.1). Proteins destined for secretion as well as membrane proteins of the plasma membrane and vacuolar apparatus are de novo synthesized on ribosomes and cotranslationally translocated into the lumen of the endoplasmic reticulum (ER). Once correctly folded by lumenal chaperones, they are transported through the different stacks of the Golgi apparatus, where they acquire their mature sugar composition. Finally, upon arrival in the *trans*-Golgi network (TGN), they are sorted, packaged into specific vesicles, and forwarded to their final destination, endosomes or the plasma membrane. Fusion of TGN-derived vesicles with the plasma membrane occurs during a process called exocytosis. It results in either the insertion of transmembrane proteins into the plasma membrane or the secretion of soluble proteins into the extracellular space. Exocytosis can be either constitutive or regulated. All eukaryotic cells carry out constitutive exocytosis, but only a small subset of cells, specialized in the secretion of hormones, neurotransmitters, or digestive enzymes, display a regulated secretory pathway.

A consequence of this subcellular organization is that the lumen of each organelle of the vacuolar apparatus is topologically equivalent with the ex-

203

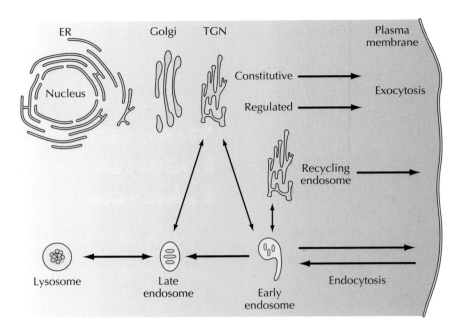

Figure 9.1 Intracellular compartments involved in endocytic and biosynthetic membrane trafficking. Newly synthesized molecules are transported from the ER through the Golgi apparatus to the plasma membrane. In the endocytic pathway, molecules are internalized at the plasma membrane and transported first to early endosomes, then to late endosomes, and finally to lysosomes. The endocytic and biosynthetic pathways are interconnected. Transport intermediates and most recycling routes are not depicted.

tracellular space. It is generally accepted that transport from one organelle to the next in the same pathway occurs by vesicular transport, which ensures that macromolecules are selectively transferred without ever being in contact with the cytoplasm. Patches of membrane-containing lipids and proteins that are to be transported first invaginate and then pinch off, yielding free transport vesicles. Upon reaching their specific destination, these vesicles dock onto and then fuse with the target membrane, releasing their content into it. Some endocytic and biosynthetic transport intermediates have been identified and characterized at the molecular level, using in vitro transport assays. Moreover, tubular intermediates are also involved in at least some transport steps. Recycling or retrieval pathways back to the donor organelle balance forward transport by replenishing donor membranes in lipids and proteins, including in components regulating transport itself. The existence of these recycling routes has become evident in experiments showing that some toxins internalized at the cell surface, like Shiga toxin, can be transported retrogradely to the ER via endosomes and the Golgi complex.

One unwanted consequence for communication with the extracellular space comes from the capacity that some parasites have evolved to mediate their own uptake by endocytosis. Some bacteria, such as *Listeria* or *Shigella*, induce their own internalization by the host cell and then escape from the vacuole into the cytoplasm where they multiply. Others, such as *Mycobacterium* or *Leishmania*, after being internalized, modify the dangerous environment of the vacuole to make themselves a decent home.

Organelles and Membrane Dynamics

Endocytic and biosynthetic organelles exhibit a wide variety of shapes and structures, which can be easily visualized by classical electron microscopy (Figure 9.2). They range from the network organization of the ER extending throughout the cytoplasm, to the pancake-like structure of the Golgi complex; from the clusters of thin, long tubules of recycling endosomes to late endosomes that contain onion-like sheets of internal membranes, tubules, or vesicles (multivesicular or multilamellar endosomes). At present, little is known about the molecular mechanisms controlling organelle shape and biogenesis, or the functional significance of such diversity. A snapshot image of the cell reveals that organelles of the biosynthetic pathway, such as the Golgi apparatus or the ER, appear as single copies, whereas endosomes are present in multiple copies. On the other hand, video microscopy observations in living cells, together with in vitro transport studies, have shown that both endocytic and biosynthetic organelles exhibit a high degree of plasticity at interphase or during mitosis, which is regulated, at least in part, by homotypic fusion and fission events. The equilibrium between fission and fusion reactions probably determines the steady-state number of apparent copies for each organelle. Thus, individual elements of both early and late endosomes may, in fact, form single functional compartments through homotypic membrane flow.

Beyond clear shape differences, the precise boundaries between different compartments are often blurred at the molecular level. In the biosynthetic pathway, the nature of the compartment between the ER and the Golgi complex—referred to as the intermediate compartment—remains ill defined and controversial. Equally under debate is the distinction between different endosome populations in the endocytic pathway. Two views have been proposed to account for transport in the endocytic pathway (Box 9.1). One model is that early endosomes mature into and become late endosomes, through the selective acquisition and retrieval of proteins by vesicular transport, whereas the other view is that early and late endosomes are stable compartments, which communicate through vesicular intermediates. Similarly, transport through the Golgi complex has been proposed to occur either via vesicular intermediates or via cisternal progression. This situation is largely due to the high dynamics of membrane flow, which results in the localization of certain molecules, in particular, cargo proteins, to more than one organelle at steady state, thus making it difficult to use the distribution of these proteins to define precise boundaries between compartments.

Ways of Entry into the Cell

Endocytic Pathways

The term "endocytosis" refers to the cellular uptake of solutes, lipids, and membrane proteins, including receptor-ligand complexes. It occurs through different molecular mechanisms, of which the clathrin-mediated endocytic pathway is the predominant one in most cell types. Clathrin and adapter proteins assemble from a cytoplasmic pool to form specialized regions on the plasma membrane called clathrin-coated pits (Figure 9.3), where receptors are concentrated by a direct interaction between endocytic signals in their cytoplasmic tails and adapter proteins. The most frequent internalization signal is tyrosine (Y) based, in the form FxNPxY or Yxxφ

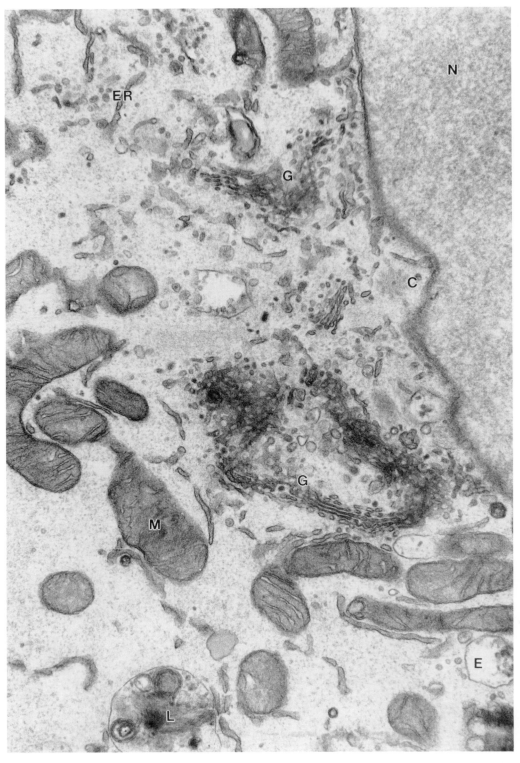

Figure 9.2   Intracellular compartments of the higher eukaryotic cell. This electron micrograph shows a perinuclear region of a Vero cell. C, centriole; N, nucleus; G, Golgi; E, possible early endosome; L, late endosome or lysosome; M, mitochondrion; ER, endoplasmic reticulum. Courtesy of Rob Parton.

BOX 9.1

Endocytic membrane transport: maturation or vesicular transport?

The past decade has witnessed an intense debate concerning two models for membrane trafficking in endocytosis. The discussion also applies to traffic through the Golgi apparatus, which is not addressed in this chapter. According to the vesicular transport model, internalized cargo molecules are packaged into transport vesicles that bud from preexisting early endosomes and fuse with preexisting late endosomes, thus transferring the contents from one organelle to the other. The maturation model, on the other hand, assumes that the cargo always remains in the same compartment, which is formed de novo by coa-

lescence of vesicles originating from the plasma membrane. The resulting early endosome then undergoes functional changes, as molecules that define its properties are progressively removed by recycling to the plasma membrane and molecules conferring late endosome function are added from the TGN.

It has not been possible to distinguish whether transport occurs as proposed by one or the other model by using the in vivo and in vitro transport assays developed so far. Thus, although endocytic organelles can be readily distinguished morphologically, it has not been possible to extrapolate to the mechanisms that give rise to the observed structures. This may be because the two models, although conceptually opposed, can both explain

the distribution and cycling of the known key regulators of membrane traffic (Rab proteins, SNAREs, etc.).

For example, the high selectivity of membrane transport implies that the tethering and docking of specific transport vesicles onto endosomes is tightly regulated. Evidence now shows that this selectivity is achieved by the coordinated action of different proteins, including SNAREs as well as Rab proteins and their effectors. However, the precise mechanism that confers endosomal identity, by ensuring that the correct set of these proteins is active at the right time and place, remains unclear. While it is clear that each model makes different predictions about how this selectivity may be achieved, it is not possible to discriminate between models at present.

(where ϕ is a large, hydrophobic residue and x can be any amino acid, although Y+2 is frequently an arginine). The Yxxϕ signal sequence has been shown to interact in an extended conformation with the $\mu2$ subunit of the AP2 adapter complex. Another sorting signal for endocytosis consists of two juxtaposed leucines (LL), which might interact with the $\beta2$ and/or $\mu2$ subunit of AP2. Phosphorylation of the serine-rich domain at the carboxyl terminus of many G-protein-coupled receptors (GPCRs) provides an additional sorting signal involved in the recruitment of β-arrestin adapters. Last, in both yeast and mammalian cells, the conjugation of ubiquitin to membrane proteins can trigger internalization.

It is becoming increasingly clear that there are also several clathrin-independent, alternative pathways of internalization into the cell. These pathways do not use known coat complexes for cargo recruitment and budding of transport intermediates. However, it is not always clear to what extent these entry pathways function constitutively or as a rescue mechanism induced by the disruption of the clathrin-dependent pathway.

Recent studies have revealed that these entry routes might exploit lateral heterogeneity in plasma membrane protein and lipid composition to select cargo. Lipid rafts, dynamic assemblies of cholesterol and glycosphingolipids, specifically recruit certain proteins and lipids for which they could act as endocytic platforms. The interleukin-2 receptor (IL-2R) provides the first example of endocytosis coupled to partitioning into lipid rafts for a transmembrane receptor. Indeed, despite the selective inhibition of clathrin-dependent endocytosis by a dominant negative mutant of Eps15, IL-2R was efficiently internalized in lymphocytes, even though these cells lacked caveolae. Similarly, some GPI-anchored proteins and certain lipids, such as lactosylceramide, globoside, and cholera toxin-clustered GM1, have been

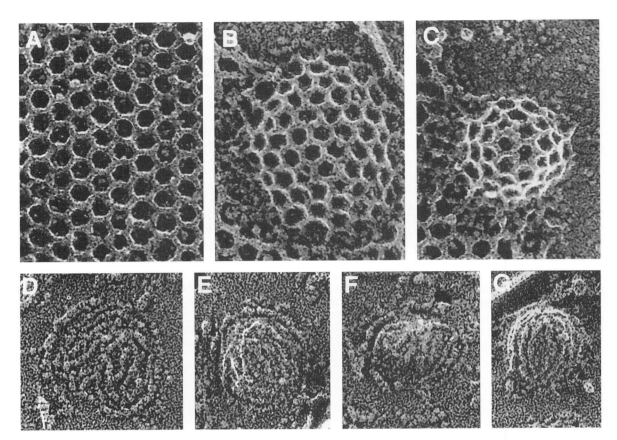

Figure 9.3   Two types of invaginations at the cytosolic side of the plasma membrane visualized by freeze-etch electron microscopy. **(A to C)** Different states of clathrin assembly, which presumably reflect different stages of vesicle formation from a flat lattice **(A)** to a deeply invaginated pit **(C)**. **(D to G)** Different types of caveolae with the typical spiraling appearance, perhaps corresponding to different stages of invagination. Courtesy of John Heuser.

shown to be internalized after association with lipid rafts. It is not clear at present to what extent rafts can be internalized through multiple pathways.

Caveolae are flask-shaped plasma membrane invaginations of 50 to 80 nm in diameter enriched in lipid rafts and the transmembrane protein caveolin (Figure 9.3). Caveolae were proposed to function as an alternative endocytic pathway, and in endothelial cells, where they are especially abundant, they have been implicated in transcytosis of molecules such as albumin. However, after disruption of the caveolin-1 gene, mice do not show a major endocytic transport defect, but trans-endothelial albumin transport in isolated cells is affected. In nonendothelial cells caveolar endocytosis might be an infrequent, possibly induced process. Recently, simian virus 40 (SV40) was shown to stimulate its own uptake by caveolae and to be delivered to the endoplasmic reticulum. It was generally believed that molecules internalized by clathrin-dependent or -independent mechanisms were all targeted to early endosomes. However, the finding that SV40 transits through caveosomes, preexisting, caveolin-1-containing organelles that do not contain endocytic markers, raises the question whether other entry routes bypass early endosomes. Despite accumulating data that caveolae can mediate endocytosis, it seems very unlikely that caveolae could account for constitutive clathrin-independent endocytosis. One might envision that caveolae represent specialized rafts stabilized by caveolin. In this context it is interesting to note that caveolin-1 was proposed to function as a negative regulator of caveolae formation.

Macropinosomes are large, noncoated vesicles of heterogeneous sizes (0.5 to 2 μm in diameter) that internalize large amounts of extracellular fluid by a clathrin-independent mechanism. They generally form by the closure of lamellipodia generated primarily at the leading edges of activated or motile cells and result in the enclosure of extracellular fluid. The function of ruffle-associated macropinocytosis is still a matter of debate. Macropinosomes might represent a side effect of membrane ruffling. However, the observed enrichment of lipid-raft markers in macropinosomes opens the possibility that selective internalization and recycling occur at the plasma membrane during cell activation and motility. Although macropinocytosis can occur constitutively in some tumor cell lines, this pathway is mainly activated by growth factors like epidermal growth factor (EGF) or mitogenic agents. Macropinocytosis contributes to antigen uptake in professional antigen-presenting cells. Moreover, bacteria such as *Legionella* or viruses such as HIV are localized in macropinosomes after their entry into macrophages. Macropinocytosis involves remodeling of the actin cytoskeleton, a process regulated by the small GTPases Ras, Rac, and Cdc42, and leads to massive membrane internalization. The GTPase ARF6 most likely regulates recycling of membranes back to the plasma membrane, because its activation leads to membrane ruffling and macropinosome accumulation.

A fourth type of noncoated vesicles has been observed. These vesicles have diameters of approximately 100 nm and have been proposed to carry fluid phase markers and toxins into the cell. Interestingly, cholera toxin can be internalized through such noncoated vesicles as well as through clathrin-coated pits. It is possible that the different routes of entry are not functionally equivalent; only nonclathrin pathways lead to delivery to the Golgi, a prerequisite for toxin activity. Indeed, treatments that affected the clathrin-dependent pathway had little effect on toxin-dependent inhibition of protein synthesis, whereas treatments that inhibited the clathrin-independent route had a strong effect.

Phagocytic Pathways

Phagocytosis constitutes the initial step for the degradation of particles larger than 0.5 μm. Professional phagocytes are specialized in the defense against pathogens and, in the case of macrophages and neutrophils, the clearance of old cells or cell debris. Phagocytosis also occurs to a lesser extent in nonprofessional phagocytes such as fibroblasts and endothelial and epithelial cells. Inert particles are captured by means of Fc receptors for Ig-opsonized particles, complement receptors for complement-opsonized particles, as well as mannose receptors, integrins, and scavenger receptors. Many pathogens can use these receptors to invade phagocytic host cells.

Several mechanisms have been described for parasite internalization. According to the zipper model, opsonized particles, which interact with surface receptors present on the phagocyte, are progressively surrounded by the plasma membrane as more and more of the opsonins distributed all around its surface bind to their receptors. To complete phagocytosis of the particle, its entire surface needs to interact with the surface of phagocytic cells, resulting in the formation of tight vacuoles, where the membrane originating from the phagocyte closely envelops the particle. Alternatively, some microorganisms such as *Salmonella* and *Shigella*, induce localized membrane ruffling and macropinocytosis, which leads to bacterial uptake. As a result, the phagosome membrane sits rather loosely around a phagocytosed particle. Other bacteria, such as *Legionella pneumophila*, enter cells by coiling phagocytosis, a phenomenon characterized by the formation of pseudopods that coil around the bacterium.

Phagocytosis requires important actin rearrangements and pseudopod extension under the control of Rho GTPases. Phagosome formation also often requires high amounts of membranes, in excess of the capacity of the plasma membrane. Recent studies highlight the fact that membranes of different organelles can be mobilized, at least in some cases, but the mechanisms involved remain poorly understood. Cell invasion by the parasite *Trypanosoma cruzi* revealed a previously unsuspected ability of conventional lysosomes to fuse with the plasma membrane. Moreover, the ER might represent the source of membrane necessary for pseudopod extension and completion of phagocytosis in macrophages.

Endosomes at the Crossroad of Membrane Traffic

Several pathways have been proposed to mediate membrane traffic between endosomes and the biosynthetic pathway. In particular, several lines of evidence support the view that direct transport routes mediate anterograde and retrograde transport between early endosomes and the TGN. Indeed, endocytosed Shiga toxin passes through early endosomes and the TGN on its way to the ER, perhaps in association with rafts.

Newly synthesized lysosomal hydrolases are sorted away from other secreted proteins at the TGN by the addition of the sugar mannose-6-phosphate onto their oligosaccharide side chains. This sugar is then recognized by the mannose-6-phosphate receptors (MPRs), which are sorted into clathrin-coated vesicles at the TGN. Next, MPRs and their bound hydrolases are transported to endosomes, but the molecular mechanisms controlling this transport are presently under debate. While the AP1 adapter complex was originally proposed to mediate this transport step, more recent studies suggest that this function is also controlled by the adapter protein GGA (Golgi localized, γ-ear containing ARF-binding). Whether AP1 is involved in anterograde TGN-to-endosome transport or retrograde traffic back to the TGN is not clear.

Once in endosomes, the receptors release the lysosomal hydrolases at the low endosomal pH. The retrograde MPR transport from endosomes back to the TGN seems to occur, at least in part, via a clathrin-independent pathway regulated by the small GTPase Rab9 and its effector TIP47, but it might also occur via an additional pathway. In yeast, vacuolar (lysosomal) enzymes do not carry the mannose-6-phosphate modification, but the corresponding pathway was shown to be controlled by the retromer complex. Homologues of these proteins exist in mammalian cells, but their functions remain to be elucidated. Lysosomal integral membrane proteins lamp1, lamp2, and CD63/lamp3 are transported from the TGN to late endosomes and lysosomes, via vesicles coated with the third adapter complex, AP3. The role of clathrin in this pathway is controversial.

Autophagy

Endosomes are also connected to the autophagic pathway. Originally described as a cellular response to cell starvation, autophagy is the main mechanism for degradation of long-lived proteins and the only mechanism for the turnover of organelles including mitochondria and peroxisomes. Upon induction, organelles and cytosol are sequestered into double-membrane vesicles with a diameter of 300 to 900 nm. The endoplasmic reticulum has long been thought to represent the origin of the sequestering membrane. Another candidate for the donor membrane is the so-called phagophore. Fusion of the outer membrane with lysosomes causes the release of the single membrane-bound inner vesicle of the autophagosome, the autophagic body, into the vacuole lumen. Because of the degradative nature of lysosomes, the internal membrane of autophagosomes is lost and they themselves become degradative. At which level the meeting of the two pathways occurs is still under debate, but the acquisition of lysosomal enzymes is believed to occur by direct fusion of either late endosomes or lysosomes with immature autophagocytic vacuoles.

PI3K activity and, hence, PI(3)P are required for autophagy in yeast, where Vps34 has been shown to coimmunoprecipitate with proteins required for autophagy. In mammalian cells it has been demonstrated that PI3K inhibitors impair autophagy and that the exogenous addition of PI(3)P increases the rate of autophagy. Recent studies in yeast have identified key components involved in the regulation of autophagy. Protein homologues exist in animal cells, and the characterization of these proteins both in yeast and in mammals reveals the existence of novel molecular mechanisms regulating protein sorting and membrane traffic along the autophagy pathway, including two ubiquitin-like systems.

Phagocytosis

Finally, complex types of interactions seem to occur between endosomes and phagosomes. Indeed, newly formed phagosomes are immature organelles that are unable to kill and degrade microorganisms. To acquire their microbicidal activity, phagosomes engage in a maturation process referred to as phagolysosome biogenesis. During this process, phagosomes sequentially fuse with early endosomes, late endosomes, and finally lysosomes. Phagosomes progressively lose plasma membrane markers and acquire first endosomal and later lysosomal markers, indicating that membrane is continuously added and removed. The pH of phagosomes decreases in parallel with the acquisition of lysosomal hydrolases, and the resulting phagolysosomes exhibit all the lysosomal characteristics required for degradation of the ingested particles or microorganisms.

The "kiss-and-run" hypothesis proposes that multiple fusion and fission events occur between endocytic organelles and phagosomes during phagolysosome biogenesis. Phagosomes and endosomes move along microtubules and interact at focal points. The outer leaflet of the two organelles can fuse and induce the formation of a fusion pore, which would permit size-limited exchange of contents without complete intermixing. Closure of the pore could be caused by changes in lipid composition. Because Rab proteins have been described as molecular switches, they were put forward as possible regulators of this kiss-and-run interaction. Recent studies support this role, in particular, for Rab5. Using a proteomic approach, over 100 proteins present on phagosomes have been identified. This database served as the starting point in the discovery that endoplasmic reticulum might represent one of the possible sources of membrane necessary for phagocytosis in macrophages, and will clearly provide new insight into phagosome functions and biogenesis.

Endocytic Pathway

Sorting in Early Endosomes

Cell surface proteins and lipids, as well as solutes, are first delivered to early endosomes (EE) (Figure 9.4). These organelles are located at the cell periphery and consist of cisternal regions from which thin tubules (60 nm in diameter) and large multilamellar vesicles (300 to 400 nm in diameter) emanate. Early endosomes are important sorting stations along the endocytic pathway. After receptor-ligand uncoupling at the mildly acidic pH (pH 6.2), housekeeping receptors are transported along the recycling route, whereas ligands follow the degradation pathway together with downregulated receptors and fluid phase markers. How sorting occurs is still poorly understood. No recycling motif has been identified, leading to the proposal that recycling to the cell surface occurs by default. But this view is difficult to reconcile with the situation in epithelial cells, in which transcytosed and recycled receptors transit through a common endosomal compartment before being transported to opposite plasma membrane domains.

Figure 9.4 Electron micrograph of an early endosome labeled with LDL-gold. LDL-gold was internalized into endosomes in vivo before the cells were fractionated. The endosomal fraction was mounted onto mica plates and processed for freeze-etch electron microscopy. The image shows the typical organization of an early endosome consisting of tubular and vesicular elements connected to a cisternal region. The large vesicular element containing LDL-gold (top right) may correspond to a forming endosomal carrier vesicle. Courtesy of John Heuser.

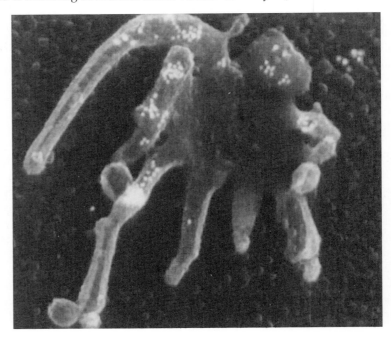

Ligands and fluid phase markers might be incorporated into forming vesicles due to a high volume-to-surface ratio. Several lysosomal targeting signals have been identified, but they bear little resemblance to each other, and a consensus has not emerged until now. However, recent studies in yeast and mammalian cells highlight the role of transient receptor monoubiquitylation in sorting to lysosomes via a complex protein machinery involving Hrs (hepatocyte growth factor-regulated tyrosine kinase substrate) and the ESCRTs-I, -II, and -III (endosomal sorting complexes required for transport) complexes (see Degradation Pathway, below). In addition, evidence is accumulating that lipid rafts might take part in early endosomal sorting. This diversity of signals might represent the need for a tightly controlled entry into the degradation pathway.

Recycling Pathway

Recycling receptors, such as the transferrin receptor, rapidly leave the early endosome and are transported to a dynamic tubular network that corresponds to recycling endosomes (RE). These do not contain ligands or receptors that are destined to be degraded. Their pH is approximately 6.4, slightly higher than that of early endosomes, and a microtubule-dependent pericentriolar accumulation of these organelles is found in some, but not all, cell types. Recycling can occur by a fast and a slow route. These could correspond to at least two separate transport steps to the plasma membrane, each with distinct molecular machinery (for example, one for early and one for recycling endosomes). Alternatively, the fast and slow routes could reflect the existence of a gradient of molecules on their way to the plasma membrane within early and recycling endosomes.

Although the boundaries between sorting and recycling endosomes are still ill defined, progress has been made in understanding the molecular mechanisms regulating protein traffic along the recycling pathway. Both Rab4 and Rab11 are involved in recycling, even if their precise functions are not clear. Two SNARE proteins, cellubrevin and syntaxin13, are localized to and function along the recycling pathway, whereas endobrevin/vamp8 is involved in the apical recycling pathway of polarized cells. Recently RME-1, a new member of the conserved family of Eps15-homology domain (EH) proteins, was found to be associated with recycling endosomes and might be involved in the exit of membrane proteins from this compartment. Transport along the recycling pathway also depends on the actin cytoskeleton and on unconventional myosin motors, which could have a mechanical role in tubule biogenesis and dynamics. Cross talk between transport, actin remodeling, and Rac-mediated signaling may depend on the small GTPase ARF6 and its partners.

Degradation Pathway

Transport toward late endosomes occurs via intermediates (endosomal carrier vesicles, ECVs), but so far it has not been possible to determine whether these vesicles change in composition during a maturation process, or whether they mediate transport between two stable compartments. These vesicular elements are relatively large and acidic (0.3 to 0.5 µm in diameter), and contain internal tubules and vesicles, thus resembling multivesicular bodies (MVBs). For these reasons they will be referred to here as ECV/MVBs. Once formed, ECV/MVBs move toward late endosomes in a microtubule- and motor-dependent manner and eventually can dock onto and fuse with late endosomes. Late endocytic transport steps, including presumably ECV/MVB-late endosome and late endosome-lysosome interactions, depend on the small GTPase Rab7 and selective subsets of SNARE proteins.

ECV/MVBs, like late endosomes, contain at least two membrane domains: internal invaginations and a limiting membrane (Figure 9.5). Whether the two membranes are physically connected, at least to some extent, or whether they interact dynamically by internal fission and fusion events is still unclear. The invaginations accumulate down-regulated receptors (for example, EGFR), reflecting a sorting mechanism for the degradation pathway. Recent evidence shows that this sorting step depends on receptor ubiquitination. Studies in yeast and mammalian cells lead to the view that the ubiquitinylated receptor is recognized by Hrs, which is itself membrane associated in part through interactions with phosphatidylinositol 3-monophosphate [PI(3)P], and which can sort the receptor into a newly identified flat clathrin coat. Then the receptor is handed over sequentially to ESCRT-I, -II, and -III and eventually translocated into endosome invaginations, where the receptor eventually accumulates. Strikingly, the membrane invagination process itself is apparently regulated by the same mechanism, involving PI(3)P signaling, Hrs, and ESCRT complexes, while the formation of ECV/MVBs seems to require annexin 2 in mammalian cells. However, this sorting mechanism does not seem to be used by all proteins destined for degradation, and the mechanism controlling the sorting of proteins present on the limiting membrane is not known, suggesting that other sorting principles may also operate at the same transport step.

The late endosome is a dynamic compartment with a complex organization that consists of cisternal, tubular, and vesicular regions with numerous membrane invaginations (Figure 9.5). Evidence now shows that limiting membranes and internal membranes (invaginations) exhibit a different protein and lipid composition. While some proteins are restricted to the limiting membrane, in particular, Lamp1 and Lamp2, others are abundant in internal membranes, including some proteins that are not destined for degradation (see below). In addition, internal membranes accumulate large amounts (~15% of total phospholipids) of lysobisphosphatidic acid (LBPA), which is a poor substrate for phospholipase and therefore resistant to degradation, and is involved in protein and lipid trafficking through this compartment. LBPA is presumably synthesized in situ and its cone-shaped structure could facilitate the formation of the invaginations.

Late endosomes play an essential role as the last sorting station before lysosomes. As discussed above, some receptors are sorted to the internal membranes for degradation. However, some proteins that are not destined for degradation also reside within internal membranes when present in late endosomes. These include, for example, members of the tetraspanin family, including CD63/lamp3 and the MHC class II complex, which is found in the late endosome-like MHC class II compartment in antigen-presenting cells (MIIC), as well as MPR in transit through late endosomes. Selective recycling out of late endosomes has been illustrated, including MPR, CD63/lamp3 in endothelial cells, and MHC class II in dendritic cells. Evidence is also accumulating that late endosomes may be connected by membrane traffic to the cell surface, at least in some cell types, perhaps accounting for the efficient MHC class II transport to the surface of maturing dendritic cells (see also Specialized Routes of Endocytic Membrane Traffic, below).

The lysosome is a hydrolase-packed organelle where degradation occurs. It is generally believed to be the end station of the endocytic, phago-

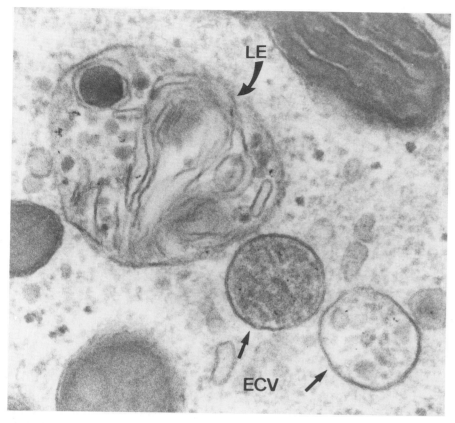

Figure 9.5 Electron micrograph of a late endosome (LE) and two endosomal carrier vesicles (ECV). Internal membranes, characteristic of organelles in the degradative pathway, are visible in both types of structures. Courtesy of Rob Parton.

cytic, and autophagocytic pathway. Lysosomes exhibit a distinct vesicular and electron-dense appearance and are generally devoid of MPR, Rab7, and Rab9 but contain high amounts of mature, dephosphorylated lysosomal enzymes. As for late endosomes, lysosomal glycoproteins are the major constituents of the lysosomal membrane and the pH is similarly acidic (5.5). In fact, no proteins or lipids have been found that would be present only in lysosomes but not in endosomes. Even the basic docking/fusion machinery seems to be shared by both compartments. Late endosomes and lysosomes can interact to form hybrid intermediates in a SNARE-, NSF-, and Ca^{2+}-dependent fashion. Lysosomes are re-formed from the hybrid organelle by a process that requires condensation of the lumenal content in the presence of ATP and lumenal Ca^{2+}.

Sorting in Polarized Cells

Polarized epithelial cells have two plasma membrane domains with a different protein and lipid composition, separated from each other by tight junctions. The apical plasma membrane faces internal cavities, whereas the basolateral surface faces underlying cells and connective tissue.

Polarity is established during biosynthesis by targeting newly synthesized molecules from the TGN to the correct plasma membrane domain, and it is maintained by sorting endocytosed molecules in early endosomes. Upon internalization from the basolateral or apical plasma membrane, macromolecules enter distinct sets of early endosomes: basolateral early endosomes or apical early endosomes. From there, endocytosed proteins destined to be degraded are delivered to a shared population of late endosomes and lysosomes. Alternatively, proteins can be recycled back or be transcytosed to the opposite surface. For this purpose, basolaterally internalized proteins are sorted to the tubular common endosomes. Subsequently, recycling proteins like transferrin and transferrin receptor are transported directly to the basolateral PM, whereas transcytosing molecules, like IgA and pIgR, are believed to be delivered to apical recycling endosomes (ARE), from where transport to the apical plasma membrane occurs. Molecules endocytosed from the apical surface probably recycle to the apical surface or transcytose to the opposite surface through the common endosome and/or the apical recycling endosome.

Basolateral sorting signals have been described and frequently appear to be similar to those involved in clathrin-mediated endocytosis. However, basolateral sorting requires the AP1 clathrin adapter containing the epithelial isoform μ1B. Clathrin/AP1 buds containing Tf/TfR are seen on the common endosome, suggesting that μ1B provides the specificity for basolateral sorting. Several mechanisms have been proposed for apical targeting, including peptide signals and lipid-based sorting (rafts hypothesis). It should be noted that some GPI-anchored proteins are sorted to the basolateral plasma membrane, even if they are lipid raft associated. These observations raise the possibility that there are different types of lipid rafts, which are sorted differently by the cell. Alternatively, lipid raft association is not sufficient to mediate apical sorting. Rather, N- and O-glycans act as the apical sorting determinant of GPI-anchored proteins as they do on transmembrane and secretory proteins.

The complex pattern of membrane traffic is subjected to numerous levels of regulation. For instance, the small GTPase rho regulates both apical and basolateral endocytosis, as well as exit from the basolateral early endosome. Rac also regulates apical and basolateral endocytosis, but it also affects the common endosome and/or the apical recycling endosome. Finally, cdc42 has been shown to control movement from both the TGN and endosomes to the basolateral plasma membrane. Recently it was found that the small GTPase ARF6 regulates both clathrin-mediated endocytosis and rearrangements of the actin cytoskeleton exclusively at the apical plasma membrane.

Specialized Routes of Endocytic Membrane Traffic

In certain cell types, endocytic membrane traffic carries out highly specialized functions. One of the best-studied examples is the recycling of synaptic vesicle components in neurons, and the molecular characterization of this process is progressing rapidly.

In contrast to professional secretory cells, which have specialized secretory granules, some cells have modified their lysosomes into a dual-function secretory-degradative compartment termed secretory lysosome. As for lysosomes, secretory lysosomes are acidic and contain degradative hydrolases. However, they are able to fuse with the plasma membrane following a defined cell stimulus. Examples of such lysosomes involved in the immune response are the lytic granules of cytotoxic T lymphocytes and NK cells. Melanosomes, which are involved in the secretion of melanin, can also be viewed as secretory lysosomes. MHC II compartments (MIIC) are involved in the antigen presentation pathway and have many of the characteristics of late endosomes. To trigger an immune response, exogenous protein antigens are endocytosed and processed into peptide fragments. These peptides are loaded onto newly synthesized MHC II in MIIC. In addition, antigen-presenting cells seem to exhibit the capacity to secrete vesicles ("exosomes") presumably derived from the internal membranes of multivesicular MIIC, a process that may contribute to modulating the immune response.

Molecular Mechanisms

In this section, we will briefly review the molecular mechanisms and proteins that regulate the transport events that we have described above.

Coat Proteins and Their Sorting Signals

Most of the budding events characterized so far involve coat proteins, which may be responsible for cargo selection and/or mechanical deformation of the membrane. Three coat complexes have been identified so far: coat protein I (COPI), which functions in the early secretory pathway and perhaps in endosomal transport; COPII, which is responsible for ER-to-Golgi transport; and clathrin, which functions in endocytosis and in transport between TGN and endosomes. Clathrin coats are highly ordered polygonal arrays as observed by electron microscopy (Figure 9.3). The main scaffold component is the 180-kDa clathrin heavy chain (CHC), which is associated with the 30-kDa clathrin light chain (CLC). Three CHCs and three CLCs form three-legged trimers, called triskelions, which oligomerize both in vitro and in vivo into hexagonal/pentagonal cages. It has been proposed that the formation of clathrin cages mechanically bends the membrane to invaginate the budding vesicles. How, and whether, the curvature of the clathrin lattice changes to form a cage out of a flat lattice is still poorly understood.

Clathrin-coated vesicle (CCV) formation in vivo requires adapters. Four types of adapter complexes (APs) have been reported. Each AP consists of two 100-kDa subunits (β and γ in AP1, α in AP2, δ in AP3, ϵ in AP4), one 50-kDa medium chain (μ), and one 20-kDa small chain (σ). Different APs are associated with specific transport pathways and confer different sorting properties. Thus, AP1 is an endosome/TGN adapter complex and AP2 is recruited to the plasma membrane. AP3 is likely to be involved in the transport of lamps glycoproteins between TGN and endosomes/lysosomes. Whether AP3 is associated with clathrin is still a matter of debate. Finally, a fourth, ubiquitously expressed adapter complex, AP4, has been discovered associated with TGN and might be part of a nonclathrin coat involved in TGN-to-endosomes transport.

GGAs (Golgi-localized, γ-ear containing, ARF-binding proteins) probably function as ARF-dependent, monomeric clathrin adapters to facilitate sorting and transport of receptors that traffic between the TGN and lysosomes, including both the cation-independent and -dependent MPRs. A new family of monomeric adapters, the stoned B family, was shown to act at the plasma membrane. Human stonin 2, a member of this family, might promote vesicle uncoating by dissociating AP2 from the plasma membrane.

Another group of proteins that function as adapters to link G-protein-coupled receptors (GPCRs) to the clathrin scaffold are nonvisual arrestins (β-arrestin 1 and β-arrestin 2).

Major advances have been performed in the characterization of molecules involved in the regulation of clathrin-mediated endocytosis, and their study has now been moving into the structural era. The assembly of CCVs is aided by an array of mostly cytosolic proteins, referred to as accessory proteins. They form a dynamic network of protein-protein and protein-phospholipid interactions by virtue of possessing multiple recruitment domains. These interactions are spatially and temporally regulated, thus coordinating the different stages of endocytosis and coupling it to signaling events. Progress has been made in understanding the nucleation, fission, and uncoating process of plasma membrane-derived CCVs in neurons. The recruitment of AP2 to the plasma membrane is probably a cooperative process involving simultaneous binding of AP2 to synaptotagmin, phosphatidylinositol-4,5-bisphosphate [$PI(4,5)P_2$], and cargo protein. Synaptotagmin, a transmembrane protein, probably acts as a docking site for AP2, while $PI(4,5)P_2$ might convert loosely membrane-associated AP2 weakly bound to cargo protein into tightly membrane-bound nucleation sites for clathrin-coated pit assembly. In many cases, the coat contains an additional protein, AP180 in neurons and CALM in other tissues, which seems to define the size of the coated vesicle. Epsin, which also binds to $PI(4,5)P_2$, interacts with Eps15, a ubiquitously expressed protein that is constitutively associated with AP2/clathrin complexes and required for the early steps of clathrin-mediated endocytosis.

Fission of the coated pit requires the action of dynamin, a GTPase homologous to the *shibire* gene product in *Drosophila*. Hydrolysis of GTP by oligomeric rings of dynamin around the neck is necessary for vesicle scission. This could occur by direct constriction, by elongation of dynamin spirals at the neck, or with the aid of downstream effectors, such as endophilin, but dynamin was also proposed to function as a regulatory GTPase. $PI(4,5)P_2$ and amphiphysin, a cytosolic protein that binds simultaneously AP2 and dynamin, have been implicated in dynamin recruitment to CCVs.

Free CCVs are rarely observed, suggesting that these vesicles undergo rapid uncoating. The polygonal clathrin lattice is disassembled by the chaperone protein hsc70 (uncoating ATPase), a member of the DnaK family, that acts in several steps of the clathrin-coated vesicle cycle. Auxilin, a DnaJ protein, aids the targeting of hsc70 to clathrin coats by associating to AP2 and clathrin, and stimulates hsc70 ATPase activity. Finally, synaptojanin seems to function both in uncoating CCVs and as a negative regulator of the interaction between the clathrin coat and the plasma membrane, probably by dephosphorylating $PI(4,5)P_2$ into $PI(4)P$.

In addition to its function in Golgi-to-ER transport, an endosomal COPI complex is believed to be involved in endosomal transport. Binding of COPI to biosynthetic and endosomal membranes is regulated by ARF1. However, endosomal and biosynthetic COPI seem to differ in their composition, as only the α, β′, δ, ε, and ζ but not the β and γ subunits are detected on endosomes. Moreover, recruitment of ARF1 and subsequent COPI binding onto endosomes, but not biosynthetic membranes, depends on an acidic lumenal pH. The precise mechanism of COPI action in endosomal transport and morphology is still unclear. However, the endosomal COP might facilitate the selective incorporation of some cargo proteins and lipids into

ECV/MVBs, since it is involved in the down-regulation of CD4 bound to the HIV-encoded Nef protein.

Ubiquitination and Sorting

Ubiquitin is a 76-amino-acid protein that becomes conjugated to lysine residues in substrates through the concerted action of three enzymes, E1, E2, and E3. It is now clear that ubiquitin can act as a regulated sorting signal at different steps of the endocytic pathway, in addition to its role to target proteins to the proteasome. One of the advantages of ubiquitination over linear peptide sorting signals is that it is reversible. In this way, individual sorting decisions can be made or redesigned at each step of the endocytic pathway.

Monoubiquitin of the cytoplasmic domain of plasma membrane proteins is sufficient to trigger endocytosis in both yeast and mammalian cells. Moreover, ubiquitin might regulate the activity of one or more component(s) of the endocytic machinery. For instance, the growth hormone receptor requires the cellular ubiquitination machinery to be efficiently internalized, even though it does not need to be ubiquitinated itself. Moreover, Eps15 becomes monoubiquitinated on stimulation of cells with EGF.

Progress has been made in our understanding of the mechanisms used by the cells to recognize and sort cargo carrying a ubiquitin signal model in that monoubiquitinated proteins can be recognized by ubiquitin-binding proteins, which link cargo to the endocytic machinery. Indeed, the E3 ligase Cbl was shown to bind to the phosphorylated EGF receptor upon EGF stimulation; to promote EGF receptor ubiquitination, which is essential for sorting to CCV; and to enhance its rate of degradation. Moreover, Cbl recruits an endophilin-CIN85 complex. Inhibition of these interactions was sufficient to block EGF receptor internalization.

Further experiments demonstrated that a 20-amino-acid sequence, ubiquitin-interacting motif (UIM), is required both for interaction with ubiquitinated proteins and for monoubiquitination (although the motif is not itself ubiquitinated), suggesting that a UIM:ubiquitin-based intracellular network might be created. Several proteins of the endocytic machinery, such as Eps15, Eps15R and Epsin, and Hrs, contain a UIM. Of particular interest is the recent finding that, after internalization, ubiquitination is also involved at the next step of the pathway, in sorting some down-regulated proteins into the multivesicular body for subsequent degradation in lysosomes. As discussed above, this sorting step depends on Hrs and the ESCRT complexes (see Degradation Pathway, above).

Cytoskeleton

In yeasts, mutations in actin and in several actin-binding proteins inhibit endocytosis, revealing a link between the actin cytoskeleton and clathrin-mediated endocytosis. In animal cells experiments using actin-disrupting drugs point to an accessory, and possibly cell-specific, role of actin.

The identification of several mammalian proteins that physically link the actin cytoskeleton with the endocytic machinery has nevertheless strengthened the notion that actin filaments participate in endocytosis. Three of these proteins, Abp1m, Pan1p, and cortactin, stimulate the F-actin nucleation activity of the Arp2/3 complex. Cortactin and Abp1m were shown to bind to dynamin, whereas Pan1p, the yeast homologue of Eps15, interacts with four clathrin-binding proteins. Two other proteins, intersectin and syndapin, bind N-WASP, an activator of actin assembly via the

Arp2/3 complex. Intersectin is a scaffold protein that regulates the formation of clathrin-coated vesicles, and syndapin interacts with dynamin, synaptojanin, and synapsin, and seems to participate in endocytosis. Hip1 and Hip1R bind to F-actin in vitro and are associated with clathrin-coated vesicles, suggesting that they may function to anchor newly forming coated pits to the cortical cytoskeleton. Finally, Myosin VI, the only myosin known to move toward the pointed ends of actin filaments, might generate actin-dependent forces for membrane invagination or for vesicle movement through the dense F-actin network that underlies the plasma membrane. Recently the budding factor dynamin has also been implicated in actin-mediated transport of endocytic vesicles.

Small GTPases of the Rho subfamily regulate the actin cytoskeleton. Evidence is emerging that they also control endocytosis. RhoD and RhoB were localized on early and late endosomes, respectively, whereas both RhoA and Rac1 seem to be involved in the formation of clathrin-coated vesicles at the plasma membrane. It has been proposed that actin participates in the formation of specific and restricted sites at the plasma membrane, referred to as hot spots, where coated-pit formation might occur. Recent data, in fact, suggest that actin assembly is coordinated with dynamin recruitment to coated pits. In addition, actin might propel endocytic vesicles along their cytosolic routes. *Listeria* and other pathogens are known to use this mechanism to drive their own intracellular transport. Actin tails have also been seen in association with newly formed endocytic vesicles in mast cells and with "rocketing endosomes."

Microtubules have precise orientations in the cell, radiating from the center to the periphery of nonpolarized cells, with their stable ends pointing at the center, and can thus provide tracks for vesicle movement over long distances, directionality being provided by specific motor proteins like cytoplasmic dynein (toward the cell center) or kinesins (toward the cell periphery). During early to late endosome transport, endosomes move from the periphery to the perinuclear region onto microtubules. This is particularly well illustrated in neurons, where ECV/MVBs move from the early endosomes located in the presynaptic region through the length of the axon back to the cell body where late endosomes are located. Furthermore, late endocytic compartments exhibit bidirectional motility along microtubules, although they tend to be clustered in the perinuclear region. The reason for this dynamic behavior is not clear, nor are the mechanisms regulating the directional switch. However, motility was shown to be controlled by the small GTPase Rab7, a candidate regulator of the motor switch. In turn, the activity of Rab7 seems to be controlled by the membrane lipid composition, a mechanism that could explain the lack of motility observed in the cholesterol storage disorder NPC.

Rab Proteins and Their Effectors

Rab proteins are a large family of small GTPases of the ras superfamily, with 63 different members in the human genome, with some being tissue specific. They exhibit a compartment-specific localization in the biosynthetic and endocytic pathways. Rab proteins can exist in a cytosolic as well as a membrane-bound form, and cycle between an active, GTP-bound state, and an inactive, GDP-bound state. The cytosolic protein GDP dissociation inhibitor (GDI) regulates membrane association. GDI can interact with most of the Rab proteins that have been tested so far in their GDP-bound state only. GDI has the dual ability to extract and load back Rabs in the correct membrane. As for most small GTPases of the ras superfamily, GTPase-activating proteins (GAPs) interact with Rabs to promote GTP hydrolysis. Similarly, guanine nucleotide exchange factors (GEFs) are responsible for their activation upon GTP binding. These cycles impose temporal and spatial regulation on membrane transport, and Rabs can therefore be viewed as molecular switches.

Activated Rabs bind to soluble factors that act as effectors to mediate membrane fusion, membrane budding, and interaction with cytoskeletal elements. Rab effectors, which interact only with the GTP forms of Rab proteins, show a high degree of heterogeneity—they are highly specialized molecules whose activities are exclusively tailored for individual organelles and transport systems.

In the early endocytic pathway, Rab5 regulates clathrin-dependent internalization, as well as homotypic fusion of early endosomes. Furthermore, Rab5 mediates the attachment of early endosomes to, and the motility along, microtubules. Both Rab4 and Rab11 have been sequentially implicated in the recycling pathway from early endosomes back to the plasma membrane, but their precise functions are still unclear. In addition, Rab11 has also been localized to the TGN and, together with a Rab6 isoform, has been proposed to function in endosome-to-TGN transport. Several other Rabs have been localized to compartments of the early endocytic pathway, such as Rab15, Rab18, Rab22, and Rab25. Rab15 may counteract the effects of Rab5 on early endocytic processes.

Rab7 is localized to late endosomes and was proposed to regulate transport from early to late endosomes, homotypic fusion of late endosomes, and late endosome motility along microtubules. Rab9 regulates recycling of MPRs from late endosomes back to the TGN. The Rab9 effector TIP47 has been proposed to function in cargo selection to ensure the incorporation of Rab9-GTP and MPRs into vesicles that mediate transport back to the TGN. TIP47 has also been proposed to distribute to lipid droplets and to share extensive amino acid sequence similarity with the lipid droplet protein ADRP (adipose-differentiation-related protein). Recently, both Rab7 and Rab9 were suggested to play a role in the endosomes-to-Golgi targeting of glycosphingolipids, since overexpression of either Rab7 or Rab9 seems to overcome the NPC-disease-associated sphingolipids and cholesterol accumulation in late endosomes.

SNARE Proteins, NSF, and SNAPs

SNARE (soluble NSF attachment protein receptor, where NSF stands for *N*-ethylmaleimide-sensitive factor) proteins constitute the conserved core protein machinery for all intracellular membrane fusion events. They are relatively small (15 to 40 kDa) proteins, and their hallmark is a heptad-repeat sequence in their membrane-proximal region that forms coiled-coil structures. Because most of the more than 30 known SNAREs are anchored to the cytoplasmic surface of distinct subcellular compartments, it was initially believed that SNARE pairing confers specificity to intracellular membrane traffic. Indeed, all of the potential v-SNAREs encoded in the yeast genome showed the predicted capacity to trigger fusion by partnering with t-SNAREs that mark the Golgi, the vacuole, and the plasma membrane. Since SNAREs inevitably spread through several compartments during vesicular transport, other mechanisms must also be involved. Current evidence indicates that Rab proteins also contribute to define specificity at the vesicle targeting and tethering step, whereas SNARE pairing is essential for membrane fusion. Functional studies of the synaptic vesicle exocytosis at the nerve terminal have provided insights into how SNAREs interact with each other to generate the driving forces needed to fuse lipid bilayers. VAMP is associated with synaptic vesicles, whereas SNAP-25 and syntaxin-1 (STX1) are anchored to the plasma membrane. The protein n-Sec1 binds to STX1 and, after a conformational change, probably opens it up. STX1 is then able to interact with the two other neuronal SNAREs, thus bringing the two membranes into close apposition. Ca^{2+} triggers the full zipping of the coiled-coil complex, which results in membrane fusion and release of vesicle contents.

NSF and its associated protein αSNAP (soluble NSF attachment protein) were originally identified as soluble factors essential for intra-Golgi transport. Since then, these proteins were shown to be involved during most transport steps in the cell, including all steps of endocytic transport that have been studied. NSF together with αSNAP carries out the dissociation of the SNARE complex into monomeric components, thus reactivating SNAREs after one round of fusion. Several studies indicate a possible model in which the ATPase activity of NSF might impart a rotational shear to dissociate the core complex.

Membrane Microheterogeneity

It is becoming clear that some proteins and lipids are not evenly distributed through the bilayer but are able to segregate and form organized and dynamic domains. These membrane domains are likely to coexist on the same organelle and to be coupled functionally in space and time. Such compartmentalization may ensure a high efficiency of processes that take place simultaneously and warrant organelle integrity.

Rab Platforms

Interference with the GTPase cycle of Rab proteins causes important morphological changes in the organelle, suggesting that Rabs contribute to membrane organization. Based on the analysis of the downstream effectors, Rab5 acts as a membrane domain organizer and can locally and temporally change the membrane environment by recruiting effectors in a regulated manner. The idea that the Rab proteins and their effectors are clustered in defined membrane domains (Rab platforms) received much support from observations that Rab5 seems to be activated by a positive feedback loop, with the exchange factor (that activates Rab5) being itself part of an effector complex. In addition, Rab5 effectors include PI(3)P-binding protein (e.g., the early endosomal antigen 1, EEA1) but also PI3-kinases, again arguing that feedback loops drive the formation of a Rab5 platform (see also Phosphoinositides, below). This notion is further supported by the fact that Rabs in early and recycling compartments form separate domains that are in dynamic equilibrium with each other. Early endosomes contain predominantly Rab5 and Rab4 domains, whereas recycling endosomes are enriched in Rab4 and Rab11 domains. Cargo progressing along the endocytic pathway traverses Rab5, Rab4, and Rab11 domains sequentially. Effectors shared by neighboring Rab domains might functionally and structurally link the domains together. Indeed, rabenosyn-5 and rabaptin-5, two proteins originally identified as Rab5 effectors, were later shown to also bind Rab4. When overexpressed, they increase the association of Rab5 and Rab4 domains. Consequently cargo recycles more efficiently from the early endosomes to the plasma membrane and accumulates less efficiently in Rab11-positive recycling endosomes.

Recent experiments revealed that Rab9 and Rab7 localize to distinct domains within late endosomal membranes, suggesting that the simultaneous presence of more than one Rab domain on the same organelle might be a more general feature.

Annexin II

Annexin proteins have long been implicated in membrane dynamics and structure. They seem to be endowed with the intrinsic ability to self-organize at the membrane surface into bidimensional ordered arrays, supporting the notion that annexins can organize membrane domains. Two members of the family, annexin I and annexin II, are present on early endosomes. Annexin II is associated with cholesterol-rich regions through a Ca^{2+}-independent mechanism, and in vivo and in vitro studies suggest that annexin II is involved in endosome dynamics, perhaps as an interface to bind elements of the cortical actin cytoskeleton. Consistently with this notion, annexin II was recently shown to play an essential role in early/recycling endosome organization and ECV/MVB biogenesis.

Lipid Raft Domains

As discussed previously, rafts and raft components can be endocytosed by more than one pathway and the distinct endocytic compartments might be connected together by transport routes. In particular, the presence of raft components in early endosomes has led to the idea that this might contribute to a lipid-based sorting mechanism for molecules desalinated to be recycled, degraded, or transported to the TGN.

The notion that lipid rafts are present on the cell surface comes from a variety of approaches, including time-lapse, light, and electron microscopy. Since such a complete analysis is not easily performed with intracellular membranes, a word of caution is needed when dealing with endosomes (and other organelles). Partial membrane resistance to solubilization with mild detergents has been used operationally to define rafts as detergent-resistant membranes (DRMs). While DRMs are undoubtedly a valuable tool, one should be careful when drawing too many conclusions from DRM experiments alone, or when comparing different detergents and different cell types. Interestingly, however, a recent study showed that BHK late endosomes contain not only raft components but also DRMs, suggesting that rafts are routed to late endosomes at least in some cell types. Consistently, cholesterol was found to be abundant within multivesicular endosomes.

LBPA Domains

LBPA is abundant in late endosomes, including in internal membranes, and this lipid is not found elsewhere in the cell. This, together with the observation that LBPA membranes can be separated from other late endosome membranes without detergent, strongly suggests that these membranes form highly specialized domains. LBPA domains are involved in protein and lipid transport through late endosomes and might play a role in some pathological conditions including the cholesterol storage disorder NPC and the antiphospholipid syndrome associated with some autoimmune diseases. In addition, in vitro studies also suggest that negatively charged phospholipids, in particular LBPA, facilitate the degradation of several glycolipids. As LBPA itself is poorly degradable, one function of LBPA membranes could be to present lipids and proteins to the hydrolytic machinery. It does appear that LBPA membranes have turnpike functions in the late endocytic machinery and that interfering with the organization of these domains leads to an endosomal traffic jam.

Phosphoinositides

Phosphorylation of the inositol ring of phosphatidylinositol in the 3, 4, and 5 positions generates seven different phosphoinositides at the cytosolic face of cellular membranes. During the past few years, phosphoinositides have emerged as regulators of membrane traffic by regulating the localization and/or activity of effector proteins. Kinases and phosphatases mediate highly localized changes in the level of phosphoinositides, providing a means for temporally and spatially regulating effectors.

As already mentioned, $PI(4,5)P_2$ seems to be a key regulator of the formation, scission, and uncoating of CCVs. This lipid also promotes actin polymerization to form comet tails on vesicles enriched in rafts.

The PI3K product PI(3)P plays a major role in endocytic traffic. PI(3)P is proposed to be generated on the early endosomes via recruitment of PI 3-kinases, including hVPS34 by the active Rab5 (see Rab Platforms, above). PI(3)P facilitates the recruitment of FYVE zinc finger domains containing proteins. Mammalian cells express more than 25 different FYVE domains containing proteins characterized by a wide range of structures and functions. One such FYVE protein, the Rab5 effector EEA1, is a tethering factor essential for homo- and heterotypic early endosome fusion. Indeed, EEA1 interacts with two SNAREs, syntaxin 13 and syntaxin 6, which mediate the homotypic early endosomal fusion and TGN to early endosomes transport, respectively. As EEA1, Rabenosyn-5 is a FYVE-containing Rab5 effector required for endosome fusion.

Recent studies have uncovered the existence of a novel, highly conserved phosphoinositide-binding domain, called the Phox homology (PX) domain. This domain was first found in $p40^{phox}$ and $p47^{phox}$, two cytosolic subunits of the phagocyte oxidase (Phox) complex. The sequential binding of $p47^{phox}$ and $p40^{phox}$ PX domains to $PI(3,4)P_2$ and PI(3)P, respectively, is crucial for the function of the Phox complex. Several PX-containing proteins belong to the family of sorting nexins, some of which have their putative functions in membrane trafficking. SNX3 is localized to early endosomes by binding to PI(3)P and when overexpressed disrupts endosome morphology. Interestingly, SNX1 interacts with the FYVE protein Hrs and is involved in EGFR degradation, whereas the SNX1 yeast homologue, Vps5p, and another PX domain-containing protein, Vps17p, are constituents of a retromer multiprotein complex that directs membrane trafficking from endosomes to the Golgi complex. In yeast, the SNARE protein Vam7p binds through its PX domain to PI(3)P on multivesicular bodies and vacuoles, and this interaction is essential for its function.

In addition, PI(3)P was also found in the internal vesicles of multivesicular endosomes, consistent with the role of PI(3)P signaling in receptor sorting (see Degradation Pathway, above). In yeast, phosphorylation of PI(3)P to $PI(3,5)P_2$ by the action of Fab1p is coupled to the regulation of sorting into the multivesicular body. PIKfyve, the mammalian Fab1p homologue, is associated with endosomes, but its precise role remains to be elucidated. Mutations in myotubularin-related proteins (MTMRs), a family of proteins recently shown to specifically dephosphorylate PI(3)P, lead to severe disorders such as myotubular myopathie and demyelinating neuropathy, perhaps suggesting that traffic defects might contribute to the pathology.

Consistent with a role of phosphoinositides in macropinocytosis and phagocytosis, inhibition of PI 3-kinase prevents both processes. PI3K might regulate the activation of phosphatidylinositol-4-phosphate-5-kinase, leading to PI(4,5)P$_2$ accumulation at the plasma membrane. Components of the actin-based cytoskeleton could then bind to PI(4,5)P$_2$ and mediate the actin rearrangements.

Selected Readings

Blott, E. J., and G. M. Griffiths. 2002. Secretory lysosomes. *Nat. Rev. Mol. Cell Biol.* **3:**122–131.

Boehm, M., and J. S. Bonifacino. 2001. Adaptins: the final recount. *Mol. Biol. Cell* **12:** 2907–2920.

Bonifacino, J. S., and B. S. Glick. 2004. The mechanisms of vesicle budding and fusion. *Cell* **116:**153–166.

Brodsky, F. M., C. Y. Chen, C. Knuehl, M. C. Towler, and D. E. Wakeham. 2001. Biological basket weaving: formation and function of clathrin-coated vesicles. *Annu. Rev. Cell Dev. Biol.* **17:**517–568.

Chen, Y. A., and R. H. Scheller. 2001. SNARE-mediated membrane fusion. *Nat. Rev. Mol. Cell Biol.* **2:**98–106.

Collins, B. M., A. J. McCoy, H. M. Kent, P. R. Evans, and D. J. Owen. 2002. Molecular architecture and functional model of the endocytic AP2 complex. *Cell* **109:**523–535.

The structure of the AP2 "core" complexed with a phosphatidylinositol headgroup analog was solved by X-ray crystallography at 2.6-A resolution. Two potential phosphatidylinositide binding sites were observed. The binding site for Yxxφ endocytic motif was buried, indicating that a conformational change must occur to allow AP2 to bind simultaneously to all of its ligands.

Desjardins, M. 2003. ER-mediated phagocytosis: a new membrane for new functions. *Nat. Rev. Immunol.* **3:**280–291.

This review reports that phagosomes are formed via direct association and fusion of the ER to the PM during early phagocytosis. ER also appears to be added in successive waves to supply membrane at various steps of phagosome maturation.

Di Fiore, P. P., and P. De Camilli. 2001. Endocytosis and signaling. an inseparable partnership. *Cell* **106:**1–4.

The authors review the molecular mechanisms that connect the machinery of receptor-mediated signaling and endocytosis.

Gaidarov, I., F. Santini, R. A. Warren, and J. H. Keen. 1999. Spatial control of coated-pit dynamics in living cells. *Nat. Cell Biol.* **1:**1–7.

Greenberg, S., and S. Grinstein. 2002. Phagocytosis and innate immunity. *Curr. Opin. Immunol.* **14:**136–145.

Gruenberg, J. 2001. The endocytic pathway: a mosaic of domains. *Nat. Rev. Mol. Cell Biol.* **2:**721–730.

Gruenberg, J., and H. Stenmark. 2004. The biogenesis of multivesicular endosomes. *Nat. Rev. Cell Mol. Biol.* **5:**317–323.

Hall, A. 1998. Rho GTPases and the actin cytoskeleton. *Science* **279:**509–514.

Hicke, L., and R. Dunn. 2003. Regulation of membrane protein transport by ubiquitin and ubiquitin-binding proteins. *Annu. Rev. Cell Dev. Biol.* **19:**141–172.

The functions of transient ubiquitination in controlling protein trafficking within the cell are reviewed. Monoubiquitination of the cytoplasmic domain of transmembrane proteins is believed to act as an internalization signal at the plasma membrane and is likely to regulate sorting to the degradative pathway.

Ikonen, E. 2001. Roles of lipid rafts in membrane transport. *Curr. Opin. Cell Biol.* **13:** 470–477.

Johannes, L., and C. Lamaze. 2002. Clathrin-dependent or not: is it still the question? *Traffic* **3**:443–451.

Knodler, L. A., J. Celli, and B. B. Finlay. 2001. Pathogenic trickery: deception of host cell processes. *Nat. Rev. Mol. Cell Biol.* **2**:578–588.

Marsh, M., and M. T. McMahon. 1999. The structural era of endocytosis. *Science* **285**:215–220.

Matsuo, H., J. Chevallier, N. Mayran, I. Le Blanc, C. Ferguson, J. Faure, N. S. Blanc, S., Matile, J. Dubochet, R. Sadoul, R. G. Parton, F. Vilbois, and J. Gruenberg. 2004. Role of LBPA and Alix in multivesicular liposome formation and endosome organization. *Science* **303**:531–534.

> Late endosomes are typically multivesicular and contain large amounts of the unusual phospholipid lysobisphosphatidic acid (LBPA). Here, it is reported that LBPA plays a crucial role in the intraluminal membrane invagination process and that the process is regulated by Alix.

Mostov, K. E., M. Verges, and Y. Altschuler. 2000. Membrane traffic in polarized epithelial cells. *Curr. Opin. Cell Biol.* **12**:483–490.

Ohsumi, Y. 2001. Molecular dissection of autophagy: two ubiquitin-like systems. *Nat. Rev. Mol. Cell Biol.* **2**:211–216.

> Authors review the molecular mechanisms of autophagy in yeast. An analysis of the genes required for autophagy has revealed the existence of two ubiquitin-like systems, indicating that ubiquitin-type conjugation is necessary for the formation of autophagosomes.

Pelham, H. R., and J. E. Rothman. 2000. The debate about transport in the Golgi: two sides of the same coin? *Cell* **102**:713–719.

> Authors discuss and review the models of biosynthetic membrane transport through the Golgi complex.

Pelkmans, L., and A. Helenius. 2002. Endocytosis via caveolae. *Traffic* **3**:311–320.

> This paper describes the caveolae-mediated entry of Simian Virus 40 and its transport to a new cellular compartment, the caveosome.

Schafer, D. A. 2002. Coupling actin dynamics and membrane dynamics during endocytosis. *Curr. Opin. Cell Biol.* **14**:76–81.

Schuck, S., M. Honsho, K. Ekroos, A. Shevchenko, and K. Simons. 2003. Resistance of cell membranes to different detergents. *Proc. Natl. Acad. Sci. USA* **100**:5795–5800.

Sever, S., H. Damke, and S. L. Schmid. 2000. Garrotes, springs, ratchets, and whips: putting dynamin models to the test. *Traffic* **1**:385–392.

Simonsen, A., and H. Stenmark. 2001. PX domains: attracted by phosphoinositides. *Nat. Cell Biol.* **3**:E179–E182.

> This paper summarizes the recent studies that uncovered the existence of a novel phosphoinositide-binding domain shared by several proteins, called the PX- or PHOX-homology domain. The interactions of some PX-domain-containing proteins with specific phosphoinositides are critical for membrane trafficking.

Simonsen, A., A. E. Wurmser, S. D. Emr, and H. Stenmark. 2001. The role of phosphoinositides in membrane transport. *Curr. Opin. Cell Biol.* **13**:485–492.

van der Goot, F. G., and J. Gruenberg. 2002. Oiling the wheels of the endocytic pathway. *Trends Cell Biol.* **12**:296–299.

Zerial, M., and H. McBride. 2001. Rab proteins as membrane organizers. *Nat. Rev. Mol. Cell Biol.* **2**:107–117.

10

Where To Stay inside the Cell: a Homesteader's Guide to Intracellular Parasitism

DAVID G. RUSSELL

The adoption of an intracellular lifestyle confers several advantages to microbial pathogens; they become inaccessible to humoral and complement-mediated attack, they no longer require a specific adherence mechanism to maintain their site of infection, and they have ready access to a range of nutrients. However, aspiring intracellular pathogens must develop specific strategies to secure and maintain their lifestyle. Some characteristics, such as the lipid-rich cell wall of mycobacteria, are obvious preadaptations that confer a head start in the acquisition of an intracellular lifestyle. However, the majority of mechanisms exhibited by intracellular pathogens are the product of evolutionary selection in response to their intracellular existence.

Successful establishment and maintenance of an intracellular infection require the resolution of a series of interconnected problems, with the solution of each frequently having profound influence on subsequent "decisions." This chapter describes these problems for a range of bacterial, protozoal, and fungal pathogens and explores, in the order in which they are encountered by the pathogen, the consequences of each decision point in the establishment of an intracellular infection.

Routes of Invasion

To infect a cell, a microbe must first adhere to it. For some pathogens, this adherence phase determines the choice of the cell to be infected, while for others, although they are capable of binding to many cell types, development of the pathogen is restricted to only certain cell lineages. Because adherence and host cell entry are tightly associated phenomena, they are discussed together as functions of the route and mechanism of invasion. There are three basic mechanisms of invasion: (i) phagocytosis, i.e., entry into professional phagocytes such as macrophages, monocytes, and neutrophils via a process dependent on the host cell contractile system; (ii) induced endocytosis and phagocytosis, i.e., entry into nonprofessional phagocytes by the active induction of internalization through the activity of the host cell contractile system; and (iii) active invasion, i.e., active entry into a passive host cell without triggering any contractile event in the host cell cytoskeleton. The different routes of host cell invasion are diagrammed in Figure 10.1.

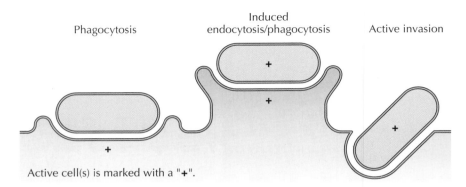

Figure 10.1   Diagrammatic representation of the three different routes of invasion of mammalian cells by intracellular pathogens. In each instance, the "active" cell or cells are labeled with a plus sign. In phagocytosis, the infecting pathogen is relatively passive in the process following ligation to host cell receptors capable of triggering internalization. This process requires little if any metabolic activity from the parasite. Examples include *Leishmania, Mycobacterium,* and *Histoplasma.* In induced endocytosis and phagocytosis, the pathogen induces a normally nonphagocytic cell to internalize the microbe. This is the least well understood route of entry and involves subversion of the host cell signaling pathways. Examples include *Salmonella* in nonprofessional phagocytes and *Trypanosoma cruzi.* In active invasion, the pathogen invades the host cell without the participation of the contractile apparatus of the host cell. In this process, the host cell is inert. Examples include all the apicomplexan parasites, *Plasmodium, Toxoplasma, Eimeria,* and microsporidia.

Phagocytosis

Many microbial pathogens are capable of either transient or sustained infection of professional phagocytes. There are two related reasons for this seemingly anomolous phenomenon. First, macrophages and other phagocytes represent the frontline defense of the host against microbial invasion. These cells migrate through tissues, internalizing and degrading foreign particles; this behavior will obviously have maximized the frequency of interaction between phagocytes and microbes. Second, the macrophage has receptors that recognize a range of ligands, including the serum opsonins antibody and complement. Pathogens that activate the alternate pathway of complement will accumulate C3b and iC3b on their surface, and because the macrophage is equipped with high-affinity receptors for these ligands, CR1 and CR3/CR4, these opsonized microbes will bind to and be internalized by the phagocyte (Box 10.1). The only specific mechanism required by the pathogen is to facilitate complement deposition while avoiding lysis by insertion of the terminal membrane attack components. Pathogens such as *Salmonella* and *Leishmania* achieve this through elongation of their surface lipopolysaccharide or lipidoglycans, respectively. These long carbohydrate chains activate complement but avoid lysis because the activating convertase is maintained some distance from the outer membrane. Complement receptors trigger phagocytosis without stimulating a strong superoxide burst from the macrophage.

Entry via phagocytic receptors activates the signaling pathways for maturation of the phagosome into an acidic, hydrolytically active compartment. Therefore, in the absence of mechanisms to subvert this process, all phagocytosed microbes will be delivered to the lysosomal system of the cell.

BOX 10.1

Complement opsonization

The complement system is one of the major defense barriers against microbial infection. It can be mobilized in both the presence (classical pathway) and absence (alternative pathway) of immunoglobulins specific to the infective agent. The direct consequence of complement activation and deposition is the polymerization of the latter components, C5 to C9, to form a membrane attack complex that creates a pore in the target membrane.

More relevant to this chapter, however, is the coating of microbes with fragments of C3, which are recognized by receptors expressed on the surface of phagocytes. C3 is the central component of both the classical and alter-

native pathways. It is a 185-Da α/β heterodimer that contains an internal thiolester bond, which becomes unstable when the protein is cleaved into its C3b form by removal of the first 77 amino acids of the α chain. The thiolester bond will interact with water or with hydroxyl (carbohydrate) or amino (protein) groups on cell surfaces, forming covalent ester or amide linkages, respectively. The alternative pathway relies on the spontaneous production of serum C3b, which will deposit on surfaces and, together with factor B, will form a C3 convertase, C3bBb. In contrast, the classical pathway relies on antibody binding to allow C1q attachment and formation of a different C3 convertase consisting of C4bC2a. Following formation of a C3

convertase, C3, which is present at more than 1 mg/ml in serum, is cleaved to form C3b, leading to massive deposition through a positive-feedback loop. C3b can be inactivated by hydrolysis or through the specific activity of factor I, which cleaves C3b into iC3b. Both C3b and iC3b are ligands for high-affinity receptors present on the surface of phagocytes and some other cell types. CR1 is expressed on erythrocytes, B lymphocytes, monocytes, granulocytes, and macrophages and recognizes both C3b and C4b. CR3 is present on granulocytes, monocytes, and macrophages and recognizes iC3b. Both these receptors, on professional phagocytes, will trigger phagocytosis under physiologic conditions.

Induced Endocytosis and Phagocytosis

Other pathogens have evolved their own ligands for adherence to host cells, an event most frequently observed in pathogens that infect nonprofessional phagocytes. Host cell entry by many of these pathogens is a facilitated process that involves the induction of an internalization response in the cell by the adherent microbe. Bacterial pathogens have been shown to use effector proteins inoculated into host cells by type III secretion systems to either stimulate invasion of nonprofessional phagocytes or, conversely, block uptake by professional phagocytes. The signaling pathways activated during induced uptake are explored in depth in chapter 6. One of the better-studied examples is *Salmonella*, which uses the type III secretion apparatus encoded in SPI1 to induce the actin cytoskeleton to form extensive membrane ruffles, or "splash," during its entry into cells. This response is accompanied by the phosphorylation of several host cell proteins as a consequence of activation of host cell signaling cascades.

Active Invasion

In contrast to both preceding mechanisms, active invasion involves invasion of the host cell without triggering any contractile events such as phagocytosis. In fact, for *Toxoplasma*, the ability to block the maturation of the entry vacuole into an acidic, lysosomal compartment is determined at the time of host cell entry. All apicomplexan parasites, including *Plasmodium*, *Toxoplasma*, and *Eimeria*, have motile invasive stages called zoites. These stages have no external locomotory organelles and have an elongate, torpedo shape determined by a spiral of subpellicular microtubules (Figure 10.2). Zoites possess an actomyosin-based motile system that mediates both gliding motility and host cell invasion. The myosin molecule is a myosin A chain that associates with components of the plasmalemma and an unusual myosin light chain and is capable of moving along actin filaments at approximately a body length a second. The external expression of this con-

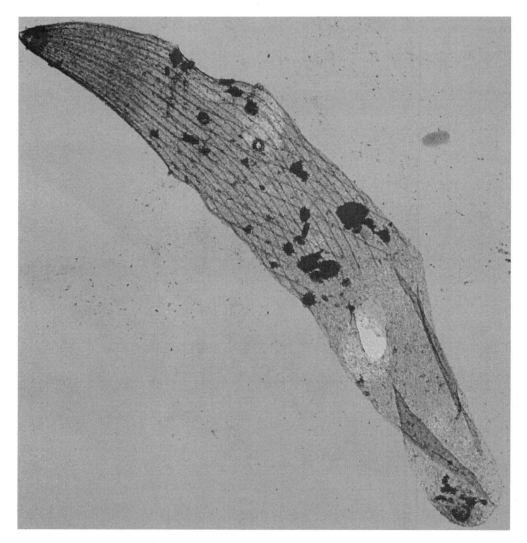

Figure 10.2 Electron micrograph of a critical-point dried, detergent-extracted, whole-cell mount of a sporozoite of *Eimeria acervulina.* The structure of this zoite is typical of that observed throughout the phyla that includes both *Plasmodium* and *Toxoplasma.* The cell adopts a spiral shape dictated by its subpellicular microtubule network. Motility and host cell invasion are achieved through an actin-myosin contractile system that caps plasmalemma constituents from the anterior to the posterior of the cell.

tractile system depends on families of membrane proteins stored in the cell's micronemes, secretory organelles located at the parasite's anterior. These proteins, called "MICs," are adhesion molecules containing a range of adherence domains like the thrombospondin domain (MIC1), epidermal growth factor (EGF) domains (MIC3), and apple domains (MIC4). Homologues of the MIC proteins of *Toxoplasma* have been found in other apicomplexa including *Eimeria, Cryptosporidium,* and *Plasmodium.*

The role of this contractile system in invasion was demonstrated by experiments with mutant host cells and *Toxoplasma* lines resistant to the antimicrofilament agent cytochalasin D in which mutant parasite lines invaded host cells in the presence of the drug irrespective of the phenotype of the host cell. If this invasion process was subverted by opsonizing the parasites with immunoglobulin G (IgG), the Fc receptors on the host cell, which were ligated during invasion, prevented the parasites from maintaining their vacuoles outside the endosomal continuum. The vacuoles acidified, and the parasites died. Conversely, if *Toxoplasma* was allowed to invade and infect host cells and was subsequently killed by treatment with pyrimethamine, an inhibitor of the parasite's dihydrofolate reductase activity, the vacuoles persisted as isolated intracellular compartments, presumably because the host cell membrane fusion apparatus could not recognize them. These data indicate an intimate link between the route of invasion and the successful establishment of an intracellular infection.

Selection of an Intracellular Niche

After invasion, intracellular microbes use many different strategies to ensure the maintenance of an intracellular infection. The niches exploited by these intracellular pathogens fall readily into three different groupings, illustrated in Figure 10.3. The first is intralysosomal, in which pathogens persist in acidic, hydrolytic compartments that interact with the endosomal network of the host. The second is intravacuolar, in which pathogens per-

Figure 10.3 Intracellular niches. Pathogens have evolved to exploit a variety of intracellular locations, which fall readily into three different groups. The first group includes those that reside in acidic, hydrolytically competent lysosomes and appear undeterred by the hostile nature of their compartment. Examples include *Leishmania, Coxiella,* and possibly *Salmonella* (in macrophages at least). The second group includes those that remain vacuolar yet avoid the normal progression of their vacuole into a lysosomal compartment. This group of pathogens is the most diverse with respect to the nature of their intracellular vacuole. Examples include *Plasmodium, Toxoplasma, Legionella, Chlamydia,* and *Mycobacterium.* The third group includes those that avoid the consequence of remaining within a phagocytic vacuole by escaping into the cytoplasm. Examples include *T. cruzi, Shigella, Rickettsia,* and *Listeria.*

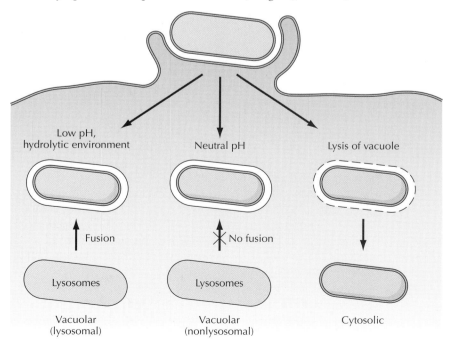

sist in nonacidic vacuoles that exhibit modified or little interaction with the endosomal system of the host. The third is cytoplasmic, in which pathogens exit the phagosome and reside within the host cell cytosol.

Intralysosomal Pathogens

In terms of modulation of host cell function and the biogenesis of phagosomes and endosomes, the pathogens that remain within the "normal" differentiation cascade of the phagolysosome appear relatively passive. *Leishmania* parasites and *Coxiella* bacilli are phagocytosed by macrophages and reside in compartments that are fully acidic, achieving a pH of 4.7 to 5.2, and contain active hydrolytic lysosomal enzymes.

Leishmania

Leishmania organisms are flagellated protozoans that are close relatives of the trypanosomes. The promastigote form found in the insect vector has a single, anteriorly orientated flagellum, which becomes vestigial in the amastigote form found within vertebrate macrophages. *Leishmania* species induce a range of diseases varying in severity from a simple lesion at the site of a fly bite to visceral leishmaniasis, where the parasite multiplies in the liver and spleen.

The vacuoles containing parasites of the *Leishmania mexicana* complex (Figure 10.4) interact freely with material internalized by the endosomal network of the host cell; in fact, the fusigenicity and access to these com-

Figure 10.4 Hoffman modulation contrast micrograph of a monolayer of murine bone marrow-derived macrophages infected with *Leishmania mexicana*. This species of *Leishmania* tends to form large fluid-filled vacuoles that contain multiple parasites, which tend to line up along the periphery of the vacuoles (arrow). The vacuoles are acidic and contain active lysosomal hydrolases.

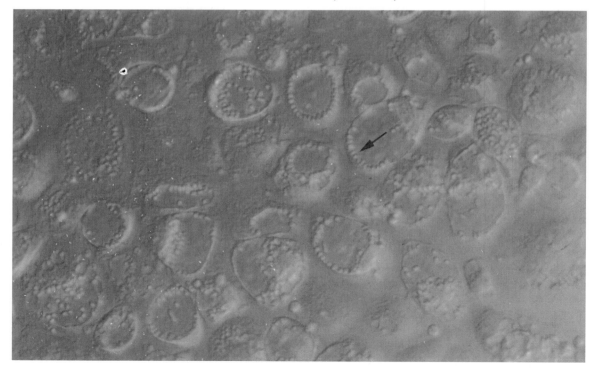

partments increase as the infection continues. The vacuoles are also competent to fuse with other particle-containing phagosomes, facilitating transfer of the particles into the parasitophorous vacuoles. The lysosomal hydrolases within the parasitophorous vacuoles of *Leishmania*, most notably cathepsins B, L, and D, are fully active and hydrolytically competent; nonetheless, the pathogens appear impervious to degradation. The vacuoles have abundant LAMP1 and LAMP2 (lgp110 and lgp120), while transferrin, which traffics through the rapid recycling pathway of mammalian cells and does not normally access the late endosome or lysosome, is not detected in *Leishmania*-infected macrophages.

Leishmania parasites also have access to cytosolic material as potential sources of nutrients through fusion with autophagosomes. These parasites are purine auxotrophs and need to salvage purines from the host, since these are probably present at low concentration in the lysosome. Access to cytosolic components sequestered in autophagosomes may provide a source of these purines. The *Leishmania* amastigote has an abundant external nucleotidase that would allow the dephosphorylation of nucleoside triphosphates, enabling uptake and use of host purines.

Existence within vacuoles that remain integral to the endosomal network carries two severe problems. The first is immediate, i.e., resistance to hydrolytic attack, while the second is more insidious, i.e., the continued sampling of parasite-derived peptides and their presentation to the host immune system. With respect to the first problem, *Leishmania* appears to adopt a strategy of avoidance. The surface of the intracellular amastigote stage of the parasite is protein poor and covered in a coat of short lipidoglycans (GIPLs). These GIPLs are relatively resistant to degradation by the host lysosomal hydrolases and thereby present minimal targets for attack. The key to the survival of *Leishmania* within the lysosome is probably its flagellar pocket. The flagellar pocket of all trypanosomatids is the main route of endocytosis and exocytosis in the cell, because the rest of the cell body is subtended by an extensive array of microtubules. Access to the pocket is partially regulated by a series of hemidesmosomal junctions that anchor the flagellum to the cell body at the mouth of the pocket, which may protect the endocytic receptors of the parasite from attack by host hydrolases.

The second problem relates to the long-term maintenance of the leishmanial infection within a host cell capable of presenting antigen via both class I and class II antigens (Box 10.2). Since it is known that both class I and class II major histocompatibility complex (MHC) molecules traffic through the parasitophorous vacuoles of *L. mexicana*- and *L. amazonensis*-infected macrophages, it is possible that they bind parasite-derived peptides which could subsequently be presented at the surface of infected cells. The induction of a T-cell response leading to the release of macrophage-activating cytokines would be detrimental to the long-term success of the parasite. Leishmanial infections are therefore an interesting system to study strategies of immune system evasion for intracellular pathogens. Examination of the ability of infected macrophages to present exogenous antigen revealed a slight diminution in the T-cell responses induced; however, this reduction varied with both T-cell lines and epitopes, suggesting that the effect was on the intracellular processing and loading rather than on the expression of class II MHC or costimulatory molecules.

Class II MHC molecules are synthesized as nascent chains that must bind to an invariant chain to be correctly processed and trafficked in cells. When the invariant chain is degraded, it facilitates the binding of appropri-

BOX 10.2

Interface between intracellular pathogens and the immune system: the postcard version

Since intracellular pathogens are sequestered beyond the ready access of the humoral immune system, it is critical that infected hosts be able to identify infection foci or the infected cells themselves. This is achieved through the antigen-processing and presentation capabilities of mammalian cells. All nucleated cells possess major histocompatibility antigens capable of binding peptides acquired inside the cell and presenting these peptides to T lymphocytes, triggering these cells to respond.

The major histocompatibility antigens fall into two groups: class I and class II antigens. Historically the roles of these two molecules were the following. Class I antigens sample pep-

tides derived predominantly from cytoplasmic antigens and therefore play a major role in the presentation of antigens from cytosolic pathogens such as *Listeria, Shigella,* and *T. cruzi,* whereas class II antigens sample peptides generated within the endosomal-lysosomal system of the host cell and play a central role in the immune response to intravacuolar pathogens such as *Mycobacterium, Salmonella,* and *Leishmania.* However, as with much immunology, this is far too simple to be completely true and class I antigens are now known to be capable of sampling antigens from the endosomal system.

Class I MHC antigens are expressed by all nucleated cell types. The class I molecules are trafficked through the cell as heterodimers complexed to α_2-microglobulin and acquire peptides predominantly in the ER. These peptides are translocated from the cytoplasm to the ER by specific transporters or TAP proteins.

Class II MHC antigens are expressed by cells of the monocyte/macrophage lineage. Class II molecules are α/β heterodimers that complex with another protein, the invariant chain, which regulates intracellular trafficking and the point of peptide acquisition.

Effective presentation to T cells is achieved when the MHC molecule with bound peptide is recognized by an appropriate T-cell receptor on a T lymphocyte. Intracellular pathogens have evolved a spectrum of mechanisms to avoid or suppress the induction or consequences of such immune system responses. These strategies include sequestration of antigens, downregulation of MHC molecule or co-stimulatory-molecule expression, suppression of T-cell proliferation, and blocking of the ability of the host cell to respond to activating cytokines. A full discussion of this interplay is provided in chapter 18.

ate peptides to the peptide-binding grove of the class II dimer. Recent analysis of *L. amazonensis*-infected macrophages showed the presence of both class II molecules and invariant chain within the parasitophorous vacuoles. The abundance of the invariant chain was enhanced by treatment of infected macrophages with inhibitors of cysteine proteinase activity. These inhibitors would block both the host cell lysosomal hydrolases cathepsins B and L and the parasite cysteine proteinases, and it was suggested that *Leishmania* degrades class II MHC molecules and thus suppresses antigen presentation by the host cell. However, this contrasts with data from another series of experiments in which the acid phosphatase gene of the parasite was engineered to express the protein in the cytosol, on the parasite surface, or secreted into the parasitophorous vacuole milieu. In live infections, the acid phosphatase was presented effectively to T cells when it was expressed as a parasite surface protein and when it was expressed as a secreted protein. Only after killing of intracellular parasites could effective presentation be detected with wild-type cells or cells overexpressing the intracellular form of the protein. These data indicate that the antigen presentation machinery of the host macrophage does interact functionally with the parasitophorous vacuole. It is likely that the parasite uses more than one strategy to avoid alerting the immune system, and minimizing the release of antigens in combination with degradation of antigen-presentation molecules would appear a rational approach.

While it had been known for a long time that intravacuolar pathogens such as *Leishmania* and *Mycobacterium* are capable of inducing strong class I-restricted CD8 responses, the site of antigen sampling by class I molecules was the subject of considerable debate. Recent data however indicate that leishmanial antigens present in phagosomes can be routed to the cytosol via the Sec61 pathway. Here they are subjected to proteasome-mediated degradation and TAP-dependent transport back into the endosomal system where the peptides are loaded onto class I molecules. Such a process could provide an explanation for class I-restricted responses to several intravacuolar pathogens.

Pathogens Sequestered in Modified or Isolated Vacuoles

Of the three different groups of intracellular pathogens, the group of pathogens sequestered in modified or isolated vacuoles is the most disparate in terms of both the properties of their intracellular compartments and the degree to which the vacuole is modified or sequestered outside the normal endocytic continuum. Some pathogens, like *Histoplasma*, survive in lysosomal compartments that fuse with endosomes yet fail to acidify; some, like *Salmonella*, survive in compartments that acidify, at least initially, but do not behave like classic lysosomes; others, like *Mycobacterium*, block the normal maturation procedure of their phagosome and fail to fuse with lysosomes; and finally, some microbes, like *Chlamydia*, *Brucella*, *Legionella*, *Toxoplasma*, *Plasmodium*, and *Cryptosporidium*, form a compartment that appears sequestered completely outside the normal endosomal trafficking pathways of their host cell.

Nonacidified Lysosomes

Histoplasma. The yeast *Histoplasma capsulatum* parasitizes macrophages in its vertebrate hosts. These vacuoles are readily accessible to endocytic tracers such as fluorescein isothiocyanate dextran and fuse with lysosomal compartments with an efficiency indistinguishable from that of *Saccharomyces*. Further examination revealed that, although *Histoplasma* vacuoles were competent to fuse with lysosomes and endosomes, they did not acidify. Analysis of the vacuolar pH up to 20 min after internalization revealed that while the vacuoles containing the yeast cell wall preparation zymosan acidified rapidly to pH 5.5, the vacuoles containing *Histoplasma* maintained a pH of 7.0.

Remodeled Lysosomal Compartments

Salmonella. *Salmonella* invades both nonprofessional phagocytes and macrophages through the induction of a macropinocytosis-like event during which the bacilli are internalized into large "spacious" phagosomes (Figure 10.5). The past few years have seen extraordinary advances in our understanding of intracellular infection by *Salmonella* emerging predominantly from our appreciation of the genes located in genetic cassettes known as "pathogenicity islands." While the pathogenicity island SPI-1 is involved in uptake into nonprofessional phagocytes, the pathogenicity island SPI-2 plays a dominant role in modulation of the intracellular compartment and is particularly important for survival in macrophages. *Salmonella*-containing vacuoles acquire LAMP1 yet do not fuse readily with vacuoles containing lysosomal cargo. *Salmonella* vacuoles in macrophages show some acidification, and, furthermore, blocking the acidification of phagosomes with bafilomycin A, a vacuolar ATPase inhibitor, reduced the viability of the bacilli. This suggests that a drop in pH is required for full induction of intracellular survival strategies in *Salmonella* that precedes the remodeling of the *Salmonella*-containing vacuole.

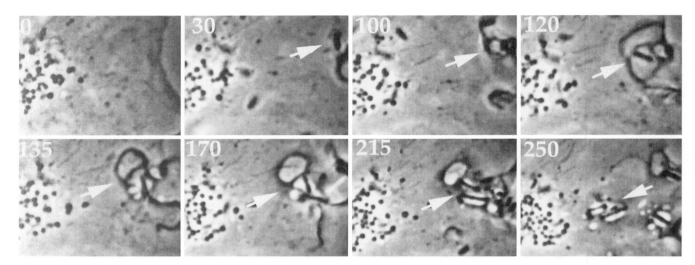

Figure 10.5 *Salmonella* induces an extreme response in mammalian cells during entry. In contrast to tight, zippering phagocytosis through which many particles are internalized, these bacteria induce a membrane "splash" or ruffle that captures the bacteria along with an appreciable volume of fluid. This phenomenon is illustrated in a series of time-lapse video frames. The point of initial contact of the bacterium is marked with an arrow in the 30-s and all subsequent time frames. The macropinosome forms (120 s), and several fluid-filled vesicles coalesce (135 and 170 s), until, finally, the phagocytosed bacilli are translocated toward the cell body (250 s). The mechanism appears analogous to the formation of macropinosomes. Courtesy of Hiroshi Morisaki, Michelle Rathman, and John Heuser.

Expression of SPI-2 genes was found to be up-regulated markedly between 1 and 6 h after infection. Induction appears to be through a two-component system encoded by *SsrAB* that can be induced in vitro by magnesium or phosphate starvation. The SPI-2 type III secretion system delivers a range of effector proteins not all of which are encoded within the pathogenicity island. SipC is thought to fulfill a chaperone-like function and aid in release of other effector molecules. SifA, SseF, and SseG are all known to each play a role in formation of the filamentous vesicles observed associated with the *Salmonella*-containing vacuoles in infected cells. Cells infected with a SifA-deficient mutant were found to have the bacteria loose in the cytosol. This indicated that SifA functions in maintenance of the *Salmonella*-containing vacuole. Intriguingly, a double mutant lacking the genes encoding both SifA and a putative acyl transferase, SseJ, did not escape into the cytosol. SseJ expression in mammalian cells induces formation of globular membrane complexes. This would suggest that SifA might promote membrane recruitment and/or fusion and SseJ could be involved in breakdown of the lipid bilayer.

In addition to regulating the morphology of the *Salmonella*-containing vacuole, SPI-2 secretion apparatus also releases effector proteins that block recruitment of the NADPH oxidase complex that is responsible for production of reactive oxygen intermediates capable of killing the bacterium. It also delivers effector proteins capable of remodeling the actin cytoskeleton of the host cell and inducing polymerization of an actin "basket" that surrounds the bacteria-containing vacuoles. Obviously *Salmonella* has developed the "cassette" approach to infection into an art form, and the biology of pathogenicity islands such as SPI-2 is discussed in chapters 15 and 16.

Developmentally Arrested Phagosomes

Some pathogenic microbes enter their host macrophages by phagocytic uptake yet prevent the normal course of differentiation of their phagosome into a phagolysosomal compartment.

Mycobacterium **spp.**   The arrest of phagosome maturation has been described for a range of pathogenic mycobacterial species capable of infecting mammals. These species include *Mycobacterium tuberculosis, M. leprae, M. microti* (a rodent pathogen), and the *M. avium* complex, which is an opportunistic pathogen of immunosuppressed individuals. Although the diseases induced by these species differ markedly, they show strong parallels with respect to the behavior of their intracellular compartments. Early work revealed that the vacuoles inhabited by pathogenic mycobacteria did not fuse with lysosomes preloaded with electron-dense colloids, in contrast to vacuoles containing nonpathogenic or dead mycobacteria. Recent analysis of *M. tuberculosis* and *M. avium* in both human monocyte-derived and murine bone marrow-derived macrophages has yielded a more complete appreciation of the properties of these vesicles. Vacuoles containing these bacteria do not acidify below pH 6.2 to 6.5 and show a paucity of the vacuolar proton-ATPase responsible for acidification of endosomal and lysosomal compartments (Figure 10.6). Despite this, the vacuoles possess LAMP1 (although it is less abundant than in neighboring lysosomes), class I and II MHC molecules, cathepsin D, and transferrin receptor.

Labeled transferrin added to the medium was internalized by infected macrophages and cycled through bacteria-containing vacuoles. In addition, surface-exposed GM_1 ganglioside, complexed with cholera toxin B subunit, was delivered rapidly to the mycobacterial vacuoles, reaching steady state within 10 min. Both observations suggest that the vacuoles are readily accessible to certain surface-derived moieties and lie within the rapid recycling pathway of the host cell.

Cathepsin D had been detected by immunoelectron microscopy in *M. tuberculosis*-containing vacuoles, but immunoblot analysis of isolated *M. avium*-containing vacuoles revealed that the enzyme was in its high-molecular-weight, immature form, and not the fully processed, cleaved form observed in IgG-bead phagolysosomes. This provides independent evidence of the limited hydrolytic capacity of the mycobacterial vacuole and, more important, suggests that endosomal constituents, such as cathepsin D and possibly LAMP1, may be delivered from the synthetic pathway rather than through fusion with existing, acidified endosomes. Analysis of IgG-bead phagosomes at very early time points indicates that this remodeling step with endosomal components delivered from the synthetic pathway is a phenomenon common to "normal" phagosome biogenesis.

All these data indicate arrest of normal phagosome biogenesis. This is supported by work demonstrating the retention of the small GTPase Rab5 on *Mycobacterium*-containing vacuoles. Rab5 is involved in modulation of homotypic fusion between vesicles of the early endosomal network. Further analysis of the vacuoles for other molecules known to be abundant in the early endosomal system showed the presence of the actin-binding coat protein coronin I, and revealed that the vacuoles were enriched for cholesterol. The cholesterol may be indicative of trafficking of lipid rafts that also contain glycosphingolipids such as GM_1. Maturation of the phagosomes containing live *Mycobacterium* spp. is reported to be arrested prior to the acquisition of the PI3P-binding protein EEA1 and the lysosome-associated

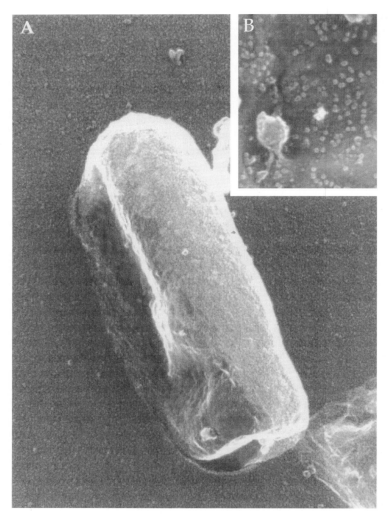

Figure 10.6 **(A)** Electron micrograph of a platinum replica from an isolated *Mycobacterium avium*-containing phagosome. The view is of the cytoplasmic face of the phagosomal membrane and reveals the atypical smooth texture of the phagosome. **(B)** Region of an *L. mexicana*-containing phagosome viewed at comparable magnification. The stud-like structures represent proton-ATPase complexes, which are rare on mycobacterial vacuoles. Proton-ATPases are responsible for the normal acidification of phagosomes. Courtesy of David G. Russell and John Heuser.

GTPase Rab7. Manipulation of the activity of the PI3 kinase VPS34 by microinjection of inhibitory antibodies produced a latex bead phagosome that expressed the same arrested maturation observed in *Mycobacterium*-containing vacuoles. These data indicate that the point of arrest lies downstream of Rab5 and PI3P acquisition yet prior to accumulation of EEA1 and Rab7. Recent data indicate that the bacterium may be suppressing the acquisition of EEA1 through direct inhibition of phosphorylation of PI by VPS34 through a Ca/calmodulin-dependent mechanism.

Treatment of infected macrophages with ATP induced both apoptosis and death of the infecting bacilli. ATP activates the purinergic receptors on the macrophage and caused increased fusion of lysosomes with the *Mycobacterium*-containing vacuole. The fusion capacity of the bacteria-containing phagosome can also be manipulated through the induction of a calcium flux by the ionophore A23187. This effect correlated with the level of phosphorylation of the calcium/calmodulin-dependent protein kinase II associated with the bacterial vacuole. These manipulations are intriguing and provide a glimpse of the level of complexity associated with the regulation of phagosome maturation.

Despite the wealth of data available on the presence or absence of a confusing range of host proteins on the bacterium-containing vacuole, the mechanism whereby mycobacteria retard the normal maturation of their phagosomes remains to be clarified. Researchers have suggested two different possibilities. First, *M. tuberculosis* and *M. bovis* BCG produce ammonia through urease activity, and it was proposed that this ammonia prevents acidification of the bacterial vacuoles. However, experiments with a urease-deficient mutant of *M. bovis* (BCG) demonstrated that, although the vacuolar pH of the wild type was higher than that of the urease-deficient mutant, there was no difference in the proportion of bacterium-containing vacuoles that fused with lysosomes preloaded with Texas red dextran. Second, mycobacterial lipids, notably cord factor (α,α'-trehalose-6,6'-dimycolate), and more recently lipoarabinomannan (LAM) are thought to have membrane fusion-blocking activities. Cord factor inhibited membrane fusion in vitro, and the presence of cord factor on *Nocardia* strains correlates with reduced fusion of vacuoles with lysosomes. In addition, phagosomes formed around LAM-coated latex beads showed a depressed maturation and acquisition of lysosomal markers in comparison with uncoated control particles. Recent data indicate that LAM may be suppressing the acquisition of EEA1 through direct inhibition of phosphorylation of PI by VPS34 through a Ca/calmodulin-dependent mechanism. Whether these lipids show the same activity on the bacterial surface and how these effects are transduced across the lipid bilayer remain to be determined.

Sequestered Compartments

The pathogens that reside in sequestered or isolated vacuoles fall into two main groups: (i) those that enter by phagocytosis but subvert the normal maturation pathway to establish a specialized compartment that no longer intersects with the endocytic network, such as *Legionella* and *Chlamydia*, and (ii) those that invade actively and build their own vacuole at the time of host cell entry, like *Toxoplasma* and *Plasmodium*. Because these compartments lie outside the normal endocytic pathway of the host cell, one of the obvious questions is how the pathogen obtains nutrients from the host. For this reason, there is particular interest in identifying pathogen-derived proteins that become constituents of the parasitophorous vacuole membrane.

Legionella. *Legionella pneumophila* is a facultative intracellular pathogen that parasitizes phagocytes. Its ability to survive within macrophages may represent a preadaptation for infection, because it is known to associate with free-living amoebae. Entry into the phagocyte can be mediated by a peculiar form of "coiling" phagocytosis, although it is unclear if this uptake pathway has any influence on the outcome of an infection. Recent work has demonstrated that the bacteria can also cause a productive infection in murine macrophages following internalization via macropinocytosis that depends on the functionality of the Dot/Icm secretion system. Recent mutational analyses have identified the *dot* or *icm* locus as a linked set of genes, or pathogenicity island, encoding an inducible secretion apparatus similar to the type III secretion systems described in chapter 15. Intriguingly, the induction of macropinocytosis in the macrophages is linked to the *Lgn*1 locus of the mouse. The phenotype of the dominant form of this host allele is partial resistance to infection, and macrophages from these mice do not show macropinocytic uptake of *Legionella*.

Within a few hours of entry, *Legionella*-containing vacuoles recruit intracellular organelles and become studded with ribosomes. The pH of these compartments has been calculated at 6.1, which is 0.8 pH unit higher than that of vacuoles formed around dead bacilli. Immunofluorescence studies on *Legionella* vacuoles demonstrated the presence of the endoplasmic reticulum (ER) protein BiP, and it has been suggested that these vacuoles represent nascent autophagous vacuoles. Treatments which enhanced autophagy (amino acid starvation of the host cells) increased the association of ER membranes with these vacuoles and enhanced bacterial division. The *dot/icm* locus is also implicated in this behavior, and *Legionella* organisms deficient in key members of this gene locus were unable to prevent delivery to lysosomes, culminating in the death of the bacilli. Recent immunofluorescence analysis of infected cells revealed that the host protein ADP-ribosylation factor 1 (ARF1) was localized to vacuoles containing wild-type bacteria but was absent from vacuoles containing *dot/icm*-deficient mutants. ARF1 is a small GTP-binding protein that regulates membrane traffic from the ER to the Golgi apparatus and would be a likely candidate to mediate the transformation of the *Legionella*-containing phagosome. ARF1 activity is regulated by nucleotide exchange factors, and a bacterial homologue of these host proteins, RalF, was identified within the *Legionella* genome. *Legionella* mutants deficient in RalF expression no longer colocalized with ARF1; however, the mutation did not have a discernible effect on bacterial survival. Recent experiments have shown that the *dot/icm* transport system actually releases a bewildering array of substrates into the host cell, suggesting extensive functional redundancy in effector molecules.

In early experiments it was shown that formalin-fixed or antibody-opsonized bacteria entered compartments that fused with secondary lysosomes, whereas treatment of cells with antibiotics postinfection did not induce fusion of bacterial compartments with lysosomes. These data suggest that a "viable" bacterial compartment, once established, cannot revert to the endosomal network. However, more recent analyses indicate that 18 h after infection the majority of the vacuoles containing live, replicating bacteria are actually acidic and contain both LAMP1 and cathepsin D. Moreover, inhibitors of proton-ATPase activity like bafilomycinA block bacterial replication, indicating that the bacteria favor an acidic environment. This ability to withstand and survive inside a more hostile environment is consistent with the absence of a survival phenotype in the RalF-deficient mutants and suggests that modulation of the vacuole is a developmentally regulated behavior in *Legionella*.

Brucella abortus.   *Brucella* is a gram-negative bacterium, related closely to the plant pathogen *Agrobacterium tumifaciens*, which is capable of entering and surviving in nonprofessional phagocytes in its host. Shortly after uptake the pathogen was found enclosed in a vacuole that acquired early endosomal markers such as Rab5, transferrin receptor, and EEA1. However, although the vacuole proceeded to become positive for the glycoprotein LAMP1, it failed to acquire other lysosomal markers, such as the aspartic proteinase cathepsin D, and did not fuse with preloaded lysosomal compartments. Morphologically the vacuoles containing viable, dividing *B.*

abortus bacilli have many of the characteristics of immature autophagosomes, and the vacuoles were shown to contain the ER marker ser61β and label with monodansylcadaverine (known to accumulate in autophagous compartments). As the infection became more established, the vacuoles lost LAMP1 labeling and associated more closely with the ER. Eventually the bacteria-containing vacuoles appear to communicate directly with the lumen of the ER. This association depends on the expression of the VirB type IV secretion system, indicating that bacterial effectors are delivered into the host cell to drive this process. This transition is opposite from the one reported for *Legionella.*

   Chlamydia.   The human pathogen *Chlamydia trachomatis* and the animal pathogen *C. psittaci* are both obligate intracellular parasites that invade nonprofessional phagocytes by activation of the host cell contractile system. Researchers on *Chlamydia* spp. are in the interesting situation of having the genome sequence for four different species but having no genetic tools available to probe function.

   Although the bacilli are actively internalized by the host cell, the vacuoles are rapidly remodeled by the pathogen to form parasitophorous vacuoles, termed inclusion bodies (Figure 10.7). Early analysis of these com-

Figure 10.7   Electron micrograph of a freeze-etch preparation of a HeLa cell infected with *Chlamydia psittaci.* The bacteria (black arrows) form an inclusion body or parasitophorous vacuole (PV) that lies within the host cell cytosol (host cell) and is excluded from the normal endocytic routes of that cell. The vacuole membrane is smooth over most of its surface; however, it is ruffled with processes (white arrows) that extend into the host cell cytoplasm in the region that subtends the host cell endoplasmic reticulum. Courtesy of David G. Russell and Ted Hackstadt.

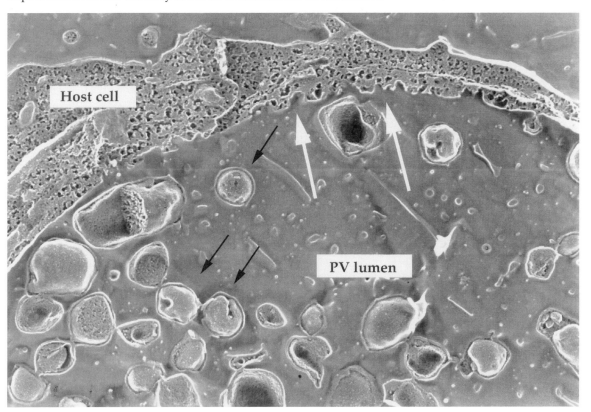

partments revealed that they did not possess acid phosphatase, nor did they fuse with ferritin-labeled secondary lysosomes. More detailed immunochemical analysis of the vacuole constituents revealed that they lacked cation-independent mannose 6-phosphate receptor (M6PR), transferrin receptor, both LAMP1 and LAMP2, cathepsin D, and vacuolar proton ATPase. Despite this clear lack of host cell proteins, the inclusion bodies showed a close association with annexins III, IV, and V. Annexins are implicated in Ca^{2+}-dependent, intracellular membrane fusion events. Indeed, although the chlamydial inclusion body did not fuse with endosomes or lysosomes, they were able to fuse with one another. Moreover, recent analysis with the fluorescent sphingomyelin precursor C_6-nitrobenzoxadiazolyl-ceramide demonstrated transport of a significant proportion of the label to inclusion body membranes and, subsequently, to the bacilli inside. These data suggest that there is at least some communication between the host cell Golgi apparatus and the parasitophorous vacuoles.

Cell fractionation techniques and the generation of monoclonal antibodies with specific staining patterns in infected cells have allowed researchers to identify proteins, termed Inc's, associated with the inclusion body membrane. Some of these proteins are species specific while others have clear homologues across species. Inc's A, F, and G have been shown to have cytoplasmically exposed carboxyl termini, and IncA is known to be phosphorylated by host cell kinases. Recent use of yeast two-hybrid screens demonstrated that IncG interacted with the host cell protein 14-3-3β that is a known contributor to many cellular signaling pathways. The degree of interaction correlated with the phosphorylation status in IncG. Additional candidates to modulate cell function include CopN, a homologue of the Yersinia type III secretion system effector protein YopN, and the CPAF proteinase that localizes to the host cell cytosol and actively degrades host cell transcription factors. Although the components of a type III secretion system have not been identified, early electron microscopy studies did report observing filamentous projections extruded from the bacterial surface, suggesting that this bacterium also exploits a type III system to modulate host cell function.

Toxoplasma. *Toxoplasma gondii* is a widespread apicomplexan parasite related to *Plasmodium*. The parasite is spread primarily through ingestion of cysts in undercooked meat or in food contaminated with cat feces. This can lead to the formation of cysts in the brain of the intermediate hosts, which include humans. Although in immunocompetent individuals this infection is almost invariably subclinical, in immunocompromised persons the parasite expansion is unrestricted and usually fatal. In some rural areas of the world where consumption of raw or rare meat is common, the sera of at least 70 to 90% of the population are positive for *Toxoplasma*.

The parasite is capable of invading and developing in an extremely diverse range of host cell types. Infection is initiated by ingestion of oocysts or tissue cysts, and the parasites form tachyzoites, which are highly motile and spread the infection. As discussed above, invasion is an active process that does not require any participation by the host cell cytoskeleton; on the contrary, ligation of receptors capable of triggering phagocytosis appears to direct the parasite into the lysosome.

▶ *For Figure 10.8, see color insert.*

Because the parasites are the sole mediators of this process, it is not surprising to find that invasion and the formation of parasitophorous vacuoles containing *Toxoplasma* require the triggered secretion of an array of proteins that are released into the nascent vacuole during the entry process (Figure 10.8). Early ultrastructural studies described an extensive tubular vesicular network in the vacuole that appeared to be contiguous with the vacuole membrane. More recent biochemical and immunolocalization studies have shown that the released proteins originate from three different secretory organelles: the micronemes, the rhoptries, and dense granules. The proteins were targeted differentially to the lumen of the parasitophorous vacuole, the tubular vesicular network, and the vacuolar membrane. One of these proteins, the nucleotide adenosine triphosphatase, has a known enzymatic function and may be involved in purine salvage, because *Toxoplasma* is a purine auxotroph. NTPase activity would yield AMP, which could be further dephosphorylated by the 5'-nucleotidase of the parasite facilitating transport of adenosine into the cell. In addition, the marked association between the membranes of the parasitophorous vacuoles and membranes of the host cell ER and mitochondria have added further significance to the localization of the rhoptry proteins ROP2, ROP3, ROP4, and ROP7 to the parasitophorous vacuole membrane. Recent experiments identified a mammalian mitochondrial targeting sequence in the amino terminus of ROP2. When expressed in HeLa cells, the protein showed partial translocation into the mitochondria. It is suggested that ROP2 acts as an anchor that ensures an intimate association between the parasitophorous vacuole membrane and the mitochondria.

The proteins from the rhoptries and dense granules must modify the parasitophorous vacuole to satisfy the requirements of the parasite and, in addition to the mitochondria tethering, the vacuole membrane is known to have pores capable of facilitating passive transport of molecules up to 1,300 Da. This would provide access to nucleotides, amino acids, and small peptides from the host cell cytoplasm. The degree of modification of this compartment is extensive and parallels that of *Chlamydia*.

Pathogens Resident in Host Cell Cytosol

A few pathogens have evolved mechanisms that allow them to avoid the potentially hostile environment of the endosomal and lysosomal network of the host cell by escaping into the host cell cytosol. This group includes both bacteria (e.g., *Shigella, Listeria,* and *Rickettsia*) and a protozoan (*Trypanosoma cruzi*). All these pathogens exhibit membrane-disruptive activities that are optimal at low pH, suggesting that there is an "activation" step during the acidification of the vacuoles formed during parasite entry. Genetic characterization of the loci involved in vacuolar escape by the bacterial pathogens within this group has revealed a battery of secreted products, many with analogous functions, delivered via type III secretion systems encoded predominantly on pathogenicity islands.

Although the adoption of a cytosolic location enables these pathogens to bypass the degradative pathway to the lysosome, exploitation of this niche carries several other problems, most specifically efficient cell-cell spread and avoidance of antigen presentation via class I MHC molecules.

Shigella. *Shigella flexneri,* one of the causative agents of bacillary dysentery, was the subject of some early experiments in the early and mid-1980s that marked the advent of the new era in microbial cell biology. It was observed that the ability of these bacilli to replicate inside cells in culture correlated not with the production of Shiga toxin but with the presence of a 140-MDa plasmid, pWR100. Moreover, introduction of the plasmid into *Escherichia coli* through transconjugation conferred the ability to survive inside HeLa cells. Ultrastructural analysis of infected cells revealed recombinant bacilli free in the cell cytosol. Coincubation of plasmid-positive and -negative strains of *Shigella* and *E. coli* together with erythrocytes demonstrated that the plasmid enabled the bacteria to induce lysis of erythrocytes.

More recent detailed analysis of the plasmid-borne genes identified a family of products, IpaA to IpaD. These proteins are secreted on contact with a host cell through a dedicated type III secretory system, encoded by the *mxi-spa* locus on the bacterial chromosome. The secreted proteins all perform functions designed to maximize the chances that the bacillus will infect the host successfully. IpaB associates with the interleukin-1β-converting enzyme and leads to the induction of apoptosis, and is discussed in depth in chapter 17. IpaC is a component of the pore in the host cell membrane but also appears to function as an F actin-nucleating element. IpaA binds to vinculin and induces cyoskeletal rearrangements reminiscent of focal adhesion plaques. These activities are also modulated by IpgD that expresses phosphatidylinositol phosphatase activity that has a dramatic effect on membrane-associated cytoskeletal elements.

The mechanism underlying *Shigella*'s escape from the vacuole is a matter of some contention and speculation. IpaD and IpaC are known to form polymers and to be required for binding and entry; however, it is still unclear if one or both of these proteins mediate vacuolar lysis. Binding of IpaC to liposomes was enhanced by acidic pH, suggesting that the intercalation of the protein into vesicles may be favored within the low pH of the endosome. *Shigella* mutants deficient in IpaC and IpaD could be complemented by the hemolysin of *E. coli,* suggesting that the molecules perform related functions. Intriguingly, however, recent characterization of the secreted proteins (Ssp) of *Salmonella enterica* serovar Typhimurium revealed that the genes encoding SspB, SspC, SspD, and SspA are homologues of those encoding IpaB, IpaC, IpaD, and IpaA and also play a role in host cell entry; however, in contrast to *Shigella, Salmonella* remains within its phagosome.

Like *Listeria, Shigella* is also capable of inducing actin-based motility in the host cytosol. This is due to the expression of IcsA that is localized preferentially at one pole of the bacteria. IcsA binds N-WASP in the host cytosol, activates the protein to bind the ARP2/3 complex, and thereby induces localized polymerization of actin.

Listeria.   *Listeria* is the causative agent of listeriosis and has emerged recently as the paradigm for defining the mechanisms of survival of an intracytosolic pathogen. Like *Shigella, Listeria monocytogenes* escapes from its vacuole into the cytoplasm shortly after entry into its host cell. However, in contrast to *Shigella,* where there is a membrane-disruptive activity but no defined molecule, *Listeria* possesses three well-characterized molecules with membrane-disrupting capabilities. The genes encoding the major players in escape from the vacuole are linked physically in a pathogenicity island that encodes LLO (listeriolysin O in *L. monocytogenes*), the metalloproteinase Mpl, the actin-binding protein ActA, and PlcB, a phosphatidylinositol-specific phospholipase C. In early experiments it was shown that expression of the listeriolysin (LLO) gene in the nonpathogenic *Bacillus subtilis* converted this bacterium into one capable of survival in tissue culture cells through its acquired ability to escape from its phagosomes. LLO is a thiol-activated cytolysin that belongs to a family of homologous proteins present in several pathogenic bacilli, most notably *Clostridium* and *Streptococcus*. Experiments conducted on LLO-negative *Listeria* complemented with the closely related perfringolysin (PFO) from *Clostridium perfringens* demonstrated that minimal alterations in the amino acid sequence of PFO rendered it capable of fulfilling the same role as LLO in vacuolar escape. However, the sequence differences between PFO and LLO render LLO pH sensitive and restrict its activity to an acidic milieu, thus limiting its toxicity to the host cell.

Although LLO appears to be sufficient to confer vacuolar escape on *Listeria*, these bacteria also possess two different phospholipases which are released during bacterial invasion. The lipases include a broad-spectrum phospholipase C (PC-PLC) and a phosphatidylinositol-specific phospholipase C (PI-PLC). The PC-PLC is released from the bacilli in an active form, while the PI-PLC is secreted as an inactive proenzyme that requires cleavage for activation. Activation can be mediated by the bacterial metalloproteinase (Mpl) or by the activity of host cell cysteine proteinases, most probably cathepsins B and/or L. The activation of PI-PLC requires a low-pH environment, suggesting that activation is concomitant with maturation of the phagosomes into acidic, hydrolytic vacuoles. Experimental data suggest that the lipases play a role in escape from the double-membrane vacuoles formed during cell-cell spread.

To survive and move within and between cells, *Listeria* exploits the ability of ActA to promote binding and polymerization of the host cell actin cytoskeleton. This activity is fundamental to the intracytosolic lifestyle of this pathogen and is detailed extensively in chapter 12.

Modulation of Intracellular Compartments by the Host Immune Response

The ability of the immune system to regulate intracellular pathogens is based on its capacity (i) to detect infection foci and (ii) to produce the appropriate response in either infected cells or bystander effector cells. Recently, much of the attention on killing of intracellular pathogens, in macrophages in particular, has focused on inducible nitric oxide synthase (iNOS). Expression of iNOS is turned on by gamma interferon and tumor necrosis factor alpha and leads to the production of NO from arginine. Inhibitors of NO production block the killing of many intracellular pathogens, and infections with *Listeria*, *Salmonella*, *Leishmania*, *Plasmodium*, and *Mycobacterium* species run uncontrolled in iNOS knockout mice. Obviously, iNOS fulfills a necessary function in regulation of these infections, but a full appreciation of its mode of action must take into account the cascade of other physiologic changes that occur during macrophage activation.

For intracellular pathogens that need to actively maintain their intracellular compartments, such as *Mycobacterium* species, the obvious question is which event comes first: the death or compromise of the infecting microbe or the differentiation of its compartment into an acidic, hydrolytically competent lysosome. If the latter is true, this translocation could alter drastically both the environment and the cofactors that would potentiate the efficacy of NO. Experiments on murine macrophages infected with *M. avium* demonstrated that activation facilitated the acidification of mycobacterial vacuoles in both de novo and established infections. The functional translocation toward more lysosomal compartments preceded any marked drop in microbial viability, suggesting that it was the product of an alteration in macrophage physiology rather than a consequence of microbial death.

The lysosomal environment of activated macrophages could potentiate NO toxicity in several ways. Oxidation of NO to nitrite and nitrate is retarded at acidic pH. NO can combine with reactive oxygen intermediates, whose production is up-regulated in activated macrophages, to make peroxynitrite ($ONOO^-$). NO can release metal ions, such as Fe^{2+}, from metalloproteins, which can combine with H_2O_2 to produce ·OH and hypervalent Fe. Furthermore, the activity of lysosomal hydrolases on the microbial cell wall will probably expose more targets to oxidative attack. The microbicidal responses of activated macrophages rely on the complex interactions of several antimicrobial phenomena, and more work on the effects of activation on the regulation of intracellular fusion within the endosomal-lysosomal continuum is required before these interactions can be appreciated.

Nutrient Acquisition: Eat In or Order Out?

The acquisition of nutrients by intracellular pathogens raises an interesting series of questions that are more accessible today than they have ever been in the past. Many virulence factors are single molecules with single functions that are readily accessible through mutant screens, but nutrient acquisition, processing, and metabolism represent a greater level of complexity and need to be dealt with in a more system-based approach, the perfect subject for microarray analysis!

The acquisition of some nutritional requirements, most notably iron, has been studied in much greater depth than any other "nutrient." Most intracellular pathogens have evolved to deal with the relatively low availability of iron and synthesize an array of iron-binding molecules to wrest the element from the host. *M. tuberculosis* synthesizes siderophores and mycobactins that are released by the bacteria to gather iron from the host. Both molecules have an affinity for iron higher than mammalian transferrin, and experiments with ^{59}Fe-loaded transferrin demonstrated that the bacteria can utilize iron delivered via transferrin to the bacteria-containing vacuoles.

Interestingly, iron can also be toxic to pathogens, either directly or through the activity of NADPH oxidase and iNOS. *Plasmodium* spp. parasitize red blood cells of their hosts and quietly digest their way through the hemoglobin inside their host cell. Hemoglobin is the oxygen-carrying pigment of the host and relies on iron to function. Excess iron is toxic to *Plasmodium* so the parasite sequesters the iron from hemoglobin in a crystalline aggregate called hemozoin that is polymerized in the food vacuole. One of the postulated mechanisms of action of the antimalarial drug chloroquine is to block polymerization of the hemozoin by binding to its nucleating agent the histidine-rich protein HRP. Parasites incubated in [^{3}H]chloroquine showed incorporation of the drug into the hemozoin (Figure 10.9), demonstrating the specific localization of the drug to its potential site of action.

Malaria is in fact the exception; very little data are available on the carbon sources utilized by most intracellular pathogens. Recent studies have shown that another apicomplexan, *Toxoplasma*, can utilize host lipids and fatty acids that it metabolized into other lipid species, although it is unclear whether this was exclusively remodeling or whether the protist

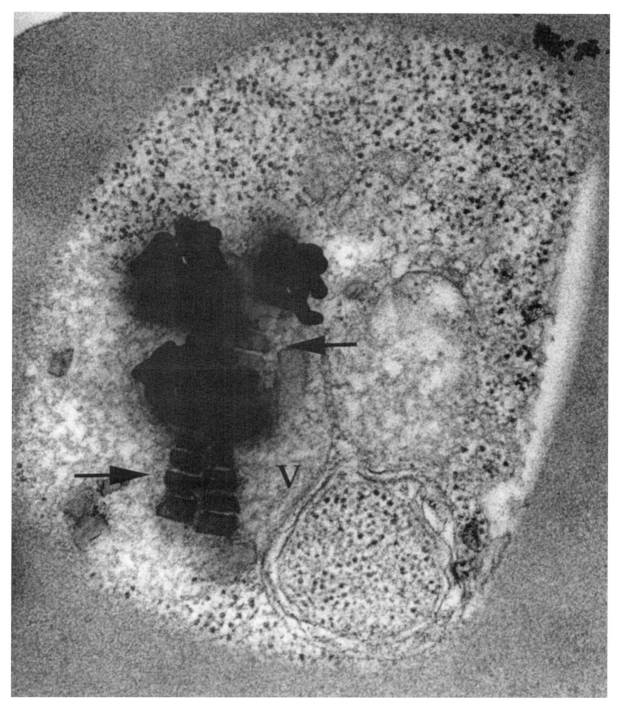

Figure 10.9 An autoradio electron micrograph of an erythrocyte (RBC) infected with *Plasmodium falciparum* (Plasmodium). The micrograph illustrates the polymerization of hemozoin (arrowed) within the degradative food vacuole (V). The parasite culture was incubated with [3H]chloroquine prior to processing, and the antimalarial drug can be seen to localize to the hemozoin polymer. This is consistent with the proposed mode of action of the drug. Courtesy of David G. Russell and Daniel Goldberg.

can actually use the fatty acids for gluconeogenesis or the de novo synthesis of amino acids. Bacterial pathogens can mobilize fatty acids for such metabolic usage. *M. tuberculosis* had long been postulated to rely on host lipids as a nutrient source. In recent experiments the enzyme isocitrate lyase was shown to be essential for maintenance of infection in an immune host, or in activated macrophages in culture. Isocitrate lyase is the gating enzyme into the glyoxylate cycle that competes with isocitrate dehydrogenase and the tricarboxylic acid (TCA) cycle for directing carbon flux into gluconeogenesis or anaplerotic metabolism versus glycolysis. It is postulated that isocitrate lyase activity correlates with an enforced switch in metabolism dictated by the changing environment inside the macrophage when it is activated by cytokines present in the immune host. The utilization of fatty acids from the host has also been demonstrated for *C. trachomatis* that acquires sphingolipids from its host via the *trans*-Golgi network. A Chinese hamster ovary cell line that is defective in sphingolipid synthesis was shown not to support growth of *Chlamydia* unless rescued genetically, or by the addition of membrane-permeable sphingolipid precursors. These data suggest a role in nutrition, although this has to be demonstrated directly.

Finally, it had long been postulated that the cell cytosol represents an environment permissive to bacterial replication; however, recent experiments in which different bacteria were microinjected into cell cytosol throw this assumption open to question. Intriguingly, *Yersinia, Salmonella, Listeria innocua,* and *B. subtilis* were all unable to replicate in the cytoplasm of viable cells, while *Shigella, L. monocytogenes,* and enteroinvasive *E. coli* were able to grow in this environment. Such an outcome could be achieved either through an active microbicidal response or through nutritional restriction. The microbicidal capacity of the cytoplasm remains unexamined, although it is known that mammalian cells can detect microbial products in the cytoplasm with specialized "Toll-like" receptors or CARD proteins. However, recent data indicate that the cytosol need not be the nutrient-rich "soup" that we imagined, and that bacteria such as *Listeria* require specific adaptations to utilize cytosolic nutrients. For *L. monocytogenes,* the rapid replication in the cell cytoplasm requires the expression of *hpt,* a gene encoding a hexose phosphate transporter, suggesting that glucose 6-phosphate is an important carbon source in the cytosol. Furthermore, the bacterium has to scavenge lipoic acid from the host as a required cofactor for pyruvate dehydrogenase activity. Bacteria defective in one of the two lipoate protein ligase genes cannot grow in the cytoplasm. The occurrence of homologues in other pathogenic bacteria indicates that this may be a common requirement.

Increasing our knowledge of the metabolic pathways important to intracellular pathogens is extremely important because it lies at the center of designing effective screens for new drugs active against these organisms. Drug screens involving organisms grown in rich medium will inevitably "miss" metabolic targets that are both unique to these organisms and essential for maintenance of infection.

Cell-Cell Spread (Metastasis) of Intracellular Pathogens

To be successful, intracellular pathogens must evolve a strategy to enable them to spread to new host cells or ultimately to new hosts. Most pathogens cause the death of their host cell at this stage. For some pathogens, such as *Plasmodium, T. gondii,* and *T. cruzi,* this has led to development of a "swarm" response whereby the host cell is lysed and there is a coordinate release of infective forms of the microbe. Other pathogens, such as *Leishmania,* just induce a progressive deterioration of their host cell, and, because they exist as amastigotes (infective forms), they infect fresh macrophages on phagocytosis following release by their degenerate host cell. These strategies appear to require minimal participation by the host cell. In contrast, both *Shigella* and *Salmonella* are capable of triggering programmed cell death (apoptotic) responses in their host cells. For *Shigella,* at least, this occurs through a very specific route, and the mechanisms behind the induction of these cell destruction pathways are discussed in chapter 17.

Finally, one of the most intriguing means of cell-cell spread is through direct transfer, which is exploited by *Listeria, Shigella,* and *Rickettsia* and is a direct product of the ability of these bacteria to subvert the actin component of the host cell cytoskeletal network to their own devices (this is discussed in chapter 12).

Conclusion

Intracellular parasitism covers the diverse lifestyles of a broad phylogenetic spectrum of pathogens. This chapter has attempted to present the major points in the biology of these pathogens within a thematic framework from the time of initial infection, through the choice of intracellular niche, avoidance or exploitation of the immune response, and culminating in the metastasis or spread of the infection. These processes function as continua that have critical points of decision which determine the fate of the microbe, and this chapter has described these decision points for different pathogens and explored the consequences of the "wrong" decision. Since the infection process is a continuum, it is equally important to read the chapters that deal with issues bordering on the establishment and maintenance of an intracellular infection; these points of intersection have been noted in the text.

Selected Readings

Brucella
Celli, J., C. de Chastellier, D. M. Franchini, J. Pizarro-Cerda, E. Moreno, and J. P. Gorvel. 2003. Brucella evades macrophage killing via VirB-dependent sustained interactions with the endoplasmic reticulum. *J. Exp. Med.* **198:**545–556.

Chlamydia
Scidmore, M. A., and T. Hackstadt. 2001. Mammalian 14-3-3β associates with the *Chlamydia trachomatis* inclusion membrane via its interaction with IncG. *Mol. Microbiol.* **39:**1638–1650.

van Ooij, C., L. Kalman, S. van Ijzendoorn, M. Nishijima, K. Hanada, K. Mostov, and J. Engel. 2000. Host cell-derived sphingolipids are required for the intracellular growth of *Chlamydia trachomatis. Cell. Microbiol.* **2:**627–637.

Legionella
Luo, Z. Q., and R. R. Isberg. 2004. Multiple substrates of the *Legionella pneumophila* Dot/Icm system identified by interbacterial protein transfer. *Proc. Natl. Acad. Sci. USA* **101:**841–846.

Nagai, H., J. C. Kagan, X. Zhu, R. A. Kahn, and C. R. Roy. 2002. A bacterial guanine nucleotide exchange factor activates ARF on *Legionella* phagosomes. *Science* **295:** 679–682.

Sturgill-Koszycki, S., and M. S. Swanson. 2000. *Legionella pneumophila* replication vacuoles mature into acidic, endocytic organelles. *J. Exp. Med.* **192:**1261–1272.

Watarai, M., I. Derre, J. Kirby, J. D. Growney, W. F. Dietrich, and R. R. Isberg. 2001. *Legionella pneumophila* is internalized by a macropinocytotic uptake pathway controlled by the Dot/Icm system and the mouse Lgn1 locus. *J. Exp. Med.* **194:** 1081–1096.

This paper describes the fascinating interplay between host and pathogen at the genetic level. Macropinocytic uptake of the bacterium depends on the expression and function of the dot/icm system. However, the bacterial TTSS is effective only in triggering macropinocytosis in mice homozygotic for the recessive form of the lgn1 locus. This suggests that the lgn1 locus encodes a product that counteracts the bacterial TTSS and renders the host less susceptible to infection.

Leishmania

Dermine, J. F., S. Scianimanico, C. Prive, A. Descoteaux, and M. Desjardins. 2000. *Leishmania* promastigotes require lipophosphoglycan to actively modulate the fusion properties of phagosomes at an early stage of phagocytosis. *Cell. Microbiol.* **2:** 115–126.

Houde, M., S. Bertholet, E. Gagnon, S. Brunet, G. Goyette, A. Laplante, M. F. Princiotta, P. Thibault, D. Sacks, and M. Desjardins. 2003. Phagosomes are competent organelles for antigen cross-presentation. *Nature* **425:**402–406.

Wolfram, M., M. Fuchs, M. Wiese, Y. D. Stierhof, and P. Overath. 1996. Antigen presentation by *Leishmania mexicana*-infected macrophages: activation of helper T cells by a model parasite antigen secreted into the parasitophorous vacuole or expressed on the amastigote surface. *Eur. J. Immunol.* **26:**3153–3162.

Listeria

Glomski, I. J., M. M. Gedde, A. W. Tsang, J. A. Swanson, and D. A. Portnoy. 2002. The *Listeria monocytogenes* hemolysin has an acidic pH optimum to compartmentalize activity and prevent damage to infected host cells. *J. Cell. Biol.* **156:**1029–1038.

Listeriolysin O is a pore-forming toxin released by *Listeria* to facilitate its escape into the host cell cytoplasm. Interesting questions are why is the activity of this toxin limited to the vacuole membrane, and why does it not kill the cell. Listeria in which LLO had been replaced with the related lysin, perfringolysin O, PFO, did kill the host cell, and it was shown that the differences in amino acid sequence between the two proteins rendered LLO pH sensitive. LLO needs an acid environment to be active and is therefore inactive in the host cell cytosol, whereas PFO retained its activity following escape into the cytosol.

Goetz, M., A. Bubert, G. Wang, I. Chico-Calero, J. A. Vazquez-Boland, M. Beck, J. Slaghuis, A. A. Szalay, and W. Goebel. 2001. Microinjection and growth of bacteria in the cytosol of mammalian cells. *Proc. Natl. Acad. Sci. USA* **9:**12221–12226.

Lecuit, M., S. Vandormael-Pourin, J. Lefort, M. Huerre, P. Gounon, C. Dupuy, C. Babinet, and P. Cossart. 2001. A transgenic model for Listeriosis: role of internalin in crossing the intestinal barrier. *Science* **292:**1722–1725.

This paper demonstrates the elegant use of transgenic mice to determine the role of a bacterial adhesin in establishment of an infection and invasion of the host. Although internalins had been shown to have a major role in binding to host cells in tissue culture systems, attempts to show their activity in invasion of animal models had met with limited success. In this paper the authors show that the infection of mice by *Listeria* through internalin activity depended on the expression of the appropriate cadherin. Mice expressing human cadherin were exquisitely sensitive to metastasis, and this invasion and spread were observed only in bacteria expressing internalin.

O'Riordan, M., M. A. Moors, and D. A. Portnoy. 2003. Listeria intracellular growth and virulence require host-derived lipoic acid. *Science* **302:**462–464.

Mycobacterium

Beatty, W. L., E. R. Rhoades, H. J. Ullrich, D. Chatterjee, J. E. Heuser, and D. G. Russell. 2000. Trafficking and release of mycobacterial lipids from infected macrophages. *Traffic* **1:**235–247.

Clemens, D. L., B. Y. Lee, and M. A. Horwitz. 2002. The *Mycobacterium tuberculosis* phagosome in human macrophages is isolated from the host cell cytoplasm. *Infect. Immun.* **70**:5800–5807.

Fratti, R. A., J. M. Backer, J. Gruenberg, S. Corvera, and V. Deretic. 2001. Role of phosphatidylinositol 3-kinase and Rab5 effectors in phagosomal biogenesis and mycobacterial phagosome maturation arrest. *J. Cell Biol.* **154**:631–644.

> This paper describes data relevant to determining the exact point of arrest of the maturation of the *Mycobacterium*-containing phagosome. Following uptake, the *Mycobacterium*-containing phagosome acquired and retained the GTPase Rab5; however, it failed to acquire the PI3P-binding protein EEA1 that is thought to facilitate maturation. PI3P is produced by the phosphatidylinositol 3'-kinase VPS34 that associates transiently with phagosome formed around inert particles and is observed around *Mycobacterium*-containing vacuoles.

McKinney, J. D., K. Honer zu Bentrup, E. J. Munoz-Elias, A. Miczak, B. Chen, W. T. Chan, D. Swenson, J. C. Sacchettini, W. R. Jacobs, Jr., and D. G. Russell. 2000. Persistence of *Mycobacterium tuberculosis* in macrophages and mice requires the glyoxylate shunt enzyme isocitrate lyase. *Nature* **406**:735–738.

> Organisms exploiting fatty acids as their primary carbon source utilize the glyoxylate cycle to facilitate retention of carbon to allow growth. In this paper *Mycobacterium* organisms deficient in the glyoxylate cycle-gating enzyme, isocitrate lyase, are shown to be defective in survival during the persistent but not the acute phase of infection. The reliance on isocitrate lyase activity can also be demonstrated in activated macrophages in culture. These data indicate that the metabolism of the bacterium changes in response to the changing environment in the activated macrophage and the immune host.

Sturgill-Koszycki, S., P. H. Schlesinger, P. Chakraborty, P. L. Haddix, H. L. Collins, A. K. Fok, R. D. Allen, S. L. Gluck, J. Heuser, and D. G. Russell. 1994. Lack of acidification in *Mycobacterium* phagosomes produced by exclusion of the vesicular proton-ATPase. *Science* **263**:678–681.

Vergne, I., J. Chua, and V. Deretic. 2003. Tuberculosis toxin blocking phagosome maturation inhibits a novel Ca2+/calmodulin-PI3K hVPS34 cascade. *J. Exp. Med.* **198**:653–659.

Plasmodium

Banerjee, R., J. Liu, W. Beatty, L. Pelosof, M. Klemba, and D. E. Goldberg. 2002. Four plasmepsins are active in the *Plasmodium falciparum* food vacuole, including a protease with an active-site histidine. *Proc. Natl. Acad. Sci. USA* **99**:990–995.

Lauer, S., J. VanWye, T. Harrison, H. McManus, B. U. Samuel, N. L. Hiller, N. Mohandas, and K. Haldar. 2000. Vacuolar uptake of host components, and a role for cholesterol and sphingomyelin in malarial infection. *EMBO J.* **19**:3556–3564.

Sullivan, D. J., Jr., I. Y. Gluzman, D. G. Russell, and D. E. Goldberg. 1996. On the molecular mechanism of chloroquine's antimalarial action. *Proc. Natl. Acad. Sci. USA* **93**:11865–11870.

Salmonella

Beuzon, C. R., S. Meresse, K. E. Unsworth, J. Ruiz-Albert, S. Garvis, S. R. Waterman, T. A. Ryder, E. Boucrot, and D. W. Holden. 2000. *Salmonella* maintains the integrity of its intracellular vacuole through the action of SifA. *EMBO J.* **19**:3235–3249.

> *Salmonella* defective in SifA expression are attenuated for survival inside macrophages. SifA is known to function in the formation of the LAMP-positive filamentous vesicles that associate with *Salmonella*-containing vacuoles. Close examination of the SifA mutants revealed that the bacteria were loose in the host cell cytosol. These data indicate that SifA function in maintenance of the bacteria-containing vacuoles, possibly through recruitment of membrane. Interestingly, these bacteria did not replicate well in the cytosol of the macrophages, which is in contrast to the study by Brumell in epithelial cells where the bacteria replicated freely.

Brumell, J. H., P. Tang, M. L. Zaharik, and B. B. Finlay. 2002. Disruption of the *Salmonella*-containing vacuole leads to increased replication of *Salmonella enterica* serovar Typhimurium in the cytosol of epithelial cells. *Infect. Immun.* **70**:3264–3270.

Ruiz-Albert, J., X. J. Yu, C. R. Beuzon, A. N. Blakey, E. E. Galyov, and D. W. Holden. 2002. Complementary activities of SseJ and SifA regulate dynamics of the *Salmonella typhimurium* vacuolar membrane. *Mol. Microbiol.* **44:**645–661.

Vazquez-Torres, A., Y. Xu, J. Jones-Carson, D. W. Holden, S. M. Lucia, M. C. Dinauer, P. Mastroeni, and F. C. Fang. 2000. *Salmonella* pathogenicity island 2-dependent evasion of the phagocyte NADPH oxidase. *Science* **287:**1655–1658.

> On ligation of certain receptors during phagocytosis the NADH oxidase complex is formed on the membrane of the nascent phagosome and is transported into the cell on the forming vacuole. This results in localized production of reactive oxygen intermediates that are extremely toxic to bacteria. *Salmonella*, defective in the expression of SPI-2 genes, are unable to block this process, and recruit the NADH oxidase complex to their phagosome, whereas the wild-type parental strain inhibits acquisition of the complex and survives.

Shigella

Niebuhr, K., S. Giuriaqto, T. Pedron, D. J. Philpott, F. Gaits, J. Sable, M. P. Sheetz, C. Parsot, P. J. Sansonetti, and B. Payrastre. 2002. Conversion of PtdIns(4,5)P(2) into PtdIns(5) by the *S. flexneri* effector IpgD reorganizes host cell morphology. *EMBO J.* **21:**5069–5078.

> One of the currently emerging themes in microbial pathogenesis is the ability of bacteria to exploit the signaling lipid moieties used by mammalian cells to regulate a range of cellular functions. In this study the bacterial effector IpgD is shown to be a phosphatidylinositol phosphatase that converts PI-4,5P into PI-5P and that this dephosphorylation event induces a profound change in the host cell cytoskeleton.

Tran, N., A. B. Serfis, J. C. Osiecki, W. L. Picking, L. Coye, R. Davis, and W. D. Picking. 2000. Interaction of *Shigella flexneri* IpaC with model membranes correlates with effects on cultured cells. *Infect. Immun.* **68:**3710–3715.

Toxoplasma

Charron, A. J., and L. D. Sibley. 2002. Host cells: mobilizable lipid resources for the intracellular parasite *Toxoplasma gondii*. *J. Cell Sci.* **115:**3049–3059.

Hakansson, S., A. J. Charron, and L. D. Sibley. 2001. *Toxoplasma* evacuoles: a two-step process of secretion and fusion forms the parasitophorous vacuole. *EMBO J.* **20:**3132–3144.

Opitz, C., and D. Soldati. 2002. 'The glideosome': a dynamic complex powering gliding motion and host cell invasion by *Toxoplasma gondii*. *Mol. Microbiol.* **45:**597–604.

> This is the only review cited in the literature section. It is cited because this laboratory has published a fascinating series of papers on the molecular basis of motility and invasion by *Toxoplasma gondii*, and it would be unfair and misleading to cite one paper without referring to several. Cell movement is pivotal to the infectious nature of this organism, and the contractile system and its expression are so novel they deserve attention.

Landmark Papers

> The landmark publications chosen define the early development of cellular microbiology and intracellular infection.

Bielecki, J., P. Youngman, P. Connelly, and D. A. Portnoy. 1990. *Bacillus subtilis* expressing a haemolysin gene from Listeria monocytogenes can grow in mammalian cells. *Nature* **345:**175–176.

> In this more recent publication Bielecki and colleagues moved the field forward by showing that the listerolysin O gene from *Listeria* contained all the information necessary to allow the bacterium to escape from the vacuole and gain access to the cytosol. The LLO gene expressed in *B. subtilis* redirected the bacillus to the host cell cytoplasm and saved it from certain death within the lysosome.

Brown, C. A., P. Draper, and P. D. Hart. 1969. Mycobacteria and lysosomes: a paradox. *Nature* **221:**658–660.

> This is the field in which I work so I may be viewed as somewhat partial; however, for me this paper was way ahead of its time. It set the tone for the cellular microbiology studies that followed it. The senior author, Philip D'Arcy Hart, renowned as the designer of the definitive clinical trial, had retired from leading the MRC Tuberculosis Unit and decided to

turn his interests to intracellular survival of *Mycobacterium* spp. In this early study he demonstrated that the bacilli are sequestered in vacuoles that fail to fuse with lysosomes and that this failure correlated with the viability of the bacteria.

Horwitz, M. A., and F. R. Maxfield. 1984. *Legionella pneumophila* inhibits acidification of its phagosome in human monocytes. *J. Cell. Biol.* **99:**1936–1943.

The next paper comes from the work of Horwitz and colleagues who were interested in analysis of the vacuole in which *Legionella* resides inside phagocytes. This laboratory had noted previously that the mode of uptake of the bacterium was unusual (coiling phagocytosis) and set out to study the vacuole itself as part of a collaboration with a bona fide cell biologist (Maxfield). The result was the accurate pH measurement within the vacuole formed around *Legionella*, which was significantly higher than the pH values of phagosomes formed around other particles. This represented a new departure for the field.

Isberg, R. R., and S. Falkow. 1985. A single genetic locus encoded by *Yersinia pseudotuberculosis* permits invasion of cultured animal cells by Escherichia coli K-12. *Nature* **317:**262–264.

This paper, and the following one by Sansonetti and colleagues, marked the next step forward, that is, the combination of bacterial genetics and cell biology in the design of a genetic screen to identify, in this instance, the invasin genes in *Yersinia*. Isberg and Falkow transformed *Escherichia coli* with the shot-gunned genome of *Yersinia*, incubated the bacteria with cells, and then exploited the hydrophilic nature of gentamycin to kill extracellular bacteria and select for clones that encoded host cell entry. The result was a classic publication exploiting and combining a range of disciplines and laying the groundwork for an emerging field.

Sansonetti, P. J., A. Ryter, P. Clerc, A. T. Maurelli, and J. Mounier. 1986. Multiplication of *Shigella flexneri* within HeLa cells: lysis of the phagocytic vacuole and plasmid-mediated contact hemolysis. *Infect. Immun.* **51:**461–469.

Similarly, Sansonetti and colleagues used classic bacterial genetics and transconjugation to transfer the "virulence" plasmid of *Shigella* into *E. coli*. On examination, the modified *E. coli* were found to be free in the host cell cytoplasm, demonstrating that the plasmid from *Shigella* encoded the capacity to allow the bacterium to escape from the phagosome and attain access to the host cell cytoplasm. Like the preceding publication, this represented a saccadic leap in the field and laid the foundation for "Cellular Microbiology."

11

The Actin Cytoskeleton: Regulation of Actin Filament Assembly and Disassembly

FREDERICK S. SOUTHWICK

Eukaryotic cells possess three kinds of cytoskeletal elements: 5- to 9-nm-diameter actin filaments, 24-nm-diameter microtubules, and 10-nm-diameter intermediate filaments. All these polymer networks are assembled reversibly from monomers. In this chapter, we will concentrate on actin. Reorganization of the actin cytoskeleton is required for leukocytes to migrate to sites of infection, for fibroblasts and endothelial cells to migrate to areas of wound healing, and for platelets to plug leaking vessels. Actin polymerization not only drives the motility of leukocytes, fibroblasts, keratocytes, and amoebae, but also provides the propulsive force for the movement of intracellular bacteria including *Listeria, Shigella,* and *Rickettsia* as well as the poxvirus, vaccinia. Thus, a working knowledge of how host cells modulate their actin filament architecture is required to fully understand cell movement, host defenses, and bacterial and viral pathogenesis.

Basic Properties of Actin Monomer and Filament

Actin exists in two interchangeable forms: monomeric globular actin (G-actin, 43 kDa) that polymerizes through reversible noncovalent interactions into a two-stranded helical polymer (F-actin). Atomic resolution structures of the actin monomer in both the ADP and ATP forms have been determined. The actin molecule consists of four domains. Domains I and II are separated from domains III and IV by a cleft containing a high-affinity nucleotide-binding site and a high-affinity divalent cation-binding site, which is usually bound by ATP and magnesium, respectively, in vivo (Figure 11.1). Actin is an enzyme that catalyzes the hydrolysis of the nucleoside triphosphate, ATP, to ADP and inorganic phosphate (P_i). G-actin has a very low ATPase activity; incorporation of the monomers into the filament significantly enhances this activity. Following incorporation of actin-ATP monomers into the polymer, ATP hydrolysis occurs, resulting in an intermediate of actin-ADP-P_i, followed by the slower release of inorganic phosphate into the surrounding medium, forming actin-ADP species (Figure 11.2). The monomers retain the ADP, and it is exchanged for ATP once the monomers have dissociated from the filament. Thus, the actin-nucleotide

255

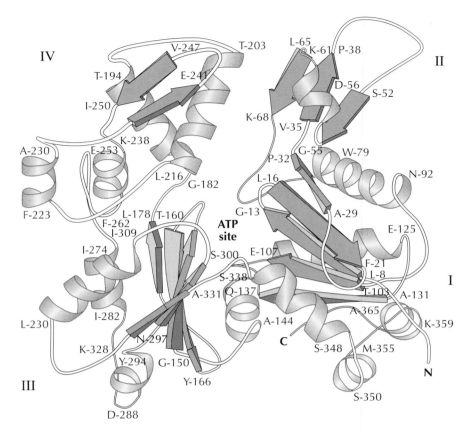

Figure 11.1 Atomic-level structure of an actin monomer showing the ATP-binding site. Based on the work of W. Kabsch, H. G. Mannherz, D. Suck, E. F. Pai, and K. C. Holmes, *Nature* **347**:37–44, 1990. The vertical axis of the monomer (as depicted in this figure) runs parallel to the long axis of the filament (see Figure 11.3). The right-hand side of the molecule is exposed to the outside of the actin filament, whereas the left-hand side is nearest the long axis of the filament. Residues 262 to 274 are thought to reach across this axis and interact with the adjacent actin monomer of the double-stranded helix. As oriented here, the polarity of the filament would correspond to the pointed end at the top and the barbed end at the bottom. The roman numerals delineate the four domains of the molecule.

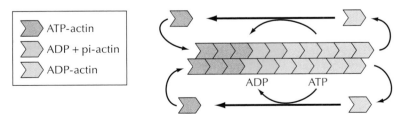

Figure 11.2 Schematic drawing of an actin filament under steady-state conditions. ATP-actin monomers add to the barbed or fast-growing end and are hydrolyzed to ADP-P$_i$, followed by the slower dissociation of P$_i$ to form ADP-actin. ADP-actin monomers dissociate from the pointed or slow-growing end. Under these conditions an actin monomer added to the barbed end will eventually treadmill through the filament and dissociate from the pointed end.

complex "matures" within the actin filament, and an individual filament contains segments carrying ATP, ADP-P_i, and ADP. A large free-energy change is associated with ATP hydrolysis; some of that energy is stored within the filament and probably as an altered state of the monomer within the filament. The ability of actin to assume different conformational as well as phosphorylation states may regulate the strength of interactions of the actin-binding proteins along the length of the filament. For example, ADF/cofilin (a filament-binding protein) has a higher affinity for ADP-actin than ATP-actin.

The crystal structure of F-actin has not yet been elucidated, and, for structural studies, one has to rely on a model of the actin filament. Holmes and coworkers constructed an atomic model of F-actin by using the best structural fit of the coordinates of G-actin to the X-ray fiber diffraction data of oriented actin gels of about 8-Å resolution. This model indicates that the large domain is located near the center of the filament axis, and the small domain is on the exterior (Figure 11.3). The filament is stabilized by both lateral and longitudinal contacts in the form of hydrophobic interactions, hydrogen bonds, and salt bridges between the monomers. This derived model of the actin filament is consistent with previously determined radial positions of selected residues. The actin filament can be thought of as two right-handed intertwined helices. Actin monomers assemble in a head-to-tail manner that accounts for the polar character of the filament. The two ends of the filament are not equivalent; the filaments display both structural and functional polarity. The structural polarity can be demonstrated by decorating F-actin with myosin heads that in the absence of ATP bind at a 45° angle forming an "arrowhead"-like pattern on electron micrographs defining a "barbed" end (also known as plus ends) and "pointed" end (also known as the minus ends).

Actin Assembly Kinetics

Actin polymerization in vitro consists of three steps: (i) nucleation (formation of actin oligomers that have a greater tendency to form filaments than to disintegrate into monomers); (ii) elongation (growth of the filament at both the barbed and pointed ends); and (iii) treadmilling (flux of monomers along the filament).

Initiation of polymerization, also known as nucleation, is an unfavorable process and, hence, a rate-limiting step during actin polymerization. It occurs by the association of actin monomers into oligomers of 3 to 4 subunits from which the actin filament grows. The size of the nucleus has been inferred by theoretical modeling. Actin nucleation is promoted by high actin-ATP concentration, as the nuclei then persist long enough for productive polymer formation. Cells contain actin monomer-sequestering proteins (see below) that suppress any spontaneous nucleation in regions where active actin polymerization is not occurring, because indiscriminate actin assembly would result in unregulated changes in shape and consistency of the cell.

After nucleation, elongation adds monomers to the two ends of the growing filament at different rates. Owing to the polar character of the actin filament, the rates of association (k_{on}) and dissociation (k_{off}) of the monomers are different at the two ends and, hence, the critical concentrations or K_Ds (k_{off}/k_{on}) at the two ends differ. When the association rate is faster than the dissociation rate, net elongation of filaments occurs. The rate

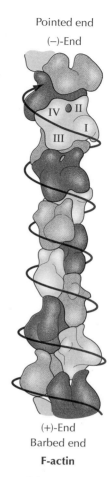

Pointed end

(−)-End

(+)-End

Barbed end

F-actin

Figure 11.3 Schematic drawing to the tertiary structure of an actin filament. Roman numerals correspond to the domains shown in Figure 11.1.

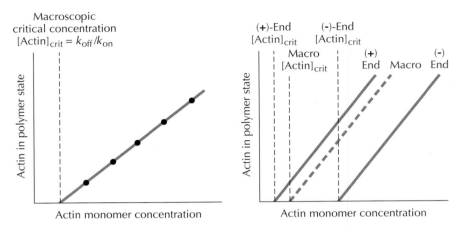

Figure 11.4   Critical-concentration behavior in actin polymerization. (Left) Plot of the steady-state actin filament concentration as a function of monomer concentration. This macroscopic behavior is often measured by the increase in fluorescence when pyrenyl-actin is incorporated into filaments. (Right) When both filament ends are uncapped, the macroscopic critical concentration lies between the microscopic critical concentrations for actin monomer interactions at the barbed end [or (+) end] and the pointed end [or (−) end]. Because the exchange rates are higher at the barbed end, the macroscopic critical concentration is closer to the microscopic critical concentration of the barbed end. At steady state, monomers will naturally dissociate from the pointed ends and will associate with the more stable barbed ends. This phenomenon is known as treadmilling. Because the exchange rates are higher at the barbed end, the macroscopic critical concentration is closer to that of the barbed end.

of elongation depends on the number of nuclei and the concentration of free actin monomers. The filament rapidly grows at both ends until a steady-state level of polymerization is reached.

At the steady state, the rate of monomer addition to ends of the filament is balanced by the rate of dissociation of the monomers, and there is no net change in the monomer concentration. The concentration of monomers at the steady state is called the critical concentration. This concentration reflects the on and off rates of both the barbed and pointed ends, and it is termed the macroscopic critical concentration of the filament. The individual ratios of the off rate over the on rate (the K_D) of the barbed end and pointed end represent the microscopic critical concentrations of the two ends (Figure 11.4). The pointed end is the slow-growing end with a higher critical concentration, while the barbed end is the fast-growing end and has a lower critical concentration. Because both the on and off rates of the barbed end are faster than those of the pointed end, the macroscopic critical concentration primarily reflects the critical concentration of the barbed end. The macroscopic critical concentration also represents the concentration at which actin filaments begin to assemble, and this concept is analogous to the critical micellar concentration, which is the minimum concentration of lipids required for micelle formation. In a plot of polymerized actin versus actin monomer concentration, no polymer will be observed until the actin concentration exceeds the critical concentration (Figure 11.4). Once steady state is reached, because the microscopic critical concentration of the barbed end is lower than that of the pointed end, there is a net growth of the filament from the barbed end and a net loss of monomers from the pointed end. Thus, the filament maintains a constant length despite a cycling of monomers through the filament. The cycling of actin monomers through the filament is called treadmilling (Figure 11.2).

Modulation of Actin Dynamics by Actin-Binding Proteins

The cytoplasm of nonmuscle cells contains concentrations of unpolymerized actin that exceed the macroscopic critical concentration of purified actin by several orders of magnitude. As cells change shape and move, they are able to quickly convert this large storage pool of actin monomers into filaments. The temporal and spatial control of actin assembly and disassembly involves a myriad of actin regulatory proteins. These proteins orchestrate the dynamic changes in the actin-cytoskeleton associated with cell movement. As listed in Table 11.1 and described in the sections below, these proteins can be classified by their mechanisms of action into six categories. It should be emphasized that no single protein can by itself explain the complex changes associated with actin-based motility. As research has advanced, it has become ever more evident that different constellations of pro-

Table 11.1   Actin-regulatory proteins

Protein	Interactions
Monomer-binding proteins	
Tβ4	Preferentially interacts with actin-ATP; weaker binding to actin-ADP
Profilin	Accelerates ATP-ADP nucleotide exchange; ushers actin monomers from Tβ4 to the barbed ends of actin filaments; binds to oligoproline sequences in VASP and other related proteins; binds PIP$_2$
Barbed-end-capping proteins	
Profilin	May bind as profilin or profilin-actin complex
CapG	Calcium-sensitive binding interaction; binding of PIP$_2$ inhibits capping function
CapZ	Calcium-independent binding interaction; binding of PIP$_2$ inhibits capping function
Gelsolin	High-affinity binding interaction associated with actin filament severing; requires calcium; binding of PIP$_2$ inhibits capping function
Filament-severing proteins	
Gelsolin	Initial side binding to filaments, followed by filament severing (also see above); recycles newly formed actin filaments
Cross-linking/bundling proteins	
α-Actinin	Creates colinear arrays of actin filaments; interacts with vinculin and phospholipids
ABP-280	Creates orthogonal networks responsible for gel formation
Actin-depolymerizing proteins	
Cofilin/actin-depolymerizing factor	Accelerates disassembly at the pointed end of actin filaments; thought to selectively interact with ADP-rich regions in actin filaments; probably weakly severs actin filaments; recycles older actin filaments
Other proteins	
Actin-related proteins (ARPs)	Bind to the sides of actin filaments, form 70° angle branches; cap pointed ends; bind profilin; nucleate actin assembly that is stimulated by free barbed ends and activated WASP proteins
Formins	Bind to the barbed ends and nucleate linear filament assembly
VASP	Binds to zyxin and vinculin; has multiple polyproline-binding sites for profilin, binds actin filaments, weakly interferes with barbed-end capping proteins
Mena and N-Mena	Similar in primary structure to VASP; bind profilin
Vinculin	Found in focal contacts; binds VASP, α-actinin, and actin filaments
Zyxin	Found in focal contacts; binds VASP; may directly bind profilin

teins combine to generate different actin structures and account for different forms of actin-based motility.

Monomer-Binding Proteins

One mechanism for regulating actin filament assembly is by binding to and altering the ability of actin monomers to polymerize. Two actin monomer-binding proteins, thymosin β4 and profilin, work in concert to maintain the high concentrations of unpolymerized actin and help to regulate actin filament assembly in nonmuscle cells.

The factor primarily responsible for preventing actin monomer assembly into filaments is the 5-kDa polypeptide known as thymosin β4 (often abbreviated as Tβ4). This small protein binds to an actin monomer to form a 1:1, or binary, complex. When bound to actin, Tβ4 sterically hinders spontaneous actin filament assembly (Figure 11.5A), blocking binding at both the pointed and barbed ends of the monomer. Tβ4 concentrations in cells are also in the 100 to 400 μM range, matching well with the concentrations of polymerization-incompetent monomeric actin in resting nonmuscle cells. Tβ4 has a higher affinity for ATP-actin monomers (the form of actin that

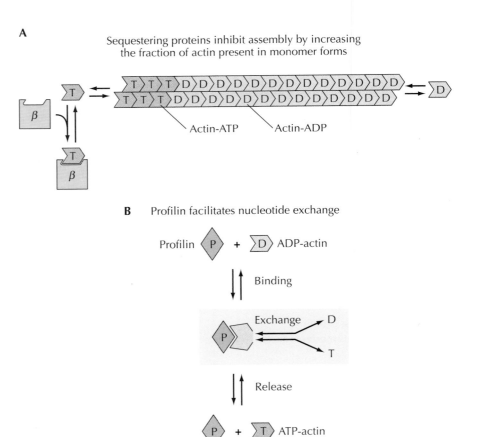

Figure 11.5 **(A)** Schematic diagram of the action of monomer-sequestering agents such as Tβ4. **(B)** Schematic diagram of how profilin may facilitate the exchange of ATP for ADP on an actin monomer. When profilin binds to an actin monomer, the central cleft of actin opens, making the nucleotide-binding site more accessible for release and exchange. Because the ATP concentration far exceeds the ADP concentration in living cells, ATP will readily replace ADP from the actin monomer.

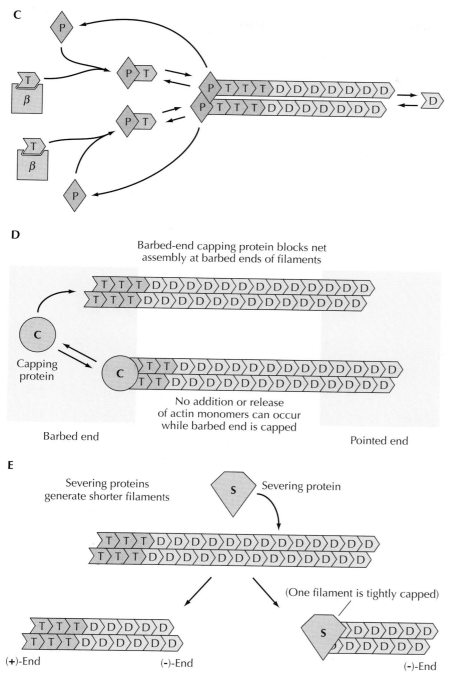

Figure 11.5 *continued* **(C)** Simplified kinetic diagram shows how free profilin can take actin monomers from the Tβ4 storage pool and usher them onto the barbed end of an actin filament. For simplicity the reverse arrows profilin-ATP to free profilin and to Tβ4-ATP actin are not shown. Once profilin-ATP binds to the barbed end, profilin rapidly dissociates. **(D)** Mechanism of a barbed-end capping protein binding to the barbed (or plus) end of an actin filament. Bound capping proteins prevent both association and dissociation of monomers; under such conditions, only the pointed end can interact with the actin monomer pool. **(E)** Model for the action filament-severing proteins. The severing protein first binds along the side of the actin filament, next interposes itself between neighboring actin subunits within the filament, and then remains tightly bound to the barbed end of one of the severed filaments.

(Figure 11.5 continues)

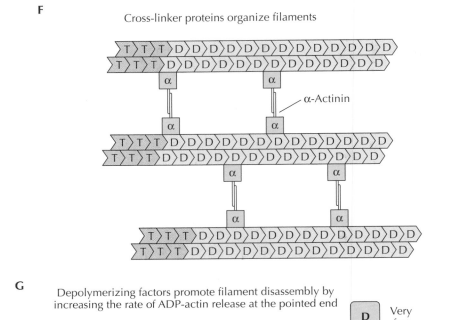

F Cross-linker proteins organize filaments

G Depolymerizing factors promote filament disassembly by increasing the rate of ADP-actin release at the pointed end

ATP-actin

ADP-actin

Figure 11.5 *continued* **(F)** Schematic diagram showing the bundling protein α-actinin, which cross-links actin filaments into parallel arrays. α-Actinin molecules form an antiparallel dimer, and each subunit contains an actin filament-binding site. **(G)** Action of a depolymerizing factor in enhancing disassembly from the pointed end of the actin filament. The dark square with a D represents an ADF or cofilin molecule binding alongside an actin filament at a site near the pointed end. Upon binding, this protein enhances the rate of the pointed end, thereby accelerating the treadmilling rate of uncapped filaments.

most readily forms actin filaments) than for ADP-actin. Binding of Tβ4 to the actin monomer changes actin's conformation and blocks the exchange of actin-bound adenine nucleotide. Because the affinity of the free barbed filament end for actin monomers is greater than that of Tβ4 (K_D for ATP-actin monomer barbed end = 0.20 μM versus 1 to 2 μM for ATP-actin monomer-Tβ4), the barbed end of an actin filament is capable of removing actin monomers from the Tβ4 complex. Addition of stable actin nuclei results in the rapid formation of filaments from the Tβ4-actin monomer. Thus, while Tβ4 is acting as an actin monomer-sequestering protein to prevent indiscriminate assembly of new filaments, its biochemical properties permit regulated growth at the barbed ends of preformed actin filaments.

The second actin monomer-binding component is the 15-kDa protein profilin. Profilin was the first actin-regulatory protein discovered, originally being isolated from spleen extracts. The intracellular concentration of profilin is approximately one-fourth that of Tβ4. Like Tβ4, profilin binds actin monomers to form a 1:1 complex and under certain conditions can form a ternary complex with Tβ4 and monomeric actin. Profilin binds to the barbed end of the actin monomer and like Tβ4 prevents the spontaneous formation of actin nuclei and indiscriminate actin assembly in the cell. However, profilin primarily functions to stimulate the rapid growth of actin filaments. Profilin binds to actin monomers and markedly enhances the rate of nucleotide exchange (Figure 11.5B). Low concentrations of profilin can catalyze nucleotide exchange, even in the presence of high Tβ4 concentrations. Therefore, profilin can readily convert ADP-actin to ATP-actin, the form of monomeric actin that favors actin assembly. Because the concentration of ATP-actin is unlikely to be rate limiting in unstimulated cells, this capacity to stimulate nucleotide exchange function may not be critical in the resting cell. However, in regions of the cell undergoing rapid actin filament turnover, the concentrations of ADP-actin should increase locally and the supply of ATP-actin would be rate limiting in the absence of profilin. In addition to increasing the concentrations of ATP-actin, profilin is able to deliver actin monomers to the barbed ends of growing actin filaments. Profilin takes actin monomers from the large pool of sequestered Tβ4-actin and ushers them onto the barbed ends of actin filaments. On binding to the actin filament, profilin quickly dissociates, leaving its actin monomer on the end of the growing filament. The resulting free profilin molecule can then compete with Tβ4 for another actin monomer and repeat the cycle (Figure 11.5C). Calculations indicate that in nonmuscle cells more than 90% of the actin monomers in filaments assemble from profilin-actin complexes rather than from free actin monomers.

In addition to being found throughout the cytoplasm, profilin concentrates in regions where new actin filaments are assembling. Profilin's unusual affinity for poly-L-proline accounts for its localization in these regions. Profilin is the only actin-regulatory protein known to bind to poly-L-proline, and a number of proteins have recently been shown to contain oligoproline repeats that are capable of binding profilin. The best studied of these profilin-docking proteins is vasodilator-stimulated phosphoprotein (VASP) (see below). Profilin also associates with the plasma membrane where it binds to the phosphoinositide phosphatidylinositol-4,5-bisphosphate (PIP_2). When micelles containing this phospholipid bind to a profilin-actin complex, actin is released from the complex.

Actin Filament-Capping Proteins

Proteins that bind to the barbed ends of actin filaments have a profound effect on filament growth (Figure 11.5D). The barbed end has a high affinity for actin monomers and in combination with profilin can readily compete with Tβ4 for sequestered ATP-actin monomers. When the barbed ends of actin filaments are capped, the critical concentration of the actin filament increases to that of the pointed end, 0.6 μM (Figure 11.6, left). Barbed-end capping also lowers the depolymerization rate of actin filaments because the off rate for the free pointed ends is considerably slower than that for the barbed end (Figure 11.6, right). In the presence of Tβ4 and profilin, the lower-affinity pointed end is unable to efficiently compete for actin monomers, and significant growth of actin filaments is unlikely to occur. Therefore, in the cell, the barbed filament end is the primary site for actin filament assembly. For unstimulated nonmuscle cells to maintain the high concentrations of unpolymerized actin, nearly all of the barbed ends must be blocked from competing with Tβ4 for actin monomers. Given the importance of regulating the barbed end to maintain high concentrations of monomeric actin in resting cells and to initiate new actin filament assembly during cell movement, it is not surprising that nonmuscle cells contain multiple proteins that are capa-

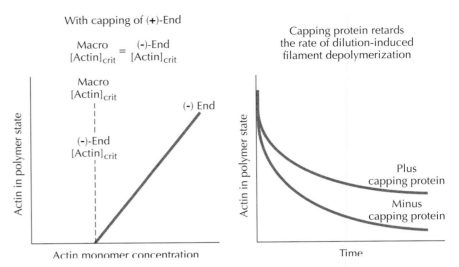

With capping of (+)-End

$$\frac{\text{Macro}}{[\text{Actin}]_{\text{crit}}} = \frac{(-)\text{-End}}{[\text{Actin}]_{\text{crit}}}$$

Figure 11.6   Effects of a barbed-end-capping protein on the critical concentration and rate of depolymerization of actin. (Left) Graph of steady-state actin filament concentration as a function of actin monomer concentration. As shown in Figure 11.5D, capping blocks all exchange at the barbed end. The free pointed end has a lower affinity for actin monomers, and this lower affinity is reflected as an increase in the critical concentration (see Figure 11.4 for comparison). (Right) Plot of the decrease in filamentous actin versus time after diluting actin filaments to below their critical concentration in the absence and presence of a barbed-end-capping protein. Capping of the barbed end retards the depolymerization, because dissociation of actin monomers occurs only at the pointed end.

ble of binding the barbed end and blocking monomer exchange. These proteins are called barbed-end-capping proteins.

CapZ is a heterodimer consisting of a 36-kDa α subunit and a 32-kDa β subunit. The protein derives its name from the observation that it localizes to the Z line in skeletal muscle. This protein has also been called β-actinin and capping protein. CapZ is found in all nonmuscle cells and is particularly abundant in neutrophils, where it represents 1% of the total cytoplasmic protein. This protein binds the barbed ends of actin filaments and prevents the release or addition of actin monomers. The dissociation constant for the CapZ filament end is 0.5 to 3 nM. The ability of CapZ to cap filaments is not affected by ionized calcium concentration but is blocked by PIP_2. In the submicromolar Ca^{2+} concentrations found in unstimulated cells, CapZ would be expected to cap the barbed ends, preventing them from competing with Tβ4 for actin monomers and allowing the cell to maintain a high concentration of unpolymerized actin (see above). Studies of CapZ in platelets reveal that stimuli that induce platelet actin assembly result in the uncapping of the ends of actin filaments by CapZ.

Another capping protein that differs structurally from CapZ is CapG, a 38-kDa protein that is closely related to the actin filament-severing proteins known as gelsolin and villin. While most abundant in macrophages and neutrophils, CapG is also found in most other cell types, with the notable exception of platelets. The affinity of CapG ($K_D = 1$ nM) for the barbed end is similar to that of CapZ. Unlike other members of the gelsolin-villin family, CapG caps only the barbed ends of actin filaments but demonstrates no actin filament-severing activity. Like other members of this family, CapG is calcium sensitive and requires 1 μM Ca^2 for half-maximal capping of actin filaments. The phosphoinositide PIP_2 inhibits barbed-end capping by CapG, suggesting that CapG responds to two intracellular signals, PIP_2 and ionized calcium. Its calcium sensitivity may permit CapG to cap and uncap filaments in response to the brief fluctuations in ionized calcium observable in the periphery of motile phagocytes. Cycles of filament capping and uncapping may represent an essential feature of peripheral membrane ruffling, and macrophages lacking CapG demonstrate a marked decrease in receptor-mediated ruffling.

Actin Filament-Severing Proteins

The most abundant severing protein in nonmuscle cells is gelsolin, an 82-kDa monomeric protein found in the cytoplasm of all nonmuscle cells. In the presence of micromolar concentrations of ionized calcium, gelsolin's tail region unfolds, exposing actin-binding sites and promoting binding to the sides of actin filaments. On binding to the filament, gelsolin interferes with monomer-monomer interactions within the filament. These actions result in the severing of actin filaments (Figure 11.5E). Gelsolin is closely related to CapG; other members of the gelsolin family include villin, a 95-kDa protein found in the intestinal brush border, and adseverin (also named scinderin), found in the adrenal medulla. With the exception of CapG, all members of this family sever actin filaments. In addition, like CapG, gelsolin, as well as all other members of the gelsolin/villin family, caps the barbed ends of actin filaments. Once gelsolin caps an actin filament, the protein binds the filament end with high affinity (K_D in the subpicomolar range) and does not dissociate from the barbed end when the ionized calcium concentration is lowered. However, as observed with both CapG and CapZ, binding to actin filaments can be inhibited by PIP_2. Experiments with increasing and decreasing gelsolin levels in cells prove that in vivo gelsolin enhances the recycling of actin filaments and facilitates actin-based motility in nonmuscle cells. The severing of actin filaments reduces the viscosity of the peripheral cytoplasm and allows the actin cytoskeleton to be rapidly remodeled. Gelsolin also plays a role in apoptosis and appears to be linked to mitochondrial porins. Reduced gelsolin levels have been noted in many cancer cells, and this condition may partly explain prolonged cancer cell survival.

Actin Filament-Bundling and Cross-Linking Proteins

Electron micrographs of nonmuscle cells reveal that actin filaments are organized into a network in which many filaments appear to cross each other at right angles, forming an orthogonal mesh. The cross-linking protein ABP-280, or filamin, is responsible for organizing these networks. This spatially extended homodimeric protein consists of two 280-kDa subunits linked at a flexible hinge region. Each subunit possesses a single actin-binding site, thereby allowing the dimer to link two actin filaments. When actin filaments are linked into a network by filamin, the solution forms a gel, giving the cell cytoplasm a firm consistency. Activation of gelsolin by calcium can abruptly shorten actin filaments and dismantle this network, causing the cytoplasm to shift from a highly viscous to a liquid consistency. Gel-sol transitions are likely to play an important role in the shape changes associated with amoeboid movement.

In addition to cross-linking proteins, nonmuscle cells possess smaller actin filament-bundling proteins, the most prominent member of this class being α-actinin (105 kDa). This protein links actin filaments into bundles of filaments in a parallel array (Figure 11.5F). Another member of this class is plastin, a 65-kDa protein (also known as fimbrin), which occurs in the so-called T and L forms.

Actin-Depolymerizing Proteins

The rate of actin filament treadmilling is much higher in vivo than in vitro. This increase in actin monomer turnover can be accounted for by a family of proteins called actin depolymerization factors (ADFs) or cofilins. These are low-molecular-weight proteins (19 to 20 kDa) that increase the dissociation rate of ADP-actin monomers from the pointed ends of actin filaments (Figure 11.5G) and may also sever actin filaments. These proteins can bind to ADP-actin monomers with considerably higher affinity (0.1 to 0.2 μM K_D) than ATP-actin monomers (1.3 μM K_D) and enhance the dissociation rate of ADP- but not ATP-actin. These proteins work in concert with gelsolin, which is able to sever ATP-actin filaments. ADF/cofilin recycles older actin filaments while gelsolin is capable of recycling newly formed filaments.

Other Actin-Regulatory Proteins

Actin Nucleation Factors

As described earlier, the monomer-binding protein profilin prevents spontaneous nucleation. Therefore, to create new actin filaments, the cell must either uncap preformed actin filaments or activate proteins that can serve as a template for the nucleation of actin assembly. The Arp 2/3 complex may nucleate actin assembly but may also amplify the growth rate of free barbed actin filament ends. First identified in *Acanthamoeba*, this complex has since been identified in mammalian cells as well as in yeast and consists of seven proteins, including actin-related protein 2 (Arp2) and actin-related protein 3 (Arp3). As their names imply, these two proteins have a high level of amino acid sequence homology to actin, and this characteristic is likely to explain the ability of the complex to bind actin monomers. Arp2/3 complex has two functions: first, this complex nucleates actin filament formation upon activation. Arp2/3 complex binds to the pointed end of the actin filament, stimulating filament growth at the fast-growing barbed end. The intrinsic ability of Arp 2/3 complex to nucleate actin filament formation is weak, and in unstimulated cells there is minimal nucleation activity. However, upon binding of agonists to the appropriate receptors, the Rho family of proteins, Rho, Rac, and Cdc42, are activated and bind to and induce conformational changes in the Wiskott-Aldrich syndrome protein (WASP) family of proteins, leading to the unmasking of binding sites for the Arp2/3 complex. The structure of the inactivated Arp2/3 complex has been solved, and Arp2 and Arp3 are not in contact with each other. Binding of activated WASP protein to the Arp2/3 complex presumably alters the conformation of the complex, bringing Arp2 and Arp3 together to form a template for the initiation of actin filament elongation. ActA from the pathogenic bacteria *Listeria* mimics an activated WASP protein and is able to directly bind to and activate the Arp2/3 complex. It has also been observed that Arp2/3 complex-mediated nucleation is greatly enhanced by the presence of preformed actin filaments with uncapped barbed ends, indicating that this complex is able to accelerate barbed-end actin assembly from preformed filaments. Thus, the Arp2/3 complex can work in concert with the uncapping of the barbed ends to expand the actin cytoskeleton.

The second function of the Arp2/3 complex is to generate Y-shaped arrays that permit directional expansion of actin filaments allowing the application of broad expansile forces to the peripheral cell membrane. The

A Branching Arp2/3 complex nucleation

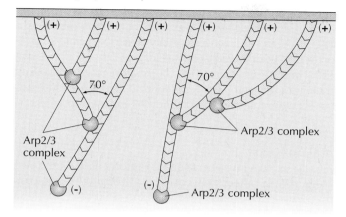

B Formin linear nucleation

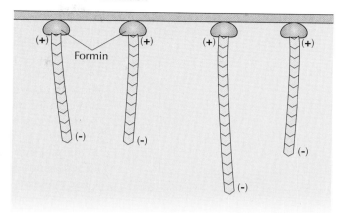

Figure 11.7   **(A)** Schematic view of Arp2/3 complex nucleated actin filament assembly. **(B)** Schematic view of formin-mediated actin filament assembly.

Arp2/3 complex is able to bind to the side of one filament and nucleate the formation of a second filament, forming a fixed angle of 70° branch (Figure 11.7A). In fibroblasts and amoeba, Arp2/3 complex is localized near the leading edge where expansion of peripheral cytoplasm takes place and is not found near stable actin-myosin bundles associated with cell adhesion and contraction. The rheological studies prove that the Arp2/3 complex is not capable of forming a gel, but rather the branched filaments are linked into a stable network by the cross-linking protein filamin.

Recently, it has also been shown that the yeast formin Bni1p is capable of nucleating actin filament formation. Formins are a group of multidomain proteins that are characterized by the presence of formin homology (FH) domains and have been implicated to play a role in regulation of cell shape and polarity. The FH1 domain is rich in oligoproline sequences capable of binding profilin. The FH2 domain is required for nucleation. Curiously, the formins bind to the barbed ends of actin filaments but do not completely cap them. Particularly in the presence of profilin, the formins nucleate the growth of actin assembly and produce linear actin filament arrays rather than a branched architecture (Figure 11.7B). Localization of formins at the peripheral membrane allows actin monomers to add to growing filaments at the membrane-actin cytoskeleton interface, and the resulting directional actin assembly could provide the force for membrane protrusion.

Tropomyosin and Myosins

Other proteins capable of binding to actin include tropomyosin and the myosins. There are multiple subtypes of tropomyosin, and all appear to bind in the groove of the actin filament in a fashion which prevents myosin binding and also blocks gelsolin side binding and severing. Myosins are another large family of proteins that produce force and movement by binding to actin filaments through the myosin head regions. The ATPase activity of the head region is activated by binding to the actin filament. The energy of hydrolysis is transduced into a structural change in the binding angle of the myosin head from 90° to 45°. This change in angle advances the filament toward its pointed end. There are two major classes of myosins. While myosin II is most abundant in muscle, this force-producing motor is also found in nonmuscle cells. The tail regions of myosin II self-associate and form filament bundles in which the myosin heads arrange at opposing ends with respect to each other. These heads can pull actin filaments toward the center of the bundle. Myosin I is a more recently described class of proteins possessing a shorter tail, which fails to self-associate. Myosin I tails can bind to actin filaments or membranes and may be responsible for moving these structures. Furthermore, myosin I proteins tend to localize in the leading edge of moving cells, whereas myosin II tends to localize toward the posterior region.

Proteins Found in Focal Contacts

When nonmuscle cells adhere to a surface, specific regions of the cell form close contacts. These regions are called focal contacts and contain high concentrations of actin filaments. Adherence receptors are linked to the actin cytoskeleton by ezrin, radixin, and moesin (the ERM proteins). In addition, talin (a 270-kDa protein) is found in these regions and binds vinculin (120 kDa). Vinculin in turn can bind α-actinin and actin filaments, and through the amino acid sequence 840-PDFPPPPPDL-849 (called an ABM-1 or FP_4 sequence, see below), it binds VASP. Vinculin contains a 90-kDa head region and a 30-kDa tail region. The ABM-1-binding site is located near the carboxy terminus of the head domain, while the F-actin-binding site is found in the tail region. The tail folds over the head, masking both the F-actin- and ABM-1-binding sites. The folding and unfolding of vinculin acts as a molecular switch. Conditions such as proteolysis, phosphorylation, and binding of PIP_2 can unfold the molecule and allow the protein to bind VASP and actin filaments. Another protein found in focal contacts is zyxin. This 84-kDa protein contains two ABM-1-binding sites and binds VASP with high affinity. VASP, a 45-kDa protein with multiple phosphorylation sites, is a central adapter protein that binds to the ABM-1-binding regions via its EVH-1 domain and concentrates profilin in regions where actin filaments are assembling. Each VASP monomer contains four amino acid sequences of the type XPPPPP, where X is glycine, alanine, lysine, or serine (called an ABM-2 sequence), and these sequences bind profilin. VASP exists as a tetramer in solution; therefore, one VASP tetramer has 16 potential profilin-binding sites. VASP is one of the founding members of the Ena/VASP protein family, which includes two other mammalian counterparts, Mena and N-mena. These proteins also contain ABM-2 sites capable of concentrating profilin. In addition to profilin-binding sequences, VASP also possesses an F-actin-binding site in its EVH-2 domain. This site may position the protein near the barbed end of the actin filament, and high concentrations of VASP can inhibit barbed-end capping. Finally VASP has an actin monomer-binding site and can weakly nucleate actin assembly. VASP-null cells demonstrate poor adherence to surfaces, abnormal filopodia, and defects in chemotaxis. The functional consequences of VASP phosphorylation are under investigation. Recent studies demonstrate that VASP phosphorylation increases as cells adhere and spread.

Polymerization Zone Model

How might cells use actin polymerization to generate force during locomotion? To form a membrane projection such as a pseudopod and filopod or to propel an intracellular bacterium, a discrete zone must be created that promotes the assembly of actin filaments. Such a polymerization zone would be expected to contain nucleating proteins, actin filaments with uncapped barbed ends, and high concentrations of profilin-ATP-actin. As new actin filaments form, they would be expected to be organized by cross-linking and bundling proteins, and expansion of this actin filament network could provide the thrust for host cell peripheral membrane and bacterial intracellular movement.

The activities within the polymerization zone can be divided into two categories. The first activity must initiate new actin filament assembly and is analogous to the starter motor in a conventional gasoline engine. In nonmuscle cells actin-based motility is initiated by the generation of new free barbed filament ends. When cells are stimulated to move and change shape, the number of free barbed filament ends markedly increases. This rapid rise in the number of free barbed ends can be accomplished in two ways. First, the barbed ends of preformed actin filaments can be uncapped and second, nucleating proteins can serve as a template to initiate the elongation of new actin filaments. Uncapping is likely to be a more energy-efficient mechanism for initiating new actin filament assembly, and considerable evidence now points to the importance of free barbed filament ends for maximizing Arp2/3 complex nucleation. The mechanisms controlling uncapping and regulating actin filament length remain to be determined, but phosphatidylinositides, VASP, and profilin are likely to play key roles. Combining severing and uncapping can further amplify the number of free barbed ends. In platelets as well as other nonmuscle cells, receptor agonists can produce a transient rise in ionized calcium and stimulate severing by gelsolin. Increased production of PIP_2 and/or the production of other uncapping activities subsequently dissociate gelsolin, resulting in the generation of multiple short actin filaments with free barbed ends. It is of interest that gelsolin-null platelets form long actin filaments and demonstrate a reduced number of free barbed ends following agonist stimulation. Furthermore, gelsolin-deficient platelets demonstrate aberrant localization of the Arp2/3 complex, indicating that in vivo the production of free barbed ends is critical for normal Arp2/3 complex localization and function. In addition to being stimulated by uncapped actin filaments, Arp2/3 complex nucleation may be activated by receptor agonists through the Rho family of proteins. In their GTP form these proteins bind to and alter the conformation of the WASP protein family, which in turn bind to and activate the Arp2/3 complex. Members of the Rho family can also activate the formins to enhance barbed-end filament growth.

The second important activity required to support actin filament growth within the polymerization zone is the efficient delivery of assembly-competent actin monomers and is analogous to the fuel delivery system of a conventional gasoline engine. Intracellular pathogen motility occurs at rates up to 1 μm/s, requiring the addition of approximately 400 monomers per second. To achieve this rate of assembly, about 200 μM actin-ATP must be immediately available for assembly. As discussed earlier, in vivo actin monomers are primarily delivered to the barbed ends as profilin-actin. However, the cytoplasmic concentration of profilin-actin, even in the most motile cell, does not exceed 60 μM. To attract the high concentrations of profilin-actin into the polymerization zone, the cell utilizes two consensus docking sequences. These sequences were originally defined by the oligo-proline modules in *Listeria monocytogenes* ActA surface protein and human platelet VASP. Analysis of the known actin regulatory proteins has led to the identification of two distinct sequences (D/E)FPPPPX(D/E) (where X is P or T) and XPPPP (where X is G, A, L, P, or S). In *Listeria*, the bacterial surface protein ActA contains a series of four sequences of the type

EFPPPPTDE. Each of these sequences may attract a VASP tetramer. Because each VASP tetramer contains 16 to 20 GPPPPP profilin-binding sites, one ActA molecule can potentially attract 64 to 80 profilin molecules. By using this protein-binding amplification cascade, extremely high concentrations of profilin can be attracted to the surface of *Listeria*. In a similar fashion, vinculin and zyxin both contain (D/E)FPPPPX(D/E) sequences that can attract VASP and profilin to the leading edge of motile cells (Figure 11.8). For actin-based *Shigella* motility, the bacterial surface protein IcsA attracts a proteolytic fragment of vinculin. Proteolysis unmasks vinculin's VASP-binding site, and unmasking of this site serves as a molecular switch that initiates the delivery of VASP and profilin to the polymerization zone. In addition, IcsA attracts N-WASP, a protein that contains multiple XPPPPP that can also concentrate profilin. In addition to ushering actin monomers onto the barbed end, the presence of profilin within the polymerization zone ensures that the weakly polymerizing actin-ADP complex will undergo facilitated exchange with ATP to produce a strongly polymerizing actin-ATP complex. The polymerization zone model also predicts that pointed-end filament depolymerization should occur in regions outside of the polymerization zone where the concentrations of profilin are lower. Such an arrangement would greatly limit the nucleotide exchange rate, and the presence of more actin-ADP is likely to prohibit filament assembly at all regions outside the active polymerization zone. Depolymerizing factors (e.g., ADF/cofilin) are apt to be selectively localized within the regions of filament depolymerization. Finally it must be recognized that the binding of high concentrations of profilin-actin to a surface will not increase the profilin-actin concentration in the solution phase; therefore, the growing filament must come in contact with the surface to allow profilin-actin to be directly transferred to the filament end. This mechanism has the advantage of not being limited by the diffusion rate and allows exquisite control of where new actin filaments are assembled.

Figure 11.8   Assembly of an ABM complex. Changes in the structure of vinculin or zyxin act as a molecular switch that exposes ABM-1-binding sequences for attracting and tethering VASP in the polymerization zone. Bound VASP then binds numerous profilin molecules, which increase the local concentrations of actin-ATP within the polymerization zone to stimulate filament assembly.

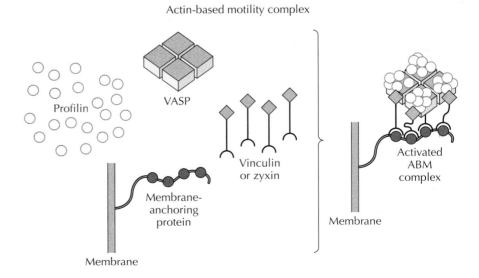

Conclusion

The precise mechanisms regulating the formation of new actin filaments in the motile cell remain to be determined. Actin filament growth can be regulated by the capping and uncapping of the barbed end. In addition, filaments can be severed by gelsolin. The combination of uncapping and severing can greatly increase the number of free barbed-filament ends available for rapid actin assembly. The generation of free barbed ends stimulates the assembly-promoting activity of nucleating proteins such as the Arp2-Arp3 complex, as does GTP-bound G proteins combined with WASP family proteins. To explain the rapid rates of actin assembly in vivo, profilin-actin must be highly concentrated on a membrane surface through a binding amplification cascade involving (D/E)FPPPPX(D/E) and XPPPP sequences. Within a discrete polymerization zone, surface-bound profilin-actin is directly transferred to growing actin filaments and this transfer bypasses the rate limitations of diffusion. ATP-actin addition to barbed-filament ends is thermodynamically favorable, and hydrolysis of filament-bound ATP-actin during filament elongation can provide the useful work needed for movement. Through the action of ABP-280 and α-actinin, newly formed filaments can be stabilized into orthogonal networks and bundles. These proteins increase the rigidity of the actin filaments, giving the peripheral cytoplasm the structure and mechanical properties required for shape change and movement. Osmotic forces and myosins may also play a role in advancing the peripheral membrane. Given the complexity of amoeboid movement, the task of determining the primary mechanisms has proved difficult.

As described in chapter 12, bacterial pathogens that utilize actin and actin-regulatory proteins for their intracellular locomotion have become model systems for identifying the minimal pathway(s) for actin-based motility.

I thank Runa Musib for her help on the actin and Arp2/3 sections of the chapter.

Selected Readings

Monomer-Binding Proteins

De La Cruz, E. M., E. M. Ostap, R. A. Brundage, K. S. Reddy, H. L. Sweeney, and D. Safer. 2000. Thymosin-beta(4) changes the conformation and dynamics of actin monomers. *Biophys. J.* **78:**2516–2527.

Dickinson, R. B., F. S. Southwick, and D. L. Purich. 2002. A direct-transfer polymerization model explains how the multiple profilin-binding sites in the actoclampin motor promote rapid actin-based motility. *Arch. Biochem. Biophys.* **406:**296–301.

Kang, F., D. L. Purich, and F. S. Southwick. 1999. Profilin promotes barbed-end actin filament assembly without lowering the critical concentration. *J. Biol. Chem.* **274:**36963–36972.

Nyman, T., R. Page, C. E. Schutt, R. Karlsson, and U. Lindberg. 2002. A cross-linked profilin-actin heterodimer interferes with elongation at the fast-growing end of F-actin. *J. Biol. Chem.* **277:**15828–15833.

  Experiments with this cross-linked complex emphasize the role of the profilin-actin heterodimer in barbed-end actin filament assembly.

Safer, D., T. R. Sosnick, and M. Elzinga. 1997. Thymosin beta 4 binds actin in an extended conformation and contacts both the barbed and pointed ends. *Biochemistry* **36:**5806–5816.

  Thymosin beta 4 is predominantly unstructured in solution. Upon binding to actin, the molecule contacts specific actin residues located at the barbed end as well as a residue in domain 2 at the pointed end, allowing thymosin beta 4 to sterically block actin polymerization at both ends of the actin monomer.

Witke, W., J. D. Sutherland, A. Sharpe, M. Arai, and D. J. Kwiatkowski. 2001. Profilin I is essential for cell survival and cell division in early mouse development. *Proc. Natl. Acad. Sci. USA* **98:**3832–3836.

Wolven, A. K., L. D. Belmont, N. M. Mahoney, S. C. Almo, and D. G. Drubin. 2000. In vivo importance of actin nucleotide exchange catalyzed by profilin. *J. Cell Biol.* **150:**895–904.

Mutations in the *Saccharomyces cerevisiae* profilin gene (PFY1) and complementation with an actin mutant having a high intrinsic nucleotide exchange rate emphasize the importance of profilin-mediated exchange of ATP for ADP for in vivo actin filament dynamics.

Capping Proteins

Schafer, D. A., P. B. Jennings, and J. A. Cooper. 1996. Dynamics of capping protein and actin assembly in vitro: uncapping barbed ends by polyphosphoinositides. *J. Cell Biol.* **135:**169–179.

CapZ has a slow dissociation rate, but rapidly dissociates when exposed to PIP_2. The relatively slow on-rate of this capping protein may explain the length distribution of actin filaments in living cells.

Witke, W., W. Li, D. J. Kwiatkowski, and F. S. Southwick. 2001. Comparisons of CapG and gelsolin-null macrophages: demonstration of a unique role for CapG in receptor-mediated ruffling, phagocytosis, and vesicle rocketing. *J. Cell Biol.* **154:**775–784.

CapG is shown to have distinct in vivo functions that differ from its family member gelsolin. This capping protein is required for receptor-mediated ruffling of macrophages and plays an important role in phagocytosis.

Severing Proteins

Lin, K. M., M. Mejillano, and H. L. Yin. 2000. Ca^{2+} regulation of gelsolin by its C-terminal tail. *J. Biol. Chem.* **275:**27746–27752.

McGrath, J. L., E. A. Osborn, Y. S. Tardy, C. F. Dewey, Jr., and J. H. Hartwig. 2000. Regulation of the actin cycle in vivo by actin filament severing. *Proc. Natl. Acad. Sci. USA* **97:**6532–6537.

These elegant studies emphasize the importance of gelsolin in actin filament turnover.

Cross-linking Proteins

Flanagan, L. A., J. Chou, H. Falet, R. Neujahr, J. H. Hartwig, and T. P. Stossel. 2001. Filamin A, the Arp2/3 complex, and the morphology and function of cortical actin filaments in human melanoma cells. *J. Cell Biol.* **155:**511–517.

Nakamura, F., E. Osborn, P. A. Janmey, and T. P. Stossel. 2002. Comparison of filamin A-induced cross-linking and Arp2/3 complex-mediated branching on the mechanics of actin filaments. *J. Biol. Chem.* **277:**9148–9154.

The rheological experiments prove convincingly that filamin, but not Arp2/3 complex, can form gels in vitro. Polymer branching in the absence of cross-linking does not lead to polymer gelation.

Actin-Recycling Proteins

Bamburg, J. R. 1999. Proteins of the ADF/cofilin family: essential regulators of actin dynamics. *Annu. Rev. Cell Dev. Biol.* **15:**185–230.

A very complete review of ADF/cofilin by the investigator who discovered ADF.

Niwa, R., K. Nagata-Ohashi, M. Takeichi, K. Mizuno, and T. Uemura. 2002. Control of actin reorganization by Slingshot, a family of phosphatases that dephosphorylate ADF/cofilin. *Cell* **108:**233–246.

Experiments with *Drosophila* emphasize the importance of ADF/cofilin dephosporylation in mediating actin dynamics.

Actin-Nucleating Proteins

Amann, K. J., and T. D. Pollard. 2001. The Arp2/3 complex nucleates actin filament branches from the sides of pre-existing filaments. *Nat. Cell Biol.* **3:**306–310.

> These experiments provide data that contradicts the findings of Pantaloni et al. (2000) and suggest that Arp2/3 complex binds to the side of actin filaments and does not require free barbed ends.

Evangelista, M., D. Pruyne, D. C. Amberg, C. Boone, and A. Bretscher. 2002. Formins direct Arp2/3-independent actin filament assembly to polarize cell growth in yeast. *Nat. Cell Biol.* **4:**260–269.

Falet, H., K. M. Hoffmeister, R. Neujahr, J. E. Italiano, Jr., T. P. Stossel, F. S. Southwick, and J. H. Hartwig. 2002. Importance of free actin filament barbed ends for Arp2/3 complex function in platelets and fibroblasts. *Proc. Natl. Acad. Sci. USA* **99:** 16782–16787.

> This third study utilizes gelsolin-null cells and CapG to prove that free barbed ends are required for efficient Arp2/3 complex nucleation and localization.

Pantaloni, D., R. Boujemaa, D. Didry, P. Gounon, and M. F. Carlier. 2000. The Arp2/3 complex branches-filament barbed ends: functional antagonism with capping proteins. *Nat. Cell Biol.* **2:**385–391.

> Demonstrates that free barbed-filament ends are required for efficient Arp2/3 complex formation.

Pruyne, D., M. Evangelista, C. Yang, E. Bi, S. Zigmond, A. Bretscher, and C. Boone. 2002. Role of formins in actin assembly: nucleation and barbed-end association. *Science* **297:**612–615.

> Formins produce an unbranched filament architecture that is distinctly different than Arp2/3 complex nucleated actin assembly. Formins remain bound to the barbed ends of growing actin filaments.

Volkmann, N., K. J. Amann, S. Stoilova-McPhie, C. Egile, D. C. Winter, L. Hazelwood, J. E. Heuser, R. Li, T. D. Pollard, and D. Hanein. 2001. Structure of Arp2/3 complex in its activated state and in actin filament branch junctions. *Science* **293:** 2456–2459.

Focal Contact Proteins

Bear, J. E., T. M. Svitkina, M. Krause, D. A. Schafer, J. J. Loureiro, G. A. Strasser, I. V. Maly, O. Y. Chaga, J. A. Cooper, G. G. Borisy, and F. B. Gertler. 2002. Antagonism between Ena/VASP proteins and actin filament capping regulates fibroblast motility. *Cell* **109:**509–521.

> VASP-transfected cells contain longer, less branched actin filaments compared with VASP-deficient cells. At μM concentrations VASP can inhibit barbed-end capping by nM concentrations of CapZ.

Castellano, F., C. Le Clainche, D. Patin, M. F. Carlier, and P. Chavrier. 2001. A WASp-VASP complex regulates actin polymerization at the plasma membrane. *EMBO J.* **20:**5603–5614.

DeMali, K. A., C. A. Barlow, and K. Burridge. 2002. Recruitment of the Arp2/3 complex to vinculin: coupling membrane protrusion to matrix adhesion. *J. Cell Biol.* **159:**881–891.

Howe, A. K., B. P. Hogan, and R. L. Juliano. 2002. Regulation of vasodilator-stimulated phosphoprotein phosphorylation and interaction with Abl by protein kinase A and cell adhesion. *J. Biol. Chem.* **277:**38121–38126.

Rottner, K., M. Krause, M. Gimona, J. V. Small, and J. Wehland. 2001. Zyxin is not colocalized with vasodilator-stimulated phosphoprotein (VASP) at lamellipodial tips and exhibits different dynamics to vinculin, paxillin, and VASP in focal adhesions. *Mol. Biol. Cell.* **12:**3103–3113.

12

Mechanisms of Bacterial Invasion

Bacterial Manipulation of the Host Cell Cytoskeleton

Jennifer R. Robbins, David N. Baldwin, Sandra J. McCallum, and Julie A. Theriot

Pathogenic bacteria have evolved many mechanisms for parasitizing the nutrient-rich environment of multicellular eukaryotic hosts, colonizing both extracellular and intracellular niches. Successful pathogens are frequently expert cell biologists, demonstrating a sophisticated practical understanding of complex cellular systems. Judicious exploitation of preexisting regulatory pathways has enabled the pathogenic bacteria to induce their eukaryotic hosts to perform energetically demanding and ultimately self-destructive behaviors. Preceding chapters describe how bacteria have evolved specific interactions with the membrane of a host cell and its proteins, including signaling mechanisms to induce (Figure 12.1A) or prevent (Figure 12.1B) internalization by phagocytosis. Of the former, some have also sought residence in the cytosol where, presumably, available nutrients and biochemical networks are plentiful (Figure 12.1C).

One host cell network that has been exploited for interactions both at the membrane and within the cytosol by a number of different bacterial species is the actin cytoskeleton. The actin cytoskeleton consists of a meshwork of protein filaments that extend throughout the cytosol and form a dense web underlying the plasma membrane. In most vertebrate cells, the actin cytoskeleton is the primary determinant of cell shape and provides the machinery for whole-cell movement. Different cell types use their actin cytoskeletons for different purposes; for example, macrophages and neutrophils use actin-dependent movement to crawl through tissues and engulf microbial invaders, while epithelial cells use actin structures to maintain strong adhesive connections with their neighbors and with their underlying extracellular matrix. At the center of actin cytoskeletal regulation is a family of monomeric GTPases, called the Rho family of GTPases. They are activated by extracellular signals and regulate a very large number of downstream effectors, which in turn biochemically regulate the nucleation, polymerization, and stabilization of actin filaments in the cytoplasm. These GTPases are regulated by guanine nucleotide exchange factors, and, as we will see in this chapter, a wide range of bacteria have chosen this step in the actin-signaling cascade as a target for manipulation. For

A Invasion

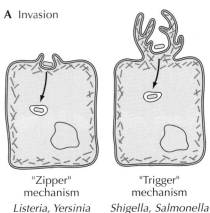 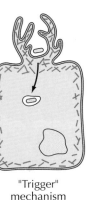

"Zipper"
mechanism

Listeria, Yersinia

"Trigger"
mechanism

Shigella, Salmonella

B Prevention of uptake

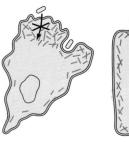 

Prevention of
phagocytosis

Yersinia

Attachment and
pedestal formation

EPEC

C Actin-based motility

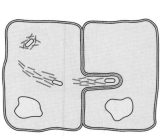

Intracellular motility
and intercellular spread

Listeria, Shigella

Figure 12.1 Schematic diagram of host cell actin rearrangements caused by bacterial pathogens. **(A)** Some pathogens induce their own uptake by nonphagocytic epithelial cells, initiating signal transduction cascades that result in polymerization of actin filaments under the host cell plasma membrane. For some pathogens, including *Listeria* and *Yersinia*, tight adhesion of the bacterium to the cell surface followed by modest actin polymerization only in the immediate neighborhood of bacterial attachment, the "zipper" form of induced phagocytosis, allows the bacterium to be taken up by the cells. For others, such as *Shigella* and *Salmonella*, the host cell throws up a large ruffling membrane that folds over and traps the bacteria in a membrane-bound compartment. This is called the "trigger" form of induced phagocytosis. **(B)** Macrophages normally phagocytose foreign particles, including bacterial invaders, but *Yersinia* species are able to prevent phagocytosis by injecting virulence factors into the host cell cytoplasm that disrupt actin cytoskeletal structures. In contrast, enteropathogenic *E. coli* colonize the surface of epithelial cells without being internalized, building a dense "pedestal" of actin filaments that raises them up from the cell surface. **(C)** *Listeria* and *Shigella* are able to escape from their vacuoles and nucleate the polymerization of actin filaments at their surfaces while growing in host cell cytoplasm. This results in the formation of a dense tail structure that pushes the bacteria through the cytoplasm. At the interface between neighboring epithelial cells, the actin-rich tail pushes the bacteria out into a membrane-bound protrusion, which can be taken up by the second cell, allowing direct cytoplasm-to-cytoplasm transmission.

more detailed information about the behavior and regulation of the actin cytoskeleton by actin-binding proteins, refer to chapter 11.

In addition to pathogens that influence cytoskeletal dynamics at close range to invade or avoid engulfment by epithelial cells and macrophages, some bacteria have developed secreted toxins with specific effects on actin filaments or the proteins that regulate the actin cytoskeleton. These toxins can act at a distance from the bacteria themselves. They are particularly interesting to cellular microbiologists because they provide important tools with which to study the actin cytoskeleton as well as yield information about microbial pathogenesis. These toxins are covered in chapter 13.

In this chapter, we will illustrate several modes of bacterial manipulation of the host cell cytoskeleton using a few well-studied examples and explore interactions of various pathogenic bacteria with the actin cytoskeleton of both phagocytic and nonphagocytic host cells. As we will see, these interactions are generally mediated by specific bacterial gene products, virulence factors, whose sole function is to mimic or interfere with normal host cell signals, in this case, those that regulate actin filament dynamics.

Mechanisms of Bacterial Invasion

The human body is not a single organism but rather a complex ecological community, most of whose members are bacteria. Adding up the bacteria found on the skin, in the nose and mouth, in the lower gastrointestinal tract, etc., there are about ten times more bacterial cells than there are human cells in the body. However, the immune system keeps them corralled in appropriate locations, and most internal tissues are normally sterile. Opportunistic pathogens must await an injury or other breach in the barriers to reach these niches, but some primary pathogens have developed mechanisms to sneak across them. Several interesting examples of host cell cytoskeleton manipulation are found among pathogens that invade epithelial cells.

Two epithelia, the lining of the intestine and of the lung, are particularly vulnerable to attack by bacterial pathogens. Since the function of these epithelia is to absorb digested nutrients and oxygen, respectively, and pass them into the bloodstream, the barrier is a cellular monolayer. Bacteria that can invade this monolayer are one step away from access to the bloodstream and the rest of the body. Epithelial cells take up proteins and other factors from the outside, using pinocytosis and receptor-mediated endocytosis, but they are generally not capable of engulfing large objects such as bacteria. To be taken up by epithelial cells, the bacteria must induce their own phagocytosis.

There are two recognized mechanisms by which pathogenic bacteria induce phagocytosis by epithelial and other nonphagocytic cells; these are termed the zipper and trigger mechanisms. Both of these mechanisms are actin dependent and require that the extracellular bacteria induce intracellular polymerization and reorganization of actin filaments at the plasma membrane. In both cases, the bacteria manipulate the host cell signal transduction apparatus to induce cytoskeletal responses similar to those that occur in response to natural stimuli. The zipper mechanism exploits pathways normally involved in cellular adhesion and motility, while the trigger mechanism resembles the membrane ruffling response to eukaryotic cell growth factors. Examples of both are described below.

The Zipper Mechanism of Bacterial Uptake by Nonphagocytic Cells

Listeria monocytogenes

One bacterial pathogen that is a recognized champion at manipulating the host cell actin cytoskeleton is *Listeria monocytogenes*. *L. monocytogenes* is a gram-positive food-borne pathogen that can cause a serious infection in pregnant women and immunocompromised people, resulting in a relatively high rate of mortality among infected adults, newborns, and fetuses. In healthy adults, the infection can be asymptomatic or result in mild, flu-like symptoms. *Listeria* is usually found in soil, in water, and on plants, but it can be isolated from animals as well and infections can be transmitted in foods such as unpasteurized dairy products. In addition, *Listeria* can contaminate stored refrigerated foods because of its unusual ability to grow at 4°C.

Once ingested, some *Listeria* pass through the upper gastrointestinal tract to the small intestine where they are thought to enter the intestinal epithelium. In tissue culture models of infection, *Listeria* can invade a wide variety of cell types including epithelial, endothelial, and fibroblastic cells by using a zipper mechanism. This term refers to the bacterium's ability to induce very local changes in actin organization that result in a close apposition of the bacteria to the host plasma membrane. The bacterium appears to sink into the cell, or pull the plasma membrane up around it, in a process that does not involve formation of membrane ruffles as in the trigger internalization process (discussed below) and does not apparently require more than a very local rearrangement of the actin cytoskeleton.

Two bacterial factors, internalins A and B, that allow *Listeria* to zipper into eukaryotic cells have been identified genetically. Internalin A (InlA) is an 88-kDa bacterial surface protein that contains leucine-rich repeats (LRRs) thought to participate in protein-protein interactions, and InlA is necessary for invasion of cultured intestinal epithelial cells. During invasion of epithelial cells, InlA interacts with a host cell surface protein, E-cadherin. E-cadherin is a cell adhesion molecule responsible in part for homotypic association of epithelial cells. It is a transmembrane protein whose cytoplasmic domain is linked to the actin cytoskeleton through a complex of proteins, which include α-, β-, and γ-catenins. While E-cadherin's ectodomain is sufficient for bacterial adhesion, the cytoplasmic domain responsible for β-catenin binding is essential for actin reorganization and zipper internalization of the bacteria. In some other cell types, InlA is not required for *Listeria* invasion; it has been shown that the 65-kDa protein InlB is responsible for the internalization of bacteria into cultured hepatocytes, Chinese hamster ovary (CHO) cells, and HeLa cells. Two cell surface molecules have been identified as receptors for InlB, the hepatocyte growth factor (HGF) receptor (also called the Met receptor tyrosine kinase) and gC1q-R (p32, a complement binding protein).

HGF is often called "scatter factor" because of its ability to make epithelial cells start crawling by actin-based motility. InlB's mimicry of this ligand allows it to direct cytoskeletal reorganization in a way that is just beginning to be understood, for its effect is not cell motility but induced internalization. The difference in the physiological responses to these two convergently evolved ligands appears to be a subtle matter of degree, one that *Listeria* has learned to exploit. The distinction is thought to arise from InlB's ability to overactivate the phosphatidylinositol-3-kinase (PI3-K)-mediated Ras-MAP kinase pathway. This ultimately results in delicately engineered actin rearrangements that cause the host to engulf the bacterium. MSF, a septin GTPase, localizes to actin filaments around the vesicles that form when InlB-coated beads are internalized, suggesting an interesting parallel between septin-mediated cytokinesis and the membrane-pinching process of *Listeria* uptake.

The recent completion of the *Listeria* genomes for both *L. monocytogenes* and the closely related nonpathogenic species *L. innocua* revealed that there are at least 19 cell surface proteins containing LRRs typical of internalins, and that 11 of these are absent from *L. innocua*. This suggests that *L. monocytogenes* has used the *inl* genes in a radiative adaptation for host cell invasion. Perhaps the diversity of this multigene family reflects the diversity of their potential cellular targets, and other gateway molecules to zipper invasion may remain to be found.

Yersinia Species

An analogous zipper-type entry mechanism is thought to be used by some species of the gram-negative pathogen *Yersinia*. *Y. enterocolitica*, *Y. pseudotuberculosis*, and *Y. pestis* are the causative agents of mild to severe bouts of gastroenteritis, animal disease similar to that seen in human patients, and bubonic plague, respectively. (Bubonic plague is also called the "black death" due to the damage to the peripheral blood vessels that gives the skin its blackish appearance in *Y. pestis* victims.) *Y. enterocolitica* and *Y. pseudotuberculosis* enter the body by the fecal-oral route and invade the mesenteric lymph nodes around the intestine through the Peyer's patch where they intentionally persist outside of the host cells (the details of this alternate strategy are discussed later in the chapter). In tissue culture models of infection, these species are capable of invading epithelial cells that are not normally phagocytic, indicating the sensitivity with which *Yersinia* discriminates between host cells and the niches they provide. A *Yersinia* outer membrane protein, invasin, binds tightly to the subset of integrins that harbor a β_1 subunit (*Y. pestis* does not apparently encode a functional invasin). Integrins, like cadherins, are cell surface adhesion proteins, but they generally mediate adhesive interactions between cells and the extracellular matrix.

Invasin has so perfected its role of binding host cell integrins that it binds with greater affinity than do their natural substrates (e.g., fibronectin), despite lacking any significant sequence similarity with host proteins. In this case, convergent evolution has generated a molecular mimic based solely on shape, for the crystal structure of invasin shows significant structural similarities to fibronectin. Invasin binding induces a signaling cascade that appears to involve the actin-regulating GTPases Rac and Cdc42, the latter of which recruits Wiskott-Aldrich syndrome protein (WASP) family proteins, which bind the actin-nucleating Arp2/3 complex. (Wiskott-Aldrich syndrome is a condition where blood cells have improperly formed cytoskeletons and is characterized by thrombocytopenia.) As with *Listeria*, inhibitors of tyrosine kinases and actin polymerization inhibit *Yersinia* invasion.

Both these examples of zipper-type invasion demonstrate bacterial exploitation of mechanisms normally involved in host cell adhesion or cell motility. Cadherins (among them the receptor for *Listeria* InlA) and integrins (the receptors for *Yersinia* invasin) mediate attachment between epithelial cells and spreading on their neighbors and on the underlying extracellular matrix, respectively. Engagement of these receptors by their normal ligands results in receptor immobilization and clustering. These events induce signaling cascades that can result in strengthening of the cell-cell and cell-matrix contacts and in cellular differentiation. When a bacterial surface protein engages the adhesion proteins on a host cell, the host cell responds as it normally would, recruiting cytoskeletal elements to the location of the attachment and attempting to strengthen the attachment. However, since the bacterium is small compared with the responding cell, the cell's attempt at spreading against the bacterial surface quickly results in bacterial engulfment. It seems a bacterium may also instruct the cell to crawl over it and similarly engulf it by posing as a motility-inducing factor, as *Listeria*'s InlB appears to do. These examples of functional convergence suggest that it may be relatively easy for a bacterial pathogen to develop the ability to invade epithelial cells by a zipper mechanism. In fact, more evidence is accumulating that many pathogens, among them *Streptococcus, Neisseriae,* and *Helicobacter,* use similar pathways to invade cells on occasion. The main requirement appears to be that the bacterium evolve a surface protein capable of binding with appropriate affinity to a host cell receptor that normally causes cell spreading or crawling.

The Trigger Mechanism of Bacterial Uptake by Nonphagocytic Cells

Salmonella Species

The trigger mechanism, the second mechanism of bacterially induced phagocytosis, is dramatically different from the zipper mechanism in mor-

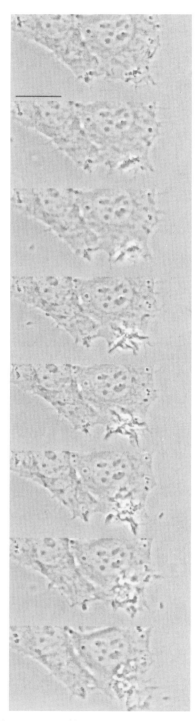

Figure 12.2 The "trigger" mechanism of bacterial invasion. Eight frames from a video sequence, recorded at 10-s intervals, show the large actin-rich ruffle formed when *S. enterica* serovar Typhimurium encounters a Henle epithelial cell in culture. Folding in of the ruffle forms numerous spacious vacuoles, visible as white circles in the later frames. Scale bar is 10 μm.

phological terms and requires a much more complicated type of bacterial machinery. In these cases, brief contact between a bacterium and the surface of a host cell results in a rapid, large-scale cytoskeletal response, where explosive actin filament polymerization under the plasma membrane pushes out huge sheets or ruffles. The extending ruffles fold over and fuse back to the cell surface, trapping large membrane-bound pockets of extracellular medium within the cell, a process termed macropinocytosis (Figure 12.2). The nearby bacteria are trapped by the infolding membrane ruffles, and the bacteria then find themselves within the membrane-bound compartments. Where zippering invasion appears to be largely a modification of cellular adhesion or motility, the triggering type of invasion bears much more resemblance to the response of cells to growth factors. Perhaps the most impressive progress with respect to understanding pathogen-host interaction at the membrane in recent years has been in the study of *Salmonella*.

Salmonella species are responsible for food-borne general gastroenteritis (*S. enterica* serovar Typhimurium, formerly called *S. typhimurium*) as well as more specific ailments such as typhoid fever (*S. enterica* serovar Typhi, formerly called *S. typhi*). They are gram-negative pathogens closely related to *Escherichia coli*. Most experimental work has been done on *Salmonella* serovar Typhimurium, which causes mild food poisoning in humans but a severe, typhoid-like systemic disease in mice. Like *Listeria* and *Yersinia*, *Salmonella* has evolved mechanisms for invading cells that are normally nonphagocytic. It has also developed specialized mechanisms to cheat death while invading phagocytic cells such as macrophages. An important aspect of *Salmonella* virulence is its ability to pass through the intestinal epithelium and invade underlying cells, including lymphocytes and macrophages; hence, it must be able to recognize and subvert a wide variety of host cells. The responses of polarized epithelial cells to *Salmonella* toxins are different from those of nonpolarized cells in culture, and many genes, which are unnecessary in tissue models, are absolutely required for bacterial survival in mice. Most that fall in this latter category are thought to help *Salmonella* to evade digestion by the macrophages it invades through macropinocytosis. Both in these normally phagocytic cells and in nonphagocytes, *Salmonella* can trigger its uptake into a specialized vacuole that it makes hospitable enough to even support bacterial replication.

Much effort has gone into exposing the clever mechanisms by which *Salmonella* tricks its host cells into providing such a nutrient-rich, protected environment. At the heart of this process is the bacterium's ability to "inject" proteins directly into the host cell before invasion, using a specialized organelle termed the type III secretion system (discussed in chapter 15). *Salmonella* carries the genes for two such complex organelles on separate pathogenicity islands within the chromosome; these are differentially regulated during the course of infection. The invasion-associated type III secretion system, encoded within the *Salmonella* pathogenicity island I (SPI-1), pierces the membrane of the host cell to direct the secretion of a sophisticated set of effector proteins into the cytoplasm. Once in the cytoplasm, these proteins have very specific targets that together induce the massive cytoskeletal rearrangements required for bacterial envelopment by membrane ruffling.

Two distinct strategies are used cooperatively by these injected effectors. Some (e.g., the Sop family) indirectly alter the cytoskeleton by modulating the behavior of actin-regulating GTPases Rac1 and/or Cdc42. Others (the Sip family) directly interact with the actin cytoskeleton.

Several members of the Sop family of effectors activate Cdc42 and Rac, members of the Rho family of GTPases that spatiotemporally regulate the actin cytoskeleton by acting on a variety of downstream targets. GTP-bound forms of the proteins are active, while GDP-bound forms are inactive, providing a useful molecular switch for the cell to control many actin-dependent processes. *Salmonella*-produced SopE and SopE2 are closely related guanine nucleotide exchange factors; the former activates both Rac-1 and Cdc42 while the latter directly activates only Cdc42 in vitro and in tissue culture cells. SopB is an inositol phosphate phosphatase that appears to activate only Cdc42. In assays on cultured fibroblasts, constitutively active Rac mutants induce broad lamellipodial formation, while expression of active Cdc42 is most often associated with spiky, actin-rich filopodia. Dominant negative forms of either protein can inhibit *Salmonella* invasion in nonpolarized epithelial cells; however, only Rac is required for apical internalization in cultured polarized epithelia, and neither is essential (although both are activated) for invasion from the basolateral surface of those cells.

Mutant forms of Rac and Cdc42 or addition of soluble growth factors have global effects throughout the cell, but the ruffling of *Salmonella* invasion is localized to the area near the invading bacterium. *Salmonella* is thought to use the actin-binding Sip proteins, which share homology with the *Shigella* Ipa proteins (discussed below), to focus and spatially limit the effects of the GTPases that Sop proteins activate. Both SipC and SipB are also necessary for delivery of at least some Sop effectors, but SipC has an additional important role. SipC localizes to the host plasma membrane during bacterial internalization where it nucleates actin filament polymerization, perhaps by a mechanism that involves SipC trimerization, thus providing a scaffold for actin polymerization analogous to the action of the cellular actin filament-nucleating complex Arp2/3. Both Arp2/3 and SipC have actin-nucleating activity when tested in a fluorescence assay that measures the polymerization of actin filaments through incorporation of pyrenated actin monomers. Arp2/3 is used both by eukaryotic cells and by bacteria moving within cells to localize the growth of new actin filaments. Presumably SipC performs a similar function at the site of *Salmonella* entry into cells.

SipA also binds to actin, but is not absolutely required for internalization. Its presence reduces the critical concentration for polymerization and enhances SipC-dependent nucleation in vitro and in cultured cells. SipA stimulates actin filament bundling by the actin-bundling host protein fimbrin, which is found abundantly in the actin-rich membrane projections near the site of uptake by *Shigella,* another pathogen which invades using the trigger mechanism (discussed in *Shigella* Species). Like invasin, neither SipA nor SipC shares any sequence similarity with the host components each mimics (in this case, actin-binding proteins), suggesting that *Salmonella* has evolved to exploit higher-order structural similarity or even a novel actin-binding domain.

After the massive actin polymerization induced to internalize *Salmonella,* the cell membrane returns rapidly to its resting state. The protein responsible for this is SptP, a *Salmonella*-produced GTPase activating protein (GAP), which is another example of structural molecular mimicry. GAPs down-regulate proteins, such as Rac and Cdc42, by accelerating GTP hydrolysis, thus returning them to their inactive state. SptP shares some sequence homology with host proteins, but its GAP domain is unlike any found in its host. The crystal structure of a SptP-Rac1 transition state shows that SptP imitates some known GAP structural motifs, but also uses novel folds to constrain certain critical catalytic sites of its target. By doing so, it effectively stops the massive, focused ruffling induced by Sop and Sip proteins and returns the normally nonphagocytic host cell to its original stoic condition.

The specificity and variety of regulatory molecules *Salmonella* has produced for manipulating the host cell cytoskeleton are indeed a remarkable testimony to the sophistication that can evolve during long-term host-pathogen interactions, and its significance does not stop at invasion. *Salmonella* SPI-2 is not required for invasion of host cells, but it is necessary for intracellular replication and systemic infection in mice and is thought to be responsible for inducing the apoptosis that macrophages undergo about one day after ingesting *Salmonella*. Virulence factors encoded by SPI-2, which has its own dedicated type III secretion system that acts in the *Salmonella* vacuole after internalization, also target actin. The SpvB protein encoded by SPI-2 has recently been shown to catalyze ADP-ribosylation of actin in cultured cells, causing the massive actin depolymerization that precedes macrophage death.

Shigella Species

Shigella is a gram-negative pathogen that is the causative agent in bacillary dysentery (bloody diarrhea). There are four species of *Shigella* that cause dysentery, *S. dysenteriae, S. sonnei, S. boydii,* and *S. flexneri.* They have similar mechanisms of virulence and, since *S. flexneri* is the best studied, the molecular descriptions presented here will be for *S. flexneri* but will presumably also apply to the other species. *Shigella*-mediated dysentery is usually a self-limiting although unpleasant diarrheal disease in healthy adults, but it kills about 500,000 children every year in developing countries, where contaminated water is all that is available. Compounding this problem is the fact that shigellae are developing resistance to the antibiotics that are safe to administer to children. *Shigella* infection can occur at a very low dose of bacteria, with a 50% infectious dose of 100 to 200 organisms, probably because the bacteria are not killed by stomach acid and find their way unharmed to the intestine. During *Shigella* infection, the bacteria can invade a variety of cell types, including M cells, enterocytes, and macrophages, and after escape from the phagocytic vacuole, shigellae can move within and between cells by actin-based motility. This intra- and intercellular motility is discussed below. In this section, we focus on the ability of *Shigella* to invade epithelial cells by a process that triggers global changes in the cellular actin cytoskeleton that result in membrane ruffles and macropinocytosis.

The trigger mechanism of *Shigella* invasion involves proteins whose genes are carried on a 220-kb virulence plasmid. Bacteria that have been cured of this plasmid are no longer invasive. As with *Salmonella* invasion, the main system involved in this process is a type III secretion system. The genes involved in this process are called the *ipa* (invasion plasmid antigens) and the *Mxi-spa* (membrane expression of *ipa*-surface presentation of antigens) genes. The *Mxi-spa* genes encode the secretory apparatus while the *ipa* genes are the secreted proteins. The protein products, IpaB, IpaC, and IpaD, have all been shown to be necessary for *Shigella* invasion.

IpaB and IpaC are secreted into the extracellular milieu and are part of a complex that is sufficient for invasion of the host cells. This "Ipa complex" does not contain IpaD but may contain other factors. Mutants in which secretion of the Ipa factors (via *Shigella*'s type III secretion system) are impeded are not internalized, and soluble Ipa factors secreted by the wild-type *Shigella* can rescue these mutants, suggesting that the Ipa proteins act as a soluble complex from the outside of the cell. Beads that are coated with the Ipa complex can be internalized into HeLa cells in a process that resembles *Shigella* invasion morphologically. These results suggest that the Ipa complex interacts with the membrane of host cells to activate the signal transduction pathways that are responsible for the generation of membrane ruffles on the host cells in much the same way as the homologous Sip proteins of *Salmonella* are thought to. IpaC can nucleate actin filaments and is a SipC homologue. It can complement a SipC deficiency in *Salmonella,* although SipC is not sufficient for *Shigella* invasion, suggesting IpaC has functionality beyond that of actin nucleation. The IpaB protein, which is homologous to and complements the *Salmonella* SipB protein, has recently been shown to bind to the cell surface protein CD44; earlier work found it associated with the $\alpha_5\beta_1$-integrin. The CD44 interaction at least is required for both *Shigella* invasion and cytoskeletal reorganization. CD44 is a receptor for hyaluronan (an extracellular matrix glycosaminoglycan) and activates changes in the cytoskeleton via the ezrin-radixin-moesin (ERM) proteins, of which ezrin is known to be required for efficient *Shigella* internalization.

ERM proteins are important cytoskeletal-membrane linkers and are thought to be regulated by Rho family GTPases, which are important for *Shigella* just as they are for *Salmonella*. Dominant negative mutant forms of Cdc42 and Rac1 severely impede *Shigella* internalization, but they do not abolish it entirely. The tyrosine kinase pp60[c-src] is thought to be a downstream target of Rho GTPases, and upon *Shigella* invasion, it phosphorylates one of its major substrates, cortactin. Cortactin, an actin-binding protein of the cortical actin cytoskeleton, is subsequently recruited to the site of bacterial attachment.

The actin-rich ruffles that extend upward around the invading *Shigella* bacterium are also enriched for the actin-bundling protein fimbrin (whose activity is enhanced by the *Salmonella* protein SipA, a homologue of *Shigella* IpaA). Cells that express a dominant negative form of fimbrin are defective for invasion of *Shigella,* supporting a role for actin bundling in this process. But IpaA has been proposed to perform another function as well. Focal adhesion proteins (such as vinculin, paxillin, α-actinin, talin, and others) have been localized to sites of *Shigella* attachment, and IpaA is able to directly recruit vinculin. A model has been proposed in which IpaA generates a focal adhesion at the site of bacterial attachment to resist being ejected by the explosive membrane ruffles that are about to erupt. This is supported by the observation that IpaA-deficient *Shigella* species are not able to maintain their contact with host cells when ruffling begins. The IpaA-vinculin complex causes depolarization of vinculin-associated actin in vitro, and if distributed in a concentration gradient descending from the point of contact, it might also aid in the smoothing of Cdc42-dependent spiky filopodia to gentler membrane sheets that ultimately engulf the bacterium.

Despite the morphological and mechanistic differences between the trigger and zipper mechanisms for bacterial invasion of nonphagocytic cells, they have certain important traits in common. Both require that extracellular bacteria send signals across the host cell plasma membrane to induce local actin polymerization. Both involve bacterial activation of signal transduction cascades that are already present in the host cell, although used for other purposes. Both are mediated by bacterial expression of specific virulence factors, whose main function is to induce these particular responses during host cell invasion. Importantly, both zipper and trigger uptake mechanisms proceed using energy derived from the host cell. The cytoskeletal and membrane rearrangements that are responsible for bacterial invasion require no energetic input from the bacterium once the type III secretion apparatus is made (for the trigger mechanism of uptake); they invade by persuasion rather than by force. This is in contrast to invasion of some eukaryotic parasites, such as *Toxoplasma gondii.* In that case, the energy for invasion comes from the parasite, which actively pushes its way into the host cell, and the host cell is a passive victim.

Preventing Phagocytosis

Yersinia Species

Thus far we have considered only cases where it is to the bacterium's advantage to be taken up by a host cell. Clearly, this is not always the case. In particular, bacteria that have gained access to normally sterile host tissues must avoid being eaten and digested by patrolling macrophages. Macrophages are highly motile and use an actin-dependent mechanism of phagocytosis to engulf any foreign particles they encounter. As we saw above, *Salmonella* avoids this fate by preemptively invading macrophages through a trigger mechanism and delivering itself into an intracellular compartment that is prevented from fusing with the macrophage lysosomes. In striking contrast, other bacteria such as *Yersinia, Pseudomonas,* and *Helicobacter* go to the opposite extreme, preventing macrophages from engulfing them at all. These bacteria disrupt the macrophage cytoskeleton by injecting factors into the host cell cytosol to block the normal rearrangements required for immune cell phagocytosis. Currently, *Yersinia* species are the best studied and we focus on them here.

Y. pestis enters the body by transmission through the skin from an infected flea bite or by aerosol transmission from an infected individual. They are then transmitted through the bloodstream to the nearest lymph nodes where they multiply. *Y. enterocolitica* and *Y. pseudotuberculosis* enter the body by the fecal-oral route and invade the Peyer's patches in the intestine and spread to the mesenteric lymph nodes. *Yersinia* species have developed a complex mechanism to evade the host immune response in lymphatic tissues by injecting factors into host cell macrophages that inhibit their ability to reorganize their cytoskeletons and engulf and kill the invaders. Injection of these factors into the host cell macrophages is mediated through a type III secretion system (discussed in chapter 15, and also described above in the context of triggered invasion by *Salmonella* and *Shigella*). These injected proteins are called Yops. Initially, the Yops were named as *Yersinia* outer membrane proteins, but now some of these factors are thought to be soluble. However, the name had already caught on, so they are still called Yops. The Yops are encoded on a 70-kb virulence plasmid, and their activation depends on interaction with the host cell plasma membrane.

The functions of the Yop effector proteins are diverse, ranging from specific inhibition of the inflammatory response to regulating the cytoskeleton. We focus on four well-characterized cytoskeletal manipulators: YopE, YopH, YopT, and YpkA/YopO. YopH is a 51-kDa protein that has homology to mammalian protein tyrosine phosphatases in its C-terminal half, a proline-rich region in its central domain that has been reported to bind to the SH3 domain of the c-Src tyrosine kinase, and an N-terminal domain which signals for bacterial secretion and translocation into host cells.

YopH is one of the most active tyrosine phosphatases ever isolated even though phosphotyrosine is very uncommon in bacteria. One might first guess that the gene for this tyrosine phosphatase was acquired by the bacterium in a horizontal transfer from a eukaryotic host. However, structural studies of the N-terminal domain responsible for substrate recognition of YopH imply otherwise. The crystal structure of the YopH phosphotyrosine domain is structurally distinct from other known eukaryotic phosphotyrosine-binding domains. It seems YopH, like previously discussed molecules such as invasin, SipA, SipC, and SptP, is another example of the bacterial ability to innovate novel functional mimics of eukaryotic proteins.

YopH acts to cause detachment of cells from the extracellular matrix. It targets focal adhesion kinase, paxillin, Fyn-binding protein, p130^{cas}, and SKAP-HOM. The inappropriate dephosphorylation of these proteins ultimately causes disruption of focal adhesions, resulting in release of the cell from the extracellular matrix and thus abolishing the cell's ability to engulf the bacteria. Not surprisingly, other actin-modulating Yops mediate their effects through the GTPases discussed earlier. YopE causes actin stress fibers to depolymerize by activiting Rho and may also have effects on Rac. YopT also acts to disrupt stress fibers but does so via RhoA. YopO binds to both RhoA and Rac. In *Y. enterolitica*, any of these four factors alone is not sufficient to resist phagocytosis by macrophages, and all are necessary for any appreciable level of survival. It seems YopH, YopE, YopT, and YopO act synergistically to render professional phagocytes harmless. Deprived of their normal cytoskeletal structure, macrophages lose control of their shape and ability to alter it to perform phagocytosis.

Bacterial Attachment without Invasion

Enteropathogenic *Escherichia coli*

A completely different type of host cell cytoskeletal manipulation has been developed by enteropathogenic *Escherichia coli* (EPEC), an important causative agent of diarrheal disease in children worldwide. Like *Listeria*, *Salmonella*, and *Shigella*, EPEC enters the human host orally and its initial site of host colonization is the intestinal epithelium. However, EPEC typically remains extracellular and forms colonies that are strongly attached to the surface of the epithelial cells. This strategy allows the bacteria to remain in the nutrient-rich environment of the intestinal lumen, while other microbial competitors are washed out by the diarrhea caused by EPEC infection. One would expect that attachment of bacteria to the epithelial cell surface via epithelial cell adhesion proteins would result in zippering and bacterial invasion, as described above for *Listeria* and *Yersinia*. EPEC has developed a mechanism to generate very strong adhesive contacts with epithelial cells, but these contacts only rarely result in uptake. Paradoxically, this anti-invasive strategy also involves recruitment of cytoskeletal elements and induction of actin filament polymerization at the site of attachment to the host cell.

In cell culture and in patient biopsies of infected intestinal tissue, epithelial cell brush borders have been observed to contain microcolonies of bacteria at sites where the microvilli have been destroyed and replaced by actin-rich structures termed "pedestals." These areas of infection are called attaching and effacing lesions because of the intimate attachment of the bacteria and consequent loss of microvilli (effacement) associated with the disease. Much speculation has surrounded their function; some have hypothesized that they are somehow required for prolonged bacterial attachment while others suggest that the pedestal structure might prevent the induction of phagocytosis observed for many other bacteria that closely adhere to the host cell surface (discussed above).

The reorganization of the membrane and cytoskeleton in this area is initiated by adhesin binding, followed by secretion of highly evolved virulence factors through a type III secretion system. Remarkably, one of the first factors the bacterium injects is actually a receptor, translocated intimin receptor (Tir), for a bacterially encoded surface ligand, intimin. When its extracellular domain is bound to intimin, the intracellular domain of Tir is a substrate for host cell protein tyrosine kinases and interacts with a series of host cytoskeletal factors to induce the host cell to retract its microvilli and create a pedestal under the attached bacterium. The EPEC bundle-forming pilus (BFP) causes the subsequent clumping of the bacteria. The BFP is a protein extension that is crucial for formation of EPEC microcolonies by allowing for the clustering of the microbes through association of their pili.

In recent years, much evidence has accumulated to suggest that Tir interacts directly with important players in cytoskeletal organization, aiding both in the construction of a new focal adhesion localized to the bacterium and in the recruitment of actin to form the pedestal shaft. When phosphorylated, Tir binds Nck, a host-cell adapter protein that is known to bind WASP family proteins. WASP, in turn, recruits the actin-nucleating Arp2/3 complex and soon, presumably, actin polymerizes to form the pedestal structure. Tir also interacts directly with focal adhesion components α-actinin, talin, and vinculin in a phosphorylation-independent manner, perhaps helping to secure the bacterium to the pedestal. Remarkably, inhibitors of the Rho family GTPases do not stop pedestal formation in cell culture (although they are required when EPEC does invade the cells). This supports the idea that EPEC has learned to govern cytoskeletal behavior more independently of its host's normal signaling pathways than many other pathogens.

On some tissue culture cell types, the pedestals themselves move laterally over the cell surface, a dramatic and unusual form of actin-based motility. The pedestal may allow EPEC to sit comfortably on its focal adhesion-like contact without being internalized by the zipper or trigger mechanisms that its cousins use, leading to the hypothesis that the more rare EPEC invasion events are "accidental" failures of the pedestal. However, evidence is accumulating that other species, among them *Neisseria gonorrhoeae* and *Helicobacter pylori*, can tightly adhere to cell surfaces, reorganize cortical actin to retract microvilli, and largely avoid internalization without necessarily forming pedestals. We will discuss the multiple host exploitation strategies of *Helicobacter* next.

Helicobacter pylori

The helical bacterium *H. pylori* is unusual in many respects. Not only does it have a relatively unusual coiled rod shape, but it thrives in the extremely hostile acidic environment of the stomach lumen and surrounds itself with a unique lipopolysaccharide (LPS) membrane. *H. pylori,* present in an estimated 50% of the world's population, is the primary causative agent of gastric and duodenal ulcers, and it has also been implicated in gastric cancer in humans. The bacteria form special contacts with the epithelium of the stomach, living in the mucosa but interacting intimately with the host cells. While of widespread clinical importance in its own right, *H. pylori* may also have much to tell us about the cytoskeleton of the eukaryotic cells it exploits.

Recent studies on the cell biology of *H. pylori* infection indicate that it exhibits many of the behaviors described for other pathogens in this chapter. Like *Yersinia*, *H. pylori* seems capable of inhibiting phagocytosis by macrophages. And like EPEC, *H. pylori* has been found to specifically adhere to epithelial cells and reorganize the actin filaments beneath their site of attachment. *H. pylori* is capable of inducing actin pedestal formation in primary gastric cells in tissue culture, although it does not always do so. Like *Listeria*, *H. pylori* may be capable of invasion by a zipper-like mechanism with a high degree of actin localization at the site of uptake. And unlike any other known bacterium, *H. pylori* has been documented as able to induce motility in at least one gastric cell line. This epithelial cell crawling is reminiscent of the effects of HGF ("scatter factor"), whose receptor the *L. monocytogenes* InlB protein stimulates to invade.

Stimulation of host cell motility depends on translocation and subsequent phosphorylation of the cytotoxin-associated gene (CagA) protein into the host cell by a type IV secretion system (discussed in chapter 16). Both CagA and the components of that secretion system are clustered in a special region of the genome called the cytotoxin-associated gene pathogenicity island (Cag PAI). *H. pylori* lacking Cag PAI are nonvirulent and live commensally with humans. So far, CagA is the only identified effector delivered by *H. pylori* into the host cell where it associates with the host membrane. CagA is phosphorylated on tyrosine residues by host cell protein kinases, which is required for at least some of *H. pylori*'s cytoskeletal effects. It has been speculated that CagA might be a functional homologue of the EPEC Tir protein. However, the two proteins share no identifiable sequence similarity. Further, both Rac-1 and Cdc42 have been found to be activated after *H. pylori* attachment in a CagA-independent manner, and shortly thereafter, stress fibers and cortical actin rings begin to degrade. Membrane ruffling has sometimes been reported. It seems likely that *H. pylori* encodes other actin-modulating factors yet to be identified.

The multitude of host cell behaviors attributed to *H. pylori* infection may represent our first glimpses of a highly evolved strategy for making the best of a hostile niche (the stomach) by mastering many host processes. An electron microscopic study of colonized stomach biopsies revealed that, while almost three-quarters of the bacteria were merely adhering to the epithelium, about 10% formed calyx-like depressions, another 10% formed pedestals, and 3% were inside of cells. This last small category may explain why most *H. pylori* infections are chronic throughout a person's lifetime. Perhaps *H. pylori* has found it best, when exploiting the cytoskeleton, to keep as many options open as possible.

Intracellular Motility and Intercellular Spread

Pathogenic bacteria that have chosen to exploit an intracellular growth niche in the host are quickly faced with a unique problem. To continue to replicate, they must find some mechanism for moving from one host cell to another. If they replicate unabated in a single cell, that cell will inevitably collapse under the pressure of nutrient deprivation and may lyse or induce apoptosis under stress. While lysis of the host cell does provide an opportunity for the liberated bacteria to reinfect a new cell, this is a time-consuming and costly process for the bacteria since they must expose themselves to the immune system. In most cases, including *Salmonella* and other well-known intracellular pathogens such as *Mycobacterium*, the bacteria replicate within a membrane-bound compartment, never entering the cytosol, until the host cell lyses and the replicated bacteria must repeat the cycle of invasion or be engulfed by resident or recruited phagocytes. A small number of pathogens have taken an alternative approach and have developed mechanisms for undergoing intercellular spread prior to the lysis of the originally infected host cell. These bacteria leave the phagocytic vacuole by secreting membrane-degrading enzymes and then, in the cytosol, exploit the host cell actin cytoskeleton for their own motility. This allows them to divide and move within the cell they have invaded as well as pass into neighboring cells via the engulfment of bacteria-containing membrane extensions. This unique mode of spread allows these crafty pathogens to move from cell to cell in a host epithelium without ever leaving the cytoplasmic environment, and they are therefore not exposed to soluble antibodies. Three bacterial genera, *Listeria, Shigella,* and *Rickettsia,* have been identified with this ability so far, and recent reports indicate that several other organisms may also form actin tails. In this discussion, we will focus on *Listeria* and *Shigella* species, for which extensive molecular characterization has been performed.

Actin-Based Motility: *Listeria monocytogenes*

As discussed earlier in this chapter, *Listeria* invades epithelial cells through a zipper mechanism. Shortly after entering the host cell, the microbe secretes a pore-forming hemolysin, listeriolysin O (LLO), which degrades the initial bacterium-containing vacuole, releasing the bacterium free into the host cell cytoplasm. After entering the host cell cytosol, *Listeria* is quickly surrounded by a "cloud" of short actin filaments, which localizes to one end of the bacterium upon cell division. Subsequently, an actin-rich tail forms and the bacteria can be observed to move at speeds ranging from 6 to 90 μm/min depending on the host cell type. The tail, which can be tens of microns in length, is made up of many thousands of short actin filaments cross-linked into a dense meshwork. Electron microscopic analysis of the actin tail of *Listeria* demonstrates that the actin filaments are arranged in a dendritically branched network and oriented with the majority of their barbed (rapidly growing) ends toward the surface of the bacterium. The actin filaments within the tail remain stationary in the host cell cytoplasm. Videomicroscopic studies of the incorporation of fluorescently labeled actin into *Listeria*'s actin tail have shown that the filaments behind the bacteria are growing only near the bacterial surface and are depolymerizing at a constant rate throughout the tail, suggesting that the propulsive force is derived from the actin polymerization at the bacterial cell surface. As we saw in the case of cellular invasion, the energy for this form of bacterial movement is provided entirely by the host cell.

A single bacterial protein, ActA, has been identified as necessary and sufficient for the subversion of the host cell actin cytoskeleton for motility by *Listeria*. ActA is a 67-kDa protein that is expressed on the surface of the bacterium. Attachment of purified ActA to small latex beads allows those beads to move by actin-based motility in cellular extracts, and expression of the ActA gene in a nonpathogenic *Listeria* species, *L. innocua,* confers motility on this species. A different gram-positive bacterium, *Streptococcus pneumoniae,* was also able to move after being coated with an ActA-derived fusion protein. Interestingly, motility was observed only after bacterial division, consistent with the observation in *Listeria* that ActA must be polarized to one end of the bacterium to support motility. Polarization of ActA by *Listeria* also appears to be regulated by bacterial division.

ActA is not homologous to any known proteins outside of the *Listeria* genus but has a number of interesting features that are important for its function. It is a surface-bound protein that has an N-terminal signal sequence and a single C-terminal membrane-spanning domain. Actin filament nucleation at the bacterial surface requires the presence of the N-terminal third of the protein. The central one-third of the protein includes four proline-rich repeats. Mutational analysis has shown that the proline-rich repeats contribute to *Listeria* motility by affecting its speed. Removal of any or all of the proline-rich repeats causes the mutant bacteria to move more slowly, with about a 20% decrease in speed for each repeat removed. Thus, ActA mediates three separable steps in *Listeria* actin-based motility. First, it catalyzes nucleation of actin filament polymerization at the bacterial surface. Second, it mediates rearrangement of these actin filaments into a polarized tail structure that can support unidirectional movement. Third, it uses specific sequences to accelerate the movement rate of the bacteria. Interestingly, purified ActA has no effect on actin polymerization or dynamics in vitro. All of its functions are mediated through association with factors supplied by the host cell.

Much of our understanding of how these factors interact has come from studies of cell-free systems. *Listeria* can undergo actin-based motility in cellular extracts from the oocytes of the clawed toad *Xenopus laevis*, human platelets, or bovine or mouse brain. When these extracts are supplemented with energy in the form of ATP and with fluorescently labeled actin monomer, *Listeria* can be observed by video microscopy to move in a manner that is very similar to that seen in cultured cells (Figure 12.3). More recently, motility has been reconstituted in a system of purified proteins. From the reconstituted system, we know that the only host proteins required for unidirectional motility are actin, the Arp2/3 complex, ADF/cofilin, and capping protein; other proteins, including vasodilator-stimulated phosphoprotein (VASP), profilin, and α-actinin, play secondary roles to enhance the speed or stability of movement. The putative roles of these proteins are summarized in Figure 12.4 and discussed below.

The Arp2/3 complex is named for the actin-related proteins (Arps) that are two of the seven polypeptides of the complex. This complex is homologous to one originally identified in *Acanthamoeba castellani* and appears to have an evolutionarily conserved role in regulation of the actin cytoskeleton in eukaryotes. The actin filament-nucleating activity of the seven-protein Arp2/3 complex relies on its two actin-like subunits (Arp2 and Arp3), which are thought to structurally mimic the end of an actin filament, thus providing a scaffold for filament growth. The complex also binds the sides of already-formed actin filaments to engender actin branches at an angle of ~70°. Those branches can be found both in the lamellipodia of locomoting cells and in the tails of bacteria or particles moving by actin-based motility.

The nucleating activity of Arp2/3 is synergistically activated by the N-terminal domain of ActA. This same domain of ActA also contains an actin monomer binding site, which may enhance the efficiency of the nucleation process. Thus, new actin filaments are continuously nucleated close to the bacterial surface as branches off the sides of preexisting filaments. The rate of polymerization and the organization of these newly nucleated filaments are modulated by other actin-associated proteins recruited to the nascent actin comet tail.

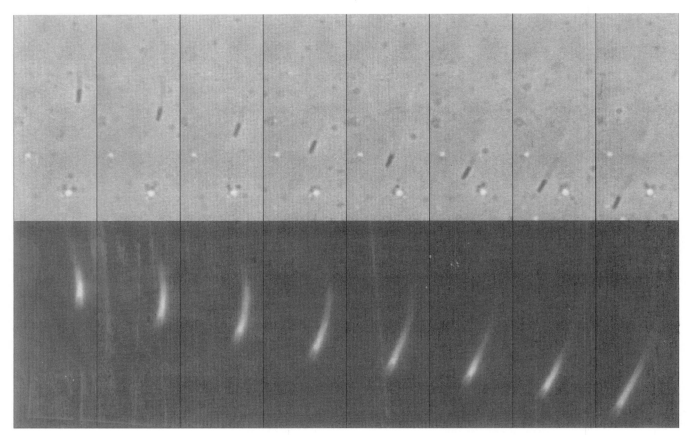

Figure 12.3 Actin tail formation and movement of *L. monocytogenes* in cytoplasmic extract. Eight frames from a video sequence are shown, recorded at 10-s intervals with phase-contrast images on the top and direct actin fluorescence on the bottom.

The ligand for the central, proline-rich region of ActA is the focal adhesion protein VASP; other members of the VASP family including Mena (mammalian enabled homologue) and Evl (Ena/VASP-like protein) can also bind to this domain of ActA and perform similar functions. Ena/VASP family proteins are localized to the end of motile bacteria associated with the actin-rich tail, and this association has been reported to occur before the formation of the actin clouds around the bacteria. VASP binds directly to ActA in vitro in a blot overlay assay using radiolabeled VASP as a probe, and its consensus binding site is the oligopeptide FPPPP found within the proline-rich domain of ActA. Mutation of the proline-rich repeats of ActA abolishes VASP binding; these bacteria can still undergo actin-based motility, but in a very slow and inefficient manner. VASP also binds directly to actin filaments, strengthening the mechanical connection between the bacterium and the tail, and modulates the association of the Arp2/3 complex and capping protein with the elongating filaments. In addition, VASP has been shown to bind to profilin and has several other putative binding partners that may be involved in cytoskeletal regulation.

The observation that deletion of the proline-rich repeats results in a slowing of bacterial motility suggests that the acceleration function of ActA is mediated through its association with VASP and profilin. Profilin's known function in accelerating the rate of actin filament elongation is the most likely cause of this change in bacterial speed. Profilin is a 14-kDa protein that binds to actin monomers. Profilin acts as a nucleotide exchange factor for actin, enhancing the dissociation of ADP and the binding of ATP to monomeric actin. In vitro, profilin can accelerate the elongation of actin filaments by acting as a chaperone to shuttle actin monomers from the thymosin β_4-bound monomer pool to the barbed ends of preexisting filaments. Its association with actin can be disrupted by the inositol lipid phosphatidylinositol 4,5-bisphosphate (PIP$_2$), and this interaction is thought to be one way that actin monomers are activated for rapid polymerization of the actin cytoskeleton in response to stimuli that generate PIP$_2$, such as growth factors. Profilin localizes to moving bacteria in a manner dependent on the presence of the proline-rich repeat domain of ActA on the bacterial surface and the presence of Ena/VASP family proteins, and an increase in profilin recruitment is correlated with faster bacterial speed.

Another protein necessary to give optimal motility in the reconstituted system, α-actinin, is a 100-kDa actin-cross-linking protein that acts by forming a homodimer that can associate with an actin filament at each end. α-Actinin has been localized to focal adhesions and filopodia, as well as the

Figure 12.4 Schematic diagram of the interactions between *Listeria monocytogenes* ActA protein and host cell cytoskeletal factors. The C terminus is anchored in the bacterial membrane. The N-terminal third of the protein mediates dimerization and also mediates nucleation of actin filaments by the Arp2/3 complex. The activated Arp2/3 complex binds to the sides of preexisting filaments to nucleate branches. Filaments nucleated close together can be efficiently cross-linked by α-actinin. A second host protein complex that includes VASP tetramers and profilin associates with the proline-rich repeats in the central region of ActA. This interaction serves to accelerate the rate of actin filament elongation.

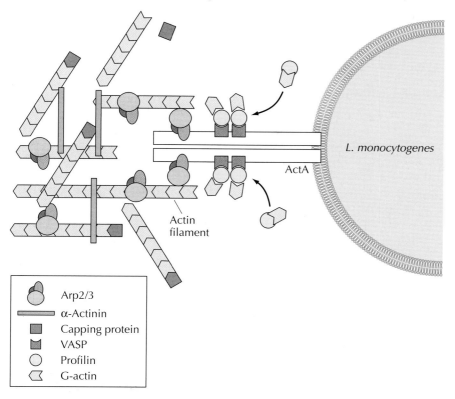

actin pedestals formed by association with EPEC (discussed above), where it is thought to be important in maintaining the bundling of the microfilaments in those structures. Microinjection of a 53-kDa piece of the α-actinin molecule that contains the actin-binding domain but not the dimerization domain causes disruption of actin bundles because they are not cross-linked. This treatment disrupts the tail of *Listeria* and abrogates motility, indicating that α-actinin is important for cross-linking actin microfilaments in the actin-rich tail of *Listeria*. A cloud of actin filaments remains associated with the bacteria after this treatment, suggesting that the cross-linking function of α-actinin is required primarily for establishing the mechanical integrity of the tail that enables it to push the bacterium forward.

Thus, the spatial organization of actin polymerization at the bacterial surface is mediated by the bacterial protein ActA, but the filament nucleation, rapid elongation, structure, and dynamics of the tail itself are governed by the host cell. First, the filaments nucleated by Arp2/3, and elongated by VASP and profilin, are cross-linked by α-actinin to form a sound structure that can push the bacterium. Then other critical host factors are required to maintain the structure, even though they do not interact with ActA. Capping protein and cofilin (ADF) are thought to play important roles in the maintenance of a pool of monomeric actin sufficient to continue polymerization at the bacterial surface. As the bacterium moves away from filaments, capping protein presumably hurries in to block the barbed ends, encouraging polymerization only where it will be useful, at the bacterial surface. Cofilin (ADF), meanwhile, accelerates depolymerization from the pointed ends, making more free monomer available. Thus, elongation can continue at steady state at the bacterial surface, resulting in a constant flux of actin subunits through the tail. Small local biochemical interactions within a few nanometers of the bacterial surface initiate a cascade of cytoskeletal rearrangements, resulting in formation of the tail that can be many micrometers in length. This thousandfold enhancement of scale is reminiscent of the host cell responses induced during bacterial invasion. Once again, the bacteria harness the power of the host cell cytoskeleton by small-scale signals, turning the host cell itself into a powerful and energetic ally.

Actin-Based Motility: *Shigella flexneri*

In contrast to the zipper-type invasion of *Listeria*, *Shigella* invades host epithelial cells using a trigger mechanism and, thereby, ends up in a different type of intracellular membrane-bound compartment. But after this point, the intracellular lifestyles of *Shigella* and *Listeria* are quite similar. *Shigella* also escapes from its membrane-bound compartment, in a process that requires IpaB. Once it has left the phagocytic vacuole, *Shigella* can usurp the host cell's actin cytoskeleton for its own motility and forms an actin-rich tail that is very similar to that formed by *L. monocytogenes* but is due to the expression of a different bacterial protein called IcsA (intercellular spread A) or VirG. IcsA (VirG) has no sequence similarity to the ActA protein, but it too has been shown to be necessary and sufficient for generation of motility: *E. coli* expressing IcsA moves by actin-based motility in host cells, cytoplasmic extracts, and a reconstituted protein system.

IcsA is a 120-kDa protein that has an N-terminal signal sequence for delivery to the periplasmic space and a C-terminal sequence of 344 amino acids (known as the β domain) that delivers and anchors the protein to the outer membrane of the cell. The remaining 706 amino acids are known as the α domain and are oriented with the N terminus of the protein extended away from the cell. This domain is predicted to be important for interactions with host cell proteins. The α domain contains five glycine-rich repeats that may be important for binding to proteins, which are responsible for the nucleation of actin filaments to form the actin-rich tail. Since IcsA is delivered to one bacterial pole but is free to diffuse in the membrane, it is distributed in an exponential gradient across the surface, ensuring unidirectional actin-based motility. IcsA is proteolytically cleaved at position 758 (between the α and β domains) by the SopA (IcsP) protease, and this yields a soluble 95-kDa fragment that has been localized to the actin-rich tail of *Shigella*. IcsP is thought to be uniformly distributed throughout the bacterial membrane and thus acts to make the IcsA gradient steeper, but it is not required for motility.

The host proteins necessary and sufficient for IcsA-mediated actin-based motility are similar to those used by ActA (actin, the Arp2/3 complex, profilin, cofilin [ADF], capping protein, and α-actinin), but VASP is not needed and an additional factor is required. Unlike ActA, IcsA does not recruit the Arp2/3 complex directly. Instead, it mimics Cdc42 to bind to neuronal Wiskott-Aldrich syndrome protein (N-WASP). N-WASP is the neuronal relative of the hematopoietic WASP protein implicated in Wiskott-Aldrich syndrome. Both of these WASP family proteins are thought to be regulated by Cdc42. However, only N-WASP has been shown to be important for the formation of actin-rich filopodial membrane projections, and WASP does not interact with IcsA. N-WASP has been shown to localize to the front of the actin-rich tail in locomoting bacteria in cultured cells, just as it does in the leading edge of lamellipodia. It also has been shown to bind to the glycine-rich repeat region of IcsA in vitro, stimulating increased Arp2/3 complex-mediated actin nucleation and accelerating the rate of actin filament growth.

Another actin-binding protein, vinculin, also binds to IcsA. Vinculin is found in focal adhesions. However, vinculin is not required for motility, since IcsA-mediated motility is still possible both in cells lacking vinculin and in a reconstituted system of purified proteins that do not include vinculin. The role of vinculin binding remains unclear, but microinjecting it into cells increases *Shigella* speed. Vinculin also contains conserved proline-rich motifs that may allow it to bind to VASP, which, while not required for *Shigella* motility, may enhance it.

The ActA-mediated actin-based motility of *Listeria* and the IcsA-mediated actin-based motility of *Shigella* have apparently arisen independently through convergent evolution, since there is no detectable similarity between the two bacterial proteins. Comparison of the two systems is thus particularly informative. In both cases, there appear to be several steps involved in actin-based motility. First, the bacterial surface proteins associate with host factors that cause actin filament nucleation (the Arp2/3 complex for *Listeria* and N-WASP plus possibly other unidentified factors in *Shigella*). Second, the nucleated clouds are rearranged to form the polarized tail, a step critically important in the generation of unidirectional movement that apparently involves both bacterial polarity and actin-cross-linking proteins provided by the host cell. Third, actin filament elongation at the bacterial surface is accelerated by yet another set of host factors, resulting in acceleration of overall bacterial speed. Dynamics and behavior of the tail at a distance from the bacterial surface are governed solely by host cell homeostatic mechanisms. In both cases, the bacterium provides no energy for its movement, but moves through host cell cytoplasm using host cell energy.

The observation that this ability has appeared two and possibly three or more times independently in the evolution of bacterial pathogens (and at least once among the viruses; see Box 12.1) suggests that this evolutionary step is not particularly demanding.

The main utility of actin-based motility for the bacteria appears to be the ability to reach the host cell surface and spread intercellularly. Objects the size of bacteria cannot diffuse appreciable distances in the viscous eukaryotic cytoplasm and so must use some form of active movement. The cytoplasm is also too viscous for bacterial flagella to rotate, so the bacteria must look elsewhere for a motile system. In eukaryotic cytoplasm, they are surrounded by the robust and adaptable actin cytoskeleton, which is preadapted and poised to generate large-scale protrusive structures in response to small local signals. Apparently all the bacteria must do is provide a binding site for host factors that catalyze actin filament nucleation and elongation at the bacterial surface, and the later steps of construction of the comet tail and generation of propulsive force are all taken care of by the obliging host cell.

Intercellular Spread by *Listeria monocytogenes* and *Shigella flexneri*

Intracellular motility allows pathogens to spread directly from the cytoplasm of one cell into the cytoplasm of an adjacent cell. This explains one potentially deadly feature of both shigellosis and listeriosis: not only can the bacterial infection spread without detection by macrophages, but it can cross both the placental and blood-brain barrier by moving directly through the cytoplasm of endothelial cells. Just as *Listeria* and *Shigella* convergently evolved ways to exploit their host's actin-based motility system, they appear to have discovered a way to exploit another host cell process in spreading from one cell to another.

The process of intercellular spread begins when the bacteria collide with the host cell membrane and form long membrane-bound protrusions into adjacent cells. *E. coli* can form these protrusions when expressing IcsA, and the frequency of their formation is statistically similar to that found with *Shigella*. The elongation of the protrusions proceeds with normal motility speeds, and for *Listeria* their extension has been shown to be ATP-dependent in MDCK kidney cells. Hence, it is thought that the protrusions are simply a consequence of the same actin-based motility observed within the cytoplasm. After a characteristic amount of time, the cell receiving the protrusion appears to pinch it off and the protrusion collapses into a double-membrane-bound vacuole inside its new host cell.

For *Listeria*, efficient escape from the new vacuole requires LLO and two phospholipases. *Shigella* uses the products of the *ipaB*, *ipaC*, and *ipaD* loci to similar effect. Although protrusions containing IcsA-expressing *E. coli* are taken up by the recipient cell, they remain trapped in vacuoles since they lack any factors for escape.

Initially, it was thought that the bacteria must modify the intercellular protrusion in some way to induce a normally nonphagocytic epithelial cell to "bite off" what should seem no more than a piece of its neighbor. However, the similarity of the characteristic process observed in multiple species argues otherwise. Further, it has been shown that epithelial cells do take up bits of neighboring membranes at a low frequency even in the absence of bacteria in a process termed paracytophagy, and this process may be important in tissue specification during development (where the transferred cell fragments are termed argosomes). The host cell requirements for para-

BOX 12.1

Everybody in the pool, viruses can swim too!

Bacteria are not the only pathogens that can exploit the host cell's cytoskeleton. Recent evidence suggests that at least one virus can utilize the actin cytoskeleton for its own intracytoplasmic motility as well, although internalization of the virus is not inhibited by cytochalasin D and is hence not thought to be actin dependent. Vaccinia virus is a complex virus related to the smallpox viruses and has a large genome of about 190 kilobase pairs. Initial observations of vaccinia particles by electron microscopy revealed virus-containing microvillus-like structures at the surface of infected cells and these structures were later shown to contain actin. Videomicroscopy has directly demonstrated that viruses associated with these actin structures are motile and that the movement of tail-associated viral particles is very similar to the actin-based movement of *Listeria* and *Shigella*.

Vaccinia may take a number of different forms during the course of infection. One, the intracellular mature virus (IMV), is released from the cell only upon cell lysis. A second, the intracellular enveloped virus (IEV), is IMV (about 5 to 15% of the total IMV) that wraps itself in intracellular membranes derived from the *trans*-Golgi network. The third form, extracellular enveloped virus (EEV), is released when IEV fuses with the plasma membrane. The IEV form of the virus appears to be the only one that can move by actin-based motility. Vaccinia makes a transmembrane protein, A36R, which is phosphorylated inside of cells and binds Nck, the same protein that the EPEC Tir recruits; in fact, the two proteins share some sequence homology and vaccinia tails contain most of the same molecules found in EPEC-induced actin pedestals. Nck recruits N-WASP, and N-WASP recruits the actin filament-nucleating complex Arp2/3. In vaccinia, the WASP-interacting protein (WIP) regulates some of this activity by stabilizing actin filaments. Recent data suggest that the IEV form may also utilize the host microtubule cytoskeleton in an A36R- and kinesin-dependent manner, but how this contributes to pathogenesis is not yet understood.

The finger-like projections that are formed by the viruses appear to be morphologically distinct from normal cellular microvilli and can be observed extending into neighboring cells. It is thought that vaccinia, like *Shigella* and *Listeria*, can spread to adjacent cells in membrane protrusions as well as by more classical budding.

It may seem surprising that these viruses have developed a pathogenic mechanism that bears such striking similarity to a pathogenic mechanism of bacteria. But the selective pressures driving the evolution of this dramatic type of host-pathogen interaction may be similar in the two cases. Most budding viruses assemble into particles from their constituent components at the host cell plasma membrane; the unassembled components can diffuse rapidly through the cytoplasm to the site of assembly at the surface. Vaccinia, however, replicates in viral "factories" near the nucleus, and it is the intact (and relatively large) viral particles that must make their way to the plasma membrane. Harnessing the protrusive behavior of the host cell actin cytoskeleton appears to be an energetically cheap and evolutionarily nondemanding mechanism to generate intracellular motility, for these large viruses as well as for bacteria.

cytophagy are not known, but they presumably involve cytoskeletal or cytoskeletally linked factors and may be physiologically (if not molecularly) similar to the zipper method of invasion. In MDCK cells, the cell-cell adhesion E-cadherin molecules have been localized along the length of *Listeria* intercellular protrusions. Studies of *Shigella* intercellular spread have found that host cells lacking E-cadherin show less intercellular spread than wild-type cells, and similarly the presence of connexins enhances the efficiency of spread. However, the basis for this has not been elucidated and may be due to global structural defects in the cells and/or the inefficiency of protrusion formation into poorly adherent neighbors.

During its brief time in the vacuole that results from protrusion uptake, *Listeria* uses a metalloprotease to strip its own surface of ActA. After escaping the vacuole, it must produce and polarize ActA again (presumably by cell division) before beginning motility. The biological reason for this is unknown, but it has been speculated that this better enables *Listeria* to take advantage of its new host, prolonging its period of replication instead of striking out immediately for another host cell with a new actin comet-like tail. In this way, *Listeria* demonstrates both cleverness, in exploiting its host's actin-based motility system, and wisdom, in knowing when not to. Further study of the process of intercellular spread should yield new insights into ordinary cellular processes that the bacteria exploit, just as study of bacterial invasion and actin-based motility have done.

The many processes used by bacterial pathogens of cellular invasion, inhibition of internalization, cellular adhesion, and intercellular spread may appear diverse, but they are linked by common molecules: the components of the host cell cytoskeleton. In becoming better pathogens, these bacteria have come to a highly evolved appreciation of the subtlety required to regulate it.

Selected Readings

Amieva, M. R., N. R. Salama, L. S. Tompkins, and S. Falkow. 2002. *Helicobacter pylori* enter and survive within multivesicular vacuoles of epithelial cells. *Cell Microbiol.* **4:**677–690.

> Uses time-lapse video microscopy, EM, and fluorescence to document the fate of individual *Helicobacter* cells over several hours, beginning with the actin-dependent formation of intracellular vacuoles.

Backert, S., S. Moese, M. Selbach, V. Brinkmann, and T. F. Meyer. 2001. Phosphorylation of tyrosine 972 of the *Helicobacter pylori* CagA protein is essential for induction of a scattering phenotype in gastric epithelial cells. *Mol. Microbiol.* **42:**631–644.

Black, D. S., and J. B. Bliska. 1997. Identification of p130Cas as a substrate of *Yersinia* YopH (Yop51), a bacterial protein tyrosine phosphatase that translocates into mammalian cells and targets focal adhesions. *EMBO J.* **16:**2730–2744.

Cameron, L. A., P. A. Giardini, F. S. Soo, and J. A. Theriot. 2000. Secrets of actin-based motility revealed by a bacterial pathogen. *Nat. Rev. Mol. Cell Biol.* **1:**110–119.

Cornelis, G. R., and F. Van Gijsegem. 2000. Assembly and function of type III secretory systems. *Annu. Rev. Microbiol.* **54:**735–774.

Cossart, P. 1997. Host/pathogen interactions. Subversion of the mammalian cell cytoskeleton by invasive bacteria. *J. Clin. Invest.* **99:**2307–2311.

Cossart, P., and H. Bierne. 2001. The use of host cell machinery in the pathogenesis of *Listeria monocytogenes. Curr. Opin. Immunol.* **13:**96–103.

Cudmore, S., P. Cossart, G. Griffiths, and M. Way. 1995. Actin-based motility of vaccinia virus. *Nature* **378:**636–638.

> First demonstration of actin polymerization-dependent movement of virus particles through the cytoplasm to infect adjacent cells.

Delahay, R. M., G. Frankel, and S. Knutton. 2001. Intimate interactions of enteropathogenic *Escherichia coli* at the host cell surface. *Curr. Opin. Infect. Dis.* **14:**559–565.

> Up-to-date molecular review of pedestal formation in EPEC.

Egile, C., T. P. Loisel, V. Laurent, R. Li, D. Pantaloni, P. J. Sansonetti, and M. F. Carlier. 1999. Activation of the CDC42 effector N-WASP by the *Shigella flexneri* IcsA protein promotes actin nucleation by Arp2/3 complex and bacterial actin-based motility. *J. Cell Biol.* **146:**1319–1332.

> Neatly ties together the molecules that link IcsA to actin, noting the protein domains responsible for each association.

Finlay, B. B., and P. Cossart. 1997. Exploitation of mammalian host cell functions by bacterial pathogens. *Science* **276:**718–725.

Fu, Y., and J. E. Galan. 1998. The Salmonella typhimurium tyrosine phosphatase SptP is translocated into host cells and disrupts the actin cytoskeleton. *Mol. Microbiol.* **27:**359–368.

Galan, J. E., and J. B. Bliska. 1996. Cross-talk between bacterial pathogens and their host cells. *Annu. Rev. Cell Dev. Biol.* **12:**221–255.

Galan, J. E., and D. Zhou. 2000. Striking a balance: modulation of the actin cytoskeleton by Salmonella. *Proc. Natl. Acad. Sci. USA* **97:**8754–8761.

Goldberg, M. B. 2001. Actin-based motility of intracellular microbial pathogens. *Microbiol. Mol. Biol. Rev.* **65:**595–626.

Grosdent, N., I. Maridonneau-Parini, M. P. Sory, and G. R. Cornelis. 2002. Role of Yops and adhesins in resistance of *Yersinia enterocolitica* to phagocytosis. *Infect. Immun.* **70:**4165–4176.

>   Wide-ranging test of the effects of various Yops on phagocytosis, with and without opsonization.

Hamburger, Z. A., M. S. Brown, R. R. Isberg, and P. J. Bjorkman. 1999. Crystal structure of invasin: a bacterial integrin-binding protein. *Science* **286:**291–295.

Hayward, R. D., and V. Koronakis. 1999. Direct nucleation and bundling of actin by the SipC protein of invasive *Salmonella*. *EMBO J.* **18:**4926–4934.

>   First description of the biochemical actin-regulating properties of SipC, the only known pathogen-made protein that acts directly on actin to both nucleate and bundle filaments.

Hayward, R. D., and V. Koronakiss. 2002. Direct modulation of the host cell cytoskeleton by *Salmonella* actin-binding proteins. *Trends Cell Biol.* **12:**15–20.

Ireton, K., and P. Cossart. 1998. Interaction of invasive bacteria with host signaling pathways. *Curr. Opin. Cell Biol.* **10:**276–283.

Kueltzo, L. A., J. Osiecki, J. Barker, W. L. Picking, B. Ersoy, W. D. Picking, and C. R. Middaugh. 2002. Structure-function analysis of invasion plasmid antigen C (IpaC) from shigella flexneri. *J. Biol. Chem.* **278:**2792–2798.

Kuhn, M., and W. Goebel. 1998. Host cell signalling during *Listeria monocytogenes* infection. *Trends Microbiol.* **6:**11–15.

Kwok, T., S. Backert, H. Schwarz, J. Berger, and T. F. Meyer. 2002. Specific entry of *Helicobacter pylori* into cultured gastric epithelial cells via a zipper-like mechanism. *Infect. Immun.* **70:**2108–2120.

>   First demonstration of the steps involved in *H. pylori* invasion.

Lafont, F., G. Tran Van Nhieu, K. Hanada, P. Sansonetti, and F. G. van der Goot. 2002. Initial steps of *Shigella* infection depend on the cholesterol/sphingolipid raft-mediated CD44-IpaB interaction. *EMBO J.* **21:**4449–4457.

Loisel, T. P., R. Boujemaa, D. Pantaloni, and M. F. Carlier. 1999. Reconstitution of actin-based motility of *Listeria* and *Shigella* using pure proteins. *Nature* **401:**613–616.

May, R. C., and L. M. Machesky. 2001. Plagiarism and pathogenesis: common themes in actin remodeling. *Dev. Cell.* **1:**317–318.

Monack, D. M., and J. A. Theriot. 2001. Actin-based motility is sufficient for bacterial membrane protrusion formation and host cell uptake. *Cell Microbiol.* **3:**633–647.

Stebbins, C. E., and J. E. Galan. 2001. Structural mimicry in bacterial virulence. *Nature* **412:**701–705.

>   Comprehensive review of pathogenic molecules that imitate host molecules to modulate the cytoskeleton, with special attention paid to SptP and invasin, as well as speculation as to how such proteins evolve.

Suzuki, T., H. Miki, T. Takenawa, and C. Sasakawa. 1998. Neural Wiskott-Aldrich syndrome protein is implicated in the actin-based motility of *Shigella flexneri. EMBO J.* **17:**2767–2776.

Welch, M. D., A. Iwamatsu, and T. J. Mitchison. 1997. Actin polymerization is induced by Arp2/3 protein complex at the surface of *Listeria monocytogenes. Nature* **385:**265–269.

>   Isolation of the Arp2/3 complex and demonstration of its actin-nucleating capacity on *L. monocytogenes*.

13

Bacterial Toxins

Mariagrazia Pizza, Vega Masignani, and Rino Rappuoli

Toxins were the first bacterial virulence factors to be identified and were also the first link between bacteria and cell biology. Cellular microbiology was, in fact, naturally born a long time ago with the study of toxins, and only recently, thanks to the sophisticated new technologies, has it expanded to include the study of many other aspects of the interactions between bacteria and host cells. This chapter covers mostly the molecules that have been classically known as toxins; however, the last section also mentions some recently identified molecules that cause cell intoxication and have many but not all of the properties of classical toxins. These belong to a rapidly expanding field and are perhaps among the most interesting molecules being studied today in cellular microbiology. Table 13.1 shows the known properties of all bacterial toxins described in this chapter, while Figure 13.1 shows the subunit composition and the spatial organization of toxins whose structures have been solved either by X-ray crystallography or by quick-freeze deep-etch electron microscopy.

The observation that culture supernatants free of bacteria fully reproduced the symptoms of deadly diseases, such as diphtheria, tetanus, cholera, and botulinum, made it obvious that in these instances bacterial toxins were the only factors needed by bacteria to cause disease, making the study of their pathogenesis straightforward. In many cases, immunity against the toxins is enough to prevent the disease caused by the bacterium that produces them, and therefore they have often been used as vaccines after being subjected to chemical treatment to remove their toxicity. The powerful effects of toxins, such as cell death, and the striking morphology changes that they caused in vivo in animal models and in vitro on cells (Figure 13.2) made it easy to study their interactions with eukaryotic cells. For this reason, toxins were the first link between bacteria and cell biology. The link became stronger when it was found that the discovery of any toxin target led to the discovery of a new, important pathway in cell biology. This is because toxins need targets consisting of molecules that play a key role in the most essential and vital processes of living organisms. Two very important path-

Table 13.1 Bacterial toxins and their targets: update

Class of toxin	Target	Toxin	Organism	Activity	Consequence	Three-dimensional structure
Extracellular toxins acting on the cell surface	Immune system (superantigens)	SEA-SEH, TSST-1, SPEA, SPEC, SSA, SPEL, SPEM	*S. aureus* and *S. pyogenes*	Binding to MHC class II molecules and to Vβ or Vγ of T-cell receptor	T-cell activation and cytokine secretion	SEA, SEB, SEC2, SEC3, SED, SEH, SET3, TSST-1, SPEA, SPEC
		SMEZ	*S. pyogenes*	Binding to MHC class II molecules and to Vβ or Vγ of T-cell receptor	T-cell activation and cytokine secretion	−
		SPEB	*S. pyogenes*	Cysteine protease, integrin binding	Proteolytic cleavage of fibronectin and vitronectin	+
		ETA, ETB, ETD	*S. aureus*	Trypsin-like serine proteases	T-cell proliferation, intraepidermal layer separation	ETA, ETB
	Cell surface molecules	BFT enterotoxin	*B. fragilis*	Metalloprotease	Cleavage of E-cadherin, alteration of epithelial permeability	−
		Elastase	*A. hydrophila*	Metalloprotease	Hydrolization of casein and elastine	−
		Elastase	*P. aeruginosa*	Metalloprotease	Corneal infection, inflammation and ulceration	−
		Collagenase	*C. hystolyticum*	Metalloprotease, calcium binding	Tissue destruction, gas gangrene	+
		Nhe	*B. cereus*	Metalloprotease and collagenase	Collagenolytic activity	−
	Cell membrane Large pore-forming toxins	Perfringolysin O	*C. perfringens*	Cell membrane permeabilization	Gas gangrene	+
		Streptolysin O	*S. pyogenes*	Cell membrane permeabilization	Transfer of other toxins, cell death	−
		Listeriolysin	*L. monocytogenes*	Cell membrane permeabilization	Apoptosis	−
		Pneumolysin	*S. pneumoniae*	Cell membrane permeabilization	Complement activation, cytokine production, apoptosis	−
	Small pore-forming toxins	α-Toxin	*S. aureus*	Cell membrane permeabilization	Release of cytokines, cell lysis	+
		Hemolysin II	*B. cereus*	Cell membrane permeabilization	Necrotic enteritis	−
		Panton-Valentine leucocidin (LukS-LukF)	*S. aureus*	Cell membrane permeabilization	Cell permeabilization and lysis	LukF

Toxin	Organism	Mechanism	Effect		
γ-Hemolysins (HlgA-HlgB and HlgC-HlgB)	*S. aureus*	Cell membrane permeabilization	Cell permeabilization and lysis	HlgB	−
CytK	*B. cereus*	Cell membrane permeabilization	Necrotic enteritis		−
β-Toxin	*C. perfringens*	Cell membrane permeabilization	Neurologic effects		−
Aerolysin	*A. hydrophila*	Cell membrane permeabilization	Cell permeabilization and lysis		+
α-Toxin	*C. septicum*	Cell membrane permeabilization	Rapid shock-like syndrome		−
δ-Hemolysin	*S. aureus*	Perturbation of the lipid bilayer	Cell permeabilization and lysis		−
Membrane-perturbing toxins					
RTX toxins					
HlyA	*E. coli*	Calcium-dependent formation of transmembrane pores	Cell lysis of erythrocytes, leukocytes, endothelial cells, and human T lymphocytes		−
AppA	*A. pleuropneumoniae*	Calcium-dependent formation of transmembrane pores	Lysis of erythrocytes and a variety of nucleated cells		−
AaltA	*A. actinomycetemcomitans*	Calcium-dependent formation of transmembrane pores	Apoptosis		−
LktA	*P. hemolytica*	Calcium-dependent formation of transmembrane pores	Activity-specific vs ruminant leukocytes		−
Insecticidal toxins					
CryIA, CryIIA, CryIIIA, etc.	*B. thuringiensis*	Channel formation	Destruction of the transmembrane potential, with the subsequent osmotic lysis of cells lining the midgut	CryIA, CryIIIA	
CytA, CytB	*B. thuringiensis*	Channel formation	Destruction of the transmembrane potential, with the subsequent osmotic lysis of cells lining the midgut	CytB	

(continued)

Table 13.1 Bacterial toxins and their targets: update *(continued)*

Class of toxin	Target	Toxin	Organism	Activity	Consequence	Three-dimensional structure
	Other pore-forming toxins	HlyE	*E. coli*	Channel formation	Tissue damage	+
		Protective antigen (PA)	*B. anthracis*	Channel formation	Pore formation and translocation of edema and lethal factor toxins	+
Extracellular toxins acting on intracellular targets	Protein synthesis	Diphtheria toxin (DT)	*C. diphtheriae*	ADP-ribosylation of elongation factor 2	Cell death	+
		Exotoxin A (ExoA)	*P. aeruginosa*	ADP-ribosylation of elongation factor 2	Cell death	+
		Shiga toxin	*S. dysenteriae*	N-Glycosidase activity on 28S RNA	Cell death	+
	Signal transduction	Cholera toxin (CT)	*V. cholerae*	ADP-ribosylation of G_s	cAMP increase	+
		Heat-labile enterotoxin (LT-I, LT-II)	*E. coli*	ADP-ribosylation of G_s	cAMP increase	+
		Pertussis toxin (PT)	*B. pertussis*	ADP-ribosylation of G_i	cAMP increase	+
		Adenylate cyclase CyaA	*B. pertussis*	Binding to calmodulin, ATP → cAMP conversion	cAMP increase	–
		Anthrax edema factor (EF)	*B. anthracis*	Binding to calmodulin, ATP → cAMP conversion	cAMP increase	+
		Anthrax lethal factor (LF)	*B. anthracis*	Cleavage of MAPKK1 and MAPKK2	Cell death	+
		Cytolethal distending toxin (CDT)	*H. ducreyi* and other species	Cell cycle arrest	Cytotoxicity	–
		α-Toxin	*C. perfringens*	Phospholipase C	Gas gangrene	+
		Toxins A and B	*C. difficile*	Monoglucosylation of Rho, Rac, and Cdc42	Breakdown of cellular actin stress fibers, apoptosis	–
		Cytotoxin-necrotizing factors 1 and 2 (CNF1, CNF2)	*E. coli*	Deamidation of Rho, Rac, and Cdc42	Ruffling, stress fiber formation	CNF1
		Dermonecrotic toxin (DNT)	*Bordetella* species	Transglutaminase, deamidation, or polyamination of Rho GTPase	Assembly of actin stress fibers and focal adhesions	–

	Toxin	Activity	Organism	Effect	Substrate/Notes
Intracellular trafficking	Tetanus toxin TeNT	Cleavage of VAMP/synaptobrevin	*C. tetanii*	Spastic paralysis	+ (Hc domain)
	Neurotoxins BoNTB, D, G, and F	Cleavage of VAMP/synaptobrevin	*C. botulinum*	Flaccid paralysis	BoNTB
	Neurotoxins BoNTA, E	Cleavage of SNAP-25	*C. botulinum*	Flaccid paralysis	BoNTA
	Neurotoxin BoNTC	Cleavage of syntaxin	*C. botulinum*	Flaccid paralysis	—
	Vacuolating cytotoxin VacA	Alteration in the endocytic pathway	*H. pylori*	Vacuole formation	—
	NAD+ glycohydrolase	NAD+ glycohydrolase	*S. pyogenes*	Enhancement of GAS spread	—
Cytoskeleton structure	Toxin C2	ADP-ribosylation of monomeric G actin	*C. botulinum*	Failure in actin polymerization	—
	ι-Toxin	ADP-ribosylation of monomeric G actin	*C. perfringens*	Failure in actin polymerization	+ (enzymatic component Ia)
	ι-Like toxin	ADP-ribosylation of monomeric G actin	*C. spiroforme*	Failure in actin polymerization	—
Toxins injected into eukaryotic cells — Signal transduction	YpkA	Serine/threonine protein kinase	*Yersinia* species	Inhibition of phagocytosis	—
	YopH	Tyrosine phosphatase	*Yersinia* species	Inhibition of phagocytosis	+
	Tir	Tyrosine phosphorylated	*E. coli* EPEC	Actin nucleation and pedestal formation	—
	CagA	Tyrosine phosphorylated, reported interactions with Grb-2, SHP-2, and cortactin. Disruption of the apical junctional complex (AJC) by interactions with ZO-1 and Jam	*H. pylori*	Alteration of the host signal transduction pathways. Inhibition of terminal differentiation with loss of polarity, increased motility, and anchorage independency	—
	YopM	Stimulates the activity of PRK2 and RSK-1 kinases (probable)	*Yersinia* species	Not known	+
Cytoskeleton	SptP	GTPase activity	*Salmonella* species	Not known	+
	ExoY	Adenylate cyclase	*P. aeruginosa*	cAMP increase	—
	SopE	Rac and Cdc42 activation	*S. enterica* serovar Typhimurium	Membrane ruffling, cytoskeletal reorganization, proinflammatory cytokine production	+ (catalytic domain in complex with Cdc42)

(continued)

Table 13.1 Bacterial toxins and their targets: update (*continued*)

Class of toxin	Target	Toxin	Organism	Activity	Consequence	Three-dimensional structure
		ExoS	*P. aeruginosa*	ADP-ribosylation of Ras	Collapse of cytoskeleton	+ (GAP domain)
		C3 exotoxin	*C. botulinum*	ADP-ribosylation of Rho	Breakdown of cellular actin stress fibers	+
		EDIN-A, -B, -C	*S. aureus*	ADP-ribosylation of Rho-GTPases	Modification of actin cytoskeleton	−
		C3-like toxin	*C. limosum*	ADP-ribosylation of Rho-GTPases	Modification of actin cytoskeleton	−
		YopE	*Yersinia species*	GTPase	Cytotoxicity, actin depolymerization	+
		YopT	*Yersinia species*	Cysteine protease, releases Rho, Rac, and Cdc42 from the membrane	Disruption of actin cytoskeleton	−
		SipA	*Salmonella species*	Binding of F-actin	Actin polymerization, mimics muscle nebulin	+
		VirA	*S. flexneri*	Not known	Microtubule destabilization and membrane ruffling	−
	Inositol phosphate metabolism	SopB	*Salmonella species*	Inositol phosphate phosphatase	Increased chloride secretion (diarrhea) phosphatase	−
		IpgD	*S. flexneri*	Inositol phosphate phosphatase	Modulation of host cell response	−
	Mediators of apoptosis	YopP/YopJ	*Yersinia species*	Cysteine protease	Apoptosis of macrophages	−
		AvrRxV	*X. campestris*	Not known	Necrotic lesions	−
		IpaB	*S. flexneri*	Binding to ICE	Apoptosis	−
		SipB	*Salmonella species*	Binding to ICE	Apoptosis	−
		ExoU	*P. aeruginosa*		Epithelial lung injury	−
Unknown mechanism of action		Zot	*V. cholerae*		Modification of intestinal tight junction permeability	−
		HBL	*B. cereus*		Fluid accumulation and diarrhea	−
		Bile salt hydrolase (BSH)	*L. monocytogenes*		Increased bacterial survival and intestinal colonization	−
		Metalloprotease AhyB	*A. hydrophila*		Hydrolysis of elastin and casein	−

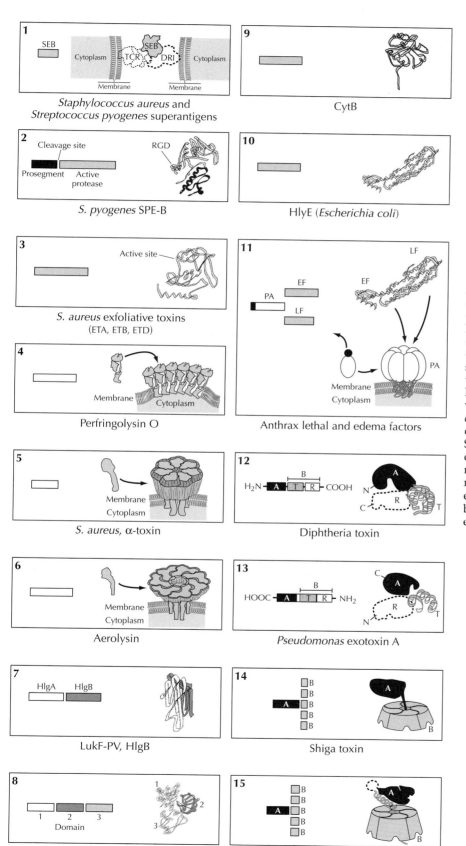

1

SEB

Cytoplasm   TCR   SEB   DR1   Cytoplasm

Membrane   Membrane

Staphylococcus aureus and
Streptococcus pyogenes superantigens

2

Cleavage site   RGD

Prosegment   Active protease

S. pyogenes SPE-B

3

Active site

S. aureus exfoliative toxins
(ETA, ETB, ETD)

4

Membrane
Cytoplasm

Perfringolysin O

5

Membrane
Cytoplasm

S. aureus, α-toxin

6

Membrane
Cytoplasm

Aerolysin

7

HlgA   HlgB

LukF-PV, HlgB

8

1   2   3
Domain

CytA

9

CytB

10

HlyE (*Escherichia coli*)

11

PA   EF   LF   EF

LF

Membrane
Cytoplasm   PA

Anthrax lethal and edema factors

12

B

H_2N — A   T   R — COOH

A
N   R   T
C

Diphtheria toxin

13

B

HOOC — A   T   R — NH_2

C
A
R   T
N

Pseudomonas exotoxin A

14

B
B
A   B   A
B
B   B

Shiga toxin

15

B
B
A   B   A
B
B   B

Cholera toxin, *E. coli* LT-I, LT-II

Figure 13.1 Structural features of bacterial toxins whose structures have been solved. (Left) Scheme of the primary structure of each toxin. For the A/B toxins, the domain composition is also shown. The A (or S1 in PT) represents the catalytic domain, whereas the B represents the receptor-binding domain. The A subunit is divided into the enzymatically active A1 domain and the A2 linker domain in Shiga toxin, CT, *E. coli* LT-I and LT-II, and PT. The B domain has either five subunits, which are identical in Shiga toxin, CT, and *E. coli* LT-I and LT-II and different in size and sequences in PT, or two subunits, the translocation (T) and the receptor-binding (R) subunits, in DT, *Pseudomonas* exotoxin A, botulinum toxin, and tetanus toxin. (Right) Schematic representation of the three-dimensional organization of each toxin. For *Staphylococcus* enterotoxin B, the protein is shown in the ternary complex with the human class II histocompatibility complex molecule (DR1) and the T-cell antigen receptor (TCR). For *Salmonella* SptP the structure is shown in the transition state complex with the small GTP-binding protein Rac1. Similarly, toxin SopE is represented in complex with its substrate Cdc42. In the case of *E. coli* CNF1 and *Pseudomonas* ExoS, only one domain has been crystallized. In the case of SipA, a three-dimensional reconstruction of SipA bound to F-actin filaments is also reported. For all toxins the schematic representation is based on the X-ray structure, except that for VacA, whose structure has been solved by quick-freeze, deep-etch electron microscopy.

(Figure 13.1 continues)

Figure 13.1 *continued*

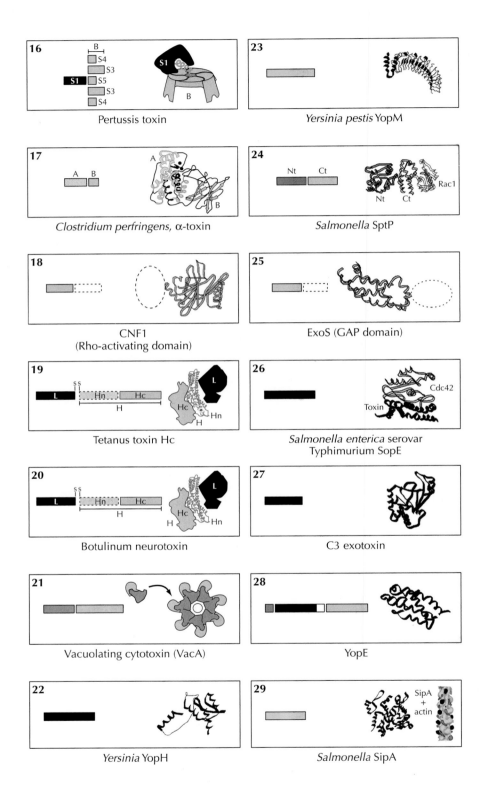

16 Pertussis toxin

17 *Clostridium perfringens*, α-toxin

18 CNF1
(Rho-activating domain)

19 Tetanus toxin Hc

20 Botulinum neurotoxin

21 Vacuolating cytotoxin (VacA)

22 *Yersinia* YopH

23 *Yersinia pestis* YopM

24 *Salmonella* SptP

25 ExoS (GAP domain)

26 *Salmonella enterica* serovar
Typhimurium SopE

27 C3 exotoxin

28 YopE

29 *Salmonella* SipA

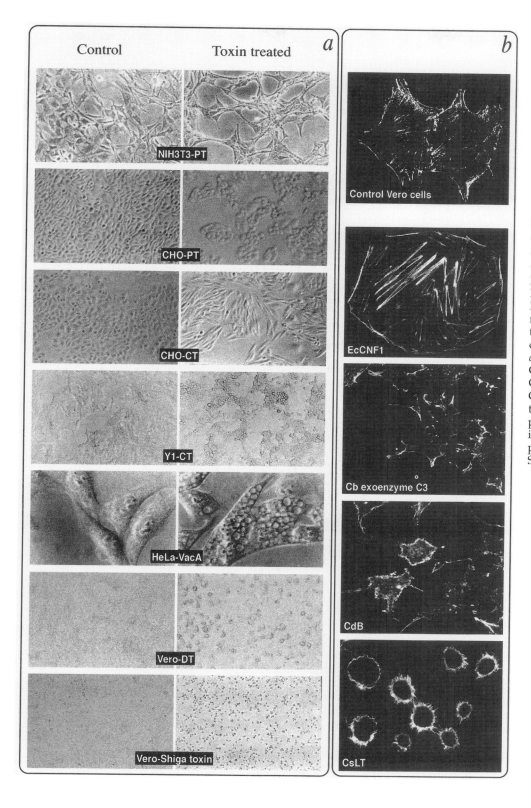

Figure 13.2 Morphology changes induced by toxins in cell lines. **(a)** Untreated cells (left) and toxin-treated cells (right). Abbreviations: NIH3T3-PT, NIH 3T3 cells treated with pertussis toxin; CHO-PT, Chinese hamster ovary cells treated with pertussis toxin; CHO-CT, Chinese hamster ovary cells treated with cholera toxin; Y1-CT, Y1 cells treated with CT; HeLa-VacA, HeLa cells treated with vacuolating cytotoxin A; Vero-DT, Vero cells treated with DT; Vero-Shiga toxin, Vero cells treated with Shiga toxin. Photographs from our laboratory or kindly provided by Ida Luzzi, Istituto Superiore di Sanità, Rome, Italy. **(b)** Immunofluorescence micrograph of untreated or toxin-treated Vero cells. Abbreviations: EcCNF1, *E. coli* cytotoxin-necrotizing factor; Cb exoenzyme C3, *C. botulinum* C3; CdB, *C. difficile* cytotoxin B; CsLT, *Clostridium sordellii* lethal toxin. Cells were stained with palloidin-fluorescein to visualize F-actin. Photographs kindly provided by Patrice Boquet, IN-SERM, Nice, France.

ways of cell biology were discovered, thanks to bacterial toxins, and they have been extensively investigated. The first is G-protein-mediated signal transduction (Box 13.1), which was discovered in the early 1980s; it could be understood only because pertussis toxin specifically blocked the inhibitory G proteins. The second pathway is the molecular mechanism of neurotransmitter release, which was discovered when tetanus toxin was found to specifically cleave vesicle-associated membrane protein, a key molecule in intracellular vesicular trafficking.

Toxins have a target in most compartments of eukaryotic cells. For simplicity, we divide the toxins into three main categories (Figure 13.3): (i) those that exert their powerful toxicity by acting on the surface of eukaryotic cells simply by touching important receptors, by cleaving surface-exposed molecules, or by punching holes in the cell membrane, thus breaking the cell permeability barrier (Figure 13.3, groups 1 and 2); (ii) those that have an intracellular target and hence need to cross the cell membrane (these toxins need at least two active domains, one to cross the eukaryotic cell membrane and the other to modify the toxin target) (Figure 13.3, group 3); and (iii) those that have an intracellular target and are directly delivered by the bacteria into eukaryotic cells (Figure 13.3, group 4).

Extracellular Toxins Acting on the Cell Surface

Some toxins deliver their toxic message just by binding to receptors on the surface of eukaryotic cells. The most common of these toxins is a family of related molecules known as superantigens. The binding domains (B domains) of the AB toxins sometimes also belong to this group. Some toxins act by cleaving important molecules exposed on the surface of eukaryotic cells. Another group of molecules that bind to the cell surface and in many cases affect cell function contains the pore-forming toxins, which act by punching holes in the cell membrane and breaking the cell permeability barrier.

BOX 13.1

Toxins act like the hero of "Independence Day"

More than 50% of bacterial toxins act on signal transduction. This suggests that perhaps the best way to poison a living organism is to interfere with cell communication. The vital importance of communication is obvious for our generation, which is dominated by multimedia and global internet connections. The importance of communication is even stressed by popular movies like "Independence Day," where the only way to defeat the extraterrestrial invaders is to interfere with their communication system, or where the most dangerous criminal combated by James Bond can get the power of the globe only by taking over communications ("Tomorrow Never Dies"). Like the hero of "Independence Day" and the baddie combated by "007," bacterial toxins have found it very convenient to attack the cell communication systems. In eukaryotic organisms, cell communication is mediated by (i) the tyrosine phosphorylation of the cytoplasmic C-terminal part of a receptor that recruits SH2-signal transducers and initiates a cascade of intracellular signaling events or (ii) modification of a receptor-coupled GTP-binding protein that transduces the signal to enzymes releasing secondary messengers, such as cyclic AMP, inositol tryphosphate, and diacylglycerol, that also initiate a cascade of intracellular signaling events. Interestingly, bacterial toxins have targeted almost exclusively signal transduction mediated by GTP-binding proteins and only in rare cases attack receptors that transmit signals by direct tyrosine phosphorylation of the cytoplasmic portion of the receptor. Why is this? The answer is likely to reside in the fact that GTP-binding proteins function as servers that transmit signals coming from many different receptors, while tyrosine phosphorylation transmits the signal coming from one receptor only. Therefore, a toxin targeting tyrosine phosphorylation would be like a criminal taking over the local radio station of a village. Nothing compared to the baddies of the Bond movies.

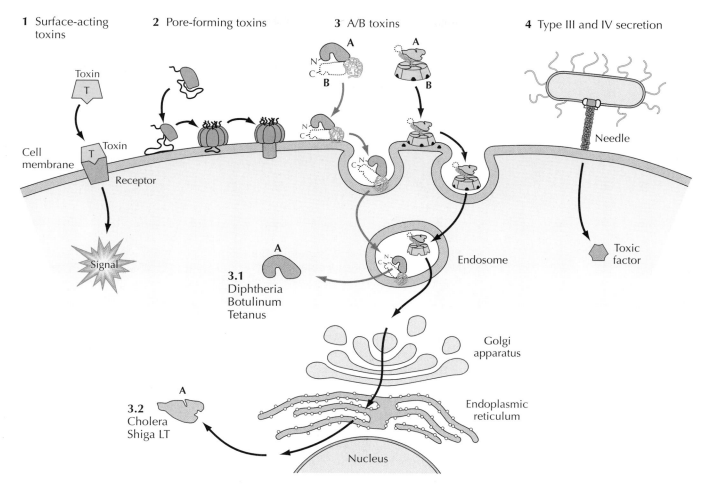

1 Surface-acting toxins

2 Pore-forming toxins

3 A/B toxins

4 Type III and IV secretion

Toxin
T

Cell membrane

Toxin
T

Receptor

Signal

A
N
C
B

A
B

N
C

A

3.1
Diphtheria
Botulinum
Tetanus

N
C

Endosome

Golgi
apparatus

A

3.2
Cholera
Shiga LT

Endoplasmic
reticulum

Nucleus

Needle

Toxic
factor

Figure 13.3   Schematic representation of the four groups of bacterial toxins. Group 1 toxins act by binding receptors on the cell membrane and sending a signal to the cell. Group 2 toxins act by forming pores in the cell membrane, perturbing the cell permeability barrier. Group 3 toxins are A/B toxins, composed of a binding domain (B subunit) and an enzymatically active effector domain (A subunit). Following receptor binding, the toxins are internalized and located in endosomes, from which the A subunit can be transferred directly to the cytoplasm by using a pH-dependent conformational change (3.1) or can be transported to the Golgi and the ER (sometimes driven by the KDEL ER retention sequence), from which the A subunit is finally transferred to the cytoplasm (3.2). Group 4 toxins are injected directly from the bacterium into the cell by a specialized secretion apparatus (type III or type IV secretion system).

Superantigens

Superantigens are produced mostly by *Staphylococcus aureus* and *Streptococcus pyogenes*. They are bivalent molecules that bind two distinct molecules (Figure 13.4), the major histocompatibility complex (MHC) class II molecule and the variable part of the T-cell receptor (Vβ or Vγ). The cross-linking between the MHC and the T-cell receptor is able to activate T cells even in the absence of a specific peptide. Therefore, superantigens are potent polyclonal activators of T cells both in vitro and in vivo. In vivo, the potent polyclonal activation results in a massive release of cytokines, such as interleukin-1 and tumor necrosis factor, which are believed to play an important role in diseases, such as the toxic shock syndrome induced by toxic shock

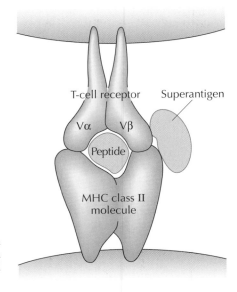

Figure 13.4 Schematic representation of the interaction of superantigens with the MHC class II molecule and T-cell receptor.

syndrome toxin 1 (TSST-1), vomiting and diarrhea caused by staphylococcal enterotoxins, and the exanthemas caused by the pyrogenic streptococcal exotoxins. Furthermore, these toxins have also been implicated in the pathogenesis of several acute or chronic human disease states, such as the Kawasaki syndrome, which is the leading cause of acquired heart disease among children in the United States, and also in other life-threatening events such as food poisoning.

The staphylococcal enterotoxins (SE) comprise a family of homologous proteins of approximately 230 amino acids, with one central disulfide bond, named staphylococcal enterotoxins A through H (SEA, SEB, SEC1, SEC2, SEC3, SED, SEF, SEG, and SEH). TSST-1 also belongs to the same family but is shorter (194 amino acids) and has no cysteines. The streptococcal erythrogenic toxins include streptococcal pyrogenic enterotoxin A and C (SPEA and SPEC) and streptococcal superantigen A (SSA). The genes for these toxins are generally carried on plasmids, bacteriophages, or other heterologous genetic elements, and all of them are translated into a precursor protein containing an N-terminal signal sequence, which is cleaved during export from the cell. The overall homology detected among the staphylococcal enterotoxins has been suggested to result from gene duplication of a common "ancestral" toxin. Crystallographic structures are currently available for most of the described staphylococcal and streptococcal superantigens. All of these toxins have a characteristic two-domain fold composed of a β barrel at the N terminus and a β grasp at the C terminus connected by a long α-helix that diagonally spans the center of the molecule (Figure 13.1). Moreover, they are characterized by a central disulfide bond (except for TSST-1, which has no cysteines) and by a Zn^{2+} coordination site, which is believed to be involved in MHC class II binding. The presence of two zinc-binding sites in SPEC indicates different modes in the assembly of the MHC-superantigen-TcR trimolecular complex.

Computer analysis of the *S. pyogenes* genome has revealed the presence of novel superantigen genes, and among them the one coding for the mitogenic exotoxin Z (SMEZ). This toxin is particularly similar to the SPEC group of superantigens and, although present in all GAS strains, it shows extensive allelic variation. Further genetic characterization has shown that SMEZ is the most potent bacterial superantigen so far discovered and that it strongly contributes to the immunological effects of GAS both in vitro and in vivo by eliciting a robust cytokine production.

Other pyrogenic toxin superantigens recently discovered by genome mining include proteins SPEL and SPEM produced by several isolates of *S. pyogenes* of the M18 serotype. The corresponding genes are contiguous and coded within a bacteriophage. Both toxins were shown to be lethal in different animal models and to directly participate in the host-pathogen interaction in some acute rheumatic fever (ARF) patients.

Other toxins that have long been known as superantigens are the streptococcal pyrogenic exotoxin B (SPEB), a virulence factor with cysteine protease activity produced by all isolates of group A streptococci, and the exfoliative toxins A and B produced by *S. aureus*. Although these proteins strongly contribute to the virulence of the corresponding microorganism, their role as mitogenic factors was disproved when it was shown that all the nonrecombinant forms were in fact contaminated with trace amounts of the SMEZ superantigen.

SPEB is expressed as a 40-kDa protein and is converted to a 28-kDa active protease by proteolytic cleavage. It causes a cytopathic effect on human endothelial cells and represents a critical virulence factor in human infection and in mouse models of invasive disease. The recently solved three-dimensional structure reveals the presence of a surface-exposed integrin-binding Arg-Gly-Asp (RGD) motif, devoted to human integrin binding. This feature is unique to SPEB among cysteine proteases and is linked to the pathogenesis of the most invasive strains of *S. pyogenes*. SPEB appears to contribute to *S. pyogenes* pathogenesis in several ways, including proteolytic cleavage of human fibronectin and vitronectin, two abundant extracellular matrix proteins involved in maintaining host tissue integrity.

The exfoliative toxins ETA and ETB of *S. aureus* are produced during the exponential phase of growth and excreted from colonizing staphylococci before being absorbed into the systemic circulation. They have been recognized as the causative agents in staphylococcal scalded skin syndrome, an illness characterized by specific intraepidermal separation of the layers of skin between the stratum spinosum and the stratum granulosum. Both superantigens are serine proteases, and this enzymatic activity could be one of the mechanisms hypothesized as the cause of epidermal separation. In fact, substitution of the active site serine residue with cysteine abolishes their ability to produce the characteristic separation of epidermal layers, but not their ability to induce T-cell proliferation. The two ETs are about 40% identical at the primary structure level and with no apparent sequence homology to other bacterial toxins. Their X-ray structures have been obtained showing an overall good conservation of the active sites.

Recently, a novel member of the exfoliative group of toxins was discovered in *S. aureus*. This protein, termed ETD, is encoded within a pathogenicity island, which also contains the genes for a serine protease and the *edin-B* gene. When injected in neonatal mice as recombinant protein, ETD has been shown to induce exfoliation of the skin with loss of cell-to-cell adhesion in the upper part of the epidermis.

Binding Domains of A/B Toxins

A/B toxins (discussed in detail in Soluble Toxins with an Intracellular Target, below) are composed of a toxic A domain and a carrier-binding domain

(B domain), which binds the toxin receptor on the cell surface and aids in translocation of the A domain across the cell membrane. Since the primary toxicity of A/B toxins is mediated by their A subunits, the role of the B domain is often neglected. However, there are several instances where receptor binding is sufficient to affect cell function. The amount of toxin necessary to achieve these effects is usually orders of magnitude greater than that required for the intoxication mediated by the A subunit, and therefore these effects are unlikely to play any major role in vivo. Nevertheless, they may be very useful for in vitro studies. The B subunits for which a cell function is known are those of pertussis toxin (PT), cholera toxin (CT), and the related *Escherichia coli* heat-labile enterotoxin (LT).

The B subunit of PT is a polyclonal mitogen for T cells. In vitro the effect requires at least 0.3 μg of PT or its B subunit. The effect has not been observed in vivo, possibly because the dose required for activity is never achieved. It also has hemagglutination activity. In vitro, the effect requires 0.3 μg of wild-type PT or its B subunit per ml. In addition, the B oligomer is able to induce signal transduction through the inositol phosphate pathway. CT and LT B subunits are systemic and mucosal immunogens, and this property depends on the binding to the GM_1 receptor. Moreover, they are strong inhibitors of T-cell activation and are able to induce apoptosis of $CD8^+$ T cells and, to a lesser extent, of $CD4^+$ T cells. Also, this effect has been described only in vitro, and there is no evidence so far of a role in vivo.

Toxins Cleaving Cell Surface Molecules

Bacteroides fragilis enterotoxin (BFT) is a protein of 186 residues that is secreted into the culture medium. The toxin has a zinc-binding consensus motif (HEXXH), characteristic of metalloproteases and other toxins such as tetanus and botulinum toxins. In vitro, the purified enterotoxin undergoes autodigestion and can cleave a number of substrates, including gelatin, actin, tropomyosin, and fibrinogen. When added to cells in tissue culture, the toxin cleaves the 33-kDa extracellular portion of E-cadherin, a 120-kDa transmembrane glycoprotein responsible for calcium-dependent cell-cell adhesion in epithelial cells, that also serves as a receptor for *Listeria monocytogenes* (for details see chapters 6 and 1, respectively). In vitro, BFT does not cleave E-cadherin, suggesting that the membrane-embedded form of E-cadherin is necessary for cleavage.

BFT causes diarrhea and fluid accumulation in ligated ileal loops. In vitro, it is nonlethal but causes morphological changes such as cell rounding and dissolution of tight clusters of cells. The morphological changes are associated with F-actin redistribution. In polarized cells, BFT is more active from the basolateral side than from the apical side, decreases the monolayer resistance, and causes dissolution of some tight junctions and rounding of some of the epithelial cells, which can separate from the epithelium. In monolayers of enterocytes, BFT increases the internalization of many enteric bacteria such as *Salmonella, Proteus, E. coli,* and *Enterococcus* but decreases the internalization of *L. monocytogenes.*

BFT belongs to a large family of bacterial metalloproteases that usually cleave proteins of the extracellular matrix. *Pseudomonas aeruginosa* and *Aeromonas hydrophila* elastases and *Clostridium histolyticum* collagenase are the best-known examples. The crystal structure of *C. histolyticum* collagenase has been solved. Recently, a novel member of this family of protein toxins was identified in *Bacillus cereus*. The protein, termed Nhe (nonhemolytic enterotoxin), is a 105-kDa metalloprotease, which shares homologies to the above-mentioned elastases and collagenases. Biochemical characterization has shown that Nhe possesses both gelatinolytic and collagenolytic activities.

Pore-Forming Toxins

Pore-forming toxins work by punching holes in the plasma membrane of eukaryotic cells, thus breaking the permeability barrier that keeps macromolecules and small solutes selectively within the cell. These toxins are often identified as lytic factors (lysins). Since erythrocytes have often been used to test the activity of these toxins, many of them are known as hemolysins. However, it should be kept in mind that the hemolytic activity was only a convenient way to measure their ability to punch holes in the cell membrane and release the intracellular hemoglobin, whose red color is easy to detect even with the naked eye. Therefore, while erythrocytes are a convenient target of hemolysins in vitro, they are never the main physiological target of this class of toxins in vivo, since the toxins exert their virulence effect mostly by permeabilizing other cell types. To generate channels and holes in the cell membrane, this class of proteins must be amphipathic, with one part interacting with the hydrophilic cavity filled with water and the other interacting with the lipid chains or the nonpolar segments of integral membrane proteins. The ultimate consequence of cell permeabilization is usually death, whereas the early consequences of losing the permeability barrier are often the release of cytokines, activation of intracellular proteases, and sometimes induction of apoptosis.

These proteins can be distinguished in large-pore-forming and small-pore-forming toxins on the basis of the dimension of the holes produced on the plasma membrane and also the kind of interaction that they establish with the eukaryotic receptor.

In addition, the pore-forming, RTX family of toxins includes a large group of Ca^{2+}-dependent hemolysins secreted by both gram-positive and gram-negative bacteria which are characterized by a conserved glycine- and aspartate-rich motif of nine amino acids (RTX, repeat in toxins).

Large-Pore-Forming Toxins

This class of cytolysins comprises more than 20 family members, which are generally secreted by taxonomically diverse species of gram-positive bacteria and which have the common property of binding selectively to cholesterol on the eukaryotic cell membrane. Large-pore-forming toxins consist of a single polypeptide chain with molecular weights ranging from 50 kDa to 80 kDa and are haracterized by a remarkable sequence similarity. Toxins contain a common motif (ECTGLAWEWWR) located approximately 40 amino acids from the carboxy terminus. Oxidation of the cysteine residue, included in this motif, abolishes the lytic activity; this activity can be restored by adding reducing agents such as thiols.

These proteins, listed in Table 13.2, are produced by *S. pyogenes*, *S. pneumoniae*, *Bacillus*, a variety of clostridia, including *Clostridium tetani*, and *Listeria*. To date, the best characterized are listeriolysin O, an essential virulence factor of *Listeria monocytogenes* that causes meningitis and abortion; perfringolysin O (PFO), a virulence factor of *Clostridium perfringens* that causes gas gangrene; pneumolysin, the major causative agent of streptococcal pneumonia and meningitis; and streptolysin O (SLO) secreted by *S. pyogenes*. SLO is widely used as a tool in cell biology to produce in vitro large pores in cell membranes, thus allowing the introduction of large molecules into the cell cytoplasm. Furthermore, recent works have demonstrated that

Table 13.2 Large-pore-forming toxins

Bacterial genus	Species	Toxin name
Streptococcus	S. pyogenes	Streptolysin O
	S. pneumoniae	Pneumolysin
	S. suis	Suilysin
Bacillus	B. cereus	Cereolysin O
	B. alvei	Alveolysin
	B. thuringiensis	Thuringiolysin O
	B. laterosporus	Laterosporolysin
Clostridium	C. tetani	Tetanolysin
	C. botulinum	Botulinolysin
	C. perfringens	Perfringolysin O
	C. septicum	Septicolysin O
	C. histolyticum	Histolyticolysin O
	C. novyi A (oedematiens)	Novyilysin
	C. chauvoei	Chauveolysin
	C. bifermentans	Bifermentolysin
	C. sordellii	Sordellilysin
Listeria	L. monocytogenes	Listeriolysin O
	L. ivanovii	Ivanolysin
	L. seeligeri	Seeligerolysin

SLO acts as a functional equivalent of type III secretion systems in *S. pyogenes* in that it is required for transfer of NAD^+-glycohydrolase into epithelial cells. The process of SLO-mediated delivery of NAD^+-glycohydrolase to the cytoplasm of human keratinocytes ultimately results in major changes in host cell biology that enhance GAS pathogenicity.

Large-pore-forming toxins share a similar mechanism of action, which consists of a former interaction, as monomers, with target cells via their receptor, cholesterol, followed by a subsequent oligomerization and insertion into the host cell membrane. This process ultimately results in serious membrane damage, with formation of large pores with diameters exceeding 150 Å that makes the cell membrane permeable to small solutes and large macromolecules (Figure 13.5A), thus leading to rapid cell death. It is interesting to note that the membrane-bound receptor, cholesterol, plays an important role in the oligomerization step as well as in membrane insertion and pore formation. However, recent data obtained for PFO would suggest a novel paradigm for explaining the mechanism of pore formation: according to this model, previous cooperative interactions between PFO monomers (prepore) would be required to promote transmembrane insertion and formation of the final β-barrel pore. Crystallographic data are available only for the thiol-activated cytolysin PFO of *C. perfringens*. Nevertheless, given the high degree of sequence conservation detected within this class of protein toxins (ranging from 43% identity in the case of PFO versus listeriolysin, to 72% identity for PFO versus alveolysin), we can consider this structure as the prototype of the entire family. Very recently, fluorescence labeling and spectroscopic techniques have been used to shed light on the mechanism of interaction of PFO within the membrane bilayer. From these studies it has emerged that only the tip of the β-barrel domain is responsible for membrane insertion and that a major conformational rearrangement takes place during pore formation.

Small-Pore-Forming Toxins

The small-pore-forming family of toxins forms very small pores (1 to 1.5 nm in diameter) in the membrane, allowing the selective permeabilization of cells to solutes that have a molecular mass lower than 2,000 Da. This group includes single-component toxins, such as *S. aureus* α-toxin and its structural homologue hemolysin II, produced by *B. cereus*, and bicomponent toxins, such as the Panton-Valentine leucocidin (LukS-PV-LukF-PV) and the γ-hemolysins (HlgA-HlgB and HlgC-HlgB). Since the initial discovery of the first small-pore-forming toxins, the number of these proteins has grown to include several members, among which is the recently identified cytotoxin K (CytK) of *Bacillus cereus*, implicated in necrotic enteritis, and the β-toxin of *C. perfringens*. These proteins are secreted as water-soluble monomers and assemble on the surface of susceptible cells to form heptameric transmembrane channels of approximately 1 nm in diameter.

The X-ray structure of the transmembrane pore of α-toxin has been solved and has confirmed the heptameric structure of the oligomer and the self-assembly of the glycine-rich region (H domain in Figure 13.5B) to form the pore.

Cells become permeable only to molecules smaller than the pores (up to 2,000 Da). Ions like Ca^{2+} are able to enter the cells, so that electrical permeability is established, while large molecules such as all cytoplasmic enzymes are retained within the cell. At this stage, the toxin induces a number of changes in the cell, such as production of eicosanoids, activation of

Figure 13.5 Schematic representation of membrane interaction and oligomerization of large-pore-forming toxins **(A)** and small-pore-forming toxins **(B).** In the case of large-pore-forming toxins, the prepore intermediate state is also shown. The dimension of the final pore can reach 35 nm, and up to 50 monomers can be involved.

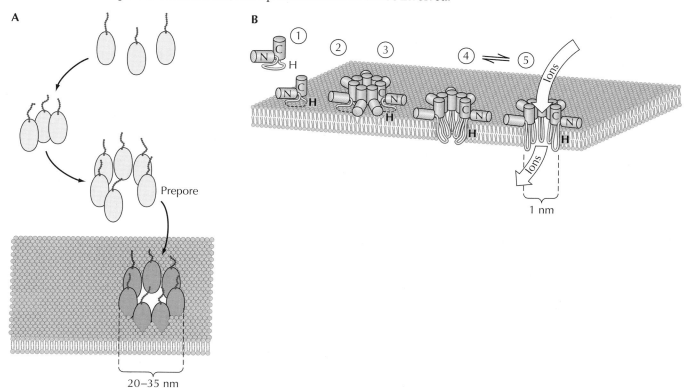

endonucleases, release of cytokines, and early apoptotic events. At high concentrations, the toxin can cause membrane rupture and cell lysis, thus killing the cells. The toxin causes membrane damage in a variety of cells, including erythrocytes, platelets, and leukocytes. In erythrocytes, for instance, the intoxication proceeds in distinct steps: binding to cell membrane, ion leakage, and eventually rupture of the cell membrane with release of larger molecules.

Leukotoxins and γ-hemolysins should be grouped together, since they form two types of bicomponent complexes that exhibit leukotoxic and hemolytic activity, respectively. Panton-Valentine leucocidin (PVL) is a closely related toxin carried by 2% of clinically isolated *S. aureus* strains and is also composed of type F and S components. The components of each protein class are produced as nonassociated, water-soluble proteins that undergo conformational changes and form oligomeric complexes after recognition of their cell targets, a process leading to transmembrane-pore formation and, ultimately, to cell death. Cell binding of bicomponent toxins is sequential. The S component is the first to adhere, thus allowing the subsequent binding of the F component. The transmembrane channels that they form are of an estimated diameter of 8 Å, thus being mainly permeable to divalent cations. The superimposition of Luk-F and HlgB monomers with the structure of α-toxin shows that the core structures of these pore-forming toxins are very similar despite the relatively low primary sequence identity (32%). A similar level of sequence similarity can be appreciated for the *C. perfringens* β-toxin, a protein that has been found to increase the conductance of bilayer lipid membranes (BLMs) by inducing channel activity.

From a structural point of view, in contrast to a wide range of bacterial and insect toxins that utilize α-helices to perturb or penetrate the bilayer, these pore-forming toxins can be defined as members of an emerging family of proteins that use bilayer-spanning antiparallel β barrels instead. Additional members of this class of β-barrel, channel-forming toxins include aerolysin of *A. hydrophila* and α-toxin of *Clostridium septicum*.

Aerolysin is secreted by *A. hydrophila* as a 52-kDa protoxin that is activated by proteolytic cleavage of the C-terminal peptide to yield a 48-kDa active toxin. The toxin binds to a family of GPI-anchored specific receptor on the surface of target cells, including the T-cell receptor RT6. Following binding, the protein oligomerizes and forms heptameric structures that insert into the cell membrane, forming pores approximately 1.5 nm in diameter. Finally, *C. septicum* α-toxin is a channel-forming protein that is an important contributor to the virulence of the organism. Recent data have proved that this toxin, like aerolysin, binds to GPI-anchored protein receptors. Furthermore, α-toxin is also active against *Toxoplasma gondii* tachyzoites. Toxin treatment causes swelling of the parasite endoplasmic reticulum, thus providing the first direct evidence that α-toxin is a vacuolating toxin.

Membrane-Perturbing Toxins

δ-Toxin or δ-hemolysin is secreted into the medium by *S. aureus* strains at the end of the exponential phase of growth. It is a 26-amino-acid peptide (MAQDIISTIGDLVKWIIDTVNKFTKK) that has the general structure of soap with a nonpolar segment followed by a strongly basic C-terminal peptide. The peptide has no structure in aqueous buffers but acquires an α-helical structure in low-dielectric-constant organic solvents and membranes. The α-helix has a typical amphipathic structure, which is necessary for the toxin to interact with membranes. The toxin binds nonspecifically parallel to the surface of any membrane without forming transmembrane channels. At high concentration, the peptide self-associates and increases the perturbation of the lipid bilayer that eventually breaks into discoidal or micellar structures. It is very interesting that mellitin, which is also a 26-amino-acid lytic peptide produced by *S. aureus,* has no sequence homology with δ-toxin but has identical distribution of charged and nonpolar amino acids. These toxins are active in most eukaryotic cells. Cells first become permeable to small solutes and eventually swell and lyse, releasing cell intracellular content.

Recent data have demonstrated that δ-hemolysin insertion is strongly dependent on the peptide-to-lipid ratio, suggesting that association of a critical number of monomers on the membrane is required for activity. The peptide appears to cross the membrane rapidly and reversibly and cause release of the lipid vesicle contents during this process.

RTX Toxins

E. coli hemolysin is a 110-kDa protein secreted into the culture supernatant by some pathogenic *E. coli* strains during exponential growth. The protein belongs to a large family of Ca^{2+}-dependent hemolysins known as RTX toxins, produced by different genera of *Enterobacteriaceae* and *Pasteurellaceae* and which are characterized by the presence of a conserved repeated glycine- and aspartate-rich motif of nine amino acids (repeat in toxins), containing multiple calcium-binding sites. The *E. coli* hemolysin is encoded by four genes, one of which, *hlyA*, encodes the 110-kDa hemolysin, while the others are required for its posttranslational modification (*hlyC*) and secretion (*hlyB* and *hlyD*). The four genes are found in a very limited number of *E. coli* clonal types and can often be located on transmittable plasmids. The activation process ultimately results in the acquired capacity of HlyA to bind target cells; this activation involves proteolytic processing and posttranslational acylation, as well as binding of Ca^{2+} ions to the repeated domain, to produce the functional toxin. Following binding, the N-terminal domain formed by hydrophobic and amphipathic transmembrane sequences is inserted into the membrane and forms transmembrane pores that are small (1 nm in diameter), unstable, hydrophilic, and cation specific. The toxin lyses erythrocytes, leukocytes, endothelial cells, renal epithelial cells, granulocytes, monocytes, and human T lymphocytes. As for other small-pore-forming toxins, the initial permeabilization causes the loss of small solutes, uptake of water, and osmotic cell lysis. To give an idea of the level of toxicity associated with *hlyA* gene product, when nonhemolytic strains of *E. coli* are transformed with recombinant plasmids encoding the hemolysin, they become 10- to a 1,000-fold more virulent than the parental strains when tested in rodent models of peritonitis.

Other members of this class of RTX proteins include the *Actinobacillus pleuropneumoniae* hemolysin (AppA), the leukotoxins of *A. actinomycetemcomitans* (AaltA) and *Pasteurella haemolytica* (LktA), and the bifunctional adenylate cyclase/hemolysin of *Bordetella pertussis* (CyaA), which will be described in Toxins Acting on Signal Transduction. In the case of HlyA and AaltA, recent studies have shown that the receptor molecule on the eukaryotic host cell membrane is represented by LFA-1, a member of the β_2-integrin family. Although a remarkable level of primary structure similarity can be detected among this group of toxins, they differ in host cell specificity and seem to adopt diverse mechanisms for cellular damage.

Insecticidal Toxins

Insecticidal proteins, also known as δ-endotoxins, include a number of toxins produced by species of *Bacillus thuringiensis* and that exert their toxic activity by making pores in the epithelial cell membrane of the insect midgut.

δ-Endotoxins form two multigenic families, *cry* and *cyt*; members of the *cry* family are toxic to insects of *Lepidoptera, Diptera,* and *Coleoptera*, whereas members of the *cyt* family are specifically lethal to the larvae of *Dipteran* insects. The insecticidal toxins of the *cry* family are synthesized by the bacterium as protoxins of molecular masses of 70 to 135 kDa; after ingestion by the susceptible insect, the protoxin is cleaved by gut proteases to release the active toxin of 60 to 70 kDa. In this form, they bind specifically and with high affinity to protein receptors and create channels of 10 to 20 Å in the cell membrane. This subgroup includes several toxins, such as CryIA, CryIIA, CryIIIA, CryIV, and CryV, whereas the only proteins so far characterized that belong to the *cyt* family are CytA and CytB.

Three-dimensional structures determined for members of the two families show that the folding of these toxins is entirely different. CryIA is a globular protein composed of three distinct domains connected by single linkers, while CytB is still a globular protein, but is composed by a single domain with a molecular mass of only 30 kDa. The region of CryIA that has been associated with receptor binding maps within a loop of domain 2, while domain 1 has been shown to be responsible for membrane insertion and pore formation. Conversely, in the case of the CytB/A, the model that has been proposed for the channel formation is based on a β-barrel structure.

Given the toxic activity exerted on several species of insects, δ-endotoxins have been formulated into commercial insecticides for more than three decades; recently, *Lepidoptera*-specific toxin genes have also been used to engineer insect-resistant plants.

Other Pore-Forming Toxins

HlyE is a novel pore-forming toxin produced by *E. coli*, which is completely unrelated to the *E. coli* hemolysin HlyA of the RTX family. Nevertheless, sequence comparison studies confirm the presence of highly homologous toxins in other pathogenic organisms such as *Salmonella enterica* serovar typhi and *Shigella flexneri* (these orthologs display 92 to 98% identity to HlyE). This observation suggests that HlyE could be the prototype of a new family of HlyE-like hemolysins specific for gram-negative bacteria. This new class of pore-forming toxins forms cation-selective water-permeable pores of 25 to 30 Å in diameter; the channel formation either could be part of a mechanism for iron acquisition by the bacterial cell or may promote bacterial infection by killing immune cells and causing tissue damage.

The precise mechanism of HlyE oligomerization to form the final transmembrane pore is at the moment unknown; nevertheless, the first step involves a process of dimerization of two HlyE molecules which pack in a head-to-tail fashion, burying the two hydrophobic patches against each other. Electron microscopy experiments have led to a model of channel formation in which the possible oligomer topology is that of an octameric complex and the β-tongue domain is primarily responsible for interaction with the membrane.

Anthrax Protective Antigen

Anthrax protective antigen (PA) is one of the three proteins of the anthrax toxin complex secreted by *Bacillus anthracis*; this complex comprises the receptor-binding PA, the edema factor (EF), and the lethal factor (LF) (Figure 13.6). PA can be considered as a B domain for two distinct A subunits such as EF and LF. The three subunits, encoded by a large plasmid, are synthesized and secreted independently. PA is a protein of 735 amino acids that binds the receptors on the cell surface. After binding, PA is activated by proteolytic removal of the 20-kDa N terminus, generating the PA C-terminal

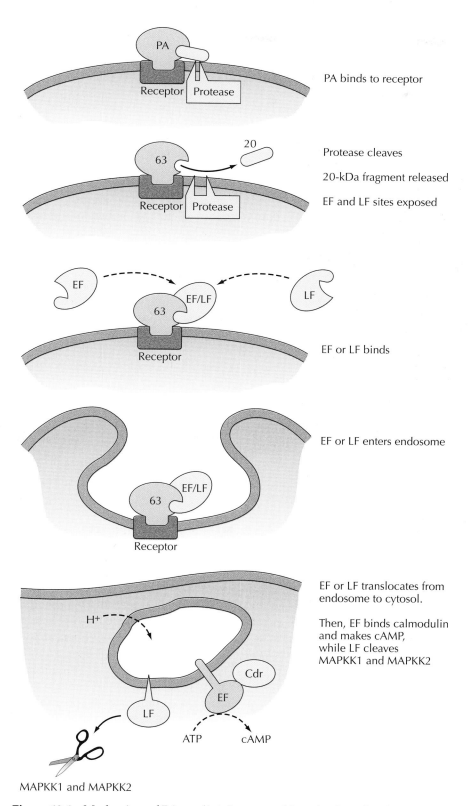

PA binds to receptor

Protease cleaves

20-kDa fragment released

EF and LF sites exposed

EF or LF binds

EF or LF enters endosome

EF or LF translocates from
endosome to cytosol.

Then, EF binds calmodulin
and makes cAMP,
while LF cleaves
MAPKK1 and MAPKK2

Figure 13.6   Mechanism of PA-mediated entry and intoxication of anthrax LF and
EF toxins.

63-kDa peptide (PA_{63}), which remains associated with the membrane. PA_{63} binds the N terminus of EF or LF. Following receptor-mediated endocytosis, the low pH causes a conformational change in PA that allows the translocation of EF or LF from the extracellular milieu into the host cell cytosol to exert toxicity. During this delivery, PA undergoes multiple structural changes, from a monomer to a heptameric prepore to a membrane-spanning heptameric pore. The catalytic factors also undergo dramatic structural changes as they unfold to allow for their translocation across the endosomal membrane and refold to preserve their catalytic activity within the cytosol. Very recently, the receptor molecule for PA was identified and cloned. The receptor, termed ATR (anthrax toxin receptor), is a membrane protein with an extracellular domain implicated in the binding to PA. The identification of ATR will allow a more detailed investigation of the mechanism of uptake of anthrax toxin by cells. EF and LF toxins will be described in the next section.

Extracellular Toxins Acting on Intracellular Targets

The group of toxins with an intracellular target (A/B toxins) contains many toxins with different structures that have only one general feature in common: they are composed of two domains generally identified as A and B. A is the active portion of the toxin; it usually has enzymatic activity and can recognize and modify a target molecule within the cytosol of eukaryotic cells. B is usually the carrier for the A subunit; it binds the receptor on the cell surface and facilitates the translocation of A across the cytoplasmic membrane. Depending on their target, these toxins can be divided into different groups that act on protein synthesis, signal transduction, actin polymerization, and vesicle trafficking within eukaryotic cells.

Toxins Acting on Protein Synthesis

The toxins that inhibit protein synthesis, causing rapid cell death, at extremely low concentrations are diphtheria toxin, *P. aeruginosa* exotoxin A, and Shiga toxin.

Diphtheria Toxin

Diphtheria toxin (DT) is a 535-amino-acid polypeptide encoded by a bacteriophage that lysogenizes *Corynebacterium diphtheriae*. The gene is regulated by an iron-binding protein and, therefore, the toxin is expressed only in the absence of iron. Following cleavage at a protease-sensitive site, the toxin is divided into two fragments (A and B) that are held together by a disulfide bridge. The A fragment (DTA) is an enzyme with ADP-ribosylating activity that binds NAD and transfers the ADP-ribose group to elongation factor 2 (EF2) according to the reaction shown below:

$$NAD + EF2 \xrightarrow{DTA} ADPR\text{-}EF2 + nicotinamide + H^+$$

The target site in EF2 is a unique amino acid resulting from the posttranslational modification of histidine at position 715; it is named diphthamide. The region containing diphthamide 715 is very close to the anticodon recognition domain of EF2. This suggests that ADP ribosylation interferes with EF2 binding to the tRNA. The ADP-ribosylated EF2 is no longer able to support protein synthesis, and the cells die. The lethal dose of DT is extremely low: 100 ng/kg of body weight. The B fragment can be further divided into two domains: the R (receptor-binding) and T (transmembrane) domains. R binds to the heparin-binding epidermal growth factor-like precursor and is internalized by receptor-mediated endocytosis. The toxin receptor is present in most mammalian cells; however, rodents are not susceptible to DT because the receptor has a few amino acid changes that abolish binding. Following acidification of the endosomes, the T domain, which is composed mostly of hydrophobic α-helices, changes conformation and penetrates the membrane and somehow facilitates the translocation of the A subunit to the cytoplasm. DT is used to make the diphtheria vaccine, where it is present after chemical detoxification by formaldehyde treatment. Furthermore, DT as well as the enzymatically inactive, nontoxic mutant CRM197 (cross-reacting material 197) is used as a carrier molecule for glycoconjugate vaccines.

Another activity described for DT is apoptosis of target cells. This activity is apparently mediated by the A fragment but is not linked to the enzymatic activity. In fact, apoptosis has also been described for the nontoxic mutant CRM197. Whether apoptosis plays a role in toxicity in vivo is unclear; however, it cannot have a major role because the mutants, which are active in apoptosis but enzymatically inactive, are nontoxic in vitro and in vivo.

Pseudomonas Exotoxin A

Pseudomonas exotoxin A (ExoA) is a 66-kDa single-chain protein encoded by a chromosomal gene. ExoA has a mechanism of action (ADP-ribosylation of EF2) identical to that of DT. However, the two proteins are totally unrelated and have no primary sequence homology. Nevertheless, the folding and three-dimensional structure of the catalytic site are conserved and superimposable on those of all mono-ADP-ribosylating enzymes known, including the A fragments of DT, CT, LT, and PT (Figure 13.7), suggesting that

Figure 13.7   Catalytic and NAD-binding site of ADP-ribosylating toxins. **(A)** Diagram of the three-dimensional structure of the catalytic site of DT and LT showing the NAD molecule (gold) inside the cavity and the two amino acids, Glu and Arg/His, important for catalysis. **(B)** Schematic representation of the catalytic site of PT showing the NAD and the two catalytic amino acids which have been mutagenized to generate the genetically detoxified PT9K/129G mutant.

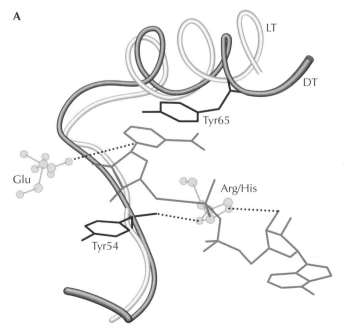

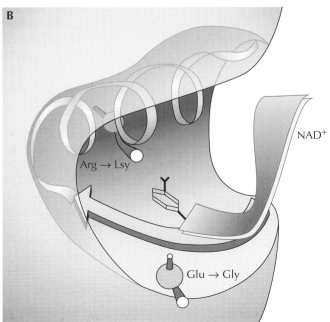

these enzymes either evolved from a common ancestor or had a convergent evolution.

The toxin is secreted into the supernatant by *P. aeruginosa* and can be divided into three functionally different domains: the N-terminal part, R, binds the α_2-microglobulin receptor on the surface of target cells; the central portion, T, is composed mostly of hydrophobic α-helix and mediates translocation of the catalytic domain to the cytosol; and the C-terminal part has ADP-ribosylating activity. Following receptor-mediated endocytosis, the toxin undergoes retrograde transport to the endoplasmic reticulum (possibly thanks to the C-terminal REDLK sequence, which resembles the endoplasmic reticulum retention signal KDEL). To become active, the ExoA toxin requires an intracellular furin-mediated proteolytic cleavage to generate a 37-kDa C-terminal fragment that is then translocated to the cytoplasm to reach the EF2 target. By using a fluorescence resonance energy transfer approach, the mechanism of interaction between ExoA and its substrate EF has been studied, showing that the binding is strongly dependent on the pH. Furthermore, the finding that EF-2 bound to GDP or GTP is still recognized by ExoA shows how adaptable this toxin is in ADP-ribosylating its substrate.

Due to the potent lethal activity, the catalytic domains of DT and *Pseudomonas* toxin have been widely used to construct fusion proteins that are able to specifically bind and kill tumor cells or other types of dangerous cells. The first of its kind, fusion between interleukin-2 and the A subunit of DT, was able to specifically kill activated T cells and has been tested for the treatment of cutaneous T-cell lymphoma.

Shiga Toxin

Shiga toxin (also known as verotoxin) is a prototype of a number of related toxins produced by the causative agents of dysentery (*Shigella dysenteriae*) and hemorrhagic colitis (*E. coli* producing Shiga toxin types 1 and 2). The toxins are encoded by an operon that carries the genes for the A and B subunits and can be located in the chromosome or on bacteriophages. Shiga toxin is a typical A/B toxin, with an enzymatically active A peptide of 35 kDa that is responsible for the toxicity. This has an *N*-glycosidase activity, which cleaves an adenine residue from the 28S RNA, altering the function of the ribosomes, which are no longer able to interact with elongation factors EF1 and EF2. This results in an inability to carry out protein synthesis and thus cell death. Interestingly, the plant toxin ricin has an identical mode of action. The B subunit is a pentamer composed of five identical monomers of 7,700 Da that bind to the globotriaosylceramide (Gb) eukaryotic cell receptor. The organization of the B pentamer is remarkably similar to that found in CT and LT, suggesting that this type of pentameric structure is favorable and has been independently and recurrently adopted during evolution. After binding to Gb$_3$, the toxin is internalized by receptor-mediated endocytosis and is transported to the Golgi and the endoplasmic reticulum, from which the A subunit is translocated to the cytoplasm, where it can gain access to the ribosomal target. The B subunit of Shiga toxin has been demonstrated as a powerful vector for carrying attached peptides into cells for intracellular transport studies and for medical research.

Toxins Acting on Signal Transduction

Communication between and within cells is essential for any living organism. In eukaryotic cells, signals from outside stimulate receptors on the surface of the cells that transmit the signal across the cell membrane mainly by two mechanisms: (i) tyrosine phosphorylation of the cytoplasmic C-terminal part of the receptor that recruits SH$_2$-signal transducers and initiates a cascade of intracellular signaling events; and (ii) modification of a receptor-coupled GTP-binding protein that transduces the signal to enzymes, releasing secondary messengers such as cyclic AMP (cAMP), inositol triphosphate, and diacylglycerol, which also initiate a cascade of intracellular signaling events. Bacterial toxins act at all levels of signal transduction: they modify GTP-binding proteins (CT, LT, and PT), produce secondary messengers such as cAMP (CyaA and EF), and act on the intracellular signal cascade either by modifying small GTP-binding proteins such as Rho, Rac, and Cdc42 (*Clostridium difficile* toxins A and B and *E. coli* CNF1 and CNF2) or by modifying kinases such as mitogen-activated MAPKK1 (LF). Other toxins (described in the next section) that are directly injected by the bacteria into cells also work on signal transduction; these can act on the intracellular signal cascade, modifying small GTP-binding proteins such as Ras and Rho (ExoS and C3) or modifying receptors by phosphorylating or dephosphorylating them (YopH and YopO).

Toxins Acting on G Proteins

CT and *E. coli* LT-I and LT-II.   CT and *E. coli* LT-I and LT-II have an identical mechanism of action and homologous primary and three-dimensional structures. CT is produced by *Vibrio cholerae,* the causative agent of cholera, and LT-I and LT-II are produced by enterotoxigenic *E. coli* (ETEC) isolated from humans with traveler's diarrhea, from pigs (LT-I), from animals with no evident disease, or from food (LT-II). The toxins are generally described as having an AB$_5$ composition, meaning that they contain one enzymatically active A subunit and a pentameric B subunit composed of five identical monomers. The toxins are encoded by a bicistronic operon located on a filamentous bacteriophage (CT) or a plasmid (LT). CT is secreted into the culture medium by *V. cholerae,* while LT is secreted across the outer membrane through the general secretory pathway and remains associated with the outer membrane bound to LPS.

The A subunit is a 27-kDa enzyme with ADP-ribosyltransferase activity, an enzymatic activity shared with DT, ExoA, PT, and *P. aeruginosa* ExoS. The enzyme binds NAD and transfers the ADP-ribose group to an arginine residue present in an -LRX**R**VXT- motif located in the central part of all G proteins, thus modifying their ability to transduce the signals from the coupled receptors. To be fully active, the A subunit has to bind to an intracellular small GTP-binding protein that is involved in vesicular trafficking (ADP-ribosylation factor [ARF]). The main targets of the enzyme are G$_s$, G$_t$, and G$_{olf}$. The effects on G$_s$ are the best known: the ADP-ribosylation causes permanent activation of the G$_s$-regulated adenylate cyclase, inducing an increase in the intracellular content of the secondary messenger cAMP. This induces an alteration of ion transport, with an increase in chloride secretion and inhibition of sodium absorption. When the toxins are released in the intestine during infection, the ultimate consequence is intestinal fluid accumulation and the watery diarrhea typical of the diseases. If the toxins are experimentally introduced into other tissues, they usually cause massive release of fluids and edema. The A subunits of CT, LT-I, and LT-II have quite a high sequence and three-dimensional homology, with LT-II being the least closely related. They also have the common fold and structure of the active site of ADP-ribosyltransferases described for DT and ExoA (Figure 13.7A).

The B subunit has a ring-like structure composed of five monomers of 11,500 Da each, with a central hole that houses the C-terminal portion of the A subunit. Each monomer binds a receptor molecule that for CT is mostly ganglioside GM_1, while LT also binds other galactose-containing structures such as other glycosphingolipids, glycoprotein receptors, polyglycosylceramides, and paragloboside.

Following receptor binding, the toxins are internalized by receptor-mediated endocytosis and undergo retrograde transport through the Golgi to the ER. Recent studies show that both A and B subunits move together from the cell surface into the ER, and this depends on the B subunit binding to ganglioside GM_1. The KDEL motif in the A2 chain does not appear to affect retrograde transport, but slows recycling of the B subunit from ER to distal Golgi stacks. Specificity for GM_1 in this trafficking pathway is shown by the failure of the *E. coli* type II toxin LTIIb that binds ganglioside GD1a to concentrate in lipid rafts, enter the ER, or induce toxicity. These results show that the B subunit carries the A1 chain from the cell surface into the ER where they dissociate, and that a ganglioside with high affinity for lipid rafts may provide a general vehicle for the transport of toxin to the ER. At this stage, the A subunits cross the membrane and end up in the cytoplasm, where they are activated by binding ARF and can finally reach and modify their targets.

The B subunits of CT and LT-I have a high degree of homology, although it is lower than that of the A subunit. Interestingly, the B subunit of LT-II, although having an identical structure, has no homology to the B subunit of CT and LT-I, indicating that the pentamer is a good evolutionary solution that can be achieved with different primary structures. In addition to carrying the A subunits, the pentameric B subunits have other biological activities on their own, such as induction of apoptosis of $CD8^+$ and $CD4^+$ T cells (as described above). CT and LT are perhaps the best mucosal immunogens and adjuvants known to date. This property can be exploited to develop mucosal vaccines against cholera and ETEC infection and to induce a mucosal response against the antigens that are coadministered. To develop safe vaccines that can be used in humans without carrying the toxic features of CT and LT, enzymatically inactive mutants have been developed by site-directed mutagenesis. These mutants are very promising candidates as mucosal adjuvants and vaccines.

PT.   PT is a chromosomally encoded, 105-kDa virulence factor secreted into the culture supernatants by *B. pertussis*, the bacterium that causes whooping cough. The toxin is composed of five noncovalently linked subunits named S1 through S5, which are organized into two functionally different domains called A and B. The five subunits are individually secreted to the periplasmic space, where the toxin is assembled and then released to the culture medium by a specialized type IV secretion apparatus (see chapter 16). The A domain, which is composed of the S1 subunit, is an enzyme that shares the active-site enzymatic activity and structure with DT, ExoA, CT, and LT (Figure 13.7A) and intoxicates eukaryotic cells by ADP-ribosylating their GTP-binding proteins. The protein transfers the ADP-ribose group to a cysteine located in the C-terminal-X**C**GLX motif of the α-subunit of many G proteins such as G_i, G_o, G_t, and G_{gust}. G_s and G_{olf} have a tyrosine instead of the cysteine and are not substrates for PT. The consequence of ADP-ribosylation is the uncoupling of G proteins from their receptors, with alteration of all signals that are transduced by them.

This may have different consequences in different tissues, the most common of which are lymphocytosis, increased insulin secretion, and sensitization to histamine.

The B domain is a nontoxic oligomer formed by subunits S2, S3, S4, and S5 in a 1:1:2:1 ratio. This domain binds to glycoproteins having a branched mannose core with attached *N*-acetylglucosamine. The toxin is then internalized by receptor-mediated endocytosis and is likely to follow the retrograde transport through the Golgi as in the case of CT; however, supportive data are not available. The importance of the Golgi localization of pertussis toxin for the S1-dependent ADP-ribosylation of G proteins was investigated using Brefeldin A (BFA) treatment to disrupt Golgi structures. This treatment completely blocked the pertussis toxin ADP-ribosylation activity of cellular G proteins, therefore indicating that retrograde transport to the Golgi network is a necessary prerequisite for cellular intoxication. In addition to the intoxication mediated by the enzymatic activity, the toxin has other biological activities, such as mitogenic activity on T cells, which are mediated only by the binding of the B domain to the receptors. These are described above. PT is one of the main components of acellular vaccines against whooping cough. In these vaccines the toxin is chemically detoxified (by formaldehyde or hydrogen peroxide) or genetically detoxified (by changing the two catalytic amino acids [Figure 13.7B] by site-directed mutagenesis). The nontoxic mutant obtained by site-directed mutagenesis (PT-9K/129G) was found in clinical trials to be more immunogenic than any other form of detoxified PT and to induce protection from disease in infants.

Toxins Generating cAMP

Adenylate Cyclase.   The adenylate cyclase (CyaA) produced by *B. pertussis, B. parapertussis,* and *B. bronchiseptica* has a molecular mass of 177.7 kDa. Unlike most of the other members of the RTX family, which are secreted into the supernatant, CyaA remains associated with the bacterial surface, through interactions with filamentous hemagglutinin (FHA). CyaA is a bifunctional protein composed of a cell-invasive and calmodulin-dependent adenylate cyclase domain (400 N-terminal residues) fused to a pore-forming hemolysin consisting of 1,306 residues. The C-terminal hemolytic domain (described as a hemolysin above) binds the receptors and delivers the N-terminal catalytic domain into the cell, where it binds calmodulin and catalyzes high-level synthesis of the second messenger cAMP, thereby disrupting cellular functions. The target cells are believed to be mainly alveolar macrophages and leukocytes.

A very similar function and mechanism of action is that of ExoY, an adenylate cyclase produced by *P. aeruginosa* and injected into the cytoplasm of eukaryotic cells by the type III secretion apparatus. However, differently from CyaA, ExoY is not activated by calmodulin. In vivo, following infection with ExoY-expressing strains, CHO cells showed a rounded morphology, which correlated with increased cAMP levels.

Anthrax Edema Factor.   Edema factor (EF) and lethal factor (LF) are the two major exotoxins produced by *B. anthracis*. These proteins are transported into host cells by the pore-forming PA (anthrax protective antigen), which is discussed above. The three subunits, encoded by a large plasmid, are synthesized and secreted independently. EF is secreted as a precursor of 800 amino acids, and cleavage of the 33-amino-acid signal peptide produces the 767-amino-acid mature protein. The N-terminal domain of EF is responsible for the binding to PA, whereas the C-terminal region contains the catalytic domain. EF binds to proteolytically activated and receptor-bound PA (Figure 13.6). The binding site of EF on PA has been recently mapped. Following receptor-mediated endocytosis, the low pH causes a conformational change in PA, allowing the translocation of EF across the cell membrane. Once inside the cells, it binds calmodulin and catalyzes the synthesis of the second messenger cAMP, thereby perturbing the cell regulatory mechanisms. The three-dimensional structure of EF alone, EF in complex with calmodulin, and EF in complex with both calmodulin and 3'-deoxy-ATP has been solved. The catalytic portion of EF is made by three globular domains. The active site is located at the interface of two domains that together form the catalytic core. The differences between the structure of EF alone and EF-calmodulin are induced by calmodulin. Following binding, an α-helix domain moves from the EF catalytic core and interacts with a disordered loop, thus leading to enzyme activation. Therefore, binding to calmodulin stabilizes the conformation of the substrate-binding site of EF. The histidine in position 351 acts as catalytic residue. Interestingly, a remarkable level of primary sequence similarity can be detected between EF and the N-terminal, calmodulin-binding domain of *Bordetella* adenylate cyclase CyaA. In particular, His351 is conserved between the two proteins.

Toxins Inactivating Cellular Kinases
Anthrax LF is a 90-kDa factor, which causes the shock-like symptoms observed in systemic anthrax infection, by inducing macrophages to overexpress proinflammatory cytokines. In animals, it induces sudden death. LF binds to proteolytically activated and receptor-bound PA. Following receptor-mediated endocytosis, the low pH causes a conformational change in PA that allows the translocation of LF across the cell membrane. Once in the cytoplasm, LF (a metalloprotease that, like clostridial toxins, contains the consensus sequence -HEXXH-) cleaves the N terminus of the mitogen-activated protein kinase kinases, MAPKK1 and MAPKK2 (Figure 13.6), which are so far the only known cellular substrates for LF. The cleavage inactivates MAPKK1 and inhibits the MAPK signal transduction pathway, a conserved pathway that controls cell proliferation, signal transduction, and gene expression. The ability of LF to induce apoptosis of activated macrophages depends on the inhibition of p38 MAPKs activation. Inhibition of p38 is sufficient to sensitize macrophages to activation-induced death by preventing induction of an anti-apoptotic factor. This mechanism may explain how *B. anthracis* paralyzes the innate immune system and is able to spread undetected through systemic infection. Crystal structure of the LF protease in complex with the N-terminal portion of its natural substrate MAPKK2 is now available.

Toxins Inducing Arrest in the G_2 Phase of the Cell Cycle
The cytolethal distending toxin produced by *Haemophilus ducreyi* (HdCDT) induces cell enlargement followed by cell death. This effect is similar to that induced by CDT produced by *E. coli*, *Shigella*, *Campylobacter*, *A. actinomycetem-comitans*, and *Helicobacter hepaticus*. HdCDT is a complex of three proteins, CdtA, CdtB, and CdtC, encoded by three genes that are part of an operon. The overall sequence similarity varies among the different members of this family of toxins. HdCDT intoxicates eukaryotic cells by causing a three- to fivefold gradual distension and induces cell cycle arrest in the G_2 phase. Transition of cells from G_2 into mitosis requires activation of the cyclin-dependent kinase p34^{cdc2} by dephosphorylation. Treatment of cells with

HdCDT induces an increased level of the tyrosine-phosphorylated form (inactive) of p34^{cdc2} and hence leads to cell cycle arrest. Whether the lack of activation of p34^{cdc2} is a direct effect of CDT activity needs to be determined. It has been recently shown that CdtB is the active subunit of the CDT toxin. CdtB has a structural homology with the DNase I family of enzymes. This homology has been supported by experimental data showing that expression of CtdB in eukaryotic cells induces consistent changes in the chromatin of transfected cells and is sufficient to induce cell cycle arrest. All the amino acids predicted to be important for nuclease activity are conserved in the CdtB of different bacteria, suggesting that the mechanism of action is the same for all CDT toxins. The role of CdtA and CdtC in delivery to target cells, in processing and/or secretion, is still unknown.

Toxins with Phospholipase C Activity

α-Toxin is the key virulence factor produced by *C. perfringens*, which is responsible for gas gangrene and is involved in the pathogenesis of sudden-death syndrome in young animals. The toxin is a zinc metalloenzyme that has phospholipase C (PLC) activity. *C. perfringens* knockout mutants are unable to cause disease, and vaccination with a genetically detoxified toxoid induces protection against disease. Originally, the toxin was identified as an enzyme with PLC activity, but subsequently it was discovered that not all toxins with PLC activity were toxic (*B. cereus* phosphatidylcholine-specific PLC [PC-PLC] is not toxic), leading to the conclusion that enzymatic activity alone was not sufficient for toxicity.

The three-dimensional structure of *C. perfringens* α-toxin has been solved, showing that the 370 residues are organized in two domains: an α-helical N-terminal domain that contains the active site and is highly homologous to *B. cereus* PC-PLC toxin, and a β-stranded C-terminal domain that is involved in membrane binding and shows high structural homology to eukaryotic calcium-binding C2 domains. This C-terminal domain has recently been discovered as the component necessary for the observed hemolytic activity of the α-toxin and that this is due to the interaction of this region with fatty acyl residues of phosphatidylcholine in the membrane. The three-dimensional structure of the toxin at acidic pH shows that the active site is closed by a small α-helix, which is not present in the open structure, with the loss of one of the three zinc ions.

Recently, other bacterial PLCs, like those from *L. monocytogenes* and *Mycobacterium tuberculosis*, have been implicated in the pathogenesis of several diseases.

Toxins Modifying Small GTP-Binding Proteins

Clostridium difficile **Toxins A and B.**   Enterotoxin A (308 kDa) and cytotoxin B (270 kDa) are secreted into the culture supernatant by *C. difficile*, bind to eukaryotic cells, and are taken up by receptor-mediated endocytosis. While toxin B has potent cytotoxic activity in vitro, the enterotoxic activity of *C. difficile* in animals has been mainly attributed to toxin A. Intracellularly, both toxins monoglucosylate small GTP-binding proteins such as Rho, Rac, and Cdc42. The site of modification is threonine 37 in Rho and threonine 35 in Rac and Cdc42. Other small GTPases, such as Ras, Rab, Arf, and Ran, are not glucosylated. The monoglucosylation results in breakdown of the cellular actin stress fibers. Recently, toxin A has been reported to induce apoptosis in T84 intestinal epithelial cells, in a time- and dose-dependent manner. These toxins are described in more detail in chapter 14.

***E. coli* CNF.**   Cytotoxin-necrotizing factors (CNF) are single-chain proteins of 1,014 amino acids that are produced by a number of uropathogenic *E. coli* and neonatal meningitis-causing *E. coli* strains. Two forms are known: CNF1 is chromosomally encoded, while CNF2 is located on a large, transmissible F-like plasmid called Vir. The two factors have 85% identical amino acid sequences and show similarities to the dermonecrotic toxins of *Pasteurella multocida* and *B. pertussis*. More recently, a CNF1-like toxin has been identified also in *Yersinia pseudotuberculosis*. These toxins cause alteration of the host cell actin cytoskeleton, induce ruffling, stress fiber formation, and promote cell spreading in cultured cells by activating the small GTP-binding proteins Rho, Rac, and Cdc42, which control assembly of actin stress fibers. Activation is achieved by deamidation of glutamine 63 of Rho and glutamine 61 of Cdc42. CNF1 induces only a transient activation of Rho GTPase and a depletion of Rac by inducing the addition of a ubiquitin chain which is known to drive to specific degradation by the proteasome. Reduction of Rac GTPase levels induces cell motility and cellular junction dynamics, allowing efficient cell invasion by uropathogenic bacteria (see chapter 14 for details). The catalytic region of CNF1 has been crystallized. The active site contains a catalytic triad, which is positioned in a deep pocket, thus explaining the restricted access to unspecific substrates and therefore its specificity. Very likely, some type of conformational rearrangement is also required to accommodate Rho in this narrow cavity.

***Bordetella* Dermonecrotic Toxin.**   Dermonecrotic toxin (DNT) is produced by *Bordetella* species as a single-chain polypeptide chain of 1,464 amino acids, which is composed of a C-terminal portion that contains the catalytic site and by an N-terminal receptor-binding domain. DNT is a transglutaminase, which catalyzes the deamidation or polyamination at Gln63 of Rho and of the corresponding residues of Rac and Cdc42. This activity causes alteration of cell morphology, reorganization of stress fibers, and focal adhesions on a variety of animal models. Recently, it has been demonstrated that the initial 54 amino acids of DNT are sufficient for cell surface recognition. However, the receptor is still unknown.

Toxins Acting on Intracellular Trafficking

Vital cellular processes, such as receptor-mediated endocytosis and exocytosis, use specialized vesicles either to internalize portions of the plasma membrane and address them to different destinations by using specialized sorting compartments or to transport molecules synthesized in the ER and modified in the Golgi apparatus to the cell surface. In some tissues, dedicated vesicle transport can be used for the local release of special cargo. A typical example of this is the local release of neurotransmitters at the presynaptic membrane. Recently, great progress in the understanding of this field has been made by the discovery that key molecules involved in vesicle docking and membrane fusion, such as VAMP/synaptobrevin, SNAP-25, and syntaxin, are specific targets of neurotoxins produced by *Clostridium tetanii* and *Clostridium botulinum*, which are specific zinc-dependent proteases. The vacuolating cytotoxin of *Helicobacter pylori* has recently been discovered to act on vesicle traffic at the level of late endosomes, a compartment whose molecular mechanisms are still totally unknown. Hopefully, the study of this toxin will help to elucidate this pathway.

Proteases of VAMP/Synaptobrevin

Tetanus Toxin. Tetanus toxin is a 150-kDa protease, produced by toxigenic strains of *C. tetani*, that is responsible for the spastic paralysis typical of clinical tetanus. The toxin is extremely potent, with a 50% lethal dose (LD_{50}) of 0.2 ng/kg. Following protease cleavage, it is divided into two fragments (H [heavy] and L [light]), held together by a disulfide bridge, a structure typical of A/B toxins. The heavy chain is composed of two fragments, H_C (or fragment C), which has recently been found to bind to di- and trisialylgangliosides on neuronal cell membranes, and H_N, which is involved in the transmembrane translocation of the L chain to the cytosol. The L chain is a 447-amino-acid fragment containing the -HEXXH- motif typical of metalloproteases. It binds zinc and specifically cleaves VAMP/synaptobrevin, a molecule essential for the docking of the neurotransmitter-containing vesicles to the cell membrane (Figure 13.8). The specific metal-dependent proteolytic activity is shared with the neu-

Figure 13.8 Scheme of a postsynaptic membrane showing the normal process of neurotransmitter (NT) release (left) and the mechanism of action of the neurotoxins (right). Vesicles containing the neurotransmitter have a transmembrane protein (synaptobrevin or v-SNARE) that binds specifically two proteins enclosed on the cell membrane (syntaxin, SNAP-25, or t-SNARE). The initial interaction becomes increasingly stronger, forcing the vesicle and cellular membranes to become in close contact and finally to fuse, thus releasing the neurotransmitters into the intercellular space. The mechanism of action of tetanus and botulinum neurotoxins is shown on the right. They cleave the v-SNARE and/or the t-SNARE, thus preventing the docking and fusion of neurotransmitter-containing vesicles.

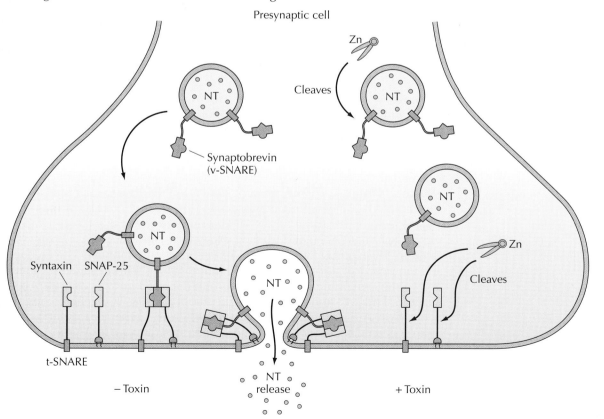

rotoxins of *C. botulinum* and *B. anthracis* LF. During intoxication, tetanus toxin is internalized at the presynaptic terminal of the neuromuscular junction and migrates retroaxonally within the motoneuron to the spinal cord, where it is released into the intersynaptic space located between the motoneuron and the inhibitory interneuron cell, penetrates the latter, is translocated to the cytosol following acid-induced conformational change, and blocks neuroexocytosis by cleaving VAMP. To determine which amino acids in tetanus toxin are involved in ganglioside binding, homology modeling was performed using the recently resolved X-ray crystallographic structures of the H_C fragment. These studies have indicated that the amino acids tryptophan 1288, histidine 1270, and aspartate 1221 are components of the GT1b-binding site on the tetanus toxin H_C fragment. Following inactivation of the toxicity by formaldehyde treatment, tetanus toxin is used for vaccination against tetanus, a practice applied for more than 60 years. Today, very promising results have been obtained by using the nontoxic recombinant fragment C for vaccination.

C. botulinum **neurotoxin.**   *C. botulinum* neurotoxin serotypes A to G are usually ingested with food and are responsible for the flaccid paralysis typical of clinical botulinum intoxication. Like tetanus toxin, they are synthesized as 150-kDa precursors that are activated by proteolytic cleavage and divided into two fragments, H and L, of 100 and 50 kDa, respectively, held together by a disulfide bridge. H binds to receptors present in the motor neurons at the neuromuscular junction and translocates the active L form into the cytosol. Recent data show that the H-L complex is embedded in the membrane in the form of a transmembrane chaperone that acts as a dynamic structure to achieve final translocation of the L chain. L chains of neurotoxins B, D, F, and G are zinc-dependent proteases that specifically cleave VAMP/synaptobrevin; they are composed of approximately 450 amino acids, with an overall homology of 30 to 60%, and contain the -HEXXH- motif typical of metalloproteases.

Proteases of SNAP-25 and Synaptobrevin
Like tetanus toxin, botulinum neurotoxins A, E, and C are synthesized as 150-kDa precursors that are activated by proteolytic cleavage and divided into two fragments, H and L, of 100 and 50 kDa, respectively, held together by a disulfide bridge. H binds to receptors present in the motor neurons at the neuromuscular junction and translocates the active L form into the cytosol. L chains are zinc-dependent proteases that specifically cleave SNAP-25 (synaptosome-associated protein of 25 kDa), at positions 197 (type A) or 180 (type E). SNAP-25 is located in the cytoplasmic face of the plasma membrane and, during exocytosis, binds VAMP, a protein located in the external face of the secretory vesicle membrane. The L chains are composed of approximately 450 amino acids, with an overall homology to the remaining botulinum neurotoxins of 30 to 60%, and they contain the -HEXXH- motif typical of metalloproteases. The gene coding for neurotoxin A is located on a plasmid, while the gene coding for neurotoxin E is on a bacteriophage. Type C toxin has the general H and L structure of the other botulinum neurotoxins; the only difference is that it cleaves syntaxin, another protein involved in the exocytotic machinery. Type C is also unique because it can also cleave SNAP-25, thus being the only one able to cleave two substrates.

Toxin Acting on Late Endosomes/Vacuoles

Helicobacter pylori **VacA.**   *H. pylori,* a pathogen living in the stomach of humans and responsible for peptic ulcers, produces vacuolating cytotoxin (VacA), a toxin that causes massive growth of vacuoles within epithelial cells. VacA also forms water-soluble oligomeric complexes and can somehow insert into lipid bilayers to produce anion-selective channels. The toxin, which is synthesized as a 140-kDa precursor, is secreted from the bacterium by the 45-kDa C-terminal region, using a secretory mechanism similar to that of immunoglobulin A (IgA) proteases of *Neisseria* (see chapter 15). The toxin purified from the supernatant has a flower-like structure consisting of seven monomers of 95 kDa, each of which can be cleaved at a protease-sensitive site into two fragments of 37 and 58 kDa (p37 and p58 moieties).

The gene coding for the N-terminal 40-kDa portion is approximately 90% homologous in the different isolates and induces vacuoles when the gene is placed under the control of a strong eukaryotic promoter and transfected into epithelial cells, suggesting that the active site must be located in this region (the mechanism of action is unknown). The C-terminal part binds the receptors on eukaryotic cells. The toxin has N-terminal and C-terminal portions with 90% identity across strains and a central portion present mainly in two different genotypes (M1 and M2), which have an overall identity of only 55%. M1 and M2 target the toxin to different cells. The toxin induces an alteration of the endocytic pathway that results in selective swelling of the vesicles having the typical markers of late endosomes. The small GTP-binding protein Rab7 is necessary for vacuole formation. It is believed that VacA must modify an unknown but fundamental effector of membrane traffic. The identification of the VacA target may help to dissect the molecular mechanisms of this still obscure part of intracellular trafficking. It cannot be excluded that the VacA toxicity is mediated only by the channel-forming activity of the toxin on cell membranes.

S. pyogenes **NAD$^+$ glycohydrolase.**   NAD$^+$ glycohydrolase is an important virulence factor produced by group A streptococci, which is thought to enhance pathogenicity by facilitating the spread of the microorganism through host tissues. This enzyme catalyzes the hydrolysis of the nicotinamide-ribose bond of NAD to yield nicotinamide and ADP ribose. Differently from ADP-ribosylating toxins, NAD$^+$ glycohydrolases possess a much higher rate of NADase activity and do not require an ADP-ribose acceptor. Interestingly, this GAS virulence factor is functionally linked to streptolysin O (SLO), a pore-forming toxin, which has been shown to be required for efficient translocation of NAD$^+$ glycohydrolase into epithelial cells. Isogenic mutants deficient in the expression of SLO, NAD$^+$ glycohydrolase, or both proteins resulted in reduced cytotoxicity and keratinocyte apoptosis with respect to the wild-type GAS. These results suggest that NAD$^+$ glycohydrolase modulates host cell signaling pathways and contributes to the enhancement of streptolysin O cytotoxicity.

Toxins Acting on the Cytoskeleton

The cytoskeleton, a structure composed of microfilaments, microtubules, and intermediate filaments, controls the shape and spatial organization of the cells and is involved in cell movement, endo- and exocytosis, vesicle transport, cell contact, and mitosis (see chapter 12). Rapid structural changes of these cytoskeletal proteins are based on their ability to polymerize and depolymerize. Bacterial toxins act mostly on microfilaments that are 7 to 9 nm in diameter and are built by polymerized actin. Many of the toxins somehow affect the structure of the cytoskeleton. Most of them do it by modifying the regulatory, small G proteins, such as Ras, Rho, Rac, and Cdc42, which control cell shape. These toxins, which have a dramatic but indirect effect on the cytoskeleton and are described in Toxins Acting on Signal Transduction, are *E. coli* CNF and *C. difficile* enterotoxins A and B (also see chapter 14). Other toxins acting on regulatory G proteins are exoenzyme S, C3, and YopE, which are described below as toxins that are directly injected into the eukaryotic cells. Other bacterial molecules that cannot be strictly considered toxins but that have a powerful ability to polymerize actin are ActA and IcsA of *Listeria* and *Shigella*, respectively. These are described elsewhere in this volume (see chapter 12). Another toxin acting indirectly on the cytoskeleton is the zonula occludens toxin (Zot) produced by *V. cholerae*, a toxin with an unknown mechanism of action that modifies the permeability of tight junctions. It has recently been shown that the Zot protein has homologies to a phage structural protein. This evidence leads to the question whether Zot is a phage protein that evolved into a toxin or whether it is just a phage protein that happens to have some biological activities (Zot is described below). Other bacterial factors, including those that induce tyrosine phosphorylation and actin polymerization at the bacterium-cell contact site, ruffling of the cell surface, and bacterial internalization and formation of pseudopodia, are discussed earlier in this volume (see chapters 11 and 12). Here we consider only toxins that have the cytoskeleton as a direct target. The only toxin shown to affect directly the cytoskeleton is the C2 toxin of *C. botulinum*, which ADP-ribosylates monomeric actin, making it unable to polymerize. A second protein that has recently been described as being able to bind actin and stabilize the fibers supporting the ruffles induced by the *Salmonella* type III secretion system is SipA (see chapter 15).

C. botulinum C2 toxin is a member of a family of "binary" cytotoxins that ADP-ribosylate monomeric G-actin at arginine 177. Since the arginine is a contact site between actin monomers, the binding of the ADP-ribose makes actin unable to polymerize. C2 is composed of two separate molecules, the 50-kDa enzymatically active toxin and the binding component, which is synthesized as a 100-kDa precursor and is proteolytically cleaved to generate a 75-kDa fragment that binds the cell receptor on the cell surface. This organization closely resembles that of the EF, LF, and PA of *B. anthracis*. Toxins related to C2 are *C. perfringens* iota toxin, and a similar molecule produced by *C. spiroforme*.

Toxins Directly Delivered by Bacteria into the Cytoplasm of Eukaryotic Cells

Most of the molecules described in this section are usually not found in discussions of bacterial toxins. Here we describe molecules that, in general, have been discovered recently and are part of a fascinating, rapidly expanding field. In the classical view, toxins were believed to be molecules that cause intoxication when released by bacteria into the body fluids of multicellular organisms (these are described above). This definition of toxins provided a rationale for the pathogenicity of the so-called "toxinogenic bacteria" but failed to explain the pathogenicity of many other virulent bacteria such as *Salmonella*, *Shigella*, and *Yersinia*, which did not release toxic proteins into the culture supernatant. Today we know that these bacteria also intoxicate their hosts by using proteinaceous weapons. The difference is that the confrontation does not take place between the whole bacterial population and the whole eukaryotic organism. Instead, each bacterium engages in an individual fight with one eukaryotic cell. The final result is not different from what happens after the bite of a poisonous insect or snake or after the release of a potent bacterial toxin into the infected host. However, in this case the battle involves one eukaryotic cell and one prokaryotic cell. These bacteria intoxicate individual eukaryotic cells by using a contact-dependent secretion system to inject or deliver toxic proteins into the cytoplasm of eukaryotic cells. This is done by using specialized secretion systems that in gram-negative bacteria are called type III or type IV, depending on whether they use a transmembrane structure similar to flagella or conjugative pili, respectively (see chapters 15 and 16).

Toxins Acting on Signal Transduction

Proteins Acting on Phosphorylation

Yersinia **YpkA and YopH.**   Phosphorylation is central to many regulatory functions associated with the growth and proliferation of eukaryotic cells. Bacteria have learned to interfere with these key functions in several ways. The best-known system is that of *Yersinia*, where a protein kinase (YpkA) and a protein tyrosine phosphatase (YopH) are injected into the cytoplasm of eukaryotic cells by a type III secretion system to paralyze the macrophages before they can kill the bacterium.

YpkA is a Ser/Thr protein kinase that also displays autophosphorylating activity in vitro. In vivo experiments have shown that this protein is essential for virulence; in fact, challenge with a *YpkA* knockout mutant causes a nonlethal infection, whereas all mice challenged with wild-type *Y. pseudotuberculosis* die. Recently, natural eukaryotic substrates of YpkA have been identified by using a two-hybrid assay. These belong to the class of small GTPases and comprise RhoA and Rac-1, but not Cdc42.

YopH is a modular protein where the tyrosine phosphatase domain shows a structure and catalytic mechanism very similar to those of eukaryotic enzymes. YopH acts by dephosphorylating cytoskeletal proteins, thus disrupting phosphotyrosine-dependent signaling pathways necessary for phagocytosis. Host protein targets include Crk-associated substrate, paxillin, and focal adhesion kinase.

EPEC Tir.   A unique protein acting on signal transduction in eukaryotic cells is Tir (translocated intimin receptor) of enteropathogenic *E. coli* (EPEC). A chaperone for Tir, called CesT, is required for stability of Tir in the EPEC cytoplasm. This is a protein containing two predicted transmembrane domains and six tyrosines. The 78-kDa protein is transferred (by a type III secretion system-dependent mechanism) to eukaryotic cells, where it becomes an integral part of the eukaryotic cell membrane and functions as a receptor for the bacterial adhesin, intimin. It is believed that, once in the host, Tir adopts a hairpin-like structure using its two putative transmembrane domains (TMDs) to span the host cell membrane. The region between the two TMDs constitutes the extracellular loop that functions as the intimin-binding domain. At this stage, the protein mediates attachment of bacteria to the eukaryotic cells and is tyrosine phosphorylated, resulting in an apparent molecular mass of 90 kDa. Following tyrosine phosphorylation, the protein mediates actin nucleation, resulting in pedestal formation and triggering tyrosine phosphorylation of additional host proteins, including phospholipase C-γ. Tir is essential for EPEC virulence and was the first bacterial protein described to be tyrosine phosphorylated by host cells.

H. pylori **CagA.** CagA is a 128-KDa protein characterized by a central region containing an EPIYA motif, which can be repeated up to six times, increasing the molecular weight of the protein. The gene is encoded within a pathogenicity island, which also encodes the type IV secretion system necessary to inject the protein into eukaryotic cells (chapter 16). Once injected into the host cell, the protein is tyrosine phosphorylated at the EPIYA motif by the kinase c-Src and Lyn. The signal is proportional to the number of EPIYA motives present. The tyrosine-phosphorylated CagA (CagA-P) activates SHP-2 and inactivates c-Src, leading to cortactin dephosphorylation triggering a signal transduction cascade (which results in cellular scattering proliferation, a phenotype indistinguishable from that induced by the hepatocyte growth factor [HGF]). Recent data indicate that the toxic activity of CagA is due to the disruption of the apical junctional complex (AJC) of cells by interactions with ZO-1 and Jam. The long-term chronic infection and the continuous stimulation increase the risk of cancer of people infected by CagA$^+$ *H. pylori*. CagA is the first bacterial protein linked to cancer in humans, and the *cagA* gene can be considered the first bacterial oncogene.

Proteins Acting on Small G Proteins

YopM. YopM is an effector protein delivered to the cytoplasm of infected cells by the type III secretion mechanism of *Yersinia pestis.* YopM is a highly acidic protein, which is essential for virulence, but whose mechanism of action is still elusive. Differently from other effectors, this toxin has been shown to accumulate not only in the cytoplasm but also in the nucleus of mammalian cells. The X-ray structure determined for YopM has shown a modular architecture constituted by leucine-rich repeats, mainly organized in an extended β-sheet structure. This organization is very similar to that found for other important proteins, such as rab geranylgeranyltransferase and internalin B produced by *Listeria.*

SptP. *Salmonella* protein tyrosine phosphatase (SptP) is a effector protein secreted by the type III secretion apparatus of *Salmonella enterica.* SptP is a modular protein composed of two functional domains, a C-terminal region with sequence similarity to *Yersinia* tyrosine phosphatase YopH, and an N-terminal domain showing homology to bacterial cytotoxins such as *Yersinia* YopE and *Pseudomonas* ExoS. Recently, it was demonstrated that this domain possesses strong GTPase-activating domain protein (GAP) activity for Cdc42 and Rac1. The crystal structure of SptP-Rac1 complex has shown that SptP is strongly stabilized by this interaction.

Proteins Acting on the Cytoskeleton

Salmonella enterica Serovar Typhimurium SopE

SopE is injected into eukaryotic cells by a type III secretion system of *Salmonella* serovar Typhimurium, and it binds and activates the small G proteins Rac and Cdc42. Transfection of eukaryotic cells with the *sopE* gene, under the control of a eukaryotic cell promoter, induces profuse membrane ruffling and actin cytoskeleton reorganization, similarly to *E. coli* CNF (see above). In vivo, the activation of Rac and Cdc42 causes membrane ruffling and cytoskeletal rearrangements that mediate the uptake and internalization of the bacterium within the host eukaryotic cells. In addition, SopE stimulates nuclear responses that induce the synthesis of proinflammatory cytokines that contribute to the induction of diarrhea. The interaction between the catalytic fragment of SopE and its host cellular target Cdc42 has been investigated by X-ray crystallography. Interestingly, the structural and functional properties of SopE are in deep contrast to those of SptP and other similar toxins.

P. aeruginosa Exoenzyme S

P. aeruginosa exoenzyme S is a 49-kDa protein with ADP-ribosylating activity that ADP-ribosylates the small G protein Ras at arginine 41. The protein contains the motifs typical of the catalytic site of ADP-ribosylating enzymes, as described above for DT (Figure 13.7). To become enzymatically active, ExoS requires an interaction with a cytoplasmic factor named factor-activating exoenzyme S, or 14-3-3 protein. The toxin is injected into eukaryotic cells by a type III secretion system. When cells are transfected with the *exoS* gene under the control of a eukaryotic cell promoter, the cytoskeleton collapses and the cell morphology changes, resulting in rounding of the cells.

C. botulinum Exoenzyme C3

Exoenzyme C3 is not a true toxin and does not necessarily belong in this discussion. It is a protein of 211 amino acids that is produced by *C. botulinum* and that in vitro ADP-ribosylates the small regulatory protein Rho at asparagine 41, inactivating its function. The enzymatic activity is identical with that of all ADP-ribosylating enzymes (described above for DT) (Figure 13.7); however, the recently solved three-dimensional structure has shown that the C3 exoenzyme structure can be distinguished by the absence of the elongated α-helix, which generally constitutes the ceiling of the active-site cleft in the ADP-ribosylating toxins crystallized so far. Seemingly, this feature does not impair the ability of C3 either to accommodate the NAD substrate or to carry out the enzymatic reaction. If the protein is microinjected into cells or if the cells are transfected with the C3 gene under a eukaryotic promoter, actin stress fibers are disrupted, the cells are rounded, and arborescent protrusions are formed. However, we do not know whether C3 alone is able to enter cells and intoxicate them, because no mechanism of cell entry has been found. It is described here in the hope that a new mechanism of this type will be discovered for it in the near future. Toxins with activity similar to C3 have been identified in *S. aureus* (EDIN), *Clostridium limosum*, *B. cereus*, and *Legionella pneumophila*.

YopE

YopE, a protein encoded by the *Yersinia* pathogenicity island and secreted by the type III secretion system, is known to paralyze macrophage phagocytosis by causing actin depolymerization and disrupting the cell stress fibers, with consequent rounding of the cells and loss of cell shape. The effector YopE was recently shown to possess GAP activity toward the Rho GTPases RhoA, Rac, and Cdc42 in vitro. Further experimentation has shown that in vivo YopE is able to inhibit Rac- but not Rho- or Cdc42-regulated actin structures. Furthermore, the structure of this toxin in complex with its chaperone SycE has been solved, showing that this interaction does not prevent catalytic activity but instead can facilitate the formation of secondary structure elements in YopE.

YopT

YopT is the prototype of a new family of 19 proteases with potent effects on host cells. These include the AVr protein of the plant pathogen *Pseudomonas* and possibly YopJ of *Yersinia*. YopT cleaves the posttranslationally modified cysteine located at the C-terminal end of Rho GTPases (DKGCASS), causing the loss of the prenyl group from RhoA, Rac, and Cdc42, and releasing them from the membrane. The inability of Rho to be located to the membrane causes disruption of the cytoskeleton. If introduced in yeast cells, YopT is lethal, and lethality can be rescued by a multiple expression of Cdc42.

SipA

SipA is a *Salmonella* protein, which is delivered into host cells to promote efficient bacterial entry. This toxin is essential for pathogenicity and exerts its function by binding F-actin, finally resulting in the stabilization of F-actin. Interestingly, despite the lack of any detectable sequence homology, it has been suggested that SipA is a bacterial structural mimic of muscle nebulin and nebulin-like proteins that are involved in the regulation of the actin-based cytoskeleton.

VirA

The invasiveness of *Shigella* is an essential pathogenic step and a prerequisite of bacillary dysentery. VirA is a *Shigella* effector protein, which is delivered into the host cell by a specialized type III secretion system. This protein can interact with tubulin to promote microtubule destabilization and membrane ruffling. Upon this mechanism, *Shigella* is able to remodel the cell surface and thus promote its entry into the host. Recent data have shown that *VirA* deletion mutants displayed decreased invasiveness and were unable to stimulate Rac1.

Proteins Altering Inositol Phosphate Metabolism

SopB, a protein secreted by a type III secretion system (see chapter 15) of *Salmonella dublin,* has sequence homology to mammalian inositol phosphate 4-phosphatase and has inositol phosphate phosphatase activity in vitro. The enzyme hydrolyzes phosphatidylinositol triphosphate (PIP$_3$), which is a messenger molecule that inhibits chloride secretion, thus favoring fluid accumulation and diarrhea. Furthermore, SopB mediates actin cytoskeleton rearrangements and bacterial entry in a Rac-1 and Cdc42-dependent manner. SopB exhibits overlapping functions with two other effectors of bacterial entry, the Rho family GTPase exchange factors SopE and SopE2. Consistent with an important role for inositol phosphate metabolism in *Salmonella*-induced cellular responses, a catalytically defective mutant of SopB failed to stimulate actin cytoskeleton rearrangements and bacterial entry. These mutants are still invasive but are unable to cause fluid secretion and neutrophil accumulation into calf ileal loops. SopB is homologous to the *S. flexneri* virulence factor IpgD, suggesting that a similar mechanism of virulence is also present in *Shigella.* Recent studies have shown that IpgD is involved in the modulation of host cell response after contact of the bacterium with epithelial cells, thus suggesting a prominent role of this effector in the manipulation of cellular processes during infection.

Apoptosis-Inducing Proteins

YopP is a 288-amino-acid (30-kDa) protein with cysteine protease activity that is encoded by a large plasmid of *Y. enterocolitica* and is injected into eukaryotic cells by a type III secretion system. The protein is called YopJ in *Y. pseudotuberculosis* and is homologous to the SipB protein encoded by the chromosome of *Salmonella,* to IpaB of *Shigella,* to AvrA of *S. enterica,* and to the AvrRxv protein of *Xanthomonas campestris.* It has been shown that *Y. enterocolitica* induces apoptosis in infected macrophages and that the gene coding for YopP is the only one necessary for this activity, suggesting that YopP is indeed the effector protein. YopP is sufficient to cause down-regulation of multiple mitogen-activated protein kinases in host cells. Recent studies have shown that the mechanism of action of YopP involves the disruption of the NF-κb signaling pathway and therefore the occurrence of apoptosis in macrophages. Furthermore, it has been demonstrated that this effect was strongly enhanced by initiation of LPS signaling. The homologous AvrRxv protein of *X. campestris* is also translocated into host plant cells, where it activates a plant defense mechanism leading to the formation of local necrotic lesions due to programmed cell death. This is the first example where animal- and plant-pathogenic bacteria share a type III secretion-dependent effector that elicits apoptosis in their hosts.

IpaB, a protein secreted by *Shigella,* also induces macrophage apoptosis by binding to the interleukin-1-converting enzyme precursor (ICE or caspase I), that induces the processing to the active caspase, a well-known mediator of apoptosis. The *Salmonella* invasion protein SipB is a 593-residue-long protein required for the delivery of bacterial effector proteins into target eukaryotic cells, which subvert signal transduction pathways and cytoskeletal dynamics. SipB has a predominant α-helical structure and contains two helical transmembrane domains, which insert deeply into the bilayer. Similarly to IpaB, SipB also induces apoptosis by binding ICE.

Cytotoxins with Unknown Mechanisms of Action

Several extracellular products secreted by the *P. aeruginosa* type III secretion system (see chapter 15) are responsible for virulence. Among these, the 70-kDa protein ExoU is responsible for causing acute cytotoxicity in vitro and in epithelial lung injury. The mechanism of action is not known; however, it has recently been proposed that injection of ExoU into eukaryotic cells is necessary for the *Pseudomonas*-mediated increase of intracellular calcium concentrations during airway epithelial cell binding.

The zonula occludens toxin (Zot) is produced by bacteriophages present in toxinogenic strains of *V. cholerae.* Zot is a single polypeptide chain of 44.8 kDa, which localizes in the outer membranes. After internal cleavage, a C-terminal fragment of 12 kDa is excreted, and this is probably responsible for the biologic effect. Zot has the ability to reversibly alter the tight junctions of intestinal epithelium, thus facilitating the passage of macromolecules through mucosal barriers. Zot has been shown to act as mucosal adjuvant in the animal model.

HBL is another enterotoxin produced by *B. cereus,* which is composed of three proteins, B, L1, and L2, each with a molecular mass of 40 kDa, and whose corresponding genes are located on the same operon. HBL has hemolytic as well as dermonecrotic and vascular permeability activities and is able to cause fluid accumulation in ligated rabbit ileal loops.

The bile-salt hydrolase (BSH) is a protein elaborated by *L. monocytogenes,* which is absent from the genome of the nonpathogenic *L. innocua.* The *bsh* gene encodes an intracellular enzyme and is positively regulated by PrfA, the transcriptional activator of known *L. monocytogenes* virulence genes. Furthermore, *bsh* deletion mutants have a reduced persistence in the intestine and reduced liver colonization, thus demonstrating the role that BSH plays in the intestinal and hepatic phases of listeriolisis.

Finally, a novel protein toxin with predicted metalloprotease activity (AhyB) has been cloned from *A. hydrophila*, the leading cause of fatal hemorrhagic septicemia in rainbow trout. AhyB has homologies to *P. aeruginosa* elastase and *H. pylori* zinc-metalloprotease and possesses an activity specifically directed toward elastin. Insertional inactivation of the corresponding gene shows a strong reduction in pathogenesis.

Novel Bacterial Toxins Detected by Genome Mining

With the advent of the Genomic Era, identification of bacterial factors possibly involved in virulence is an easier challenge. In fact, given the vast amount of information that we now possess on toxins, including sequence data, and thanks to the growing number of sequenced bacterial genomes, it is possible to proceed by homology criteria to predict novel members of important classes of bacterial toxins. Several examples exist where computer-based methodologies have been instrumental in the identification of novel potential bacterial toxins in sequenced genomes. Among them, we mention here the case of mono-ADP-ribosyltransferases.

Mono-ADP-ribosyltransferases (mADPRTs) constitute a class of potent toxins in bacteria, which generally play an important role in the pathogenesis of related microorganisms (examples are diphtheria toxin, cholera toxin, pertussis toxin, and *Pseudomonas* exotoxin A) (Figure 13.7). Despite the poor overall conservation at the primary structure level, the catalytic subunits of these toxins show a remarkable similarity within the enzymatic cavity, so that these portions of the proteins are quite well conserved. For these reasons, and encouraged by the availability of a growing number of sequenced bacterial genomes, a series of studies have been directed toward the computer-based identification of novel members of this family of enzymes by means of sequence-homology criteria in finished and unfinished genome sequences. As a result, more than 20 novel putative ADP-ribosyltransferases have been identified both in gram-positive and gram-negative organisms, including five from *Pseudomonas syringae*, five from *Burkholderia cepacea*, two from *Enterococcus faecalis*, and one each from *Salmonella enterica* serovar Typhi, *S. pyogenes*, *Mycoplasma pneumoniae*, *Streptomyces coelicolor*, *Bacillus halodurans*, and *Vibrio parahaemolyticus*. With the exception of the protein detected in *Salmonella*, which is adjacent to an open reading frame similar to the S2 subunit of pertussis toxin, all the other genome-derived putative ADPRTs lack a predicted translocation domain. So far, none of these bacterial proteins has been tested either for their ADP-ribosyltransferase activity or for the capability of entering eukaryotic cells; however, sequence data indicate a possible role of these proteins in the pathogenesis of the corresponding microorganisms.

Very recently, a new protein has been added to the list of ADP-ribosyltransferases detected by computer analysis. This novel factor has been identified by means of primary and secondary structure analysis in the genomic sequence of a virulent isolate of *Neisseria meningitidis* and has been named NarE (*Neisseria* ADP-ribosylating enzyme). As predicted by "in silico" studies, biochemical analysis has demonstrated that NarE is capable of transferring an ADP-ribose moiety to a synthetic substrate. The identification of possible natural substrates for this protein and the characterization of its possible role in the virulence of *Meningococcus* are still unknown.

Selected Readings

Alouf, J. E., and J. R. Freer (ed.). 1999. *The Comprehensive Sourcebook of Bacterial Protein Toxins*. Academic Press, London, United Kingdom.

A fundamental reading for a comprehensive knowledge of major bacterial toxins and their implications in cell biology and vaccine design.

Bradley, K. A., J. Mogridge, M. Mourez, R. J. Collier, and J. A. Young. 2001. Identification of the cellular receptor for anthrax toxin. *Nature* **414**:225–229.

A breakthrough article reporting the identification of type 1 membrane protein ATR as the cellular receptor for the protective antigen (PA) subunit of anthrax toxin. This valuable information will allow a detailed investigation of the mechanism of uptake by cells and will hopefully lead to the development of new approaches for the treatment of anthrax.

Collier, R. J. 1967. Effect of diphtheria toxin on protein synthesis: inactivation of one of the transfer factors. *J. Mol. Biol.* **25**:83–98.

Duesbery, N. S., C. P. Webb, S. H. Leppla, V. M. Gordon, K. R. Klimpel, T. D. Copeland, N. G. Ahn, M. K. Oskarsson, K. Fukasawa, K. D. Paull, and G. F. Vande Woude. 1998. Proteolytic inactivation of MAP-kinase-kinase by anthrax lethal factor. *Science* **280**:734–737.

Flatau, G., E. Lemichez, M. Gauthier, P. Chardin, S. Paris, C. Fiorentini, and P. Boquet. 1997. Toxin-induced activation of the G protein p21 Rho by deamidation of glutamine. *Nature* **387**:729–733.

Hotze, E. M., A. P. Heuck, D. M. Czajkowsky, Z. Shao, A. E. Johnson, and R. K. Tweten. 2002. Monomer-monomer interactions drive the prepore to pore conversion of a beta-barrel-forming cholesterol-dependent cytolysin. *J. Biol. Chem.* **277**:11597–11605.

Just, I., J. Selzer, M. Wilm, C. von Eichel-Streiber, M. Mann, and K. Aktories. 1995. Glucosylation of Rho proteins by *Clostridium difficile* toxin B. *Nature* **375**:500–503.

Li, H., A. Liera, and R. A. Mariuzza. 1998. Structure-function studies of T-cell receptor-superantigen interactions. *Immunol. Rev.* **163**:177–186.

Llewelyn, M., and J. Cohen. 2002. Superantigens: microbial agents that corrupt immunity. *Lancet Infect. Dis.* **2**:156–162.

Masignani, V., E. Calducci, F. Di Marcello, S. Savino, D. Serrato, D. Veggi, S. Bambini, M. Scarselli, B. Aricò, M. Comanducci, J. Adu-Bobie, M. M. Giuliani, R. Rappuoli, and M. Pizza. 2003. NarE: a novel ADP-ribosyltransferase from Neisseria meningitidis. *Mol. Microbiol.* **50**:1055–1067.

Pallen, M. J., A. C. Lam, N. J. Loman, and A. McBride. 2001 An abundance of bacterial ADP-ribosyltransferases: implications for the origin of exotoxins and their human homologues. *Trends Microbiol.* **9**:302–307.

Pellizzari, R., C. Guidi-Rontani, G. Vitale, M. Mock, and C. Montecucco. 2000. Lethal factor of Bacillus anthracis cleaves the N-terminus of MAPKKs: analysis of the intracellular consequences in macrophages. *Int. J. Med. Microbiol.* **290**:421–427.

Pizza, M., A. Covacci, A. Bartoloni, M. Perugini, L. Nencioni, M. T. De Magistris, L. Villa, D. Nucci, R. Manetti, M. Bugnoli, F. Giovannoni, R. Olivieri, J. T. Barbieri, H. Sato, and R. Rappuoli. 1989. Mutants of pertussis toxin suitable for vaccine development. *Science* **246**:497–500.

This publication describes the genetic inactivation of pertussis toxin (PT), which will be used as a principal component of the first acellular vaccine against pertussis.

Rappuoli, R., and C. Montecucco (ed.). 1997. *Guidebook to Protein Toxins and Their Use in Cell Biology*. Sambrook and Tooze Publication, Oxford University Press, Oxford, United Kingdom.

Schiavo, G., B. Poulain, O. Rossetto, F. Benfenati, L. Tauc, and C. Montecucco. 1992. Tetanus toxin is a zinc protein and its inhibition of neurotransmitter release and protease activity depend on zinc. *EMBO J.* **11**:3577–3583.

This work describes the functional characterization of one of the most potent toxins known, tetanus toxin (TeNT). The authors show that TeNT and possibly also botulinum neurotoxins are zinc metalloprotease and that they block neurotransmitter release via this protease activity.

Schmidt, G., P. Sehr, M. Wilm, J. Selzer, M. Mann, and K. Aktories. 1997. Gln63 of Rho is deamidated by *Escherichia coli* cytotoxic necrotizing factor-1. *Nature* **387:**725–729.

Shao, F., P. M. Merritt, Z. Bao, R. W. Innes, and J. E. Dixon. 2002. A Yersinia effector and a Pseudomonas avirulence protein define a family of cysteine proteases functioning in bacterial pathogenesis. *Cell* **109:**575–588.

Silhavy, T. J. 1997. Death by lethal injection. *Science* **278:**1085–1086.

Stein, M., F. Bagnoli, R. Halenbeck, R. Rappuoli, W. J. Fantl, and A. Covacci. 2002. c-Src/Lyn kinases activate Helicobacter pylori CagA through tyrosine phosphorylation of the EPIYA motifs. *Mol. Microbiol.* **43:**971–980.

  This work describes the posttranslational modification occurring to CagA, one of the major toxins produced by *H. pylori*. After injection into eukaryotic cells by a specialized type IV secretion apparatus, the toxin is tyrosine-phosphorylated at specific sequence repeats. This modification is required to induce host cell signaling events and to promote the dramatic changes that occur in the morphology of cells growing in culture.

Wallace, A. J., T. J. Stillman, A. Atkins, S. J. Jamieson, P. A. Bullough, J. Green, and P. J. Artymiuk. 2000. X-ray crystal structure of the toxin and observation of membrane pores by electron microscopy. *Cell* **100:**265–276.

14

Bacterial Protein Toxins as Tools in Cell Biology and Pharmacology

Klaus Aktories

At least four properties of bacterial protein toxins make them suitable as cell biological and pharmacological tools (Tables 14.1 and 14.2). First, the toxins enter cells without damaging the cell integrity. This ability is typical of intracellularly acting protein toxins and depends on very complex uptake mechanisms, which are described in detail elsewhere. In general, uptake of toxins depends on at least three steps: receptor binding, endocytosis of the receptor-bound toxin, and translocation of the toxin through the vesicle/compartment membrane into the cytosol. Of course, this does not apply to toxins that explicitly act on the cell surface and form membrane pores.

Second, the toxins possess high specificity. A high cell specificity is most often based on a toxin-specific membrane-binding domain and on specific receptors present on the surface of eukaryotic target cells. This is, for example, the reason for the cell specificity of neurotoxins. The cell surface receptors for botulinum neurotoxins, which are still not well defined (and this holds true also for tetanus toxin), are found exclusively on neuronal cells. Other cell types that may contain the protein target are not affected, because the toxins are not able to enter these cells. Most important, many bacterial protein toxins are characterized by a very specific target recognition. Two examples illustrate this fact. First, diphtheria toxin and *Pseudomonas* exotoxin A selectively ADP-ribosylate diphthamide, a post-translationally modified histidine residue found exclusively in eukaryotic elongation factor 2 but not in other proteins. Second, *Clostridium botulinum* C2 toxin ADP-ribosylates nonmuscle actin but not skeletal muscle actin, although these two actin isoforms are more than 95% identical in their amino acid sequence and share the acceptor amino acid arginine 177.

Third, many bacterial protein toxins are remarkably potent and extremely efficient. It has been estimated that one molecule of diphtheria toxin per cell is sufficient to kill the cell by ADP-ribosylation of elongation factor 2 and subsequent inhibition of protein synthesis. Moreover, botulinum neurotoxins are by far the most potent agents known. A high affinity of the toxins for their specific cell receptor may contribute to this high potency. However, in most cases, the high potency is believed to be

341

Table 14.1   Bacterial toxins frequently used as tools

Study of signal transduction involving heterotrimeric G proteins
  G_s protein activation by cholera toxin and related toxins
  $G_{i,o}$ protein inhibition by pertussis toxin
  G_q protein activation by *Pasteurella multocida* toxin
Study of signal transduction involving low-molecular-mass GTPases
  Rho subfamily protein activation by CNF and
  RhoA, RhoB, and RhoC inactivation by C3-like transferases
  Rho/Ras subfamily protein inhibition by large clostridial cytotoxins
Study of the involvement of the actin cytoskeleton in cellular processes
  Depolymerization of actin by *C. botulinum* C2 toxin and *C. perfringens* iota toxin
  Modulation of actin regulation via Rho proteins (see above)
Study of the involvement of synaptic peptides in exocytosis
  Cleavage of synaptic peptides by botulinum neurotoxins and tetanus toxin
Construction of cellular protein delivery systems
  Single- or dichain transport systems consisting of the binding-translocation subunits of diphtheria toxin and *Pseudomonas* exotoxin A
  Binary transport systems consisting of the separate components of anthrax toxin, *C. botulinum* C2 toxin, or *C. perfringens* iota toxin
Selectively killing of target cells
  Inactivation of elongation factor 2 by fusion toxins or immunotoxins consisting of diphtheria toxin or *Pseudomonas* exotoxin A
Permeabilization of eukaryotic cells
  Streptolysin O
  S. aureus α-toxin
Tracing of neurons
  By retrograde transport of CT
  By retrograde transport of neurotoxins

Table 14.2   Enzyme activities and targets of the bacterial protein toxins most frequently used as tools

Toxin	Activity or target
ADP-ribosyltransferases	
Diphtheria toxin	Elongation factor 2
Pseudomonas exotoxin A	Elongation factor 2
CT	G_s proteins
PT	$G_{i,o}$ proteins
C. botulinum C2 toxin	Nonmuscle β/γ-actin, smooth muscle γ-actin
C. perfringens iota toxin	All actin isoforms
C3-like transferases (e.g., *C. botulinum* C3 exoenzyme)	RhoA, RhoB, and RhoC
Glycosyltransferases	
C. difficile toxins A and B	Rho, Rac, and Cdc42
C. sordellii lethal toxin	Rac, Cdc42, Ral, Ras, and Rap
Deamidases	
CNF1 and CNF2	Rho, Rac, and Cdc42
Metalloendoproteases	
Botulinum neurotoxins A, C, and E	SNAP25
Botulinum neurotoxins B, D, F, and G	Synaptobrevin (VAMP)
Botulinum neurotoxin C	Syntaxin
Tetanus toxin	Synaptobrevin (VAMP)
Anthrax toxin	MAPKK

due to the enzyme activities of the toxins. Therefore, toxins are endoproteases, *N*-glycosidases, ADP-ribosyltransferases, glucosyltransferases, or deamidases (see also Table 14.2), which covalently modify their protein substrates with high efficiency. Although most toxin-induced enzymatic reactions are reversible in vitro and can be driven back under specific experimental conditions, the toxin-induced modifications are essentially irreversible and complete in intact cells and under physiological conditions. Moreover, the covalent modification by toxins occurs even in protein mixtures, in which the target protein is present only in small amounts. Therefore, the reaction is exploited to selectively radiolabel target proteins by [^{32}P]ADP-ribosylation or ^{14}C-glycosylation to facilitate the purification of these targets or to allow analysis by sodium dodecyl sulfate-polyacrylamide gel electrophoresis.

Fourth, the remarkable cell biological efficiency of bacterial toxins is not based simply on the specific kinetics of the enzyme reactions catalyzed by the toxins. A pivotal physiological role of the eukaryotic target is also important. Thus, very often the toxins encounter cellular "master" regulators, thereby inhibiting or stimulating crucial signal pathways absolutely necessary for the normal physiological function of the cells targeted. This aspect is of particular importance for the use of toxins as pharmacological and cell biological tools and has been extensively exploited during recent years. In fact, research on the action of novel bacterial toxins or on the molecular mechanisms of toxins which were hitherto obscure has been motivated greatly by the prospect of unraveling crucial biological pathways or novel pivotal regulators of the target cells.

Toxins as Tools To Study Nucleotide-Binding Proteins

For unknown reasons, GTP-binding proteins are the preferred substrates for bacterial toxins. For example, heterotrimeric G proteins, which couple to heptahelical membrane receptors and are involved in various signaling processes, are ADP-ribosylated by cholera toxin (CT), the related heat-labile *Escherichia coli* enterotoxins, and pertussis toxin (PT). Various evidently unrelated bacterial toxins modify low-molecular-mass GTPases (e.g., Rho proteins) by either ADP-ribosylation, glycosylation, or deamidation. Finally, elongation factor 2, the eukaryotic protein substrate of diphtheria toxin and *Pseudomonas* exotoxin A, is a GTP-binding protein. Actin, another important eukaryotic substrate for ADP-ribosylation by bacterial toxins, is not a GTP-binding protein but an ATP-binding protein. Because all these nucleotide-binding proteins are functionally important cellular proteins, the toxins, which allow their selective covalent modification, are widely used as tools.

Cholera Toxin, Pertussis Toxin, and *Pasteurella multocida* Toxin as Tools To Study G-Protein-Mediated Signaling

The G-protein ADP-ribosylating toxins CT and PT are the "classical" protein toxins which are used as pharmacological tools. Both toxins are used to study signal pathways involving heterotrimeric G proteins (Figure 14.1 and Box 14.1).

Cholera Toxin

CT, which is produced by *Vibrio cholerae*, is the principal cause of the watery diarrhea of cholera. The toxin (85 kDa) is composed of an enzyme compo-

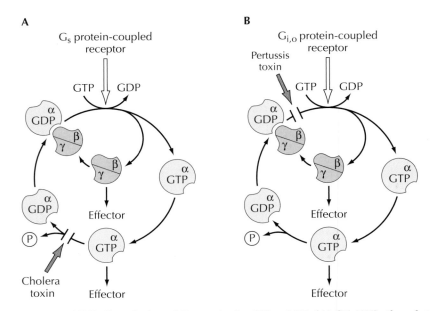

Figure 14.1 ADP-ribosylation of G proteins by CT and PT. **(A)** CT ADP-ribosylates the α subunits of G_s proteins. ADP-ribosylation blocks the GTPase activity of $G_{s\alpha}$ and activates the G protein persistently. CT is used as a tool to radiolabel G_s proteins and to manipulate the signaling via G_s. **(B)** PT ADP-ribosylates the α subunits of $G_{i,o}$ proteins, thereby blocking the receptor-mediated activation of the G protein. Thus, inhibition of a specific signal transduction process by PT indicates the involvement of "PT-sensitive" G proteins in this signaling pathway.

nent (A subunit, ~28 kDa) and a binding component, which is formed by five identical B subunits (each ~11.6 kDa). The enzyme component consists of an A1 subunit (~21 kDa) and an A2 subunit (~6 kDa), which are linked by a disulfide bridge. A1 possesses ADP-ribosyltransferase activity, whereas A2 mediates the interaction with the B component. The physiological target

---

BOX 14.1

G proteins, targets of CT, PT, and PMT

Whereas CT activates G_s protein, PT inhibits the activation of $G_{i,o}$ proteins (Figure 14.1). PMT activates G_q protein by an unknown mechanism. The functional consequences of the modification of G proteins depend on the signal transduction systems of the specific cell targeted by the toxins.

G_s proteins are, for example, coupled to adrenergic (β_{1-3}), adenosine (A2), dopamine (D1 and D5), histamine (H2), serotonin (5-HT4), and prostaglandin (DP, EP2, EP4, and IP) receptors. $G_{i,o}$ proteins are coupled to adrenergic (α_2), muscarinic (M2 and

M4), dopamine (D2, D3, and D4), serotonin (5-HT1), opioid (OP1), and somatostatin (SST1) receptors. Gq couple to adrenergic (α_1), muscarinic (M1 and M3), serotonin (5-HT2), angiotensin (AT1), histamine (H1), and vasopressin (V1) receptors.

The α subunit of G_s activates adenylyl cyclase (all isoforms) and causes an increase in cAMP, which stimulates protein kinase A, cyclic nucleotide-dependent cation channels, and an activator (Epac) of the small GTPase Rap1. The α subunit of $G_{i,o}$ induces inhibition of adenylyl cyclase and stimulation of src kinase. The main effect of the α subunit of G_q, which couples to classical calcium-mobilizing receptors, is the ac-

tivation of phospholipase Cβ. PLCβ produces diacylglycerol (DAG) and inositoltrisphosphate (IP3), which releases calcium from internal stores. Moreover, the free β/γ subunits of heterotrimeric G proteins interact with effectors. They are released by CT-induced activation of G_s and remain associated with $G_{i,o}$ after inactivation of the G protein by PT. The β/γ subunits activate phospholipase Cβ and adenylyl cyclase isoforms II, IV, and VII, but inhibit adenylyl cyclase isoform I. They activate phophatidylinositol-3-kinase (PI3-kinase) and increase K^+ currents (Neves et al., 2002). Thus, all these signal transduction pathways are affected by CT, PT, and PMT.

of CT (the same holds true for the related *E. coli* heat-labile enterotoxins) is the $G_{s\alpha}$ subunit of heterotrimeric G proteins. The toxin ADP-ribosylates arginine 201 (or equivalent arginine residues depending on the splice variants) of $G_{s\alpha}$. ADP-ribosylation inhibits the intrinsic GTPase of $G_{s\alpha}$, induces dissociation from $\beta\gamma$ subunits, and renders the α subunit persistently active. In enterocytes, stimulation of adenylyl cyclase and increase in intracellular cyclic AMP levels caused by the persistently active $G_{s\alpha}$ subunits contribute to the fluid and electrolyte loss that occurs in the course of cholera. At least in vitro, CT also ADP-ribosylates other G proteins, such as transducin and even $G_{o/i}$ (under specific conditions). Substrates also include cell proteins with a suitable arginine residue, as well as small arginine derivatives like agmatine, and, finally, the toxin itself is modified by an auto-ADP-ribosylation reaction. Moreover, like all other ADP-ribosyltransferases, CT possesses NAD glycohydrolase activity and splits NAD into ADP-ribose and nicotinamide in the absence of a suitable substrate. However, this activity is very low compared with the ADP-ribosyltransferase activity. In vitro, ADP-ribosylation by CT is stimulated largely by the presence of cofactors called ADP-ribosylation factors, which have been identified as members of a subfamily of small GTPases (ARF subfamily of GTPases). Although the specificity of CT in in vitro experiments is not very impressive, in intact cells $G_{s\alpha}$ appears to be the preferred substrate of the toxin.

In addition to its usage as a tool to manipulate signal transduction pathways involving G_s protein or to label $G_{s\alpha}$ protein, CT has been successfully used as an adjuvant applied with an unrelated antigen to stimulate mucosal immune responses. Moreover, CT is used as a tracer for retrograde transport in neuronal cells.

Pertussis Toxin

PT (formerly called islet-activating protein) is one of the exotoxins of *Bordetella pertussis*, the causative agent of whooping cough. It is a hexameric toxin of 105 kDa and consists of the enzyme subunit S1 ($\sim$26 kDa, A subunit) and the binding pentamer (B subunit) formed of S2 ($\sim$22 kDa), S3 ($\sim$22 kDa), two S4 subunits ($\sim$12 kDa), and S5 ($\sim$12 kDa). The enzyme subunit (S1) of PT ADP-ribosylates G_α isoforms of the G_i subfamily of heterotrimeric G proteins in the presence of $\beta\gamma$ subunits (Figure 14.1). The toxin catalyzes the ADP-ribosylation of a cysteine residue located 4 amino acid residues from the carboxy-terminal end of the α subunit of sensitive G proteins (e.g., G_{i1-3}, $G_{o1,2}$, and G_t). More than 20 years ago, PT was most important for the identification of the G_i protein as an additional G protein besides G_s. The consequence of toxin-catalyzed ADP-ribosylation is a functional uncoupling of G proteins from their membrane receptors with subsequent blockade of the G-protein-transduced signal pathways. Importantly, G proteins of the G_s, G_q, G_{12}, and G_z subfamilies and a splice variant of G_{i2} [$G_{i2(L)}$] are not modified by pertussis toxin. Thus, the toxin is widely used to test whether a signal transduction pathway involves so-called "PT-sensitive" G proteins.

P. multocida Toxin

P. multocida toxin is an $\sim$150-kDa protein, which is responsible for turbinate bone atrophy associated with porcine atrophic rhinitis observed after *Pasteurella* infections. It is an extremely potent mitogen in many cells and causes an extensive increase in phospholipase Cβ activity. Although the precise molecular mode of action is not known so far, a large amount of evidence exists that PMT activates $G_{\alpha q}$ proteins. For this purpose it is increasingly used in signal transduction studies. Importantly, it stimulates $G_{\alpha q}$, but not the highly related $G_{\alpha 11}$. However, in studies with PMT one has to consider that RhoA is also activated by PMT and that other G proteins like $G_{12/13}$ might be targets of the toxin.

Toxins as Tools To Study the Regulation of the Actin Cytoskeleton

Actin, which is one of the most abundant proteins in eukaryotic cells, is the major component of the microfilament system. It regulates the architecture of cells and is involved in various motile processes. Besides its function in skeletal muscle contraction, it plays important roles in migration, phagocytosis, endocytosis, secretion, and intracellular transport. The actin cytoskeleton is the target of various bacterial toxins that affect the microfilament protein either directly by ADP-ribosylation or indirectly by modifying the regulatory mechanisms involved in the organization of the actin cytoskeleton. In the latter case, Rho proteins, which are regulators of the cytoskeleton, are especially important targets of bacterial protein toxins.

Actin ADP-Ribosylating Toxins

Actin is the specific substrate for the family of actin ADP-ribosylating toxins. This family has four members: *Clostridium botulinum* C2 toxin, *C. perfringens* iota toxin, *C. spiroforme* toxin, and the ADP-ribosylating toxin from *C. difficile.* All these toxins are binary in structure and consist of separate enzyme and cell-binding/translocation subunits, which are not linked by either covalent or noncovalent bonds. (Recently, an actin-ADP-ribosylating toxin [SpvB] from *Salmonella enterica,* which is not binary and most likely transported by direct bacterial contact, has been reported.) After attachment of the binding component to the cell surface and induction of a binding site for the enzyme component take place, the toxin subunits will interact with each other (Figure 14.2).

    C. botulinum C2 toxin and *C. perfringens* iota toxin are used as cell biological tools. The binding component (C2II) of *C. botulinum* C2 toxin is about an 80-kDa protein that has to be activated by trypsin to release a ca. 65-kDa active fragment. The molecular mass of the enzyme component (C2I) is about 50 kDa. C2I ADP-ribosylates nonmuscle β/γ actin and γ-smooth muscle actin but not α-actin isoforms at arginine 177. In contrast, *C. perfringens* iota toxin catalyzes the ADP-ribosylation of all mammalian actin isoforms known. Interestingly, the difference between the various mammalian isoforms is less than 5%. The modification by the toxins is highly specific. Neither G proteins, which are substrates for the arginine-modifying CT, nor other cytoskeletal elements such as tubulin are ADP-ribosylated. ADP-ribosylation of actin has several functional consequences. Most important is the inhibition of the actin polymerization (Figure 14.2). The acceptor amino acid of actin is located at an actin-actin contact site, and it is believed that the attachment of the ADP-ribose moiety inhibits the interaction of the actin monomers. This is also the reason why G-actin but not polymerized actin is a substrate for ADP-ribosylation. Second, ADP-ribosylated actin behaves like a capping protein that binds to the barbed ends of actin filaments, thereby inhibiting the polymerization of unmodified G-actin at the plus ends of filaments. In contrast, ADP-ribosylated actin does not interact with the minus ends of filaments. Therefore, the critical actin concentration for polymerization increases to values that correspond to the critical actin con-

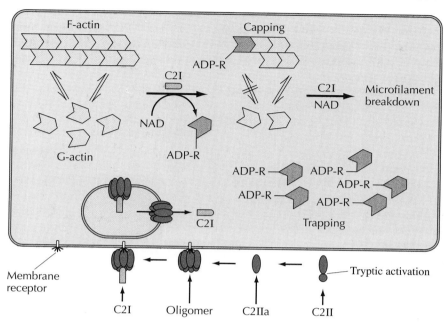

Figure 14.2   Model of the action of the actin-ADP-ribosylating *C. botulinum* C2 toxin. The trypsin-activated binding component (C2IIa) forms heptamers, binds to the cell surface receptor of the target cell, and is the binding site for the enzyme component C2I. The toxin-receptor complex is internalized. At the acidic pH of endosomes, the binding component inserts into the membrane and allows the translocation of C2I into the cytosol, where monomeric actin is ADP-ribosylated. ADP-ribosylation of actin blocks its polymerization ("trapping" of the monomeric form). Moreover, ADP-ribosylated actin monomers bind like capping proteins to the plus ends of actin filaments, thereby inhibiting polymerization of unmodified actin monomers ("capping"). Because the minus ends of filaments are free, actin can be released at this site. The released actin is immediately ADP-ribosylated and trapped. Thus, the major consequence of ADP-ribosylation is breakdown of the microfilaments.

centration at the minus end of actin filaments. This means that actin filaments depolymerize. However, the released monomeric actin is immediately ADP-ribosylated by the toxin and trapped in the nonpolymerizable form. Third, ADP-ribosylation of actin completely blocks actin ATPase activity and increases the ATP exchange. Finally, ADP-ribosylation affects the interaction of actin with actin-binding proteins, which are essential for regulation of actin polymerization. For example, the nucleation activity of the gelsolin-actin complex is blocked after ADP-ribosylation of actin. All these functional consequences of the ADP-ribosylation of actin finally result in morphological changes of targeted cells (rounding up) (Figure 14.3), redistribution of the microfilaments, depolymerization of F-actin, and an increase in the amount of G-actin. Thus, the actin ADP-ribosylating toxins are the most powerful tools to induce depolymerization of microfilaments in intact cells. Actually, the complete actin cytoskeleton of cells can be depolymerized. Therefore, the toxins have been used to study the role of actin in migration, secretion, endocytosis, superoxide anion production, and intracellular vesicle transport.

C. botulinum C2 toxin was used to investigate the activation of neutrophils by chemotactic agents. In neutrophils, complement C5a, *N*-formyl peptides, or leukotriene B_4 induce cellular responses such as cell shape

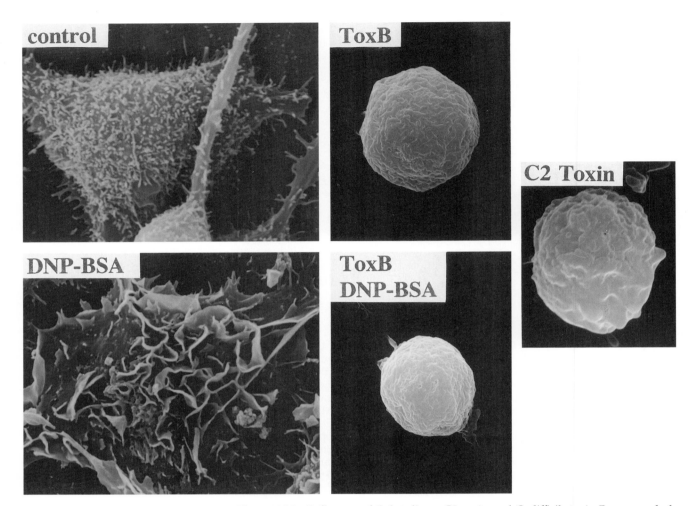

Figure 14.3 Influence of *C. botulinum* C2 toxin and *C. difficile* toxin B on morphology of RBL cells. Stimulation of immunoglobulin E-primed RBL cells by antigen (2,4-dinitrophenyl-bovine serum albumin [DNP-BSA]) via the high-affinity antigen receptor (FcεRI) induces membrane ruffling. Pretreatment of cells with *C. botulinum* C2 toxin, which depolymerizes actin by ADP-ribosylation, causes dramatic morphological changes but increases regulated serotonin release. *C. difficile* toxin B, which inactivates Rho family proteins by glucosylation, induces a similar morphology of RBL cells but completely blocks antigen-induced serotonin release. The experiment shows that inhibition of serotonin release by toxin B is not simply caused by an action on the actin cytoskeleton and indicates that Rho proteins are essentially involved in the signal transduction of the FcεRI receptor. Scanning electron micrographs courtesy of J. Wilting, Freiburg, Germany.

change, adhesion, migration, degranulation, and phagocytosis. It was shown by using the toxin that all these events depend on the redistribution of the cytoskeleton and on changes in actin polymerization. For many years, the mycotoxins phalloidin, which induces polymerization, and cytochalasin, which causes depolymerization of F-actin, were used as tools to elucidate the physiological functions of actin. Whereas phalloidin is hampered by its poor cell accessibility, the shortcomings of cytochalasins are the incomplete depolymerization of microfilaments and additional nonspecific effects. Therefore, actin-ADP-ribosylating toxins are used as tools for selective and complete depolymerization of actin. (Note: Recently, jasplakinolide and latrunculin, both low-molecular-mass toxins from marine sponges, were used to polymerize actin and depolymerize actin, respectively.) By using C2 toxin, it was shown that redistribution of the actin cytoskeleton largely affects the activity of superoxide anion-producing NADPH oxidase of neutrophils, thereby confirming previous results obtained with cytochalasins.

C2 toxin has been used in studies on exocytosis. Actin is suggested to function as a subcortical barrier, preventing vesicle fusion with the cell membrane. In fact, in various cell types C2 toxin increases the regulated exocytosis. This was shown for the *N*-formyl peptide-stimulated release of *N*-acetylglucosaminidase and of vitamin B_{12}-binding protein from neutrophils, for steroids in Y-1 adrenal cells, for serotonin in RBL cells, and for insulin in isolated rat islets. In general, similar results were obtained with cytochalasin. In all these cases, C2 toxin did not affect the basal release of mediators. However, biphasic effects of C2 toxin were observed on the release of noradrenaline from PC12 cells. Treatment of the cells with the toxin for up to 1 h increased noradrenaline release, whereas longer incubation reduced the mediator release induced by depolarization or by carbachol. In hamster insulinoma HIT-T15 cells, C2 toxin treatment inhibited insulin release by about 50%. Inhibition of insulin release was more pronounced in the second phase of the biphasic insulin release. These findings can be interpreted to indicate that actin filaments are involved in the recruitment of vesicles to the releasable pool and that C2 toxin blocks this action. The toxin was used to show the involvement of actin microfilaments in the retrograde transport from the Golgi to the endoplasmic reticulum. Also, in suspended mast cells, C2 toxin inhibited regulated mediator release. In these cells, however, the inhibitory effect of C2 toxin turned into a stimulatory effect after adhesion of the mast cells, suggesting a complex role of actin, which cannot be explained simply by a barrier function or vesicle recruitment.

C2 toxin was used to investigate the regulation of ileal smooth muscle contraction. It was shown that C2 toxin inhibited the contraction in ileal longitudinal smooth muscle preparations induced by electrical stimulation, by agonists like bradykinin or carbachol, and even by Ba^{2+} ions, which directly stimulate smooth muscle contraction. Since the toxin modifies G- but not F-actin, these data were taken as an indication that G/F-actin transition is important in smooth muscle contraction. Moreover, C2 toxin was used to investigate the autoregulation of the actin synthesis that appears to depend on the levels of G-actin and on the G/F-actin ratio. In all these studies, proper control of the specific action of C2 toxin was necessary. For example, the single toxin components (C2I or C2II) applied should not induce any effect in the system studied. The toxin effect should occur with some delay of at least 15 to 30 min. This time is necessary for the translocation of the toxin. Moreover, it should be tested whether actin is in fact ADP-ribosylated by the toxin. Finally, it is also recommended to compare the actions of C2 toxin with other actin-depolymerizing agents such as cytochalasins and latrunculin.

Toxins That Affect Small GTPases

In recent years, various toxins that modify low-molecular-mass GTP-binding proteins have been identified (Figure 14.4). Among these are the ADP-ribosylating C3-like transferases, the glycosylating large clostridial toxins, and a rather novel group of deamidating and/or transglutaminating toxins. Because these toxins (especially C3-like transferases) were most

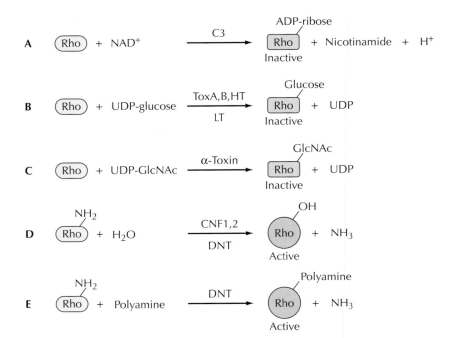

Figure 14.4 Covalent modification of Rho proteins by bacterial protein toxins. **(A)** C3-like transferases ADP-ribosylate RhoA, RhoB, and RhoC by using NAD as a cosubstrate. **(B)** Rho family proteins are glucosylated by *C. difficile* toxin A and B and by the hemorrhagic (HT) and lethal (LT) toxins of *C. sordellii*. The cosubstrate is UDP-glucose. While *C. difficile* toxins A and B and HT glucosylate all Rho family members, LT modifies Rac and Cdc42 (depending on the producer strain) but not Rho. Additionally, Ras family proteins (e.g., Ras, Rap, and Ral) are substrates for glucosylation by LT. **(C)** The α-toxin from *C. novyi* catalyzes an *N*-acetylglucosaminylation. **(D)** Rho proteins (Rho, Rac, and Cdc42) are activated by cytotoxic necrotizing factors (CNF1 and CNF2) from *E. coli* and by the dermonecrotic toxin (DNT) from *Bordetella* species. **(E)** DNT is also able to activate Rho GTPases by transglutamination, e.g., the attachment of polyamines onto the GTPases.

important for the functional characterization of small GTPases, these proteins are briefly reviewed.

At least five families of small GTPases can be described; these are the Ras, Rab, Arf, Ran, and Rho subfamilies, which have about 30% sequence identity. Members within a specific GTPase subfamily exhibit at least 50% identity in the amino acid sequence. Like heterotrimeric G proteins, these low-molecular-mass or "small" GTP-binding proteins are regulated by a GTPase cycle. They are inactive with GDP bound and activated after GDP-GTP exchange. Hydrolysis of bound GTP terminates the active state of the GTPases. It has been shown that especially Rho subfamily GTPases are targets for bacterial protein toxins. This family comprises more than 15 GTPases (including RhoA, RhoB, RhoC, RhoD, RhoE, RhoG, Rac1, Rac2, Rac3, Cdc42Hs, G25K, and RnD/RhoE). Rho proteins are controlled by three groups of regulatory proteins. Guanine nucleotide exchange factors (GEFs) induce the activation of Rho GTPases by facilitating the GDP-GTP exchange. GTPase-activating proteins (GAPs) stimulate and catalyze the GTP hydrolysis, thereby inactivating the GTPases, and, finally, guanine nucleotide dissociation inhibitors (GDIs) block nucleotide exchange, keep the inactive form of Rho proteins in the cytosol, and may be involved in presentation of the GTPases to other proteins (Figure 14.5).

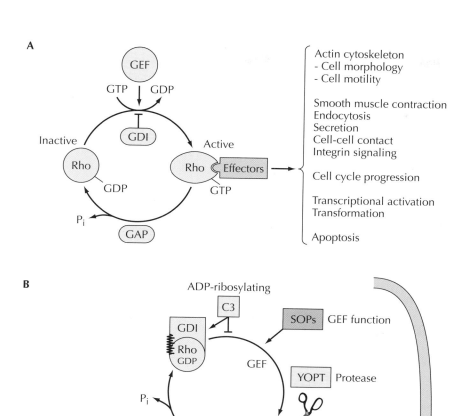

Figure 14.5 **(A)** Regulatory GTPase cycle of Rho proteins. Rho GTPases are inactive in the GDP-bound form and active after GTP-GDP exchange. The nucleotide exchange is stimulated by guanine nucleotide exchange factors (GEF) and inhibited by guanine nucleotide dissociation inhibitors (GDI). The active state of the GTPase is terminated by hydrolysis of bound GTP. GTP hydrolysis is stimulated by GTPase-activating proteins (GAP). Some of the various processes regulated by Rho GTPases are indicated. **(B)** Effects of Rho-modifying toxins. C3 exoenzyme and C3-like transferases ADP-ribosylate Rho at Asn 41, inhibiting the activation of Rho by blocking GEF-induced activation and increasing the stability of the GDI/Rho complex. Glucosylation of Rho GTPases by large clostridial cytotoxins blocks the interaction of active Rho with effectors and causes inactivation of Rho GTPase-dependent processes. The cytotoxic necrotizing factor CNF and the dermonecrotic toxin DNT inhibit the switch-off mechanism of activated Rho GTPases by deamidation and transglutamination of a glutamine residue (Gln 63 of Rho), which is essential for GTP hydrolysis resulting in constitutively activated Rho. In addition, bacterial proteins, which are introduced into target cells by a type III system, are indicated. *Yersinia* YopT possesses protease activity to cleave the isoprenylated C terminus of Rho GTPases, thereby inactivating the GTPases. SOP is a *Salmonella* protein with GEF-like function to activate Rho. The *Yersinia* YopE, the *Salmonella* SptP, and the *Pseudomonas aeruginosa* ExoS are bacterial GAPs, which inactivate Rho GTPases.

The small GTPases of the Rho family (Rho, Rac, and Cdc42) are involved in the regulation of the actin cytoskeleton. Rho induces the formation of stress fibers and adhesion complexes. Rac is involved in lamellipodium formation and induces adhesion complexes, which appear to be different from those induced by Rho. Finally, Cdc42 induces the formation of microspikes. Beside their roles in the organization of the actin cytoskeleton, Rho proteins act as molecular switches in various signal transduction processes (see below) and may play essential roles in invasion of eukaryotic cells by various bacteria (e.g., *Salmonella* and *Shigella*).

Rho-Inactivating Toxins

C3-Like ADP-Ribosyltransferases

Rho GTPases are the specific substrates of the C3-like ADP-ribosyltransferases. The C3-like transferases are all basic proteins (IP > 9) with molecular masses of about 25 kDa, having a sequence identity of 30 to 70% (at least 7 isoforms from *C. botulinum*, *Clostridium limosum*, *Bacillus cereus*, and *Staphylococcus aureus* are known). ADP-ribosylation of Rho by C3 transferases needs no additional factors than NAD. Rho ADP-ribosylation is affected by guanine nucleotides, divalent cations, detergents, and temperature. Purified endogenous Rho, recombinant Rho proteins, and the membranous Rho protein are better substrates for ADP-ribosylation when bound to GDP than when bound to GTP. By contrast, the ADP-ribosylation of cytosolic Rho by C3 is increased by GTP or stable GTP analogs. Similarly, detergents and phospholipids increase ADP-ribosylation most probably by dissociating the GDI-Rho complex.

C3 transferases ADP-ribosylate RhoA, RhoB, and RhoC at asparagine 41, thereby blocking the biological activity of Rho (Figure 14.5). Although asparagine 41 is located in the so-called effector region of the Rho protein, the precise mechanism of Rho inhibition by ADP-ribosylation is not entirely clear. For example, ADP-ribosylated Rho is still able to bind to its effectors protein kinase N and Rho kinase. Whether other Rho effectors are also able to interact with ADP-ribosylated Rho remains to be clarified. Although C3 modifies Rac by up to 5% in vitro, all experimental evidence indicates that in intact cells C3-like transferases ADP-ribosylate RhoA, RhoB, and RhoC but not Rac and Cdc42. This is most important because C3 has been widely used to study the cell biological role of Rho. However, there is one exception: C3stau2 (also called EDIN B), which modifies in addition the GTPases RhoE/(Rnd3).

C3 transferases appear to lack any specific binding and translocation subunit. Therefore, the cell accessibility of these ADP-ribosyltransferases is rather poor. Most probably this is also the reason why C3 shows low toxicity compared with other toxins. For example, intraperitoneal injection of 100 μg of C3 into mice has no obvious consequences. Thus, it appears questionable to describe C3 as a "real" toxin. To circumvent the problem of poor cell entry, several approaches have been used. First, C3 was introduced into cells by osmotic shock. In fact, all methods used to permeabilize cells can be applied to allow C3 to enter cells. Mostly, pore-forming toxins like streptolysin O were used. Many excellent studies, especially from the laboratory of Alan Hall, used microinjection techniques. Even more sophisticated approaches involve the use of chimeric toxins. To this end, a fusion toxin that consists of C3 and the receptor-binding and translocation subunits of diphtheria toxin was constructed. Although this fusion toxin is able to enter all diphtheria toxin-sensitive cells, the uptake pathway of the fusion toxin differs from that of diphtheria toxin. Another chimeric toxin was constructed of *C. botulinum* C2 toxin and C3 transferase. C3 was fused to the N-terminal part of the enzyme component of C2I, which is apparently involved in the interaction of C2I with C2II. This C3-C2IN fusion toxin could be introduced into cells via the binding and translocation component C2II. Similarly, as observed with the diphtheria toxin-C3 fusion protein, in intact cells the Rho-ADP-ribosylating activity of C3-C2IN was increased several hundred-fold compared with that of the native C3 transferase. Recently, cell accessibility of C3 was also increased by attachment of transport peptides to C3, including the peptide Tat from the human immunodeficiency virus, Antp from the *Drosophila* antennapedia homeodomain, and proline- or arginine-rich peptides. However, most studies took advantage of the fact that C3 enters most cells when used at high concentrations for a long time (e.g., 24 to 48 h). For comparison, the effect of the C3-C2IN fusion protein was observed 1 to 2 h after addition to the culture medium.

As mentioned above, the use of C3 was most important in elucidating the functions of Rho proteins. Because the C3-induced effects on Rho appear to be very specific, the inhibition of a specified signaling pathway by using C3 can be taken as an indication that Rho is involved in the signaling process. Thereby, it was shown that RhoA specifically control stress fiber formation and induction of cell adhesions. C3 was used in studies to show that Rho participates in the control of cell aggregation, integrin signaling, control of phosphatidylinositol 3-kinase, phosphatidylinositol-4-phosphate 5-kinase, and phospholipase D. Moreover, C3 transferases were used to characterize the role of Rho proteins in endocytosis, secretion, and control of transcription, cell cycle progression, and cell transformation. Finally, studies on the involvement of Rho proteins in neurite outgrowth and nerve growth were largely supported by the usage of C3. Because Rho GTPase inhibits neurite outgrowth, inhibition of its biological activity by C3 facilitates nerve growth and regeneration. Thus, it is proposed to use C3 as a drug for treatment of nerve damage repair (Figure 14.5).

Large Clostridial Cytotoxins

Recently, it has been shown that Rho proteins are the targets of various clostridial cytotoxins, which modify the GTPases by glycosylation. These toxins have molecular masses between 250 and 308 kDa and are therefore called "large" clostridial cytotoxins. From a medical point of view, *C. difficile* toxins A and B, which are the major virulence factors in antibiotic-associated diarrhea, are the most important. Toxin B is several hundredfold more potent in inducing cytotoxic effects in cell culture than is toxin A and is therefore designated a cytotoxin. However, toxin A also induces cytotoxic effects. Whether the difference in potency is mainly related to the use of different cell membrane receptors is still a matter of debate. After parenteral administration, both toxins are lethal at an identical dose (minimum lethal dose in mice ~50 ng i.p.).

In most cells, *C. difficile* toxins induce depolymerization of the actin cytoskeleton, leading to a morphology similar to that induced by C3-like transferases. It has been shown that Rho proteins are also substrates for *C. difficile* toxins (Figures 14.2 and 14.5). However, these large clostridial toxins catalyze the glucosylation of Rho GTPases by using UDP-glucose as a cosubstrate. Notably, not only RhoA, RhoB, and RhoC but also other members of the Rho subfamily like Cdc42 and Rac are substrates. Other small GTPases like Ras, Rab, Arf, or Ran are not glucosylated. Glucosylation occurs at Thr37 of Rho and Thr35 of Rac and Cdc42. Threonine 35/37 is conserved in all small GTPases and is involved in the coordination of the divalent cations and of nucleotides. Moreover, this residue is located in the switch-I region of the proteins. This region undergoes major conformational changes upon nucleotide binding and participates in effector coupling. It has been demonstrated that glucosylation inhibits the RhoA-effector interaction, explaining the functional inactivation of the GTPase after toxin treatment. Moreover, the acceptor amino acid residue for glucosylation is located very close to the site of ADP-ribosylation (Asn41 in RhoA). This explains why glucosylated Rho is no longer ADP-ribosylated by C3 and vice versa. Thus, failure to label Rho in lysates of toxin B-treated cells by C3-induced [^{32}P]ADP-ribosylation indicates that all Rho was already modified by toxin B in intact cells.

Because *C. difficile* toxins A and B glucosylate all Rho subfamily GTPases, they are used to screen whether Rho GTPases are involved in certain regulatory pathways. For example, toxin B was used to study the involvement of Rho proteins in signal transduction of the FcεRI receptor and in histamine or serotonin secretion from mast cells or RBL cells (Prepens et al., 1996) (Figure 14.5). Toxin B inhibited regulated secretion induced by antigen, compound 48/80, and even the calcium ionophore A23187. Because Rho proteins are involved in various diverse regulatory pathways, including regulation of the actin cytoskeleton, it is important to clarify whether the inhibitory effect of the toxin on secretion is secondary and induced by depolymerization of actin. In this case, another toxin that selectively inhibits actin polymerization but has no effect on Rho proteins, such as *C. botulinum* C2 toxin, is useful. In the above-mentioned case of histamine or serotonin secretion from RBL cells, it was shown that C2 toxin largely increased regulated secretion, excluding the possibility that the effect by toxin B is caused merely by disturbing the actin cytoskeleton.

Other members of the family of large clostridial cytotoxins are also glycosyltransferases but differ in cosubstrate or protein substrate specificity; however, the conserved Thr35/37 residue of the GTPases is modified in each case. *C. novyi* α-toxin modifies Rho subfamily proteins like *C. difficile* toxin but catalyzes *N*-acetylglucosaminylation by using UDP-GlcNAc as a cosubstrate (Figure 14.2). *C. sordellii* lethal toxin, which is about 90% similar to *C. difficile* toxin B, also uses UDP-glucose as a cosubstrate but differs in its protein targets. Whereas Rac is a very good substrate for this toxin, Rho is poorly modified (most probably not at all in intact cells). In addition, lethal toxin glucosylates and inactivates Ras subfamily proteins like Ras, Rap, and Ral. Thus, the lethal toxin interferes with the Ras-signaling pathway and blocks activation of the mitogen-activated protein kinase (MAPK) cascade by growth factors.

Other Rho-Inactivating Bacterial Proteins

Rho GTPases are also inactivated by bacterial effectors, which are introduced into the eukaryotic target cells by type C3 secretion. Among these are *Yersinia* YopT, a protease, which cleaves the isoprenylated C terminus of Rho GTPases, thereby inactivating the GTPases. Moreover, *Yersinia* YopE, *Pseudomonas aeruginosa* ExoS, and *Salmonella* SptP are bacterial GAPs (note, ExoS and SptP are multidomain proteins possessing also ADP-ribosyltransferase and phosphatase activities, respectively. Because these bacterial effectors do not possess translocation units, they are not extensively used as tools.

Rho-Activating Toxins

Constitutively active Rho GTPases can be constructed by exchange of amino acid residues, which are essential for GTP-hydrolyzing activity. The exchanges of glycine 14 and glutamine 63 of Rho for valine and leucine, respectively, to obtain Rho with very low GTP-hydrolyzing activity are well known. These mutant proteins were often used in microinjection studies to elucidate the role of Rho in organization of the actin cytoskeleton. Rho proteins are also activated by bacterial protein toxins. Recently, it has been shown that Rho GTPases are activated by the cytotoxic necrotizing factors CNF1 and CNF2 from *E. coli* and by the dermonecrotic toxin produced by various *Bordetella* species (Figure 14.4). CNF1 and CNF2 are 115-kDa proteins, which are 99% similar. The toxins cause formation of multinucleated cells and induce the polymerization of actin and the formation of stress fibers, microspikes, and lamellipodia. CNF catalyzes the deamidation of glutamine 63 of RhoA to glutamic acid. Because glutamic acid is not able to fulfill the function of glutamine, the intrinsic and GAP-stimulated GTPase activity of Rho is blocked. Most probably, CNF not only deamidates Rho but also deamidates Cdc42 and Rac, explaining the formation of microspikes, lamellipodia, and membrane ruffles after toxin treatment. Similarly to CNFs, dermonecrotic toxin (DNT; 154 kDa) causes deamidation of Rho proteins. However, DNT not only deamidates the crucial glutamine (e.g., Gln63 of Rho) residue but also causes attachment of polyamines onto this amino acid residue, also resulting in Rho activation. Thus, it acts like a transglutaminase. Both CNF and DNT can be used to activate Rho proteins, for example, in cell culture. The effective activation of Rho-GTPases by the toxins is tested in pull-down experiments performed with Rho-GTPase effectors (or the GTPase-binding domains of the effectors), which interact only with the activated forms of the GTPases.

In addition to CNFs and DNT, Rho is activated by *Salmonella* SOPs, which are GEFs for Rho GTPases. These bacterial effectors are introduced into target cells by type III systems.

Toxins as Part of a Transmembrane Carrier System

Many toxins affecting eukaryotic cells must be translocated into the cytosol to find their target. Some of these translocated toxins are among the biggest toxins known (e.g., large clostridial cytotoxins with molecular masses of 250 to 308 kDa). The ability of these toxins to cross the cell membrane and, in whole or in part, to enter the cytosol has been exploited for use as a carrier system for proteins not related to the toxins themselves. As mentioned above, these toxins are A/B toxins and consist of the biologically active component (A, enzyme component) and the binding and translocation component (B). Actually, the binding and translocation domains are functionally and structurally separate entities, and therefore the toxins are composed of three major domains. In most toxins, these functional domains are located on a single toxin chain (e.g., *Pseudomonas* exotoxin A), are positioned on different chains linked by disulfide bonds (e.g., diphtheria toxin, botulinum neurotoxins), or are located on specific components which are noncovalently associated (e.g., CT and PT). Another group of toxins is characterized by nonlinkage of the enzyme and binding-translocation components. These toxins are also called binary toxins. To construct a protein delivery system, all three toxin components can be used. If it is desired to target a particular cell type, the toxin receptor-binding part must be changed (see below). On the other hand, it might be desirable to transfer into the cytosol an enzyme which otherwise is not able to cross the cell membrane. In this case, the receptor and translocation domains of the toxins are fused to the enzyme. This type of fusion toxin is exemplified by the above-mentioned constructs of C3 with diphtheria toxin or with C2 toxin. Also, the light chain of botulinum neurotoxin has been introduced into cells by means of an anthrax toxin fusion protein.

Clostridial Neurotoxins as Tools To Study Exocytosis

Tetanus toxin and botulinum toxins consist of a heavy chain (~100 kDa), which is responsible for binding and membrane translocation, and a light chain (~50 kDa), which is biologically active and harbors the enzyme activity. The chains are linked by a disulfide bond. The clostridial neurotoxins are metalloendoproteases. They selectively cleave proteins (synaptobrevin [VAMP] syntaxin and SNAP-25) which are involved in exocytosis of synaptic vesicles, thereby blocking neurotransmitter release from presynaptic nerve endings. Whereas tetanus toxin occurs in a single serotype, at least seven different serotypes of botulinum neurotoxins (types A, B, C1, D, E, F, and G) have been identified. The toxins differ by their protein substrate specificity; e.g., synaptobrevin is cleaved by tetanus toxin and botulinum toxins B, D, F, and G; SNAP25 is cleaved by botulinum toxins A, C, and E; and syntaxin is cleaved by botulinum toxin C. Moreover, most toxins cleave the synaptic proteins at different sites. Although tetanus toxin and botulinum neurotoxins act on the molecular level in a very similar manner, their actions differ on the anatomic level, a fact that causes completely different symptoms of intoxication. Whereas botulinum neurotoxins act on peripheral nerves and cause flaccid paralysis, tetanus toxin acts on the central nervous system and induces spastic paralysis (tetanus).

The delineation of the molecular mechanisms and the identification of the targets of the clostridial neurotoxins have promoted the use of these extremely potent agents as cell biological tools to study the involvement of their target proteins in exocytosis from neuronal cells or synaptosomes. Moreover, in combination with permeabilization methods (e.g., usage of pore-forming toxins), clostridial neurotoxins are now also used in studies on signal secretion coupling of nonneuronal cells (e.g., insulin-secreting cells). These cells are otherwise insensitive toward neurotoxins because they lack the specific membrane receptor. Notably, botulinum neurotoxins, which are not only the most potent toxins but also the most potent biologically active substances ever identified, are now used as therapeutic agents (Box 14.2).

Toxins as Tools for Permeabilization of Eukaryotic Cells

Pore-forming toxins are widely used as biological tools to permeabilize eukaryotic cells. The aim of this approach is to manipulate the intracellular ionic milieu, to introduce small molecules, which are otherwise membrane impermeable (e.g., nucleotides) into the cytosol, or even to transfer peptides and proteins such as antibodies or toxin fragments (e.g., the light chain of tetanus toxin) into the cell. Most often used are *S. aureus* α-toxin and streptolysin O from *Streptococcus pyogenes.* However, several other pore-forming toxins have been described (e.g., the RTX family, with *E. coli* hemolysin being one of the best-studied pore-forming toxins), which are of potential importance as cell biological tools. Note, Madden and coworkers (2001) proposed that cytolysins like streptolysin O can act functionally equivalent to type III secretion systems, suggesting that the transport properties of these

BOX 14.2

Toxins as therapeutic agents

Bacterial protein toxins are used to produce toxoids for vaccination. A most effective vaccination is that by tetanus toxoid, which might have prevented the deaths of hundreds of thousands of people. For other diseases, complex toxoid preparations were used, often containing several components (e.g., antipertussis vaccination), some of which might be responsible for the side effects of vaccination. Therefore, recombinant mutant toxins (e.g., PT), which are inactivated by site-directed mutagenesis, are clinically tested, with promising results.

For several years, botulinum neurotoxins, which are the most potent toxins known (in mice, the 50% lethal dose is 100 pg/kg), have been approved as therapeutic agents in a variety of ophthalmological and neurologic disorders. The first published report concerned the use of these toxins in strabismus, but their usage has been extended to many diseases, including blepharospasm, hemifacial spasm, and several types of dystonias like spasmodic dysphonia and cervical dystonia. Other potential applications are certain types of tremor, urinary retention, and anismus.

Toxin fusion proteins exploit the potent cytotoxic effects of bacterial protein toxins in efforts to kill cancer cells. Therefore, the enzymatic portion of a toxin and the toxin domain that causes membrane translocation are fused to a receptor-binding protein, which specifically binds to the target cell. When antibodies directed against a specific cell surface receptor are used as the receptor-binding domain of the fusion toxin, these toxins are called immunotoxins. The enzyme and translocation domain of diphtheria toxin and *Pseudomonas* exotoxin A are most often used in these fusion toxins. Problems with these immunotoxins are large size, instability, inadequate biological specificity, insufficient penetration into solid tumors, and antigenicity of the constructs. Another promising but still experimental approach is the use of these fusion toxins to purge stem cell preparations of cancer cells ex vivo before autologous transplantation.

---

toxins are not only relevant for their use as pharmacological tools but are also important for their function as virulence factors.

S. aureus α-toxin is a hydrophilic polypeptide of 293 amino acids, whose crystal structure was analyzed recently. The model derived from the crystallographic data shows a mushroom-shaped homo-oligomeric heptamer with a transmembrane domain composed of a 14-strand antiparallel β-barrel with a pore diameter of 2.6 nm. The pore formation caused by α-toxin is limited to the cell membrane because the toxin monomers (~33 kDa) are not able to pass through the pore.

Streptolysin O (~60 kDa) interacts with membrane cholesterol, oligomerizes, and forms much larger pores (up to 30 nm), which are permeable for large molecules (>150 kDa). Therefore, the toxin can also interact with internal membranes. To avoid damage to intracellular membranes, the binding of and pore formation by streptolysin O can be dissociated by a temperature shift. Thus, binding is carried out at low temperature (0°C). Thereafter, surplus streptolysin O is removed by washing and the permeabilization is initiated by increasing the incubation temperature to 37°C. Recently, it was shown that even reversible membrane permeabilization is possible with streptolysin O depending on calcium-calmodulin and intact microtubules and allowing the transport of up to 100-kDa proteins into the cytosol (Walev et al., 2001). Pore-forming toxins have been widely used in studies on exocytosis from various secretory cells, on calcium regulation, on signal-contraction coupling in smooth muscle cells, and on intracellular membrane trafficking. Researchers using pore-forming toxins must consider that these toxins induce a wide spectrum of biological effects, many of which are explained by changes in cellular ion fluxes and appear to be triggered by monovalent ion fluxes and by Ca^{2+} influx (e.g., secretion, generation of lipid mediators, and cytoskeletal redistribution). However, some of the effects of pore-forming toxins cannot be explained in this way. Examples of the latter effects are activation of G-protein signaling and proteolytic shedding of membrane proteins. All these effects must be considered when interpreting data obtained with pore-forming toxins as tools.

Finally, it should be mentioned that pore-forming toxins play an immense commercial role as pesticides against insects. These members of a large family of pore-forming toxins (Cry families) are produced by *Bacillus thuringiensis*. The various related toxins specifically kill insect larvae but are apparently not harmful to mammals. Therefore, spraying plants with spores of *B. thuringiensis* seems to be environmentally safe. Importantly, the various toxins are specific for certain insects only. Recently, the genes for toxins were introduced into some crop plants in an effort to protect them from insect attack.

Selected Readings

Ahnert-Hilger, G., I. Pahner, and M. Höltje. 2000. Pore-forming toxins as cell-biological and pharmacological tools, p. 557–575. *In* K. Aktories and I. Just (ed.), *Handbook of Experimental Pharmacology*. Springer, Heidelberg.

Reviews the methods to use pore-forming toxins and gives important hints for the bench work.

Aktories, K., M. Bärmann, L. Ohishi, S. Tsuyama, K. H. Jakobs, and E. Habermann. 1986. Botulinum C2 toxin ADP-ribosylates actin. *Nature* **322:**390–392.

The classical report on the action of C2 toxin.

Aronson, A. I., and Y. Shai. 2001. Why *Bacillus thuringiensis* insecticidal toxins are so effective: unique features of their mode of action. *FEMS Microbiol. Lett.* **195:**1–8.

Barth, H., F. Hofmann, C. Olenik, I. Just, and K. Aktories. 1998. The N-terminal part of the enzyme component (C2I) of the binary *Clostridium botulinum* C2 toxin interacts with the binding component C2II and functions as a carrier system for a Rho ADP-ribosylating C3-like fusion toxin. *Infect. Immun.* **66:**1364–1369.

Gives the construction and effects of C3 fusion toxin on the basis of C2 toxin.

Bishop, A. L., and A. Hall. 2000. Rho GTPases and their effector proteins. *Biochem. J.* **348:**241–255.

Comprehensive review on Rho GTPases and their effectors.

Cassel, D., and T. Pfeuffer. 1978. Mechanism of cholera toxin action: covalent modification of the guanyl nucleotide-binding protein of the adenylate cyclase system. *Proc. Natl. Acad. Sci. USA* **75:**2669–2673.

Classical work which identified G proteins as a target of cholera toxin.

Chardin, P., P. Boquet, P. Madaule, M. R. Popoff, E. J. Rubin, and D. M. Gill. 1989. The mammalian G protein rho C is ADP- ribosylated by Clostridium botulinum exoenzyme C3 and affects actin microfilament in Vero cells. *EMBO J.* **8:**1087–1092.

Flatau, G., E. Lemichez, M. Gauthier, P. Chardin, S. Paris, C. Fiorentini, and P. Boquet. 1997. Toxin-induced activation of the G protein p21 Rho by deamidation of glutamine. *Nature* **387:**729–733.

Excellent study on the identification of the molecular mechanism of CNF.

Fu, Y., and J. E. Galán. 1999. A *Salmonella* protein antagonizes Rac-1 and Cdc42 to mediate host-cell recovery after bacterial invasion. *Nature* **401:**293–297.

Just, I., J. Selzer, M. Wilm, C. Von Eichel-Streiber, M. Mann, and K. Aktories. 1995. Glucosylation of Rho proteins by *Clostridium difficile* toxin B. *Nature* **375:**500–503.

Gives the story on the identification of the molecular mechanism of toxin B. The usage of toxin B as a pharmacological tool is based on this study.

Kreitman, R. J. 1999. Immunotoxins in cancer therapy. *Curr. Opin. Immunol.* **11:**570–578.

Lehmann, M., A. Fournier, I. Selles-Navarro, P. Dergham, A. Sebock, N. Leclerc, G. Tigyi, and L. McKerracher. 1999. Inactivation of Rho signaling pathway promotes CNS axon regeneration. *J. Neurosci.* **19:**7537–7547.

Li, G., E. Rungger-Brändle, I. Just, J.-C. Jonas, K. Aktories, and C. B. Wollheim. 1994. Effect of disruption of actin filaments by *Clostridium botulinum* C2 toxin on insulin secretion in HIT-T15 cells and pancreatic islets. *Mol. Biol. Cell* **5:**1199–1213.

Madden, J. C., N. Ruiz, and M. Caparon. 2001. Cytolysin-mediated translocation (CMT): a functional equivalent type III secretion in gram-positive bacteria. *Cell* **104:**143–152.

Exciting work on a new hypothesis of the action of pore-forming toxins as pathophysiological transporters.

Masuda, M., L. Betancourt, T. Matsuzawa, T. Kashimoto, T. Takao, Y. Shimonishi, and Y. Horiguchi. 2000. Activation of Rho through a cross-link with polyamines catalyzed by *Bordetella* dermonecrotizing toxin. *EMBO J.* **19:**521–530.

Neves, S. R., T. R. Prahlad, and R. Iyengar. 2002. G protein pathways. *Science* **296:**1636–1639.

Gives a nice review on the complex signaling of G proteins. Many links and internet-based schemes for detailed information on the signaling of G protein-coupled receptors.

Nobes, C. D., and A. Hall. 1995. Rho, Rac, and Cdc42 GTPases regulate the assembly of multimolecular focal complexes associated with actin stress fibers, lamellipodia, and filopodia. *Cell* **81:**53–62.

Classical work on the role of Rho GTPases in regulation of the actin cytoskeleton.

Nürnberg, B. 1997. Pertussis toxin as a cell biological tool, p. 47–60. *In* K. Aktories (ed.), *Bacterial Toxins: Tools in Cell Biology and Pharmacology.* Chapman & Hall, Weinheim, Germany.

Very helpful review with many hints for the use of PT as a tool.

Popoff, M. R., O. E. Chaves, E. Lemichez, C. Von Eichel-Streiber, M. Thelestam, P. Chardin, D. Cussac, P. Chavrier, G. Flatau, M. Giry, J. Gunzburg, and P. Boquet. 1996. Ras, Rap, and Rac small GTP-binding proteins are targets for Clostridium sordellii lethal toxin glucosylation. *J. Biol. Chem.* **271:**10217–10224.

Prepens, U., I. Just, C. Von Eichel-Streiber, and K. Aktories. 1996. Inhibition of FcεRI-mediated activation of rat basophilic leukemia cells by *Clostridium difficile* toxin B (monoglucosyltransferase). *J. Biol. Chem.* **271:**7324–7329.

Ridley, A. J., and A. Hall. 1992. The small GTP-binding protein rho regulates the assembly of focal adhesions and actin stress fibers in response to growth factors. *Cell* **70:**389–399.

Classical work which elucidated the role of Rho GTPases in actin cytoskeleton regulation by using C3.

Rozengurt, E., T. Higgins, N. Chanter, A. J. Lax, and J. M. Staddon. 1990. *Pasteurella multocida* toxin: potent mitogen for cultured fibroblasts. *Proc. Natl. Acad. Sci. USA* **87:**123–127.

Sadoul, K., J. Lang, C. Montecucco, U. Weller, R. Regazzi, S. Catsicas, C. B. Wollheim, and P. A. Halban. 1995. SNAP-25 is expressed in islets of Langerhans and is involved in insulin release. *J. Cell Biol.* **128:**1019–1028.

Schiavo, G., F. Benfenati, B. Poulain, O. Rossetto, P. Polverino de Laureto, B. R. DasGupta, and C. Montecucco. 1992. Tetanus and botulinum-B neurotoxins block neurotransmitter release by proteolytic cleavage of synaptobrevin. *Nature* **359:**832–835.

Schiavo, G., M. Matteoli, and C. Montecucco. 2000. Neurotoxins affecting neuroexocytosis. *Physiol. Rev.* **80:**717–766.

Excellent review on the action of neurotoxins with important implication for their usage as tools and drugs.

Schmidt, G., P. Sehr, M. Wilm, J. Selzer, M. Mann, and K. Aktories. 1997. Gln63 of Rho is deamidated by *Escherichia coli* cytotoxic necrotizing factor 1. *Nature* **387:**725–729.

Van Aelst, L., and C. D'Souza-Schorey. 1997. Rho GTPases and signaling networks. *Genes Dev.* **11:**2295–2322.

Van den Akker, F., F. A. Merritt, and W. G. J. Hol. 2000. Structure and function of cholera toxin and related enterotoxins, p. 109–131. *In* K. Aktories and I. Just (ed.), *Bacterial Protein Toxins.* Springer, Berlin.

van der Goot, F. G. (ed.) 2001. Pore-forming toxins. *Curr. Top. Microbiol. Immunol.* **257:**1–166.

A recent review booklet on pore-forming toxins with excellent chapters on various toxins.

Wahl, S., H. Barth, T. Ciossek, K. Aktories, and B. K. Mueller. 2000. Ephrin-A5 induces collapse of growth cones by activating Rho and Rho kinase. *J. Cell Biol.* **149:** 263–270.

Walev, I., S. C. Bhakdi, F. Hofmann, N. Djouder, A. Valeva, K. Aktories, and S. Bhakdi. 2001. Delivery of proteins into living cells by reversible membrane permeabilization with streptolysin-O. *Proc. Natl. Acad. Sci. USA* **98:**3185–3190.

Wilde, C., and K. Aktories. 2001. The Rho-ADP-ribosylating C3 exoenzyme from Clostridium botulinum and related C3-like transferases. *Toxicon* **39:**1647–1660.

Zywietz, A., A. Gohla, M. Schmelz, G. Schultz, and S. Offermanns. 2001. Pleiotropic effects of *Pasteurella multocida* toxin are mediated by Gq-dependent and -independent mechanisms. Involvement of G_q but not G_{11}. *J. Biol. Chem.* **276:**3840–3845.

Important contribution to the target of *Pasteurella multocida* toxin showing the high specificity for G_q as compared with the very similar G_{11} protein, however, also indicating other targets than G_q.

15

Type III Secretion Systems in Animal- and Plant-Interacting Bacteria

MATTHEW S. FRANCIS, KURT SCHESSER, ÅKE FORSBERG, AND HANS WOLF-WATZ

A multitude of activities performed by bacteria, as well as eukaryotic cells, depend on the transport of proteins through membranes. In bacteria, these activities include the synthesis of extracellular organelles such as flagella and the delivery of toxins to their site of action. Selectively transporting a protein through a lipid bilayer is not a trivial task, considering the relative physical dimensions of proteins compared with small solutes such as water and sodium ions. This task is further complicated by the fact that most proteins predestined for transport through a membrane must be prevented from assuming their three-dimensional structure prior to transport. This chapter discusses the type III secretion system (TTSS) possessed by several gram-negative bacteria that live, for at least part of their life cycle, in close association with eukaryotic cells. What makes type III secretion unique is that it functions to inject (also termed translocate) bacterial proteins not just across the entire bacterial envelope (consisting of the inner and outer membranes) but also across the eukaryotic membrane. In this context, secretion is referred to as the transport of proteins across the bacterial envelope, while injection/translocation refers to the passage of proteins through the eukaryotic cell plasma membrane and then entry into the cytosol.

Gram-negative bacteria possess several protein secretion systems (Figure 15.1). The best-studied systems at the mechanistic level are the *sec*-dependent type II and V secretion systems, which likely represent evolutionarily ancient systems since some components also have homologues in gram-positive bacteria, the archaebacteria, and the eukaryotes. Proteins secreted by these systems possess N-terminal signal sequences of 16 to 26 residues (the signal sequences in gram-positive bacteria can be a little longer) that consist of a basic N-terminal domain, a hydrophobic central core segment, and a distal domain that contains a cleavage site in which the signal sequence is removed during the transport across the inner membrane. The basic and hydrophobic domains are essential for the recognition event between the protein to be transported and components of the secretion complex located on the cytoplasmic face of the inner membrane. Following this initial recognition event, this protein is transported across the inner membrane by

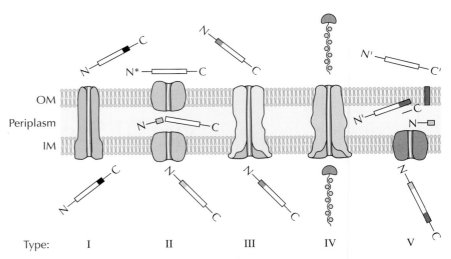

OM

Periplasm

IM

Type: I   II   III   IV   V

Figure 15.1   Schematic diagram of bacterial type I to V secretion systems. Recognition sequences that target the proteins to the respective secretion complexes are indicated either in light or dark gold (for N-terminal signal sequences of type II and IV and type III systems, respectively) or grey and black (for C-terminal signal sequences of type V and type I systems, respectively). Proteins secreted by the type I and III pathways traverse the inner membrane (IM) and outer membrane (OM) in one step, whereas proteins secreted by the type II and V pathways cross the inner membrane and outer membrane in separate steps. The N-terminal signal sequences of proteins secreted by the type II and V systems are enzymatically removed upon crossing the inner membrane, in contrast to proteins secreted by the type I and III systems, which are exported intact. Although proteins secreted by the type II and V systems are similar in the mechanism by which they cross the inner membrane, differences exist in how they traverse the outer membrane. Proteins secreted by the type II system are transported across the outer membrane by a multiprotein complex, whereas those secreted by the type V system autotransport across the outer membrane by virtue of a C-terminal sequence which is enzymatically removed upon release of the protein from the outer membrane. Type IV pathways secrete either polypeptide toxins (directed against eukaryotic cells) or protein-DNA complexes between either two bacterial cells or a bacterial and eukaryotic cell. Shown in the figure is the protein-DNA complex delivered by *Agrobacterium tumefaciens* into a plant cell that consists of a 20-kb single-stranded DNA molecule and the VirD2 (grey) and VirE2 (gold) proteins.

a mechanism that has been extensively studied and is fairly well understood. After transport across the inner membrane, a protein may remain in the periplasm, integrate in the outer membrane, or continue its journey across the outer membrane. Proteins secreted by the type II and V pathways differ from each other in how they traverse the outer membrane. Proteins secreted by the type V secretion system, which include the immunoglobulin A proteases of *Neisseria gonorrhoeae* and the *Helicobacter pylori* vacuolating cytotoxin, mediate their own secretion ("autotransport") across the outer membrane. In contrast, proteins secreted by the type II secretion system, such as the extracellular degradative enzyme pullulanase of *Klebsiella oxytoca*, require additional proteins for their secretion across the outer membrane.

Proteins secreted by the type I secretion system are mainly toxins such as the hemolysin of *Escherichia coli* and the cyclolysin produced by *Bordetella pertussis*. Unlike proteins secreted by the type II secretion system, these proteins contain no signal sequence at their N termini but instead contain domains at their C termini that are necessary for recognition by the type I secretion complex. This complex is relatively simple, consisting of only three proteins: an ATPase that provides the energy for transport as well as forms a pore through the inner membrane, a protein that spans the periplasmic space, and a protein located in the outer membrane. Both the nature of the type I secretion signal and the transport mechanism are currently unclear. Furthermore, it is unknown why gram-negative bacteria evolved a dedicated system for the secretion of these toxins.

Type IV secretion systems in gram-negative bacteria are specialized for the cell-to-cell transfer of a variety of biomolecules (see chapter 17 for details). This secretion system is used for the transfer of protein-DNA complexes either between two bacterial cells (e.g., during bacterial conjugation) or between a bacterial and eukaryotic cell (e.g., *Agrobacterium tumefaciens*-mediated DNA transfer into plant cells). Additionally, *B. pertussis* and *H. pylori* possess type IV secretion systems that mediate the secretion of pertussis toxin and interleukin-8-inducing factor, respectively. Approximately 10 *A. tumefaciens* proteins are required to transfer a single-stranded 20-kb DNA molecule (referred to as the T complex) into plant cells. Based on several different types of experiments, these proteins are thought to form a transmembrane pore spanning the bacterial envelope (and perhaps even the plant membrane as well). The structural basis of the type IV secretion system that allows it to secrete polynucleotide and polypeptide molecules is not known.

Several gram-negative bacteria that interact with either plant or animal cells possess a secretion system, referred to as the type III secretion system (TTSS), that functions to translocate proteins directly from the bacterial cell into the cytosol of the eukaryotic cell. This secretion system was first described and characterized in *Yersinia* spp. that are pathogenic for humans. In the mid-1950s it was noted that the ability of *Y. pestis*, the causative agent of bubonic plague, to grow at 37°C in media lacking calcium correlated with a loss of virulence. Subsequently, it was found that virulence depended on the presence of a 70.5-kb extrachromosomal plasmid (pCD1). This plasmid in turn was shown to direct the massive secretion of about 10 proteins into the culture supernatant by bacteria incubated under the "nongrowing" conditions (i.e., 37°C in media lacking calcium). Genetic analysis of the 70.5-kb virulence plasmid revealed that it encoded both the proteins that were secreted and approximately 20 proteins that comprised the secretion complex (Figure 15.2). For a long time the function of these secreted proteins was unknown. Almost all of them were shown to be required for full virulence in the mouse model system, but in soluble form they had no observable effects on cultured eukaryotic cells or whole animals. However, in the mid-1990s it was demonstrated that these secreted proteins are in fact injected by bacteria directly into the eukaryotic cytosol, from which they exert their effects. These observations instigated a terminology shift: the type III secreted proteins are now commonly referred to as type III "effector proteins," which distinguishes them from the proteins that comprise the type III secreton. Since the type III effector proteins pass directly from the bacterium into the eukaryotic cell, the "secretion" of these proteins by bacteria grown in pure culture could be artifactual.

It also turned out that TTSSs were involved in another phenomenon that was first described in the mid-1950s: the gene-for-gene hypothesis underlying the plant hypersensitivity response toward bacterial pathogens (see Box 15.2). A major breakthrough in the study of host-parasite interactions came with the discovery that several gram-negative bacteria that infect plants possess secretion systems that are clearly evolutionarily related to the TTSSs possessed by bacteria that infect animals. This key finding

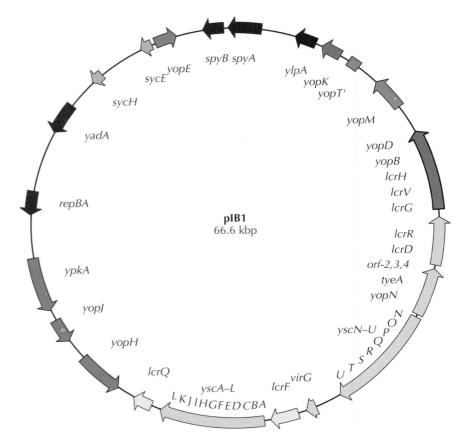

Figure 15.2 The TTSS of *Y. pseudotuberculosis* is encoded on the virulence plasmid pIB1. The genes encoding components of the type III secretion system and those involved in plasmid replication (*repAB*) and plasmid partitioning (*spyAB*) are depicted in the map as arrows showing the direction of transcription. A nonfunctional, truncated gene has no arrowhead and is indicated by an apostrophe to denote the truncation. Genes in light grey (*yscA* to *yscU*, *virG*, *lcrDR*, *yopN*, and *tyeA*) encode either components of the secretion apparatus or those involved in controlling protein secretion. The genes shown in dark grey (*lcrGVH yopBD* and *yopK*) encode proteins involved in controlling translocation of effectors into target cells. The genes encoding the secreted effector proteins (*yopE, yopH, yopJ, yopM, yopT´*, and *ypkA*) are shown in dark gold. The genes encoding the chaperones required for efficient secretion of effector Yops (*sycE* and *sycH*) are shown in light gold, and important regulatory genes are depicted in soft gold. Note that some proteins may have dual functions. For example, several proteins encoded by the *lcrGVH yopBD* operon are involved in both controlling translocation and regulating the system. Moreover, TyeA is involved in control of Yop secretion but is also required for translocation of the YopE and YopH effector molecules. The DNA sequence of pIB1 on which this map is based is unpublished data provided by Peter Cherepanov and Thomas Svensson, Department of Microbiology, FOI NBC-Defence, Umeå, Sweden.

immediately suggested that plant- and animal-infecting bacteria have common strategies at the cellular and molecular levels in how they deal with eukaryotic cells. This view has recently been strengthened by observations that, in addition to the structural similarities of the type III secreton, some of the type III effector proteins of animal- and plant-interacting bacteria possess similar biochemical activity (see Box 15.2).

In the past few years TTSSs have been discovered in numerous genera of bacterial animal and plant pathogens (Table 15.1). Somewhat unexpect-

Table 15.1 Type III secretion systems of prominent animal- and plant-interacting bacteria

Organism	Location	Type III system	Typical substrates	Function in virulence
Animal-interacting bacteria				
Yersinia spp.	Virulence plasmid	Ysc (*Yersinia* secretion)	Yop (*Yersinia* outer proteins)	Blocks phagocytosis and induction of cytokine expression, induces apoptosis
Pseudomonas aeruginosa	Chromosome	Psc (*Pseudomonas* secretion)	Exo (exoenzyme toxins) Pop (*Pseudomonas* outer proteins)	Blocks phagocytosis, induces cytotoxicity
Shigella flexneri	Virulence plasmid	Spa/Mxi (surface presentation of Ipa-antigens/membrane expression of Ipa)	Ipa (Invasion protein antigens)	Induces uptake into eukaryotic cells, apoptosis, and vacuole membrane lysis
Salmonella serovar Typhimurium, *Salmonella* serovar Dublin	Chromosome (SPI-1)[a]	Inv/Spa (invasion/see *Shigella*)	Sip (Secreted invasion proteins) Sop (*Salmonella* outer proteins)	Induces uptake into epithelial cells, apoptosis, cytokine expression, and fluid secretion in the intestine
Salmonella serovar Typhimurium, *Salmonella* serovar Dublin	Chromosome (SPI-2)[a]	Ssa (secretion system apparatus)	Sse (Secretion system effectors) Sif (*Salmonella* induced filament) Ssp (*Salmonella* secreted proteins) SpiC (?), PipB (?)	Essential for virulence and intracellular replication
EPEC[b]	Chromosome (LEE)[c]	Sep/Esc (secretion of *E. coli* proteins/*E. coli* secretion proteins)	Esp (*E. coli* secreted proteins) Tir (transmembrane intimin receptor)	Induces tight adherance between the bacterium and its target cell
Chlamydia spp.	Chromosome	Cds/Sct (contact-dependent protein secretion/secretion and cellular translocation)[d]	Cop (*Chlamydia* outer proteins) Inc (Inclusions)?	Survival and replication in intracellular inclusions
Plant-interacting bacteria **Pathogens**				
Erwinia amylovara	Chromosome or plasmid[e]	Hrc (Hrp-conserved)[f]	Harpin	Virulence-associated proteins
Ralstonia solanacearum *Pseudomonas syringae* *Xanthomonas campestris*			Avr (avirulence proteins)	Elicits hypersensitive response in resistant plants
Symbionts				
Rhizobium spp.	Plasmid	Rhc (*Rhizobium* conserved)	Nop (nodulation outer proteins)	Induces nodule formation in legumes

[a] SPI-1 and SPI-2, *Salmonella* pathogenicity islands 1 and 2.
[b] EPEC, enteropathogenic *Escherichia coli*.
[c] LEE, locus of enterocyte effacement.
[d] Sct is the recommended unified nomenclature for type III systems.
[e] *Ralstonia solanacearum*.
[f] Hrp, hypersensitive reaction and pathogenicity.

edly, TTSSs have also been found in bacteria that peacefully coexist with multicellular organisms. The nitrogen-fixing plant symbiont *Rhizobium* uses a TTSS to secrete effector proteins involved in the formation and maintenance of root nodules (where nitrogen fixation occurs). Perhaps it should not come as a surprise that *Rhizobium* would possess a TTSS, since in many respects the way in which *Rhizobium* induces legumes to form nodules resembles an infection process. In addition to *Rhizobium*, TTSSs have also been described in *Sodalis glossinidius* and *Photorhabdus luminescens*, endosymbionts of the tsetse fly and nematodes, respectively. Although it remains to be determined how these various TTSSs are utilized within the context of a mutualistic bacterium-host relationship, it is tempting to speculate that these few examples represent just the tip of the iceberg of the extent of this form of interkingdom communication.

Among the described bacterial species possessing TTSSs, there is a high degree of similarity between the various TTSSs both in terms of genetic organization and at the level of the individual genes. The genes encoding the TTSS are clustered in blocks that largely display a conserved genetic order, suggesting that these genes have spread in toto by horizontal gene transfer. This hypothesis is further strengthened by the fact that in many cases these gene clusters either are contained on plasmids (for example, see Figure 15.2) or, for chromosomally located gene clusters, are flanked by insertion elements (the latter are usually referred to as pathogenicity islands). Among the individual genes themselves, which encode the type III secretion apparatus, some are found in virtually all species while others are more restricted in their distribution. Thus, TTSSs from different species can be grouped into subfamilies based on sequence similarities and genetic organization.

Divergence of type III gene sequences between species could be due, at least in part, to different selection pressures that are imposed on the bacteria during their coevolution with their respective hosts. In this scenario, a particular bacterial species would first acquire a type III-like secretion system and then modify that system, through natural selection, to fit its specific needs. These needs would be shaped by numerous factors, such as the host defense system and the strategies used by the bacterium to grow at the expense of the host organism. The intestinal pathogen *Salmonella* provides an illustrative example of the adaptation of type III gene clusters. This pathogen possesses two entirely separate TTSSs, encoded by different gene clusters, which, based on genetic evidence, apparently were acquired at different times during evolution. These two systems are required for different stages of the *Salmonella* infection process. One system secretes proteins involved in the invasion of eukaryotic cells, while the other system is required for the bacterium to survive once it is inside the eukaryotic cell.

Unlike the clustered blocks of genes that encode the TTSS, most, but not all, of the secreted effector proteins are genetically unlinked. This suggests that type III effector-encoding genes can be acquired or lost independently of the genes encoding the type III secretion complex. This may enhance the bacteria to quickly adapt to host countermeasures and/or to an entirely new host. The bacterial adaptation process may even have involved acquiring (or "capturing") genes from the host organism and modifying those genes through natural selection into those encoding type III effector proteins (see Box 15.1). Under this scenario, such type III effectors are in fact the products of two evolutionary processes: one occurring within the eukaryotic host followed by refinement within the bacterium to streamline each individual virulence strategy.

BOX 15.1

"Eukaryotic-like" type III effectors

Several of the type III effectors "look" like eukaryotic proteins. This is most obvious when one examines their sequences; several type III effectors contain domains that bear sequence similarities to eukaryotic-like phosphotases, kinases, GTPases, proteases, and cellular localization signals. Although several type III effectors have been physically demonstrated to be injected into eukaryotic cells (see the text), the presence of these eukaryotic-like sequences supports the idea that these effector proteins are in fact active within the host cell. In several cases tested so far, type III effectors that have been experimentally mutated in their eukaryotic-like sequences have lost their biological activity. For example, the type III effector AvrBs3 of the plant pathogen *Xanthomonas campestris* contains a nuclear localization sequence that, following its injection into the host cell by the *Xanthomonas* type III secretion system, is recognized by the plant cell machinery that transports proteins from the cytoplasm to

the nucleus. The animal pathogens *Yersinia* and *Salmonella* each encode type III effectors, YopH and SptP, respectively, that are protein tyrosine phosphotases (PTPases) whose catalytic domains contain invariant residues present in all eukaryotic PTPases. *Yersinia* mutant strains containing a catalytically inactive YopH-encoding gene are no longer virulent in the mouse model, indicating that tyrosine dephosphorylation of host proteins plays an important role in the virulence of *Yersinia*.

How did these bacterial species come to possess type III effector proteins with eukaryotic-like domains? There are at least two possibilities. These effectors could have evolved from "normal" bacterial proteins through a process of convergent evolution. Alternatively, these animal- and plant-interacting bacteria could have acquired (or "captured") genetic material from their host and modified it to encode our present-day type III effectors (this being an example of divergent evolution). Recently the three-dimensional structures of a number of type III effector proteins have been determined and have shed light as to

their possible origins. A particularly interesting structure is that of SptP, which actually comprises two independent modules. In addition to a PTPase domain-containing module noted above, SptP also has a module that possesses a biochemical activity (GTPase) toward host regulatory proteins. Due to its structural similarity to eukaryotic PTPases, it is thought that the PTPase portion of SptP divergently evolved from a "captured" host PTPase. If this is true, this module of SptP can then be thought of as the product of two sequential and opposing lines of evolution; the first occurring within and serving the host followed by selection occurring within *Salmonella* and serving the needs of the bacterium. This contrasts to the GTPase portion of SptP, which structurally differs substantially when compared with host GTPases, suggesting that this portion of SptP convergently evolved from a preexisting bacterial protein. As more structural, biochemical, and cellular data of type III effectors become available, we should be able to get a better idea as to how bacterial pathogens have evolved these proteins to manipulate the host cell.

It is becoming increasingly clear that several of the proteins secreted by the TTSS inactivate or at least retard the host response. Apparently, the injected proteins target key processes within eukaryotic cells that are normally activated upon exposure to gram-negative bacteria. The working hypothesis is that the type III effector proteins have evolved to exploit chinks in the eukaryotic host defense armor. A substantial amount of research is currently focused on identifying the eukaryotic processes that are targeted by the type III effector proteins. By determining how these bacterial proteins function within eukaryotic cells, much will be learned about animal and plant defense systems.

Type III Secretion Systems

TTSSs of animal and plant pathogens and of the flagellum-specific secretion pathway have a number of features in common. Since flagellar motility is an ancient system that probably existed before the divergence of archaebacteria and prokaryotes, it is likely that TTSSs evolved from the flagellum-specific secretion pathway as a means of secreting and translocating antihost factors into host cells. Since flagellar motility was probably present in most free-living bacteria, this could explain why the TTSSs are so widespread between gram-negative animal and plant pathogens. The secretion and assembly of flagellar components have been reasonably well characterized. The flagellum-specific secretion pathway mediates the secretion of all proteins that assemble into a continuous axial structure, from proximal rod to hook to filament and, finally, to the filamentous cap. A similar picture is emerging for the assembly of the so-called needle complex of the TTSS of animal pathogens. Secretion and assembly of the external needle only occurs once the bacterial envelope-spanning components, including the outer membrane-associated parts of the needle complex, are assembled. Another feature common to both systems is that protein secretion appears to occur by a continuous process without any detectable periplasmic protein intermediates. Flagellar components are assembled at the distal end of the nascent structure, with evidence suggesting that assembly is mediated by protein traffic through a channel within the growing structure. A central channel is observed in the filament and most probably also in the rod. The needle components of needle complexes isolated from animal pathogens also appear as hollow structures, suggesting that all TTSSs serve as conduits in protein secretion. Moreover, neither the flagellar proteins secreted during biosynthesis nor the virulence proteins secreted by a type III-dependent mechanism are processed during secretion. In both systems, the recognition sequence targeting proteins for secretion resides in the N terminus of the individual proteins, but no common motif has been identified.

The proteins required for flagellar secretion and assembly localize mainly to the cytoplasmic membrane or remain within the cytoplasm. This suggests a mechanism where the secretion apparatus identifies and targets proteins to the secretion channel in the proximal rod and provides the required energy to transport the proteins out to the bacterial surface. Since homologues to the proteins that constitute the flagellum-specific secretion pathway are present in all TTSSs so far identified, it is considered that secretion of virulence effectors by type III pathways occurs by an analogous mechanism. Additional proteins unique to subfamilies of TTSSs are also known. These proteins may collectively participate to form a rod-like structure extending through the periplasm to the pore in the outer membrane. Proteins destined for secretion are identified and then targeted to the secretion channel to permit secretion. The type III secreton also provides the energy required to drive the channeling of proteins to the exterior. In addition, the channel is gated to prevent accidental leakage through this structure.

Secreton Assembly

Flagella are extracellular appendages that mediate bacterial motility. A flagellum is composed of three structural components: a thin helical filament of variable length, a hook, and a basal body. The basal body is embedded in the bacterial envelope and serves to anchor the flagellum filament to the bacterial surface via the hook protein linkage (a probable universal joint). The basal body consists of a central continuous rod surrounded by several ring structures (Figure 15.3). The L ring (for lipopolysaccharide) resides in the outer membrane, and the P ring (for peptidoglycan) is embedded in the peptidoglycan layer. The MS ring (for membrane-supramembrane) is situated in the inner membrane, and the bell-shaped C ring (for cytoplasm), harboring the motor/switch proteins (FliG, FliM, and FliN), is associated with the MS ring in the inner membrane but projects into the cytoplasm. Another component of the flagellar biosynthetic machinery consists of a family of proteins located in either the inner membrane or the cytoplasm, which collectively

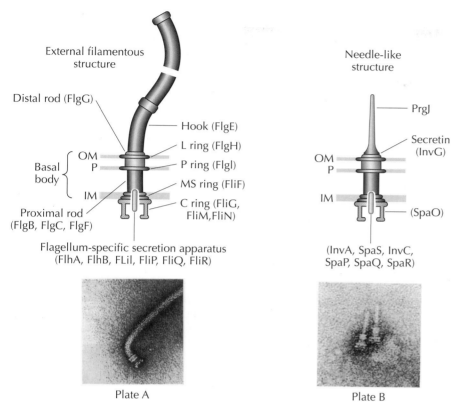

Figure 15.3   Structural comparison between the enteric bacterial flagellum and the *Salmonella* needle complex. Three major components of the flagellum structure are shown: external filamentous structure, basal body, and flagellum-specific secretion apparatus. The structure of the *Salmonella* TTSS spanning the bacterial envelope is represented like the basal body, as deduced from genetic, electron microscopic, and scanning electron microscopic studies (see the micrograph inset of a purified flagellum [plate A] and the *Salmonella* SPI-1 needle complex [plate B]). Nevertheless, these structures are shown in black to indicate limited sequence similarity among the corresponding protein components. Conversely, some proteins of the needle complex that have extensive similarities to components of the flagellum-specific secretion apparatus, located within the bell-shaped C ring (proteins indicated in parentheses; see Table 15.2), are shown in light grey. Plate A is shown with copyright permission from ASM Press, and plate B is reprinted from Kubori et al., *Science* **280**:602–605, 1998, with permission of AAAS. OM, outer membrane; P, peptidoglycan; IM, inner membrane.

function as the flagellum-specific type III secretion apparatus responsible for the sequential secretion of proteins forming the rod, hook, and filaments.

Interestingly, for several TTSSs, structures resembling basal bodies have been isolated. The best characterized is that encoded by SPI-1 (for *Salmonella* pathogenicity island 1) of *Salmonella* (Figure 15.3), but structures of similar dimensions have also been isolated from *Shigella* and enteropathogenic *E. coli* (EPEC). In particular, the part embedded in the bacterial envelope is physically similar to the ring structures in the inner and outer membranes of the basal body. The proteins constituting the MS and C rings in the basal body have been identified, but these display only limited similarity to particular type III secretion proteins of pathogenic bacteria. Moreover, a significant feature of the *Salmonella* type III secreton is that instead of the extracellular flagellar appendage, a hollow needle-like structure encoded by PrgJ (80 nm long and 13 nm wide) extends from the bacterial surface. Similar extensions composed of *prgJ* homologues were also isolated from *Shigella* and EPEC. Interestingly, *prgJ* homologues exist in TTSS of all animal pathogens. This fact, together with the similar physical observations from studies on *Salmonella, Shigella,* and EPEC, indicates that other animal pathogens will have an analogous type III secreton composed of a basal-body-like structure in association with an extracellular needle protrusion. As such, this secreton has also been denoted the "needle complex."

The assembly of the needle complex encoded by SPI-1 of *Salmonella* has been studied in some detail. First, PrgH is inserted in the inner membrane to form the inner-ring structures. Thereafter, a *sec*-dependent process assembles the putative periplasmic spanning protein PrgK and the outer-ring structure formed by invG/InvH. Once the ring structures are assembled, the proteins encoding the secretion system InvA, InvC, SpaO, SpaP, SpaQ, SpaR, and SpaS (see discussion below and Table 15.2) are located in or around the inner membrane presumably linking with the cytoplasmic face of the PrgH-encoded inner ring. Only after completion of this structure is the major needle component, PrgJ, finally exported and assembled at the bacterial surface by a *sec*-independent process. The needle part of the structure is required for functional TTSS and may function to bridge the gap between the bacterium and the host cell to establish translocation of virulence proteins across the host cell plasma membrane. In support of this, the secreted EspA protein required for translocation of effector proteins by EPEC forms a filamentous structure that appears to link the needle appendage of

Table 15.2   Proteins common to the flagellum-specific secretion apparatus and the *Salmonella* type III secreton[a]

SPI-1	Flagella	Location	Function/structure
InvC	FliI	Cytoplasm	F_0F_1 proton-translocating ATPase: energizer of flagellar secretion and assembly
SpaO	FliN/Y[b]	Cytoplasmic face of the inner membrane	Probable component of the C ring
SpaP	FliP	Inner membrane	Secretion apparatus
SpaQ	FliQ	Inner membrane	Secretion apparatus
SpaR	FliR	Inner membrane	Secretion apparatus
SpaS	FlhB	Inner membrane	Secretion apparatus, suppresses formation of hook multimers
InvA	FlhA	Inner membrane	Secretion apparatus, forms a channel?
PrgK	FliF	Transmembrane/ periplasmic spanning	Structural protein (basal body)
InvJ[c]	FliK	Exported	Sensor (substrate switch mechanism)

[a] While components of the SPI-1 type III secretion system of *Salmonella* are illustrated, all type III secretion systems encoded by animal- and plant-interacting bacteria contain proteins with extensive similarity to these components of the flagellum-specific secretion apparatus.
[b] FliN (*Escherichia coli* and *Salmonella* serovar Typhimurium); FliY (*Bacillus subtilis*).
[c] Although sharing only limited genetic similarity, members of the InvJ family are clearly functionally related to FliK.

the needle complex to the infected eukaryotic cell (see Figure 15.11). Presumably, this allows effector proteins to be channeled via the needle complex directly into the host cell. For the other animal pathogens, however, it has not yet been possible to physically link the secreted proteins essential for effector translocation to the needle complex.

By analogy, there is a substantial amount of indirect evidence suggesting that proteins secreted by TTSSs of plant-interacting bacteria, like bacterium-animal cell systems, are injected directly into the plant cell cytoplasm. However, no *prgJ* homologue encoding a needle component has been identified in the gene clusters encoding plant TTSSs. However, the Hrp (for "hypersensitive reaction and pathogenicity") TTSS of the plant pathogen *Pseudomonas syringae* produces a novel pilus-like surface appendage, termed Hrp pilus. The Hrp pilus has a diameter of 8 nm and is at least 2 μm long. It is likely that this structure has evolved to mediate the penetration of plant cells through pores that naturally exist in the usually impermeable polysaccharide cell wall that surrounds individual cells. (The needle structures of animal pathogens are not adapted for this task.) This would enable the direct injection of Avr (for "avirulence") (Box 15.2) proteins into the plant cy-

---

BOX 15.2

Type III secretion system and plant defense responses

Like animals, plants are under constant attack by microbial pathogens and have elaborate defensive systems to prevent viral, bacterial, and fungal pathogens from establishing infections. Two possible outcomes can occur when a bacterial pathogen enters a plant through a stomata or wound. In a so-called "susceptible" plant host, the bacterial pathogen is able to colonize the plant and cause disease. If, on the other hand, the bacterium is unfortunate enough to find itself within a "resistant" plant host, it is rapidly recognized and killed by any one of a number of plant defense responses. These responses include the hypersensitivity response, which entails a programmed cell death localized to the area of the plant surrounding the initial site of bacterial penetration, that serves to prevent the bacterium from spreading and establishing a systemic infection. (Hypersensitivity responses are commonly observed as brown spots on the leaves of houseplants.)

What is the basis for the recognition of the bacterium by the plant, and why is the recognition often restricted to specific plant species or lines? For 50 years it has been known that the recognition event requires the expression of complementary plant and bacterial genes designated the resistance (*R*) and avirulence (*avr*) genes, respectively. This phenomenon is commonly referred to as the "gene-for-gene" hypothesis. If either the plant possesses a mutated *R* gene (i.e., an *r* allele) or the bacterium lacks the corresponding *avr* gene, the plant does not recognize the bacterium, and consequently the bacterium is able to spread and cause disease. Thus, the term "avirulence" refers to the fact that if the *avr* gene (or a specific allele) expressed by the bacterium is matched with its plant *R* gene counterpart, the bacterium is rendered avirulent due to activation of the plant defense responses. Several indirect lines of evidence suggest that bacterial plant pathogens, analogous to bacterial animal pathogens, inject Avr proteins directly into plant cells via a type III secretion system. It was once thought that Avr and R proteins physically interacted with one another, but recent evidence suggests that R proteins in fact serve as "guards" for still other plant proteins that are involved in plant defense responses. The latter are believed to be the targets of the injected Avr proteins and a hypersensitivity response is induced if a cell senses, through the R protein guard, that these defense proteins have been targeted by the Avr proteins.

Recently it has been found that plant and animal pathogens encode type III effectors with similar biochemical activities. For example, the animal pathogen *Yersinia pseudotuberculosis* encodes a type III effector, YopT, which possesses a cysteine protease activity. The plant pathogen *Pseudomonas syringae* also encodes a type III effector, AvrPphB, possessing cysteine protease activity. Inactivating the protease activity of either YopT or AvrPphB abolishes their biological activities (entailing either disrupting the eukaryotic cytoskeleton or eliciting a hypersensitivity response, respectively). These and other examples, along with the fact that plant R proteins have been found to be structurally and functionally related to proteins involved in immune signaling in animal cells, have bolstered the view that animal- and plant-interacting bacteria face similar challenges in coping with, and in some cases prevailing over, their respective host's defensive systems.

A **B** **C**

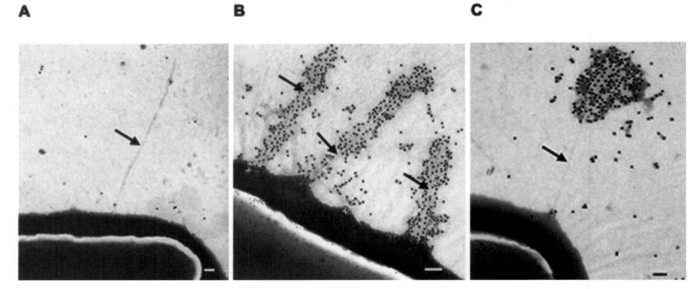

Figure 15.4 Secretion of effector proteins occurs through the Hrp pilus of the plant pathogen *P. syringae*. A strain of *P. syringae* was engineered so that Hrp pilus assembly was uncoupled from effector synthesis by placing the *avrPto* effector gene under the control of a heterologous inducible promoter. Shown are electron microscopy images of recombinant *P. syringae* grown in Hrp-inducing media with no *avrPto* induction (**A**), simultaneous induction of AvrPto synthesis (**B**), and AvrPto synthesis induced long after Hrp pilus assembly (**C**). Arrows indicate the Hrp pili extending from the bacterial surface. Secreted AvrPto is detected by a specific immunogold labeling technique that is reflected by the appearance of black spots on the micrographs. When *avrPto* was coinduced with Hrp-pili, secreted effector was uniformly localized along the entire structure (**B**). However, when pili were allowed to assemble prior to induction of AvrPto synthesis, secreted effector was routinely localized at the tip portion of pili (**C**). This strongly supports a conduit model whereby effectors travel through a hollow pilus and are secreted at the tip. Modified from Figure 2 in Jin et al., *Science* **294:**2556–2558, 2001, with permission of AAAS and provided by Sheng-Yang He, Michigan State University, East Lansing.

tosol. Injection of Avr proteins into plant cells through such structures remains to be shown. However, using an elegant pulse-labeling technique combined with electron microscopy, secretion of the effector protein AvrPto through the Hrp pilus of *P. syringae* was demonstrated (Figure 15.4). In particular, the continuously expressed Avr effector protein localized along the entire length of the pilus, while newly synthesized protein localized only at the distal part of the pilus. This finding provides strong evidence for a common conduit mechanism for type III secretion.

Homologous Components

While there is no significant similarity among proteins comprising the flagellar basal body and TTSSs, some proteins that constitute the flagellum-specific type III secretion apparatus show similarity to particular proteins of TTSSs. The properties of these proteins are summarized in Table 15.2. Seven of these flagellum-specific type III secretion proteins localize in or around the inner membrane and appear to be linked to the basal-body ring structures, possibly localizing to the central pore inside the MS ring. As discussed above and also suggested in Figure 15.3, it is very likely that a similar structure is formed by protein components of the type III secreton specific for secretion of virulence proteins. Importantly, such a structure is needed for secretion of the needle component, so formation must occur prior to assembly of the outer needle appendage. Additionally, based on the observation of continuous secretion (no periplasmic protein intermediates) by type III secretons, it has been postulated that a flagellar rod-like structure is associated with the ring structures in the inner membrane and extends through the periplasm to the outer membrane. However, no proteins with significant sequence similarity to the flagellar rod proteins FlgB, FlgC, FlgF, and FlgG have been observed for any component of TTSSs. Nevertheless, as discussed above, it is possible that PrgK and other proteins unique to TTSSs are involved in forming a structure that connects the postulated rings in the inner and outer membranes.

Although most proteins of TTSSs appear to be located in the bacterial inner membrane or cytoplasm, a subfamily of proteins, including InvG (from *Salmonella*), YscC (*Yersinia*), and MxiD (*Shigella*), are clearly located in the outer membrane and are required for type III secretion. These proteins display extensive similarity to a family of outer membrane proteins denoted secretins. Secretins mediate the transport across the outer membrane of various large molecules, including extracellular enzymes via the type II secretion system (PulD of *Klebsiella* spp.), filamentous bacteriophages (the pIV family), and type IV pili (PilQ of *Pseudomonas aeruginosa*). Interestingly, a common feature among this diverse protein family is that they contain an N-terminal signal sequence allowing their transport to the outer membrane via a *sec*-dependent mechanism. In addition to the *sec* system, the PulS lipoprotein is required to prevent periplasmic degradation of PulD and to ensure correct outer membrane positioning of this enzyme. While no proteins specific to TTSSs appear similar to the PulS chaperone, the VirG lipoprotein from *Yersinia* performs an analogous function to PulS, being responsible for localizing YscC to the outer membrane. Significantly, the YscC protein family is not related to the proteins that form the L and P rings in the outer membrane of the flagellar basal body. This is not surprising considering that different proteins are assembled on the L ring of the flagellar basal body than the secretin-like proteins in the outer membrane of TTSSs. Secretin multimers from several systems form pore-like structures with external diameters of about 200 Å and internal diameters of between 50 and 80 Å, sufficient for passage of large proteins and bacteriophages through the outer membrane. A similar-sized pore in the outer membrane of *Yersinia* is formed by the multimerization of the YscC protein. It is believed that the outer membrane pore structure formed by TTSSs permits the channeling of virulence effectors through the outer membrane to the exterior. Since this pore contains a large aqueous channel, it would be expected that a protein(s) would be required to gate this channel, preventing nonspecific passage of components in or out of the cell. Assuming that the needle protrusion assembles from this secretin-pore base, it is possible that the narrow-channeled needle structure not only allows active secretion, but also gates the fully assembled system.

Substrate Specificity Switch

One critical step in the secretion assembly of flagella is the switch from secretion of hook proteins to flagellar filament proteins once the hook structure is completed. The secreted FliK protein likely senses when the hook is completed, since *fliK* mutants are unable to secrete flagellin, but continue to assemble polyhook structures. Interestingly, certain mutations in the cytoplasmic part of *flhB*, an inner membrane protein required for secretion, can suppress the phenotype of *fliK* mutants and at least partly restore flagellin secretion. This indicates that after FliK senses completion of the hook structure, an interaction with FlhB is permitted. In turn, this promotes a switch in substrate specificity, so that the system goes from secreting hook protein to filament protein. While the assembly of a complex structure like a flagellum appears very different from a system simply devoted to protein secretion, the FlhB and FliK proteins are actually well conserved between the flagellar secretion system and the systems devoted to protein secretion in animal pathogens (see Table 15.2). Mutations in the *fliK* homologue found in the type III secreton of the animal pathogens *Salmonella, Shigella,* and *Yersinia,* all result in loss of effector protein secretion. Presumably, this is a secondary consequence of hyperextended needle filament formation, which poorly secrete substrates. Thus, it appears that these FliK homologues sense when the needle structure is completed. They then convey a specific signal(s) to the secretion system so that needle component secretion can be terminated, concomitant with a switch to translocator and effector protein secretion. By analogy with the flagellar system, some mutations in the FlhB homologue YscU of *Yersinia* can partly suppress the phenotype of *yscP* (*fliK* homologue) with respect to needle length and protein secretion.

In summary, the emerging picture for the mechanism of type III secretion is that proteins destined for secretion are identified and targeted to the secreton at the inner membrane via a common mechanism for both effector proteins of pathogens and flagellar proteins involved in motility (Figure 15.3). Indeed, this recognition signal resides in the very N terminus of the secreted proteins (see below for details). In addition, another common mechanism (involving one membrane protein and one secreted protein) controls a substrate specificity switch from early secretion of needle/hook protein to later secretion of effector/filament proteins, respectively. Yet additional common proteins of the secretion apparatus likely gate the pores present in both the inner and outer membranes (these are connected by a channel extending through the periplasm), while others would provide the energy required to channel the proteins to the exterior.

Secreted Proteins

Type III secretion is a specific mechanism used by numerous bacteria to establish an infection and/or symbiotic relationship with eukaryotic host cells by mediating the targeted injection of effector proteins into the host cell cytosol. The secretion signal of proteins targeted either by the type III mechanism or by the flagellum-specific secretion apparatus resides in the N-terminal coding region of the respective genes. For example, the first 11 amino acids of the YopE cytotoxin from pathogenic *Yersinia* spp. are sufficient to ensure secretion. Interestingly, no common sequence motif or structure has been identified for the secreted proteins, and when the N-terminal coding region of the secreted YopE and YopN proteins of *Yersinia* was systematically mutagenized, no mutants with abolished secretion were observed. This led to the suggestion that putative structures within the mRNA form the secretion signal for YopE and YopN. In each case, stem-loop structures that incorporated the start codon and the ribosome-binding site within a computer-simulated mRNA base-paired duplex were predicted to control secretion. This indicated that translation and secretion might be intimately coupled at the very site of the secretion apparatus. However, for YopE and InvJ, two type III secretion substrates of *Yersinia* and *Salmonella,* respectively, extensive mutagenesis which altered the mRNA sequence while maintaining the amino acid sequence did not influence secretion. At least in these cases, this suggested the secretion signal does not reside in the mRNA structure but rather in the amino acid sequence of the very N termini and possibly is composed of a sequence of alternating polar and hydrophobic residues. Clearly, further scrutiny of the "real" signal that predestines a substrate for secretion is still required.

Some of the secreted proteins are not only secreted through the two bacterial membranes but also translocated across the plasma membrane of the eukaryotic target cell (for details, see below). This appears to be a universal feature of TTSSs. The translocation process requires that all proteins contain a distinct translocation motif independent of the secretion signal and additional translocator proteins also secreted by TTSSs. The recognition sequence for translocation of Yop effectors of *Yersinia* resides in the region immediately downstream of the N-terminal secretion signal. It is important to stress that although translocation requires protein secretion, these are mutually exclusive events, and it follows that different regions of the translocated protein encode the information necessary for each event.

Thus, virulence effector proteins injected into the interior of the host cell by TTSSs of pathogenic bacteria are modular proteins usually segregated into three functionally distinct domains (Figure 15.5). For example, the extreme N-terminal coding region of Yops produced by *Yersinia* contains the specific signal that determines a protein for type III secretion. Immediately downstream of the secretion domain resides a larger signature domain (usually between 50 and 80 amino acids) that identifies a protein targeted

Figure 15.5 Schematic diagram of the modular organization of effector proteins injected into eukaryotic cells by TTSSs. The *Yersinia* YopE and YopH antiphagocytic proteins, SptP from *Salmonella* serovar Typhimurium and ExoS from *P. aeruginosa*, are given as examples. The minimal N-terminal domain required for secretion of each protein is shown in grey. The secretion signal is immediately followed by a minimal region required for the proteins to be injected into eukaryotic cells (corresponding to "Translocation and chaperone binding"). These translocation domains are also overlapped by high-affinity chaperone-binding sites. The C-terminal region of YopE (amino acids 99 to 215) is essential for the cytotoxic effector function. In particular, this domain functions as a Rho GTPase-activating domain. Both the SptP and ExoS proteins also display a modular organization, having regions with extensive similarity to the cytotoxic domain of YopE. An essential arginine (Arg) finger motif defines this functional domain. SptP possesses a second effector domain with similarity to the PTPase activity domain of YopH. The cysteine (Cys) residue is essential for tyrosine dephosphorylation of host proteins. The second functional domain of ExoS located at the C terminus exhibits ADP-ribosyltransferase activity, which blocks receptor-stimulated Ras activation through a modification of Ras in vivo.

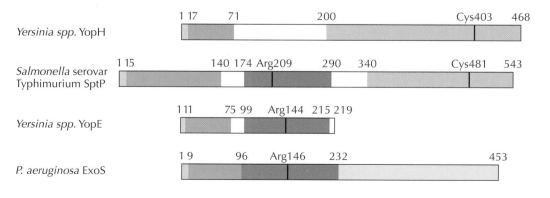
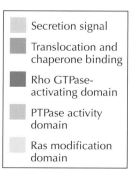

for translocation across the eukaryotic plasma membrane. For several effector proteins, this region is protected or capped by specific chaperones in the cytoplasm of the bacterium (see below). Finally, the C-terminal regions of the proteins contain effector domains, which in some cases display enzymatic activities normally associated with eukaryotic cell function (Box 15.1).

Chaperone Proteins

Small cytoplasmic accessory proteins (chaperones) specifically interact with their cognate type III secreted protein to ensure presecretory stabilization and efficient secretion. Unlike the universal role of SecB in *sec*-dependent secretion, TTSSs encode several dedicated chaperones that specifically interact with one or at most a few partner proteins destined for secretion. In *Yersinia,* the region in which chaperones interact with their partner Yop also overlaps with the motif necessary for translocation (see Figure 15.5). Thus, chaperone-Yop interactions in the cytoplasm of bacteria presumably "caps" the region required for translocation to prevent premature interaction with other proteins of the type III secretion apparatus and/or self-aggregation prior to secretion. In several TTSSs, the lack of a specific chaperone reduces secretion of the partner protein due to degradation of the aggregated protein in the bacterial cytoplasm. This is best exemplified by the IpgC chaperone from *Shigella* that independently binds to IpaB and IpaC in the cytoplasm. These specific interactions serve to inhibit the premature association and subsequent degradation of an IpaBC protein complex before secretion. Hence, the term "bodyguards" has also been used to describe the function of type III specific chaperones while the cognate proteins still reside in the cytoplasm. However, it is intriguing that not all secreted substrates require a chaperone. This observation has led to the notion that chaperones may introduce a secretion hierarchy; a chaperone-substrate combination promotes more rapid secretion, while substrates without a chaperone are secreted secondarily. This would be consistent with an infection scenario, as it is envisaged that some secreted substrates would be required immediately to establish bacterial colonization while other proteins would be required later to short circuit the host's attempts at bacterial clearance. The implication of chaperones in establishing spatial and temporal regulatory patterns is further discussed in Common Regulatory Mechanisms.

Even with the widespread existence of these customized chaperones in all TTSSs of both animal- and plant-interacting bacteria, their concise role in protein secretion is debated. It follows that several key questions still remain unanswered. (i) Does the chaperone dock at the cytoplasmic face of the secretion apparatus to facilitate efficient protein secretion? (ii) Are substrates secreted in a folded or unfolded state being influenced by direct chaperone binding? (iii) What is the mechanism for dissociation of the chaperone-substrate complex? These questions have captured the imagination of biologists from diverse backgrounds and prompted the resolution of crystal structures for some chaperones, either alone or in complex with their secreted partner. Although chaperones are generally dissimilar at the amino acid level, the overall chaperone structure of CesT from EPEC, SycE from *Yersinia* spp., and SigE and SicP from *Salmonella* appear remarkably conserved. These chaperones all exist as a dimer with conserved substrate-binding groves that complex with their cognate substrate in a 2:1 binding ratio (Figure 15.6). These interactions are predominately hydrophobic, interspersed with polar interactions. Thus, in the absence of a common signature sequence, the surface distribution of hydrophobic and polar residues likely confers unique

▶ *For Figure 15.6, see color insert.*

substrate specificity among each individual chaperone. Clearly, the region of substrate bound by the cognate chaperone is unfolded and resistant to proteolysis. On the other hand, the unbound C-terminal enzymatic region of each substrate still remains functionally active, indicative of these effector domains still being fully folded (Figure 15.6). Therefore, the partial unfolding of a substrate induced by chaperone binding likely primes this substrate for secretion. Since protein unfolding requires significant energy input, perhaps chaperone binding triggers a more energetically favorable unfolding pathway to ensure a secretion-competent protein. Even though many questions still remain, novel structural information is providing valuable insights into understanding the complex mechanism of type III secretion.

Delivery and Effectors

Injection System

The proteins secreted by the TTSS interact with eukaryotic cells in several ways. In most systems studied so far, the effector proteins are translocated across the eukaryotic plasma membrane into the cytoplasm by extracellular bacteria (Figure 15.7). For *Yersinia* and EPEC, it is thought that one function of the effector proteins is to either prevent phagocytosis (*Yersinia*) or form a tight adherence between the bacterium and the target cell that induces the formation of so-called pedestal structures (EPEC). In contrast, *Salmonella* translocates effector proteins into the eukaryotic cytoplasm that actually induce the eukaryotic cell to internalize the bacterium. In addition, *Salmonella* utilizes the type III needle complex to inject effector proteins following their internalization within the eukaryotic cell. These examples illustrate that gram-negative bacteria have evolved different effector proteins that induce and/or modulate eukaryotic host responses that occur during various stages of the infection process. This gives each individual pathogen their specific lifestyle.

For Figure 15.7, see color insert.

In all cases studied so far, the translocation of effector proteins depends on physical contact between the bacterium and the eukaryotic cell. The bacterial adhesion molecule mediating this tight interaction may not necessarily be an integrated component of the TTSS. In *Yersinia,* following the intimate binding between the bacterium and the surface of the eukaryotic cell, Yop effector proteins traverse the eukaryotic cell membrane only at the interface between the two cells. Thus, the translocation process appears to be polarized. Additionally, eukaryotic cell contact somehow also induces *Yersinia* to increase the expression of Yop effector proteins (Box 15.3). Similarly, *Salmonella* also has a cell contact-induced injection system of the Sip and Sop proteins. In contrast to *Yersinia*, however, eukaryotic cell contact does not result in increased *sip/sop* gene expression, nor does it appear that the TTSS of *Salmonella* injects proteins into eukaryotic cells in a polarized fashion. (Sip/Sop proteins are also found in the tissue culture media during cell culture infections.)

Most of what is known about how the TTSSs translocate proteins across the eukaryotic membrane has been derived from studies on *Yersinia*. The injection process in *Yersinia* depends on at least three translocator proteins, YopB, YopD, and LcrV, which are themselves secreted to the exterior of the bacterial cell by the type III needle complex. Bacterial mutants deficient in any of these proteins are still able to secrete the other Yops from the bacterial cell but are unable to inject these proteins into eukaryotic cells; instead, these proteins accumulate at the zone

BOX 15.3

How does *Yersinia* utilize target cell contact to induce Yop synthesis and promote their polarized translocation? A model

Contact of *Yersinia* with a HeLa cell results in the induction of *yop* transcription and the subsequent polarized injection of Yops across the target cell plasma membrane at the zone of contact between the interacting cells. Presumably, to facilitate the induction of Yop synthesis, a signal must be sensed by the bacterium and then transmitted from the surface into the bacterial cytoplasm. With this in mind, how does the bacterium sense the interaction with a eukaryotic cell surface ligand? The type III secretion system per se is essential not only to secrete Yops but also to restrict Yop synthesis to concise periods during bacterial infection. It is proposed that the Ysc secretion apparatus transports a protein to the bacterial surface that specifically functions to sense physical contact between the pathogen and the host cell. This protein may also block the secretion channel such that only during physical contact would the

blockage be released. This would limit Yop translocation to the zone of contact between the interacting cells—a process termed polarized translocation.

Since Yop synthesis depends on a functional Ysc secretion apparatus, how is the signal transmitted through the secretion apparatus to facilitate this tight regulatory control? Significantly, a mutation in the *lcrQ* gene is always derepressed for Yop synthesis and when overexpressed in *trans*, LcrQ down-regulates *yop* expression. Furthermore, secretion of LcrQ into the extracellular medium, via the Ysc secretion apparatus, correlates with induction of Yop synthesis and secretion. Moreover, in *ysc* secretion mutants normally repressed for Yop synthesis, LcrQ is found only in the bacterial cytoplasm. However, on introduction of an *lcrQ* mutation into these *ysc* secretion mutants, Yop synthesis is derepressed. Based on these phenotypes, the following model for contact-dependent Yop regulation by *Yersinia* has been proposed. Contact between the pathogen and its target cell results in the opening of the gated secretion apparatus. This allows the rapid secretion of LcrQ, significantly

reducing its intracellular concentration. This releases the LcrQ-mediated repression of *yop* expression and results in increased Yop synthesis and secretion. This model explains how the bacterium coordinates up-regulation of *yop* expression and polarized secretion and also suggests that the same sensing and gating mechanism controls these two regulatory events.

Interestingly, the parallel between the flagellum-specific secretion pathway and type III secretion system can be extended to how each system regulates gene expression. Just as Yop synthesis and secretion is induced by the rapid and specific secretion of the *yop* regulatory protein LcrQ through the Ysc secretion apparatus, expression of flagellum-specific genes is induced upon the secretion of the anti-σ factor, FlgM, through the flagellum itself. The release of the antagonist FlgM permits the flagellum-specific σ factor (FliA) to interact with DNA to induce gene transcription. While LcrQ and FlgM have no obvious genetic or functional similarities, it is interesting how two systems involved in the biosynthesis of cell surface organelles have adopted a similar regulatory mechanism to coordinate gene expression.

of contact between the bacterial and eukaryotic cells (Figure 15.7). As LcrV, YopB, and YopD cooperate to cause erythrocyte lysis and associate with lipid membranes, it has been proposed that a complex of these proteins forms a pore in eukaryotic cell membranes through which the other Yops are injected into the eukaryotic cytosol. The size of the translocation pore is estimated to be about 3.5 nm in diameter, i.e., enough to allow non-folded proteins to pass through. Interestingly, YopD itself is also injected into the cytosol of cultured eukaryotic cells infected with *Yersinia*. This may indicate a further role for YopD within eukaryotic cells.

As discussed above, the ability to inject effector proteins is functionally conserved in different species. Thus, not only the secretion system but also the ability to inject the effectors across the plasma membrane of the eukaryotic cell is mediated through a similar process in different species. Functional conservation of the injection process was first shown for the *Yersinia* YopE protein, which was secreted and translocated into the eukaryotic cytosol by the TTSS of *Salmonella*. Surprisingly, *Salmonella* does not possess proteins that bear any obvious similarities to the *Yersinia* LcrV, YopB, and YopD, although the SipB protein of *Salmonella* does have a limited degree of similarity to YopB, which may indicate a similar function. SipB is encoded by the *sipBCDA* operon; all proteins encoded by this operon except SipA are involved in translocation of effector proteins. Interestingly, like YopD, the Sips are injected themselves, which could indicate that they are also active within eukaryotic cells. As these proteins associate with the eukaryotic plasma membrane, however, it is also possible that their apparent localization inside eukaryotic cells is not through direct injection, but rather comes about because a part of the membrane-associated protein is exposed on the cytosolic face of the membrane. Nonetheless, even in this location, a virulence role inside the cell is still feasible.

The injection systems of *P. aeruginosa* and *Yersinia* are probably similar since *P. aeruginosa* encodes proteins that are very similar to LcrV, YopB, and YopD. It follows that the corresponding *pcrV* and *popB* genes of *Pseudomonas* can functionally complement *lcrV* and *yopB* mutants of *Yersinia*. It is generally accepted that the overall mechanism used by various gram-negative bacteria to translocate effector proteins into eukaryotic cells via the TTSS will turn out to be highly conserved. Nevertheless, it should not come as a surprise that variations would exist between the various species, since the needs of each species, in terms of a protein translocation system, would be influenced by such factors as the preferred site of infection and the local host defense system.

Effectors

As discussed above, gram-negative pathogens possessing TTSSs secrete proteins that directly affect eukaryotic cells. These bacterial proteins probably exert their activities by physically interacting with key eukaryotic proteins involved in host defense functions. Our knowledge of what these proteins are doing varies greatly, depending on the protein. For some of the injected proteins, nothing is really known about their effects on eukaryotic cells, although in many cases it is known that the protein is required to cause disease in the animal model system. At the other extreme, an increasing number of proteins have well-defined cellular functions and the eukaryotic proteins with which they physically interact are known. Some of the better-characterized effector proteins are discussed in this section.

Yersinia

The three species of *Yersinia* that are recognized as human pathogens (*Y. pestis, Y. pseudotuberculosis*, and *Y. enterocolitica*) have a common tropism for lymphatic tissue and are able to resist the naive immune system by blocking both phagocytosis and the induction of proinflammatory cytokines. Both these processes depend on the TTSS and the directional translocation of the Yop effector proteins into the cytosol of the infected cell (Figure 15.8). Since both these activities could be assayed by infecting cultured eukaryotic cells with mutant strains of *Yersinia,* it was relatively simple to determine which of the Yops were required to block phagocytosis and inductive cytokine expression.

The full antiphagocytic activity of *Yersinia* requires the additive effects of the YopH and YopE proteins. To engulf a bacterial cell, eukaryotic cells must reorganize their cytoskeletal system (the actin-based component that confers the structural integrity of a cell). This reorganization encloses the bacterium in a membrane-bound vacuole. It appears that YopH and YopE are able to interfere with the structural rearrangements required for the normal phagocytic process to occur. YopH is a protein tyrosine phosphatase

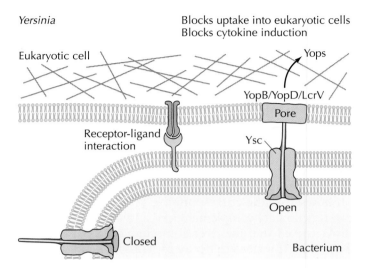

Yersinia

Blocks uptake into eukaryotic cells
Blocks cytokine induction

Eukaryotic cell

Yops

YopB/YopD/LcrV

Pore

Receptor-ligand
interaction

Ysc

Open

Closed

Bacterium

Figure 15.8 A proposed model for *Yersinia* interaction with target cells. Intimate association between the pathogen and its target cell is established by a receptor-ligand interaction. In phagocytic cells, the basis of this interaction is not known. In epithelial cells, however, β_1-integrins bind to the bacterial surface-located protein, invasin. This target cell contact opens the TTSS, allowing Yop secretion, some of which forms a pore in the target cell plasma membrane, while others are injected through the pore into the target cell. This results in blockage of bacterial uptake and suppression of cytokine expression.

(PTPase) that is similar to eukaryotic PTPases (see Figure 15.5 and Box 15.1). *Yersinia* strains mutated in YopH are avirulent in the mouse model and are rapidly phagocytosed by cultured eukaryotic cells. Following injection into the cytosol, YopH rapidly (less than 1 min) dephosphorylates specific target cell proteins. In cultured HeLa cells, these dephosphorylated proteins include p130^{Cas} and focal adhesion kinase (FAK). Both of these proteins are involved in the regulation and assembly of new focal adhesion complexes, a cellular process that accompanies phagocytosis. Focal adhesion complexes, located on the inner surface of the plasma membrane, are thought to link the cellular cytoskeletal system to extracellular structures such as matrix proteins. Interestingly, YopH was found to localize to focal adhesion structures after infection of HeLa cells, and YopH activity was associated with the disassembly of these structures. Additionally, YopH was found to physically interact with FAK and p130^{Cas} in vivo, further suggesting that YopH targets these proteins during the infection cycle. Thus, these observations have led to the suggestion that YopH specifically targets and destroys focal complexes as a means of blocking phagocytosis. It is also reasonable to expect that p130^{Cas} and FAK, whose transient tyrosine phosphorylation is required for normal phagocytosis, belong to a group of proteins that constitutes the actual in vivo substrates for the YopH PTPase activity.

The Fyn-binding protein (Fyb) and the protein SKAP-HOM are also associated with YopH function. Fyb and SKAP-HOM can interact and are phosphorylated during normal macrophage function but are rapidly dephosphorylated by YopH. In this way, YopH interferes with an essential signal transduction pathway in macrophages. In addition, YopH also interferes with T- and B-cell proliferation and blocks the synthesis of monocyte chemoattractant protein 1 (MCP1) in cultured macrophages infected with

Yersinia. Collectively, interference at any of these levels would disrupt the normal immune response to *Yersinia* infection. However, it must be stressed that YopH is extremely enzymatically active, indiscriminately dephosphorylating a vast majority of the tyrosine-phosphorylated proteins in an infected cell. Therefore, results obtained from YopH studies must be treated with some care.

Both the phagocytic process and cell motility require a substantial amount of cytoskeletal reorganization, which involves a balance of actin polymerization and depolymerization. After being injected into the cytosol of the infected cell, YopE disrupts the cellular cytoskeletal system by inducing a massive depolymerization of actin stress fibers. Members of the Rho family of small G proteins, RhoA, Rac, and Cdc42, act as molecular switches ("on" when bound by GTP and "off" when the GTP is hydrolyzed to GDP) in coordinating the reorganization of the actin cytoskeleton and regulation of the actin meshwork. Therefore, it came as no surprise to learn that YopE affects the activity of these three small G proteins. Each G protein exhibits an intrinsic GTPase activity (hydrolysis of GTP to GDP) that is stimulated by specific GTPase-activating proteins (GAP). All GAPs have a arginine residue (termed the arginine finger) that is critical for stimulating GTP hydrolysis. Interestingly, YopE possesses an arginine finger at position 144 and displays GAP activity toward RhoA, Rac, and Cdc42. Exchange of this arginine with an alanine results in loss of GAP activity, and the corresponding *yopE* mutant is neither able to block phagocytosis nor establish an infection in an animal model. Thus, the GAP activity of YopE is biologically relevant. Interestingly, YopE shows high homology with two other type III effector proteins, ExoS of *P. aeruginosa* and StpP of *Salmonella* (see Figure 15.5). These three proteins are GAPs interacting with the Rho family of small G proteins. Although they all possess an arginine finger motif, indicating that their mode of function is similar to the eukaryotic counterparts, they show no structural similarity to eukaryotic GAPs. Thus, in contrast to YopH that clearly has eukaryotic homologues and likely arose by horizontal gene transfer, the YopE family of bacterial GAPs has likely evolved via convergent evolution (see Box 15.1).

The activity of yet another effector, YopT, is also coupled to actin depolymerization. However, YopT might be a redundant virulence determinant since some pathogenic strains of *Y. pseudotuberculosis* lack this protein. Nevertheless, YopT does possess a very interesting enzymatic function, being responsible for the specific cleavage of RhoA, Rac, and Cdc42 near their C termini. Inactivation subsequently occurs as the cleaved G proteins detach from the inner face of the plasma membrane (the proteins are normally membrane anchored via a covalently bound lipid moiety). Why *Yersinia* possesses two different effectors (YopE and YopT) to inactivate the same proteins is not known. However, there are about 20 different proteins belonging to the Rho family of small G proteins. Therefore, it is possible that the "real" target of both YopE and YopT is to be found among these less-studied proteins.

In addition to extracellular growth in eukaryotic tissues (due to the antiphagocytic activity of Yops), another characteristic of *Yersinia* infections is that the pathogen is able to proliferate in tissues without provoking an inflammatory-like response. For all three human pathogenic *Yersinia* species, blocking of the host inflammatory response depends on the 70.5-kb virulence plasmid encoding the type III-associated genes (see Figure 15.2). For example, in mice infected with *Y. pestis*, visceral organs such as the liver become heavily colonized with the pathogen. In contrast, in tissues of mice infected with a plasmid-cured strain of *Y. pestis*, inflammatory cells are recruited to the site of infection, leading to the formation of protective granulomas. The lack of an inflammatory response at the tissue level is likely due to the ability of *Yersinia* to repress, at the cellular level, the expression of proinflammatory cytokines. Inducible cytokine secretion is a result of a multistep process: a danger signal is perceived at the cell surface; this signal has to be transmitted to the nucleus, where the appropriate gene must be transcribed; the transcript must be exported to the cytoplasm and translated; and, finally, the resulting protein must be secreted from the cell. Recent work has indicated that *Yersinia* most probably blocks inducible cytokine secretion by blocking the activation of signaling molecules involved in some of the early events that occur when eukaryotic cells encounter bacteria and other forms of stress.

The *yopJ* locus of *Y. pseudotuberculosis* (and the homologous locus in *Y. enterocolitica, yopP*) is required to block inducible expression of the cytokines tumor necrosis factor alpha (TNF-α) and interleukin-8 (IL-8). The TNF-α and IL-8 gene promoter regions contain a number of binding sites for various families of transcription factors that control their basal and/or inductive expression levels. Binding sites for one such family, NF-κB, are important for the inducible expression of these genes. The NF-κB transcription factors are retained in the cytoplasm of uninduced cells by forming a complex with anchor proteins like the inhibitor IκB. On the appropriate stimulus, such as lipopolysaccharide, phosphorylation of IκB and its subsequent degradation by the proteasome enables the translocation of NF-κB transcription factors to the nucleus. Here, they bind to the promoter regions of several genes such as those encoding for TNF-α and IL-8 and increase their expression. Not surprisingly, it was found that *Yersinia* blocked the activation of NF-κB and that this activity depends on YopJ/YopP. This activity was also required for the *Yersinia*-mediated blockage of the mitogen-activated protein kinase (MAPK) superfamily of signaling proteins involved in eukaryotic stress responses. As the initial phosphorylation of the anchor protein is conducted by the IκB kinase complex (IKKB), a member of the MAPK superfamily, phosphorylation of IκB is presumably prevented by YopJ/YopP-mediated inhibition of the IKKB kinase. In this scenario, NF-κB is not activated and cytokine expression is suppressed during a *Yersinia* infection. In accordance with the ability to block MAPKs, YopJ/YopP also blocks the activation of the CREB transcription factor by targeting p38 (another kinase belonging to the MAPK superfamily). CREB is a key regulator of the immune defense response. Thus, through the broad activity range of YopJ/YopP, *Yersinia* is capable of shutting down very sensitive stress response signaling pathways in the eukaryotic cell during a time of extreme distress.

Yersinia can also induce or trigger apoptosis (programmed cell death) in macrophage cell lines. Whether *Yersinia*-mediated apoptosis occurs during an actual infection has not been determined. Apoptosis of cultured macrophages, like that of repressing cytokine expression, depends on the *yopJ/yopP* locus. Triggering apoptosis also depends on the *Yersinia* TTSS, suggesting that, like apoptosis-inducing proteins of plant-interacting bacteria, YopJ/YopP is recognized, or its activity occurs, inside the eukaryotic cell and that this recognition or activity triggers apoptosis. The effects of translocated YopJ/YopP are seen by the release of cytochrome *c* from mitochondria, a signal that triggers the caspase-signaling cascade resulting in apoptotic cell death. However, the precise series of events is still unclear. It is also possible that the apoptotic response is enhanced by the YopJ/YopP-dependent suppression of NF-κB activation, since the latter is required for the synthesis of antiapoptotic factors.

What is the significance of apoptosis in animals infected by bacterial pathogens? Currently, there is disagreement about whether apoptosis is in the best interest of the bacterium or, alternatively, occurs as a host defense strategy. For plants, bacterium-induced apoptosis by proteins translocated by type III secretons is clearly a host defense response that serves to cordon off the bacterial pathogen to the initial site of infection (Box 15.2). Alternatively, inducing apoptosis in macrophage cells (and perhaps other cell types) may benefit *Yersinia* by preventing those cells from performing important immune functions.

Salmonella

Salmonella spp. causes a broad spectrum of different diseases, usually originating from oral infections. These infections can be life threatening, such as typhoid fever, or relatively mild, such as gastroenteritis. As mentioned earlier, *Salmonella* possesses two different gene clusters, designated SPI-1 and SPI-2, each encoding apparently independent TTSSs. *Salmonella* organisms grown in culture secrete more than 20 proteins. The SPI-1 secreton secretes most of these proteins, and the molecular functions of some of these effector proteins have been dissected. Less is known about the *Salmonella* proteins secreted by the SPI-2-encoding pathogenicity island. Clearly, however, the SPI-1-encoded TTSS actively induces uptake of *Salmonella* into epithelial cells, and the SPI-2-encoded TTSS supports intracellular survival of *Salmonella* in eukaryotic cells. Remarkably, neither bacteria with mutations in the SPI-1-region nor those with mutations in the associated SPI-1-secreted effector proteins are drastically attenuated in virulence in the mouse model. In contrast, bacteria with mutations in the SPI-2 region are severely virulence attenuated. Nevertheless, the *inv/spa* secretion system (as the SPI-1 is also known) of *Salmonella* is essential for the ability of the pathogen to invade epithelial cells and to induce cellular cytotoxicity and apoptosis (Figure 15.9). This clearly indicates the importance of the SPI-1-associated secretion system in virulence.

When wild-type strains of *Salmonella* are grown in culture, the SipA, SipB, SipC, and SipD proteins, encoded by the *sipBCDA* operon, are secreted to high levels in the culture supernatant. Bacteria with mutations in *sipB*, *sipC*, or *SipD* are all unable to translocate other effector proteins into target cells. Thus, Sips are essential for the actual translocation process, suggesting that they perform functions similar to LcrV, YopB, and YopD of *Yersinia*. Additionally, extracellular and intracellular *Salmonella* organisms translocate SipB and SipC into the cytosol of eukaryotic cells, and SipB, SipC, and SipD are all associated with *Salmonella*-induced morphological effects on cultured eukaryotic cells. Collectively, these findings may indicate that Sip proteins per se have effector functions, although this is not experimentally proven. In contrast, the last protein expressed by the *sip* operon, SipA, does not possess any obvious role in effector protein translocation. Rather, SipA binds directly to actin and inhibits depolymerization of actin stress fibers. This demonstrates that SipA is a translocated effector protein and an interaction with actin facilitates the *Salmonella*-induced uptake process into eukaryotic cells.

In addition to the Sips, *Salmonella* secretes at least five other effector proteins denoted SopA, SopB (SigD), SopC, SopD (SptP), and SopE. Specif-

Salmonella

A

Induces uptake
Growth inside host phagosomes

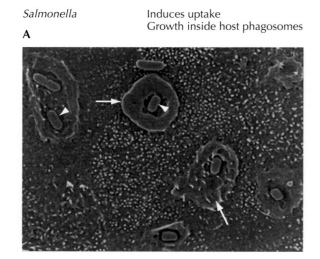

Figure 15.9 A proposed model for invasion and intracellular survival of *Salmonella* (arrowheads) within epithelial cells. **(A and B)** Upon contact with the eukaryotic epithelial cell (mucosal cells or M cells), *Salmonella* injects a number of effector proteins (Sops) into the cytosol of the target cell via the SPI-1-encoded TTSS. The injected effector proteins induce membrane ruffling (arrows) and subsequent uptake of the pathogen into a membrane-bound phagosome. **(C)** After uptake into phagosomal compartments within eukaryotic cells, *Salmonella* activates the SPI-2-encoded TTSS, including the associated effector proteins. Thereafter, the effector proteins are translocated across the phagosomal membrane into the lumen of the target cell. These effectors apparently inhibit a variety of functions such as phagosome-lysosome fusion (SpiC), vesicular trafficking, and avoidance of killing by the respiratory burst. Thus, the SPI-2-encoded TTSS supports the intracellular survival and proliferation of *Salmonella*. Panel A was provided by Jorge Galán, Yale University School of Medicine, New Haven, Conn.

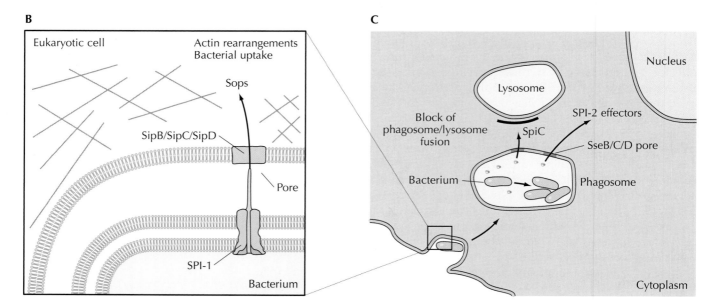

ically, when SopE is introduced into eukaryotic cells either by the *Salmonella* TTSS or artificially by microinjection, cytoskeletal rearrangements are induced that accompany membrane ruffling and eventual internalization of the bacterium (Figure 15.9). This is simply not an artifact of cell culture systems, because *Salmonella* also induces similar ruffles in M cells of the Peyer's patches, the portal through which this bacterium induces uptake during a natural infection. SopE physically interacts with and activates small GTP-binding signaling proteins of the Rho family. Rho-type signaling proteins are normally activated when eukaryotic cells are exposed to external signals. This activates cellular responses that typically include membrane ruffling that occasionally leads to bacterial uptake. Apparently, *Salmonella* has evolved a way to manipulate a cellular signaling pathway to serve its own means (to invade the eukaryotic cell).

Another interesting type III secreted and translocated protein of *Salmonella* is SopD/SptP. The N-terminal portion of this protein is very similar to YopE and the N-terminal domain of ExoS from *P. aeruginosa,* while the C-terminal portion is very similar to the PTPase of *Yersinia* YopH, including the tyrosine phosphatase catalytic domain (see Figure 15.5). As expected from the sequence, SptP possesses a protein tyrosine phosphatase activity and induces a "YopE-like" cytotoxic effect on cultured eukaryotic cells (disruption of the actin cytoskeletal system). Thus, SptP exhibits two apparently independent activities. It is therefore surprising that *sptP* mutants, in contrast to *yopE* and *yopH* mutants of *Yersinia,* are only moderately attenuated in virulence in the mouse model. It is important, however, to stress that the mouse model does not reflect all aspects of virulence, and it seems very likely that SptP plays a significant role in the overall pathogenicity of *Salmonella* disease.

The sophisticated ability of *Salmonella* to manipulate host cell responses is elegantly demonstrated by the opposing actions of SopE and SptP. These two SPI-1-translocated effectors are intimately involved in actin cytoskeletal rearrangement facilitating bacterial uptake. SopE is a potent guanine nucleotide GTPase exchange factor (GEF) that, when translocated, activates the small G proteins Cdc42 and Rac-1. Their activation (when G proteins are in a GTP-bound form) via SopE induces dramatic actin rearrangements leading to membrane ruffling and bacterial uptake. Only after a short delay, however, the infected cell surface returns to normal due to the opposing GAP effect mediated by the "YopE domain" of translocated StpP. This protein specifically inactivates (when G proteins are in a GDP-bound form) both Cdc42 and Rac-1 (see description of YopE in *Yersina,* above). This SopE and SptP synergy facilitates healing of the host cell and guards against the potential harmful side effects caused by the induction of Cdc42/Rac-1 signaling.

Another SPI-1 TTSS effector, SopB (also known as SigD), is an inositol phosphatase that specifically dephosphorylates inositol (1,3,4,5,6)-pentakisphosphate (Ins1,3,4,56P$_5$) both in vitro and in vivo. This produces Ins(1,4,5,6)P$_4$, which is an indirect activator of Cdc42. Therefore, SopB function complements the action of SopE. Why *Salmonella* harbors two partially overlapping functions is a mystery, although these functions might be cell type specific (the proteins are used at different stages of the infection process). This is evidenced by the ability of SopB, but not SopE, to induce fluid secretion (diarrhea) in an animal infection model. However, more work is needed to address this question.

Not surprisingly, cells infected with *Salmonella* show a stress response that is manifested by increased tyrosine phosphorylation of host proteins. This stress response is counteracted by the phosphatase activity exerted by StpP. Although the molecular target of this phosphatase activity has not been identified, the role of StpP is likely to down-regulate cellular defense reactions. Therefore, the activity of StpP is different from that of the *Yersinia* YopH phosphatase that is actively involved in antiphagocytosis. This is surprising given the high homology between YopH and StpP. This observation is a clear example that simple conclusions drawn from basic structural similarities between proteins can be misleading.

As mentioned above, *Salmonella enterica* invades and replicates within host cells that include both epithelial cells and macrophages. Intracellular replicating *Salmonella* is contained within membrane-bound phagosomes that were generated during the uptake process. Specific mutations of *Salmonella,* which render the bacterium incapable of growing in murine macrophages, are avirulent. These mutations generally cause defects in the TTSS encoded by SPI-2. Thus, this TTSS is essential for the survival of the pathogen inside these phagosomes. By analogy with other TTSS, some of the SPI-2-secreted proteins form a translocation pore in the phagosomal membrane containing *Salmonella*. The pore-forming proteins (SseB, SseC, and SseD) show homology with the corresponding proteins (such as SipB, SipC, and SipD) of other systems. Pore formation precedes the translocation of SPI-2-secreted effector proteins into the host cell cytosol. However, the cellular targets of the SPI-2 effector proteins are unknown, although some of their effects on cells have been described. For example, the SpiC effector may interfere with important cellular trafficking events, such as phago-some-lysosome fusion, normal vesicular trafficking, and endocytosis. Thus, the SPI-2 pathogenicity island seems to have a role in the cascade of events leading to intracellular survival of the pathogen in macrophages.

EPEC

EPEC is one of several different *E. coli* strains that cause diarrhea after oral infection. At the tissue level, an EPEC infection results in localized destruction of intestinal brush-border microvilli. Interestingly, bacterial binding to intestinal epithelial cells results in the formation of an actin-rich, pedestal-like structure, known as the attaching and effacing (A/E) lesion, underneath the bound bacterial cell (Figure 15.10). EPEC resides on this structure during an infection. This feature of EPEC is associated with a 35-kb chromosomal pathogenicity island called the locus of enterocyte effacement (LEE). LEE encodes a TTSS and several type III substrates (EspA, EspB, EspD, EspF, and Tir). With the exception of EspF, all these secreted proteins are associated with the formation of A/E lesions.

Figure 15.10 The proposed stages in the generation of tight adherence between EPEC and its target cell. After initial interaction between EPEC and the eukaryotic cell has been established (independent of the TTSS), a type III-dependent EspA organelle that extends from the needle complex bridges the small gap between the pathogen and the eukaryotic cell surface (see Figure 15.11). This facilitates the secretion of EspB and EspD, which subsequently form a translocation pore in the target cell plasma membrane through which the Tir protein is translocated. Tir then integrates into the host cell plasma membrane. Tight association between EPEC and the eukaryotic cell is then established via the interaction between the bacterial outer membrane protein intimin and Tir, which in this case works as a bacterially induced "eukaryotic cell receptor." Concomitantly, pedestal-like structures are formed as a consequence of dramatic actin rearrangements within the host cell. EPEC bacteria reside at the top of these pedestal structures. During this process the ability to visualize EspA is lost.

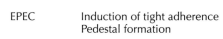

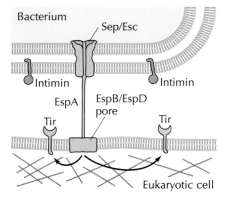

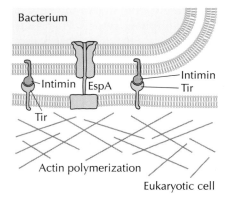

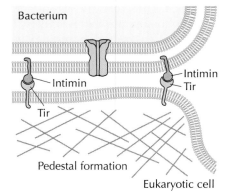

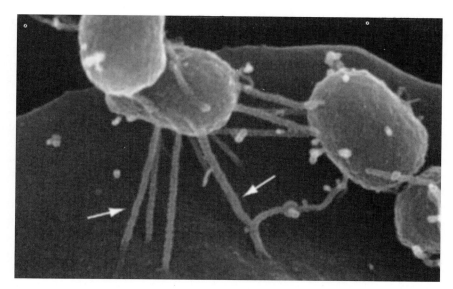

Figure 15.11   The EspA surface appendage bridges the gap between EPEC and the eukaryotic cell. This scanning electron micrograph captures the EspA filamentous structures (indicated by arrows) that promote attachment between EPEC and an infected red blood cell. The EspA filament is required for the translocation of the Tir effector into eukaryotic cells. Provided by Stuart Knutton, University of Birmingham, Birmingham, United Kingdom.

The intimate interaction between EPEC and the target cell is initiated by surface-localized, bundle-forming pili (BFP), which mediate loose adherence of the bacterium to the eukaryotic cell. This step is independent of the LEE-encoded TTSS. Rather, BFP are plasmid encoded and belong to the class of type IV pili. After initial attachment, the TTSS is actively engaged, although the type III substrate EspA is already secreted prior to target cell contact. Secreted EspA forms a filamentous, pilus-like organelle on the surface of the bacterium that protrudes from the tip of the type III needle complex. This organelle constitutes a conduit that likely connects the TTSS to the host cell plasma membrane (Figure 15.11). At least three other secreted proteins, Tir, EspB, and EspD, are translocated into the target host cells apparently through this hollow EspA organelle. To date, no effector function has been assigned to EspB and EspD. However, these two proteins contribute to pore formation in the target cell through which effector proteins are translocated. This is consistent with their sequence similarity to the YopB and YopD translocators of *Yersinia*.

Intriguingly, only one effector protein, Tir (for translocation intimin receptor), has been identified within the LEE locus. However, this protein is the "cornerstone" in the remarkable ability of EPEC to tightly adhere to and induce pedestal formation in the target eukaryotic cell. Translocated Tir serves as a receptor for intimin, a bacterial outer membrane protein that is localized by a type III-independent mechanism. Intimin is a powerful adhesin, such that the Tir-intimin interaction triggers intimate attachment between the target cell and the pathogen. This is then the cue for active pedestal formation. However, while it is easy to understand how intimin and Tir cooperate to mediate tight binding between EPEC and enterocytes, less is known about the subsequent signaling steps leading to pedestal formation. Given that Tir is localized to the tip of the pedestal, Tir may somehow interact with cytoskeletal components that regulate actin stress fiber formation at the cell surface. Interestingly, Tir is modified by tyrosine phosphorylation (likely mediated by host tyrosine kinases), which is an essential mediator of the subsequent induction of F-actin nucleation underneath the pathogen. Therefore, Tir is clearly important in the process of pedestal formation. Tir may recruit WASP and the Arp 2/3 complex (two crucial regulators of F-actin nucleation) to the tip of naive pedestals. This would be sufficient to induce the actin polymerization process directly underneath the pathogen. However, this Tir property has not been satisfactorily proven and the actin rearrangements could be explained by other means. Perhaps other undiscovered effector proteins also have vital functions in this process. Nevertheless, an interaction between the two EPEC proteins, intimin and translocated Tir, to establish tight bacterial-target cell contact, is yet another beautiful example of the extensive sophistication pathogens have acquired to interfere with and (mis)use the host system for their own benefit.

Common Regulatory Mechanisms

Because TTSSs are intrinsically complex organelles, the ability to coordinate appropriate synthesis of each individual component requires delicate cross talk between several positive and negative regulatory control loops. Bacteria interpret their host microenvironment either by physical contact with target cells or by monitoring physiological (temperature, acidity, osmolarity, population density) and/or nutritional (iron and cation concentration, amino acid composition, carbon source) conditions. Virulent bacteria establish colonization of specific niches that are each characterized by a unique set of environmental signals. These signals are then transmitted across the bacterial envelope into the cytoplasm where they are decoded to determine spatial and temporal expression of type III genes. This is most simply exemplified by the different colonization niches of animal pathogens; *Salmonella* spp., *Shigella* spp., and *Chlamydia* spp. seek an intracellular lifestyle, while EPEC, *Yersinia* spp., and *P. aeruginosa* all establish extracellular niches. No doubt plant-interacting bacteria interpret yet another set of diverse environmental cues. This section is designed to briefly mention key examples that may represent universal modes of type III regulatory control.

At least all the animal pathogens utilize an AraC-like transcriptional activator for up-regulation of type III genes. While a conserved C-terminal domain thought to be important for DNA-binding characterizes the AraC family, those involved in regulating virulence are routinely more conserved throughout the protein, possibly reflecting a common means of interpreting environmental stimuli. In addition, small ubiquitous DNA-binding proteins such as H-NS and FIS also cooperate with the transcriptional activators to modulate gene transcription. In fact, in the absence of H-NS, expression of the *Shigella* AraC-like VirF transcriptional activator is constitutive, resulting in uncontrolled type III gene expression even in the absence of inducing signals.

As mentioned in an earlier section, type III chaperones perform a unique customized function in ensuring the stabilization and efficient secretion of a specific substrate(s). However, this role has been complicated by the increasing awareness that a distinct subset of these chaperones, those that bind two substrates, act as important regulatory cofactors. The SicA chaperone of *Salmonella* serovar Typhimurium exerts its function on the secreted substrates SipB and SipC. However, free SicA acts in concert with the AraC-like InvF transcriptional activator, being responsible for transcription from antihost effector gene promoters. As SicA does not bind directly to DNA, but does copurify with InvF, this positive regulatory loop likely involves an InvF-SicA complex. Similarly, free IpgC chaperone of *Shigella* spp. is visualized to act in concert with the AraC-like MxiE transcriptional activator to turn on a subset of virulence genes. However, presumably gene transcription is inhibited when SicA or IpgC is in prese-cretory complexes with either of their substrates SipB and SipC or IpaB and IpaC, respectively, since this would inhibit binding to the transcriptional activators. This model is not unprecedented as the weakly homologous LcrH chaperone of *Yersinia* spp, when in complex with the YopD substrate, down-regulates synthesis of secreted components of the TTSS. Therefore, type III chaperones clearly represent an ingenious mechanism by which bacteria tightly couple regulated protein synthesis with their ultimate secretion. It is noteworthy that such a mechanism instills a secretion hierarchy to guarantee a synchronous bacterial infection. This is ensured in *Shigella* spp. by utilizing the IpgC-MxiE complex to induce transcription of effector virulence genes only after proteins important for initial bacterial entry (notably IpaB and IpaC) have been synthesized and secreted. Importantly, the full gamut of versatility within this intriguing chaperone family remains to be uncovered.

Cell contact by most animal and plant pathogens expressing a functional TTSS is a prerequisite for increased production and subsequent delivery of antihost effector proteins across host cell plasma membranes. An example of *Yersinia*'s response to target cell contact is given in Box 15.3. Importantly however, synthesis and assembly of the entire type III secreton are also induced on cell contact in the plant pathogen *Ralstonia solanacearum*. In particular, transcription of the HrpB transcriptional activator, which controls type III gene expression, is significantly induced on cell contact regardless of a functional TTSS. Rather, this response relies on three type III independent proteins, PrhA, PrhI, and PrhR, which are required for transmitting the cell contact signal across the bacterial envelope. PrhA, located in the outer membrane, senses an unidentified, nondiffusible, cell component during contact, resulting in the transmission of a signal across the outer membrane to the membrane-spanning component PrhR. In turn, the signal is transmitted across the inner membrane to activate the alternate sigma factor PrhI, probably prompting its release from the inner membrane. Active PrhI is then involved in the transcription of the regulatory genes necessary to initiate transcription of additional TTSS-encoding operons. Thus, *R. solanacearum* has evolved a three-component signaling module that spans the bacterial envelope to activate pathogenicity determinants in response to a nondiffusible plant cell wall signal.

It seems that every type III regulatory network is unique to a given bacterial species. In essence, by fine-tuning this common virulence mechanism, each bacterial species has created a device tailor-made for either a host-adapted or strictly pathogenic life cycle. Throughout evolution, the choice has been theirs.

Conclusion

Only recently have we become aware of the subtle ways in which bacteria interact with eukaryotic cells. As discussed in this chapter, one very sophisticated approach that gram-negative bacteria have evolved to manipulate eukaryotic cells involves direct injection of proteins into the cytosol of the infected cell. These bacterial proteins injected by the TTSS either impede or activate particular eukaryotic cellular responses depending on what is beneficial for the bacterium. The variety of species that possess TTSSs, ranging from animal pathogens to plant symbionts, suggest that animal- and plant-interacting bacteria have common strategies to overcome, or at least survive, the eukaryotic defense system.

Knowledge of the TTSS and the proteins it injects will undoubtedly allow the design of novel therapeutic strategies for use against microbial pathogens. For example, the *Salmonella* type III system has been used to inject heterologous proteins (from both viral and bacterial origin) into eukaryotic cells that are eventually displayed on the eukaryotic cell surface in association with major histocompatibility complex (MHC) proteins. This surface presentation of foreign antigens promotes the development of the adaptive immune response that serves to protect the animal from subsequent microbial challenge. Similarly, understanding how the injected effector proteins function within eukaryotic cells will certainly give us a better grasp of the animal and plant defense systems.

Selected Readings

Anderson, D. M., and O. Schneewind. 1997. A mRNA signal for the type III secretion of Yop proteins by *Yersinia enterocolitica*. *Science* **278:**1140–1143.

Blocker, A., P. Gounon, E. Larquet, K. Niebuhr, V. Cabiaux, C. Parsot, and P. Sansonetti. 1999. The tripartite type III secreton of *Shigella flexneri* inserts IpaB and IpaC into host membranes. *J. Cell Biol.* **147:**683–693.

Brito, B., D. Aldon, P. Barberis, C. Boucher, and S. Genin. 2002. A signal transfer system through three compartments transduces the plant cell contact-dependent signal controlling *Ralstonia solanacearum hrp* genes. *Mol. Plant. Microb. Interact.* **15:**109–119.

Darwin, K. H., and V. L. Miller. 2001. Type III secretion chaperone-dependent regulation: activation of virulence genes by SicA and InvF in *Salmonella typhimurium*. *EMBO J.* **20:**1850–1862.

Goguen J. D., J. Yother, and S. C. Straley. 1984. Genetic analysis of the low calcium response in *Yersinia pestis* mud1(Ap *lac*) insertion mutants. *J. Bacteriol.* **160:**842–848.

Guan, K. L., and J. E. Dixon. 1990. Protein tyrosine phosphatase activity of an essential virulence determinant in *Yersinia*. *Science* **249:**553–556.

Håkansson, S., K. Schesser, C. Persson, E. E. Galyov, R. Rosqvist, F. Homble, and H. Wolf-Watz. 1996. The YopB protein of *Yersinia pseudotuberculosis* is essential for the translocation of Yop effector proteins across the target cell plasma membrane and displays a contact-dependent membrane disrupting activity. *EMBO J.* **15:**5812–5823.

Heesemann, J., B. Algermissen, and R. Laufs. 1984. Genetically manipulated virulence of *Yersinia enterocolitica*. *Infect. Immun.* **46:**105–110.

Jin, Q., and S. Y. He. 2001. Role of the Hrp pilus in type III protein secretion in *Pseudomonas syringae*. *Science* **294:**2556–2558.

Herein, the authors use an in situ immunogold labeling procedure to visualize the secretion of an effector protein, AvrPto, from the tip of the Hrp-pilus, a surface organelle assembled by the type III secretion system of the plant pathogen *P. syringe*. While the Hrp-pilus is morphologically distinct from the type III-associated needle complexes of animal pathogens, this study is the first to physically demonstrate that type III secretion organelles function as a conduit for protein delivery across the two bacterial membranes. It may also indicate that proteins translocated into target cells occur via the tip of these organelles.

Kenny, B., R. DeVinney, M. Stein, D. J. Reinsheid, E. A. Frey, and B. B. Finlay. 1997. Enteropathogenic *E. coli* (EPEC) transfers its receptor for intimate adherence into mammalian cells. *Cell* **91:**511–520.

This study highlights a remarkable event in which EPEC uses a type III secretion system to insert its own bacterial receptor (Tir) into target cells. Translocated Tir becomes tyrosine phosphorylated and integrates into mammalian plasma membranes, where surface-exposed motifs engage the bacterial ligand, intimin. This intimate attachment initiates pronounced actin accumulation beneath adherent bacteria that results in pedestal formation characteristic of this group of pathogenic organisms.

Knutton, S., L. Rosenshine, M. J. Pallen, L. Nisan, B. C. Neves, C. Bain, C. Wolff, G. Dougan, and G. Frankel. 1998. A novel EspA-associated surface organelle of enteropathogenic *Escherichia coli* involved in protein translocation into epithelial cells. *EMBO J.* **17:**2166–2176.

Kubori, T., and J. E. Galán. 2003. Temporal regulation of salmonella virulence effector function by proteasome-dependent protein degradation. *Cell* **115:**333–342.

Kubori, T., Y. Matsushima, D. Nakamura, J. Uralil, J. M. Lara-Tejero, A. Sukhan, J. E. Galán, and S.-I. Aizawa. 1998. Supramolecular structure of the *Salmonella typhimurium* type III protein secretion system. *Science* **280:**602–605.

This microscopy study provided the first visualization of the membrane-spanning supramolecular structure of a type III secretion system from an animal pathogen. It revealed a "needle complex" structure with architecture remarkably similar to that which is involved in flagella assembly, further strengthening their evolutionary link. The authors suggest the needle complex is the functional equivalent of the flagella basal body, operating as a conduit through which substrates transit the bacterial envelope.

Lloyd, S. A., M. Norman, R. Rosqvist, and H. Wolf-Watz. 2001. *Yersinia* YopE is targeted for type III secretion by N-terminal, not mRNA, signals. *Mol. Microbiol.* **39:**520–531.

Luo, Y., M. G. Bertero, E. A. Frey, R. A. Pfuetzner, M. R. Wenk, L. Creagh, S. L. Marcus, D. Lim, F. Sicheri, C. Kay, C. Haynes, B. B. Finlay, and N. C. Strynadka. 2001. Structural and biochemical characterization of the type III secretion chaperones CesT and SigE. *Nat. Struct. Biol.* **8:**1031–1036.

McDaniel, T. K., and J. B. Kaper. 1997. A cloned pathogenicity island from enteropathogenic *Escherichia coli* confers the attaching and effacing phenotype on *E. coli* K-12. *Mol. Microbiol.* **23:**399–407.

Minamino, T., and R. M. Macnab. 2000. Domain structure of *Salmonella* FlhB, a flagellar export component responsible for substrate specificity switching. *J. Bacteriol.* **182:**4906–4914.

Ménard, R., P. Sansonetti, C. Parsot, and T. Vasselon. 1994. Extracellular association and cytoplasmic partitioning of the IpaB and IpaC invasins of *S. flexneri*. *Cell* **79:**515–525.

Pettersson, J., R. Nordfelth, E. Dubinina, T. Bergman, M. Gustafsson, K.-E. Magnusson, and H. Wolf-Watz. 1996. Modulation of virulence factor expression by pathogen target cell contact. *Science* **273:**1231–1233.

Rosqvist, R., K.-E. Magnusson, and H. Wolf-Watz. 1994. Target cell contact triggers expression and polarized transfer of *Yersinia* YopE cytotoxin into mammalian cells. *EMBO J.* **13:**964–972.

This investigation pioneered the concept that some gram-negative pathogens utilize a dedicated secretion mechanism to specifically direct antihost effector translocation into target cells. Using *Yersinia* as a model, the authors demonstrated that upon target cell contact, Yop effector expression was specifically derepressed and then the type III secretion system, precisely at the zone of bacterial-cell contact, translocated these effectors into eukaryotic cells.

Russmann, H., H. Shams, F. Poblete, Y. Fu, J. E. Galan, and R. O. Donis. 1998. Delivery of epitopes by the *Salmonella* type III secretion system for vaccine development. *Science* **281:**565–568.

Shea, J. E., M. Hensel, C. Gleeson, and D. W. Holden. 1996. Identification of a virulence locus encoding a second type III secretion system in *Salmonella typhimurium*. *Proc. Natl. Acad. Sci. USA* **93:**2593–2597.

Sory, M. P., and G. R. Cornelis. 1994. Translocation of a hybrid YopE-adenylate cyclase from *Yersinia enterocolitica* into HeLa cells. *Mol. Microbiol.* **14:**583–594.

Stebbins, C. E., and J. E. Galán. 2000. Modulation of host signaling by a bacterial mimic: structure of the Salmonella effector SptP bound to Rac1. *Mol. Cell* **6**:1449–1460.

> Several bacterial proteins secreted by type III secretion systems possess domains common to eukaryotic proteins. Presented herein are X-ray crystal structures of the *Salmonella* effector SptP, a GAP toward the Rho family of small G proteins, complexed with a molecular target, Rac1. They provide a high-resolution glimpse at the efficiency of bacterial pathogens to evolve molecular mimics to components of host cell signal transduction cascades.

Stebbins, C. E., and J. E. Galán. 2001. Maintenance of an unfolded polypeptide by a cognate chaperone in bacterial type III secretion. *Nature* **414**:77–81.

Sukhan, A., T. Kubori, J. Wilson, and J. E. Galán. 2001. Genetic analysis of assembly of the *Salmonella enterica* serovar Typhimurium secretion-associated needle complex. *J. Bacteriol.* **183**:1159–1167.

Tamano, K., S. Aizawa, E. Katayama, T. Nonaka, S. Imajoh-Ohmi, A. Kuwae, S. Nagai, and C. Sasakawa. 2000. Supramolecular structure of the *Shigella* type III secretion machinery: the needle part is changeable in length and essential for delivery of effectors. *EMBO J.* **19**:3876–3887.

Van den Ackerveken, G., E. Marois, and U. Bonas. 1996. Recognition of the bacterial avirulence protein AvrBs3 occurs inside the host plant cell. *Cell* **87**:1307–1316.

> This paper documented the finding that the product of the plant gene *Bs3* specifically recognized the avirulence protein AvrBs3 from a bacterial plant pathogen, when transiently produced inside the plant cell, to elicit hypersensitive cell death. This work, coupled with observations that intact *hrp* (hypersensitive reaction and pathogenicity) genes are essential for recognition of the bacterium by a plant cell, led these authors to introduce the notion that plant pathogens utilize a type III secretion system composed of products of the *hrp* genes to directly translocate bacterial Avr proteins into plant cells.

Wattiau, P., and G. R. Cornelis. 1993. SycE, a chaperone-like protein of *Yersinia enterocolitica* involved in the secretion of YopE. *Mol. Microbiol.* **8**:123–131.

> This breakthrough report introduced the concept that specific accessory proteins (chaperones) are required for presecretory stability and/or targeting of one, or at most a few, cognate substrates to the type III secretion machinery. In this study, the SycE chaperone was shown to specifically bind to the YopE cytotoxin and promote its type III-dependent secretion in the model bacterium *Y. enterocolitica.* The presence of dedicated chaperones is now accepted as a defining feature of type III secretion systems.

16

Bacterial Type IV Secretion Systems: DNA Conjugation Machines Adapted for Export of Virulence Factors

PETER J. CHRISTIE AND ANTONELLO COVACCI

Bacteria have evolved transport systems that until recently were thought to function specifically for the conjugative transfer of plasmid DNA between cells. Our early understanding of conjugation was developed with studies of the F plasmid of *Escherichia coli*. Extensive genetic experiments showed that transfer initiates upon pilus-mediated contact between donor and recipient cells. The pilus then retracts, bringing donor and recipient cells into direct physical contact. Stabilization of the mating pair triggers the processing of F plasmid into a translocation complex with a single-stranded DNA intermediate generated by the DNA transfer and replication proteins (Dtr complex) acting at the origin of the transfer sequence (*oriT*). The relaxase introduces a strand-specific nick to the 5' end where it binds, covalently forming with the Dtr proteins a substrate "competent" for uptake. This nucleoprotein particle is delivered across the donor cell envelope via a conjugal or "mating" pore to the recipient cell, and this step is followed by second-strand synthesis and dissociation of the mating pair. By definition, conjugation is a contact-dependent process. Thus, conjugation bears a strong mechanistic resemblance to the type III secretion systems (see chapter 15). Pathogenic bacteria of humans and plants have coopted conjugation systems to export virulence factors to eukaryotic host cells. Although this is a functionally diverse family, there are some unifying themes: (i) exporters are assembled at least in part from subunits of DNA conjugation systems, and (ii) the known substrates recognized by these transporters are large macromolecules such as nucleoprotein particles, scaffolding proteins, guanine nucleotide exchange factors, or multisubunit toxins.

An Expanding Family of Type IV Injectosomes

The best-characterized type IV system is that used by *Agrobacterium tumefaciens* for exporting oncogenic DNA (T-DNA) to plant cells. The result of DNA transfer is the proliferation of plant cells and ultimately the formation of tumorous tissues termed crown galls (Figure 16.1). T-DNA transfer requires approximately 20 virulence (Vir) proteins, which are encoded by six

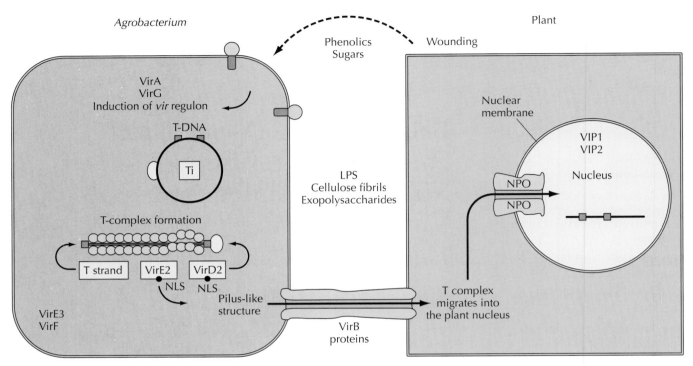

Figure 16.1   Model of the *Agrobacterium* infection process. Specific classes of phenolic and sugar molecules released from wounded plant cells induce the Vir regulon, resulting in the interkingdom transfer of the oncogenic T complex to plant nuclei.

operons. Some of the Vir proteins, notably the VirD2 relaxase and the VirE2 single-stranded DNA-binding protein (SSB), interact with single-stranded T-DNA (the T strand) to form the transfer intermediate (the T complex). In addition, the VirE3 and the VirF proteins are required for plant infection. Other Vir proteins, including the 11 *virB* gene products and the *virD4* gene product, assemble into a gated channel-pilus complex. The channel serves as a transenvelope conduit through which the T complex passes, while the pilus is proposed to mediate the physical contact between *A. tumefaciens* and recipient plant cells.

Both VirD2 and VirE factors possess nuclear-localization sequences (NLS) that mediate the nuclear import. Within the host cells the VirD2 protein is cyclically phosphorylated and dephosphorylated on serine/threonine residues then recognized by the carrier protein Kariopherin alpha for nuclear homing. VirE2 also interacts with the nuclear proteins VIP1 and VIP2 and with the Ran GTPase, suggesting that all are part of a system for targeting the nucleus and for the integration process. Figure 16.2 shows that the VirB components of the T-DNA transport system are highly related to Tra protein components of transfer systems encoded by several broad-host-range plasmids. The most extensive similarities are to Tra proteins encoded by the IncN plasmid, pKM101, and by the IncW plasmid, R388. Both of these Tra systems code for homologues of all 11 VirB proteins. In addition, the genes coding for related proteins are colinear in the respective *tra* and *virB* operons, providing further evidence of a common ancestry. A subset of the Tra proteins comprising the transfer systems of other broad-host-range plasmids, including RP4 (IncP) and the narrow-host-range plasmid F (IncF), also are related to VirB proteins.

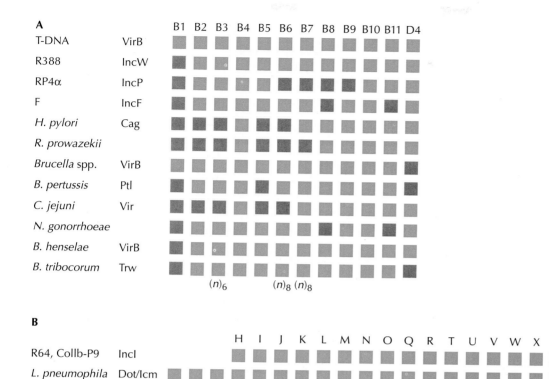

Figure 16.2 Alignment of genes encoding related components of the type IV transport systems. **(A)** Of the 11 VirB proteins, those encoded by *virB2* through *virB11* are essential for T-complex transport to plant cells (T-DNA). The broad-host-range plasmids and the narrow-host-range (NHR) plasmids code for Tra proteins related to most or all of the VirB genes. Type IV transporters found in bacterial pathogens of humans export toxins or other protein effectors to human cells show regions that are highly conserved among species. **(B)** Schematic representation of *Legionella pneumophila* Dot/Icm region and R64 plasmid.

The type IV secretion family also is composed of toxin exporters used by several bacterial human pathogens (Table 16.1). *Bordetella pertussis,* the causative agent of whooping cough, uses the Ptl transporter to export the A-B type pertussis toxin across the bacterial envelope. All nine Ptl proteins are related to VirB proteins, and, as with several of the plasmid Tra systems, *ptl* genes are colinear with the corresponding *virB* genes in the respective operons (Figure 16.2). Type 1 strains of *Helicobacter pylori,* the causative agent of peptic ulcer disease and a risk factor for the development of gastric adenocarcinoma, contain the 40-kb *cag* pathogenicity island that encodes for 31 different genes, and among these there are six Cag proteins related to VirB proteins and one Cag protein related to VirD4 (coupling protein or CP) essential for the translocation of the effector CagA oncoprotein within host cells. Two homologues of VirB4 and one homologue each of VirB8, VirB9, VirB10, VirB11, and VirD4 were detected in the complete nucleotide sequence of *Rickettsia prowazekii,* suggesting that this microorganism encodes a type IV transport system similar to *H. pylori.* Mutational studies aimed at identifying genes involved in intracellular growth of *Legionella pneumophila* and in macrophage killing resulted in the identification of the *dot/icm* (defective for organelle trafficking and/or intracellular multiplication) locus

Table 16.1   Biological effects mediated by type IV secretion systems in various bacterial pathogens

Bacterium	Effects
Helicobacter pylori	IL-8 induction on epithelial cells
	Tyrosine phosphorylation of the 145-kDa CagA molecule translocated in gastric epithelial cells
	Cell shape elongation effect
	Anchorage independency
	Disruption of the AJC and terminal differentiation process
	NF-κB activation after exposure of epithelial cells to type I strains
	Loss of polarity and increased motility interactions with ZO–1 and JAM
	Colonization and bacterial density in animal models and human biopsy specimens
Agrobacterium tumefaciens	Export of T-DNA-protein complex to plant cells
Escherichia coli	Intra- and interspecies conjugative DNA transfer to other bacteria and yeasts
Bordetella pertussis	Pertussis toxin export in the extracellular compartment
Legionella pneumophila	Phagolysosome fusion in macrophages
	Host cytotoxicity by pore formation via DotA
	Reduction of viability by LidA
	Ralf1-mediated activation of ARFs
	DNA transfer
Rickettsia prowazekii	Phagolysosome fusion in eukaryotic cells
	N-Wasp-Arp2/3 activation
	DNA transfer

sharing limited homologies with the VirB10 and VirB11 proteins but extended homologies with the IncI family of plasmids like R64 and ColI-P9. The *dot/icm* proteins act as an exporter for virulence factor(s) targeted to the intracellular environment since the causative agent of Legionnaires' disease and Pontiac fever resides into a vacuole that support replication. *Coxiella burnetii* encodes for 21 components mechanistically related to the Legionella *dot/icm* apparatus, suggesting that the virulent agents of Q fever in humans create specialized endocytic vacuoles using a functional type IV secretion system (TFSS) similar to *Legionella*. Recent studies on *Bartonella tribocorum* have identified two TFSSs: one with similarities to plasmid R388 (IncW) and one resident on a PAI that has duplications scattered around the VirB2, VirB5, VirB6, and VirB7 homologues. The magnitude of the phenomenon suggests that the variants could generate a repertoire useful to circumvent the immune response. Several effector proteins are under characterization, and we will soon learn about factors acting at the level of erythrocytes and at the level of small vessels.

The growing list of pathogens that utilize TFSS for delivery of effector molecules into the host cell environment, comprising species like *Brucella, Actinobacillus, Ehrlichia, Wolbachia,* and *Xilella*, is under continous revision. The chapter includes a description of the effects associated with intracellular delivery of well-characterized effectors of virulence in the hope that *Bordetella, Helicobacter,* and *Legionella* will expand our knowledge about TFSS dissemination, diversification, and evolution and that new experimental organisms will document, at the end, many additional ways to circumvent basic cellular processes.

The *A. tumefaciens* T-Complex Transfer System as a Paradigm for Assembly of Type IV Secretion Systems

Extensive studies of the *A. tumefaciens* T-complex transporter components have generated the following information. There is compelling evidence that VirB2 is the major pilin subunit for the pilus. VirB2 localizes at the inner membrane: after cleavage by the prepilin peptidase and cyclization, via cyclic peptidase, it is recruited to the outer membrane for pilus assembly. The growth of the pilus depends on the coordinate action of the VirB1 transglycosilase for lysis of the peptidoglycan layer. The VirB proteins are required for pilus assembly and pilus morphogenesis, providing an anchor at the membrane for pilus attachment. VirB4 and VirB11 contain consensus Walker "A" nucleotide-binding motifs, and each hydrolyzes ATP in vitro. These proteins utilize the energy of ATP hydrolysis to drive transporter assembly (VirB11) and substrate translocation (VirB4). The structure of the *Helicobacter* homologue of VirB11 (HP0525) has recently been solved and has validated the model that predicted that the size of the pore at the center of the hexamer can be regulated by the binding of ATP (close) and its release (open). Each of these ATPases forms homohexameric structures, and they are likely to interact to form a pore with a variable size tightly associated with the inner membrane. The VirD4-coupling protein is also associated with the inner membrane, aggregating into hexamers and connecting the cytoplasm to the periplasm. This particular component is involved in association with the substrate, acting as an adapter for specific protein-protein interactions with secreted components. VirB4 is an integral membrane protein with two domains either embedded into or through the membrane into the periplasm. VirB11 is tightly bound to the membrane and most probably is localized exclusively on the cytoplasmic face of the membrane. The functions of the remaining transporter components have not been defined in detail. Several integral inner membrane protein components such as VirB6 and VirB8 are likely candidates as structural components of an inner membrane pore, with VirB10 forming a bridge between the inner membrane and the outer membrane as a part of periplasmic channel. Other outer membrane components such as VirB7 and VirB9 are likely candidates as components of an outer membrane pore. The remaining VirB proteins may be structural subunits of the putative transenvelope channel, or they may transiently assist in the assembly of this structure. Figure 16.3 depicts a model of the T-complex transporter-pilus with the likely positions of the VirB and VirD4 proteins (Figure 16.4).

The possible roles of individual VirB proteins in transporter assembly have been analyzed in part by construction of nonpolar virB mutants. Studies of these mutants revealed that some of the VirB proteins, most notably VirB6, VirB7, and VirB9, provide important stabilizing functions for other VirB proteins. VirB7, an outer membrane lipoprotein, was shown to stabilize itself as well as VirB9 through the assembly of a disulfide cross-linked VirB7-VirB9 heterodimer. Several lines of evidence suggest that the correct positioning of the VirB7-VirB9 heterodimer at the outer membrane is required for assembly of a functional transporter. Thus, it has been proposed that this heterodimer functions as a nucleation center for transporter assembly. Of further interest, VirB6, a polytopic inner membrane protein, recently has been shown to facilitate VirB7 dimer formation or stabilization, as well as the stabilization of other VirB proteins, including the VirB4 and VirB11 ATPases. Taken together, these biochemical results have led to the

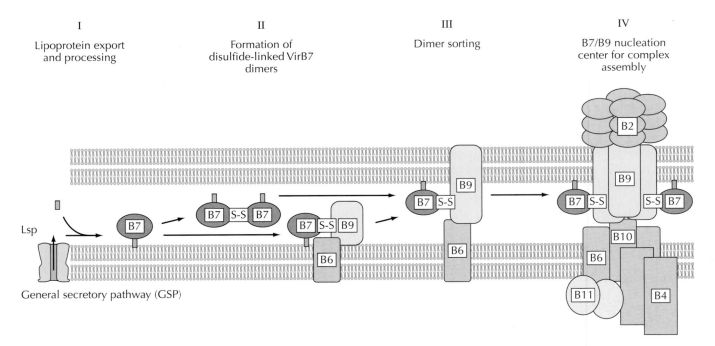

I

Lipoprotein export
and processing

II

Formation of
disulfide-linked VirB7
dimers

III

Dimer sorting

IV

B7/B9 nucleation
center for complex
assembly

Lsp

General secretory pathway (GSP)

Figure 16.3 Early steps in a proposed assembly pathway for the T-complex transporter. VirB proteins, including VirB7 and VirB9, are exported across the cytoplasmic membrane via the general secretory pathway. Lsp, signal peptidase II, processes the prelipoproteins. The VirB7 lipoprotein (fatty acid modification denoted by lollipop stick) assembles as homodimers and heterodimers with VirB9, possibly facilitated by VirB6. The VirB6-VirB7-VirB9 protein subcomplex is a proposed nucleation center for recruitment and stabilization of other VirB proteins, leading to assembly of the proposed gated channel-T pilus shown on the right.

hypothetical model shown in Figure 16.3, depicting several early stabilizing interactions that must occur before the biogenesis of a functional T-complex transporter can take place. Interestingly, the PtlI and PtlF homologues of VirB7 and VirB9 also assemble as disulfide cross-linked dimers, and dimer formation also is important for Ptl protein stabilization. Furthermore, the PtlD homologue of VirB6 has a stabilizing effect on other Ptl transporter components. Finally, the PtlH homologue of VirB11 requires an intact ATP-binding domain for function, supporting the sequence-based prediction that PtlH provides energy from ATP hydrolysis for transporter assembly or function. Thus, even at this early stage of analysis of these systems, it is clear that several common mechanistic principles are applied to build functional T-complex and pertussis toxin exporters.

The existence of a subset of VirB homologues in the *H. pylori cag* and *L. pneumophila icm/dot* systems underscores the functional importance of these types of proteins in macromolecular export. Of particular note, the *cag* PAI codes for homologues of the two VirB ATPases, VirB4 and VirB11, and proposed structural components of the transfer channel, VirB7, VirB9, VirB10, and VirD4. Together, these findings raise the intriguing possibility that this subset of proteins corresponds to a minimal ancestral protein subassembly that bacteria have built upon to construct transporters designed for novel purposes ranging from intercellular DNA transfer, toxin export, and, possibly, direct injection of virulence factors into eukaryotic cells.

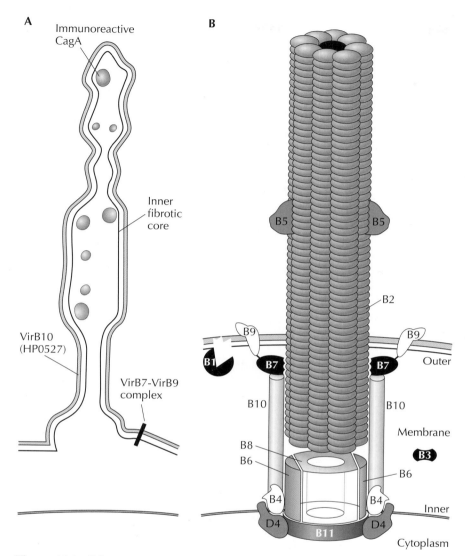

Figure 16.4 Schematic prototype model for a type IV secretion injectisome. **(A)** Schematic illustration of the structure of the *cag* organelle featuring an inner core of fibrotic structures surrounded by a membranous sheet containing the VirB10 (HP0527) protein and the protruding basal complex of the VirB7-B9 homologues. **(B)** After the formation of a nucleation center, the transglycosilase VirB1 lyses the petidoglycan layer, facilitating the growth of the pilus structure formed by the assembly of aggregates of the VirB2 cyclic monomers. Immunoreactive CagA is represented as black bodies.

Substrate Recognition and the Export Route

During conjugation, the donor cell generates a transfer intermediate consisting of a single strand of DNA covalently bound at its 5′ end with the processing endonuclease and coated along its length with an SSB. How the conjugation machine recognizes this nucleoprotein particle is a subject of great interest. There are several lines of evidence suggesting that substrate recognition is conferred not by the DNA but rather by proteins associated with the DNA. Although the nature of these signals is unknown, it is likely the exported protein components contain a conserved peptide sequence or a recognition motif generated on protein folding. It is therefore tempting to speculate that conjugation machines are in fact protein exporters that evolved to accommodate "hitchhiker" DNA during transport.

The *A. tumefaciens* T-DNA-processing reaction resembles the conjugative DNA-processing reaction, resulting in formation of the T-strand/VirD2/VirE2 nucleoprotein particle or T complex. Perhaps the most compelling evidence that conjugation machines recognize proteins as translocation-competent substrates is that VirE2 SSB can be exported to plant cells independently of the T-strand/VirD2 complex. Conjugal DNA transfer intermediates are thought to be delivered across the donor cell envelope in a single step through a proteinaceous channel (mating pore). In striking contrast, the highly related Ptl exporter is proposed to function quite differently. Pertussis toxin is a multisubunit toxin composed of five different subunits, each possessing characteristic signal sequences. These subunits are proposed to be delivered across the inner membrane via the general secretory pathway. Pertussis toxin assembles in the periplasm, where it is exported by the Ptl system across the outer membrane.

In view of the proposed one-step translocation model for conjugal DNA transfer and the two-step model for export of pertussis toxin, it is of considerable interest to recall that the two transporters are assembled from almost identical sets of protein homologues (Figure 16.1). Among the essential VirB proteins, only VirB5 has no counterpart in the Ptl system. This observation raises the interesting possibility that VirB5 and related Tra proteins supply a function(s) specifically for assembly of DNA transporters. In addition, in contrast to the contact dependence of conjugal DNA transfer, *B. pertussis* excretes functional pertussis toxin into the extracellular milieu, which has led to the hypothesis that the Ptl proteins do not elaborate a pilus. However, it is notable that PtlA is related to the VirB2 pilin subunit. This could mean that in fact the Ptl system does assemble at least a vestige of a pilus or that VirB2-type proteins supply a function, which is critical to transporter assembly but unrelated to pilus assembly.

The substrates exported by the *L. pneumophila icm/dot* systems have recently been identified. However, one known substrate for the *icm/dot* system is the IncQ plasmid RSF1010. This is a non-self-transmissible plasmid that possesses the origin of transfer (*oriT*) and cognate mobilization (*mob*) proteins for processing the transfer intermediate (here referred to as the R complex). The proposed R complex resembles the T complex as a linearized single strand of RSF1010 DNA covalently associated at its 5' end with MobA endonuclease and coated along its length with SSB. Many conjugative plasmid transfer systems, as well as the T-complex transporter, recognize and export the R complex to recipient cells. Indeed, wild-type *A. tumefaciens* cells carrying RSF1010 show a reduced ability to export T-DNA, possibly as a result of substrate competition for a limited number of transporters.

Similarly, the *L. pneumophila icm/dot* system directs the movement of plasmid RSF1010 between bacteria. Cells carrying RSF1010 display a reduction in intracellular multiplication and human macrophage killing. Both of these features of the infection process are thought to result from export of a toxin effector via the *icm/dot* system. These observations suggest that the *icm/dot* system has retained a functional vestige of the ancestral DNA conjugation system from which it evolved. Whether other type IV toxin export systems, including the *B. pertussis* Ptl system and the *H. pylori cag* system, also are capable of directing conjugative DNA transfer to recipient bacteria or even to eukaryotic cells remains to be tested.

Molecular Biology of Type IV Secretion Systems

Genomes increase in complexity through processes like gene duplication. The flagellar apparatus, or a simplified version of it, was proposed as a common ancestor of the type III secretion machines. These systems eventually specialized as extracellular tubular protrusions or as an intracellular gated complex involved in substrate transfer between different subcellular compartments. This analogy can be extended to the type IV family, which was proposed to originate from a conjugative apparatus that later associated with other classes of genes, providing, over the evolutionary timescale, an entirely new range of functions. We speculate that both the types III and IV systems have a progenitor in the filamentous-phage assembly and secretion processes.

Type IV systems are involved in a variety of physiopathological effects (Figure 16.4). These systems deliver toxins and nucleoprotein particles that interact with discrete receptors and, for all of the known systems, are internalized into cells, where they mediate a variety of intracellular responses. Within the eukaryotic cell, the effector molecules may perturb cellular functions through direct structural interactions. The liberation of pertussis toxin is a well-known example: the toxin has an ADP-ribosylating activity specific for GTP-binding proteins, causing cell intoxication. Alternatively, the transferred molecule may be DNA, in which case the expression of genes alters the cellular physiology. The example highlighted in this article is *Agrobacterium*-mediated T-DNA transfer, which ultimately disturbs plant hormone balances, leading to loss of cell growth control and to tumor formation. Similarly, the conjugal transfer of plasmids between bacterial cells often induces changes in cell physiology that aid in cell survival in harsh environments or increased virulence.

In the other type IV secretion systems, the variety of effects is extremely rich even through the exported substrate. In *Helicobacter*, the *cag* system induces the epithelial secretion of IL-8, a mediator of chronic inflammation, and this process depends on NF-κB activation. In addition, the bacterial CagA protein is tyrosine phosphorylated by the host, and changes in cell shape and junctional activity are disrupting the integrity of the epithelium, reducing the wound-healing mechanisms and producing a basolateralization of the target cells. Interestingly, a similar event of basolateralization is common in the attaching-effacing lesions induced by the type III secretion system of enteropathogenic strains of *E. coli*. The *Legionella icm/dot* system transfers DNA, although DNA is not thought to be the substrate involved in *Legionella* virulence. Rather, in mammalian hosts, the *dot/icm* products export at least three different factors that promote intracellular multiplication and killing of human macrophages and prevention of phagosome-lysosome fusion. This latter response suggests the *dot/icm* system plays a direct role in altering communication processes between distinct subcellular compartments.

The Cell Biology of Effector Molecules

Bordetella pertussis ptl System

B. pertussis is the causative agent of whooping cough. Most of the symptoms depend on the release of a potent toxin of the A-B family that consists of enzymatic S1 subunits responsible for ADP-ribosylation of the inhibitory Gα subunit of the heterotrimeric G proteins and a pentameric B component composed by four proteins (S2, S3, S4, and S5) at a molar ratio of 1:1:2:1, respectively. Subunits are translocated in the periplasmic space by the general secretory pathway after cleavage of the signal sequence. The holotoxin assembly is then secreted in the periplasmic milieu by the coordinate action of a TFSS called *ptl* (pertussis toxin liberation). This hybrid system of translocation and secretion utilizes the general secretory pathway (GSP) coupled to a contact-dependent secretion system for cargo delivery. The cell-binding activity of the B-pentamer for surface glycoproteins allows the internalization of the catalytic subunit by receptor-mediated endocytosis. ADP-ribosylation uncouples the G protein from their receptors, causing deregulation of the signal transduction pathways and resulting in increased insulin secretion and histamine sensitization.

Helicobacter pylori cag System

H. pylori colonizes the human gastric mucosa for the entire lifespan of the host, causing strong local inflammation, continuous epithelial damage, and destruction of the mucosal surface well beyond the vascularized layer. Due to the deepness of crateriformic lesions, ulcerations usually are repaired by a fibrotic scar that will never be covered by the regenerating epithelium. This absence of *restitutio ad integrum* could explain why the risk of tumorigenic conversion is increased, since damage and partial healing are inflicted endlessly at the site of infection. Homing and cell damage are achieved by intracellular injection of the CagA protein with *cag*, a TFSS spanning 31 ORFs of a pathogenicity island associated with the genomic locus of glutamate racemase. CagA is a large molecule of 128 to 142 kDa with distinct functional and binding domains that mimic a scaffolding structure. Once translocated it is tyrosine phosphorylated at the EPIYA motifs by the src/lyn host cell kinases and was found to be membrane associated. Two EPIYA target sequences are part of all CagA protein of western and eastern origin, but additional EPIYA motifs originate by tandem duplication of 102-bp modules, each identical and composed by three segments of the CagA gene that encodes the EPIYA and hydrophilic amino acid linkers (called D1, D2, and D3). This "variable region" of CagA can be defined as a "recombinogenic region." Repeats from Western and Asian strains are just variants, simply arising by a specific reassortment of well-conserved sequences of CagA that always include the canonical motif for tyrosine phosphorylation (Figure 16.5). Tyrosine phosphorylation is a stoichiometric reaction where all sites are progressively modified and the biological activity is a function of the number of activated sites. The major phenotypic effect is a massive cell elongation followed by increased motility. This phenomenon reflects the strong involvement of the EPIYA sequences in the elongation processes mediated by reorganization of the filamentous actin and a parallel reduction in cell-to-cell contacts. The coordinate use of CagA mutants in experiments based on transfection and infection of polarized MDCK cells has revealed that the carboxy-terminal region (EPIYA region) is responsible only for cell-shape elongation and depends on tyrosine phosphorylation. The amino-terminal region disrupts the maturation of the apical junctional complex (AJC) after binding ZO-1 and JAM proteins, causing basolateral leakage. Interestingly, CagA$^+$ strains induce aberrant and ectopic protojunctions on precancerous AGS cells, by recruitment of the scaffolding protein ZO-1. Foremost, Δ*cagA* strains are impaired in apical-junction association, and adhere to the apical surface of the host epithelial cells without any particular preference for the intercellular regions. Finally, CagA association with the junctional components is phosphorylation independent and spatially colo-

A

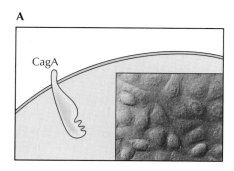

B

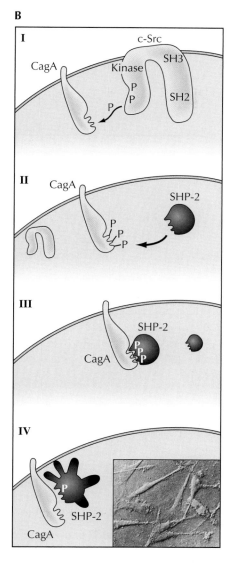

Figure 16.5 Suggested model for CagA action. Translocated CagA associates with cell membrane **(A)** and is tyrosine phosphorylated **(B)** by members of c-Src family (c-Src and c-Lyn) at the level of the EPIYA motifs (I). The presence of phosphate residues favors the recruitment of the cytoplasmic form of the phosphatase SHP-2 (II) and the transition from a closed into an open structure (III). Cell elongation depends on the number of CagA-phosphorylated sites and on the formation of a complex with the phosphatase (IV).

calized with the apical-junctional components ZO-1, JAM, and E-cadherin (Figure 16.6).

A plethora of effects related to CagA have been described, especially in the field of signal transduction pathways where possible interactions with cellular partners have been outlined. CagA has been reported to interact

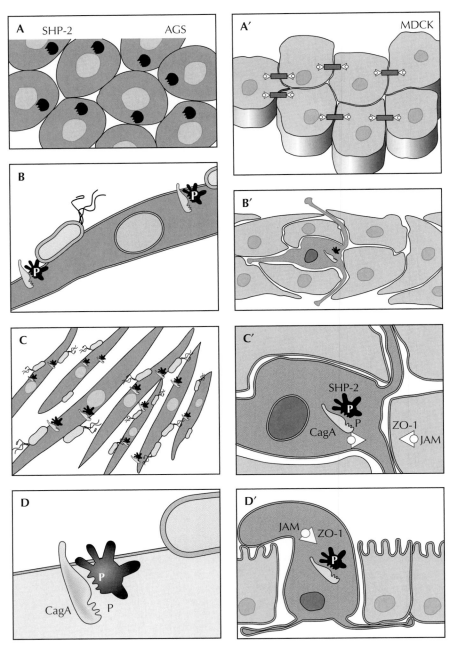

Figure 16.6 A revised model based on the different behavior of nonpolarized AGS cells and polarized MDCK cells. **(A to D)** The absence of junction formation is typical of AGS cells. CagA translocation induces ectopic synthesis of the ZO-1 and Jam proteins that are associated with the early stage of junctional assembly. Transfection with the N-terminal fragment of CagA is associated with elongation phenotype resulting from the phosphorylation activity. **(A′ to D′)** During monolayer formation of individual MDCK cells and assembly of junctional components, a single transfectant for the C-terminal portion of CagA exhibits dominant negative interference with the core complex of the apical junctional complex (AJC) and lacks of junctional activity. The cell is characterized by increased motility **(B′)**, condensation activity for ZO-1 and Jam, and recruitment of SHP-2 at the level of junction **(C′)**. In the absence of the AJC controls, the cell protrusions look for inhibitory contacts and the cell is marked by stratified growth **(D′)**.

with the tyrosine phosphatase SHP-2. This phosphatase is a Src homology 2 (SH2) domain-containing protein. SHP2-CagA binding requires tyrosine phosphorylation. SHP-2 is a cytoplasmic protein in a close conformation status. It is after binding CagA that SHP-2 opens its structure and associates with CagA in the membrane fraction, participating in regulation of cell morphology and motility. CagA is also capable of interacting with C-terminal Src kinase (Csk) via its SH2 domain. Upon complex formation, CagA stimulates Csk, which in turn inactivates the Src family of protein tyrosine kinases. Binding of CagA with the adapter protein Grb-2 and the hepatocyte growth factor (HGF) receptor c-Met has been reported. CagA/Grb-2 interaction has been proposed to be crucial to induce the growth factor-like morphological changes in the epithelial host cell (Figure 16.7).

Ectopic expression of ZO-1 induced by infection with CagA$^+$ strains and ZO-1 colocalization with the transfected CagA-N terminus, strongly suggest that the CagA protein binds selectively to the junctions. Tight junctions are continuous circumferential intercellular contacts at the apical poles of lateral cell membranes, appearing in electron micrographs as a series of discrete contacts between the plasma membranes of adjacent cells. Tight junctions form a barrier that regulates the paracellular transit of water, solutes, and immune cells across an epithelium, and are essential for establishing cell polarity by separating the apical and basolateral

Figure 16.7   Schematic summary of effects depending on CagA translocation. Grb-2 interactions trigger signal transduction pathways. Tyosine phosphorylation is responsible for the elongating effect and disruption of the AJC by ZO-1 and Jam recruitment and SHP-2 delocalization alters terminal differentiation processes.

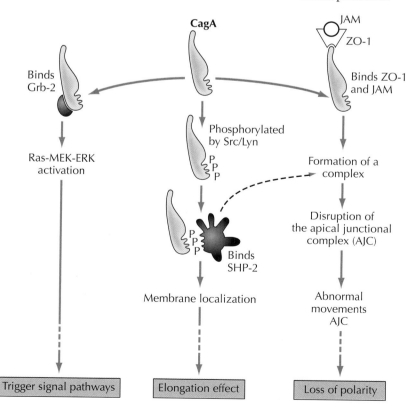

domains of polarized epithelial cells. Besides these "mechanical" functions, tight junctions are also involved in signal transduction pathways to modulate gene expression, cell proliferation, and differentiation. Tight junctions form a complex of transmembrane, scaffolding, and signaling proteins. ZO-1, in particular, has been implicated in recruiting structural and signaling protein to the apical junctional complex. Tight junction regulation can occur indirectly through changes in the perijunctional actomyosin ring, or directly through changes in specific junctional proteins. Bacterial pathogens, such as Clostridia, *Vibrio cholerae*, Enteropathogenic *E. coli* (EPEC), and *Listeria monocytogenes,* have learned how to disrupt tight junctions of epithelial cells in several different ways. Disruption of specific tight-junction proteins can result from selective degradation by bacterial derived proteases or by biochemical alterations such as phosphorylation or dephosphorylation. MDCK monolayers infected for 3 days with a CagA$^+$ *H. pylori* strain alter the tight junctions simply through association or by recruitment of a biologically active molecule in close proximity to them. The phosphatase SHP-2 cofractionates with CagA and with junctional proteins in iodixanol density gradients (Optiprep). The recruitment of SHP-2 at the tight junctions may alter apical-junctional complex function by dephosphorylating junctional proteins or substrates located close to them. Moreover, SHP-2 activated by CagA is likely to interfere with the normal signal pathways starting and arriving at the AJC (Figure 16.6).

Two reports have recently suggested that the pilus structure of *H. pylori* can deviate from the classical structure of a conjugative pilus. Transmission electron microscopy (TEM) images showed a core of fibers surrounded by a membranous sheet with large engulfments. The sheet was partially composed by the homologue of VirB10 (HP0527), a protein with 74 segments that resembles the fiber-forming channel used by ditters to feed larvae. In addition, the VirB7-B9 complex (HP0532 and HP0538) was observed to extend from the cell surface (Figure 16.4A). Further studies are necessary to complete the dissection, but it is worth mentioning that appendages of the *cag* TFSS have been described as required during the early phase of colonization in animal models of infection, in inducing Raft aggregates (patches) in Jurkat T-cell lines, and in triggering IL-8 induction on monolayers.

Legionella pneumophila dot/icm System

Legionella pneumophila is a facultative intracellular pathogen that internalizes and multiplies within specialized phagosomes that escaped the fusion with lysosomes by interception of vesicles from the endoplasmic reticulum (ER). The secretory vesicles fuse with the membrane of the replicative vacuole, forming a coat that mimics the rough ER membrane. During replication *Legionella* secretes the DotA protein by using the *dot/icm* system. The protein makes, after oligomerization, channels in the plasma membrane, which could also be used for substrate delivery directly into the cytosol (Figure 16.8).

Another factor that affects viability and replication of bacterial cells has been discovered by mutagenesis and called LidA (lowered viability in the presence of Dot). The protein localizes at the vacuolar surfaces after TFSS-dependent secretion and is essential for bacterial survival and for vesicle recruitment. Among the effectors the discovery of the ARF1 protein has provided important insight in intracellular vesicle transport. ADP-ribosylating factors (ARFs) are GTP-binding proteins that control

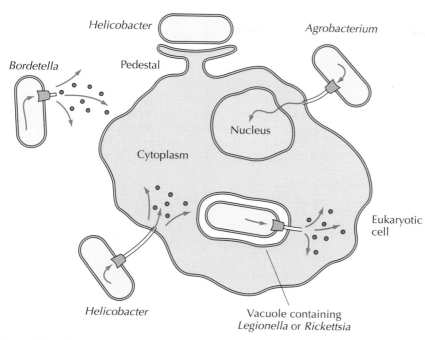

Figure 16.8 As with the type III systems, type IV systems are present in bacteria with different pathogenic behaviors. *Helicobacter* attachment to the eukaryotic cell surface is mediated by cell skeleton remodeling. *Agrobacterium* and *Helicobacter* likely establish contact with eukaryotic host cells via a pilus structure prior to intracellular delivery of effector molecules. *Bordetella* secretes toxin via the Ptl system. Intracellular microorganisms, like *Legionella* and *Rickettsia,* confined in vacuoles, cross talk with the cytoplasmic compartment with a Ptl-like system, presumably deprived of piliated protrusions.

the vesicular traffic between the ER and Golgi apparatus. Selective inhibition of ARFs by brefeldin A blocks the targeting of ER-associated proteins to the phagosomes, inhibiting the biogenesis of replication-permissive vacuoles where *Legionella* multiplies. ARF1 accumulation in the phagosomes is *dot/icm* dependent. GDP-bound ARF1 is localized in the cytosol while the GTP-bound form is concentrated in the cytosolic surface of vesicle membranes. ARF1 activation depends on active secretion of RalF, a guanine nucleotide exchange factor that stimulates GDP to GTP exchange in ARF proteins. The discovery of RalF implicates that the *dot/icm* system is involved in secretion and translocation of RalF across the phagosomal membrane. RalF secretion could promote the concentration of the ARF1 protein on the phagosomes, stimulates vesicle transport between ER and Golgi, favoring phagosome remodeling, and stimulates replicative vacuole biogenesis.

Conclusion

The evolution of a family of secretion systems from ancestral DNA conjugation machines raises many interesting questions and exciting new research directions. For example, it is of considerable biomedical importance to identify all the effectors exported by the *H. pylori* Cag and *L. pneumophila* Icm/Dot transporters. Equally important are detailed mechanistic and comparative studies of the type IV transporters with respect to the definition of (i) the basis for substrate recognition, (ii) the architectural arrangement and assembly pathway(s), (iii) the energy requirements for transporter assembly and function, and (iv) molecular interactions between these systems and eukaryotic host cells. Such information ultimately will be useful for applied studies aimed, for example, at vaccine development, the design of drugs for selective inactivation of type IV secretion systems, or, even more enticing, testing the potential of these systems for delivery of therapeutic protein or DNA macromolecules directly to eukaryotic cells.

Selected Readings

Amieva, M. R., R. Vogelmann, A. Covacci, L. S. Tompkins, W. J. Nelson, and S. Falkow. 2003. Disruption of the epithelial apical-junctional complex by Helicobacter pylori CagA. *Science* **300:**1430–1434.

Cascales, E., and P. J. Christie. 2003. The versatile bacterial Type IV secretion systems. *Nat. Rev. Microbiol.* **1:**137–149.

Covacci, A., J. L. Telford, G. Del Giudice, J. Parsonnet, and R. Rappuoli. 1999. Helicobacter pylori virulence and genetic geography. *Science* **284:**1328–1333.

Ding, Z., K. Atmakuri, and P. J. Christie. 2003. The outs and ins of bacterial type IV secretion substrates. *Trends Microbiol.* **11:**527–535.

Hatakeyama, M. 2003. *Helicobacter pylori* CagA: a potential bacterial oncoprotein that functionally mimics the mammalian Gab family of adaptor proteins. *Microb. Infect.* **5:**143–150.

Higashi, H., R. Tsutsumi, S. Muto, T. Sugiyama, T. Azuma, M. Asaka, and M. Hatakeyama. 2002. SHP-2 tyrosine phosphatase as an intracellular target of Helicobacter pylori CagA protein. *Science* **295:**683–686.

Matter, K., and M. S. Balda. 2003. Signalling to and from tight junctions. *Nat. Rev. Mol. Cell Biol.* **4:**225–236.

Montecucco, C., and R. Rappuoli. 2001. Living dangerously: how Helicobacter pylori survives in the human stomach. *Nat. Rev. Mol. Cell Biol.* **2:**457–466.

Nagai, H., and C. R. Roy. 2003. Show me the substrates: modulation of host cell function by type IV secretion systems. *Cell Microbiol.* **5:**373–383.

Rohde, M., J. Puls, R. Buhrdorf, W. Fischer, and R. Haas. 2003. A novel sheathed surface organelle of the Helicobacter pylori cag type IV secretion system. *Mol. Microbiol.* **49:**219–234.

Schaeper, U., N. H. Gehring, K. P. Fuchs, M. Sachs, B. Kempkes, and W. Birchmeier. 2000. Coupling of Gab1 to c-Met, Grb2, and Shp2 mediates biological responses. *J. Cell Biol.* **149:**1419–1432.

Stein, M., F. Bagnoli, R. Halenbeck, R. Rappuoli, W. J. Fantl, and A. Covacci. 2002. c-Src/Lyn kinases activate Helicobacter pylori CagA through tyrosine phosphorylation of the EPIYA motifs. *Mol. Microbiol.* **43:**971–980.

17

Induction of Apoptosis by Microbial Pathogens

Jeremy E. Moss, Ilona Idanpaan-Heikkila,
and Arturo Zychlinsky

There are two distinct mechanisms of cell death: apoptosis and necrosis. The most significant difference between them is that in apoptosis the cell's own molecules are ultimately responsible for its death, while in necrosis the cell is a victim of molecules synthesized by other cells. For example, apoptosis can be initiated by a cytokine binding to its receptor, which initiates a signaling cascade that activates a cell death program. Necrosis results from the effects of toxic molecules on a cell. Activation of complement results in the formation of a pore on the membrane of the target cell called the membrane attack complex. This pore disrupts the integrity of the cell membrane, causing a rapid exchange of ions and, eventually, osmotic lysis.

In multicellular organisms both development and homeostasis are achieved through a balance between cell death and cell growth. Programmed cell death, or apoptosis, is activated in multicellular organisms when death is desirable for the well-being of the whole organism. A classic and graphic example of the importance of apoptosis in development can be seen during the morphogenesis of the limb. During embryogenesis, the digits are interconnected by a web of cells. At a precise time point, the cells that constitute the interdigital web undergo programmed cell death, allowing the formation of independent fingers.

There are two criteria to distinguish apoptosis from necrosis: morphology and DNA fragmentation. Typical morphological changes (Figure 17.1) that occur during apoptosis include cell shrinkage and loss of normal cell-to-cell contacts, blebbing at the cell surface, and intense cytoplasmic vacuolization. Conservation of organelle structure, condensation of the chromatin (often at the perinuclear region), and loss of normal nuclear architecture also occur. In contrast, during necrotic cell death, the organelles are critically damaged, the plasma membrane is ruptured, the cytoplasmic elements are dispersed into the extracellular space, and the shape of the nucleus is normally conserved, although flocculation of chromatin may be detected.

DNA fragmentation is a biochemical characteristic that was detected early in the study of cell death. Apoptotic cells break up their DNA into

409

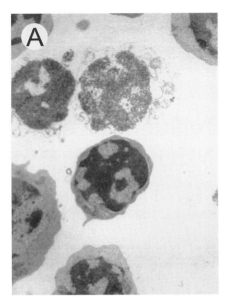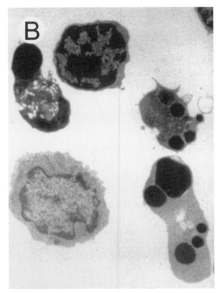

Figure 17.1 Morphology of thymocytes undergoing necrosis and apoptosis. **(A)** Thymocytes undergoing necrosis. Organelles are critically damaged, the plasma membrane is ruptured, the cytoplasmic elements are dispersed, the shape of the nucleus is normally conserved, but flocculation of chromatin can be detected. **(B)** Thymocytes undergoing apoptosis. Cell shrinkage, blebbing at the cell surface, intense cytoplasmic vacuolization, conservation of organelle structure, condensation of the chromatin at the perinuclear region, and loss of nuclear morphology, which appears like segmentation of the nucleus, are all apparent.

multimers of approximately 200 bp, correlating roughly with the size of a nucleosome. This fragmentation of DNA can be quantified and used as a marker for programmed cell death. In contrast, during necrosis, the DNA of the dying cell remains intact initially, although it is degraded eventually.

Another characteristic of apoptotic cells is that they express specific markers on their surface, such as vitronectin and phosphatidylserine, which can be recognized by professional phagocytes. These cells can rapidly engulf the apoptotic bodies before their intracellular contents are spilled into the tissue, and therefore they prevent inflammation.

Interestingly, the morphological changes, the fragmentation of DNA, and the expression of markers for recognition by phagocytes are very similar across different cell types and species. This reflects remarkable conservation of the mechanism of apoptosis, which is further underscored by the homology across species of the genes involved in this process.

Apoptosis Is Genetically Programmed

Initially, the demonstration that there was genetic information for cell death, i.e., that cell death was "programmed," came from now classical studies of the nematode *Caenorhabditis elegans*. Two features of *C. elegans* make it an ideal organism for the study of development and consequently of cell death. First, these worms are transparent, and so the fate of each cell can be microscopically observed in living animals. Second, of 1,090 somatic cells generated during hermaphrodite development, 131 naturally undergo programmed cell death. Furthermore, each cell in the worm has been mapped in both genealogy and position.

Table 17.1   *C. elegans* genes involved in apoptosis

Type of gene	Gene in:	
	C. elegans	**Mammals**
Proapoptotic	*ced-3*	Caspases
	ced-4	*apaf-1* family
Antiapoptotic	*ced-9*	*bcl-2* family

In mutagenesis studies, two cell death (*ced*) genes, *ced-3* and *ced-4*, were shown to be required for cell death. If either of these genes was inactivated, death failed to occur and mutant adult worms ended up with "extra" cells (Table 17.1). A third gene involved in the regulation of cell death is *ced-9*. The product of this gene inhibits cell death and functions upstream of *ced-3* and *ced-4*. The discovery of these genes demonstrated unequivocally that cell death was genetically programmed. Amazingly, the apoptotic program described in *C. elegans* is conserved in mammalian cells since not only are there mammalian homologues of *ced-3, ced-4*, and *ced-9*, but also, in some cases, mammalian genes can complement mutations in worms and worm genes are functional in mammalian cells.

In mammals, the program for cell death is more complex than that in worms and the induction of apoptosis is highly regulated. Furthermore, there are many homologues of certain cell death genes, suggesting both redundancy and multiple pathways of apoptosis. There has been a recent explosion in the identification of genes that are involved in apoptosis; however, a clear picture of how all of the pieces to the puzzle fit together has not been established.

Triggering of Apoptosis

Signal Transduction of a Cell Death Signal

Many apoptotic signals are received by the cell through surface receptors. Several mechanisms exist to transduce either the pro- or the antiapoptotic signal from the cell surface to the cell death machinery. The initial signal frequently comes from the binding of a ligand to a specific receptor at the cell membrane. This signal is transduced within the cell and can initiate several pathways, including generation of second messengers like an increase in intracellular cyclic AMP (cAMP) or calcium concentrations, activation of specific kinases that start the apoptotic program, and interaction with molecular adapters that directly connect with cell death effectors.

An illustrative and relatively simple example of this process is the case of Fas ligand-induced apoptosis. Fas ligand is a molecule related to the cytokine tumor necrosis factor alpha (TNF-α) and is important in T-lymphocyte cytotoxicity. Fas ligand binds to Fas, its receptor, on the target cell. This binding leads to an active receptor complex which transduces a death signal via the cytosolic adapter molecule Fas-associating protein with death domain (FADD). FADD connects the receptor and an effector protease of the caspase family. The function of caspases is discussed below.

Apoptosis and the Cell Cycle

Apoptosis can also be triggered from within, when cells receive conflicting signals for cell cycle progression and arrest or after irreparable damage to DNA. p53 is proposed to be instrumental in blocking cell cycle progression and activating apoptosis. p53 is a DNA-binding protein that can transactivate genes. Inactivation of p53 prevents cells from undergoing apoptosis after specific stimuli, supposedly because it controls the expression of cell death effector genes at a crucial cell cycle checkpoint. This tumor suppressor gene might prevent malignant transformation by activating programmed cell death. In fact, it is the most commonly mutated gene in human cancer.

Effector Molecules of Apoptosis

Caspases

Caspases are a family of cysteine proteases that play a central role in the apoptotic pathway. Among cysteine proteases, caspases are unique in requiring an aspartate at the cleavage site. The first caspase to be isolated, interleukin-1β (IL-1β)-converting enzyme (ICE; caspase 1), was identified by classical biochemical studies, using the limited proteolysis of IL-1β as an assay. IL-1β is a cytokine, i.e., a protein that signals to other cells, with significant proinflammatory activity. IL-1β is synthesized as a biologically inactive 30-kDa protein, which is cleaved to a mature form of 17 kDa. The initial link between ICE and apoptosis was made by sequence comparison; the *ICE* gene is homologous to the *C. elegans* cell death gene *ced-3.* Furthermore, both *ICE* and *ced-3* induce apoptosis when overexpressed in mammalian cells in tissue culture, and *ICE* complements the *ced-3* mutation in *C. elegans.*

More than 10 caspases have been identified in humans. Most of these induce apoptosis when overexpressed in tissue culture cells, but not all are required for different apoptosis pathways, as shown in gene disruption experiments with mice. For example, ICE knockout mice develop normally and have no overt phenotype in apoptosis. This suggests either that ICE plays no role in developmental or homeostatic cell death or that there is redundancy in caspase function. Caspase 3 is one of the most commonly activated caspases in apoptosis. In contrast to ICE knockout mice, caspase 3 knockout mice have profound developmental alterations, although some of the mice are viable. These mice usually die young due to severe brain malformations.

The regulation of caspase activation is one of the key steps in the control of the apoptotic process (Figure 17.2). The caspase precursors are abundant in the cell cytosol. In response to diverse apoptotic stimuli, they can be proteolytically cleaved into active enzymes. Caspase precursors contain an amino-terminal end that varies in size depending on the caspase and is thought to regulate the cleavage of the precursor. The activation of some caspase precursors is autocatalytic and can lead to cleavage and activation of other members of the family. It has not been clearly established whether caspases are organized in a proteolytic cascade or whether there are caspases that can cleave their relevant substrates and induce apoptosis independently of all other caspases.

The key caspase substrates are beginning to be discovered. Caspase targets appear to be crucial in maintaining the cell architecture, RNA splicing, and DNA repair. These substrates include nuclear lamins, gelsolin, poly(ADP-ribose) polymerase, and the retinoblastoma protein. The hallmark of apoptosis, degradation of DNA into nucleosomal fragments, results from caspase cleavage of a substrate called DNA fragmentation factor. It is still unclear whether all the aforementioned substrates are required for the induction of apoptosis or whether a coordinate cleavage of substrates is needed. Therefore, although caspase activation results in apoptosis, the cas-

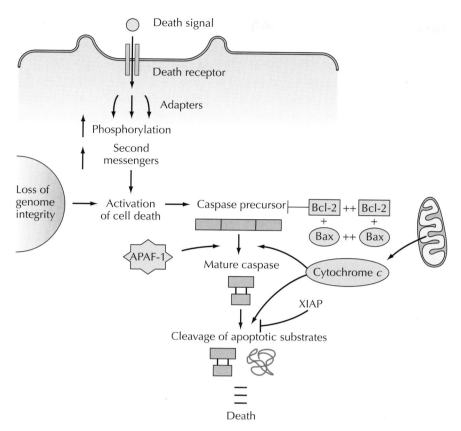

Figure 17.2   Models for induction of apoptosis. Receptor binding or DNA damage activates a signal cascade that culminates in caspase activation and/or disruption of binding of members of the Bcl-2 family and release of mitochondrial proteins required for cell death.

cade of programmed cell death that leads from these proteases to apoptosis is not yet understood. It is interesting that although all caspases require an aspartate at the cleavage site, each caspase recognizes a different amino acid sequence, suggesting that they cleave different substrates.

Caspase activity can be blocked by specific inhibitors, like the cellular inhibitors of apoptosis (IAPs). Viruses, like baculovirus and poxvirus, which are strict intracellular pathogens and can benefit from preventing host death, encode other caspase inhibitors.

Bcl-2 Family

bcl-2, the first gene identified in the Bcl-2 family, maps to a conserved chromosomal translocation site in B-cell lymphomas. Several Bcl-2 family members have been identified in mammals and viruses. Members of the *bcl-2* family encode proteins that either inhibit (like Bcl-2 and Bcl-xL) or activate (like Bax, Bad, Bik, and Bak) apoptosis. Bcl-2 is homologous to the *C. elegans* apoptosis inhibitor gene *ced-9*. *bcl-2* complements the *ced-9* mutation in worms and inhibits apoptosis in many different instances when overexpressed in mammalian cells. The highest homology between members of this large family is found in two specific regions called Bcl-2 homology domains 1 and 2 (BH-1 and BH-2). Both of these domains are crucial for binding to other Bcl-2 homologues.

The Bcl-2 family of proteins forms homo- and heterodimers, which antagonize or enhance the functions of the two partners (Figure 17.2). For example, overexpression of Bad, which results in apoptosis, leads to both formation of Bad-Bad homodimers and disruption of Bcl-2–Bcl-2 homodimers to form Bad–Bcl-2 heterodimers. In this case, it is an open question whether the proapoptotic activity of Bad is the result of the activity of Bad-Bad homodimers or of the disruption of the protective Bcl-2–Bcl-2 homodimer or both. It appears that the fate of the cell depends on the relative amounts of Bcl-2 inhibitors and activators present.

Mutating both the BH-1 and BH-2 domains of Bcl-xL does not abrogate the antiapoptotic activity of this molecule, indicating that it has important functions other than binding Bcl-2 family members. Bcl-xL also binds to Ced-4, the product of a gene important in the induction of apoptosis in *C. elegans*. Ced-4 can bind to and activate caspases. Recently, Apaf-1, the first human protein with sequence similarity to Ced-4, was isolated and shown to activate caspase 3 in a cytochrome *c*-dependent manner. Thus, it appears that members of the Bcl-2 family might modulate caspase activity through Ced-4 homologues.

Bcl-2 also prevents cytochrome *c* release from mitochondria. Cytochrome *c* is a mitochondrial protein localized in the intermembrane space and is involved in cellular respiration. Intact cells undergo apoptosis after release of cytochrome *c* in the cytosol, indicating that this protein, in addition to its function in respiration, has a function in apoptosis. Thus, it has been postulated that Bcl-2 could act in situ on mitochondria by inhibiting cytochrome *c* release.

The mechanism of action of this family of proteins remains elusive. Bcl-2 localizes to the outer mitochondrial membrane and the endoplasmic reticulum and might regulate the redox potential of the cell. Interestingly, the three-dimensional structure of Bcl-xL has revealed striking similarity to the pore-forming subunits of bacterial proteins, such as diphtheria toxin and colicins. The structural similarity led to experiments that indicate that Bcl-xL forms channels in lipid membranes. This activity allows for alterations in mitochondrial permeability by the Bcl-2 family of proteins.

Apoptosis in Diseases

The inappropriate induction of cell death is involved in the pathogenesis of a number of diseases. The untimely activation or inhibition of apoptosis plays a role in cancer, viral latency, autoimmune diseases, and microbial infections.

Infectious agents have evolved ways to affect the host cell suicide program. Some viruses and bacteria are able to activate the apoptotic program of the infected cell as part of their means of causing disease. Several microbes modulate apoptotic pathways to survive host defense mechanisms or otherwise guarantee optimal living conditions. Another goal might be to facilitate the initiation of infection and inflammation, thus securing efficient microbial spread. Finally, other microbes might inhibit apoptosis to guarantee the life of a host cell that is useful for their survival or persistence.

Induction of Apoptosis by Microorganisms

Many bacterial pathogens induce apoptosis in host cells (Table 17.2). This review groups microbial pathogens by the mechanisms they use to induce

Table 17.2   Bacteria that cause apoptosis

Actinobacillus actinomycetemcomitans	*Mycobacterium* spp.
Bordetella pertussis	*Pasteurella haemolytica*
Clostridium difficile	*Pseudomonas aeruginosa*
Corynebacterium diphtheriae	*Salmonella* spp.
Escherichia coli	*Shigella* spp.
Helicobacter pylori	*Staphylococcus aureus*
Legionella pneumophila	*Streptococcus pyogenes*
Leptospira interrogans	*Yersinia* spp.
Listeria monocytogenes	

apoptosis. The proapoptotic strategies used include activation of cell surface receptors, mimicry of second messengers, regulation of caspase function, inhibition of protein synthesis, disruption of the host cell membrane, and, finally, unknown mechanisms.

Activation of Host Cell Receptors That Signal for Apoptosis

Staphylococcus aureus is a gram-positive coccus that can be part of the normal flora of the skin and mucosa. This microorganism is also the etiological agent of a variety of diseases, including skin lesions, food poisoning, toxic shock syndrome, endocarditis, and osteomyelitis. Many virulence factors implicated in staphylococcal virulence, including superantigens, can cause apoptosis (discussed below). *Streptococcus pyogenes* is another gram-positive coccus; it is an important cause of pharyngitis and is sometimes associated with serious sequelae. It has also been associated with toxic shock-like syndrome through exotoxins that act as superantigens. Superantigens are proteins that activate T cells by directly binding both to major histocompatibility complex class II molecules on antigen-presenting cells and to the T-cell receptor (TCR) on T cells. Normally, T cells are activated only when the TCR recognizes a peptide in the context of the major histocompatibility complex. The superantigens of streptococci and staphylococci recognize specific TCRs of the Vβ family.

Engaging the TCR on T cells can activate programmed cell death. Activation of the TCR through superantigens like staphylococcus exotoxin B (SEB), SEA, SED, and SEE induces apoptosis in T cells and thymocytes both in vitro and in vivo. Although it is likely that superantigens activate apoptosis by engaging the TCR, the precise pathway by which superantigen activation results in programmed cell death is not yet known. It is clear, however, that T cells have to be active and proliferating to die. SEB-induced cell death is mediated by both protein kinase C and an ATP-gated ion channel.

Vβ cells are depleted in patients infected with toxigenic *S. pyogenes*. Furthermore, peripheral blood mononuclear cells isolated from these toxic shock patients undergo apoptosis in vitro. This suggests that induction of apoptosis plays an active role in superantigen-induced toxic shock syndrome.

It has yet to be established whether superantigen induction of T-cell apoptosis plays a role favorable or harmful to the host in the pathogenesis of microbial infections. On the one hand, it can be argued that the reduction in the number of T cells after superantigen exposure and the consequent decrease of immune system function may be important for pathogen survival. On the other hand, the deletion of T cells by superantigens might control the unregulated proliferation and activation of T cells during an infection and thus prevent an exaggerated immune response.

Induction of Second Messengers

Bordetella pertussis, a gram-negative rod, is the etiological agent of whooping cough, an upper respiratory tract infection characterized by a cough with an inspiratory "whoop." *Bordetella* is highly contagious and is transmitted through air droplets. In the course of the infection, the bacteria remain localized to the respiratory tract, where they evoke an acute inflammation.

B. pertussis kills macrophages by apoptosis in vitro by secreting adenylate cyclase-hemolysin (AcHly) toxin. This toxin has two domains: (i) a potent adenylate cyclase activity, which is activated by calmodulin, and (ii) a hemolysin activity, which is a pore-forming protein that is thought to allow the translocation of the cyclase into the host cell cytoplasm. AcHly kills macrophages only when both parts of the toxin are functional. This indicates that the disruption of the host cell cytoplasmic membrane through the pore-forming domain of this toxin is not sufficient to kill the cell.

An increase in the intracellular concentration of cAMP triggers pathways that lead to apoptosis, and curtailing the production of this second messenger can prevent programmed cell death. Hence, it is interesting to speculate that *Bordetella* initiates apoptosis by sharply increasing the intracellular concentration of cAMP to activate a program for cell death.

Bordetella encodes another toxin, pertussis toxin (PT), which also increases the intracellular cAMP concentration. PT is a member of the A/B family of toxins, in which the B subunits allow the translocation of the enzymatically active A subunit into the cytoplasm of the target cell. PT ADP-ribosylates a G protein, inhibiting the inhibitory subunit, which acts on the host adenylate cyclase. Thus, indirectly, by inhibiting the down-regulator of the host cell endogenous cyclase, PT activity leads to an increase in the intracellular cAMP concentration. Interestingly, PT is not necessary for *Bordetella*-mediated induction of macrophage cytotoxicity. Thus, it appears that the kinetics and/or magnitude of the increase in the cAMP concentration may be the key determinant of apoptosis activation.

The importance of *Bordetella*-induced apoptosis is still not understood. AcHly is produced early in the infection and is then down-regulated during the later, chronic phase of the disease. Therefore, it is possible that *Bordetella* kills alveolar macrophages to eliminate the first line of defense that it encounters. Alternatively, early macrophage apoptosis might be important in triggering an inflammatory response (see the next section).

Regulation of Caspase Activity

Activation of Caspases

Shigella is a gram-negative rod that causes dysentery, a severe form of diarrhea that often contains blood and mucus. *Shigella* is transmitted through the fecal-oral route and is an extremely infectious agent.

Shigella is an invasive organism that induces macrophage apoptosis. This pathogen is phagocytosed by macrophages but escapes from the phagocytic vacuole. Inside the cytoplasm of the macrophage, *Shigella* secretes, among other proteins, invasion plasmid antigen B (IpaB). IpaB disseminates throughout the cytoplasm, binding to and activating ICE (caspase 1 [see above]). The activity of mature ICE is responsible for both macrophage apoptosis and maturation of IL-1β. *Shigella* induces classical apoptosis, as determined by morphological changes and the fragmentation of the host cell DNA. Since IpaB has to be delivered by the bacterium into the compartment where ICE resides, *Shigella* is able to induce apoptosis only from within the cytoplasm.

All clinical isolates of *Shigella* tested thus far induce macrophage cell death. Apoptosis is up-regulated in tissues of animals experimentally infected with *Shigella* and in patients suffering from shigellosis. Taken together, these data indicate that apoptosis is activated in *Shigella* infections and is not an in vitro artifact.

Recent findings, including infections of mice with a targeted deletion in ILE, IL-1β, and IL-18, strongly suggest that macrophage apoptosis is an important step in *Shigella* pathogenesis. This bacterium has a very specific tropism for the colon, where it penetrates through M cells (Figure 17.3) into lymphoid follicles. After M-cell translocation, *Shigella* encounters resident macrophages. Colonic macrophages are usually activated, probably because of the constant sampling of bacterial products from the lumen of the colon. The infected macrophages undergo apoptosis, and because *Shigella*-induced apoptosis depends on ICE activation, the proinflammatory cytokine IL-1β and another ICE substrate IL-18 are processed to their mature forms. IL-1 is likely to be the first signal to initiate the inflammatory response to this invasive organism. An acute inflammation, rich in polymorphonuclear lymphocytes (PMNs), ensues. These inflammatory cells compromise the integrity of the intestinal barrier as they migrate toward the lumen of the intestine and allow for the entry of more bacteria. The disruption of the epithelial cell barrier is a necessary step in the pathogenesis of *Shigella*.

Salmonella, like *Shigella,* is a gram-negative enteric pathogen that is transmitted orally. Depending on the bacterial serovar and host specificity, *Salmonella* either remains localized in the gut and produces gastroenteritis or is dispersed hematogenously to other organs, such as the spleen and the liver, as in typhoid fever. Interestingly, *Salmonella* activates an ICE-dependent cell death through SipB (Box 17.1). SipB is a virulence factor with high homology to IpaB. Furthermore, ICE-knockout animals are extremely resistant to an oral challenge with *Salmonella*.

Inhibition of Caspases

Several viral proteins are known to inhibit caspase activity, preventing the host cell from undergoing apoptosis. An interesting example is that of the poxviruses, a family of DNA viruses that replicate in the eukaryotic cytoplasm. Cowpox virus, a member of this family, causes skin lesions that are the result of both viral replication and the host inflammatory response. This virus contains the *crmA* gene, encoding cytokine response modifier A (CrmA). CrmA is a caspase competitive inhibitor of the serpin family. This inhibitor is quite selective in its ability to block caspases, showing the highest affinity for ICE. In vitro, CrmA inhibits the enzymatic activity of ICE, and when it is ectopically expressed in mammalian cells, it prevents programmed cell death initiated by a variety of stimuli.

Infections with poxvirus *crmA* mutants result in larger skin lesions than those from wild-type viruses. This suggests that in infections with wild-type virus, inhibition of ICE results in a lower production of IL-β and that this cytokine is a key mediator of the inflammatory response to cowpox virus. The viral inhibition of ICE prevents the intense inflammation, possibly allowing further viral replication. As with activation of caspase in

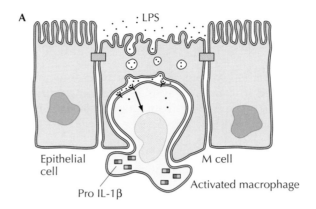

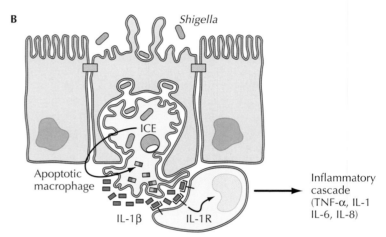

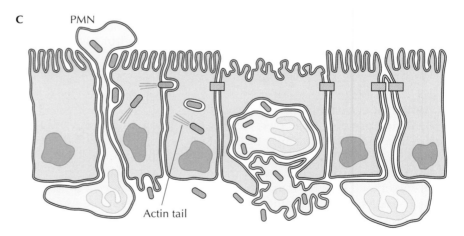

Figure 17.3 Model of *Shigella* pathogenesis. **(A)** Colonic macrophages are activated to synthesize IL-1β by exposure to bacterial products of the normal microflora, like lipopolysaccharide (LPS). **(B)** *Shigella* translocates through M cells, infects resident macrophages, and escapes the phagocytic vacuoles. It secretes IpaB, which binds to ICE, which is activated. ICE activation leads to macrophage apoptosis and IL-1β maturation. IL-1β initiates an acute inflammatory response. **(C)** PMNs break the epithelial barrier, allowing further bacterial invasion. Bacteria invade epithelial cells through the basolateral side and then spread from cell to cell. Redrawn from A. Zychlinsky and P. J. Sansonetti, *Trends Microbiol.* **5:**201–204, 1997.

BOX 17.1

Why does *Salmonella* have two mechanisms to induce apoptosis?

*S*almonella carries virulence genes in pathogenicity islands. *Salmonella* pathogenicity island 1 (SPI1) is important in the intestinal phase of the disease and carries the effector SipB, which, like IpaB (see text), induces an ICE-dependent cell death. ICE is essential for the intestinal phase of *Salmonella* infections. SPI2 is necessary in the later phases of the infection, and it orchestrates the intracellular life of this bacterium. SPI2 is also important in inducing an apoptotic process that is significantly slower than SPI1-dependent cell death, but at least in part requires ICE. It is unclear why *Salmonella* acquired two mechanisms to induce apoptosis.

Shigella infections, it remains to be determined whether the poxvirus modulation of inflammation through ICE is linked to the proapoptotic potential of this enzyme.

Inhibition of Apoptosis via Bcl-2

Until recently, except for *Rickettsia*, there has been no description of a bacterium that inhibits apoptosis of its host cell. This is surprising since some intracellular microbial pathogens could benefit from prolonging the life of their host. Many viruses, however, use homologues of Bcl-2 or other members of that gene family to prolong the life of their host cell. For example, adenovirus encodes the protein E1B19-kD, which is a Bcl-2 homologue. Infections with adenovirus carrying mutations in E1B-19kD induce apoptosis in infected cells. Bcl-2 can complement this mutation, suggesting that Bcl-2 and E1B-19kD have an analogous function. Although the sequence similarity between E1B-19kD and Bcl-2 is not significantly high over the entire protein, the domains involved in the inhibition of apoptosis are conserved. Furthermore, it has been shown that E1B-19kD and Bcl-2 interact with a similar set of cell death proteins, explaining the analogy in function between these two proteins.

A number of herpesviruses, including Epstein-Barr virus, herpesvirus saimiri, and Kaposi's sarcoma-associated virus, contain homologues of *bcl-2*. Most of these viruses can cause persistent infections, and the products of the *bcl-2* homologues might allow the extended survival of the host cell. Epstein-Barr virus encodes another gene, latent membrane protein 1 (LMP1), which regulates the expression of genes that control apoptosis, including *bcl-2*.

Inhibition of Protein Synthesis

Corynebacterium diphtheriae, a non-spore-forming gram-positive rod, and the gram-negative bacteria *Pseudomonas aeruginosa, Shigella dysenteriae,* and enterohemorrhagic *Escherichia coli* (EHEC) secrete toxins that inhibit eukaryotic translation and activate apoptosis in host cells. All these toxins are of the A/B type, like PT. They inhibit translation through different mechanisms. Inhibition of eukaryotic translation or transcription by other substances, like cycloheximide or actinomycin, also eventually activates apoptosis in many cell types.

The symptoms of diphtheria are sore throat, fever, and a characteristic gray, adherent pseudomembrane at the back of the pharynx. This fibrillar pseudomembrane contains bacteria and infiltrating inflammatory cells. A dangerous complication of diphtheria is obstruction of the airways by an enlarged pseudomembrane.

C. diphtheriae lives extracellularly and secretes diphtheria toxin (DT). The B subunit of DT binds to an extracellular glycoprotein receptor, allowing the toxin to be endocytosed by several different cell types including epithelial and myeloid lines. The acidic environment of the phagolysosome exposes a specific domain and creates an aqueous pore that allows the translocation of the A subunit into the host cell cytoplasm. The A subunit is an enzyme that ADP-ribosylates elongation factor 2 (EF-2), an essential component of the translation machinery. The ribosylation inhibits EF-2 activity and blocks protein synthesis, eventually leading to apoptosis. Three lines of evidence strongly support the hypothesis that DT induces apoptosis by inhibiting translation: (i) the levels of inhibition of protein synthesis and cytotoxicity correlate very tightly in DTX-treated cells; (ii) host cells carrying mutations in EF-2 which prevent ADP ribosylation are insensitive to DTX-induced apoptosis; and (iii) inhibition of ADP-ribosylation by DT blocks both cytotoxicity and the inhibition of protein synthesis.

P. aeruginosa is found in soil and water and sometimes in the flora of the gut. It can cause sepsis, urinary tract infection, and pneumonia, particularly in immunocompromised persons and cystic fibrosis patients. One of the key virulence factors of *P. aeruginosa* is exotoxin A (ExoA), which, like DT, ADP-ribosylates EF-2 and inhibits protein synthesis. ExoA and DT do not have significant sequence homology. However, ExoA also induces apoptosis in a human monoblastoid cell line. Mutations in EF-2 confer resistance to ExoA, suggesting that, as with DT, inhibition of translation is necessary for ExoA-mediated apoptosis. Interestingly, it has been reported that ExoA, like DT, has a nuclease activity. The role of ExoA in *Pseudomonas* infections has not been thoroughly analyzed.

S. dysenteriae and EHEC cause a dysenteric syndrome and can be associated with sequelae involving the kidneys and the central nervous system (CNS). *S. dysenteriae* and EHEC both produce toxins, Shiga toxin (ST) or Shiga-like toxins (SLT), respectively, which are almost identical. The B subunits of these A/B toxins mediate binding to globotriaosylceramide (Gb$_3$), the receptor on the host cell membrane. The A subunit cleaves eukaryotic rRNA and disrupts ribosomal function and hence inhibits protein synthesis. Induction of apoptosis by these toxins has been documented both in vitro and in vivo. The purified B subunit of the toxin is sufficient to kill epithelial cells, suggesting that Gb$_3$ binding might activate a signal transduction cascade that culminates in apoptosis. The inhibition of protein synthesis might synergize with the Gb$_3$-dependent pathway, preventing the synthesis of inhibitors of apoptosis.

Although these toxins are cytotoxic in vitro, the significance of the induction of apoptosis in vivo is still unclear. Cells in both the kidneys and the CNS are enriched in Gb$_3$ receptors. This distribution might explain the specific localization of the sequelae of infections with bacteria that secrete either ST or SLTs. Significantly, macrophages do not express the Gb$_3$ receptor and are not susceptible to ST or SLT cytotoxicity.

Disruption of the Cytoplasmic Membrane

Several pore-forming proteins (PFP) are made by microbes. These toxins include *S. aureus* α-toxin, *Actinobacillus actinomycetemcomitans* leukotoxin, *Listeria monocytogenes* listeriolysin, and *E. coli* hemolysin. It appears that large doses of PFP massively disrupt the integrity of the cell membrane and cause necrosis. At lower doses, a more delicate modification of the host cell membrane results in apoptosis. It is still unclear how disruption of the cell membrane can lead to apoptosis. Nevertheless, mutations in PFP that lower their hemolytic activity also decrease their apoptotic potential, demonstrating that the pore-forming ability is directly linked to apoptosis.

It is still not known whether bacterial pore-forming proteins bind to a specific host cell receptor. It has been postulated that low doses of *S. aureus* α-toxin bind to a host cell receptor and allow the influx of Na^+. This Na^+ influx might be cytotoxic because of the ionic imbalance it generates. Hyperpolarization of the cytoplasmic membrane has been previously shown to lead to apoptosis. Alternatively, the Na^+ influx might indirectly lead to an increase in the intracellular concentration of Ca^{2+}, which serves as a second messenger to activate apoptosis. High doses of α-toxin generate larger holes in the host cell membrane. These larger holes allow the free flow of Ca^{2+} and are rapidly cytotoxic to cells. α-Toxin appears to be an important staphylococcal virulence factor; however, the relevance of α-toxin-induced apoptosis in any of the diseases that *S. aureus* can cause (see Activation of Host Cell Receptors That Signal for Apoptosis, above) is not clear.

A. actinomycetemcomitans is a gram-negative coccobacillus that is associated with periodontitis as well as meningitis and endocarditis. It produces leukotoxin which causes cytotoxicity in lymphoid cells but spares fibroblasts, platelets, and endothelial and epithelial cells. Destruction of immune system cells may be important in the development of disease. In addition, *Actinobacillus* induces apoptosis independently of leukotoxin. There are *Actinobacillus* strains that do not make leukotoxin but still kill macrophages. This second pathway requires intracellular bacteria and involves a protein kinase C pathway but not a cAMP-dependent protein kinase pathway.

L. monocytogenes is a gram-positive rod that can cause meningitis and sepsis in infants and immunosuppressed patients. It is transmitted orally, and after invading the gastrointestinal tract, it can spread systemically. *Listeria* invades cells and escapes from the phagolysosome into the cytoplasm of the host cell. It can kill cells, however, without invading them. *Listeria* induces apoptosis in dendritic cells and hepatocytes, but it is not cytotoxic to macrophages. Dendritic cells are antigen-presenting cells located in lymphoid aggregates throughout the body including the gut. The killing of dendritic cells might be an important way of preventing the immune system from mounting a timely immune response against this pathogen.

The liver receives much of the blood drainage from the lower gastrointestinal tract and is one of the first organs to be infected with *Listeria* in a systemic infection. When *Listeria* infects the liver, it induces hepatocyte apoptosis. Hepatocyte apoptosis causes the release of a yet-to-be-defined PMN chemoattractant. Consequently, these important immune system cells are brought into an infected liver, probably to promote bacterial clearance. Thus, it is likely that, like *Shigella* infections of macrophages (see above), *Listeria* infections of hepatocytes are proinflammatory.

E. coli hemolysin can kill stimulated but not unstimulated T lymphocytes, suggesting that a factor that allows apoptosis is up-regulated (or an inhibitor is down-regulated) upon stimulation. Hemolysin is important in the pathogenesis of *E. coli* infections of the urinary tract.

Alternative Mechanisms

Yersinia

Yersinia, like *Salmonella* and *Shigella,* to which it is closely related, also causes apoptosis of murine and human macrophages. *Y. enterocolitica* and *Y. pseudotuberculosis* are acquired by ingestion of contaminated food.

Yersinia gains access to the intestinal epithelium and replicates in Peyer's patches. Unlike *Salmonella* and *Shigella,* however, *Yersinia* does not persist in macrophages; instead, it exerts cytotoxic effects from outside the cell. Several bacterial genes are associated with the ability of *Yersinia* to induce macrophage apoptosis. YopJ or its homologue yopP inactivates kinases in the cascade that eventually allows translocation of the transcriptional vector NF-κB. NF-κB transcribes antiapoptotic genes. Thus, *Yersinia* can block an important "antiapoptotic" pathway. The "pro"-apoptotic effector in *Yersinia* is still not clearly identified.

Mycobacterium

Mycobacterium tuberculosis is an acid-fast bacterium that causes the respiratory tract disease tuberculosis. *M. tuberculosis* is transmitted by respiratory droplets and invades the lungs, where it infects alveolar macrophages, which serve as its residence for the duration of the infection. The macrophages elicit an immune response against the bacteria, leading to the characteristic histological caseating granuloma of tuberculosis. *M. avium-M. intracellulare* causes a similar disease but only in immunocompromised individuals. *M. avium-M. intracellulare,* like *M. tuberculosis,* also resides in alveolar macrophages. Mycobacteria induce apoptosis in macrophages in vitro by an as-yet-undefined mechanism that possibly involves TNF-α. In tissue sections of mycobacterially infected lungs, apoptotic cells have been detected by electron microscopy and by in situ demonstration of DNA fragmentation. Since mycobacteria require the host cell to replicate and survive, induction of macrophage apoptosis would probably be beneficial to the host, since it would deprive the pathogen of a place to live. Indeed, it has been shown that more virulent *Mycobacterium* strains induce less apoptosis than avirulent strains do.

Helicobacter pylori

The gram-negative rod *Helicobacter pylori* causes gastritis, peptic ulcers, gastric atrophy, and carcinoma. This bacterium encodes VacA, a toxin that disrupts the mitochondrial membrane, allowing the release of cytC and Apaf-1, two proteins that recruit caspases and activate apoptosis in the cytoplasm. There is a significant increase in the number of apoptotic cells in patients infected with *H. pylori* compared with controls, and eradication of the infection after treatment reduces the number of apoptotic cells to control levels. The apoptosis-inducing ability of *H. pylori* may be the key to understanding the gastric atrophy in infected persons.

Selected Readings

Ellis, H. M., and H. R. Horvitz. 1986. Genetic control of programmed cell death in the nematode *C. elegans. Cell* **44:**817–829.

A landmark paper on the discovery of the genetic basis of cell death.

Monack, D. M., C. S. Detweiler, and S. Falkow. 2001. Salmonella pathogenicity island 2-dependent macrophage death is mediated in part by the host cysteine protease caspase-1. *Cell Microbiol.* **3:**825–837.

Orth, K., Z. Xu, M. B. Mudgett, Z. Q. Bao, L. E. Palmer, J. B. Bliska, W. F. Mangel, B. Staskawicz, and J. E. Dixon. 2000. Disruption of signaling by Yersinia effector YopJ, a ubiquitin-like protein protease. *Science* **290:**1594–1597.

A combinatorial approach to understanding a cell death pathway activated by bacterial pathogens in different systems.

Sansonetti, P. J., A. Phalipon, J. Arondel, K. Thirumalai, S. Banerjee, S. Akira, K. Takeda, and A. Zychlinsky. 2000. Caspase-1 activation of IL-1beta and IL-18 are essential for Shigella flexneri-induced inflammation. *Immunity* **12**:581–590.

    In vivo validation of the caspase-1-dependent pathway of pathogenesis.

Thornberry, N. A., and Y. Lazebnik. 1998. Caspases: enemies within. *Science* **281**:1312–1316.

    Thorough review of caspases and their intricacies.

Weinrauch, Y., and A. Zychlinsky. 1999. The induction of apoptosis by bacterial pathogens. *Annu. Rev. Microbiol.* **53**:155–187.

    Comprehensive review of the induction of apoptosis by bacteria and bacterial products.

Wyllie, A. H. 1980. Glucocorticoid-induced thymocyte apoptosis is associated with endogenous endonuclease activation. *Nature* **284**:555–556.

    Landmark paper on the discovery of DNA fragmentation during apoptosis.

Yuan, J., S. Shaham, S. Ledoux, H. M. Ellis, and H. R. Horvitz. 1993. The *C. elegans* cell death gene ced-3 encodes a protein similar to mammalian interleukin-1β-converting enzyme. *Cell* **75**:641–652.

    Key contribution on the genetic identification of proteases as mediators of cell death.

Zychlinsky, A., M. C. Prevost, and P. J. Sansonetti. 1992. *Shigella flexneri* induces apoptosis in infected macrophages. *Nature* **358**:167–169.

    Initial report on the induction of apoptosis by bacterial pathogens.

18

Interaction of Pathogens with the Innate and Adaptive Immune System

Emil R. Unanue and Ennio De Gregorio

The host response to pathogenic microorganisms is extraordinarily diverse. The extent and degree of the host response depend on the nature of the pathogen itself and the route and extent of the infection. Some general features of host-pathogen interactions are discussed in this chapter.

Innate and Adaptive Immunity

Microbial infections can be controlled by one or more effector systems that are brought into play during the infection. Two broad kinds of responses or interactions, distinguished by the extent of involvement of the lymphocyte, take place between the host and the pathogen. The responses that are independent of lymphocytes have been termed "natural immunity," "innate immunity," or "T-independent" responses. The responses that involve lymphocytes are the adaptive responses. The innate response involves various leukocytes, in particular, the cells of the mononuclear phagocyte system (macrophages), the cells of the dendritic cell lineage, the granulocytes, and the natural killer (NK) cells, which can be mobilized and activated without the direct participation of the T cell. Their response to pathogens is fast and immediate. These findings have led to the concept that the innate cellular response is the initial step in the host response. The cells of the innate response also cooperate with the lymphocytes, in part by presenting antigens in the form of peptide fragments and in part by releasing a number of modulatory molecules. It is safe to say that the cells of the innate system serve to regulate and support the adaptive response, when it comes into operation.

Much of the adaptive response centrally involves the T cell. T cells are the cells derived from the thymus, which form part of the recirculating pool of lymphocytes (those that migrate from the blood to the lymphoid tissues, to the lymph, and back to the blood). Upon activation, T cells rapidly produce and release mediators, particularly cytokines that regulate and activate the cellular response. T cells also respond by direct cellular interactions. Thus, T cells are involved not only in activation of macrophages, for example, but also in killing of infected cells and in regulating B cells for antibody formation. All these different cellular expressions are centrally focused on an activated T lymphocyte, i.e., T lymphocytes of either the CD4 and/or CD8 subset that have responded to antigen. Finally, the B cell plays a major role in the antimicrobial response by producing antibodies. The antibody response plays a major role by neutralizing many microbes and/or their products. The growth of many microbes is controlled by antibody molecules, while the growth of others is resistant. (Pus-forming gram-positive bacteria are rapidly eliminated if bound to antibodies and engulfed by neutrophils, but this does not happen with intracellular facultative bacteria.) Antibody molecules are usually produced by B cells interacting with T cells, both recognizing microbial antigens (Box 18.1).

Innate Response

Probably the best experimental approach to the analysis of the innate response is to examine mutant mice that can use only the innate system, by virtue of the absence of lymphocytes. The first strains of mice used for this purpose were the SCID mice, discovered by the Bosmas when examining the antibody response of the CB.17 strain of inbred mice. SCID mice have a defect in the formation of the T-cell and B-cell antigen receptor, resulting from a defect in the enzyme DNA-PK involved in the recombination of the V gene segments required to form the antigen receptor of T and B cells. Mice which lack other enzymes in this recombination process, like Rag and Ku proteins, and which, like the spontaneous SCID strain, exhibit a selective

---

BOX 18.1

Functions of B cells and T cells

B Cells

1. Recognition of antigen molecules occurs by way of surface Ig (IgM and IgD). The Igs assemble by recombination of the different gene segments (V, D, and J). Each B cell has a unique receptor (clonal selection).
2. Recognition of multimeric repeating epitopes can partially trigger B-cell activation. (This is the response to bacterial polysaccharides, the "T-independent response.")
3. Recognition of proteins requires the helper function of T cells.
4. B-cell–T-cell interaction involves binding of protein by B cells, internalization, and processing with generation of peptides bound to class II MHC molecules. CD4 T cells recognize the peptide-MHC complex and activate B cells.

5. Activation involves switching of Ig constant heavy-chain genes, point mutations of Ig genes resulting in selection of high-affinity Ig molecules, and generation of B-cell memory.

T Cells

1. There are two major subsets of T cells, the CD4 and CD8 T cells, each with receptors for peptide-MHC complexes displayed by APC; i.e., their receptors have dual specificity (self-MHC and peptides).
2. The thymus is the main source of T cells. These undergo a selection process in the thymus in which the thymocytes first express both CD4 and CD8 molecules together with the TCR, maturing to express either one or the other. For positive selection, if the TCR has affinity for self-MHC but weak reactivity to the self-peptides, the T cells mature to CD4 or CD8 and exit the thymus. Each T cell will react with a unique

peptide-MHC combination usually produced by a "foreign" peptide. For negative selection, if the TCR also recognizes self-peptides at high affinity, the T cell dies. This results in the death of many self-reactive T cells. For death by negligence, if the TCR does not show specificity for self-MHC, the T cells die by apoptosis.
3. CD4 T cells that recognize peptide-MHC complexes are activated, secrete cytokines, proliferate, and differentiate into distinct sets of cytokine-secreting cells. CD4 T cells are involved in macrophage activation and in B-cell–T-cell interaction. Each CD4 T cell recognizes a unique complex of peptide with class II MHC molecules.
4. CD8 T cells are activated after recognition of peptides presented by class I MHC molecules. CD8 T cells can kill cells presenting the peptide-MHC complex.

absence of lymphocytes have now been produced by gene ablation techniques.

The examination of SCID mice infected with a variety of viruses, bacteria, and parasites has allowed the response to be examined, unencumbered by lymphocytes. By reconstituting these mice with lymphocytes, the influence of the adaptive response on the infection could be compared with that in SCID mice. SCID mice exhibited a surprising first line of defense toward many pathogens. Leukocytes were mobilized, some sets of cytokines were produced, and the infection was restricted or delayed. The three cell types that permitted the innate response in the SCID mice to operate were the macrophages, neutrophils, and NK cells. Two results need to be emphasized regarding the response of SCID mice. First, SCID mice showed no sterilizing immunity; thus, the growth of the pathogens could be partially controlled but they were not eliminated. Second, as predicted, SCID mice did not develop an anamnestic or secondary response following a primary infection. The state of cellular activation was finite and could not be perpetuated unless lymphocytes were present.

Macrophage Response—the Activated Macrophage

The mononuclear phagocyte system comprises the circulating monocytes and their products of differentiation, the various tissue macrophages, and the dendritic cell lineage. Macrophages are found widely among tissues, usually near epithelial surfaces and blood vessels. Such macrophages constitute a first barrier to the dissemination of exogenous material. Mononuclear phagocytes are also involved in the reorganization of tissue during inflammation (i.e., in wound healing) and in the removal of tissue debris and apoptotic cells. Macrophages function in part by releasing mediators that affect the surrounding cells. The early release of cytokines by macrophages is an important response to microbes. Macrophages also express histocompatibility molecules and participate in interactions with CD4 and CD8 T cells (Table 18.1).

The production and differentiation of macrophages are controlled by colony-stimulating factors (CSF), of which CSF-1 is the major regulatory factor. Indeed, the number of circulating monocytes is much influenced by CSF-1, a protein elaborated by many cells including mesenchymal cells. The absence of CSF-1 translates into a deficit of mature macrophages in some organs.

The interaction between pathogens and macrophages that results in their uptake and internalization is mediated through a variety of cell surface receptors, which recognize a range of proteins and polysaccharides of bacteria, viruses, and parasites. Among the cell surface receptors identified are the family of scavenger receptors, several of which have been cloned (types A, I, II, and B, CD36, and others). These receptors are of broad specificity and interact with gram-positive and gram-negative bacteria, participating in their clearance from the circulation. Scavenger receptors bind to negatively charged molecules, low-density lipoproteins, polynucleotides, and anionic polysaccharides and phospholipids. Scavenger receptors have homologies to structures found in hemocytes of insects, presumably the early evolutionary counterparts of phagocytes. Macrophages also have a variety of other receptors for polysaccharides, including macrosialin (CD68) and the mannose receptor, for microbial lipoproteins (CD14), and for immunoglobulin G (IgG) and complement (C) proteins. The Fc and C

Table 18.1 Properties of macrophages

Membrane receptors for diverse chemical structures
    Scavenger receptor
    Complement receptors
    Fcγ receptors
    Sialoadhesin
    Mannose receptors
    Macrosialin
    Cytokine receptors (IFN-γ and TNF)
    CD14-LPS receptor

Production of cytokines
    IL-1α and β
    TNF-α
    IL-12
    IL-10
    IL-6
    Fibroblast growth factor

Antigen presentation

Production of enzymes involved in antimicrobial responses and/or acting on
    connective tissue proteins or cells
    Collagenase
    Elastase
    Lysozyme
    Lysosomal enzymes

Production of bioactive lipid and small radicals
    Prostaglandins
    Platelet-activating factor
    Reactive oxygen intermediates
    Reactive nitrogen intermediates

receptors are of major importance in that they promote, by several thousandfold, the uptake of microbes containing bound antibody and/or C proteins. Various Fc receptors for IgG (Fc-RI, Fc-RII, and Fc-RIII) have been identified and cloned. The same holds true for CR1, CR2, and CR3, the receptors for degradation intermediates of the C3 protein, the major C opsonin in blood. These receptors are also expressed in neutrophils, B cells, and the follicular dendritic cells of germinal centers. Since the original description of its phenomenon, opsonization is recognized as a major step in the clearance and elimination from the blood and extracellular fluids of many microbes and proteins. Some further comments on general features of this phenomenon of opsonization should be made.

1. C proteins can directly interact with many microbial surfaces in the absence of antibodies (and without the binding of C1, C4, and C2, the first three components of the C activation cascade). This alternative pathway of complement activation is thought to be vital in the clearance of many encapsulated organisms. Indeed, genetic deficiency of C3, the key opsonic protein deposited on microbe surfaces, results in pronounced infections with diverse microorganisms but particularly with extracellular bacteria (Box 18.2).

2. Blood contains a pool of antibodies that arise spontaneously, in the absence of overt antigenic stimulation. Many of these natural antibodies arise from a subset of B cells called B-1 and have low affinity and broad binding specificities, some of which are directed to polysaccharides. It is still unknown whether these natural antibodies act in an early recognition step in infections.

3. Other proteins have been identified that interact with polysaccharides and help in the clearance of abnormal glycoproteins. Collectins are proteins that also constitute a first defense element by interacting with polysaccharides and glycoproteins. These proteins have general structural features: they are polymers made of subunits, each with a carbohydrate recognition unit and a collagen-like stalk. Among the collectins are the mannose-binding protein found in blood, the serum bovine protein conglutinin, and the lung surfactant proteins A and D. The surfactant proteins are presumed to be involved in the rapid clearance of inhaled bacteria at the level of lung alveoli.

Macrophages can respond to external stimuli and become activated. An activated macrophage exhibits distinct features. Morphologically it is a larger cell than a normal macrophage and has more vacuoles and more pinocytic activity. The activated macrophage expresses a number of cytocidal molecules, including reactive oxygen and nitrogen metabolites, which enable them to control the growth of intracellular pathogens. Highly reactive oxygen derivatives are formed during the consumption of oxygen by the macrophages. The reactions involve the assembly of a phagocyte oxidase that utilizes NADPH. The oxidase contains three cytosolic proteins and a unique membrane cytochrome *b*, which utilizes NADPH to transfer

---

BOX 18.2

Complement cascade

Classical Pathway
Antibody binds to antigen, to which C1, the first component of C, binds to initiate the cascade of interactions. C1 is composed of three subunits: C1q, C1r, and C1s. C1q binds to the Fc portion of Ig molecules, C1r is a serine proteinase that cleaves C1s, and C1s is another serine protease that cleaves C4 and C2. C4 binds to the activating surfaces after its partial cleavage by C1 (C4b is the cleavage product). C2, upon cleavage, assembles with C4 to form the "C3 convertase," a complex that cleaves C3. C3, the major blood C protein, is the major serum opsonin. It binds to activating surfaces upon cleavage by the C3 convertase. The complex of C4bC2b cleaves C5.

Alternative Pathway
C3 binding to activating molecules, including those from microbes, results in cleavage. Unique proteins of the alternative pathway are factor D, which cleaves factor B to generate an active fragment (Db); factor B, which is a serine protease that upon its cleavage by factor D assembles with C3 to form the "C3 alternative pathway convertase," which cleaves C3; and properdin, a protein that assembles with the C3 convertase to produce a more stable enzyme.

Terminal Components
C5, C6, C7, C8, and C9 form the terminal components. After the cleavage of C5, the C5b fragment induces the assembly of the membrane attack complex. C6 to C9 assemble on membranes and form a transmembrane pore. This box outlines the major components of the classical and alternative C activation pathways. (This summary is based on information in A. K. Abbas, A. M. Lichtman, and J. S. Pober, *Cellular and Molecular Immunology*, 1994, The W. B. Saunders Co.) Aside from the proteins involved in the cascade, other soluble and membrane proteins participate in regulating the activity of the C proteins. Among the soluble regulatory proteins are the C1 inhibitor (which controls the activity of C1), factor I (an enzyme that cleaves the C4b and C3b components of the C3 convertase), and factor H (a cofactor for factor I). Among the membrane proteins are CR1, a receptor-cleaved C3b, which results in the opsonization of cells and can also act to dissociate the C3 convertase. CD46 and decay accelerating factor are two proteins that control the activity of the C3 convertase.

oxygen. Oxidants that are formed include the superoxide anion (O^{2-}), the perhydroxyl radical (HO_2), and the hydroxyl radical ($OH^{.}$). The production of $NO^{.}$ by the activated macrophage results from the expression of an inducible enzyme, the nitric oxide synthetase. (There are two constitutive isoforms of the enzyme, the endothelial and neuronal forms.) $NO^{.}$ is produced from the metabolism of arginine to citrulline. Nitration of a number of enzymes, including RNase reductase, results in the impaired growth of cells. Activated macrophages that are infected with pathogens have restricted growth, and part or all of this restriction can be attributed to $NO^{.}$ production. For example, inhibition of $NO^{.}$ production in mice infected with a number of intracellular pathogenic bacteria results in uncontrolled infection. The same effects were found in mice lacking the gene for the inducible nitric oxide synthetase.

Activated macrophages are the hallmark of the cellular response to intracellular pathogens. They were first detected in the initial clinical and experimental studies of the tuberculous granuloma by Koch. The tuberculous granuloma consists of a mass of activated macrophages, some of which contain the bacilli. Some of the macrophages fuse to form multinucleated giant cells. (This capacity of macrophages to organize in infective foci and restrict the spread of microbes may be a primitive evolutionary behavior. A similar behavior is noted in invertebrates as a response to exogenous stimuli, for example, an accumulation of hemocytes that restrict the dissemination of the inflammation-inciting material.)

There is a relationship between activated macrophages and control of infection with intracellular pathogenic bacteria. Lurie and collaborators first showed that activated macrophages curbed the growth of the tubercle bacilli whereas serum antibodies did not. This relationship was subsequently established by Mackaness and coworkers in studies of murine infections with *Listeria monocytogenes*. The results of their observations established that antibody molecules were not the major effector molecules that restricted the growth of intracellular pathogens but, rather, that a particular cellular change in the phagocytes, characteristic of the activated phagocyte, restricted the growth. In marked contrast, the antibody molecules controlled the changes induced by extracellular toxins, the infections with many extracellular pyogenic bacteria, and the blood and extracellular stages of some viral infections.

The activation of macrophages results, to a major extent, from the interaction of macrophages with the cytokine gamma interferon (IFN-γ). IFN-γ binds to specific receptors found in many different cells, including those of the mononuclear phagocyte lineage. Neutralizing IFN-γ with monoclonal antibodies or infecting mice that do not produce IFN-γ or do not respond to it because of lack of the IFN-γ receptor (i.e., due to gene ablation techniques) results in overwhelming infection, with the absence of activated macrophages. IFN-γ regulates the number of cellular effector pathways by activating a number of important genes (including the LMP2 and LMP7 genes, involved in proteasome activation, and the NO synthase gene, involved in NO production). The effect of IFN-γ on macrophages is markedly potentiated by second stimuli that include interactions with bacteria or their products (lipopolysaccharides are the most notable) or with cytokines, particularly tumor necrosis factor (TNF). Macrophage activation, on the other hand, can be inhibited, in particular, by cytokines like interleukin-4 (IL-4), IL-10, and transforming growth factor β (TGF-β), which are anti-inflammatory. The way in which the balance of IFN-γ versus

Table 18.2   Generation of activated macrophages

Pathway I (T-cell independent)
  Microbe activates macrophages
  Macrophages release early cytokines: IL-12, TNF-α, IL-1β, IL-6
  IL-12 + TNF activates NK cells
  NK cells produce IFN-γ
Pathway II (T-cell dependent)
  Microbe activates macrophages
  Macrophages release cytokines, as above
  Macrophages present microbial peptides to T cells
  Macrophages up-regulate B7-1/B7-2
  T-cell activation takes place
    TCR is engaged by the peptide-MHC complex of the macrophage
    T-cell–APC interaction is fostered by adhesion/costimulatory
      molecules/CD40-CD40L molecules
  T cells differentiate to Th1 pattern of differentiation through IL-12 release

IL-4/TGF/IL-10 takes place during an infectious process can be critical and needs to be evaluated, particularly for chronic infections.

The activated macrophage is found as a result of activation of either the innate immune system or the T-cell system (Table 18.2). The innate system is activated when macrophages interact with microbes and release early cytokines that induce NK cells to produce IFN-γ. The adaptive T-cell system is activated in a clonal manner when the antigen-reactive cloned T cells recognize the specific peptides from pathogens presented by histocompatibility molecules.

Interactions Involving Neutrophils

Neutrophils are the other sets of essential effector cells that control the growth of a number of infectious organisms. Their importance is clearly indicated by the susceptibility to infections in the neutropenic individual. Neutrophils are essential in infections with extracellular bacteria, which are rapidly eliminated by the oxidative and nonoxidative neutrophil microbicidal mechanisms. The oxidative burst, as described for monocytes, involves activation of the NADPH oxidase with the generation of superoxide anion. In the neutrophil a great part of O_2^- is converted to hydrogen peroxide, which in turn reacts with Cl^- ions to generate hypochlorous acid (HOCl) in a reaction catalyzed by the enzyme myeloperoxidase. HOCl is a short-lived compound but can also react with amines to form the more stable microbicidal N-chloramines. The importance of the NADPH oxidase became evident in studies of the clinical disease chronic granulomatous disease, where defects in the elimination of gram-positive organisms are evident. The nonoxidative mechanisms involve cationic peptides of the neutrophil granule. The defensins comprise a family of antimicrobial peptides, some of which are located in the primary granules of neutrophils. (However, defensins and related molecules are also produced in epithelial cells and could serve as major microbicidal molecules in infections of respiratory and gastrointestinal epithelia. Similar molecules have been identified in invertebrates and could represent the most primitive line of defense.)

The migration of neutrophils to sites of inflammation is critical. Directed migration or chemotaxis takes place when the neutrophil recognizes a gradient of the chemoattractant. Among the chemoattractants are products derived from C activation (C5a), small lipids (LTB$_4$, from the leukotriene cascade of mediators), microbial products, and a large family of small polypeptides, the chemokine family. Chemokines comprise about a dozen or more small proteins, now classified into three groups, CXC, CC, and C chemokines, depending on the presence of terminal cysteine residues. The CXC chemokines (i.e., two terminal cysteines with a residue in between) are powerful neutrophil chemoattractants, while the CC and C chemokines attract monocytes and lymphocytes. Chemokines are induced strongly by interaction with a range of bacteria. Particularly prominent is the release of CXC family molecules by streptococci and staphylococci.

Neutrophils are activated by interactions with microbes, particularly if these are opsonized. Cytokines also contribute to part of the activation. Neutrophils are also found in tuberculous granulomas, but their role is not clear. For some intracellular pathogens, neutrophils reduce the microbial load.

In summary, antibody molecules opsonize extracellular pyogenic bacteria, together with complement protein C3; the bacteria opsonized by neutrophils in particular are killed in a process involving oxygen intermediates. In contrast, the many facultative and obligate intracellular pathogens are not entirely or partially eliminated by neutrophils, even if opsonized. They are eliminated by macrophages activated by IFN-γ released by NK cells during innate stages of infections and/or CD4 and CD8 T cells during the adaptive phase.

Interaction Involving the NK-Cell Response

NK cells represent about 10% of circulating leukocytes, are derived from stem cells of the bone marrow, and mature in the absence of lymphocytes. NK cells participate in the early response to viruses and some bacteria (e.g., *Listeria*) and in the control of some parasitic infections (i.e., *Toxoplasma gondii*). Tumor growth and tumor metastasis can be influenced by NK cells.

For many of these responses, NK cells function through their cytolytic properties, involving the pore-forming protein perforin. In this process, NK cells establish contact with the target cell, and this contact results in their degranulation and the release of perforin. Other granule-associated proteins, like the granzymes, are involved. The granzymes enter the target cell through the pore made by perforin and cause cell death via apoptosis. However, NK cells are also involved in the response to intracellular pathogens by virtue of their production of large amounts of cytokines, particularly IFN-γ, as described above. The production of IFN-γ can be extensive, placing the NK cell as a central cell in the regulation of the early stages of an infectious process, particularly with intracellular pathogens.

Although NK cells can produce cytokines after their interaction with target cells, the main stimuli for their cytokine production derive from macrophages. A pathway of stimulation from macrophages to NK cells has been identified whereby interaction of macrophages with pathogenic organisms results in release of the cytokines IL-12 and TNF, which in turn result in activation of NK cells for IFN-γ release (Table 18.2). Other cytokines (IL-1α, IL-1β, and IL-6) are also released from the macrophages, and each has a predominant target of action. This cytokine cascade pathway (i.e., pathogen → macrophages → IL-12 + TNF → NK cells → IFN-γ) operates in a variety of infections, particularly with intracellular pathogenic bacteria, parasites, and some viruses. Upon production of IFN-γ, the macrophage system is activated and primed to become a highly cytocidal cell. NK cells are also regulated by a number of cytokines, including IL-2 and IL-15 molecules, as well as IFN-α/β that promote their growth and activation.

NK cells recognize their target cells by using two sets of surface receptors: the activation receptors, which permit recognition and killing of the target cells, and, importantly, the inhibitory receptors, which inhibit the activation of the NK cell. The nature of all the activating receptors is not entirely known. One activating receptor is FcγRIII (CD16), which allows the interaction of NK cells with target cells bound to antibody molecules. There are examples of receptors (see below) that can be inhibitory or activating depending on the presence or absence, respectively, of inhibitory signals in their cytosolic domains. The inhibitory receptors interact with various forms of class I major histocompatibility complex (MHC) molecules. An NK cell that interacts with a target cell bearing class I MHC molecules may be inhibited from killing the target cell. This inhibition can be released by blocking and/or removing the class I MHC molecules (this is the "missing-self hypothesis" proposed by Karre). Thus, the NK cells may favor the recognition of target cells bearing low levels or abnormal forms of class I MHC molecules, as can occur with some virally infected or tumor cells. The inhibitory receptors are diverse and are represented by two sets of molecules, C-type lectin receptors and Ig-like receptors. The former are disulfide-linked type 2 integral membrane dimers (e.g., human CD94 and mouse Ly49A). The Ig-like receptors, which are termed KIR, are diverse membrane proteins that vary in their expression of Ig domains. NK cells appear to vary in the extent and diversity of NK-cell-inhibiting receptors. Different allelic forms of receptors exist. All these express cytoplasmic tyrosine inhibitory products (ITIM), which upon phosphorylation associate with tyrosine phosphatases, antagonizing the activation of tyrosine protein kinases. It is noteworthy that some viruses express a class I-like MHC molecule which can have biological consequences by engaging and neutralizing the NK-cell receptors. This has been documented for murine cytomegalovirus.

Macrophage–NK-Cell Interaction: Lessons from Listeriosis

The most extensively studied infection that has led to insights into the activation of the innate immune system is *L. monocytogenes* infection in the SCID mouse (Figure 18.1; Table 18.3). Results similar to those initially found in *Listeria* were also found in infections involving other intracellular pathogens. In *Listeria* infection in SCID mice, exponential growth of *Listeria* takes place after systemic infection. After a few days, growth of *Listeria* stops but the numbers of *Listeria* organisms are maintained at a steady state for prolonged periods. Examination of the macrophage discloses a heightened expression of class II MHC molecules and of their antigen-presenting capabilities. The macrophages show enhanced microbicidal activity and all the properties of activated macrophages. The activated macrophages restrict the growth of *Listeria* in the granulomas, particularly from the release of NO.

In *Listeria* infection of normal mice, the macrophages are activated by both CD4 and CD8 T cells, which recognize antigens from the microorganism. However, interestingly, both normal and SCID mice exhibit the same number and distribution of activated macrophages after infection. The difference in the infection between the two strains is marked: the normal

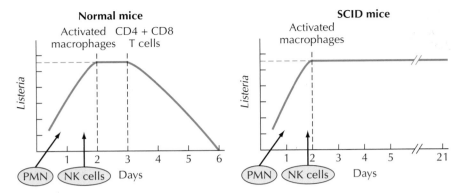

Figure 18.1 Dynamics of *Listeria* infection in normal and SCID mice. In normal mice there is exponential growth and the infection is controlled. The cells involved at different stages are indicated. In SCID mice, the absence of lymphocytes results in persistence of the infection. The early components of the infection involve neutrophils and NK cells, while lymphocytes are essential for clearance of the infection.

strain develops sterilizing immunity, whereas the SCID strain becomes a chronic carrier of the infection. Thus, activated macrophages by themselves are not capable of entirely eliminating an infection with intracellular pathogens, and so lymphocytes must play a role in producing sterilizing immunity beyond that of activating macrophages by producing IFN-γ. Indeed, sterilizing immunity in *Listeria* infection results from CD4 and/or CD8 T cells that are generated during presentation of *Listeria* antigens by the macrophages or dendritic cells. *Listeria* organisms can reach the cytosol as a consequence of their production of a pore-forming enzyme, listeriolysin O. This cytosolic stage allows *Listeria* to disseminate within cells in tissues with an extracellular phase. The vacuolar and cytosolic stages of *Listeria* generate peptides for either class I or II MHC molecules, which trigger T-cell activation (see the next section).

The crux in *Listeria* infection is that this organism is a powerful agonist for the release of cytokines by macrophages. The elements in *Listeria* that induce this response probably involve more than one chemical entity. Not all microbes induce this powerful macrophage response, and thus the extent to

Table 18.3 Properties of SCID mice

Selective absence of B and T lymphocytes

Acellular thymus with only stromal cells; small lymph nodes and spleen

Normal number of granulocytes, macrophages, and NK cells

Leukocytes can be mobilized upon sterile or infectious inflammation.

Normal antigen-presenting function: in culture, their antigen-presenting cells stimulate allogeneic or syngeneic T-cell responses.

Spleen cells do not release cytokines upon addition of mitogens (like concanavalin A); cytokines are released by addition of some microbes; macrophage-NK interaction can induce IFN-γ release from NK cells.

Partial resistance to microbial infections; reduced microbial growth, development of a carrier state of resistance, no sterilizing immunity

Different cytokines produced from macrophages and NK cells; these early cytokines include IL-1α, 1L-1β, TNF-α, TNF-β, IL-6, IL-10, IL-12 (all from macrophages), and IFN-γ (from NK cells); no production of IL-2, IL-4, or IL-5.

Addition of T cells (or bone marrow stem cells) reconstitutes normal immune system function.

---

which SCID mice mobilize the innate system will vary. The SCID mouse responds to *Listeria* uptake via the release of IL-12 and TNF by macrophages that have taken up the microbe. Both of these cytokines play seminal roles in the regulation of the innate system, as well as the lymphocytes. IL-12 and TNF both bind to specific receptors and drive NK cells to express IFN-γ. As with IFN-γ, neutralization of these cytokines by specific antibodies or in mice with ablation of the IL-12 gene or TNF p55 receptors results in uncontrolled infection. The role of TNF is, however, much broader than that of IL-12. TNF acts not only on various leukocytes but also in the vascular endothelium. TNF-activated endothelia express adhesion molecules (E-selectin, intercellular cell adhesion molecule [ICAM]) that promote the adhesion of leukocytes and their migration to the extracellular milieu.

Another issue that has become very prominent in *Listeria* infection is the influence of different effector mechanisms during the various stages of the infective process. The early-exponential growth of *Listeria* requires control by an early infiltration by neutrophils within a few hours of the infection. The activated macrophages do not become apparent until 2 or 3 days later. The neutrophil role is not only to curb the dissemination of the microbe but also to control the liver infection. *Listeria* infects hepatocytes, where it grows exuberantly unless neutrophils restrict it. The last stage of the infection is that ultimately mediated by CD4 and CD8 T cells, producing sterilizing immunity. The CD8 T-cell response can be very prominent and persists for long periods, ensuring a state of immunological memory.

Symbiosis between the Innate Cellular System and the T-Cell Response—the MHC System

T-cell responses depend on the recognition by T cells of peptides derived from the intracellular processing of protein antigens. These peptides are bound to the MHC molecules. Initially discovered as transplantation antigens, the MHC molecules were later shown to regulate all the cellular interactions involving T cells. MHC molecules are peptide-binding molecules that rescue peptides from intracellular digestion and transport them to the cell surface. The peptide-MHC molecular complex represents the molecular substrate that engages the T-cell receptor (TCR) for antigen. The composition of the peptide-MHC complex reflects the intracellular milieu of the antigen-presenting cells (APC). Thus, it is via the MHC molecules that the T-cell system is informed whether an abnormal or previously unrecognized molecule has entered the APC. The MHC molecules thus connect the innate cellular response to the T-cell response.

The presentation of a peptide-MHC complex by the APC initiates activation of the T cells; with T-cell activation, a series of cellular events profoundly change the cellular environment. These events result from the different expressions of activated T cells: first, the release of cytokines like IL-2, IL-4, IL-6, lymphotoxin, and IFNs that alter the environment, and second, direct cellular interactions that can activate or kill cells. Prominent among these interactions are those of CD4 T cells with B cells, resulting in antibody formation; those of CD4 or CD8 T cells with APC, releasing IFN-γ and activating the macrophage system; and those of CD8 T cells with APC or other MHC-bearing cells, causing the death of these cells.

The large MHC gene segments encode a series of proteins involved in host defense, particularly two major sets of proteins, the class I and II MHC molecules. Several loci encode distinct class I and II MHC molecules with common structural and biological features. The class I MHC molecules are made of a heavy chain (of ~44 kDa) and a small (12-kDa) polypeptide, β_2-microglobulin. (The gene encoding β_2-microglobulin is not included in the MHC gene segment.) The class II MHC molecules are made up of two chains (α and β), which are noncovalently linked. Both class I and II MHC molecules are transmembrane proteins with a small segment in the cytosol (Figure 18.2).

The MHC genes are highly polymorphic. The sites of allelic differences are the amino acid residues located in the peptide-binding site. Thus, the evolutionary selective pressure for allelic diversity resides in the peptide-binding property of these molecules. The greater the allelic diversity, the more capable is the species of binding peptides from different pathogens.

Both class I and II MHC molecules bind peptides: their binding site lies at the most distal end of the molecule. The overall structure of the peptide-binding site is similar in both sets of molecules. In the class I MHC

Figure 18.2   General features of a class I MHC molecule. The α_1 and α_2 domains create the peptide-binding site, which is on top of the molecule. The α_3 domain is an Ig-like domain. A small polypeptide, β_2-microglobulin, forms part of the complex. (In the class II MHC molecule, the combining site is formed by the two most external domains of the α and β chains, which make up a structure similar to the combining site of class I molecules.)

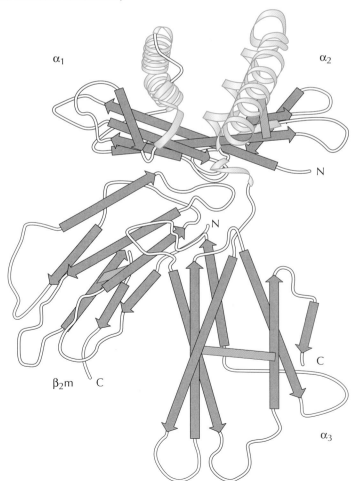

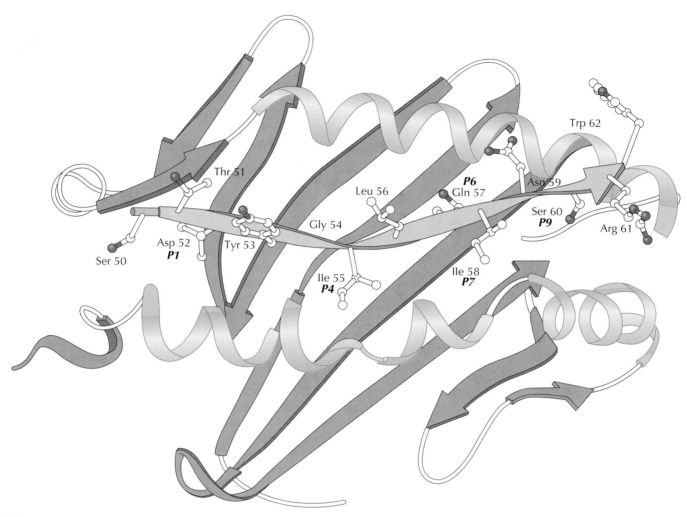

Figure 18.3   Structure of a peptide bound to the combining site of a class II MHC molecule. This specific example shows a lysozyme peptide bound to the murine *I-A^k* molecule. The peptide is stretched out and slightly twisted. The TCR contacts some of the solvent-exposed residues as well as the helices of the α_1 (top) and β_1 (below) domains. Adapted from D. H. Fremont, D. Monnaie, C. A. Nelson, W. A. Hendrickson, and E. R. Unanue, *Immunity* **8**:305, 1998.

molecules, both the α_1 and α_2 domains, the two most distal domains, at the amino end of the protein, contribute to the binding site (the third domain, the α_3 domain, is an Ig-like domain). In the class II MHC molecules, each of the two distal domains of each chain contributes to peptide binding (α_1 and β_1 domains) (Figures 18.2 and 18.3; Table 18.4).

The combining site is made of a platform of pleated chains bounded by two helices, which leave an open cleft or groove where the peptide binds (Figure 18.3). Although both class I and II binding sites are similar, there are some important differences. At a given time, the combining site holds a single peptide, usually of 8 to 10 residues for class I MHC molecules but 10 or more for the class II molecules. The binding specificity for peptides by MHC molecules tends to be broad. A given MHC allele can interact with a wide range of peptides. This broad, "promiscuous" binding ensures that many different peptides can be recognized. Peptides bind to MHC molecules with

Table 18.4   Properties of class I and II MHC molecules

Property	Class I	Class II
Chemistry	Two chains, a heavy chain that forms the binding site and α_2-microglobulin	α and β chains that pair to form the binding site
Peptide binding	Peptides of 8–10 residues	Peptides usually of >10 residues
Site of peptide binding	Primarily in ER	Vesicular system
Derivation of peptides	From cytosolic proteins, but not exclusively	From proteins in the vesicular system
Assembly pathway	Proteins in cytosol → proteasome catabolism → ER transport via TAP → assembly involving auxiliary molecules (tapasin, calnexin)	α and β chains transported from ER with invariant chain → release of invariant chain → peptide binding
Expression	All cells; high in hematopoietic cells	Mostly in hematopoietic cells

different affinities. For a given protein, some segments are preferentially displayed on the plasma membrane; these give rise to the dominant peptides that preferentially stimulate T-cell activation.

Several features responsible for peptide binding are well recognized since the initial analysis of the X-ray crystal structure. Peptides interact via amino acid side chains that establish hydrophobic and polar interactions with sites or pockets in the combining site. These sites are usually those having allelic specificity. Contributing to the binding affinity is an extensive network of hydrogen bonding between the peptide main carbon chain and many of the conserved residues of the helices and platform of the combining site. Some of the amino acid side chains are solvent exposed and serve to contact the TCR. The receptor establishes contact both with peptide residues and with the residues in the α-helices. This dual specificity is the basis for the phenomenon of MHC restriction; i.e., the recognition of a foreign peptide is always linked to recognition of self-MHC molecules.

From the very early studies, it became clearly apparent that class I MHC molecules were sampling peptides from proteins that localized to the cytosol. In contrast, the peptides that bound to class II MHC molecules were derived mainly from proteins taken by the vesicular system of the APC. Regardless of the source, it is important to note that the MHC system does not discriminate between autologous and foreign proteins. Peptides derived from foreign proteins or from self-proteins bind similarly; i.e., some bind well, others bind weakly, and some do not bind at all, depending on their sequence of amino acids. This issue is vital both for our understanding of how T-cell differentiation takes place and for placing the recognition of foreign and autoimmunity in the correct perspective.

We now understand that because the MHC binds to self-peptides, this allows, in the thymus, the elimination of T cells that spontaneously express an autoreactive receptor. The T cells go through a process of selection during their maturation in the thymus gland, as they interact with epithelial cells and thymic APC, both of which express MHC molecules. Only developing T cells that express a receptor that interacts with self-MHC molecules but poorly with the self-peptides contained in them are selected to mature, and they eventually spread to seed the lymphoid tissues. In contrast, the T cells that have a TCR directed to self-MHC containing an autologous peptide are eliminated and will die. As a result, part of the self-reactive repertoire of receptors is purged in the thymus.

The result of thymic selection is that the TCR from T cells that peripheralize to secondary lymphoid organs will recognize many non-self (foreign) peptides if they are presented by their own syngeneic APC. This "MHC restriction" was discovered, to a great extent, in the context of microbial peptides. Thus, in the APC–T-cell interaction, it was first noted by Rosenthal and Shevach in inbred guinea pigs immunized with dead tubercule bacilli. Importantly, the antiviral response to lymphocytic choriomeningitis virus was found by Zinkernagel and Doherty to be restricted by the class I MHC alleles. Thus, in the latter experiments, CD8 T cells from mice from the H-2^k strain would lyse target cells infected with this virus but only if they were of the H-2^k strain. Either the virus was modifying the structure of the H-2 molecules (altered self) or the virus and the H-2 protein in some way contributed to an interaction between these molecules and the T cells.

The class I MHC molecules are central for presentation of peptides derived from viruses or bacteria that reach the cytosol. The CD8 T-cell response is the major cellular system responsible for the antiviral response. CD8 T cells have cytolytic properties and will lyse infected cells. Thus, by lysing infected cells and releasing cytokines with antiviral effects, like IFN-α/β and IFN-γ, the CD8 T cell restricts the growth and spread of the virus. An important point is that class I MHC molecules are expressed in most cells. Thus, most infected cells can signal their infection through class I MHC molecules.

The pathway of presentation of peptides from cytosolic proteins is complex, involving a number of critical steps. The catabolism of the protein is carried out by the multicatalytic proteasome, a structure of about 1,500 kDa, made up of several protein subunits. The proteasome is a barrel-shaped structure with a central channel where the unfolded peptide interacts. Two of these proteins, LMP2 and LMP7, are encoded within the MHC gene complex. The expression of both proteins is also enhanced by IFN-γ. The presence of the different subunits influences the subsets of peptides that are generated. Indeed, mice engineered to have mutations in either LMP2 or LMP7 have susceptibilities to different viruses. The importance of proteasomal catabolism in the generation of peptides has been shown by the use of aldehyde inhibitors like lactacystine.

The peptides generated from proteasomal catabolism are then transported into the endoplasmic reticulum (ER) by two peptide transporters, which form a bimolecular molecular complex, the TAP-1 and TAP-2 molecules (for "transports of antigen processing"). There is some degree of specificity in the peptides transported by way of TAP-1 and TAP-2. Once in the ER, the peptide itself forms part of the assembly of the complex. The heavy chain associates with several accessory molecules like tapasin, calnexin, and calreticulin, which maintain it in an unfolded state. The exact sequence of assembly is still under evaluation. Once assembled, the peptide-MHC complex moves out of the ER into the Golgi and the plasma membrane.

The class II MHC system samples proteins found in the vesicular system, i.e., proteins that are internalized from the extracellular milieu by either receptor-mediated or fluid-phase endocytosis. Thus, the class II molecules are essential for presentation of many of the microbes that are taken into the APC through phagocytosis. CD4 T cells recognize the class II

MHC-peptide complex and are the central cells in the response to intracellular pathogens.

The assembly of the class II MHC-peptide complex is also intricate. The class II MHC molecules are synthesized in the ER as separate chains which pair and associate with a third chain, the invariant chain (called Ii because it does not show allelic polymorphism like the MHC molecules). Ii, also a transmembrane protein, plays two distinct roles: it has cytosolic residues that govern its transport out of the ER and Golgi, and it assembles with the class II MHC molecules, covering the combining site and thus blocking it from binding peptides. Once out of the ER-Golgi the complex reaches a proteolytic environment, where the Ii is degraded by cathepsins. A small peptide from the Ii chain is left in the combining site. At this stage, another auxiliary molecule, HLA-DM/H2M, comes into play; it favors the release of the Ii-derived peptide and the association of peptides from surrounding proteins. The new peptide-MHC complex is now free to be transported to the plasma membrane. The foreign proteins in the vesicular compartment are transported to the proteolytically rich vesicles bearing the MHC molecules, where they are denatured and partially proteolysed. How the opened polypeptide chain binds to the class II MHC molecule is not entirely settled. The point is, however, that once a segment of the polypeptide chain becomes bound, it becomes protected from catabolism by the cathepsins that surround it.

Recent studies have indicated an important facet of the interaction between microbes and the host, i.e., interference with antigen processing and presentation. This interference has been found for several DNA viruses. An important finding has involved the inhibition of presentation of one of the nuclear antigens of Epstein-Barr virus (EBV). For example, EBNA-1, despite displaying a class I MHC epitope, is not presented. This is as a result of having a protein domain made up of a series of Gly-Ala repeats that inhibits processing. Manipulation of the EBNA-1 gene by deleting this domain results in the expression of the class I MHC epitope. This cycle could be important in the maintenance of the viral latency in EBV-infected B cells, where viral gene expression is restricted to EBNA-1.

Cytomegaloviruses are important for their effect on immunocompromised individuals. They contain a gene, U6, which inhibits class I MHC assembly, also by targeting the TAP transporters. Another gene, US11, also inhibits class I MHC expression by channeling these molecules from ER to the cytosol, where they are rapidly degraded. Herpes simplex viruses also interfere with antigen presentation; they do this through the expression of a protein, ICP47, which inhibits the assembly of class I MHC molecules by binding to TAP molecules and inhibiting their function of peptide transport.

Another example of how microbes alter the interaction with MHC molecules is that of superantigens. Superantigens are proteins, derived from bacteria or viruses, that are capable of binding to class II MHC molecules outside the combining site and to TCR of particular sets of T cells. This can result in extensive activation of T cells. Much of the pathology of infection with toxic shock syndrome toxin type A of staphylococci is caused by the sudden release of cytokines like TNF.

APC–T-Cell Interaction

The interaction of T cells with APC bearing the peptide-MHC complex involves two components. The first component consists of the TCR binding to the MHC-peptide complex. The second component involves auxiliary molecules in or from the APC and from the T cell that foster and modulate the cellular interactions and the differentiation of each cell. There are four sets of interactions dominated by auxiliary or non-antigen-specific proteins and cytokines.

Interactions with Adhesion Molecules

As the name implies, pairs of adhesion molecules foster the cell-to-cell contact. A number of coreceptors involving integrins and molecules of the Ig superfamily have been defined. (The Ig superfamily contains proteins with an Ig fold or domain, usually of 90 to 110 residues, made up of two antiparallel β-strands connected by a critical disulfide bond. Many molecules involved in cellular interactions bear Ig domains. Proteins of the Ig superfamily are probably derived from an early gene that diversified as required for regulation of multiple cellular interactions in higher vertebrates.) The integrins belong to various families that foster not only cell-cell interactions but also interactions with connective tissue proteins and between lymphocytes and various cells including endothelial, epithelial, and connective tissue cells. Important to cite here are the interactions between ICAM molecules on APC or target cells and the family of integrins on T cells (i.e., LFA-1 or CD11/CD18).

Interaction of CD40 with CD40L

CD40L is a molecule found on T cells that pairs with CD40 on APC including B cells. This interacting pair is crucial for B-cell activation and differentiation and is also important for activation of APC. The importance of CD40-CD40L was first appreciated during analysis of infants with the hyper-IgM syndrome. These infants show marked susceptibility to infections caused by mutations of the gene encoding CD40L. In these infants, the B cells produce IgM but do not class switch to produce IgG antibodies. The IgG is required for interaction with the opsonic receptors. CD40-CD40L interaction also has a profound effect on the biology of APC. APC will produce cytokines like IL-1, TNF, and IL-12 and also produce a burst of NO˙ production. Thus, APC stimulation can develop not only directly as a result of microbes, as described above, but also during the stage of interaction with T cells. It follows that interactions with microorganisms or proteins that do not directly stimulate the macrophages depend highly on the CD40-CD40L system to induce this state of activation. The CD40-CD40L interaction importantly enhances the expression of B7-1 and B7-2 molecules involved in the third set of interactions (see below).

Pairing of B7-1/B7-2 with CD28/CTLA-4

B7-1 and B7-2 are a pair of molecules of the Ig superfamily that are expressed under basal conditions at low levels on APC. B7-1 and B7-2 have complementary molecules on T cells, i.e., CD28 and CTLA-4. Both are disulfide-linked homodimers. B7-1 and B7-2 are up-regulated during interactions of APC with a variety of microorganisms. CD28 is expressed in a constitutive way, while CTLA-4 is up-regulated during T-cell activation. The engagement of B7-1/B7-2 with CD28 favors T-cell activation and the expression of a number of T-cell cytokines. Also important is the up-regulation of a number of antiapoptotic molecules of the Bcl-2 family. Lack of CD28 engagement results in poor or limited T-cell stimulation. The CTLA-4 molecule appears to inhibit T-cell activation, an issue made very apparent by the phenotype of mice having null mutations of it, which show massive lymphoproliferation and activation of T cells. Thus, there is a balance between the two molecules, one favoring activation (CD28) and the other dampening it.

Role of Cytokines

Several conditions during the APC–T-cell interaction have a profound influence on the differentiation of T cells. T cells polarize into two subsets depending on their activation of particular sets of cytokine genes. The Th1 subset produces the cytokine IFN-γ as well as IL-2. Thus, Th1 cells, via IFN-γ, markedly influence the cellular response that eliminates many intracellular pathogens. The Th2 subset produces little if any IFN-γ but makes many of the B-cell-activating molecules, such as IL-4, IL-5, IL-6, and IL-10. Some of these cytokines have a negative effect on macrophage activation. Thus, the Th2 response favors antibody responses and is much less favorable for the macrophage activation pathway.

The decision by the T cells to polarize in a Th1 or Th2 direction rests on the early stage of interaction with APC and is influenced by many factors, but a major influence is the cytokine environment. Production of IL-12 is a major and important signal for Th1 differentiation, while IL-4 (resulting for mast cells or lymphocytes) favors Th2 polarization (Figure 18.4).

Many of the cellular responses are of mixed type at the start of the response, but persistence of antigenic stimulation will skew the response toward one or the other, and these are of major importance in determining the outcome of an infection. The importance of Th1-Th2 differentiation has

Figure 18.4 Regulation of CD4 T-cell differentiation. Reprinted from A. O'Garra, *Immunity* **8**:275–283, 1998.

Regulatory T cells control Th1 and Th2 responses

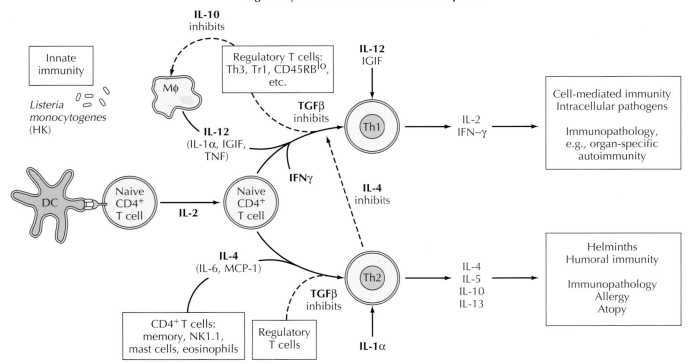

Table 18.5   Experimental manipulations that influence resistance to *Leishmania major* in the mouse

Favor resistance	Favor susceptibility
Injecting IFN-γ early in the infection	Neutralizing IFN-γ, IL-12, or TNF with antibodies
Neutralizing IL-4 with antibodies or infecting IL-4 null mice	Using mice with ablation of IFN-γ, IL-12, or TNF genes
Injecting IL-12 early in the infection	Neutralizing NO˙ or infecting mice lacking inducible nitric oxide
Abolishing early immune response to dominant epitope of *L. major* for CD4 T cells	Neutralizing early production of IFN-αβ

been strikingly shown in the model of *Leishmania major* infection of mice. *Leishmania* infection is controlled by activated macrophages produced primarily by IFN-γ secreted particularly by CD4 Th1 cells. The activated macrophages restrict *Leishmania* growth because of the release of NO˙. Indeed, infections where IFN-γ is not produced or is neutralized by antibodies or where NO˙ is likewise not produced because of ablation of the inducible NO synthase gene or NO˙ is neutralized by drugs all result in dissemination of the infection (Table 18.5). In brief, because of as yet unidentified susceptibility genes, the BALB/c strain responds early to the infection by showing a predominant IL-4 response over IL-12 response, resulting in a Th2 differentiation pattern. The IL-4 and IL-10 cytokines do not favor macrophage activation, resulting in dissemination of the parasite. Manipulations that favor a Th1 pattern (i.e., neutralization of IL-4 or addition of IL-12) produce the contrasting response, with macrophage activation and elimination of the parasite. One of the manipulations summarized in Table 18.5 is the one involving the early response to a *Leishmania* peptide, which is very fast. In the BALB/c susceptible strain, the early response has a high IL-4 component, which immediately skews the response to a Th2 pattern. If such a response is neutralized, this lack of IL-4 response can allow the IL-2 stimulation to the protective Th1 phenotype. The differences in Th1 and Th2 responses may be important in chronic infections and may have been indicated in the two polar extremes of leprosy infection: tuberculoid leprosy, with activated granulomas controlling the bacterial load, and lepromatous leprosy, with widespread bacillary infiltration.

Apoptosis of Lymphocytes

Finally, during an ongoing response to infection, lymphocyte growth is controlled largely by mechanisms that involve apoptosis. Apoptosis of lymphocytes is a component of the normal response as lymphocytes become activated. Activation-induced cell death involves a number of cell surface ligand pairs, the most extensively studied of which involves Fas and Fas ligand, both of which are expressed in lymphocytes and up-regulated during cell activation. The levels of these molecules also increase during infection, particularly with intracellular pathogens. During diverse viral infection, virally infected cells die; this process is mediated by CD8 T cells (or NK cells) but also not infrequently by direct viral infestation. Interestingly, a number of viral products influence apoptosis. These have been identified, particularly in DNA viruses, and include inhibitors of caspases, the enzymes involved in the apoptotic response, which can be produced by poxviruses (the CrmA inhibitor); the viral FLICE proteins, which interrupt the signaling pathways involved with Fas proteins; and the AIA protein from adenoviruses, which affects the DNA cycle of the infected cell.

Pathogen Detection by the Innate Immune System: the Toll-Like Receptors

Recognition of pathogens by the adaptive immune system relies on the presence of a wide repertoire of protein complexes (soluble antibodies, B-cell receptors, and T-cell receptors) encoded in lymphocytes by genes resulting from somatic DNA rearrangement events (Box 18.1). These adaptive immune sensors bind with high-affinity foreign antigens derived from individual species of microorganisms. By contrast, the innate immune system detects foreign organisms through a defined number of germ line-encoded proteins that recognize conserved pathogen-associated molecular patterns (PAMPs) common to several species of microorganisms but not present on self-tissues. These proteins are generally extracellular or associated with the plasma membrane of the immune cells. One example of an extracellular innate immune sensor is the mannan-binding lectin, which binds mannose on pathogen surfaces and directly activates the complement cascade. One of the membrane-associated proteins that recognize PAMPs is the already mentioned mannose receptor, which promotes phagocytosis of pathogens by macrophages. The identification of two mammalian intracellular bacterial sensors, NOD1 and NOD2, has shown that PAMP recognition can also occur, in the case of intracellular pathogens, in the cytoplasm of infected cells. NOD1 and NOD2 are characterized by two main functional domains: a leucine-rich repeat (LRR) domain, believed to associate with bacterial peptidoglycan, and a caspase activation and recruitment (CARD) domain, which triggers an intracellular signaling cascade leading to nuclear factor kappa B (NF-κB) activation.

Recently, a new family of transmembrane proteins was discovered, the Toll-like receptors (TLRs) that play a fundamental role in non-self-detection by the innate immune system. Mammalian Toll-like receptors have first been identified by homology with *Drosophila* Toll that serves two central functions in the fruit fly biology: in early embryos it controls the formation of the dorso-ventral pattern, while in larvae and adults it is implicated in the immune response against fungal and gram-positive bacterial pathogens. It is important to note that *Drosophila*, like all invertebrates, does not possess an adaptive immune system and relies exclusively on innate immune reactions to fight microbial infections. The most important effector mechanism of *Drosophila* immunity is the production of several families of potent antimicrobial peptides that make the flies very resistant to infection by a wide spectrum of microorganisms. Strikingly, flies that do not have a functional Toll protein fail to produce antimicrobial peptides and die a few days after microbial infection.

Mammalian TLRs share with *Drosophila* Toll conserved structural features. They are characterized by an extracellular LRR domain, believed to interact directly with PAMPs, similarly to the LRR of NOD proteins, and a cytoplasmic domain similar to that of interleukin 1 (IL1) receptor, called Toll-IL1 receptor (TIR) domain, which initiates intracellular signaling. Several studies conducted mainly in mice by gene targeting have demonstrated that individual TLRs are activated by distinct elicitors (Table 18.6). Mice lacking TLR4 are resistant to LPS-induced septic shock, suggesting that TLR4 is the main LPS sensor and pointing to the fundamental role played

Table 18.6 PAMPs from various microorganisms that elicit mammalian Toll-like receptors

TLR	Ligands	Microorganism
TLR1 (heterodimers with TLR2)	Lipoproteins	Bacteria
TLR2	Peptidoglycan	Bacteria
	Lipoteichoic acid	Bacteria
	Porins	*N. meningitides*
	Zymosan	Yeast
TLR3	dsRNA	RNA viruses
TLR4	LPS	Bacteria
	Heat shock proteins	Endogenous
	Fibrinogen	Endogenous
	Fibronectin	Endogenous
TLR5	Flagellin	Bacteria
TLR6 (heterodimer with TLR2)	Lipoproteins	Mycoplasma
TLR7	Imidazoquinoline	Antiviral
TLR8	Imidazoquinoline	Antiviral
TLR9	CpG	Bacteria
TLR10	Unknown	

by TLRs during the inflammatory response. Interestingly, it has been reported that naturally occurring mutations in TLR4 are associated with LPS hyporesponsiveness also in humans. Subsequent studies have identified additional TLR4 elicitors including other pathogen-associated structures and, more surprisingly, endogenous molecules like heat shock proteins, fibrinogen and fibronectin. TLR2 recognizes peptidoglycan, lipoteichoic acid, and lipopeptides from bacteria and zymosan from the yeast cell wall. In addition, TLR2, combined with TLR1 or TLR6, can also bind to lipoproteins from bacteria and mycoplasma. TLR5 binds to flagellin, a structural component of bacterial flagella. TLR3 is activated by double-strand RNA, which is produced in high amounts during replication of RNA viruses, while TLR9 is activated by unmethylated CpG motifs that are enriched in bacterial DNA. Although the natural ligands for TLR7 and TLR8 remain unknown, it has been shown that some antiviral compounds of the imidazoquinoline class (resiquimod R-848) activate both receptors. Most of the TLRs are localized on the cell surface with the exception of TLR9 that associates with the cell endosomes and probably recognizes DNA fragments derived from bacterial lysis. For their direct binding with PAMPs, mammalian TLRs are defined as pattern recognition receptors (PRRs). With regard to this feature, mammalian TLRs differ from *Drosophila* Toll, which is not a direct microbial sensor. In the flies, in fact, gram-positive bacteria are recognized by two soluble PRRs called peptidoglycan recognition protein (PGRP)-SA and gram-negative binding protein (GNBP) 1, which trigger an extracellular proteolytic cascade culminating in the activation of Spaetzle, the sole Toll protein ligand. However, functional evidence for a direct association of the mammalian TLRs with PAMPs is often weak or missing. In addition, the X-ray structures of the TLR leucine-rich domains are not currently available; thus, the molecular basis of the LRR-PAMP interaction is still unknown. Due to the lack of data on the direct interactions between TLRs and PAMPs, it has been suggested that accessory soluble proteins or coreceptors might contribute to TLR activation. At least for LPS-mediated

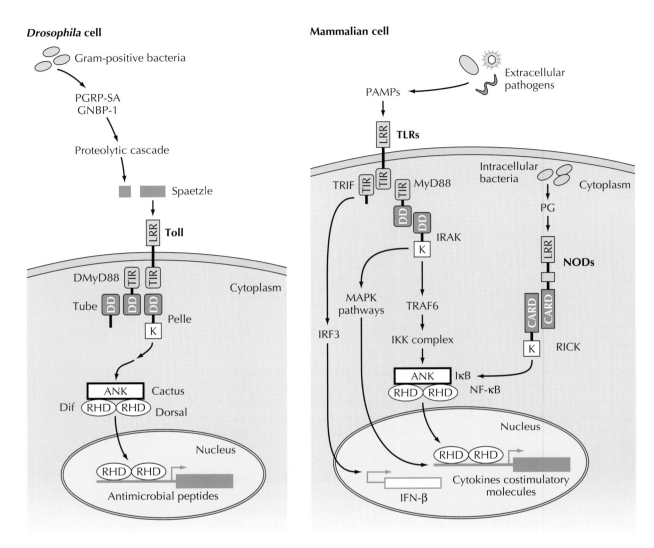

Figure 18.5 Pathogen-activated innate immune pathways in *Drosophila* and mammalian cells. In *Drosophila* (left panel) gram-positive bacteria are recognized by two PRRs, PGRP-SA and GNBP1, which trigger the activation of a proteolytic cascade that activates Spaetzle. A cleaved form of Spaetzle directly activates Toll. The signal is transmitted through a *Drosophila* homologue of MyD88 (DMyD88) and an IRAK-like protein kinase (Pelle), leading to phosphorylation of the IκB-homologue Cactus, the nuclear translocation of the NF-κB homologues Dif and Dorsal and the activation of *Drosophila* immune-responsive genes, including antimicrobial peptide genes. In mammals (right panel) PAMPs associated with various extracellular pathogens elicit the activation of TLRs, which can transmit the signal through two TIR containing proteins: MyD88, which activates NF-κB and MAPK pathways leading to up-regulation of several immune responsive genes including proinflammatory cytokines and costimulatory molecules, and TRIF, triggering the IRF-3/interferon β response. Peptidoglycan (PG) associated with intracellular bacteria is recognized by NOD1 and 2, which activate NF-κB pathway through RICK. Abbreviations: PAMPs, pathogen-associated molecular patterns; TLRs, Toll-like receptors; LRR, leucine-rich repeat domain; TIR, Toll-IL1 receptor domain; DD, death domain; CARD, caspase activation and recruitment domain; RHD, Rel homology domain; ANK, ankyrin repeats domain; K, kinase domain; IRAK, IL1-receptor-associated kinase; TRAF, TNF-receptor-associated kinase; NF-κB: nuclear factor κB; IκB, inhibitor of κB; IKK: IκB kinase; MAPK, mitogen-activated protein kinase; IFN, interferon; PGRP, peptidoglycan recognition protein; GNBP, gram-negative binding protein.

activation of TLR4, the involvement of two extracellular proteins (LPS-binding protein and MD2) and one coreceptor (CD14) has been demonstrated. The presence of extracellular adapter proteins interacting with TLRs might explain how several elicitors with little or no structural similarities can activate the same TLR. In addition to that, the identification of some endogenous molecules like heat shock proteins and fibrinogen, capable of activating TLRs, has suggested that TLRs might play a dual function in sensing non-self-PAMPs and self-danger signals derived from tissue damage or cell lysis following microbial infection. Interestingly, some cellular products generated by cell lysis, like DNA and RNA, might contain small amounts of known TLR-inducing motifs (CpG and dsRNA).

The activation of TLRs is believed to occur through formation of homo- or heterodimers followed by recruitment of TIR-containing proteins by the TLR cytoplasmic TIR domain. One of these TIR-containing proteins, MyD88, binds to all TLRs and mediates the recruitment of IRAK family kinases leading to the activation of NF-κB and mitogen-activated protein kinase (MAPK) pathways. Interestingly, most of the intracellular proteins participating in the MyD88-dependent activation of the NF-κB pathway in mammals are remarkably conserved in the *Drosophila* Toll pathway (Figure 18.5). Unlike MyD88, a second mammalian TIR-containing protein, TRIF, associates only to a subset of TLRs (TLR3 and 4) and triggers the activation of interferon regulative factor-3 (IRF3), which in turn induces the transcription of the genes coding for interferon α and β. This finding contributes to the explanation of how different TLR elicitors can trigger different transcriptional programs in the same immune system cell type.

Although TLR expression has been detected in several cell types, most of the studies on the biological function of TLRs have focused on APCs, in particular, macrophages and dendritic cells. TLRs activation induces in these cells the expression of pro-inflammatory cytokines such as TNF-α, IL-1, IL6, and IL12 and the up-regulation of MHC and costimulatory molecules like B7.1 and B7.2. As a result of this maturation program, APCs are significantly more efficient in priming naive T cells. This mechanism contributes to ensure that T cells are primed only when antigen presentation is associated with microbial infection, limiting the danger derived from the activation of autoreactive T cells. Recent studies have extended the function of TLRs on B cells. It has been shown that TLR9 activation through CpG induces both maturation and proliferation of the human memory B-cell subset, suggesting that a TLR-mediated mechanism might contribute to the maintenance of the serological memory.

In conclusion, it is interesting to note that Toll, a central element in the control of invertebrate innate immunity, has been conserved during evolution and has assumed in mammals many different roles in the control of both innate and adaptive immune reactions.

Selected Readings

Akira, S. 2003. Mammalian Toll-like receptors. *Curr Opin. Immunol.* **15:**5–11.

Arbour N. C., E. Lorenz, B. C. Schutte, J. Zabner, J. N. Kline, M. Jones, K. Frees, J. L. Watt, and D. A. Schwartz. 2000. TLR4 mutations are associated with endotoxin hyporesponsiveness in humans. *Nat. Genet.* **25:**187–191.

Babbitt, B. P., P. M. Allen, G. Matsueda, E. Haber, and E. R. Unanue. 1985. Binding of immunogenic peptides to Ia histocompatibility molecules. *Nature* **317:**359–363.

Bernasconi, N.L., E. Traggiai, and A. Lanzavecchia. 2002. Maintenance of serological memory by polyclonal activation of human memory B-cells. *Science* **298**:2199–2202.

Bjorkman, P. J., B. Saper, B. Samraoui, W. S. Bennett, J. L. Strominger, and D. C. Wiley. 1987. The foreign antigen binding site and T cell recognition regions of class I histocompatibility regions. *Nature* **329**:512–515.

Buchmeier, N. A., and R. D. Schreiber. 1985. Requirement of endogenous interferon-γ production for resolution of Listeria monocytogenes infection. *Proc. Natl. Acad. Sci. USA* **82**:7404–7409.

Epstein, J., Q. Eichbaum, S. Sheriff, and A. B. Ezekowitz. 1996. The collectins in innate immunity. *Curr. Opin. Immunol.* **8**:29–35.

Freeman, M., J. Ashkenas, D. J. Rees, D. M. Kingsley, N. G. Copeland, N. A. Jenkins, and M. Krieger. 1991. An ancient, highly conserved family of cysteine-rich protein domains revealed by cloning type I and II murine macrophage scavenger receptors. *Proc. Natl. Acad. Sci. USA* **88**:4931–4935.

Girardin, S. E., P. J. Sansonetti, and D. Philpott. 2002. Intracellular versus extracellular recognition of pathogens: common concepts in mammals and flies. *Trends Microbiol.* **10**:193–199.

Girardin, S. E., J. Hugot, and P. J. Sansonetti. 2003. Lessons from Nod2 studies: towards a link between Crohn's disease and bacterial sensing. *Trends Immunol.* **24**:652–658.

Gobert, V., M. Gottar, A. A. Matskevich, S. Rutschmann, J. Royet, M. Belvin, J. A. Hoffmann, and D. Ferrandon. 2003. Dual activation of the Drosophila Toll pathway by two pattern recognition receptors. *Science* **302**:2126–2129.

Gordon, S., S. Clark, D. Greaves, and A. Doyle. 1995. Molecular immunobiology of macrophages: recent progress. *Curr. Opin. Immunol.* **7**:24–33.

Grewal, I. G., and R. A. Flavell. 1998. CD40 and CD154 in cell-mediated immunity. *Annu. Rev. Immunol.* **16**:111–135.

Hengel, H., and U. H. Koszinowski. 1997. Interference with antigen processing by viruses. *Curr. Opin. Immunol.* **9**:470–476.

Hsieh, C. S., S. E. Macatonia, C. S. Tripp, S. F. Wolf, A. O'Garra, and A. M. Murphy. 1993. Development of Th1 CD4 T cells through IL-12 produced by *Listeria*-induced macrophages. *Science* **260**:547–549.

Janeway, C. A., Jr., and R. Medzhitov. 2002. Innate immune recognition. *Annu. Rev. Immunol.* **20**:197–216.

Lanier, L. L. 1997. Natural killer cells—from no receptor to too many. *Immunity* **6**:371–378.

Lemaitre, B., E. Nicolas, L. Michaut, J. M. Reichhart, and J. A. Hoffmann. 1996. The dorsoventral regulatory gene cassette spaetzle/Toll/cactus controls the potent antifungal response in Drosophila adults. *Cell* **86**:973–983.

Lurie, M. B. 1964. *Resistance to Tuberculosis: Experimental Studies in Native and Acquired Defensive Mechanisms*. Harvard University Press, Cambridge, Mass.

Mackaness, G. B. 1964. The immunological basis of acquired cellular resistance. *J. Exp. Med.* **120**:105–113.

Medzhitov, R., P. Preston-Hurlburt, and C. A. Janeway, Jr. 1997. A human homologue of the Drosophila Toll protein signals activation of adaptive immunity. *Nature* **388**:394–396.

Moore, K. V., A. O'Garra, R. de Waal Malefyt, P. Vieira, and T. R. Mosman. 1993. Interleukin 10. *Annu. Rev. Immunol.* **11**:165–184.

Nelson, C. A., N. J. Viner, and E. R. Unanue. 1996. Appreciating the complexity of MHC class II peptide binding: lysozyme peptide and I-Ak. *Immunol. Rev.* **151**:81–105.

Pasare, C., and R. Medzhitov. 2003. Toll-like receptors: balancing host resistance with immune tolerance. *Curr. Opin. Immunol.* **15**:677–682.

Ravetch, J. V., and R. A. Clynes. 1998. Divergent roles for Fc receptors and complement *in vivo. Annu. Rev. Immunol.* **16:**421–432.

Rollins, B. J. 1997. Chemokines. *Blood* **90:**909–928.

Unanue, E. R. 1997. Studies in Listeriosis show the strong symbiosis between the innate cellular system and the T cell response. *Immunol. Rev.* **158:**11–25.

Yamamoto, M., S. Sato, H. Hemmi, K. Hoshino, T. Kaisho, H. Sanjo, O. Takeuchi, M. Sugiyama, M. Okabe, K. Takeda, and S. Akira. 2003. Role of adaptor TRIF in the MyD88-independent Toll-like receptor signaling pathway. *Science* **301:**640–643.

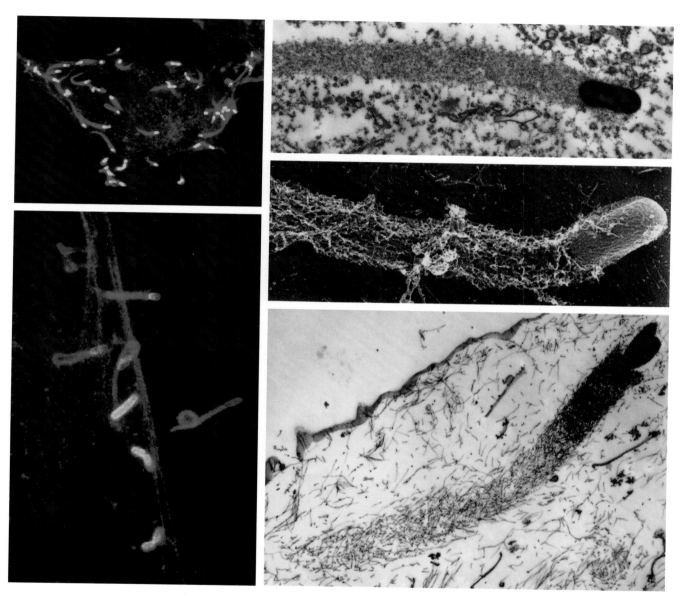

Figure 1.3 *L. monocytogenes* action tails. (Left top and bottom) Actin tails visualized by immunoflourescence. *L. monocytogenes*-infected cells were labeled with flourescein isothiocyanate-conjugated phalloidin (to label F-actin) and anti-*Listeria* antibodies. (Right) Actin tails visualized by electron microscopy. (Top) Thin-section electron micrograph from an infected tissue culture cell showing a moving bacterium associated with an F-actin comet tail (Tilney technique). (Middle) Three-dimensional visualization of the actin comet tail by the quick-freeze/deep-etch technique (courtesy of J. Heuser). (Bottom) Thin section through an *L. monocytogenes* cell with a tail whose actin filaments have been decorated with subfragment 1 (S1) of myosin.

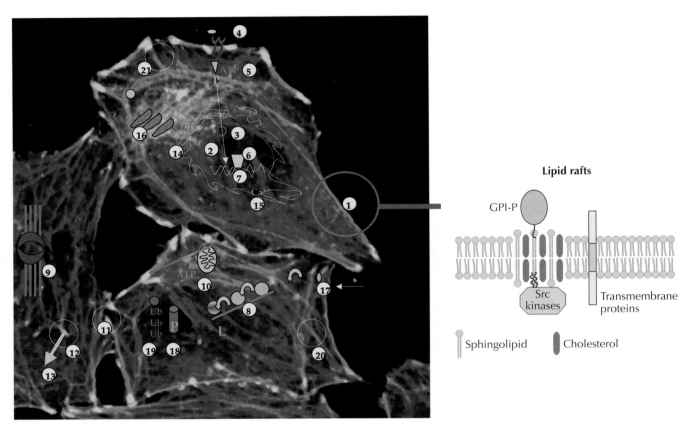

Figure 3.1 General cell organization (see text for details).

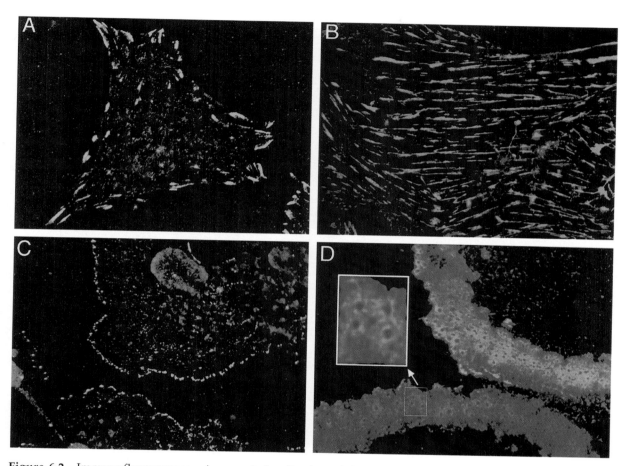

Figure 6.2 Immunofluorescence microscopic localization of the main forms of integrin-mediated matrix adhesions. **(A)** Human foreskin fibroblast, labeled for phosphotyrosine, displaying mainly "classical" focal adhesions, located primarily at the cell periphery. **(B)** Fibrillar adhesions of human foreskin fibroblast labeled for tensin. These adhesions are typically associated with fibronectin fibrils and are enriched in central regions of the cells. **(C)** Human fibroblasts (SV80 line) treated with the Rho-kinase inhibitor Y-27632 and immunolabeled for phosphotyrosine. The labeling is associated primarily with small dot-like structures associated with the lamellipodium, which are identified morphologically as focal complexes. **(D)** Paxillin-labeled podosomes formed by a primary rat osteoclast. Individual podosomes consist of a ring containing several "plaque proteins" (see insert), and an actin-rich central domain. As seen in this picture, podosomes often tend to cluster into large arrays. Reproduced with permission from *Nature Reviews Molecular Cell Biology* (Geiger et al., 2001), Macmillan Magazines Ltd.

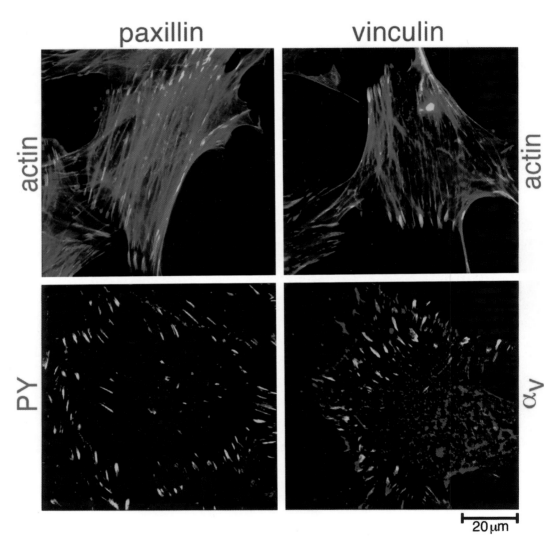

Figure 6.4 (Top) Double fluorescent labeling of fibroblasts for actin (green) and vinculin or paxillin (red). (Bottom) Double immunofluorescence labeling (right) for vinculin (red) and α_v integrin (green) or (left) for paxillin (red) and phosphotyrosine (PY) (green). Both vinculin and paxillin are associated with the termini of actin-containing stress fibers. Vinculin and phosphotyrosine are also associated with cell-cell adherens junctions, while paxillin and integrin are present only in matrix adhesions.

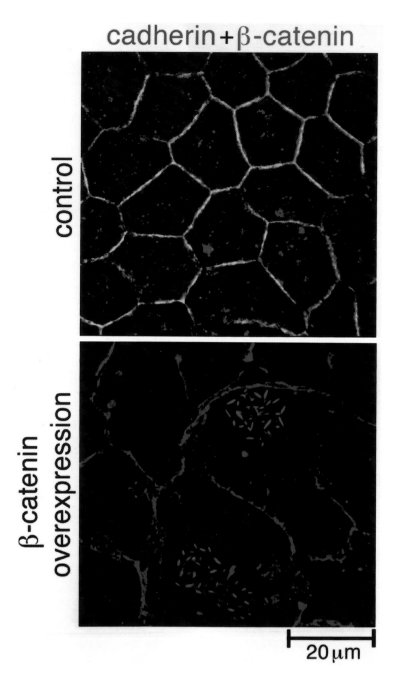

Figure 6.5 Components of cell-cell adherens junctions. (Top) Double immuno-fluorescence labeling of epithelial MDCK cells for cadherin (red) and β-catenin (green). Cell-cell junctions are seen as yellow lines due to superposition of red and green colors, showing nearly complete overlap between cadherin and β-catenin at cell-cell adhesion. (Bottom) Overexpression of a chimeric molecule consisting of β-catenin and green fluorescent protein (which allows visualization of the molecule) in MDCK cells by transient transfection followed by immunolabeling for cadherin (red). The fluorescent β-catenin appears green. Note that when overexpressed, β-catenin accumulates in the nucleus, forming aggregates of different shapes.

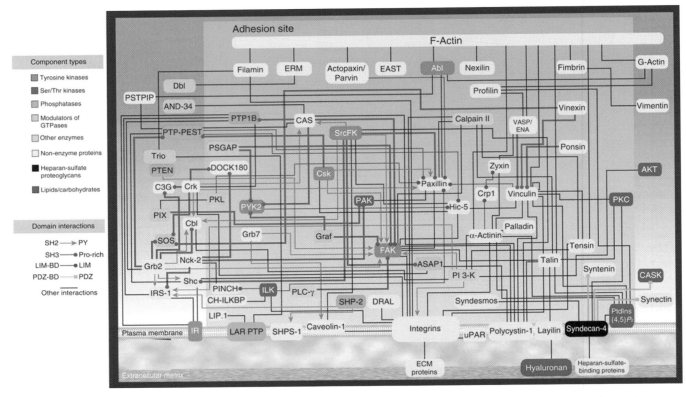

Figure 6.6 A scheme summarizing known interactions between the various constituents of cell-matrix adhesions. Components that were found to be associated with cell-matrix adhesion sites are placed inside the internal green box, whereas additional selected proteins that affect matrix adhesions but were not reported to stably associate with them are placed in the external blue frame. The general property of each component is indicated by the color of its box, and the type of interaction between the components is indicated by the style and color of the interconnecting lines, as indicated in the legend. From Zamir and Geiger (2001).

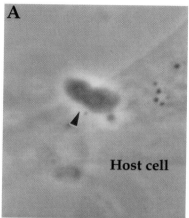

Host cell

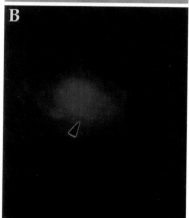

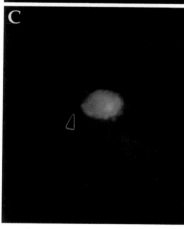

Figure 10.8 Phase-contrast **(A)** and fluorescence **(B and C)** micrographs of a tachyzoite of *T. gondii* caught in the act of invading a mammalian cell. The preparation was reached with antibody against the parasite surface protein SAG1 **(B)**. permeabilized, and incubated with antibody against the microneme protein MIC2 **(C)**. The fluorescence shows the capping of the surface protein SAG1 concomitant with the secretion of the microneme protein MIC2 into the nascent parasitophorous vacuole during the entry process. The junction point between the parasite and the host cell is indicated by an arrow. The parasite is known to remodel this vacuole extensively during establishment of intracellular infection. Courtesy of L. David Sibley.

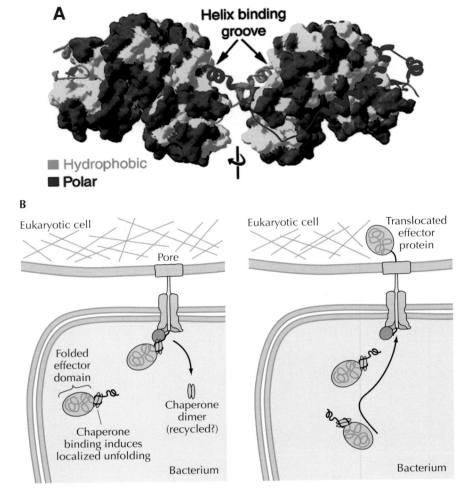

Figure 15.6 Structural analysis of type III secretion chaperones. **(A)** Molecular architecture of the SicP chaperone from *Salmonella* consists of two sets of tightly bound homodimers. Shown is the surface distribution of polar regions (grey) and hydrophobic regions (yellow). The SicP homodimer pairs are encased in a complex with two SptP effector molecules (simplistically shown as a red and blue ribbon) that lie within a helix-binding grove. The effector-interacting surfaces of SicP are predominantly hydrophobic. Modified from Figure 4 in C. E. Stebbins and J. E. Galán, *Nature* **414:** 77–81, 2001, with permission and provided by Jorge Galán, Yale University School of Medicine, New Haven, Conn. **(B)** A proposed model for the chaperone-mediated secretion of type III substrates. As a prerequisite for translocation into a target eukaryotic cell, an effector substrate binds to a cognate chaperone homodimer. At this site of binding (near the N-terminus), localized effector unfolding occurs, while the C-terminal enzymatic domain remains folded and functionally active (depicted by a gold globular shape). This effector – chaperone complex probably docks at the inner face of the type III secreton. Localized unfolding may 'catalyze' a general unfolding along the entire effector as it is secreted through the secreton. The chaperone homodimer is released and might be recycled to form a complex with another newly synthesized cognate effector molecule pre-destined for secretion.

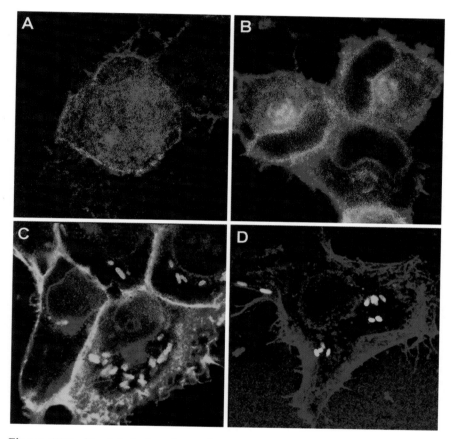

Figure 15.7 *Yersinia*-induced translocation of Yops into cultured HeLa cells. Translocation of Yops is visualized by confocal laser-scanning microscopy. **(A to C)** Yops are represented by green fluorescence: YopH (A), YopE (B), and YpkA (C). HeLa cell plasma membranes are illustrated by the red fluorescence. Note that the individual Yops have different locations within the eukaryotic cell: YopH is widely distributed in both the cytosolic and nuclear compartments, YopE is enriched in the perinuclear region, and YpkA is localized at the inner surface of the plasma membrane. Yellow indicates colocalization of YpkA and the plasma membrane. **(D)** Mutants in any of *lcrV, yopB,* or *yopD* are translocation deficient.

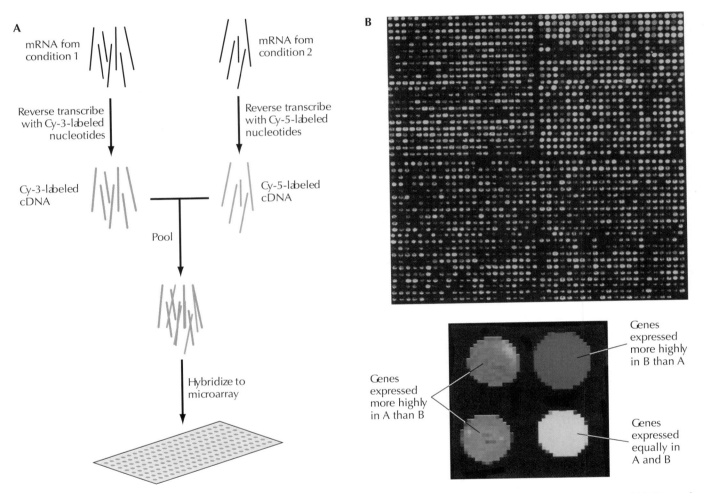

Figure 20.2   Gene expression experiment performed on a microarray. **(A)** Flow of a microarray experiment. **(B)** Scanned image of hybridized microarray and close-up image showing three possible outcomes of hybridization.

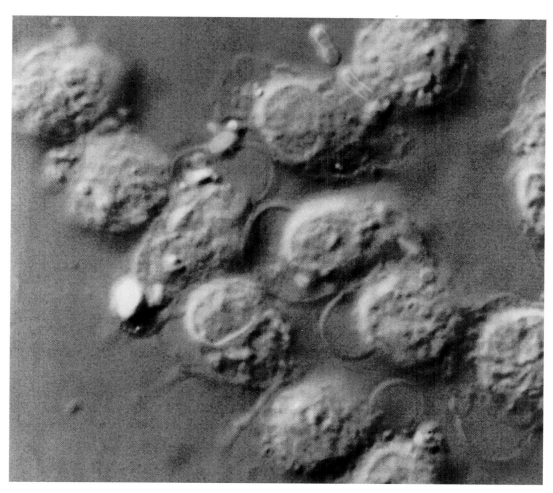

Figure 20.6 GFP as a tool to detect gene expression in intracellular bacteria. Mouse macrophages were infected with *Salmonella* serovar Typhimurium containing a *gfp* reporter gene fused to a promoter expressed only when the bacteria reside within host cells. Bacteria inside cells appear as green rods while bacteria outside (stained with an anti-*Salmonella* antibody conjugated to Texas Red) do not express GFP.

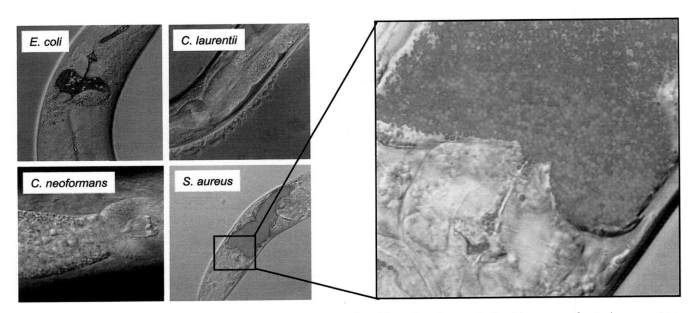

Figure 23.5 Bacterial and fungal pathogens but not innocuous bacteria or yeast accumulate in the *C. elegans* intestine. In most, but not all cases, killing of *C. elegans* by microbial pathogens involves the accumulation of the pathogenic microbe in the *C. elegans* intestine. Normally, when *C. elegans* is feeding on a relatively innocuous bacterium such as *E. coli* expressing green fluorescent protein (GFP) as in this photomicrograph, almost all the ingested bacteria are ground up in the pharyngeal grinder organ and very few intact bacteria enter the intestinal lumen. Similarly, when feeding on the innocuous yeast *Cryptococcus laurentii,* no intact yeast cells can be observed in the intestine. In this latter case, the yeast cells are large enough to be readily seen without being labeled by GFP. In contrast to the results obtained with innocuous microbes, when feeding on most pathogenic bacteria or yeast, such as *S. aureus* or *C. neoformans* pictured here, high titers of the pathogenic microbes accumulate in the intestine.

19

Electron Microscopy

CHANTAL DE CHASTELLIER

Pathogens have evolved a wide variety of strategies to circumvent the host microbicidal activities and to use the cellular machinery to their own advantage. When pathogens invade the host, they immediately encounter professional phagocytes, i.e., macrophages and neutrophils. Depending on their surface properties, they will either be or not be phagocytosed by macrophages. In the latter case, they will either remain as extracellularly growing bacteria or invade nonphagocytic cells, i.e., fibroblasts, endothelial cells, or epithelial cells.

Each intracellular pathogen has evolved distinct strategies to manipulate host cell organelles and/or constituents, thus enabling it to find favorable conditions for survival and multiplication. After binding to cell surface receptors, microorganisms and particles are internalized by phagocytosis into membrane-bound compartments called phagosomes. The latter immediately become part of the organelles of the endocytic pathway. Under normal conditions, the newly formed phagosomes intermingle contents and membrane with the successive compartments of the endocytic pathway (early endosomes, late endosomes, lysosomes) through a complex series of fusion and fission events. As they are processed into phagolysosomes, they undergo gradual modifications by specific addition and removal of membrane constituents. In addition, they become acidified due to the vacuolar proton pump ATPase located in the membrane and acquire toxic constituents, including hydrolases, which will ultimately destroy bacteria. One of the major strategies used by endoparasites, but by no means the only one, is to modulate these interactions. A wide variety of situations have been described. Pathogens can (i) use the acidic and hydrolase-rich phagolysosomal environment to survive and multiply, or (ii) avoid the cytolytic environment of the phagolysosome by preventing phagosome maturation and fusion with lysosomes at different steps of the endocytic pathway. Mycobacteria, for example, do not inhibit phagosome-lysosome fusions directly but rather affect the preceding step of phagosome maturation. Other pathogens escape the endocytic pathway. They can either (i) escape from the phagosome, after lysis of the phagosome membrane, and invade the cytoplasm, in which they multiply; (ii) exclude, from the phagosome mem-

451

brane, plasma membrane-derived constituents and/or non-plasma membrane-derived fusion-mediating factors, thereby depriving the phagosome membrane of recognition signals required for fusion with the successive compartments of the endocytic pathway; or (iii) even segregate from the endocytic pathway to interact with the endoplasmic reticulum in which they multiply. Whatever the strategy used to modulate or prevent interactions with compartments of the endocytic pathway, it is of the utmost importance since it will profoundly affect drug targeting to the intracellular site of replication of pathogens and also antigen presentation.

Although some of the survival strategies have been identified for several pathogens in past years, the underlying molecular mechanisms remain largely unknown. For cells, several molecules have been reported to be involved in the successive events of phagosome formation and processing. Among these, one can cite cell surface receptors involved in recognition and adhesion of bacteria; membrane constituents necessary for recognition, docking, fusion, and fission of endocytic compartments among themselves and with the maturing phagosomes; and the cytoskeletal network, i.e., actin filaments, microtubules, and associated proteins involved in phagocytosis, organellar movement, and fusion events. For pathogens, several constituents important for their survival have been identified. However, our knowledge of the interactions between pathogen and cell constituents remains very restricted for most pathogens.

Our understanding of the molecular mechanisms of pathogen survival depends a great deal on the tools and techniques we use to obtain information about the pathogen and the cellular machinery. Morphological methods, at the light and especially electron microscopic levels, especially when combined with molecular biology tools (mutants, knockouts) or with drugs that modify the cellular machinery, are extremely valuable tools to unravel the cellular and molecular mechanisms that enable pathogens to survive and multiply within the host cells.

This chapter is devoted to electron microscopy, which is the approach of choice for optimal resolution and precision. A wide variety of cytochemical and immunoelectron microscopy methods can be used to characterize pathogens, analyze the intracellular compartment in which they reside, and localize bacterial virulence factors or cell components involved in their survival. Some of the major morphological methods, recent and less recent, of special interest to host-pathogen interplay are reviewed here. The detailed protocols are not given because they are documented and explained at length in excellent handbooks and recent review articles. The emphasis is on some of the important aspects of the methods and their advantages, drawbacks, and necessary compromises. It is important to realize that no standard recipes can be given and that trials and errors with different methods are often necessary. This is particularly true when one wants to localize molecules, as explained in the sections on immunolocalization of bacterial and/or cell constituents by electron microscopy. Given the wide range of methods and applications, it is not possible to cite the many references related to the methods and/or applications described here. The most pertinent ones are cited (and the others can be found in these references), and a few articles have been selected as examples to illustrate the methods.

Basic Conventional Electron Microscopy

Before going into complicated and sophisticated methods, very basic conventional electron microscopy methods should be used to characterize the survival strategies of pathogens. These techniques give very valuable information, provided that optimal fixation and embedding protocols are used. It is vital to choose fixation conditions that preserve ultrastructure and avoid shrinkage of pathogens or swelling of organelles. Maintenance of membrane integrity is of the utmost importance, especially if it is desired to determine whether a pathogen resides permanently in a phagosome or whether it escapes from the phagosome after membrane lysis. The quality of the products, the method of preparation and storage of fixatives, and the fixation procedure itself (temperature, incubation time, buffer, osmolarity, and added ions) are most important. A very satisfactory procedure consists of a three-step fixation procedure with glutaraldehyde followed by osmium tetroxide and, finally, en bloc staining with uranyl acetate.

The choice of the embedding resin is also important. The most widely used resin is Epon. However, upon sectioning of Epon-embedded samples, tearing can occur at the interface between the bacterial wall and the resin. This can be overcome by choosing more fluid resins such as Spurr. However, depending on the method used for embedding with Spurr, membranes can become nearly impossible to visualize. This can be very misleading, and it is advisable to compare other resins, such as Epon, with Spurr before drawing any conclusions about bacterial escape from the phagosome. Other particles such as latex beads, for which there is no lysis of the phagosome membrane, can be used as a control.

These simple methods will provide definite and precise answers to the following questions: (i) What is the morphological state of bacteria, i.e., are they intact or damaged (Figure 19.1A and B) (intact bacteria are usually live, and damaged ones are dead)? (ii) If CFU counts remain stationary, is it because bacteria have become dormant or because they are being degraded at the same rate as they multiply? (iii) Do bacteria replicate in a single, increasingly larger phagosome or does bacterial division induce separation of the phagosome? (iv) Is there any specific interaction between the pathogen and the phagosomal membrane? For example, is there a tight apposition between the bacterial surface and the phagosome membrane all around the bacterium or at discrete sites? (v) Do the phagosomes contain vesicular or tubular structures of either bacterial or cellular origin that might be important for bacterial replication? (vi) What is the spatial organization of the phagosome with respect to other cellular compartments, such as the endocytic organelles, the endoplasmic reticulum, or mitochondria, and is there any visible interaction between the two? (vii) Do bacteria lyse the phagosome membrane (Figure 19.1C)? (viii) Do bacteria induce a reorganization of the cytoskeletal network (Figure 19.1C and D)? To answer this final question, one possibility is to use other fixation conditions before and especially after cell permeabilization and decoration with subfragment 1 (S1) of myosin.

One must be aware that conventional resin embedding is prone to aggregation artifacts. Electron microscopists have made considerable efforts to minimize the aggregation of cell components through the development of low temperature embedding and cryoelectron microscopy of vitreous sections (these methods are discussed in Immunoelectron Microscopy, below, as they are mostly used for such a purpose). In addition, labile bacterial wall and cytoplasmic constituents may be partially lost during conventional embedding and, more particularly, during the dehydration step. This is especially true for bacteria, such as mycobacteria, that have a high lipid content, which is particularly susceptible to extraction by organic solvents. It has been shown that in such cases, freeze substitution provides a more

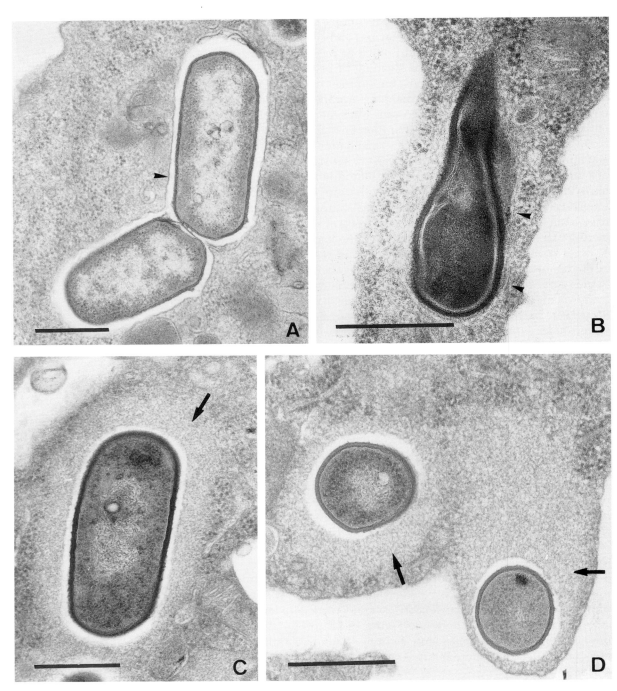

Figure 19.1   Morphological appearance of bacteria in conventional electron microscopy. Macrophages were infected with live *L. monocytogenes* LO-28 for 45 min. At 2 h after infection, the bacteria were found within phagosomes **(A and B)**, in which they were morphologically intact **(A)** or damaged **(B)**, or had escaped from the phagosome **(C and D)** after lysis of the membrane (arrowheads). In the cytoplasm, the bacteria are surrounded by a thick network of actin filaments (arrows in **C and D**). Some bacteria are being extruded from the cell **(D)**. Bar, 0.5 μm.

accurate image of structural organization than that achieved by conventional procedures.

Cytochemical Methods To Characterize Bacteria or Host Cells

Contact and binding of bacteria to the cell surface, which is the initial and obligatory event in host-pathogen interaction, depends on the surface properties of bacteria and its host cell. Surface negative charges can promote adhesion of bacteria to the cell surface, and hydrophobic interactions seem to favor phagocytic uptake. Specific bacterial carbohydrates that bind to lectin-like receptors at the cell surface can also promote bacterial binding and subsequent ingestion as well as the release of mediators by host cells. The nature and distribution of bacterial or cell surface components can be analyzed by a variety of cytochemical methods, as follows. (i) Negatively charged residues are visualized by labeling with cationized ferritin or colloidal ferric hydroxide. Pretreatment with alcian blue allows discrimination between strongly and weakly acidic groups. (ii) Lectins, which are proteins or glycoproteins with binding sites that recognize a specific sequence of sugar residues (e.g., concanavalin A with a sugar specificity for α-D-mannose and α-D-glucose residues), can be used for localizing sugar residues, glycoproteins, and glycolipids. (iii) In a more general manner, ruthenium red can be used to stain cell surface acidic polysaccharides and all polysaccharides can be stained with the methods described by Thierry or Rambourg. Although these methods provide interesting information on the overall composition of the outer surface of bacteria as well as the cell surface or intracellular compartments, they do not allow investigators to obtain quantitative information about the labeled molecular species.

Methods To Study Intersection with the Endocytic Pathway

Analysis of the intracellular compartment in which pathogens are harbored and how it intersects with the compartments of the endocytic pathway is a necessary component of any research on the survival strategies of intracellular pathogens. Several methods can be used. The first set of approaches consists of analyzing the fusogenic properties of phagosomes toward early endosomes versus late compartments of the endocytic pathway (late endosomes, lysosomes). For such studies one can use electron-dense endocytic content tracers, content or membrane markers that are tagged with electron-dense particles such as gold, or radioactively labeled. One can also stain cells for the presence of lysosomal enzymes by cytochemical methods. The second approach consists of labeling cells for content or membrane markers considered to be specific for the different organelles of the endocytic pathway by immunoelectron microscopy methods (see corresponding paragraph).

Newly formed phagosomes first intermingle contents and membrane with early endosomes. During this process, they mature to a state where they no longer fuse with early endosomes. Only then do they become able to fuse with lysosomes. When designing experiments, one must keep in mind, first, that fusion and intermingling events are highly dynamic and, second, that some pathogens might only slow down and thereby delay the successive events along the endocytic pathway rather than inhibit them.

Acquisition of Content Markers

Phagosomes that retain intermingling characteristics of early endosomes are considered to be immature; those that become mature lose their ability to fuse with early endosomes and fuse with lysosomes to become phagolysosomes. Because immature and mature phagosomes cannot be distinguished directly, implicit parameters must be used. These are based on direct morphological evidence of fusion events with either early endosomes or lysosomes labeled with endocytic tracers and on the kinetics of acquisition by phagosomes of such tracers added to cells either at selected intervals after infection with pathogens or chased to lysosomes prior to phagocytic uptake of pathogens.

Acquisition of Newly Internalized Content Markers by Phagosomes
Horseradish peroxidase (HRP) has been widely used by endocytologists as a content marker because it is easily stained by cytochemical methods, it is resistant to standard fixation conditions, and the reaction product forms a very dense, insoluble deposit that is easy to visualize under the electron microscope. The great advantage of this marker is that early endosomes can be distinguished from prelysosomes and lysosomes in terms of two parameters. First, they differ in their cytochemical staining pattern after HRP uptake. In early endosomes, the HRP reaction product lines only the inner face of the membrane (Figure 19.2A). In prelysosomes and lysosomes, the entire lumen is filled with the HRP reaction product (Figure 19.2B). Second, early endosomes acquire newly internalized endocytic marker (HRP) immediately, whereas lysosomes display the marker only after a characteristic lag of 5 to 10 min after uptake in macrophages, and this time can reach 15 to 30 min for other cell types. The first parameter is used to observe whether phagosomes fuse with early endosomes or with lysosomes (Figure 19.3), and the second is used to classify phagosomes as resembling either early endosomes or lysosomes.

Figure 19.2 Morphological appearance of early endosomes **(A)** and lysosomes (and prelysosomes) **(B)**. Macrophages were given HRP (25 μg/ml) for 30 min, fixed, and stained for the endocytic content marker, HRP. In early endosomes (E), the reaction product lines only the inner face of the membrane, whereas in lysosomes (L), it entirely fills the lumen **(B)**. Bar, 0.25 μm.

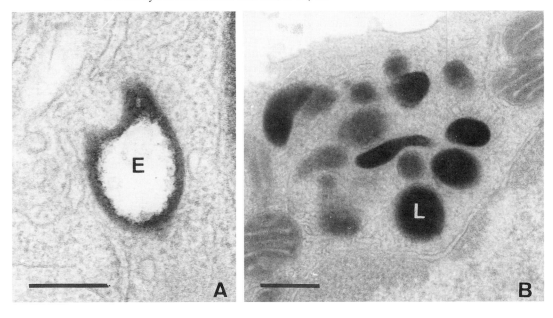

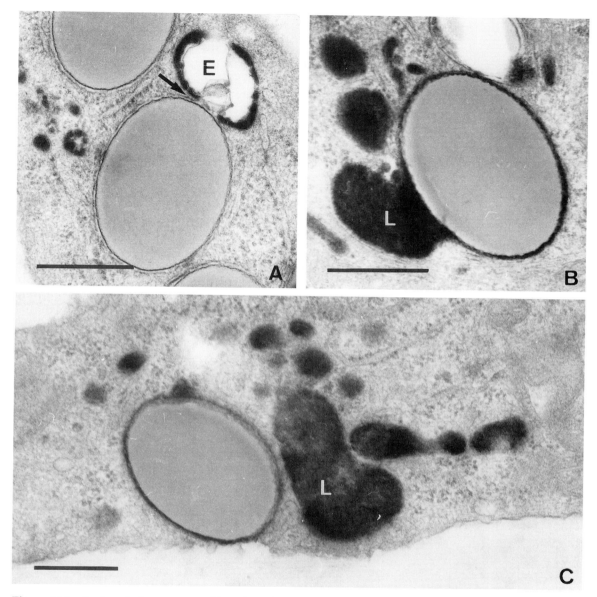

Figure 19.3   Fusion of phagosomes with early endosomes or lysosomes stained with HRP, as observed by electron microscopy. The cells underwent phagocytosis of different types of latex beads. At 2 h after phagocytic uptake, cells were given HRP as an endocytic content marker. They were then fixed and stained for HRP. The 1-μm-diameter hydrophobic bead-containing phagosomes fuse with early endosomes (E) (arrow in **A**) but not with lysosomes (L) (**C**). The 1-μm-diameter hydrophilic bead-containing phagosomes have matured and fuse with lysosomes (**B**). Bar, 0.5 μm.

Many other endocytic content markers can be used, provided that they can be tagged with dense probes. One possibility is to use biotinylated ligands (dextran or albumin) as content markers and then exploit the biotin-streptavidin interaction, with streptavidin coupled to HRP or to gold particles, to localize the ligands intracellularly. A much simpler method consists of tagging ligands with gold particles (bovine serum albumin [BSA] tagged with gold is the most widely used). This method is quite useful to analyze interactions between compartments of the endocytic pathway by successive addition of ligands tagged with gold particles of different diameters (5, 10, and 15 nm). However, with all these ligands, it is not possible to morphologically distinguish early endosomes from late compartments, as with HRP. If one opts for BSA tagged with gold, it is advisable to prepare one's own by the method of Slot and Geuze. Although it is tedious and somewhat tricky to prepare, the commercially available BSA tagged with gold does not have an adequate concentration of marker for studies of the acquisition of endocytosed material by phagosomes.

Acquisition of Content Markers Chased to Lysosomes Prior to Particle Uptake

The most widely used method to study phagosome processing into phagolysosomes involves chasing an endocytic content marker to lysosomes prior to phagocytosis and then analyzing acquisition of the marker by phagosomes at selected intervals after phagocytosis. Several markers have been used, such as ferritin and thorotrast, to cite a few, but the most frequently used at present is BSA coupled to gold particles. This method is illustrated in Figure 19.4 to show fusion or no fusion of phagosomes with lysosomes. To obtain the most reliable data, the marker must be rapidly endocytosed by the cell and in sufficient amounts so that it will label the entire lysosomal compartment when chased after uptake. However, the marker must not be packed too tightly within the lysosome, or it might form a sort of rigid gel or network. Under such conditions, the lysosomal contents

Figure 19.4   Acquisition of lysosomal marker by phagosomes. Macrophages were incubated for 30 min with BSA tagged with gold, washed, and incubated for 2 h in medium devoid of BSA tagged with gold to chase the marker to the lysosomes (L). The cells were then incubated in the presence of different types of latex beads. Phagosomes containing 1-μm-diameter hydrophobic beads do not fuse with lysosomes **(A)**, but phagosomes containing smaller beads (0.1–0.5-μm diameter) do **(B).** Bar, 0.5 μm.

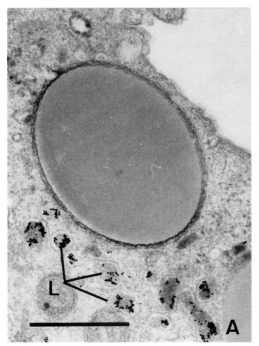
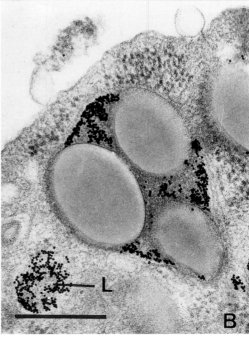

would not be transferred to phagosomes upon fusion of lysosomes with phagosomes, and this would be misinterpreted as a "no-fusion" event. It is advisable to try different concentrations of the marker and different incubation times in the presence of the marker and to use control particles, such as *Bacillus subtilis,* hydrophilic latex beads, or dead bacteria, that do not inhibit phagosome-lysosome fusion events when defining the optimal conditions.

Finally, it is important to recall that lysosomal contents, especially nondegradable ones, are recycled out of the lysosomal compartment via small recycling vesicles. This will eventually lead to excretion of the marker via the normal recycling pathway. The recycling vesicles can also fuse with and deliver contents to early endosomes. This phenomenon is particularly important in the case of macrophages where BSA tagged with gold chased to lysosomes reappears as tiny patches in early endosomes within a 2-h chase period. The latter can then be transferred to immature phagosomes upon fusion of early endosomes with phagosomes. These small patches of gold, however, can usually be distinguished from the large amounts of gold particles transferred to phagosomes upon phagosome-lysosome fusion. In macrophages, about 50% of the label is secreted within a 20-h chase period.

Enzyme Cytochemistry Methods: Acquisition of Lysosomal Enzymes by Phagosomes

To determine whether phagosomes have been processed into phagolysosomes, it is possible to stain cells for the presence of lysosomal enzymes at selected intervals after infection.

Enzymes are not electron dense and so are not visible in electron microscopy unless enzyme cytochemistry or immunoelectron microscopy methods are applied. Cytochemistry does not visualize the enzyme itself but the product of the enzymatic reaction. The fact that it must be electron dense to be visualized has limited the number of enzymes that can be localized. Acid phosphatase, aryl sulfatase, and trimetaphosphatase are enzymes of choice, because the phosphate liberated during the enzymatic reaction in the presence of substrate will react with the lead citrate used as a capture agent to form insoluble and electron-dense lead phosphate precipitates. These precipitates remain in the organelles containing the enzyme, provided that the ultrastructure, and more especially the phagosome membrane, is well preserved.

For this to be achieved, the cells must be fixed before the enzymatic reaction occurs, and a compromise must be reached between preservation of ultrastructure and of enzymatic activity. Most enzymes are inactivated by chemical fixatives and, below a threshold concentration, they will no longer be detected. When studying the acquisition of lysosomal enzymes by phagosomes, it is advisable to try different fixation conditions and include controls such as phagocytosis of particles that do not inhibit phagosome-lysosome fusion and, therefore, acquisition of hydrolases by phagosomes. One must keep in mind that hydrolases, which are concentrated in prelysosomes and lysosomes, are also present in small amounts in early compartments and can, therefore, be acquired by phagosomes via multiple fusion events with early endosomes. The amount of enzyme transferred to phagosomes in such conditions is, however, usually below the threshold level of detection. Staining for acid phosphatase is illustrated for *Mycobacterium avium*-containing phagosomes in Figure 19.5A.

The immunoelectron microscopy methods, see below, can be used to localize any enzyme, provided that antibodies are available.

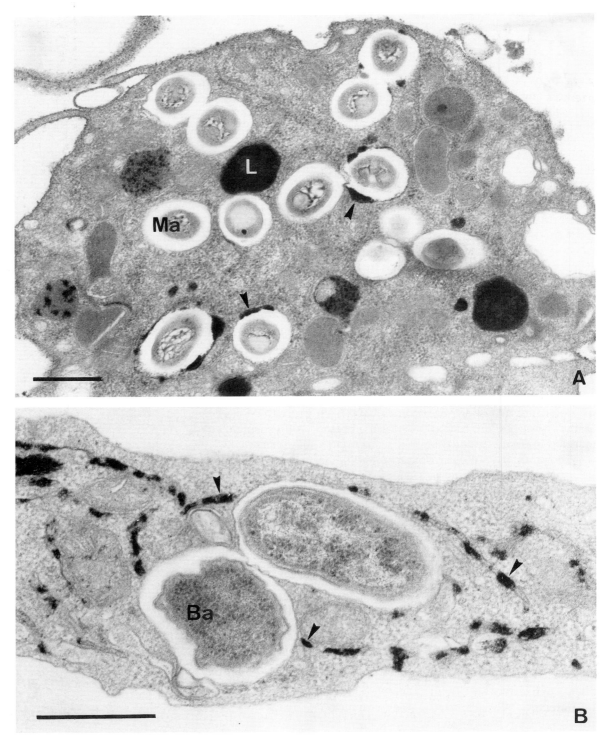

Figure 19.5 Examples of enzyme cytochemistry. **(A)** Staining of *M. avium*-infected macrophages for the hydrolytic enzyme acid phosphatase. Bone marrow-derived macrophages from BCG-susceptible mice (BALB/c) were infected with *M. avium* TMC 724 (Ma). Seven days later, the cells were fixed, stained for acid phosphatase, and processed for electron microscopy. Lysosomes (L) were strongly labeled, but most phagosomes were not stained. Only a few of them displayed small deposits (arrowhead). **(B)** Staining of *Brucella abortus* (Ba)-infected macrophages for glucose-6-phosphatase (arrowheads), an enzyme specific to the endoplasmic reticulum. Although the ER has been recruited in the vicinity of the two phagosomes, it has not yet fused with them. As a result, the two phagosomes are not stained for the enzyme. Bar, 0.5 μm.

Acquisition of Membrane Markers and Analysis of Phagosome Membrane Composition

Localization of Membrane Markers Specific to Endomembrane Compartments

Immunoelectron microscopy localization of membrane proteins (see Methods in Immunoelectron Microscopy, below), known to be located predominantly in the membrane of a specific compartment of the endocytic pathway, is a widely used method to characterize the phagosome membrane as resembling the membrane of early endosomes, prelysosomes, or lysosomes. Some of the proteins most commonly used as markers are listed in Table 19.1. Proteins of the Rab family have become very useful, because these small GTPases, which are involved in the regulation of intracellular membrane traffic, are often associated with only one compartment. In contrast, localization of membrane markers considered to be specific to the late compartments can be misleading because the pathways for targeting membrane proteins to lysosomes involve several compartments. Newly synthesized proteins in the endoplasmic reticulum are transported to the Golgi complex and upon arrival at the *trans*-Golgi network can be first targeted either to early endosomes or to the cell surface, from where they can be rapidly internalized into early endosomes for final delivery to lysosomes. The presence of lysosomal markers, therefore, does not necessarily mean that a phagosome has been processed into a phagolysosome. While remaining immature, it may acquire membrane constituents that are considered to be lysosomal as a result of extensive fusion events with early endosomes, especially if recycling is impaired or directly from the *trans*-Golgi network. A typical example is the *Mycobacterium*-containing phagosome. It has been extensively shown that *Mycobacterium* inhibits phagosome maturation and therefore fusion with lysosomes, and yet the phagosome membrane has been shown to be LAMP-1 positive. As a consequence, it is advisable to also label cells for specific early endosome membrane markers, such as the transferrin receptor or Rab5, which are never found in the membrane of late compartments, and especially to use other methods, as described above, before drawing any conclusions about the nature of the compartment in

Table 19.1 Membrane or content markers used to identify phagosomes as resembling a given endocytic compartment or the ER

Compartment	Markers[a]
Early endosome	Membrane markers: TfR, Rab5, annexins I, II, and III Proteases: Immature cathepsin D
Late endosome	Membrane markers: M6PR, Rab7, LAMP1, LAMP2, CD63, CD68 Hydrolases: acid phosphatase, aryl sulfatase, trimetaphosphatase Proteases: cathepsin B, D, H, dipeptidyl peptidase I and II Phospholipid: LBPA
Lysosome	Membrane markers: LAMP1, LAMP2, CD63 Hydrolases: acid phosphatase, aryl sulfatase, trimetaphosphatase Proteases: cathepsin B, D, H, dipeptidyl peptidase I and II
ER	Membrane markers: calnexin, calreticulin Enzyme: glucose-6-phosphatase

[a] TfR, transferrin receptor; M6PR, mannose-6-phosphate receptor; LAMP, lysosome-associated membrane protein; LBPA, lysobisphosphatidic acid.

which a pathogen resides and how it modulates the interactions with the endocytic pathway.

Trafficking of Membrane Markers and Composition of the Phagosome Membrane in Terms of Plasma Membrane-Derived Membrane Markers
Another approach involves analyzing the acquisition, by phagosomes, of newly internalized plasma membrane markers by morphometry and autoradiography approaches. After labeling of cell surface glycoconjugates by enzymatic glycosylation with radioactive galactose, this membrane marker (i.e., [³H]galactose-labeled glycoconjugates) redistributes to membranes of the different endocytic organelles with distinct kinetics. Early endosomes acquire the membrane marker immediately, whereas lysosomes display the marker only after a characteristic lag of 5 to 10 min after uptake. These parameters can be used to determine whether the phagosome membrane resembles the membrane of early endosomes or that of lysosomes in terms of kinetics of acquisition of marker. This approach can also be used to study the composition of the phagosome membrane in terms of plasma membrane-derived glycoconjugates according to the following rationale. When cell surface glycoconjugates are labeled with [³H]galactose, subsequent endocytic membrane traffic leads to a redistribution of label to all membrane compartments that lie on the path of this traffic. When steady-state distribution has been reached, the fraction of label that ends up in a given compartment reflects the relative pool size of this compartment for the types of molecules that are susceptible to this labeling technique. When this pool size is related to the amount of membrane of a given compartment (here, the number of autoradiographic silver grains per membrane area measured by morphometric methods), the obtained value corresponds to the concentration of labeled molecular species in that membrane compartment. This concentration is a measure for the relative membrane composition in terms of the type of molecules that carry the label. In macrophages, for example, the relative membrane composition in terms of plasma membrane-derived glycoconjugates is about twofold lower on early endosomes than on the plasma membrane, and lysosomes display a twofold lower concentration of label than early endosomes do. These parameters can be used to determine whether the phagosome membrane has the same membrane composition as the compartment with which it interacts or whether the membrane composition is modified while the pathogen establishes its replicative niche.

Such methods are extremely informative because very precise information can be obtained about the phagosomal membrane composition in relation to early endosomes and prelysosomes/lysosomes as well as on the kinetics of membrane trafficking to and from the phagosome compartment. They are, however, difficult to apply without some expertise in autoradiography and morphometry.

Surface labeling with sulfobiotin has also been used, but it is much less informative, because it can be used neither for a quantitative analysis of the membrane composition of phagosomes nor for comparative studies with respect to the different endomembrane compartments.

Phagosome Acidification
3-(2,4-Dinitroanilino)-3′-amino-N-methyldipropylamine (DAMP) is a probe that is widely used to study phagosome acidification at the electron microscopy level. This weak base accumulates by diffusion within acidic compartments. Once protonated, it can no longer diffuse out. During chemical fixation with aldehydes, it becomes covalently linked to proteins, which

allows it to be retained in acidic organelles during processing of samples for electron microscopy. DAMP is then localized with appropriate postembedding immunolabeling methods (anti-dinitrophenol [DNP] antibodies followed by protein A coupled to gold) on Lowicryl or LR White thin sections.

An important advantage of this method is that the number of gold particles per phagosome can be converted to values for intraphagosomal acidification by the method described by Orci provided that the phagosome contains sufficient protein for all the protonated DAMP molecules to be retained in the phagosome after fixation. An example of this labeling method is illustrated in Figure 19.6 for macrophages infected with *Listeria monocytogenes.*

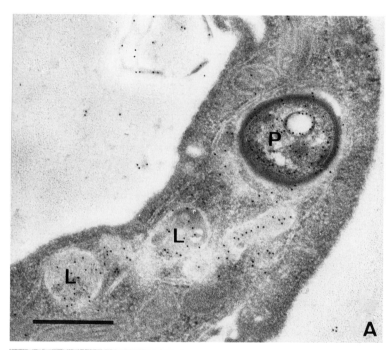

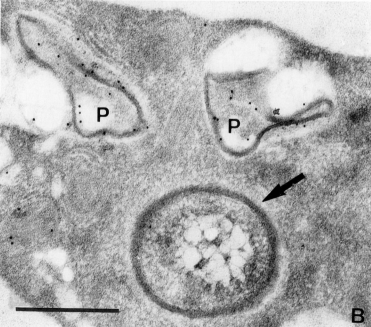

Figure 19.6 Morphological assessment of phagosome acidification. Macrophages were infected for 45 min with *L. monocytogenes,* washed, and reincubated in fresh medium. One hour later, DAMP (30 min, 60 μM) was added for passive accumulation in acidic organelles. The cells were fixed and embedded in Lowicryl. The probe was localized by a postembedding labeling method as follows. Lowicryl thin sections were sequentially incubated with rabbit anti-DNP (dinitrophenol) antibodies and with protein A coupled to 10-nm-diameter gold particles. **(A)** A high abundance of gold particles was observed within lysosomes (L) and phagosomes (P), thereby indicating that they are acidic. **(B)** Intracytoplasmic bacteria (arrow) were not labeled. Bar, 0.5 μm. From such pictures, one can estimate the intraphagosomal pH by the method described by Orci.

Methods To Study Intersection with the Endoplasmic Reticulum

Endocytosis and phagocytosis both involve an extensive transfer of membrane in both directions between the cell surface and intracellular membrane compartments. Quantitative analysis of this membrane traffic has shown that it is a homeostatic process, which means that the relative amounts of membrane in the participating compartments remain constant. Recycling to the cell surface of membrane internalized during endocytosis has been directly demonstrated in macrophages and other mammalian cells. Major recycling occurs at a prelysosomal stage within minutes after membrane has been internalized. Membrane is also retrieved from the lysosome compartment. In phagocytic cells, the phagosome also constitutes a site for membrane recycling to the cell surface. Given the large amounts of membrane required during uptake of particles and also during replication of endoparasites within their phagocytic vacuole, one of the questions that has been raised recently was whether other cell compartments, and more especially the endoplasmic reticulum (ER), could be an additional source of membrane for forming and/or dividing phagosomes. Conventional electron microscopy will allow us to visualize recruitment of ER in the close vicinity of phagosomes. In contrast, fusion events between the two compartments are extremely rapid and therefore difficult to visualize. In some cases, e.g., *Brucella*- or *Legionella*-containing vacuoles, it is obvious that the phagosome membrane has acquired ER-derived membrane because it becomes studded with ribosomes. In most cases, the interaction between the phagosome membrane and the ER is less obvious and conventional electron microscopy (EM) used alone is insufficient to obtain precise information on ER-phagosome interactions. EM cytochemical staining for glucose-6-phosphatase, an enzyme that is specific to the ER, and/or immunoelectron microscopy approaches that allow for labeling of specific ER membrane constituents (Table 19.1) are good methods to determine whether the ER fuses either with the macrophage plasma membrane during formation of phagocytic cups or with fully formed phagosomes during their processing into a replicative niche. Staining for glucose-6-phosphatase is illustrated for bone marrow-derived mouse macrophages infected with *Brucella abortus* (Figure 19.5B). At present, however, there are no methods allowing for a quantitative evaluation of the concentration of ER constituents within a given phagosome membrane.

Immunoelectron Microscopy

Methods

Immunoelectron microscopy is the best choice to study the distribution of antigens at high resolution. Immunolabeling specialists face a perpetual dilemma, i.e., to find the best conditions for sample processing and labeling to achieve optimal preservation of ultrastructural details, maintenance of antigenic determinants, and immunoreactivity and precise immunolabeling with negligible background. It is equally important to ensure equal accessibility of all antigens, especially when they are located at different sites, and to find conditions where antigens and the markers used to localize them remain at their natural intracellular site.

A variety of different methods have been developed to detect antigens at the EM level. Most of them are depicted schematically in Figure 19.7. Because some antigens will be readily localized by one method and not by another, it is advisable to try several methods to optimize the labeling effi-

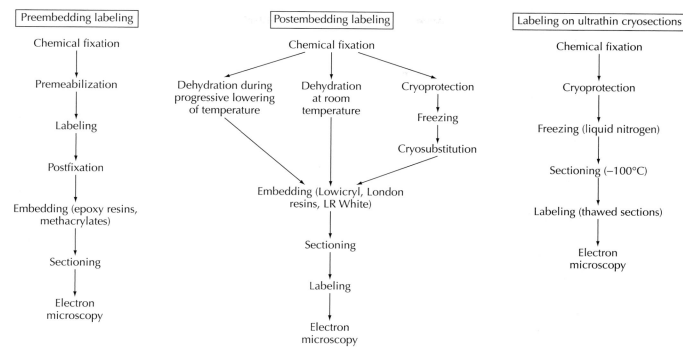

Figure 19.7   Schematic depiction of methods used for immunoelectron microscopy. See the text for details.

ciency for a given antigen. The method of choice will also depend on the material under investigation, the kind of information needed (qualitative versus quantitative), and the available equipment.

As shown in Figure 19.7, there are three main options: (i) preembedding ultrastructural immunocytochemistry, in which labeling is performed after fixation of the cell sample but before tissue dehydration, embedding, and sectioning; (ii) postembedding immunolabeling in which labeling is performed on thin sections of fixed, dehydrated, and embedded (Epon, Lowicryl, LR Gold, or LR White) material; and (iii) immunolabeling of ultrathin cryosections, in which samples are fixed, cryoprotected, and frozen in liquid nitrogen. In the third case, the frozen samples are cryosectioned (−95 to −120°C, usually), and labeling is performed on thawed cryosections, which are then contrasted and stabilized before observation by EM. Postembedding immunolabeling and immunolabeling of ultrathin cryosections are illustrated in Figures 19.6 and 19.8, respectively. The detailed procedures, including recent improvements in ultracryomicrotomy, are not given here because they can be found in the handbook by Griffiths, reviews, and articles as recommended at the end of this chapter. However, some of the crucial points are discussed here.

Crucial Points

Fixation

Whatever the method used to localize a given molecule, samples must first be fixed unless one performs cryoelectron microscopy of vitreous sections (see below). In the absence of fixation, antibodies are internalized by endocytosis, sequestered, and eventually degraded in the lysosomes.

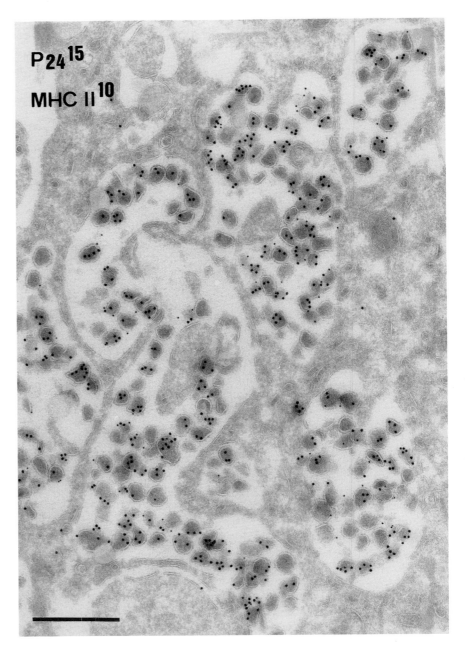

Figure 19.8   Localization of constituents by immunolabeling of ultrathin cryosections. Human macrophages were infected with HIV-1. Fourteen days later, they were fixed and processed for immunolabeling of ultrathin cryosections. Sections were double-immunogold labeled for p24 (gold particles 15 nm in diameter) to localize HIV-1 and for MHC II (major histocompatibility class II antigens) (gold particles 10 nm in diameter). Note the accumulation of HIV particles positive for p24 in enlarged compartments. These compartments contain MHCII. In addition, MHCII also colocalizes with the p24-positive virions. Bar, 0.5 μm. From Raposo et al. (2002), reprinted with permission.

The choice of fixative is crucial because chemical fixatives often introduce artifacts and morphological changes. Chemical fixations are performed with paraformaldehyde or glutaraldehyde (and sometimes acrolein) either alone or in combination. Paraformaldehyde is often preferred to glutaraldehyde. Although the latter permits much better preservation of ultrastructure, it often causes irreversible changes in the antigenic determinants due to efficient cross-linking. Antibodies are then unable to bind to the specific epitopes, especially when monoclonal antibodies are used. When deciding on the fixative of choice for a given method and even for an unknown antigen-antibody combination, different fixation procedures should be tested to achieve optimal labeling.

Some antigens are extremely sensitive to chemical fixatives. In this case, cryoelectron microscopy of vitreous sections is the method of choice. This method consists of cryoimmobilizing, by high-pressure freezing (HPF), plunge-, slam-, or jet and spray freezing, the native biological material so rapidly that water does not have the time to crystallize. In principle, cryoelectron microscopy of vitreous sections is free of any aggregation artifact (which is a common event during either the chemical fixation or dehydration steps) and it also allows for better preservation of the antigenicity. HPF is followed by low-temperature cryosubstitution in organic solvents to ensure protective stabilization by forming stable cross-links between macromolecules by hydrophobic interactions. Cryosubstitution without the use of a chemical fixative allows for good preservation of cellular ultrastructure. Although the method is still technically difficult, recent progress has made it available to practical application in EM laboratories specializing in such methods.

Permeabilization

One of the important obstacles in the detection of intracellular antigens by immunolabeling methods (both fluorescence and electron microscopy) is the low penetration of antibodies through cell membranes, a problem that is enhanced by cell fixation. To improve the permeability of the cell membrane for preembedding labeling, methods include freeze-thawing, mechanical sectioning, and partial permeabilization with detergent. Saponin, which interacts with cholesterol in membranes, is generally preferred to Triton because it is a mild detergent and hence avoids extraction of antigens and can be used after mild fixation with aldehydes without causing damage to the ultrastructure. The necessary concentration must be adjusted to the cell type and fixative and must be added throughout the labeling procedure since permeabilization with saponin is reversible. If labelings are instead performed on ultrathin resin-embedded or frozen sections, the difficulties of antigen accessibility are removed and samples need not be permeabilized.

One must also keep in mind that membrane antigens are more difficult to localize than antigens in the lumen of cell organelles, because the former are less abundant and less accessible to antibodies.

Labeling

With preembedding methods, only one antigen can be labeled at a time. The most common procedure is to first incubate cells with a specific antibody and then with immunoglobulin G (IgG) from another species coupled to HRP. Cells are then stained for HRP by the usual procedure. For postembedding labeling or immunolabeling of cryosections, gold probes can be used instead of HRP. The most commonly used single-labeling procedures are (i) specific antibody followed by protein A coupled to gold particles (PAG), (ii) specific antibody followed by bridging antibody and then by PAG, and (iii) specific antibody followed by IgG of a different species coupled to gold particles. The specific antibody can be a monoclonal or polyclonal antibody; the bridging antibody and the IgG conjugated to the gold probe originate from different species. A bridging antibody can be used either to amplify the signal or in situations where specific antibodies do not react with PAG. The two-step PAG labeling (specific antibody followed by PAG) gives the best result in terms of resolution and low background. Moreover, the use of PAG permits double and even triple labelings with antibodies from the same species, which is not as readily achievable when specific IgG-gold complexes are used.

In performing double and triple labelings with antibodies from the same species, it is necessary to beware of possible pitfalls due to cross-reactions of secondary antibody and/or gold probes. The size of gold particles (5, 10, or 15 nm) and the sequence of antibodies should be chosen carefully. There are no general rules, but it is advisable to do the most critical reactions first and with the largest gold particles to facilitate detection. In multiple-labeling procedures, when antigens are present in the same intracellular location, the yield of subsequent labeling steps is lower than in a single-labeling experiment, probably due to steric hindrance. Another problem is false-positive reactions (colocalizations) due to the reaction of the gold probe or bridging antibody with previously used immunoreagents. To overrule this problem, a glutaraldehyde cross-linking step is used between successive labelings.

Quantitative Immunocytochemistry

The use of particulate markers such as colloidal gold has the advantage that the immunolabeling can be quantitated. However, quantitation in immunocytochemistry is usually not absolute. This method generally cannot reveal the quantities of antigen present, because it is not known to what extent a measured labeling density reflects the actual concentration of antigen. Some of the variables that affect the labeling efficiency, such as fixation, dehydration, and embedding, can be standardized. Variables due to the antigen itself are difficult to control. Differences in the intracellular matrix densities will also introduce variables in the labeling efficiency. This generates differences in accessibility of antigens to the immunoreagents due to differential penetration into the section. This can be avoided by embedding cells in Lowicryl or LR White, prior to sectioning. In this way, the surface of the section is labeled and no variations in penetration can occur. However, the labeling efficiency will still decrease since only antigens at the thin-section surface are accessible. Simple methods that consist of counting gold particles and determining surface areas by morphometric methods will give very meaningful results, provided that the gold particles are uniformly distributed over organelles. This method has been successfully used to estimate the intraphagosomal pH of *M. avium*-containing phagosomes.

Conclusion

A wide variety of morphological methods at the electron microscopy level can be used to analyze the molecular mechanisms involved in the survival of pathogens within the host cell. From the different illustrations and comments, it is clear that electron microscopy techniques are the most suitable methods to obtain maximum resolution and precision when studying the pathogen's intracellular replicative niche and how it interacts with cell compartments of importance for the pathogen's survival, such as those of the endocytic and of the antigen presentation pathway, to cite a few, and defining the specific cell constituents and virulence factors involved in survival of pathogens in general.

Concerning the immunolocalization of pathogen or cell constituents, there is no standard procedure. Each method has its advantages and limitations, and each antigen-antibody combination may require a different approach to achieve optimal labeling. Before using the more complicated electron microscopy methods, immunofluorescence should be performed as a first assay. Nevertheless, a positive result does not guarantee a satisfactory reaction at the electron microscopy level. All three immunoelectron microscopy methods permit the study of the distribution of antigens with good resolution. However, ultracryomicrotomy and immunogold labeling are the techniques of first choice since they have proven to combine excellent preservation of ultrastructure and antigenicity and a very precise localization of antigens.

I am grateful to Graça Raposo (Institut Curie, CNRS UMR 144, Paris, France) and Hans J. Geuze (University of Utrecht Medical School, Utrecht, The Netherlands) for kindly providing the print for Figure 19.8 (reprinted from Raposo et al., 2002, with permission).

Selected Readings

Anderson, R. G. W., J. R. Falck, J. L. Goldstein, and M. S. Brown. 1984. Visualization of acidic organelles in intact cells by electron microscopy. *Proc. Natl. Acad. Sci. USA* **81:**4838–4842.

Bohrmann, B., and E. Kellenberger. 2001. Cryosubstitution of frozen biological specimens in electron microscopy: use and application as an alternative to chemical fixation. *Micron* **32:**11–19.

> This paper describes and discusses cryofixation and cryosubstitution methods applied to bacteria for ultrastructural studies and for immunolocalization of bacterial antigens that are sensitive to chemical fixation.

Carlemalm, E., W. Villiger, J. D. Acetarin, and E. Kellenberger. 1984. Low temperature-embedding, p. 147–154. *In* O. Johari (ed), *Science of Biological Specimen Preparation.* SEM Inc. O'Hare, Chicago, Ill.

> This chapter emphasizes and discusses the use of low-temperature embedding methods for immunoelectron microscopy labelings.

Clemens, D. L., and M. A. Horwitz. 1996. The *Mycobacterium tuberculosis* phagosome interacts with early endosomes and is accessible to exogenously administered transferrin. *J. Exp. Med.* **184:**1349–1355.

de Chastellier, C., and L. Thilo. 1997. Phagosome maturation and fusion with lysosomes in relation to surface property and size of the phagocytic particle. *Eur. J. Cell Biol.* **74:**49–62.

> This paper describes and illustrates the methods discussed in Methods To Study Intersection with the Endocytic Pathway on the acquisition of content markers added before or after phagocytic uptake and how they were used to propose a model for how mycobacteria prevent phagosome maturation.

de Chastellier, C., and L. Thilo. 2002. Pathogenic *Mycobacterium avium* remodels the phagosome membrane in macrophages within days after infection. *Eur. J. Cell Biol.* **81:**17–25.

> This paper describes and emphasizes the use of autoradiography and morphometry methods as discussed in Methods To Study Intersection with the Endocytic Pathway to study membrane trafficking and to analyze the phagosome membrane composition in terms of plasma-membrane-derived glycoconjugates. These methods were used to show how mycobacteria remodel the phagosome membrane in the course of infection.

de Chastellier, C., T. Lang, and L. Thilo. 1995. Phagocytic processing of the macrophage endoparasite *Mycobacterium avium,* in comparison to phagosomes which contain *Bacillus subtilis* or latex beads. *Eur. J. Cell Biol.* **68**:167–182.

de Chastellier, C., C. Fréhel, C. Offredo, and E. Skamene. 1993. Implication of phagosome-lysosome fusion in restriction of *Mycobacterium avium* growth in bone marrow macrophages from genetically resistant mice. *Infect. Immun.* **61**:3775–3784.

This paper illustrates and describes the method used to stain cells for acid phosphatase.

Dubochet, J., and N. Sartori Blanc. 2001. The cell in absence of aggregation artefacts. *Micron* **32**:91–99.

This paper compares and discusses different EM methods, i.e., (i) conventional resin-embedding and sectioning, (ii) low-temperature embedding and sectioning of freeze-substituted samples, and (iii) cryosections of vitrified samples, in terms of cell structure and organization.

Gagnon, E., S. Duclos, C. Rondeau, E. Chevet, P. H. Cameron, O. Steele-Mortimer, P. Paiement, J. M. Bergeron, and M. Desjardins. 2002. Endoplasmic reticulum-mediated phagocytosis is a mechanism of entry into macrophages. *Cell* **110**:119–131.

Griffiths, G. (ed.). 1993. *Fine Structure Immunocytochemistry.* Springer-Verlag, Berlin, Germany.

Excellent book about immunoelectron microscopy methods.

Griffiths, G., J. M. Lucocq, and T. M. Mayhew. 2001. Electron microscopy applications for quantitative cellular microbiology. *Cell. Microbiol.* **3**:659–668.

This paper emphasizes the power of EM approaches in conjunction with stereology approaches to analyze intracellular membrane trafficking.

Griffiths, G., P. Quinn, and G. Warren. 1983. Dissection of the Golgi complex. I. Monensin inhibits the transport of viral membrane proteins from *medial* to *trans* Golgi cisternae in baby hamster kidney cells infected with Semliki forest virus. *J. Cell Biol.* **96**:835–850.

This paper indicates how to stain cells for glucose-6-phosphatase.

Harding, C. V., and H. J. Geuze. 1992. Class II MHC molecules are present in macrophage lysosomes and phagolysosomes that function in the phagocytic processing of Listeria monocytogenes for presentation to T cells. *J. Cell Biol.* **119**:531–542.

Liou, W., H. J. Geuze, and J. W. Slot. 1996. Improving structural integrity of cryosections for immunogold labeling. *Histochem. Cell Biol.* **106**:41–58.

Mayhew, T. M., J. M. Lucocq, and G. Griffiths. 2002. Relative labelling index: a novel stereological approach to test for non-random immunogold labelling of organelles and membranes on transmission electron microscopy thin sections. *J. Microsc.* **205**:153–164.

Murk, J. L., G. Posthuma, A. J. Koster, H. J. Geuze, A. J. Verkleij, M. J. Kleijmeer, and B. M. Humbel. 2003. Influence of aldehyde fixation on the morphology of endosomes and lysosomes: quantitative analysis and electron tomography. *J. Microsc.* **212**:81–90.

Orci, L., P. Halban, A. Perrelet, M. Amherdt, M. Ravazzola, and R. G. W. Anderson. 1994. PH-independent and -dependent cleavage of proinsulin in the same secretory vesicle. *J. Cell Biol.* **126**:1149–1156.

Ozawa, H., R. Picart, A. Barret, and C. Tougard. 1994 Heterogeneity in the pattern of distribution of the specific hormonal product and secretogranins within the secretory granules of rat prolactin cells. *J. Histochem. Cytochem.* **42**:1097–1107.

This paper describes and illustrates postembedding immunolabelings on LR White thin sections.

Paul, T. R., and T. J. Beveridge. 1992. Re-evaluation of envelope profiles and cytoplasmic ultrastructure of Mycobacteria processed by conventional embedding and freeze-substitution protocols. *J. Bacteriol.* **174**:6508–6517.

This article is a good example of how freeze-substitution protocols provide a more accurate image of structural organization (wall architecture, distribution of nuclear material) than that achieved by conventional procedures.

Raposo, G., M. J. Kleijmeer, G. Posthuma, J. W. Slot, and H. G. Geuze. 1997. Immunogold labeling of ultrathin cryosections: application in immunology, p. 208.1 –208.11. *In* L. A. Herzenberg and D. M. Weir (ed.), *Weir's Handbook of Experimental Immunology*, vol. 4. Blackwell Science Inc., Oxford, United Kingdom.

> This article gives an excellent and detailed account of immunolabeling methods of cryosections (from sample preparation to immunolabeling) and discusses some of the pitfalls/problems that might be encountered.

Raposo, G., M. Moore, D. Innes, R. Leijendekker, A. Leigh-Brown, P. Benroch, and H. Geuze. 2002. Human macrophages accumulate HIV-1 particles in MHC II compartments. *Traffic* **3:**718–729.

Rastogi, N., C. Fréhel, A. Ryter, H. Ohayon, M. Lesourd, and H. L. David. 1981. Multiple drug resistance in Mycobacterium avium: is the wall architecture responsible for the exclusion of antimicrobial agents? *Antimicrobiol. Agents Chemother.* **20:**666–677.

Roth, J., J. M. Lucocq, and D. J. Taatjes. 1988. Light and electron microscopical detection of sugar residues in tissue sections by gold labeled lectins and glycoproteins. I. Methodological aspects. *Acta Histochem. Suppl.* **36:**81–99.

Russell, D. G., J. Dant, and S. Sturgill-Koszycki. 1996. *Mycobacterium avium* and *Mycobacterium tuberculosis*-containing vacuoles are dynamic, fusion-competent vesicles that are accessible to glycosphingolipids from the host cell plasmalemma. *J. Immunol.* **156:**4764–4773.

Slot, J. W., and H. J. Geuze. 1985. A new method of preparing gold probes for multiple-labelling cytochemistry. *Eur. J. Cell Biol.* **38:**87–93.

Slot, J. W., H. J. Geuze, and A. J. Weerkamp. 1988. Localization of macromolecular components by application of the immunogold technique on cryosectioned bacteria. *Methods Microbiol.* **20:**211–236.

Studer, D., W. Grabber, A. Al-Amoudi, and P. Eggli. 2001. A new approach for cryofixation by high-pressure freezing. *J. Microsc.* **203:**285–294.

Tilney, L. G., D. J. DeRosier, and M. S. Tilney. 1992. How Listeria exploits host cell actin to form its own cytoskeleton. I. Formation of a tail and how that tail might be involved in movement. *J. Cell Biol.* **118:**71–81.

Tougard, C., and R. Picart. 1986. Use of pre-embedding ultrastructural immunocytochemistry in the localization of a secretory product and membrane proteins in cultured prolactin cells. *Am. J. Anat.* **175:**161–177.

> This article gives an excellent and detailed account of preembedding immunolabeling methods (from sample fixation to immunolabeling) and discusses some of the pitfalls/problems that might be encountered.

20

New Tools for Virulence Gene Discovery

TIMOTHY K. MCDANIEL AND RAPHAEL H. VALDIVIA

As the preceding chapters have demonstrated, the starting point of many cellular microbiology studies is the identification of the microbial genes necessary for virulence. The logic underlying the identification of such virulence genes is codified into a set of rules known as molecular Koch's postulates. These rules establish the criteria to prove that a suspected virulence gene is necessary for a given pathogenic phenotype, just as the original Koch's postulates established the criteria to prove that a suspected pathogen causes its associated disease. The basis of these molecular postulates is that the relationship between a gene and a functional phenotype, such as virulence, may be established by investigating the phenotypes of strains differing only in a defined genetic lesion. While this may be an oversimplification for a very complex phenotype (virulence), it has nonetheless provided the framework for the discovery of most of the virulence factors described to date. The molecular Koch's postulates are fulfilled if all three of the following conditions are met:

1. A gene is found in strains with a certain virulence phenotype.
2. Mutating the gene abolishes the virulence phenotype.
3. Reintroducing the gene reconstitutes the virulence phenotype in the mutant strain.

Obviously, the conditions required to fulfill the molecular Koch's postulates are attainable only for organisms that can be genetically manipulated. Beyond this requirement, a candidate virulence gene is needed. As the information from sequenced microbial genomes rapidly outpaces our ability to characterize these microbes, cellular microbiologists require progressively more sophisticated methods to screen bacterial genomes for virulence genes. Such methods have been developed in recent years and are the subject of this chapter. Although these new methods use a variety of technologies, they all identify virulence gene candidates by one of two general approaches: mutagenesis or analysis of differential gene expression.

Identifying Candidate Virulence Genes by Mutation

Early forays into identifying virulence genes used the basic tools of bacterial genetics: mutation and complementation. In such screens, DNA-damaging agents or transposons (Tn) are typically used to mutate a pathogen, generating a collection of bacteria ("mutant library") where each member of the collection contains a unique mutation.

The most direct way to screen such libraries is to test individual library members for the ability to infect and cause disease in laboratory animals. However, such a screen would require testing thousands of mutants and animals, which is unrealistic given that animal tests are expensive and labor-intensive. The development of in vitro infection models with cultured mammalian cells has permitted more rapid screening of mutant libraries for loss of phenotypes relevant to the host-pathogen interaction, such as adherence, invasion, and intracellular survival. Although more convenient than assays with whole animals, assays based on cultured cell models are still time-consuming and cumbersome, especially when tens of thousands of library members must be individually tested to ensure complete mutational coverage of a typical bacterial genome. Despite some successes, the time and expense associated with such brute-force screens have limited their use. In the future, however, as information from genome sequences and new techniques for the rapid generation of gene replacements in pathogenic bacteria become available, collections of strains bearing defined deletions in all nonessential genes will be undoubtedly constructed and tested in various virulence assays.

Enrichment Strategies

The overwhelming difficulty of individually screening members of a mutant library can be avoided if the investigator devises a genetic selection in which avirulent mutants are enriched. Where such selections can be devised, whole mutant libraries can be tested simultaneously. For example, a decades-old strategy for isolating nonreplicative mutants has been adapted to isolate *Salmonella enterica* serovar Typhimurium mutants defective for growth during intracellular residence. This strategy relies on the fact that β-lactam antibiotics, such as penicillin, kill only bacteria that are actively dividing. To select the desired mutants, bacteria residing in cultured mammalian cells are exposed to a β-lactamase antibiotic. Bacteria that replicate within the cell are selectively killed by the antibiotic, thus enriching for nonreplicating mutants. This approach revealed a number of mutants with altered cell envelope components, nutritional requirements, and susceptibility to host cell antibacterial compounds.

Signature-Tagged Mutagenesis

In most cases, an investigator cannot devise an enrichment strategy for desired mutants and brute-force testing of individual mutants is impractical. A technique that overcomes these limitations, known as signature-tagged mutagenesis (STM), allows the simultaneous screening of large numbers of distinct mutants for those that fail to survive a challenge of the investigator's choosing. This is achieved by tagging each bacterium with a unique DNA sequence, called a signature tag, which allows individual mutants to be tracked within a large pool of bacteria. The original STM procedure was a Tn mutagenesis-based screen used to identify *Salmonella* serovar Typhimurium genes necessary for survival in the mouse (Figure 20.1). A collection of Tn is engineered such that each contains a unique 40-bp DNA

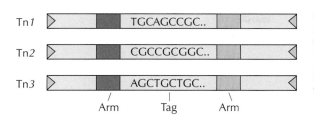

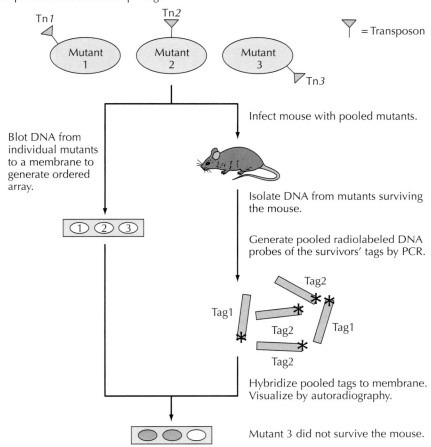

Figure 20.1   Signature-tagged mutagenesis. This diagram shows the procedure for a pool of three mutants. In practice, pools of up to 96 mutants have been used to infect a single animal.

sequence (tag). The delivery of a Tn to the *Salmonella* genome thus provides the means of mutating, as well as tagging, each library member. A pool of tagged mutants is used to infect a mouse, and the bacteria that survive to colonize the spleen are harvested and regrown in vitro. The DNA is isolated from the survivors, and the tags are amplified by PCR to generate a pooled collection of radiolabeled DNA probes representing only the tags present in the surviving pool of bacteria. This pooled probe is then hybridized to an ordered array of DNA from all bacteria used in the initial infection, and individuals that failed to colonize the spleen are identified by the failure of their signature tags to hybridize to the pooled probes. The STM technique led to the discovery of novel *Salmonella* virulence genes, such as those within the SPI-2 pathogenicity island that are required for bacterial survival in host tissues. Since its inception, the basic concept of identifying missing members from a pool of Tn-mutagenized population has been successfully applied to discover virulence genes in a variety of gram-negative, gram-positive, and fungal pathogens. Technical advances in the original STM methodology include the preoptimization of tags prior to mutagenesis, the use of new generations of hyperactive Tn with little sequence specificity, and in vitro transposition systems which can be used to rapidly mutate to saturation the genomes of many pathogenic bacteria. Similarly, significant advances have been made to increase the numbers of mutants that can be screened simultaneously by taking advantage of the information obtained from genome sequencing and PCR amplification. As with STM, where the hybridization signal obtained from a particular signature-tagged Tn mutation depends on the fitness of the bacteria bearing such a mutation, the abundance of a PCR-product amplified with gene "X" and Tn-specific primers will also reflect the fitness of an "X" Tn mutant during an environmental challenge. By choosing sets of primers specific for different bacterial genes, multiple PCR products can be simultaneously resolved in a single gel to determine the relative abundance of amplification products from a mutant library before and after a selective pressure. These types of approaches have led to development of systems, such as genome analysis and mapping by in vitro transposition (GAMBIT) and essential gene tests (EGT), where a complete Tn mutant library pool can be analyzed in a single tube. In GAMBIT, genomic map information is used to design a set of "anchor" primers to be used with Tn-specific primers in long polymerase amplification reaction of DNA isolated from a Tn library. Since Tn mutagenesis is random, a ladder of evenly spaced products is obtained from these DNA amplifications. Any gaps in this ladder suggest that a Tn mutation could not be isolated in this region of the chromosome and indicate that mutations in this region are deleterious. While the application of these approaches has been limited to identifying essential genes, their use for the identification of virulence genes is likely to become routine.

The fundamental biology of some host-pathogen interactions can limit the ability to apply STM and related approaches. For some bacterial species, only a few individual "founder" bacteria establish an animal infection even when large numbers of bacteria are used for the inoculation. If the number of founders that establish an infection is smaller than the number of distinct mutants in the initial pool used for infection, tags will be randomly missing from the output pool even when their corresponding mutations are not deleterious. In the example of the successful *Salmonella* STM screen described above, 96 individual mutants could be screened simultaneously in a single animal that had been infected by intraperitoneal injection. However, when the same number of mutants is used in STM screens in oral models of *Vibrio cholerae* and *Yersinia pseudotuberculosis* infection, only a minority of the initial pool survives to seed the animal. This low seeding rate necessitates using smaller mutant pools for each infection and correspondingly increasing the number of animals needed to screen the same number of mutants.

Another potential complication of STM is that some mutants containing disruptions in essential virulence genes can survive selection in an animal if its function is complemented by other members of the mutant library. For instance, a secreted bacterial toxin may alter the host environment in a way that is necessary for survival of the bacterium, and a mutant deficient in toxin synthesis, which is expected to be cleared by the host, might survive if its environment is altered by toxin produced by the neighboring bacteria.

Identifying Candidate Virulence Genes by Differential Expression

Bacteria are frugal; therefore, if a gene is expressed only in a certain environment, it is likely that the gene encodes some function useful within that environment. As a result, knowing when and where a gene is expressed can provide important clues as to its function. For example, the bacterial genes required to colonize, survive, replicate, and cause disease are usually expressed only inside the host and at the appropriate time and place during infection. This principle underlies several strategies for isolating candidate virulence genes, where preferential expression within animals or cultured cells or under laboratory conditions that mimic the environment within the host are prerequisites for further study. Often the "host-like" conditions studied are merely laboratory media with slight recipe changes, such as the addition of acid to mimic the acidic lysosome of phagocytic cells or the omission of iron to mimic the iron-sequestered environment of host tissues.

Various techniques are available to identify these differentially expressed genes. Some entail identifying mRNAs whose expression is induced on association with the host. Other approaches involve sophisticated genetic selections that use the host environment as a selective medium to identify promoters with host-specific activity. Once the mRNAs or promoters associated with a host-induced gene are identified, the gene itself can be identified simply by comparing the isolated sequences (often only a fragment of the gene or 5' flanking sequence) to DNA sequence databases. In the increasingly rare case where one is working with an unsequenced genome, the retrieved DNA sequence can be used as a probe to isolate clones from a genomic library.

Gene Expression Profiling with DNA Microarrays

The most direct way to test whether a gene is differentially expressed is to monitor the abundance of mRNA under different conditions. The relative abundance of an mRNA transcript is determined by the quantitative hybridization of total mRNA to labeled DNA probes specific for the gene of interest (Northern hybridization). The identification of a gene that is preferentially transcribed under conditions mimicking the host environment (for example, in the presence of blood serum or acidic pH of the stomach) can provide important clues as to the function of that gene during infection. Northern hybridizations have been used to study the expression of a limited number of genes at a time, but applying quantitative hybridization methods to search for differentially regulated genes at the whole-genome level, until recently, was not possible because of two technological shortcomings: (i) the absence of a sequenced bacterial genome and (ii) the lack of tools to monitor the simultaneous hybridization of thousands of genes. Fortunately, "shotgun" cloning, automation of DNA purification, new DNA-sequencing technologies, and powerful computational tools that can process immense amounts of raw DNA sequence data have made the sequencing of bacterial genomes a relatively routine process. Similarly, a major breakthrough in the parallel processing of DNA hybridization experiments occurred with the invention of devices known as DNA microarrays

BOX 20.1

DNA microarrays

The most common type of microarray used in academic research is known as the spotted array, because its manufacture entails placing spots of DNA onto the array surface. The DNA sequences to be placed on the array are assembled in microtiter plates. A robotic device called a spotter dips tiny printing tips that work like quill pens into the wells and then deposits the DNA from the wells in nanoliter-volume droplets on the surface of a glass slide. The droplet dries to form a spot on the array. The printing heads have enough tips to handle 4 to 32 sequences at a time, and can print a few dozen to more than 100 slides with each dip. Printing a set of slides is thus a repetitive process consisting of hundreds to thousands of cycles in which the print head dips into a few wells, prints the sequences, gets flushed clean, then dips into a new set of wells to print the next positions of the array. The print heads used in this process are indifferent to the material they spot, giving the investigator flexibility in what is printed. Arrays can be made from PCR products containing entire genes or fragments of genes, as well as oligonucleotides correspond-ing to only 30 to 70 bases of a gene's sequence.

Microarrays are made by a variety of other processes, including the use of photolithographic machinery adapted from semiconductor manufacturing to synthesize the array sequences directly on the array surface, ink-jet technology similar to that used in computer printers, and the synthesis of arrays on microscopic beads (one unique sequence per bead). Microarray technology is rapidly advancing, so it is likely that the approaches most common today will be supplanted by new ones.

▶ *For Figure 20.2, see color insert.*

(Box 20.1). Microarrays (also called DNA chips) are solid platforms, such as a glass slide, on which microscopic spots (~100 μm) of DNA have been immobilized in a grid pattern. Because of the microscopic dimensions, a microarray the size of a standard microscope slide can contain tens of thousands of different gene sequences, one sequence per spot—easily enough to contain every gene of the largest bacterial genome.

In a microarray experiment, the problem of performing thousands of simultaneous DNA hybridizations is solved by reversing the order of standard Northern hybridizations (Figure 20.2). Instead of performing nucleic acid hybridizations in a solution containing a single labeled probe to an unlabeled, immobilized cellular RNA sample, one hybridizes a labeled cellular RNA sample to a collection of thousands of unlabeled, immobilized ordered probes. Microarray experiments usually use fluorescently labeled probes instead of radiolabeled ones, which allows their hybridization patterns to be read automatically by confocal laser microscopes. While various experimental designs are used, a common approach is to use two different fluorescent dyes to label RNA obtained from cells exposed to two different conditions. The samples are mixed and spots on the array that hybridize more intensely to one of the labeled RNAs, but not the other, indicate that the genes represented by those spots are preferentially expressed under one of the conditions. In this way, it is possible to rapidly and simultaneously read the expression levels of every gene of a sequenced genome in response to conditions of interest.

Applied to cellular microbiology, DNA microarrays to date have been used to identify genes preferentially expressed in a variety of conditions mimicking the host environment, including low iron, body temperature, or in vitro conditions that induce biofilm formation. As of this writing, microarrays have not been used to directly examine genes expressed by bacteria infecting animals, although they have been used to examine genes expressed by *Neisseria meningitidis* during infection of cultured mammalian cells. It is especially difficult to isolate sufficient quantities of high-quality bacterial mRNA in animal tissues, where one faces the additional challenge of separating bacterial RNA from that of the host. However, this is a topic of intense interest and investigation, and it seems likely that more sensitive microarray techniques will enable such studies in the years to come.

It should be noted, however, that any methodology that pools transcripts from a population will, by definition, provide an average of the responses of each member of the population. Such an average is an accurate reflection of individual bacterial responses when the population is subjected to a relatively homogeneous stimulus. This may not be the case during infections where distinct microenvironments, even within a single organ, provide heterogeneous stimuli that vary spatially and temporally. By pooling these RNAs, a typical DNA microarray experiment may significantly underrepresent the true gene expression levels of individual bacteria residing in these specialized niches.

Promoter Traps

The most common method to identify differentially expressed genes is a genetic strategy known as a promoter trap. In this approach, genes are identified on the basis of the strengths of their promoters rather than the levels of their mRNA transcripts. Once conditionally active promoters are isolated, they can be sequenced and the genes or operons under their control can be determined by sequencing downstream regions or by comparison with sequence databases.

In a promoter trap, promoter activity is measured by fusing promoters to a reporter gene (a gene whose product is easily measured and thus permits the investigator to distinguish bacteria that are expressing the gene product from those that are not). For example, the *lacZ* gene is widely used as a reporter because the activity of its product, β-galactosidase, is easily detected by using a chromogenic substrate of this enzyme. Alternatively, the reporter gene can encode for a selectable marker (e.g., drug resistance), allowing for the positive selection of bacteria bearing activated promoters. A promoter trap begins with the construction of a promoter library, a collection of bacteria genetically engineered such that each bacterium contains the same reporter gene inserted downstream of a different promoter. The reporter gene contains no promoter of its own, so the product is made only if the gene is fused to an active promoter. This allows one to identify conditionally active promoters by screening for library members exhibiting reporter activity preferentially under the conditions of interest. The promoter fusions can be created in a variety of different ways. Gene fusions can be constructed in plasmids and maintained episomally or integrated at homologous sites in the chromosomes to create partial diploids. Alternatively, Tn can be modified to randomly insert a reporter gene in the bacterial chromosome. This latter approach has the added advantage of generating a mutation that can be directly tested in virulence assays.

The advantage of promoter traps over the biochemical isolation of RNA transcripts as a means of measuring differential gene expression is that promoter traps generate a collection of live biological reporters of gene activity. These reporters can then be used to probe different environments or in further genetic screens to identify extragenic regulators of the trapped promoter.

The activities of enzyme-based reporter genes can be readily measured only in relatively simple environments such as laboratory media or agar plates. As a consequence, promoter traps originally had limited utility when applied to probing bacterial interactions with host cells or tissue. In the past decade, selection schemes using different combinations of se-

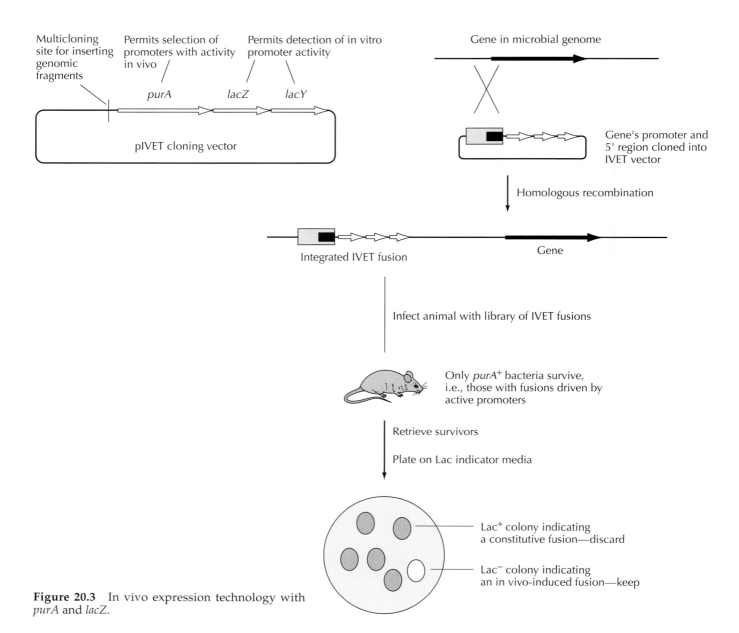

Figure 20.3 In vivo expression technology with *purA* and *lacZ*.

lectable markers and recently developed fluorescent reporters have expanded the use of promoter traps to environments of interest to the cellular microbiologist, including the surface and interior of cultured mammalian cells and even the interior of organs of living animals.

In Vivo Expression Technology

A variety of promoter traps permit the direct selection of bacteria bearing gene fusions induced during animal infections; this type of trap is termed in vivo expression technology (IVET). The original IVET made use of an *Salmonella* serovar Typhimurium library consisting of a promoterless dual reporter operon fused randomly in the bacterial chromosome (Figure 20.3). One of the two reporters, *purA*, encodes an enzyme involved in purine biosynthesis whose expression is necessary for survival in mice. A mouse is infected with *Salmonella* bearing a promoter library, and only bacteria containing in vivo-induced fusions survive, since only they express *purA*. Bacteria are then harvested from the animal and grown on laboratory media where the activity of the second gene in the IVET operon, *lacZ*, is assessed. The investigator determines the *lacZ* activity of bacteria grown on indicator media, and only those with low *lacZ* activity (i.e., those with promoters active in the mouse but inactive in vitro) are kept for further study.

A bias in this IVET strategy is that the *purA* gene used to assess in vivo promoter activation must be continuously active for survival in the mouse. As a result, the original IVET screen identified promoters that are continuously induced during infection but not those that are only transiently induced. This bias meant that the screen was liable to miss promoters driving genes needed only during restricted times or in restricted sites of infection. This problem has been addressed in more recent IVET strategies involving alternative reporters to *purA*, which allow the investigator to choose the time after infection when selection is applied or to avoid lethal selections within the animal altogether. For example, genetic recombination in concert with antibiotic selection has been used as a reporter of in vivo gene transcription. In recombination-based IVET screens (Figure 20.4), the background strain contains an antibiotic resistance gene flanked by copies of a DNA sequence that is the substrate of a DNA resolvase enzyme. The resolvase catalyzes recombination at the repeat sequences, resulting in a deletion of the DNA flanked by these sequences. The promoter library is made by constructing promoter fusions with a gene encoding a DNA resolvase. If the fused promoter is even transiently activated, the resolvase is made and catalyzes the deletion of the antibiotic resistance gene. This deletion permanently records the fact that the promoter was activated in the infected animal.

Regardless of the gene reporters used, the growth conditions used outside the host during the screen can strongly influence which genes are retrieved in an IVET screen. For example, if the counterscreen for in vitro-activated promoters is performed in nutrient-rich laboratory media, a high proportion of "in vivo"-induced promoters will be isolated that regulate metabolism and nutrient biosynthesis genes. This trend is probably seen because such genes are needed for survival in host compartments, in which

Figure 20.4   Resolvase as an IVET reporter. Promoter fragments are inserted before a promoterless *tnpR* gene, which encodes a resolvase enzyme. If the gene is fused to an active promoter, the resolvase is produced and deletes an antibiotic resistance gene that has been inserted into the bacterial chromosome flanked by two resolvase recognition sites. Loss of antibiotic resistance therefore permanently records that the promoter was activated.

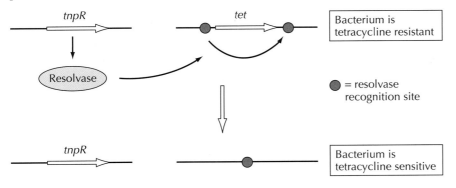

essential nutrients are depleted or sequestered, whereas they are not needed in nutrient-rich laboratory media. While this indicates that the host environment places demands on bacterial metabolism, we have learned relatively little about bacterial genes responsible for modulating the function of host cells. A recent IVET-based screen for *Yersinia* virulence genes avoided this problem by using defined minimal media to grow the bacteria during the counterscreen against genes that are active in vitro. This design ensured that promoters for biosynthetic genes would be induced during both phases of the selection and that these promoters would thus be eliminated from further study.

Differential Fluorescence Induction

One of the most versatile tools to probe host-pathogen interactions is the green fluorescent protein (GFP). GFP is readily expressed in bacteria and fluoresces green upon excitation with blue light. GFP fluorescence is readily detected by fluorescence microscopy, fluorimetry, or flow cytometry and allows the monitoring of gene expression and/or protein localization in single bacteria. The fluorescent signal from individual bacteria can be detected with a fluorescence-activated cell sorter (FACS), an instrument that sorts living cells into pure populations on the basis of their light-scattering and fluorescence properties. As a result, FACS can be used to screen, in an automated fashion, for activated promoters from among a library of promoters fused to *gfp*. The FACS can also sort eukaryotic cells in association with GFP-expressing bacteria (adhered or internalized), making possible the detection of bacterial genes activated by interactions with host cells.

The use of GFP as a tool for the identification of bacterial genes expressed during infection was first realized with the development of a promoter trap methodology known as differential fluorescence induction (DFI). In DFI, conditionally active promoters are isolated from a GFP-promoter library by selecting library members whose fluorescence is stimulus dependent (Figure 20.5). While the original DFI methodology was limited to cell culture systems, more recently the methodology has been extended to the isolation of bacteria expressing *gfp* directly from infected tissues. In a particularly striking application, bacteria have been collected by FACS from granulomas to identify mycobacterial promoters that specifically activated during chronic infections.

Figure 20.5 Isolation of host cell-induced microbial genes by differential fluorescence induction.

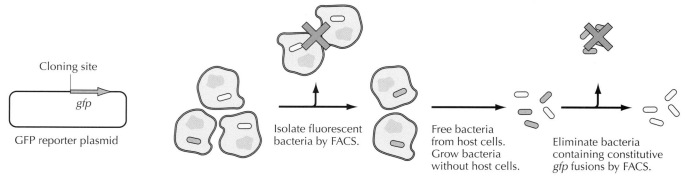

Cloning site

gfp

GFP reporter plasmid

Isolate fluorescent bacteria by FACS.

Free bacteria from host cells. Grow bacteria without host cells.

Eliminate bacteria containing constitutive *gfp* fusions by FACS.

Infect cultured cells with *gfp* library.

BOX 20.2

Beyond genetic screens: the many uses of GFP

GFP is a tool with many uses. Because it can be observed in live cells that have not been fixed or stained, it can be used to observe bacterium-specific proteins that interact with host cells and tissues. Examples of specific applications are given below.

Identifying Subsets of Host Cells Infected within Animal Tissues
Salmonella serovar Typhimurium targets, among other sites, the spleen, an organ containing a mixture of different immune system cells. To find which of the spleen cells become infected, investigators infected mice with *Salmonella* serovar Typhimurium constitutively expressing GFP, excised their spleens, and stained the spleen cells with a mixture of fluorescently tagged antibodies, where each antibody was a different color and was specific for a different type of cell. These stained cells were then analyzed by a FACS machine: the green signal of the bacteria was associated with phagocytes, such as neutrophils and macrophages, but not with B or T lymphocytes.

Tracking Bacterial Gene Expression in Host Cells
Several *Salmonella* serovar Typhimurium genes are activated in association with macrophages, but which step of the interaction confers the inducing signal? *Salmonella* serovar Typhimurium organisms containing *gfp* driven by the promoter of the macrophage-induced *aas* gene were observed by time-lapse fluorescence microscopy as they infected macrophages. The bacteria fluoresced only after entering the cells and associating with host vacuoles. In this experiment, the gene was never induced in bacteria outside macrophages or in those merely adhering to macrophages. (Figure 20.6 [see color insert]).

Monitoring Subcellular Localization of Bacterial Toxins within Host Cells
Genes for bacterial toxins can be fused to the *gfp* gene to create toxin-GFP hybrid proteins. Such hybrids can retain both the activity of the toxin and the fluorescence of GFP. With these hybrids and a fluorescence microscope, an experimenter can record in real time where the toxin accumulates in the host cell and can identify probable sites of action.

In addition to its amenability to automation, other features of the GFP reporter contribute to the versatility of DFI. The fluorescent signal of GFP requires no enzymatic substrate, and so promoter activity can be monitored in environments such as living tissues and subcellular compartments of living cells, which cannot be probed by enzymatic reporter genes that require artificial substrates for their detection. GFP is a stable molecule, and so its signal is maintained even after multiple steps of processing, such as those needed to harvest bacteria from animal tissues. The ability to visualize GFP in situ by fluorescence microscopy means that the promoter fusions retrieved from a DFI screen provide the researcher with a set of ready-made tools for detailed cellular microbiology studies. Indeed, the true power of DFI and new generations of IVET-like systems (e.g., recombinase-based IVET-RIVET) is not in their application to virulence gene discovery but their use to generate versatile reporters of gene expression. For example, one can analyze the temporal expression of different bacterial virulence factors within different anatomical sites in infected animals. The availability of GFP variants with distinct spectral properties in conjunction with bioluminescent reporters will increase the number of bacterial genes that can be monitored simultaneously and significantly further our understanding of the complexity of the genetic program activated by both the pathogen and the host during infection (Box 20.2).

The properties of GFP also place some limitations on the utility of DFI. Unlike enzymatic reporters, which amplify their signal by converting multiple substrate molecules into signal-generating products, the fluorescent signal of GFP comes directly from the protein itself. This lack of enzymatic amplification means that the GFP signal is weak compared with the best enzymatic reporters. Hyperfluorescent variants of GFP that considerably improve the sensitivity of the reporter have been developed, but identification of weak-to-medium-strength promoters still requires the use of libraries based on plasmids, which amplify the fluorescent signal via an increased copy number of the *gfp* gene. When cloned on multicopy plasmids, promoters can behave differently from when they are found on the bacterial chromosome, so that results generated from plasmid-based screens should be viewed cautiously until confirmed by other means.

Despite this theoretical problem, DFI-based screens have yielded a high proportion of virulence genes relative to housekeeping genes. Part of this success results not from any particular advantage of the DFI technique per se but from a principle that is important when designing any screen based on analysis of differential expression, be it DNA microarrays or IVET. This principle is that the investigator should minimize the differences between inducing and noninducing conditions. The most successful gene hunts have been those designed around narrow changes in the environment of the bacterium. For instance, successful DFI-based screens for macrophage-induced *Salmonella* and *Mycobacterium* genes compared gene expression of bacteria grown in the presence and absence of macrophages but otherwise in identical growth medium. When many variables are changed at once, a plethora of genes responding to the various environmental cues will be retrieved, only some of which will be relevant to the response of the pathogen to the host.

Functional Approaches

Promoter traps and microarray methods cannot reflect changes in protein levels, protein phosphorylation, or other posttranslational processing steps that may occur in response to the host environment. The increasing resolving power of commercial two-dimensional polyacrylamide gel electrophoresis systems and the development of novel electrophoresis methods coupled with mass spectrometry allow for up to 10^4 proteins to be resolved in a single gel. With these tools investigators can monitor en masse protein levels and posttranslational responses that are modulated by the bacterium in response to the host environment. Techniques for rapid, proteome-wide analysis are only in their infancy and have not achieved the ease of use offered by many of the above-described genetic screens. However, proteomics is an area of rapid development and will be a field to watch in the coming years.

There is a substantial gap between the information available from genome sequences and our understanding of the functions of the proteins encoded by these genomes. To address this gap, investigators are increasingly developing new methods to either assign a biochemical function to bacterial proteins or link them functionally to a disease-related process. For example, recent screens in *Pseudomonas syringae* have focused on the identification of proteins secreted by this plant pathogen's type III secretion system. By taking advantage of a *Pseudomonas* protein (AvrRpt2) that is normally translocated into plants and has a visible phenotype, investigators have created a "signal sequence" trap to identify sequences that permitted the translocation of a truncated version of AvrRpt2 lacking the signals required for transport via type III secretion systems. This approach, much like a promoter trap, identified fragments that can functionally substitute for AvrRpt2 translocation signals. This screen led to the identification of signature amino acid sequences that are sufficient to export *Pseudomonas* proteins through a type III secretion system and the discovery of >15 new proteins that are likely translocated into plant cells during infection.

It has also become apparent that even though we have identified hundreds of virulence factors, we still know remarkably little about what these proteins do. This has led to a new push to identify host targets of these virulence factors. DNA microarrays, for example, are being increasingly used to monitor the host's transcriptional profiles in response to challenges with defined bacterial mutants. In this manner, investigators want to link virulence factors to the activation of particular host cellular responses. Another approach that is becoming increasingly popular is to use alternative host systems to study the function of virulence factors. For example, recent reports have shown that the nematode *Caenorhabditis elegans* is colonized and killed by a variety of human pathogens including *Pseudomonas* and *Salmonella* species. Furthermore, many of the same virulence factors that are required for mammalian infections are also required to infect nematodes. *C. elegans* is one of the prevalent model systems used to study the development of multicellular organisms. It has a short generation time and mutants can be rapidly isolated and characterized. This allows for the potential identification of nematode mutants that are either resistant or hypersensitive to bacterial infections. These mutants will likely lead to the identification of host proteins that interact with pathogenic bacteria. Other model eukaryotes (e.g., the yeast *Saccharomyces cerevisiae*, the fruit fly *Drosophila melanogaster,* and the soil amoeba *Dictyostelium discoidem*) that have been extensively characterized biochemically and are amenable to genetic manipulation are also currently being explored as surrogate infection models. These new approaches will undoubtedly simplify the identification of the eukaryotic targets of various virulence proteins and lay the foundation for a biochemical analysis of host-pathogen interactions.

Comparative Genomics

The availability of whole genomic sequences presents another method of virulence gene discovery: bioinformatics. By comparing the sequences of pathogens to closely related species that are nonpathogens or pathogens that cause disease by different mechanisms, one can identify the collection of genes unique to the pathogen under study. The emerging discipline of comparing sequences at the whole genome level to infer biological function is known as comparative genomics.

An example of the application of comparative genomics to the study of bacterial pathogenesis came out of the sequencing of the *Salmonella* serovar Typhimurium genome. Computer programs were used to compare the sequence to the complete or nearly complete genome sequences of *Salmonella* serovar Typhi, *S. paratyphi* A, and to the closely related enterics *Escherichia coli, E. coli* O157:H7, and *Klebsiella pneumoniae.* These studies showed that of the 4,597 genes of *Salmonella* serovar Typhimurium genome, fewer than one quarter of the genes are unique to the *Salmonella* genus, fewer than one in eight are unique to *Salmonella enterica* subspecies I (the group of *Salmonella* serovars, including serovar Typhimurium, that infect warm-blooded animals), and fewer than 3% (121 genes) are unique to *Salmonella* serovar Typhimurium (Figure 20.7). Thus, an investigator who is interested in what encodes the unique traits of *Salmonella* serovar Typhimurium can focus on 121 genes of more than 4,597 in the genome.

The investigators in the *Salmonella* study complemented their direct sequencing comparisons with comparative genomic microarray studies. To compare genome contents, total DNA from one bacterial strain is labeled with a fluorescent dye and hybridized to an ordered array made from a se-

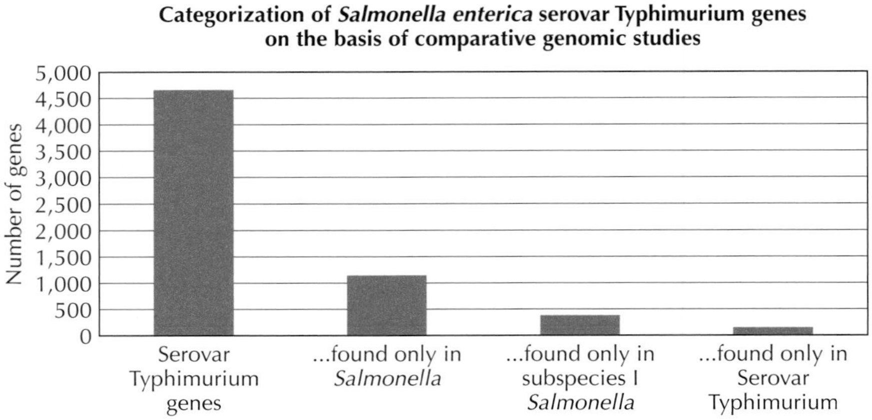

Categorization of *Salmonella enterica* serovar Typhimurium genes on the basis of comparative genomic studies

Figure 20.7 Categorization of *Salmonella* serovar Typhimurium genes based on comparative genomic studies of *Salmonella* serovar Typhimurium strain LT2 to five other strains of *Salmonella* and three other enteric bacterial strains. Note the small fraction of genes unique to *Salmonella* serovar Typhimurium. The precise number of genes in any given category is not static since the numbers shown depend on the limited set of strains examined.

quenced strain (see Gene Expression Profiling with DNA Microarrays). Positions in the array that provide a positive fluorescence signal indicate a gene that is conserved between strains, while those that fail to give a signal indicate a gene that is either absent from the strain or so divergent that it cannot hybridize. In the *Salmonella* study, the investigators made a microarray based on the *Salmonella* serovar Typhimurium sequence and hybridized it to genomic DNA from *S. paratyphi* B, *S. enterica* subsp. *arizonae*, and *S. bongeri*, none of which had been sequenced. By using microarrays, the investigators were able to expand their comparative genomic survey to more species, even in the absence of genomic sequences. This approach to studying pathogens is becoming widespread, as it can be used to broadly survey genomic variation in bacterial species or groups of closely related species even when only one representative has been sequenced.

As with other methods, however, one should be careful not to let comparative genomic approaches oversimplify pathogenesis. Many studies have shown that complex traits, such as virulence, are the result of the interplay of multiple genetic loci (multilocus traits) and that the function of a particular allele of a gene can be heavily influenced by its genetic context. Therefore, the mere absence or presence of a gene may be insufficient to define the pathogenic properties of *Salmonella* serovar Typhimurium. Rather, it is likely that many genes that are common to both pathogenic and nonpathogenic species are present as a combination of distinct alleles that can significantly modulate *Salmonella* virulence.

Conclusion

The preceding sections have outlined a variety of approaches used to identify candidate virulence genes. Each technique has its weaknesses and strengths, which may leave the reader wondering which approach is best. The answer is simple: none of them is. All have limitations that bias toward the isolation of certain genes and not others. In cases where more than one technique has been applied to the same pathogen, each approach has yielded genes missed by the others. In many cases, the choice of technique will be dictated by the availability of genetic tools for use with the pathogen in question (e.g., availability of plasmid vectors, sequenced genomes, transposition system, etc.).

It is important to remember that the techniques described in this chapter identify virulence gene candidates. These cannot be considered true virulence genes until they have satisfied the requirements of the molecular Koch's postulates. Mutational strategies such as STM offer the advantage that the investigator can satisfy these requirements simply by complementing the mutations in the isolated mutants. Screens based on differential gene expression (microarrays and promoter trapping) necessitate several additional steps: isolating the full gene, mutating the gene, testing the mutant in assays for pathogenicity, and showing that the defect can be corrected by complementation. Nonetheless, at the end of a well-designed differential expression screen, the investigator better understands the exact conditions under which the gene is induced, which can guide further studies into the role of this gene in disease.

The genome sequence of every major pathogen (and most minor ones) is or will soon be available. These genomic sequences, augmented by the gene discovery methods described in this chapter, should permit the discovery of novel virulence genes at an unprecedented pace. Future editions of this textbook will almost certainly describe presently unknown mechanisms of microbial infection that will have been discovered by the approaches described in this chapter.

Selected Readings

Camilli, A., and J. J. Mekalanos. 1995. Use of recombinase gene fusions to identify *Vibrio cholerae* genes induced during infection. *Mol. Microbiol.* **18:**671–683.

Falkow, S. 1988. Molecular Koch's postulates applied to microbial pathogenicity. *Rev. Infect. Dis.* **10:**S274–S276.

Grifantini R., E. Bartolini, A. Muzzi, M. Draghi, E. Frigimelica, J. Berger, G. Ratti, R. Petracca, G. Galli, M. Agnusdei, M. Monica Giuliani, L. Santini, B. Brunelli, H. Tettelin, R. Rappuoli, F. Randazzo, and G. Grandi. 2002. Previously unrecognized vaccine candidates against group B meningococcus identified by DNA microarrays. *Nat. Biotechnol.* **20:**914–921.

One of the first examples describing how DNA microarray data can be used to identify novel vaccine targets.

Guttman, D. S., B. A Vinatzer, S. F. Sarkar, M. V. Ranall, G. Kettler, and J. T. Greenberg. 2002. A functional screen for the type III (Hrp) secretome of the plant pathogen *Pseudomonas syringae*. *Science* **295:**1722–1726.

A clever approach to identify targets of the TTSS system.

Hensel, M., J. E. Shea, C. Gleeson, M. D. Jones, E. Dalton, and D. W. Holden. 1995. Simultaneous identification of bacterial virulence genes by negative selection. *Science* **269:**400–403.

This classic paper laid the foundation for the concept of parallel screening of large mutant libraries to identify bacterial genes required for colonization and survival in animal hosts. This methodology has rapidly become the most widely used approach to identify virulence factors.

Mahan, M. J., J. M. Slauch, and J. J. Mekalanos. 1993. Selection of bacterial virulence genes that are specifically induced in host tissues. *Science* **259:**686–688.

The "original" in vivo expression technology (IVET) paper. This work introduced the concept that bacterial genes required for survival in host tissues are most likely preferentially expressed in host tissues.

McClelland, M., K. E. Sanderson, J. Spieth, S. W. Clifton, P. Latreille, L. Courtney, S. Porwollik, J. Ali, M. Dante, F. Du, S. Hou, D. Layman, S. Leonard, C. Nguyen, K. Scott, A. Holmes, N. Grewal, E. Mulvaney, E. Ryan, H. Sun, L. Florea, W. Miller, T. Stoneking, M. Nhan, R. Waterston, and R. K. Wilson. 2001. Complete genome sequence of Salmonella enterica serovar Typhimurium LT2. *Nature* **413:**852–856.

Valdivia, R. H., and S. Falkow. 1997. Fluorescence-based isolation of bacterial genes expressed within host cells. *Science* **277:**2007–2011.

This paper describes the use of *gfp* to identify bacterial genes expressed during infection. It also details the advantages of monitoring bacterial gene expression with single-cell resolution in complex host environments.

21

DNA Arrays
Microarray Design and Production
Applications
Proteomics
Mass Spectrometry
Genetic Analysis of Protein Interactions Using "Two-Hybrid" Screens
Protein Arrays
Conclusion
Selected Readings

Genome-Wide Approaches to Studying Prokaryotic Biology

Su L. Chiang and Stephen Lory

Rapid progress in large-scale sequencing of complete genomes of living organisms, including the human genome, has ushered in a new era in scientific enterprise. It is estimated that over 100 microbial genome sequences will be completed by 2004, including several sequences of individual species. Because of its important role in genome-sequencing efforts, computational biology has emerged as a central component of modern biomedical research and will continue to be the dominant technology supporting all disciplines of postgenomic research.

The major challenge now faced by the research community is to understand the biological function of the thousands of newly discovered gene products, specifically within the context of the parallel activity of all the components that make up a living cell and that are defined by the cell's genetic repertoire. This postgenomic phase of research is termed "functional genomics," and its primary goal is the systematic identification of the function of entirely novel proteins deduced from genomic sequence analysis. This field is driven by technological developments in different areas, including classical genetics, physics, engineering, molecular biology, and biochemistry.

Almost 50% of all identified open reading frames encode proteins of unknown function. These proteins are likely to play roles in "new biology" cellular processes that are fundamentally different from those characterized by extensive pregenomic research on proteins of known function. This chapter will summarize the development and use of two broad experimental approaches directed toward defining the biological function of unknown proteins. The first approach is based on the well-known biological phenomenon of coordinate gene expression, which can be analyzed through transcriptional profiling. The second approach requires the identification and analysis of protein complexes in which unknown proteins participate. These strategies typically rely on mass spectrometry methods and interactive genomic databases. Finally, many standard genetic and biochemical tools, such as two-hybrid or coimmunoprecipitation analysis, are now being exploited in a truly comprehensive fashion, using entire sets of expressed proteins of living organisms.

DNA Arrays

Of all the postgenomic technologies developed during the past 5 years, the most significant scientific impact has been due to the use of high-density DNA arrays for monitoring transcriptional activities of cells or for determining the gene content of specific genomes. Extensions of DNA microarray technology are currently being developed to detect the activities of individual proteins.

A DNA microarray consists of multiple unique DNA fragments (probes) attached to a solid support in a specific pattern. Each distinct probe location in the pattern on the array is termed a feature. The number of features on an array is determined by various parameters, including the feature size, synthesis area, and the number of sequences to be represented. It is possible to place as many as 500,000 features on an array that is less than 1.5 cm^2 in area, but most arrays have a much lower number of features.

Most microarray analyses are comparative hybridization experiments that compare the DNA or RNA content of two samples. The basic outline of microarray construction and a single-array experiment is shown in Figure 21.1. Total RNA or DNA is isolated from two different samples. These target samples may be cells obtained under two different conditions, or they may be different cells grown under the same conditions. The RNA or DNA from one sample is labeled with a fluorescent dye, and the RNA or DNA from the other sample is labeled with a different fluorescent dye. The two differentially labeled target pools are mixed together and cohybridized to the same array. Fluorescence detection is then used to determine the relative signal, or fluorescence intensity ratio, from the dyes at each spot on the array. This ratio provides a quantitative measurement of the relative amount of that sequence in the two samples.

Microarray Design and Production

Today, virtually all microarrays are addressable. That is, the identity and location of each element on the array are known. This is in contrast to libraries, where the array elements consist of a comprehensive collection of unknown gene sequences.

The basis of array construction and analysis is the complete genome sequence. There are currently over 90 complete microbial genome sequences available in public databases and approximately 180 genome-sequencing projects in progress. A list of complete genome sequences and those in progress can be found at http://www.tigr.org/tdb/mdb/mdbcomplete.html.

The most important property of a probe is its unique sequence. The DNA should hybridize only with the complementary sequence from the same gene. Due to large numbers of homologous sequences within each genome (paralogs), potential cross-hybridizing sequences are eliminated computationally by finding unique segments within the coding regions of the genome. In general, the shorter the probe sequence, the more likely it will represent a sequence that will hybridize only to its cognate target.

Array Production

Two DNA microarray platforms are currently in use. (i) Platform I. For spotted arrays, the probe DNA is first prepared and placed into multiwell containers. The probes can be entire clones, partial open reading frames, PCR products, or synthetic oligonucleotides. Using manual or robotic spot-

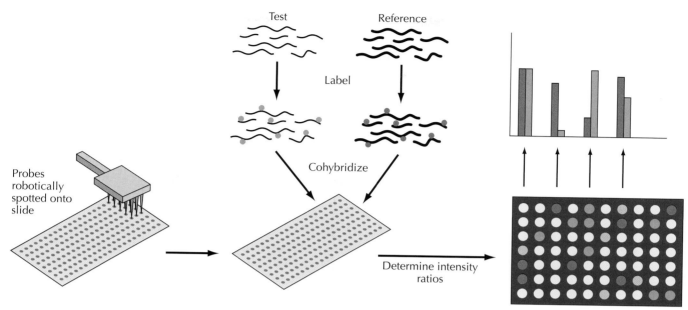

Figure 21.1 Transcriptional profiling using DNA microarrays. Shown is an outline of a two-color labeling scheme, with the objective to determine the transcriptome of a microorganism grown under two different conditions. A spotted microarray is first generated by PCR-amplifying all open reading frames in the organism's annotated genome, and spotting them onto derivatized microscope slides using a robotic arrayer. The bacteria are grown under two different conditions, and total RNA is isolated from each culture. These are then converted to labeled target by first-strand cDNA synthesis, with simultaneous incorporation of Cy3 or Cy5 nucleotides. The two labeled targets are combined and cohybridized to the probes on the microarray. Following hybridization and washing, the amount of target bound to each probe is determined by scanning the array and recording the fluorescence of Cy3 and Cy5 after excitation at their respective wavelength, which gives maximal emission. The relative intensities reflect the abundance of the mRNAs present in each sample, and the relative amount of bound target is usually displayed by pseudocolor representation. Red represents excess binding of Cy5 over Cy3, while green indicates that this probe bound more Cy3 than Cy5. Yellow indicates that equal amounts of Cy3 and Cy5 bound to each probe, and that the corresponding gene is not differentially regulated.

ting tools or ink-jet dispersion processes, these DNAs are then transferred to and immobilized on a solid support, which most often is derivatized glass or plastic. (ii) Platform II. Oligonucleotide arrays consist of 20- to 80-mer single-stranded oligonucleotides that are synthesized in situ on a solid support using a variety of synthetic methods, including photolithotrophy or ink-jet synthesizers.

Regardless of the manufacturing format, the use of all arrays is to detect sequence abundance by quantitating the extent of its hybridization to the array.

Manual spotting. DNA probes are manually transferred to nylon membranes using 96- or 384-pin replicators, after which the DNA is cross-linked to the membrane with ultraviolet light. Target binding can be measured using radioactive detection methods. Manual spotting is inexpensive and produces arrays that are reusable and flexible (i.e., individual DNA sequences can easily be added to or omitted from the array during manufacture). However, spot volume is difficult to control, the arrays are of low density, and a significant amount of probe DNA is required.

Robotic spotting. With this method, a small volume of probe (1 to 2 nl of 100 to 500 μg of probe per ml) is spotted onto a derivatized surface such as glass, and the DNA is UV-cross-linked to the support. Since such small volumes of probe are used, a typical PCR reaction can provide 700 to 1,000 spots. The relatively high-density arrays produced by this method contain a maximum of ~10,000 spots per slide and are flexible and highly sensitive. The disadvantage is that specialized printing and analysis equipment are required, thus significantly increasing the cost of array manufacture. Also, this type of array is not reusable.

In situ synthesis of probes. Spotted arrays are made by printing presynthesized DNA molecules onto slides, but arrays can also be manufactured by synthesizing oligomers directly on slides. One such approach uses photolithotrophic methods similar to those used in the semiconductor industry (Figure 21.2). Linker molecules are first attached to a derivatized surface. These molecules must undergo deprotection by UV light before a nucleotide can be attached. Photolithotrophic masks are then used to light-activate specific positions on the array for nucleotide addition. The newly added nucleotides must also be light-deprotected before subsequent addi-

Figure 21.2 Construction of a DNA microarray by photolithographic synthesis of oligonucleotide probes. Probes corresponding to small (~25 nucleotide) portions of genes can be synthesized by sequential light-induced deprotection of nucleotides and subsequent addition of nucleotides. Each deprotection step necessitates the use of a mask, which exposes only those probes that need to be deprotected and serve as acceptors of the next nucleotide. This method allows high-density synthesis of up to a million probes per array, with each gene represented by 10 to 20 probes.

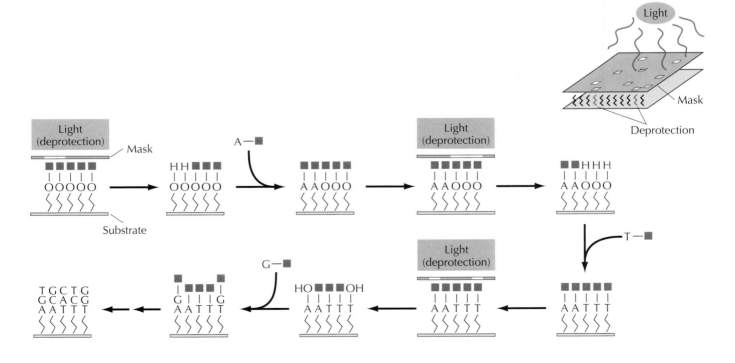

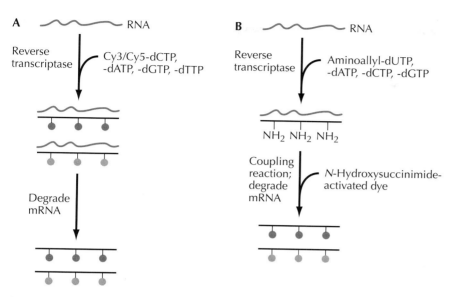

Figure 21.3   Methods of labeling targets for DNA microarray. **(A)** Incorporation of Cy3 or Cy5 nucleotides during DNA synthesis. **(B)** A two-step labeling procedure, whereby amine-modified nucleotides are incorporated during cDNA synthesis. The cDNA is then labeled by covalent linkage of *N*-hydroxysuccinimide Cy3 or Cy5.

tions can occur. Sequential rounds of masking, deprotection, and nucleotide coupling result in addressed synthesis of oligomers. The feature density of these arrays can be as high as 390,000/cm^2.

Additional technologies have been recently developed for miniaturized in situ synthesis of oligonucleotides that do not require the use of masks, such as the use of ink-jet oligonucleotide synthesizers and maskless photolithography.

In situ-synthesized arrays can possess a very high feature density, and they are extremely sensitive. However, they are expensive, not reusable, and in some cases not flexible. Also, unlike spotted arrays, the ability to manufacture in situ-synthesized arrays is completely beyond the capability of individual laboratories.

Target Labeling

Most labeling methods involve synthesis of a complementary DNA strand, using reverse transcriptase, DNA polymerase, or T7 RNA polymerase (Figure 21.3). In some methods, nucleotides labeled with cyanin dyes or radioisotopes are incorporated directly into the cDNA molecule. Indirect labeling approaches use incorporation of amine-modified nucleotides followed by labeling with a reactive fluorescent dye. Alternatively, target molecules can be synthesized using biotinylated nucleotides, and hybridization to the array is subsequently detected with streptavidin conjugates (Figure 21.4).

Applications

DNA microarray technology is most often used for two applications: the determination of gene expression levels (i.e., steady-state RNA levels) and the identification of specific sequences in genomes, including the detection of mutations. Two newer applications are genome-wide location analysis and transposon site hybridization. These approaches have been applied exten-

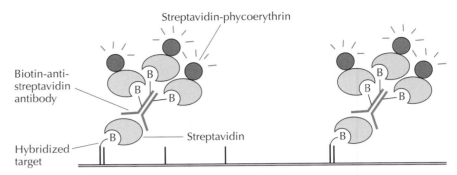

Figure 21.4   Amplification of target signal on the microarray. Biotin-labeled cDNA target is generated by incorporating biotin-modified nucleotides during cDNA synthesis or by attaching a biotinylated nucleotide to the 5′ end of the cDNA using terminal transferase. The bound cDNA is detected by sequential application of streptavidin, biotinylated antistreptavidin antibody, and fluorescently labeled streptavidin.

sively to address basic biological processes in many microorganisms. Moreover, transcriptional profiling is now a major tool for studying host-pathogen interactions, and some examples are reviewed here.

Transcriptional Profiling
Transcriptional, or expression, profiling attempts to quantitate the steady-state RNA levels in cell populations grown under defined conditions. By comparing different transcriptional profiles, it is possible to obtain information about the bacterium's genome-wide response to specific changes or signals. This method has been used to examine bacterial responses to inactivation of certain genes and to conditions such as nutrient limitation, growth phase, oxidative stress, temperature, light, mechanical stress, or growth within a host.

In bacteria, genes with related function are often expressed at the same time and only when needed. The simultaneous control of multiple genes is termed coordinate regulation, and researchers have attempted to define regulatory networks by undertaking searches for genes that respond to the same signals. Microarrays offer unprecedented opportunities for defining regulatory pathways (i.e., the group of genes controlled by a common signal, often acting through a transcriptional factor). Since the expression of the entire genetic repertoire of an organism can be monitored simultaneously, it is possible in a single experiment to identify all the genes that are induced or repressed by a given stimulus.

Understanding the flow of genetic information encoded in mRNA is important for all biological processes. However, it should not be assumed that the relative changes in mRNA levels, as detected by microarrays, reflect corresponding changes in protein levels or activities. It is well known that there are regulatory mechanisms that control protein function at several posttranscriptional levels, including protein synthesis, folding, degradation, covalent modification, and compartmentalization. The connection between transcription and translation is likely to be more direct for bacteria, in which mRNA is rapidly turned over, and the primary mechanism of controlling the levels of specific intracellular gene products occurs at the level of transcriptional initiation of the corresponding genes.

Profiling Responses to In Vitro Conditions

Pathogenic bacteria use a variety of virulence factors to infect their hosts and to cause damage. The expression of these virulence factors in many cases responds to experimental parameters (e.g., temperature, pH, and nutrient limitation) that presumably reflect some aspect of the host environment. At present, in vitro conditions cannot fully replicate the host environment, but in vitro studies offer the advantages of permitting strict control of the experimental conditions and isolation of large numbers of axenic bacteria for RNA purification.

Examples of microarray profiling of pathogenic bacteria exposed to different in vitro conditions include profiling of *Pasteurella multocida* in response to iron limitation, group A *Streptococcus* in response to temperature shift, *Mycobacterium tuberculosis* in response to nutrient limitation, and the *Escherichia coli* O157:H7 acetate-induced acid tolerance response. Other studies profiled responses to acid in *Helicobacter pylori* and *M. tuberculosis*. In *M. tuberculosis*, the genes most highly induced by acid (Rv3083 to 3089) resembled genes involved in nonribosomal peptide synthesis. These experiments also detected acid-enhanced expression of the isocitrate lyase (*icl*) gene, which is induced in macrophages and required for fatty acid metabolism and survival within the host.

Microarrays were also used to investigate *M. tuberculosis* responses to the antituberculous drug isoniazid and to low oxygen conditions. In the former study, both previously known and new INH-induced genes were identified. The latter study identified a cluster of three hypoxia-induced genes that contained a predicted two-component response regulator pair. Disruption of two of these genes resulted in decreased expression of α-crystallin, which was known to be induced by hypoxia and required for *M. tuberculosis* growth in macrophages.

Expression profiling was used to identify *Salmonella enterica* serovar Typhimurium genes regulated by low-shear modeled microgravity (LSMMG), which had been found to enhance several virulence phenotypes. Microarray profiling detected LSMMG-induced repression of lipopolysaccharide (LPS) biosynthesis genes, which correlated with reduced LPS synthesis in LSMMG-grown cultures. Since surface structures like LPS are targets for the host immune response, it may be that decreased LPS production permits *Salmonella* serovar Typhimurium to evade the immune response more effectively, thus accounting at least in part for the LSMMG-induced increase in virulence.

Contact with host cells is another in vitro condition that can be examined with microarray analysis. Profiling of gene expression changes induced in *Neisseria meningitidis* by contact with human epithelial cells identified numerous genes encoding membrane-associated or periplasmic proteins. These results suggest that extensive remodeling of the *N. meningitidis* membrane occurs upon host cell contact. This study also identified five adhesion-induced antigens that were subsequently found to be capable of eliciting bacteriocidal antibodies in mice.

Several expression-profiling analyses have been performed in *Pseudomonas aeruginosa*. One of these studies found that relatively few genes were differentially expressed in *P. aeruginosa* biofilms compared with planktonic bacteria. However, one gene that was repressed in biofilms (*rpoS*) had previously been implicated in virulence and biofilm formation, and further investigation showed that biofilms formed by an *rpoS* mutant of *P. aeruginosa* possessed increased resistance to the antibiotic tobramycin.

Another study profiled the *P. aeruginosa* response to iron limitation. As expected, genes induced by iron limitation included many involved in siderophore-mediated iron acquisition. Subsequent profiling was performed on a *P. aeruginosa* strain with a mutation in *pvdS*, which is required for production of the siderophore pyoverdine. This work led to the identification of novel genes involved in pyoverdine biosynthesis.

Profiling of Mutants

Another strategy that has been used to define regulatory networks is transcriptional profiling of mutant strains. This type of experiment simply compares transcriptional profiles generated in a wild-type strain and isogenic mutants that do not express particular transcription factors. Ideally, it is thus possible to identify genes under the control of those regulators, since genes activated by the regulator will show relatively decreased expression in a mutant, whereas genes normally repressed by the regulator will show a relative increase in expression in a mutant. In actuality, microarray analysis cannot distinguish between genes directly controlled by the regulator and those that are responding indirectly to physiological changes caused by the mutation. This point will be addressed again later in this chapter.

Microarray profiling of regulatory mutants has been performed in a variety of different pathogens, some of which have been noted above. In *Staphylococcus aureus*, this strategy identified new genes regulated by the *agr* and *sarA* loci, which were already known to control virulence factor expression. A similar analysis in group A *Streptococcus* defined numerous genes whose expression was controlled both in vitro and in vivo by CovR, a known virulence regulator. Profiling of *Vibrio cholerae* strains carrying mutations in characterized virulence regulators (ToxRS, TcpPH, and ToxT) identified many loci not previously known to be controlled by these factors. Profiling of *M. tuberculosis* strains with mutations in two different sigma factor genes (*sigE* and *sigH*) identified a number of genes controlled by these sigma factors, and this information was used to establish putative consensus promoter sequences for both σ^E and σ^H. Another profiling study in *M. tuberculosis* identified a number of genes regulated by iron and IdeR.

In some instances, these microarray analyses led quickly to the discovery of new virulence factors or pathways. Transcriptional profiling of a *Streptococcus pneumoniae* strain deleted for the *ciaRH* two-component signaling system resulted in the identification of a new gene (*htrA*) required for colonization of infant rats. Profiling of an autoinducer synthetase mutant of *E. coli* O157:H7 found a role for quorum sensing in regulation of motility and toxin production, and profiling of a *V. cholerae luxO* mutant demonstrated that quorum sensing regulates virulence factor expression and biofilm formation in this bacterium. Finally, microarray analysis of a *Salmonella* serovar Typhimurium polynucleotide phosphorylase (PNPase) mutant showed that mRNA stability regulates virulence genes located on the SPI1 and SPI2 pathogenicity islands. In accordance with these findings, the PNPase mutant secreted higher than wild-type levels of SPI1 effector proteins and showed enhanced invasion and intracellular replication in tissue culture cells and mouse spleen.

Expression profiling of wild-type and *prfA* mutant strains of *Listeria monocytogenes* identified three classes of genes controlled by PrfA, a known regulator of *Listeria* virulence. Further analysis showed that a putative PrfA-binding site was located upstream of some, but not all, PrfA-regulated genes identified by microarray, indicating that PrfA may act indirectly on some genes. It was also found that PrfA can act as both an activator and a repressor, and that PrfA-mediated regulation may involve the use of alternative sigma factors.

In *P. aeruginosa,* it was known that calcium levels and the regulatory factor ExsA were involved in coordinated regulation of type III secretion system (TTSS) genes. Profiling of an *exsA* mutant under high- and low-calcium conditions led to the identification of a unique membrane-associated adenylate cyclase (*cyaB*) as a TTSS regulator. Subsequent analyses demonstrated that cyclic AMP acts in concert with the known regulator Vfr to control transcription of all known TTSS genes, components of three other virulence pathways, and several predicted chemotaxis genes.

Profiling Bacterial Gene Expression within Hosts

Much useful information is obtained by using in vitro conditions as surrogates for signals encountered by bacteria within the host, but it is clear that in vitro studies do not accurately replicate the host environment. Earlier, several genetic methods were devised to identify bacterial genes that are specifically induced "in vivo" during an actual infection (see chapter 20). However, these approaches identify only genes that are induced in vivo, not those that are repressed in vivo, and these methods are also unable to provide a quantitative measure of changes in gene expression. Microarray analysis therefore promises greatly to enhance our ability to analyze bacterial gene expression under in vivo conditions.

Transcriptional profiling analyses of bacteria recovered from infected hosts are difficult for several reasons. First, there are significant technical problems associated with preparing labeled target from bacterial RNA mixed with a much larger quantity of host RNA. Second, the host environment is more variable than an in vitro environment, and minor or uncontrollable differences in host nutritional status or immune competence may affect the course of the infection. In addition, bacteria within a host may encounter different microenvironments, and bacteria in one microenvironment may not be separable from those in another. If the transcriptional profiles of these populations differ, their superposition could lead to misleading or uninterpretable results. Finally, since all microarray experiments involve a test and a reference sample, care must be taken when choosing a reference against which to compare in vivo profiles.

Erwinia chrysanthemi. In vivo transcriptional profiling of the plant pathogen *E. chrysanthemi* compared the expression profile of bacteria recovered from African violet leaves with that of bacteria grown in rich medium. This study was performed prior to the completion of the *E. chrysanthemi* genomic sequence, using an array constructed with 5,000 randomly selected 3-kb clones from a plasmid library of *E. chrysanthemi* genomic DNA. Although the study was complicated by the use of clones containing multiple genes, 48 genes were identified as host induced. Insertion mutations were constructed in 17 of these genes, and three of these mutants showed significant virulence defects.

P. multocida. Genomic microarrays were used to examine the gene expression profile of *P. multocida* during infection of chickens. Relative to bacteria grown in laboratory medium, bacteria isolated from the blood of infected chickens showed reproducible differential expression of 40 genes. Ten of the 17 in vivo-induced genes were predicted to be involved in amino acid transport and metabolism or energy production. The majority (12) of the 23 in vivo-repressed genes had unknown functions, and no strong conclusion could be drawn based on the identity of the others.

This study documented considerable variation in the microarray data obtained from different chickens; 40 genes were differentially expressed in 3 of 3 chickens, whereas 522 genes were differentially expressed in at least 1 of 3 chickens. It should be noted that the number of bacteria recovered from the animals was quite high (10^9 to 10^{10} CFU/ml of blood), and presumably the amount of contaminating host RNA was minimal. These observations underscore the fact that even under seemingly favorable experimental conditions, the host environment is significantly more complex and difficult to control than an in vitro environment, and that multiple data sets are required for in vivo studies.

Borrelia burgdorferi. During *B. burgdorferi* infection of a mammalian host, the number of bacteria in host tissues is very low. Therefore, to obtain sufficient numbers of in vivo-grown bacteria for expression profiling, investigators grew *B. burgdorferi* in dialysis membrane chambers implanted in rats. This study compared the transcriptional profiles of *B. burgdorferi* grown thus in vivo and also in vitro under conditions thought to simulate those found in fed- and unfed-tick vectors. The results indicated that expression profiles from unfed-tick and in vivo conditions differed more from fed-tick conditions than from each other. Differentially expressed genes identified here were distributed roughly equally between the chromosome and the plasmid population, although some individual plasmids possessed no differentially expressed genes.

V. cholerae. Two separate transcriptional profiling studies were performed on *V. cholerae* present in stools obtained from cholera patients. The first study compared the gene expression pattern of stool bacteria with that of bacteria grown to stationary phase in vitro, while the second compared stool bacteria with cells grown to midexponential phase in vitro. Both studies concluded that *V. cholerae* in the human gut encounters an environment that is oxygen, iron, and nutrient limited. Similar results were obtained in microarray profiling of *V. cholerae* grown to midexponential phase in vivo in rabbit ligated intestinal loops and in vitro in rich medium.

Comparative Genomics

DNA microarrays are also often utilized to compare the gene content of different strains, with the goal of identifying genes that confer virulence properties, discovering whether and how these genes are transmitted among different strains, and tracing the evolutionary lineages of related strains.

High-variability pathogens. Microarray-based comparative genome studies have in several cases found a high degree of variability among strains of a given species. It should be kept in mind that microarrays cannot detect the presence or absence of genes not represented on the array. Since genome sequences have not been obtained for many different isolates of any bacterial species, the arrays used in comparative genome studies likely do not contain a full representation of all the genes present in the natural bacterial population. Therefore, microarray studies to date may have underestimated the genomic diversity in certain bacteria.

In *S. aureus* and pneumococcus, approximately 22 to 25% of all genes showed variation among the strains examined. Similarly, a whole-genome microarray study of 11 *Campylobacter jejuni* strains found that at least 21% of genes in the sequenced strain were absent or highly diverged in at least one other isolate. Much of the strain-specific variation involved genes associated with synthesis of surface structures such as flagella, lipooligosaccharide, or capsule.

Another enteric pathogen demonstrating high genomic variability in whole genome array analysis is *H. pylori*. Analysis of 15 clinical *H. pylori* isolates found that 22% of the genes on the array were absent in at least one isolate. Many of the strain-specific genes were located either in the PAI pathogenicity island or in the PZ plasticity zone, and most strain-specific genes with known function encoded restriction modification system components or transposases. Another study examined gene content differences among 14 *H. pylori* strains isolated from a single human host. Although both genomic and nongenomic diagnostic methods indicated that the strains were genetically closely related, only genomic microarray analysis detected a 3% variation in genomic content among the strains. These differences included both absence of sequences and the presence of additional sequences relative to the reference strain.

Mycobacterium.   Microarrays based on the genome sequence of *M. tuberculosis* H37Rv were used to compare the gene content of H37Rv, the subspecies *M. bovis*, and several strains of *M. bovis* derivative BCG, which is the current vaccine used against tuberculosis. This analysis detected 16 regions that were deleted in BCG strains. Nine of these regions were absent in BCG and all virulent *M. bovis* strains, and these potentially contain genes involved in the different virulence properties of *M. tuberculosis* and *M. bovis*. Other deletions were correlated with the known history of the BCG strains and used to establish genealogical relationships among them.

Group A *Streptococcus.*   Genomic microarray studies of the M1, M3, and M18 serotypes of group A *Streptococcus* indicated that much of the genetic difference between these serotypes is due to variability in regions containing phage or phage-like sequences. Since many of these phage regions contain known and putative virulence factors, they may contribute to the differing disease manifestations associated with the different GAS serotypes.

Salmonella **serovar Typhimurium.**   Microarrays based on the *Salmonella* serovar Typhimurium LT2 genome were used to examine the genetic differences among 22 strains representing all seven *Salmonella* subspecies and the related species *Salmonella bongori*. This analysis identified 56 "signature genes" that were present in all 22 strains, as well as 74 genes that were unique to subspecies I, which is highly adapted to mammals and birds. SPI1 pathogenicity island genes were mostly present in all *Salmonella*, while SPI2 genes were found in all *S. enterica* strains but not *S. bongori*. These data led to a predicted sequence of evolution of the Salmonellae and also suggested that *Salmonella* genomes are highly dynamic, with multiple acquisitions of many different genes by different lineages.

V. cholerae.   Genomic microarray comparison of 10 *V. cholerae* strains identified genes unique to all pandemic strains, as well as genes specific to seventh pandemic El Tor and related O139 serogroup strains. These loci may confer traits specifically associated with displacement of the preexisting pathogenic *V. cholerae* strains in South Asia and also promote the establishment of disease in previously cholera-free locations.

Genome-Wide Location Analysis

As we have seen, microarrays are commonly used to define regulatory pathways, often by comparing transcriptional profiles generated in a wild-type strain and an isogenic regulatory mutant. However, these profiles are often very complex, and they do not discriminate between direct and indirect effects of the factors. Since binding sites for the factors are often degenerate, their locations are not always identifiable. A microarray application called genome-wide location analysis (Figure 21.5) was developed to address this problem. This strategy is based on chromatin immunoprecipitation methods and the ability of transcription factors to bind DNA with sufficiently high affinity to allow chemical cross-linking of the complex in intact cells. DNA fragments cross-linked to the specific DNA-binding protein are then recovered by immunoprecipitation, the cross-link is reversed, and the DNA is amplified by PCR to generate a target for microarray analysis. The comparison of an expression profile (mutant versus wild type) and the result of the location analysis are used to identify genes that are directly

Figure 21.5 Genome-wide location analysis. Genome-wide location analysis is a DNA microarray application designed to identify transcription factor binding sites. First, protein-DNA complexes within cells are cross-linked with formaldehyde. The cross-linked complexes are isolated by immunoprecipitation with specific antibody, and the DNA is released by reversing the cross-links. The DNA is then amplified by PCR to generate targets for microarray analysis. A corresponding sample is generated from total DNA that has not been subjected to immunoprecipitation. The signal ratio derived from the two samples identifies sequences that are directly bound by specific factors, thus demonstrating direct regulation of those genes by those factors.

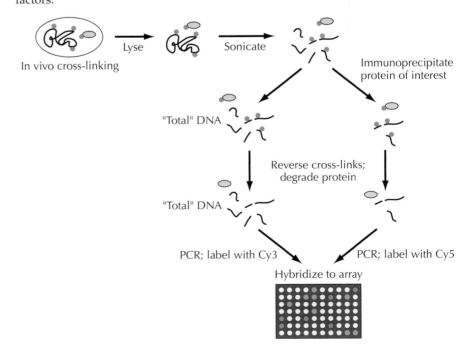

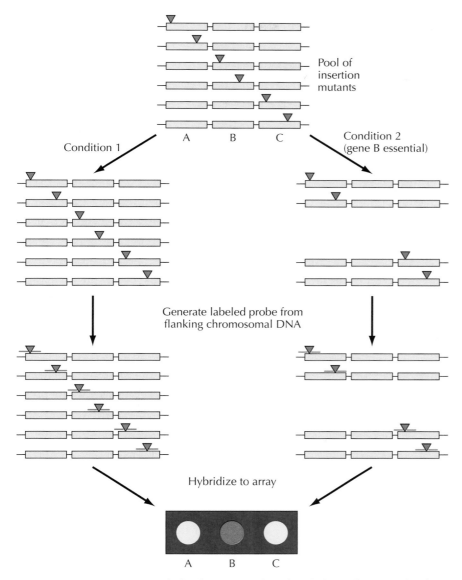

Figure 21.6   Transposon site hybridization analysis (TraSH). Replicate pools of mutants are grown under two different conditions, and differentially labeled target is produced from the chromosomal sequences flanking each transposon in each pool. The two labeled targets are cohybridized to a mircroarray, and the signal ratio at each spot quantitates the representation of that sequence in the two pools. TraSH thus provides information about which genes are required for survival under a given condition.

controlled by the regulatory protein. This approach was used to identify the binding sites for CtrA, the master regulator of *Caulobacter crescentus* that controls expression of cell cycle-regulated genes.

TraSH

Another genomic analysis method that has taken advantage of microarray technology is transposon site hybridization (TraSH), which uses microarrays simultaneously to map all the insertions present in a pool of transposon insertion mutants (Figure 21.6). Replicate pools of mutants are grown under two different conditions, and differentially labeled target RNA is produced from the chromosomal sequences flanking each transposon in each of the two pools. The labeled target from the two pools is then cohybridized to a genomic microarray. The relative fluorescence intensity at every spot on the array provides information regarding the representation of that sequence in the two pools. In this way, it is possible to determine the presence or absence of specific transposon insertion mutants under a given condition. TraSH has been used to conduct a genome-wide identification of conditionally essential genes in mycobacteria.

Proteomics

Genome-wide characterization of gene expression patterns represents the initial effort toward obtaining a complete understanding of cellular activity. However, it is proteins that carry out the vast majority of cellular processes, and protein levels do not necessarily correlate with transcript levels. In addition, since posttranslational modifications play a vital role in regulating protein activity, it is impossible based on gene expression studies alone to determine what forms of a protein are present and active within a cell. The research community is therefore already starting to identify and quantitate all the proteins in a cell and to characterize how individual proteins act in conjunction with one another.

A protein's normal function is often defined by its interactions with other proteins, and identifying all of a protein's interaction partners may provide vital information about its function within a living cell. Genetic and biochemical approaches have long been utilized to study protein complexes, but this undertaking was recently extended to the genomic level. That is, the goal is now systematically to define all potential interactions among the open reading frames encoded in a particular organism's genome, thus yielding an interaction map, or interactome.

Interactome analysis may be an important functional genomic tool. It can greatly aid in defining the function of unknown proteins by detecting their interactions with proteins of known function. If one identifies an unknown protein's interactive partner, and if the partner is sufficiently well characterized, testable hypotheses can be made regarding the function of the unknown protein. Of course, this approach is not universally applicable, as it is possible that the unknown protein interacts with other unknown proteins, which in turn have no characterized partners.

Mass Spectrometry

Mass spectrometry (MS) is a key proteomic tool that has been used to characterize the identity, abundance, and modification of individual proteins, and to analyze the individual components of multisubunit complexes. Advances in MS applications have been greatly fueled by the development of highly sophisticated instrumentation linking fractionation techniques with mass spectrometry. In addition, direct bioinformatic links between MS results and genome datebases allow the accurate identification of proteins based on fragmentation patterns predicted from genomic sequences.

Almost all techniques utilizing MS for protein identification require some separation or enrichment technique to reduce the complexity of the mixtures to be analyzed. MS is most popularly used in conjunction with two-dimensional gel electrophoresis, which separates complex mixtures into individual proteins by isoelectric focusing and sodium dodecyl sulfate-polyacrylamide gel electrophoresis. Individual protein spots are visualized (i.e., stained), excised, fragmented by proteolytic digestion, and then identified by M. The comparison of the spots in two gels, each containing extracts from cells propagated under two different conditions, can be used to quantitate the relative abundance of proteins. This method remains one of the important tools of proteomic research, despite problems associated with the relative lack of comprehensive two-dimensional gel methods. For instance, poorly expressed and hydrophobic proteins are often undetectable by visualization of spots after two-dimensional electrophoresis. Column chromatography (especially high- or fast-performance liquid chromatography) can replace two-dimensional gels, but its resolution is inferior to that of electrophoretic gels. One of the newer methods for analyzing complex mixtures is multidimensional liquid chromatography, which consists of combined ion-exchange and reverse-phase chromatography coupled to two-stage tandem mass spectrometry.

Another new technology for quantitative proteomic analysis is based on protranslational labeling of complex mixtures with tags that can be used to purify peptides and accurately quantitate their relative abundance. This method is referred to as isotope-coded affinity tag (ICAT) labeling and is outlined in Figure 21.7. One early proteome-wide application of ICAT explored the metabolic perturbations in *Saccharomyces cerevisiae* grown in the presence of glucose or galactose, resulting in accurate quantification of changes in hundreds of proteins. Recently, ICAT was used to study the magnesium response in *P. aeruginosa.* Under magnesium limitation, which mimics conditions in the lungs of cystic fibrosis patients, alterations in the levels of 145 proteins were detected. The ICAT analysis was further extended to membrane proteins and demonstrated extensive remodeling of the bacterial membrane when magnesium is limited. Magnesium stress-dependent induction was detected for quorum-sensing proteins, particularly the enzymes responsible for synthesis of the *Pseudomonas* quorum-sensing (PQS) signaling molecule, and these results were validated by demonstrating a significant increase in PQS levels in bacteria grown in magnesium-limited media.

Genetic Analysis of Protein Interactions Using "Two-Hybrid" Screens

All two-hybrid systems detect readouts generated by the interaction of two proteins or, very often, of two reporter protein domains fused to putative interacting protein domains (Table 21.1). If the putative interacting domains are able to mediate reconstitution of the reporter domains into an active complex, this interaction is detected via a transcriptional readout. The yeast-based two-hybrid system remains the main workhorse in the genetic protein-protein interaction field, although several methods have also been developed in mammalian cells and in bacteria. Since even the relatively simple genomes of microorganisms require pairwise testing of thousands of interactions, a two-hybrid method must be amenable to high-throughput formats to be useful for functional genomic studies.

In the original two-hybrid screen, the coding sequences of the modular transcriptional factor GAL4 is split between two plasmid vectors. One vector expresses the GAL4 DNA-binding domain, and the other expresses the transcriptional activation domain. Coexpression of the two domains alone does not lead to reconstitution of an active GAL4 complex. However, if the

GAL4 domains are fused to heterologous interactive domains, interaction between the heterologous domains will reconstitute the GAL4 complex by bringing the DNA-binding and transcriptional activation modules together. The active hybrid complex then directs transcription of reporter genes from GAL4-dependent promoters.

In general, two approaches can be taken toward defining an interactome, both of which have been used to define the interactomes of several model organisms. The first is simply a scaled-up version of a common two-hybrid library strategy. A single coding sequence is cloned into a "bait" vector and screened against a highly complex library of protein-coding sequences cloned into the "prey" vector. This "one clone vs. library" strategy implies that a fully comprehensive analysis requires as many individual screens as there are identified ORFs in a genome. Since each positive prey

Figure 21.7   Applications of isotope-coded affinity tag (ICAT) technology. Protein samples, such as lysates from bacteria grown under two different conditions, are chemically derivatized with the isotopically light or heavy version of the ICAT reagent. The labeled samples are combined and proteolyzed to yield small peptide fragments. The derivatized fragments are isolated by affinity chromatography, using a tag that is part of the ICAT reagent. The isolated peptides are separated, identified, and quantified following liquid chromatography (LC) and tandem mass spectrometry (MS).

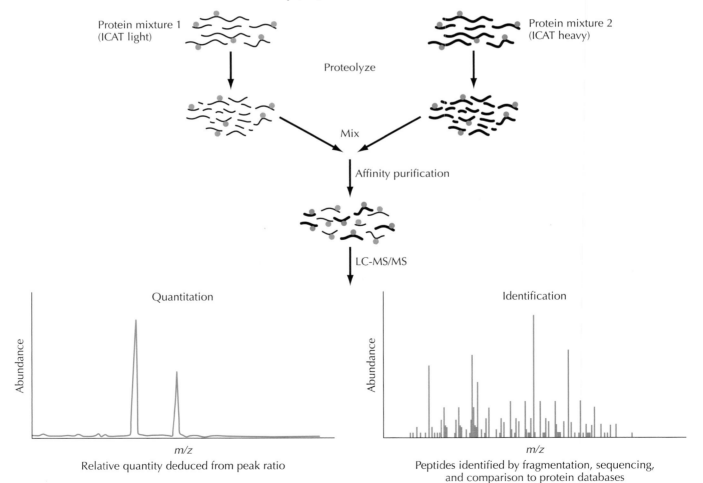

Table 21.1 Genetic methods for studying protein-protein interactions

Protein used to study interactions	Host system
Systems based on reconstitution of signaling proteins or transcriptional activators	
GAL4	Yeast
LexA-VP16	Yeast
LexA-B42	Yeast
Sos, Ras	Yeast
LexA	*E. coli*
AraC	*E. coli*
ToxR	*E. coli*
cI	*E. coli*
Zif-RpoA	*E. coli*
Systems based on reconstitution of active single-chain proteins	
Ubiquitin	Yeast
Adenylate cyclase	*E. coli*
Dihydrofolate reductase	*E. coli*

clone has to be isolated and sequenced, the labor intensiveness of this approach strongly depends on the quality of the prey library, which should contain as many random, nonredundant fragments as possible and a low background of false positives.

The second possible type of screen involves constructing complete sets of expressing clones, and this approach has been used in several model organisms. A complete set of open reading frames (ORFs) is cloned into both a bait and a prey vector, and the two libraries are introduced into *S. cerevisiae* of opposite mating types to facilitate the formation of diploid cells coexpressing the two GAL4 fusions. These clones are usually arrayed in a matrix format (e.g., into 96- or 384-well plates) and individually screened against each other. Since the array format is addressable (i.e., the identity and location of each clone in the matrix are known), any positive interaction immediately identifies an interactive partner. In theory, the comprehensive screen requires a large number of screens, each defined by a unique combination of bait and prey in the same diploid cell. For example, the comprehensive analysis of *S. cerevisiae* (6,200 ORFs) can be accomplished by analyzing 37,000,000 fusions. The magnitude of this work can be significantly reduced by pooling bait clones; positive interactions can then be deconvoluted to identify individual interactive pairs.

Viral Interactomes

Viral genomes are small and thus offer relatively simple systems for testing proteome analysis methods. Yeast two-hybrid methods were used to conduct comprehensive identification of protein-protein interactions in several viruses. The genome of bacteriophage T7 consists of 55 ORFs, and a combination of library and array-based screens identified 22 interactions, including known interactions among four proteins. A more comprehensive analysis was performed with vaccinia virus. Its 266 full-length ORFs were subjected to an array-type screen (a total of over 70,000 combinations) that led to the identification of 37 interactions. A slightly different approach was used to define interactions among hepatitis C proteins. When an array-based approach failed to identify any interactions, a set of bait plasmids was constructed from approximately 200 fragments derived from 10 predicted viral ORFs. These baits were screened against a random-fragment genomic library, and 15 interactions were identified.

S. cerevisiae Interactome Analysis

The most informative proteomic analyses to date have been performed in *S. cerevisiae*. These studies not only defined complex interactive networks but also provided a useful assessment of the reproducibility and sensitivity of the yeast two-hybrid technology, and its suitability as a major tool for proteomic analysis of other organisms. In addition, the results revealed the power and limitations of any single approach toward understanding one of the more complex problems in biology, namely, how thousands of proteins present at different levels interact in an orderly matter within living cells.

The interactome of *S. cerevisiae* was analyzed in two separate efforts. An array-based screen using 6,200 cloned full-length yeast ORFs identified 841 interactions, the majority of which were novel. The level of confidence of a particular detected interaction is based on identification of the same interaction in independent clones, and only those that are detected more than three times are considered relevant. A similar screen combining library- and array-based approaches detected 691 interactions, most of which were also not previously described. Surprisingly, only 141 interactions were identified by both of these nearly identical approaches, and this discrepancy in results raises concerns about the sensitivity of the assay. A more difficult problem lies in the inability to validate the biological significance of interactions detected in an artificial system. This is true even though these yeast proteome studies were conducted in yeast.

An independent effort used a λ repressor dimerization system to detect homotypic interactions among *S. cerevisiae* proteins. In an entirely library-based approach, *S. cerevisiae* DNA was fused to the λ phage repressor coding sequence, and several million-member libraries were analyzed for dimerization. A total of 35 interactions were detected. Interestingly, only 5 of these 35 interactions were also found in the two studies described above. However, the λ repressor method was validated by its identification of interactions that had previously been demonstrated by other means.

Two other studies used an entirely different approach to define protein complexes. These studies were based on purification of protein complexes from cell lysates, followed by mass spectrometry analysis of the complexes. Two reports described the affinity-tagging and purification of 589 and 493 complexes, respectively, and subsequent identification by mass spectrometry of the components of these complexes. Comparison of the data showed only 10% overlap in the results. Although these discrepancies have not been resolved, this approach resulted in the description of a larger fraction of previously characterized interactions than the yeast two-hybrid method, thus providing a somewhat independent validation of the approach. Moreover, the affinity purification method revealed a network of interactions involving multiple proteins, which would not be detected by binary two-hybrid analyses. However, this property may also complicate the interpretation of the results, as it is unclear whether these interactions represent true multisubunit complexes or experimental artifacts.

To date, only one report of a bacterial interaction map has been published. To identify interaction partners in *H. pylori*, a classical yeast two-hybrid screen of 261 *H. pylori* ORFs (of 1,590 total) was screened against a "prey" library of random *H. pylori* chromosomal fragments. The choice of the 261 bait ORFs was biased toward genes encoding potential virulence factor genes and randomly selected genes encoding proteins of unknown function. A total of 1,280 interactions were identified, which link together almost half of the *H. pylori* proteome. An important subsequent study demonstrated the value of this approach in assigning function to unknown ORFs. This work identified an unknown ORF as an anti-sigma factor based on its interaction with the flagellum-specific sigma factor FliA. Reciprocal screening of a library, using this hypothetical protein as bait, validated the interaction. Since random prey libraries contain overlapping fragments of each ORF, it was possible to define the domains mediating between FliA and its interactive partner. These domains were all consistent with known characteristic interactions of sigma–anti-sigma factor binding. The anti-sigma factor protein was thereafter termed FlgM, and it is analogous to similar proteins controlling flagellar expression in other bacterial species. Furthermore, analysis of *H. pylori* flagellin and FlgM expression validated the role of FlgM as a negative regulator of flagellin transcription.

Traditional two-hybrid analysis has been extremely successful, but its major limitation is its poor performance in analyzing interactions between membrane proteins. The main problem appears to be the failure of proteins with transmembrane segments to be effectively transported into the nucleus or to remain soluble during the reconstitution of the active GAL4 transcriptional factor. Since secreted or membrane proteins can make up to 30% of the proteome in most living organisms, several alternatives to the original two-hybrid system have been developed to allow analysis of interactions among membrane proteins. Particularly useful among these is the split-ubiquitin system, which is based on reconstitution of a ubiquitin-tagged substrate for ubiquitin-specific proteases (Box 21.1).

Ubiquitin can be expressed separately as amino-terminal (Nub) and carboxy-terminal (Cub) domains, which reassemble in yeast. For analysis of protein-protein interactions, the Nub portion is engineered with a mutation (termed NubG) that prevents assembly of the two ubiquitin fragments, while Cub carries a reporter fusion (CubR). Interaction mediated by the fused heterologous domains (analogues to the bait and prey constructs in the classical two-hybrid system) leads to degradation of the ubiquitin and release of an activator of reporter genes. This method is particularly useful for membrane proteins since the interaction that produces active ubiquitin takes place in the yeast membrane, and release of the active, readily transportable transcriptional factor leads to activation of reporter genes in the

---

BOX 21.1

The split ubiquitin system for membrane proteins

Two plasmids are introduced into *S. cerevisiae*. One expresses the N-terminal portion of ubiquitin with a mutation (NubG) that eliminates its natural interaction with its carboxy-terminal domain. The second plasmid directs the expression of Cub-TF, the carboxy-terminal domain of ubiquitin fused to an artificial transcriptional activator (a hybrid of the bacterial LexA protein and the herpes simplex VP16 transactivator protein). The halves of the ubiquitin can be brought together by fusion of NubG and Cub-Tf with two interactive membrane proteins, X and Y, respectively. Ubiquitin dimerization in the membrane leads to the release of the transcriptional factor, which enters the nucleus and activates the expression of reporter constructs (*lacZ* and HIS30). Productive interaction of X and Y can be monitored by the appearance of blue colonies on media containing β-galactosidase substrate, or by loss of the histidine requirement.

nucleus. The split ubiquitin assay has been used to demonstrate interaction of individual membrane proteins in a variety of eukaryotic organisms, but it has not yet been applied in a genome-wide screen.

Protein Arrays

The use of DNA microarrays in postgenomic research was wildly successful and thus became common in many branches of molecular biology. It is apparent that similar technologies are needed to allow highly parallel quantitation of individual protein levels. The most obvious application of protein arrays would be to measure the response of proteins to perturbations, analogous to gene expression profiling, but protein arrays potentially have many more applications than DNA arrays.

Ideally, protein-detecting arrays would report in a highly parallel, miniaturized format on the abundance of each protein in a proteome; this application would make it analogous to the measurements of mRNA abundance using DNA microarrays. Protein-detecting arrays consist of arrayed capture ligands (e.g., antibodies) that are specific for individual proteins, and bound protein is quantified in a subsequent step. Protein activity arrays have additional applications, such as analysis of enzymatic activities, posttranslational modifications, or interactions with small molecules or protein partners. A generalized schematic of the production and use of these basic array formats is shown in Figure 21.8.

Figure 21.8 Two protein array formats. **(A)** Protein-detecting array. A set of capture reagents, usually antibodies specific for each protein analyzed, is immobilized on a solid surface. A protein mixture is applied, and bound protein is detected using a second reagent, such as another specific antibody. **(B)** Protein function array. A set of expressed proteins is immobilized in an array format on a solid support. These can be then used in a variety of assays to study enzymatic activity or interaction with substrates, including other proteins or nucleic acids.

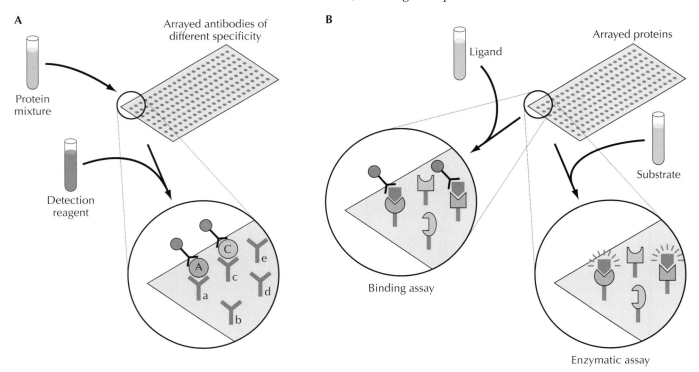

There are many inherent difficulties associated with generating complete sets of expressed proteins and with controlling their behavior in array formats. The field has thus far progressed slowly, but several key developments provide indications that protein arrays have the potential to revolutionize many area of biomedical research.

Production of Protein Arrays

The major obstacle to using whole-genome protein arrays is the requirement for extensive customization. Nucleic acid probes and targets are generally stable and can be generated in large amounts using standardized high-throughput procedures. In contrast, proteins often have specific synthesis and folding requirements, and they are often destabilized during the process of being immobilized on arrays. Uniform conditions for expression, folding, and immobilization are therefore usually achievable only for a limited group of proteins. Proteins also differ in their stability, and, in general, they cannot be made too much in advance of their use. However, these concerns vary depending on the intended use of the array. For example, tertiary structure may be of less importance for a protein array used to identify serum antibodies in infected hosts than for arrays used for miniaturized enzymatic assays of an unknown protein or the identification of an interactive partner.

The initial technical consideration in constructing protein arrays is the synthesis of sufficient amounts of protein. This is probably the most significant rate-limiting step impeding the large-scale use of protein arrays. Several recombinant host/vector systems have been adapted for generating partial, full-length, or affinity-tagged proteins. Coupled in vitro transcription/translation systems can yield comparable amounts of proteins as many recombinant systems, and these can be programmed with synthetic DNA generated by PCR, removing the need for cloning particular genes.

Immobilization on a solid surface can be accomplished by various methods. Affinity tags can facilitate both protein purification and immobilization of proteins on an array. The coding sequence of the affinity tags is generally included in the expression vector, or it is incorporated at either the amino- or carboxy-terminal coding sequence of each cloned ORF. Commonly used tags include hexa-histidine, glutathione-*S*-transferase, or streptavidin-binding peptide. Following affinity purification, these proteins can be immobilized on glass surfaces modified by chelated nickel or streptavidin. If the proteins are tagged with a specific peptide epitope, they can be attached to a surface that contains immobilized antiepitope antibody.

Proteins can also be covalently attached to derivatized surfaces via functional groups provided by their side chains (e.g., sulfhydryl, amino, carboxy, hydroxyl, and imidazole groups). For these applications, a wide range of derivatized glass surfaces are available that allow covalent cross-linking of the deposited protein. Presynthesized peptides can be directly coupled to appropriately modified surfaces to generate arrays.

Finally, several technologies for direct synthesis of peptides on modified surfaces are available. These methods are analogous to the light-directed in situ synthesis of nucleotides, including the use of photogenerated amino acids. In situ synthesis has also been adapted for fabrication of peptide microarrays.

Detection Methods

The detection method used is dictated by the particular application of the array. In protein-detecting arrays, it is necessary to identify and accurately quantify the individual proteins captured from a solution of analyte. When capture is based on a specific antibody, an obvious detection method is to use a second antibody (similar to the detection method of a sandwich enzyme-linked immunosorbent assay). However, secondary antibody may not be readily available, and without amplification, the sandwich approach is not very sensitive. Alternatively, a ratiometric approach may be used, relying on differential labeling of all proteins in the analyte (e.g., with Cy3 and Cy5 dyes) and determining the relative amounts of bound protein.

For protein function arrays, the readout is based entirely on the assay used. Although many highly sensitive enzymatic assays are available, they are often not amenable to high-throughput screening formats. Many of the enzymatic assays are not applicable to flat-structure arrays, and most successful applications have been performed in nanowells. However, it is reasonable to expect that protein function arrays will develop rapidly, in particular, with the increased availability of new, highly sensitive fluorescence reagents, improved methods of reproducible signal amplification, and improvements in the sensitivity of instruments used for analysis. One such promising technology amplifies the signal generated by binding an antibody to an immobilized protein and then generates hundreds of tandemly linked single-stranded DNA segments via a rolling-circle DNA replication mechanism. Depending on the method of detecting the in situ-generated DNA, this method can result in up to 1,000-fold amplification of signal.

Protein-Detecting Arrays

Using specific chemical reagents to capture an entire proteome is not feasible; they can be used only with specific groups of proteins possessing a high-affinity binding site, such as those for various cofactors or pseudosubstrates. In contrast, antibodies and peptide and nucleic acid aptamers provide sufficient specificity to bind thousands of proteins, some of which may have only subtle differences in their primary amino acid sequence.

Antibodies in particular offer several advantages, and protein detection arrays most commonly utilize antibody capture reagents. The diversity of antibody libraries is virtually limitless, and antibody-antigen interactions are specific and relatively stable. Antibodies can be generated as recombinant proteins (scFv) that can be modified to provide desirable properties, such as incorporation of various tags into their coding sequences. The optimal conditions for antigen-antibody binding are relatively conserved, making it possible to standardize high-throughput screens.

Other analogous capture reagents, such as peptides or nucleic acid aptamers, hold a great deal of promise over conventional antibody-based capture systems. They are relatively easy to customize for directed projects, and they possess enhanced potential for adaptation to high-density surface display formats. For instance, a directed library of small molecules can be used as capture reagents in detection array format to study interactions with specific ligands.

Protein Activity Arrays

Investigators studying the simple eukaryote *S. cerevisiae* made critical contributions to the early development of array-based proteomic analysis. In 1999, nearly all the yeast ORFs were expressed as GST fusions, purified, and used in a variety of enzymatic assays. Using pooling strategies, three unknown ORFs were assigned biochemical activity: a cytochrome *c* methyltransferase, and an enzyme with an Appr-1″-P-processing activity. Although this work was carried out in "microwells" rather than high-density nanowells on flat surfaces, the ability to express and analyze an entire proteome and use it to discover novel activities was a pioneering effort in the development of protein activity arrays.

Due to the difficulties with array customization, protein activity arrays initially contained only specialized classes of proteins. For example, analysis of the *S. cerevisiae* genome identified 122 ORFs encoding putative protein kinases. A nanowell-format microarray containing 119 of these kinases (as GST-fusion proteins) was used to assess autophosphorylation. In a complementary experiment, an array containing 17 different substrates was used to determine enzymatic specificities following addition of the individual kinases and $[\gamma\text{-}^{32}P]ATP$. These arrays performed with remarkable accuracy, verifying known enzyme-substrate combinations. Moreover, evidence of possible tyrosine phosphorylation activity was obtained, even though no sequence signature motifs for tyrosine protein kinases were found in the kinases tested by the arrays. This work therefore demonstrates another important contribution of protein arrays technology. Based on the recognition of tyrosine phosphorylation substrates by enzymes with presumed Ser/Thr specificity, key residues were identified that are shared with enzymes capable of phosphorylating tyrosine residues. Discoveries of unexpected biological activities, based on the use of comprehensive protein function arrays, will undoubtedly be exploited in developing new computational algorithms for predicting protein function from primary sequence.

The first proteome chip was constructed in 2001, and it contained approximately 5,800 yeast open reading frames (93.5% of the total) that were cloned and expressed as hybrid proteins carrying amino-terminal GST-hexahistidine tags. The GST tag was used to purify the recombinant proteins, while the hexahistidine tag allowed immobilization on nickel-coated glass slides. This array was used in several biological assays, which identified new calmodulin- and phosphatidylinositol-binding proteins. Consensus sequences resembling those found in other calmodulin-binding proteins were identified in 14 of the 39 proteins that bound calmodulin on the microarrays, and the remaining 25 proteins may contain new calmodulin-binding sequences or structural motifs.

Membrane proteins present a true challenge for the design and use of protein arrays, because the maintenance of active conformation requires their placement into a hydrophobic lipid environment. Nevertheless, there have been a few reports of successful parallel analysis of membrane protein function. G-protein-coupled receptors represent a large family of membrane proteins in eukaryotic cells, and they are involved in the transmission of a variety of cellular responses. They also represent a group of potential drug targets, which provides a further incentive to develop high-throughput microarray formats for analyzing their function and for screening compound libraries for modulators of their function. Such membrane arrays were used to demonstrate binding of a number of known ligands, with affinities comparable to those demonstrated by other methods that do not use immobilized receptors.

One of the more useful applications of protein arrays is in serodiagnosis of infectious disease. A simple antigen protein array, containing features based on proteins from *Toxoplasma gondii*, cytomegalovirus, herpes simplex virus, and rubella virus, was shown to detect immunoglobulins in human sera with sensitivity and specificity comparable to conventional enzyme-linked immunosorbent assay methods. Antigen arrays based on the entire proteome of a particular pathogen may soon be used to scan sera from infected patients, thus providing the key tool for obtaining a global understanding of host response to infection. Moreover, correlation of the immune response (or lack thereof) with the clinical condition of individual patients, may serve as an important guide for identifying vaccine candidates. In this regard, the recent construction of a carbohydrate microarray is significant, since this technology may allow production of microarrays that contain both proteins and carbohydrates that are present on the surfaces of pathogenic microorganisms, and that are often the primary molecules recognized by the host during infection. In fact, these studies have shown that antibodies to specific bacterial polysaccharides can be detected in human sera with a significant level of sensitivity. Since the basic manufacturing platform for the carbohydrate array was similar to those used for many protein arrays (nitrocellulose-coated glass slides), it is conceivable that future proteome arrays will be supplemented with glycosylated proteins and polysaccharides, thus allowing monitoring of the host response to a truly comprehensive set of potential antigens expressed by pathogens.

Conclusion

Biological science is being redefined by the availability of vast amounts of genomic sequence information and the application of associated technological and computational tools. It is obvious that we are only at the beginning of a new era in biology, in which it is possible to study biological questions in the context of whole cells, whole organisms, or microbial communities such as biofilms. These approaches are already yielding many exciting discoveries that were unattainable through traditional approaches.

Microbial systems will continue to play an important role in the system-wide approaches to biology for several reasons. The ability of microbes to provide large quantities of relatively uniform cell material may be necessary for the development or use of various functional genomics tools. Microbial genetic technologies are also well established and ideally suited for validating hypotheses generated by system-wide approaches. Finally, the combination of systems biology and pathogenic microbiology in particular could have far-reaching impact. The development of new antimicrobial therapeutics or vaccines has been relatively stagnant during the past two decades, but it would not be surprising to see a significant reversal of this trend in response to seminal discoveries regarding the activities of pathogens during interaction with their hosts.

Selected Readings

DNA Array Profiling

Clements, M. O., E. Eriksson, A. Thompson, S. Lucchini, J. C. D. Hinton, S. Normark, and M. Rhen. 2002. Polynucleotide phosphorylase is a global regulator of virulence and persistency in *Salmonella enterica*. *Proc. Natl. Acad. Sci. USA* **99:**8784–8789.

Wolfgang, M. C., V. T. Lee, M. E. Gilmore, and S. Lory. 2003. Coordinate regulation of bacterial virulence genes by a novel adenylate cyclase-dependent signaling pathway. *Dev. Cell.* **4:**253–263.

> Transcriptional profiling of *Pseudomonas* was used to identify genes that are coregulated with the genes encoding the type III secretion system of this organism. This led to the discovery of a second adenylate cyclase-cAMP signaling mechanism, which is primarily dedicated to regulating the expression of virulence factors. The paper illustrates the systematic approach, starting with the generation of a transcriptome, toward identifying genes that can be disrupted and defining the hierarchical organization of a regulatory network.

In Vivo Profiling

Merrell, D. S., S. M. Butler, F. Qadri, N. A. Dolganov, A. Alam, M. B. Cohen, S. B. Calderwood, G. K. Schoolnik, and A. Camilli. 2002. Host-induced epidemic spread of the cholera bacterium. *Nature* **417**:642–645.

Transcriptional profiling of *Vibrio cholerae* bacteria shed from human cholera patients identified genes that are differentially expressed in stool compared with growth in laboratory medium. The results indicate that vibrios in the human intestine experience oxygen-, iron-, and nutrient-deprived conditions and represent a starting point for investigating the authors' observation that human-shed bacteria appear to have enhanced infectivity in a mouse model.

Okinaka, Y., C. H. Yang, N. T. Perna, and N. T. Keen. 2002. Microarray profiling of *Erwinia chrysanthemi* 3937 genes that are regulated during plant infection. *Mol. Plant Microbe Interact.* **15**:619–629.

Comparative Genomics

Behr, M. A., M. A. Wilson, W. P. Gill, H. Salamon, G. K. Schoolnik, S. Rane, and P. M. Small. 1999. Comparative genomics of BCG vaccines by whole-genome DNA microarray. *Science* **284**:1520–1523.

Dziejman, M., E. Balon, D. Boyd, C. M. Fraser, J. F. Heidelberg, and J. J. Mekalanos. 2002. Comparative genomic analysis of *Vibrio cholerae*: genes that correlate with cholera endemic and pandemic disease. *Proc. Natl. Acad. Sci. USA* **99**:1556–1561.

Israel, D. A., N. Salama, U. Krishna, U. M. Rieger, J. C. Atherton, S. Falkow, and R. M. J. Peek. 2001. *Helicobacter pylori* genetic diversity within the gastric niche of a single human host. *Proc. Natl. Acad. Sci. USA* **98**:14625–14630.

This paper describes analysis of the gene content of a clonal isolate of *Helicobacter pylori* that was used for sequence analysis of the construction of a DNA microarray. The authors show several unexpected features of genome evolution in vivo, including deletion of genes from the chromosome and acquisition of exogenous genetic material by horizontal gene transfer.

Other DNA Array Technologies

Laub, M. T., S. L. Chen, L. Shapiro, and H. H. McAdams. 2002. Genes directly controlled by CtrA, a master regulator of the *Caulobacter* cell cycle. *Proc. Natl. Acad. Sci. USA* **99**:4632–4637.

Ren, B., F. Robert, J. J. Wyrick, O. Aparicio, E. G. Jennings, I. Simon, J. Zeitlinger, J. Schreiber, N. Hannett, E. Kanin, T. L. Volkert, C. J. Wilson, S. P. Bell, and R. A. Young. 2000. Genome-wide location and function of DNA binding proteins. *Science* **290**:2306–2309.

The authors develop a technique for using DNA microarrays to identify regulatory regions on the chromosome. The method utilizes immunoisolation of transcriptional regulators, reversibly cross-linked to their cognate regulatory site. This method in conjunction with transcriptional profiling allows discrimination between direct and indirect effects of perturbation of regulatory networks.

Sassetti, C. M., D. H. Boyd, and E. J. Rubin. 2001. Comprehensive identification of conditionally essential genes in mycobacteria. *Proc. Natl. Acad. Sci. USA* **98**:12712–12717.

This paper describes the development and use of transposon site hybridization (TraSH) to conduct genome-wide identification of conditionally essential genes in mycobacteria. Pools of transposon mutants are subjected to different growth conditions, and microarray hybridization is subsequently used to determine the presence or absence of specific insertion mutants in different pools. This rapid and simultaneous analysis of extremely large libraries of transposon mutants permits high-throughput identification of genes that are essential for survival under specified conditions.

Proteomic Studies

Colland, F., J. C. Rain, P. Gounon, A. Labigne, P. Legrain, and H. De Reuse. 2001. Identification of the *Helicobacter pylori* anti-sigma 28 factor. *Mol. Microbiol.* **41**:477–487.

In many bacteria, a specialized sigma factor (sigma 28) is required for transcription of the flagellin genes, which encode the major component of the flagellar filament. This paper describes the use of a yeast two-hybrid screen to identify proteins that interact with sigma 28, including an unknown protein that binds to sigma 28 and antagonizes its action. Transcription of the flagellin gene is greatly stimulated in mutants lacking this anti-sigma factor. The two-hybrid screen defined the interactive domains within sigma 28 and the anti-sigma factor.

Gavin, A. C., M. Bosche, R. Krause, P. Grandi, M. Marzioch, A. Bauer, J. Schultz, J. M. Rick, A. M. Michon, C. M. Cruciat, M. Remor, C. Hofert, M. Schelder, M. Brajenovic, H. Ruffner, A. Merino, K. Klein, M. Hudak, D. Dickson, T. Rudi, V. Gnau, A. Bauch, S. Bastuck, B. Huhse, C. Leutwein, M. A. Heurtier, R. R. Copley, A. Edelmann, E. Querfurth, V. Rybin, G. Drewes, M. Raida, T. Bouwmeester, P. Bork, B. Seraphin, B. Kuster, G. Neubauer, and G. Superti-Furga. 2002. Functional organization of the yeast proteome by systematic analysis of protein complexes. *Nature* **415:**141–147.

Ho, Y., A. Gruhler, A. Heilbut, G. Bader, L. Moore, S. Adams, A. Millar, P. Taylor, K. Bennett, K. Boutilier, L. Yang, C. Wolting, I. Donaldson, S. Schandorff, J. Shewnarane, M. Vo, J. Taggart, M. Goudreault, B. Muskat, C. Alfarano, D. Dewar, Z. Lin, K. Michalickova, A. Willems, H. Sassi, P. Nielsen, K. Rasmussen, J. Andersen, L. Johansen, L. Hansen, H. Jespersen, A. Podtelejnikov, E. Nielsen, J. Crawford, V. Poulsen, B. Sorensen, J. Matthiesen, R. Hendrickson, F. Gleeson, T. Pawson, M. Moran, D. Durocher, M. Mann, C. Hogue, D. Figeys, and M. Tyers. 2002. Systematic identification of protein complexes in *Saccharomyces cerevisiae* by mass spectrometry. *Nature* **415:**180–183.

Ideker, T., V. Thorsson, J. Ranish, R. Christmas, J. Buhler, J. Eng, R. Bumgarner, D. Goodlett, R. Aebersold, and L. Hood. 2001. Integrated genomic and proteomic analyses of a systematically perturbed metabolic network. *Science* **292:**929–934.

Ito, T., T. Chiba, R. Ozawa, M. Yoshida, M. Hattori, and Y. Sakaki. 2001. A comprehensive two-hybrid analysis to explore the yeast protein interactome. *Proc. Natl. Acad. Sci. USA* **98:**4569–4574.

Marino-Ramirez, L., and J. C. Hu. 2002. Isolation and mapping of self-assembling protein domains encoded by the *Saccharomyces cerevisiae* genome using lambda repressor fusions. *Yeast* **19:**641–650.

Uetz, P., L. Giot, G. Cagney, T. A. Mansfield, R. S. Judson, J. R. Knight, D. Lockshon, V. Narayan, M. Srinivasan, P. Pochart, A. Qureshi-Emili, Y. Li, B. Godwin, D. Conover, T. Kalbfleisch, G. Vijayadamodar, M. Yang, M. Johnston, S. Fields, and J. Rothberg. 2000. A comprehensive analysis of protein-protein interactions in *Saccharomyces cerevisiae. Nature* **403:**623–627.

ICAT

Guina, T., S. O. Purvine, E. C. Yi, J. Eng, D. R. Goodlett, R. Aebersold, and S. I. Miller. 2003. Quantitative proteomic analysis indicates increased synthesis of a quinolone by *Pseudomonas aeruginosa* isolates from cystic fibrosis airways. *Proc. Natl. Acad. Sci. USA* **100:**2771–2776.

Gygi, S. P., B. Rist, S. A. Gerber, F. Turecek, M. H. Gelb, and R. Aebersold. 1999. Quantitative analysis of complex protein mixtures using isotope-coded affinity tags. *Nat. Biotechnol.* **17:**994–999.

This paper describes one of the most powerful proteomic technologies for the detection and quantification of proteins: isotope-coded affinity-labeling technology (ICAT). The method combines differential labeling of proteins at cysteine residues with covalent tags carrying both density and affinity tags, followed by mass spectrometric identification of the sequences and quantities of the labeled peptides. ICAT technology was used in this paper to demonstrate changes in the expression of glucose-regulated proteins in yeast.

Protein Arrays

Houseman, B., J. Huh, S. Kron, and M. Mrksich. 2002. Peptide chips for the quantitative evaluation of protein kinase activity. *Nat. Biotechnol.* **20:**270–274.

Zhu, H., M. Bilgin, R. Bangham, D. Hall, A. Casamayor, P. Bertone, N. Lan, R. Jansen, S. Bidlingmaier, T. Houfek, T. Mitchell, P. Miller, R. Dean, M. Gerstein, and M. Snyder. 2001. Global analysis of protein activities using proteome chips. *Science* **293:**2101–2105.

New Technologies

Seetharaman, S., M. Zivarts, N. Sudarsan, and R. R. Breaker. 2001. Immobilized RNA switches for the analysis of complex chemical and biological mixtures. *Nat. Biotechnol.* **19**:336–341.

Wang, D., S. Liu, B. J. Trummer, C. Deng, and A. Wang. 2002. Carbohydrate microarrays for the recognition of cross-reactive molecular markers of microbes and host cells. *Nat. Biotechnol.* **20**:275–281.

22

Cell Biology of Virus Infection

Mark Marsh

Viruses are transmissible genetic elements that have the capacity to replicate within the confines of a cell. Most viruses exhibit a cell-free stage in their life cycle. It is during this phase that their nucleic acid can be transferred from an infected to an uninfected organism or cell, and characteristic virus particles can be isolated. These particles exhibit no metabolic activity and are incapable of replicating themselves. Only within a cell do viruses exert their activities and capacity for self-replication.

Viruses range in complexity from relatively simple agents, such as alphaviruses, which encode just 6 to 7 proteins in a small (<10 kb) positive-sense RNA genome, to herpes- and poxviruses, which contain large DNA genomes (approximately 150 and 200 kb, respectively) encoding on the order of 160 and >200 proteins, respectively. Regardless of their complexity, all viruses rely on the host cell to provide the biochemical machineries essential for their replication. They rely on cells to provide receptors and mechanisms for entry, mechanisms for ensuring delivery to a necessary location in the cell, and machineries for replication, assembly, and release of progeny virions. Different viruses use different strategies for replication and exploit different sets of cellular functions to achieve these aims. Indeed the ability of different viruses to use distinct cellular machineries has made them extraordinary tools for studying basic aspects of cell function.

This chapter focuses on events in viral replication cycles that have similarity or relevance to the interactions of bacterial pathogens with cells, and the examples discussed will draw mainly on membrane-containing enveloped viruses. A text such as *Fields' Virology* (2001) should be read for a more complete account of viral replication strategies.

Virus Organization and Structure

The genetic information in viruses can be carried as either RNA or DNA. RNA genomes can be positive or negative sense or, in retroviruses, serve as templates for reverse transcription to DNA (Table 22.1). Because eukaryotic cells do not encode machineries for replicating RNA, RNA-containing viruses must carry this machinery with them, either as enzymes or encoded

517

in their genome. Thus, for retroviruses, the reverse transcriptase carried in the virus particle generates a DNA transcript of the viral RNA soon after the virus has entered a new host cell. Alphaviruses, with positive-sense RNA genomes, encode an RNA polymerase, while rhabdoviruses, with negative-sense genomes, carry an RNA polymerase in their particles to generate a positive-sense message early after entry into a new host cell.

Viral families are defined through similarities in their genetic organization and sequences. Groups of related viruses frequently show similarity in particle structure and morphology and in many aspects of their replication and properties within an infected cell. For all viruses the nature of the particle that is formed for the cell-free stage of the life cycle, however brief, is crucial. Viruses have evolved two basic strategies for generating coats that protect the viral genome during these stages. These coats must be robust enough to resist a range of environmental insults, but sufficiently labile to allow release of the viral nucleic acid during infection of a cell.

One type of coat is found on enveloped viruses (Table 22.1). The envelope is a bilayer lipid membrane that surrounds the core containing the viral nucleic acid and associated proteins. The membrane is acquired when the core, which may itself assemble on the cytoplasmic surface of a specific membrane system, progressively wraps membrane around itself until it undergoes scission to release a completely enveloped particle with a nonleaky membrane coat.

The second type of coat is a protein shell. These coats are often icosahedral in organization and are assembled from protein subunits encoded by the virus. The capsids are assembled within the cytoplasm or nucleoplasm of an infected cell from which they must be released, often by mechanisms involving lysis of the infected cell. By contrast to enveloped viruses, these viruses are often known as nonenveloped viruses.

Some viruses incorporate both types of coats in their particles. Herpesviruses, for example, produce DNA-containing capsids in the nucleus of

Table 22.1 Genetic composition and coat structures of different viruses

Genetic composition	Segmentation	Coat structure	Virus(es)
DNA viruses			
Double-stranded		Enveloped	Herpes-, baculo-, pox-
		Nonenveloped	Adeno-, papova-, polyoma
Single-stranded		Nonenveloped	Parvo-
Double-stranded with single-stranded intermediate		Enveloped	Hepadna-
RNA viruses			
Double-stranded positive sense	Nonsegmented	Enveloped	Cysto-
	Segmented	Nonenveloped	Birna-, reo-
Double-stranded positive sense with DNA intermediate[a]		Enveloped	Retro-
Single-stranded positive sense	Nonsegmented	Enveloped	Corona-, flavi-, toga-
		Nonenveloped	Picorna-, astro-, calci-
Single-stranded negative sense	Segmented	Enveloped	Orthomyxo-
	Nonsegmented	Enveloped	Rhabdo-, filo-, paramyxo-

[a] RNA serves as a template for reverse transcription and not as mRNA.

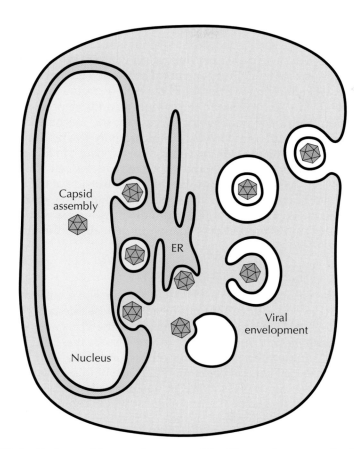

Figure 22.1   Cartoon of herpesvirus assembly. Herpesviruses use both protein coats and lipid membranes to generate free virus particles. Capsids assemble in the nucleus of the infected cell and bud through the inner nuclear membrane to generate a transient enveloped particle in the space between the inner and outer nuclear membranes (this space is continuous with the lumen of the ER). These enveloped particles are then believed to fuse with the outer nuclear/ER membrane to deliver the capsid to the cytoplasm. Subsequently the capsids acquire virally encoded tegument proteins that are available in the cytoplasm and undergo a second envelopment step into a cytoplasmic membrane-bound compartment. The tegument proteins form a layer between the protein shell of the capsid and the membrane. Subsequently, viruses are released from the cell when the virus-containing vesicles fuse with the plasma membrane.

an infected cell. These nucleocapsids then appear to translocate to the cytoplasm using a mechanism that involves envelopment at the inner nuclear membrane, a transient enveloped particle in the intermembrane space of the nuclear membrane (a compartment that is continuous with the lumen of the ER) and subsequent fusion with the outer nuclear or ER membrane. In the cytoplasm, these capsids undergo a second envelopment process by budding into membrane-bound cytoplasmic organelles. Viruses are released from cells when these compartments fuse with the plasma membrane (Figure 22.1).

Entry

In many respects it is in their entry mechanisms that viruses most resemble certain bacterial and cellular pathogens. Because their replication depends on processes taking place in the cytoplasm and/or nucleoplasm of a cell, all viruses must cross the barrier imposed by the plasma membrane. To this end viruses have evolved a variety of different entry strategies. These can be broken down into two main principles that reflect the structure of the virus. All enveloped viruses enter cells by membrane fusion, i.e., the virus fuses its membrane with a limiting membrane of the cell (see below). This fusion delivers the viral core or capsid into the cytoplasm and the processes leading to replication begin. For nonenveloped viruses, the mechanisms of entry vary somewhat for different viruses. For picornaviruses, binding of virions to specific receptors on the cell surface and/or exposure to low pH can unlock the capsid in such a way that the RNA can cross the cellular membrane to the cytoplasm, leaving an empty capsid on the outside of the membrane. By contrast, adenovirus entry involves lysis of the endosome membrane and release of the capsid into the cytosol. Subsequently, the capsid undergoes a stepwise disassembly process culminating in delivery of the viral DNA to the nucleus.

Receptors

As with many pathogens, the first event in the entry of a virus into a target cell is binding to a specific cell surface receptor. These interactions occur between binding sites displayed on the surface of the virus and components of the target cell plasma membrane. For some viruses these interactions appear relatively straightforward, but for others binding is complex and may involve multiple cell surface components. These binding interactions may just recruit viruses to the cell surface, or they may play a major role in penetration. Binding can also be the first stage at which cellular tropism is determined, i.e., permissive infection may be restricted to cells expressing a specific receptor(s). In some cases interaction with cell surface receptors may not lead to infection but may still play important roles in viral pathogenesis. Table 22.2 provides examples of some of the cellular components identified as virus receptors.

Initial Cell Surface Interactions

Many viruses are initially recruited to the surface of cells through interactions with cell surface proteins, proteoglycans, or other moieties. These interactions themselves do not lead to penetration or productive infection, but appear to serve primarily as a means of presenting virus to specific entry receptors. For example, herpesviruses (herpes simplex virus [HSV-1] and cytomegaloviruses), as well as a number of adenoviruses and the adeno-associated virus-2, bind glycosaminoglycans. The human immunodeficiency virus type 1 (HIV-1) appears to bind proteoglycans prior to interaction with CD4. Viruses such as HIV incorporate cellular proteins into their membrane during particle formation (see below); these can include adhesion proteins and ligands for various cell surface proteins. Again, interaction between these proteins on the virus particle and the cognate cell surface receptor may facilitate HIV attachment to target cells. Antibodies or other ligands that bind at least some of these proteins can inhibit virus entry. Although in many cases these interactions may lack specificity, some can be highly specific. For example, adenoviruses, types 2 and 5 (Ad2 and Ad5), bind to the coxsackie virus/adenovirus receptor (CAR) with high affinity prior to interaction with integrins. The extent to which these types of interactions are used by different viruses is unclear, but it is likely that low-affinity, somewhat nonspecific interactions are a common means for viruses to gain an initial foothold on the surface of a potential host cell.

Table 22.2   Examples of cellular receptors used by viruses

Virus family	Virus	Receptor
Enveloped viruses		
Orthomyxo-	Influenza virus	Sialic acid (glycoproteins/glycolipids)
Paramyxo-	Sendai virus	Sialic acid (glycoproteins/glycolipids)
	Measles virus	CD150 (all virus strains), CD46 (vaccine strains)
Retro-	HIV-1, HIV-2, SIV	CD4 + CCR5 or CXCR4 (DC-SIGN, DC-SIGN-R)
	Mouse mammary tumor virus	Transferrin receptor
	Murine leukemia virus (MLV-E)	Cationic amino acid transporter
	Gibbon ape leukemia virus, Feline leukemia virus, MLV-A	Phosphate transporters
Flavi-	Dengue virus	DC-SIGN and/or Fc receptors via IgG
Herpes-	Epstein-Barr virus	Complement receptor CR2
	Herpes simplex virus	Heparan sulfate proteoglycans and Nectin 1
	Human herpes virus 6	CD46
Rhabdo-	Rabies virus	Acetylcholine receptor
Corona-	Human coronavirus 229E (and some other coronaviruses including the procine TGEV)	CD13 (aminopeptidase N)
	Murine hepatitis virus	Murine biliary glycoprotein (Bgp1a, also called CEACAM1)
Nonenveloped viruses		
Adeno-	Adenovirus type-2	CAR + $\alpha_v\beta_3$ and $\alpha_v\beta_5$ integrins
Picorna-	Major group human rhinoviruses	ICAM 1
	Minor group human rhinoviruses	Low-density lipoprotein-related protein
	Foot-and-mouth disease virus	$\alpha_v\beta_1$, $\alpha_v\beta_3$, and $\alpha_v\beta_6$ integrins
	Coxsackie virus	CAR
Parvo-	Canine parvovirus	Transferrin receptor
	Feline panleukopenia virus	Transferrin receptor
Polyoma-	SV40	MHC I, gangliosides

Specific Receptor Binding

One of the most straightforward and well-understood virus receptor systems is that used by the orthomyxoviruses, which include influenza viruses. Influenza virions carry three virally encoded membrane proteins. One is the hemagglutinin (HA); the others are NA and M2 (see below). HA is remarkably well understood. It was one of the first membrane proteins to be fully sequenced and the first for which a high-resolution X-ray crystal structure was solved (Box 22.1). HA has two important roles in entry. It mediates (i) binding to cell surface receptors and (ii) membrane fusion. Influenza viruses bind to sialic acid (neuraminic acid [NANA]) carried on the terminal positions of complex N-linked oligosaccharides of membrane glycoproteins and on the glycosyl moieties of glycolipids. These negatively charged sugar moieties bind into a cavity located at the tip of the HA molecule (HA is a trimer, so there are three NANA-binding sites on each HA protein). Membrane fusion occurs following receptor-mediated endocytosis of intact virions into the cell (see below). Thus, the principal role of receptor binding is to recruit virions to the surface of a cell prior to their endocytosis. A second protein in the viral envelope is the neuraminidase (NA), which is able to remove NANA from oligosaccharides. NA plays a role in releasing virus from infected cells but may also function in entry. Influenza virus cannot discriminate NANA on en-

BOX 22.1

Selected landmark papers

1. **Helenius, A., J. Kartenbeck, K. Simons, and E. Fries.** 1980. On the entry of Semliki forest virus into BHK-21 cells. *J. Cell Biol.* **84:**404–420.
This paper first indicated how enveloped viruses internalized by clathrin-mediated endocytosis could penetrate cells by low pH-induced fusion of the viral membrane with membranes of endocytic organelles.
2. **Wilson, I. A., J. J. Skehel, and D. C. Wiley.** 1981. Structure of the haemagglutinin membrane glycoprotein of influenza virus at 3 Å resolution. *Nature* **289:**366–373.
This paper provided the first detailed insights into the structure and mode of action of a viral envelope protein involved in receptor binding and fusion. This paper together with that of Carr and Kim (*Cell* **73:**823–832, 1993) provided the first clear insights into

the molecular basis of viral membrane fusion.
3. **Pelkmans, L., J. Kartenbeck, and A. Helenius.** 2001. Caveolar endocytosis of simian virus 40 reveals a new two-step vesicular-transport pathway to the ER. *Nat. Cell Biol.* **3:**473–483.
This paper demonstrated how a non-enveloped virus, SV40, can use a lipid-raft-dependent mode of endocytosis to gain entry to cells. This paper was crucial in demonstrating that some viruses require an endocytic mechanism distinct from clathrin-mediated endocytosis for entry and productive infection. In addition it identified a novel cellular organelle now termed caveosomes.
4. **Garrus, J. E., U. K. von Schwedler, O. W. Pornillos, S. G. Morham, K. H. Zavitz, H. E. Wang, D. A. Wettstein, K. M. Stray, M. Cote, R. L. Rich, D. G. Myszka, and W. I. Sundquist.** 2001. Tsg101 and the

vacuolar protein sorting pathway are essential for HIV-1 budding. *Cell* **107:**55–65.
This paper was one of several showing a role for cellular ESCRT proteins in retroviral assembly. These proteins are normally involved in the topologically similar budding of membrane vesicles and sorting to lysosomes, but they are recruited by various enveloped viruses during assembly and release.
5. **Maddon, P. J., A. G. Dalgleish, J. S. McDougal, P. R. Clapham, R. A. Weiss, and R. Axel.** 1986. The T4 gene encodes the AIDS virus receptor and is expressed in the immune system and the brain. *Cell* **47**(3):333–348.
One of the first studies to use expression methods to identify and demonstrate a functional virus receptor. Related strategies have been used to identify key receptors for a range of different viruses.

docytically active cells from NANA on cell surface components that do not undergo endocytosis. NA can release virions from "dead-end" binding that does not result in endocytosis and productive infection.

Many enveloped and nonenveloped viruses use endocytosis and low pH-dependent mechanisms to enter cells (see below). Mouse mammary tumor virus (MMTV) is one of just a few retroviruses that appear to use this pathway. MMTV uses transferrin receptors for entry. The transferrin receptor is a well-studied cellular protein that undergoes constitutive endocytosis through clathrin-coated vesicles (CCVs). Although it seems appropriate for MMTV to use an endocytically active receptor, there is no general trend for viruses that rely on endocytosis for infection to use receptors that are known to undergo efficient endocytosis.

A key event in viral entry is penetration, i.e., transfer of the viral genetic material to the cytoplasm of the target cell. Low pH can provide a trigger for these events for many enveloped and nonenveloped viruses, but it is not a universal trigger. Many viruses do not require exposure to acidic conditions for entry. For these so-called pH-independent viruses alternate triggers exist. In some cases receptor binding can act as the trigger. A well-characterized example is HIV-1. CD4 (a coreceptor to the T-cell antigen receptor [TcR] that is expressed on subsets of T lymphocytes and monocytic cells) is a key component for productive infection by HIV. However, CD4 alone is not sufficient. Members of a subfamily of heptahelical G-protein-coupled receptors (GPCRs) for small chemotactic peptides called chemokines act together with CD4 to mediate HIV fusion. Currently, two such molecules, the CC chemokine receptor 5 (CCR5) and the CXC chemokine receptor 4 (CXCR4), have been linked to HIV infection and pathogenesis (note: these receptors are related to the Duffy antigen that functions as a cellular receptor for malaria). Of these, CCR5 is required in most cases for initial infection of individuals. Later in the course of the disease viruses with a tropism for CXCR4 can emerge in infected persons. Current models suggest that during entry, the HIV envelope glycoprotein (Env) binds the N-terminal immunoglobulin domain of CD4. This interaction initiates conformational changes that display a site in Env that binds either CCR5 or CXCR4 (or both in some cases of dual tropism). The second interaction induces further conformational changes in Env that initiate the events leading to fusion of the viral and target cell membranes (see below).

Receptor binding also plays significant roles in the penetration of a number of nonenveloped viruses. For these viruses the mechanisms involved in penetration are less well understood. Picornaviruses are a large family of small nonenveloped RNA viruses that includes the agents that cause common cold, poliomyelitis, and foot-and-mouth disease. One model for the entry of these viruses proposes that the viral RNA escapes from the capsid and is transferred across the membrane, leaving an empty capsid structure outside the cell. For polio and at least some other picornaviruses, binding of the receptor into a depression on the surface of the virion initiates some opening of the capsid that allows the RNA to escape. For other picornaviruses, this process may be enhanced by, or even fully dependent on, exposure to low pH. Thus, with both enveloped and nonenveloped viruses, receptor engagement can initiate the events leading to penetration and productive infection.

Signaling through Receptors
A number of viruses bind to receptors capable of signal transduction (Table 22.2). HIV, for example, uses the GPCRs CCR5 and CXCR4, together with CD4 (which in T cells is coupled to the nonreceptor tyrosine kinase p56^{Lck}). Binding of HIV-1 Env CCR5 can activate G-protein-mediated signal transduction, calcium mobilization, activation of the tyrosine kinases pyk2 and FAK, and chemotaxis. While these activities are not required for virus fusion per se, a role in enhancing postfusion events may exist. HIV also activates extracellular signal-regulated kinase (ERK) through CD4/p56^{Lck} together with the MEK kinase Raf-1 and the coreceptors CXCR4 and CCR5 (in some cases).

The entry of Ad2 depends on signaling. Ad2 initially binds CAR, but must engage the integrins $\alpha_v\beta_3$ and $\alpha_v\beta_5$ to mediate internalization through CCVs. Binding of Ad2 to cells activates phosphatidylinositol-3 kinase (PI3K) and the Rho GTPase downstream effectors of PI3K, Rac, and CDC42, which regulate the actin cytoskeleton. Ad2 binding to integrins causes filopodia formation, the subsequent extension of lamellipodia and macropinocytosis. Macropinocytosis appears to be essential for the release of internalized Ad2 from endosomes. How the clathrin/endosomal and macropinocytic pathways interact is unclear, though it is apparent that agents such as cytochalasin D, which disrupt actin polymerization, PI3K inhibitors, and dominant negative forms of Rac and CDC42, can all inhibit Ad2 internalization and infection. In addition, Ad2 capsids are translocated on microtubules (see below). When Ad2 capsids are injected into cells, they are immobile. Motility can be rescued when virus is added to the cells externally. Two signaling pathways from the cell surface can influence viral motility. One pathway leads to activation of protein kinase A and requires the α_v integrin receptors. A second integrin-independent pathway activates the p38/mitogen-activated protein kinase (MAPK) cascade. Knocking out the MAPK kinase

MKK6 inhibits transport of Ad2 to the nucleus. Thus, Ad2 stimulation of p38 appears to suppress plus end-directed transport.

The polyomavirus simian virus 40 (SV40) is internalized via caveolae (see below). These organelles undergo little constitutive endocytosis and usually remain linked to the plasma membrane. However, they can be induced to internalize by agents that modulate phosphorylation. SV40 can be recruited to cells by major histocompatibility class I (MHC I) antigens, but must transfer to, or also bind, a receptor capable of inducing caveolae internalization. Recent evidence suggests this second receptor is a ganglioside that can couple to tyrosine kinases as tyrosine kinase inhibitors can block SV40 uptake and infection.

Receptors, but Not for Productive Infection

Viruses can also bind to cells and undergo endocytosis without inducing a productive infection. These interactions can be important for viral pathogenesis. HIV Env, for example, can bind C-type lectins expressed on various types of dendritic cells (DCs). In particular interaction with DC-SIGN (DC-specific intercellular-adhesion-molecule [ICAM]-3-grabbing nonintegrin) appears to provide a route through which HIV can be sequestered into endocytic compartments. When DCs that have internalized HIV subsequently undergo maturation and migrate to lymph nodes, these internalized virions can be regurgitated and presented in an infectious form to T cells. In this manner DCs play an important, if not essential, role in disseminating HIV from its initial point of entry to the body, the rectal or vaginal epithelium, to sites rich in CD4-positive target cells. Other viruses, including the human cytomegalovirus (HCMV), Ebola virus, Dengue virus, and hepatitis C virus, can also bind DC-SIGN and/or its liver homologue L-SIGN, although for some of these viruses this interaction may lead to productive infection of the receptor-expressing cell. Interestingly, a number of nonviral pathogens including *Helicobacter pylori, Mycobacterium tuberculosis, Leishmania,* and *Schistosoma mansoni* also bind DC-SIGN.

Endocytosis

Endocytosis is a key feature of the entry of a number of viruses (Figure 22.2), being required to deliver pH-dependent viruses to an acidic environment. However, use of endocytic pathways may not be restricted to pH-dependent viruses. Viruses that use other triggers for penetration may also use endocytic mechanisms. Endocytosis can occur by several distinct biochemical mechanisms, each of which has unique properties and capabilities. Some of these processes, such as clathrin-mediated uptake and phagocytosis, are well understood. Others are poorly appreciated, in terms of both mechanism and functional importance. Nevertheless, there is evidence that some viruses have evolved to exploit these different pathways for entry. Although endocytosis may bring virions within the confines of the plasma membrane, a virus within an endocytic vesicle is still topologically outside the cell. The genome of an enveloped virus, for example, is still separated from the cytoplasm by its own and the vesicle membranes, and membrane fusion is still required to bring the genome into the cytoplasm.

Clathrin-Mediated Endocytosis

Clathrin-mediated endocytosis is one of the best-characterized endocytic mechanisms. CCVs are distinguished morphologically by their appearance in electron micrographs, being ~100 nm in diameter and coated on their cy-

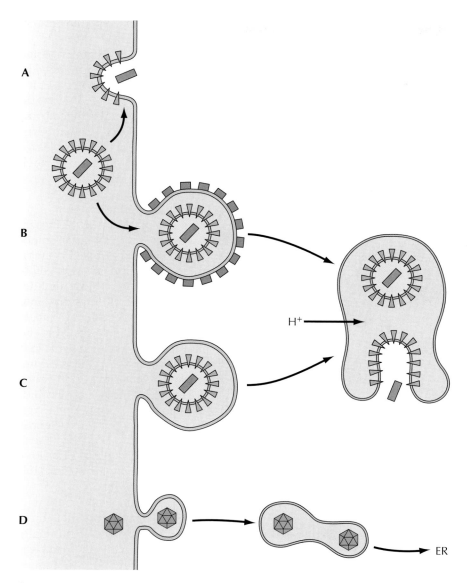

Figure 22.2 Virus entry. Entry mechanisms for animal viruses. **(A)** Direct fusion or penetration at the plasma membrane, seen for many pH-independent viruses, e.g., HIV. **(B)** Endocytosis of viruses in clathrin-coated vesicles and fusion or penetration from endosomes as seen for many pH-dependent viruses, e.g., SFV. **(C)** Endocytosis in non-clathrin-coated vesicles that also leads to delivery to endosomes and provides an alternative route for some pH-dependent viruses, e.g., influenza virus. **(D)** Uptake through caveolae and delivery to caveosomes as observed for SV40.

toplasmic aspect by a bristle coat formed of clathrin and associated proteins. CCVs are formed by invagination of special regions of the plasma membrane called coated pits, where clathrin is recruited to the membrane through interaction with AP2 adapter complexes (and possibly other proteins). Reorganization of the clathrin lattice through a series of intermediate states of shallow pits to deeply invaginated pits drives vesicle formation that is completed when the vesicle is finally pinched off and released into the cytoplasm. Clathrin-mediated endocytosis occurs in virtually all cells (with the exception of nonnucleated red blood cells) and is a highly dynamic process. In some cell lines, CCVs form rapidly (approximately 1 min) at rates of up to 2,000 vesicles/cell/min. CCV formation is constitutive, i.e., it is not receptor induced, though clathrin recruitment to the membrane can be modulated by ligation of some signaling receptors. Clathrin-mediated endocytosis is frequently linked to receptor-mediated endocytosis, the process whereby specific cell surface receptors and their ligands are efficiently internalized by recruitment of the receptor into an endocytic vesicle. While CCVs do mediate the efficient internalization of a number of receptors (the best examples are the receptors for transferrin and low-density lipoprotein, LDL), clathrin-independent endocytic mechanisms also have the ability to internalize certain receptors and their ligands.

CCVs are transport intermediates and their lifetimes are brief (approximately 1 min). Soon after their scission from the plasma membrane, the coats are removed so that the vesicles can fuse with, and deliver their membrane and content to, early endosomes. It is within these organelles that pH changes start to occur. Vacuolar proton ATPase complexes pump protons into the lumen of the organelle, acidifying the environment to about pH 6. Subsequent compartments of the pathway, late endosomes and lysosomes, are progressively more acidic (pH 5.5 and 4.8, respectively). It is these acidic environments that some viruses have learned to exploit so effectively. Because clathrin-mediated endocytosis is constitutive, viruses can use a variety of cell surface moieties for internalization (though specific receptors may be used to bring viruses to other non-clathrin-mediated pathways). Viruses such as MMTV use the transferrin receptor that is recruited efficiently to coated pits. Others, such as the alphavirus Semliki Forest virus (SFV), have poorly defined receptors but are nevertheless internalized efficiently through CCVs. Inhibitors of clathrin-mediated endocytosis, e.g., dominant negative dynamin or Eps15, are efficient inhibitors of SFV infection. SFV undergoes fusion at about pH 6.2 within early endosomes. Other enveloped viruses, e.g., influenza, fuse at more acidic pHs (pH ~5.5) from within late endosomes. Significantly, though morphological experiments show influenza viruses in CCVs, dominant negative inhibitors of clathrin-mediated endocytosis have minimal influence on influenza virus infectivity, suggesting that nonclathrin-dependent pathways can also be used. The extent to which these pathways are used under normal conditions has not been established. One criterion that may determine a virus' ability to use the clathrin-mediated pathway is size. SFV particles are 65 nm in diameter and fit easily into endocytic CCVs. Others, such as the rhabdovirus vesicular stomatitis virus (VSV), which has particles 150 nm long, are a tighter fit and may be at the limit of what the clathrin system can accommodate.

Phagocytosis

Phagocytosis is a second well-studied endocytic mechanism. By contrast with clathrin-mediated endocytosis, phagocytosis is activated by engagement of specific receptors and is dependent on actin remodeling. This process, which is most frequently seen in specialized phagocytic cells of the immune system but can occur to some extent in most cells, has the capacity to internalize large particles, including bacteria, cellular pathogens, and even whole cells. One mechanism through which phagocytosis is induced is through engagement of either Fc receptors (receptors that bind the Fc domains of immunoglobulins) or complement component receptors. Immunoglobulins and complement components are frequently deposited on the surface of invading pathogens as part of the innate and acquired immune responses. This so-called opsinization marks the invader for uptake and destruction by professional phagocytes such as neutrophils and macrophages. Engagement of Fc and complement receptors activates small GTP-binding proteins of the Rho family, leading to remodeling of cortical actin on the membrane underlying the bound pathogen and progressive envelopment of the particle within a membrane vacuole or phagosome. Once formed these vacuoles may fuse with endosomes and/or lysosomes. Although phagocytosis is used as an entry mechanism by many bacterial and cellular pathogens, there are few clear cases in which it is used for viral entry. However, its use by large DNA viruses such as poxviruses cannot be excluded. Nevertheless, phagocytosis plays a key role in the clearance of cells infected by viruses, either following antibody or cell-mediated killing of these cells or after apoptosis of the cell in response to the virus infection. Though not necessarily involved in phagocytic uptake of viruses, Fc receptors and complement component receptors have been implicated in infection by, for example, flavivirus (Dengue virus) and herpesvirus (Epstein-Barr virus [EBV]) and may have a significant role in antibody-enhanced virus infection.

Macropinocytosis

Macropinocytosis is similar to phagocytosis in that it is dependent on remodeling of cortical actin, but it is not dependent on the ligation of specific receptors. Macropinocytic vesicles tend to form when plasma membrane ruffles collapse back onto the membrane, trapping portions of the extracellular medium. As with phagosomes, macropinosomes can be large (on the order of 1 μm in diameter) and can mediate delivery of their membrane and content to endosomes and lysosomes. Macropinocytic activity is seen in antigen-presenting cells (APCs), such as immature DCs, where it is thought to provide a means for sampling large volumes of extracellular fluid for antigens. Significantly, macropinocytic activity is decreased when these cells mature and switch from acquiring antigen to presentation. Although not usually seen in other cell types, macropinocytosis can be activated by, for example, phorbol esters or growth factors. A number of viruses, including HIV, have been seen in macropinosomes, but it is not clear that this uptake has a significant role in infection. As discussed above, Ad2 enters cells in CCVs but transiently activates macropinocytosis.

Caveolin-Mediated Endocytosis

Caveolae are small (50-nm)-diameter, flask-shaped plasma membrane invaginations. The protein caveolin is associated with the cytoplasmic aspect of these invaginations, and their integrity depends on the presence of cholesterol in the membrane. Caveolae have been linked to membrane proteins and lipids associated with detergent-resistant, so-called raft domains of the plasma membrane, and to signal transduction. Under unstimulated conditions, caveolae do not exhibit endocytic activity. However, some stimuli can induce their internalization. One such stimulus is the polyomavirus SV40. These small nonenveloped DNA viruses are recruited to the plasma membrane through interaction with MHC I antigens. They subsequently transfer to a second receptor, recently identified as a ganglioside, through which signaling events lead to internalization of virion-containing caveolae. Caveolae deliver their cargo to caveosomes, a cellular compartment that is distinct from endosomes and lysosomes and which is also associated with caveolin. Internalized SV40 virions are subsequently redistributed via tubular extensions of caveosomes and eventually appear in the endoplasmic reticulum (ER) where, by means that remain to be understood, they are released to the cytoplasm. The viral DNA is finally taken into the nucleus. Whether caveolae are involved in the entry of all polyomaviruses is unclear.

Caveolae have also been implicated in the entry of filoviruses. Infection by Ebola virus can be inhibited by cholesterol-depleting agents, such as methyl-β-cyclodextrin, which disrupt caveolae. In addition, a glycophosphatidylinositol (GPI)-linked protein, the folate receptor (which exhibits a tendency to associate with lipid rafts), has been implicated as an Ebola virus receptor. However, it is unclear how these relatively large and morphologically complex viruses fit into the much smaller dimensions of a typical caveolus. Other receptors, such as DC-SIGN, have also been linked to Ebola uptake. Whether this receptor functions in productive infection or fulfills some other role akin to that seen with HIV remains to be determined.

Non-Clathrin-, Non-Caveolin-Mediated Endocytosis

Experiments using dominant negative proteins to inhibit specific endocytic pathways have led to the view that cells possess multiple mechanisms for endocytosis. Dynamin is a GTPase that has been implicated in the scission events required to pinch off or complete the formation of CCVs. Expression of dominant negative dynamin inhibits CCV formation. It also inhibits caveolin-mediated endocytosis and macropinocytosis. When dominant negative dynamin is expressed in a number of different types of cells, fluid phase endocytosis and internalization of some receptors can still be observed. Thus a pathway that is independent of clathrin, caveolin, and dynamin exists. Studies with polyomaviruses also indicate that a raft-dependent endocytic pathway that is independent of caveolin can operate in cells. Little is known of the molecular mechanisms underlying these pathways or the extent of their activity in cells.

Fusion

For enveloped viruses the limiting membrane of the cell, be it the plasma membrane or the endosome membrane that confronts a virus that has undergone endocytosis, is a barrier that must be traversed for infection to proceed. Enveloped viruses have adopted the strategy of membrane fusion to overcome this barrier (Figure 22.2). Given that cellular membranes do not fuse spontaneously, viruses have acquired fusion proteins to facilitate these reactions. Some of these viral fusion proteins share features with the SNARE proteins involved in membrane trafficking in eukaryotic cells. Model systems using enveloped viruses, such as influenza, have provided the best models for understanding membrane fusion in general. A key aspect of these reactions is that fusion is nonleaky, i.e., the integrity of the target cell membrane is maintained throughout the process.

Viral fusion proteins can be grouped into several distinct classes based on the organization of the protein. Class 1 fusion proteins (Figure 22.3), which include influenza HA (see above) and HIV Env, are synthesized as single-chain transmembrane proteins that assemble into trimers in the ER of the infected cell. En route to the cell surface, the ectodomain of each subunit is proteolytically cleaved to generate a protein with three transmembrane domains (HA2 for influenza, TM or gp41 for HIV) and three associated surface units (HA1 for influenza, SU or gp120 for HIV) (Figure 22.3). These subunits can be covalently coupled, as in the case of HA, or noncovalently associated, as in the case of HIV. The proteolytic cleavage has two key

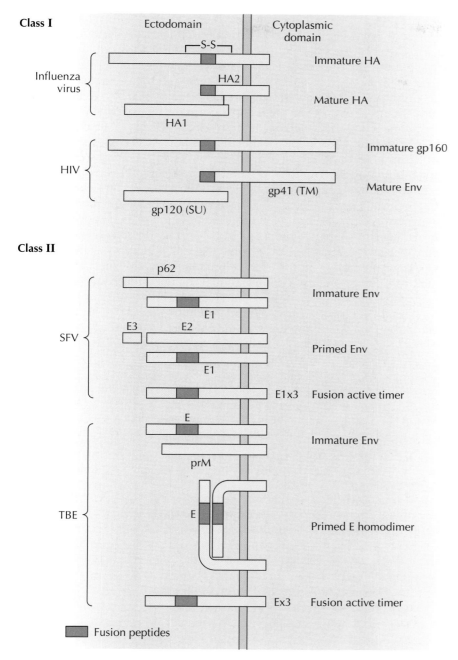

Figure 22.3   Fusion proteins. Cartoon showing the domain organization of viral fusion proteins. The shaded boxes indicate fusion peptides. HIV, human immunodeficiency virus; SU, surface unit; TM, transmembrane; SFV, Semliki Forest virus (*Alphaviridae*); TBE, tick-borne encephalitis virus (*Flaviviridae*).

functions. (i) It generates a novel N-terminal domain on the transmembrane subunits and places a stretch of hydrophobic residues at or close to this new N terminus. This short sequence of hydrophobic residues has come to be known as the "fusion peptide" and is believed to insert into the target membrane during the fusion reaction. (ii) It allows a conformational change in the protein that renders it competent for fusion.

Current understanding of class 1 fusion reactions is based largely on studies of influenza HA. Influenza requires exposure to low pH to trigger fusion. Crystal structures of the neutral and low pH forms of influenza HA have been generated. The neutral pH form shows that HA2 forms a rod-like core supported by a long α-helix. This α-helix is linked to a second α-helix that runs antiparallel to the first. The sequence linking the two helices is an extended loop, but the sequence would normally be expected to form an α-helix (Figure 22.4). At the tip of the second helix, the N-terminal fusion peptide is held in a hydrophobic pocket in the shaft of the core. Overlying this central core are the HA1 subunits. These subunits hold the complex together and act as a clamp to stop the extended loop in HA2 undergoing rearrangement to its more stable α-helical conformation. This neutral pH HA is the metastable, primed form of the protein ready to mediate fusion. Following exposure to low pH, the HA1 subunits dissociate. Removal of the clamp allows the extended loop in HA2 to form an α-helix, generating one long α-helical structure from the two antiparallel helices discussed above. As HA is a trimer, these long helices align to form a triple helical coiled-coil structure. The result of this conformational change is that the fusion peptide

Figure 22.4   pH-induced conformational changes in influenza HA. (1) Mature influenza HA is a trimer of 3 × HA1 + 3 × HA2 subunits (derived from the precursor 3 × HA by proteolytic cleavage). The HA1 subunits contain the sialic acid-binding pockets involved in receptor recognition. (2–4) Following exposure to low pH, the HA1 subunits dissociate (but remain linked to their respective HA2 subunit via and an S–S bond), releasing the clamp that maintains the metastable neutral pH form of HA2. (3) The extended loop in HA2 adopts a helical formation and in so doing moves the N-terminal helical domain (shaded) and fusion peptide (dark gold) to the top of the long α-helix (light gold). (4) Subsequently, the dark gold helix is believed to fold back on the light gold helix to form the six-helix bundle that brings the N and C termini of the protein, which are embedded in the host cell and viral membranes, respectively, close together to drive merger of the two membrane systems.

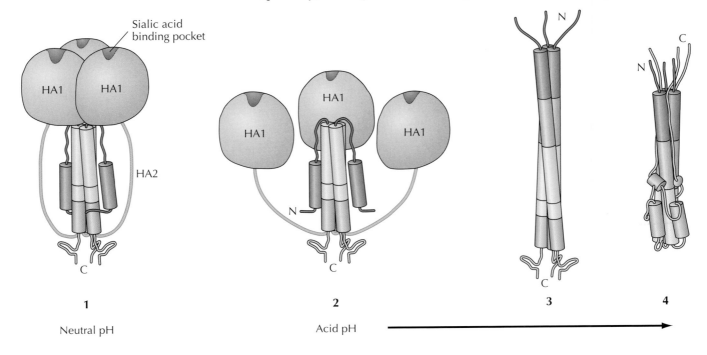

is moved several Ångstroms to the tip of the triple helical rod and is believed to be positioned for interaction with the target membrane. To then bring the viral and target membranes close enough for fusion to occur, the helical rod is believed to fold back to form a six-helix bundle with a third helical domain in HA2.

Other class 1 fusion proteins also appear to form six-helix bundles similar to that proposed for HA2, and there are likely to be similarities in the mechanisms through which these proteins induce fusion. How exactly these changes lead to the formation of a fusion pore, the nature of the fusion pore, and a clear view of how lipid rearrangement and bilayer fusion occur remain to be understood. However, the six-helix bundles seen for class 1 fusion proteins are similar to the structure that is believed to be adopted by SNARE proteins during the fusion of cellular transport vesicles with their target organelles, suggesting that the basic mechanism might be similar.

In contrast to the class 1 proteins, class 2 fusion proteins contain internal fusion peptides, i.e., these peptides are not at, or close to, the N terminus of the protein. The fusion protein itself does not undergo an activating cleavage, although an associated "chaperoning" protein may. Two well-characterized examples of class 2 fusion proteins are the flavivirus tick-borne encephalitis (TBE) virus and SFV. For these two viruses there is evidence that the fusion proteins (E and E1, respectively) exist as a homodimer and heterodimer (with E2), respectively, at neutral pH. During low pH activation, the proteins undergo a dramatic reorganization that results in them forming homotrimeric spikes that project from the virus and interact with the target membrane. How these trimeric proteins then induce membrane fusion is unclear. There is evidence that the fusion proteins must interact with sphingomyelin and/or cholesterol in the target membrane. In addition, emerging structural data are suggesting that a "fold-back"-type mechanism proposed for class 1 fusion proteins also operates for class 2 proteins.

A number of other viruses contain fusion proteins that have not yet been designated to either of these classes. Rhabdoviruses, for example, contain a single noncleaved envelope glycoprotein (G), and herpesviruses appear to use multiple envelope proteins to mediate fusion. Whether these represent fusion protein with mechanistically distinct fusion properties remains unclear.

A Role for Raft Domains?

Lipid rafts are microdomains in the plasma membrane and other membranes of the vacuolar apparatus that are enriched in cholesterol, sphingolipids, and glycolipids. These domains have been implicated in sorting, signaling, and a variety of other membrane functions. These domains also have the capacity to concentrate specific types of membrane protein, such as GPI-linked membrane proteins, as well as certain acylated transmembrane proteins. The integrity and activities of raft domains can be dissipated when cholesterol is removed from membranes using agents such as methyl-β-cyclodextrin. Raft domains have also been implicated in the entry of some viruses. With HIV it appears that raft domains may facilitate fusion when receptor expression levels are low, but these domains are not required when expression levels are high. Cholesterol and sphingomyelin are required in the target membrane for SFV fusion activity. However, experiments with reconstituted membranes indicate that cholesterol and sphingomyelin requirements do not correlate with the ability of these lipids to form lipid rafts.

Uncoating

Although membrane fusion is essential for the penetration of enveloped viruses, further reactions are required in the cytoplasm to complete uncoating and initiate replication. For SFV, binding of ribosomes to the capsid protein is crucial for uncoating the viral RNA. With influenza viruses, acidification of the viral core is required for transport of incoming ribonucleoprotein complexes (RNPs) to the nucleus. The influenza M1 protein regulates trafficking of RNPs into and out of the nucleus. During synthesis of RNPs, M1 is transported into the nucleus and facilitates the export of these complexes to the cytoplasm. During entry, M1 must be inactivated, or its activities reversed, to allow RNPs to enter the nucleus. Acidification of the viral core causes M1 to be released from the RNPs. Although the membrane of the virus normally prevents protons from reaching the viral core, influenza virus contains several copies of a small virally encoded proton channel called M2. The channel activity of this protein is activated when viruses are exposed to low pH in endosomes and permits proton translocation to the viral core. Significantly, small inhibitors such as amantidine can block the M2 channel. These agents are particularly effective in inhibiting influenza virus infection.

Analogous processes are likely to occur with most viruses. For example, the capsids of adenoviruses undergo a series of alterations, which result from exposure of the capsids to the reducing environment in the cytosol and proteolysis that allows the capsids to interact with nuclear pore complexes (NPCs), through which the viral DNA is transported into the nucleus. During assembly of herpesviruses, packaging of the viral DNA into the capsid requires ATP. Thus during entry, following transport of incoming capsids to the perinuclear region of the cell and docking to NPCs, the stored energy from the assembly reaction may drive the release of the viral DNA and its transfer into the nucleus, leaving the empty capsid outside the nucleus. By contrast, for small DNA viruses, such as parvoviruses, the entire capsid may enter the nucleus and uncoating occurs within the nucleoplasm.

pH-Dependent Entry

Exposure to low pH is a key factor in the entry of many enveloped and nonenveloped viruses (Table 22.3). All acid-dependent viruses must un-

Table 22.3 Examples of the pH dependence of entry of various virus families

pH dependent	pH independent
Enveloped viruses	
Flaviviruses (e.g., tick-borne encephalitis virus)	
Rhabdoviruses (e.g., vesicular stomatitis virus)	
Orthomyxoviruses (e.g., influenza virus)	
Retroviruses (e.g., mouse mammary tumor virus)	Retroviruses (e.g., human immunodeficiency virus)
Bunyavirus	
Filoviruses	
Nonenveloped viruses	
Adenoviruses (e.g., Ad type 2)	Polyomavirus (e.g., SV40)
Parvoviruses	
Arenaviruses	
Arteriviruses	

dergo endocytosis to ensure delivery to acidic intracellular compartments. One method that has been used extensively to determine whether viruses are acid dependent for entry is to examine infection in the presence of agents that raise the pH within acidic organelles. Three classes of drugs have been used extensively: (i) inhibitors of the vacuolar ATPase such as bafilomycin A; (ii) carboxylic ionophores, such as monensin, that exchange protons for monovalent cations; and (iii) weak bases, such as chloroquine or ammonium chloride. With each of these agents, endosomal pH can be raised above 6.0 and the entry of pH-dependent viruses inhibited. Although these agents can indicate whether a virus is pH dependent, they do not necessarily indicate whether endocytosis is used, as pH-independent viruses may use endocytic pathways

Viral Assembly and Egress

Virus Factories

Many viruses establish regions within an infected cell that have been described as "virus factories." African swine fever virus (ASFV), a large DNA virus, assembles factories within a basket of intermediate filament proteins. Within these factories, viral DNA and protein synthesis occur and progeny viral particles undergo assembly. HCMV also seems to undergo its cytoplasmic envelopment reaction within discrete regions of the cytoplasm where capsids and budding particles are frequently seen. Other areas of cytoplasm appear to be largely devoid of viral structural proteins and assembly intermediates. All positive strand RNA viruses replicate their RNA within cytoplasmic vesicles from where the products must be exported for incorporation into progeny virions, be they nonenveloped picornaviruses or enveloped alphaviruses and flaviviruses. These membranes are believed to be derived from compartments of either the endocytic and/or exocytic pathways and their sequestration for viral replication requires the expression of specific viral proteins. Although membrane components may be sequestered from exocytic organelles, the functional activities of the exocytic pathway remain intact, allowing enveloped viruses that rely on the exocytic pathway the means of release from an infected cell. The synthesis of viral components in domains within the cell may both enhance the efficiency of synthesis and contribute to the spatial and temporal control of virus assembly (see below). It may also hide or protect the viral nucleic acid from the cellular defense mechanisms that would otherwise detect and destroy the viruses.

Viral Membrane Protein Synthesis

Viral membrane proteins are synthesized using the same machineries used by the host cell to make the majority of its own membrane proteins. The proteins are synthesized on the ER and undergo cotranslational translocation and N-linked glycosylation. Proteins emerging into the lumen of the ER undergo folding reactions that are required to generate the mature functional protein. These reactions involve a set of abundant ER proteins, including chaperones, such as calnexin and calreticulin, as well as enzymes such as protein disulfide isomerase, Erp57 and others, which ensure the newly synthesized proteins are folded correctly. This is particularly important for the viral fusion proteins, which must be clamped into metastable forms capable of eliciting fusion.

For flaviviruses newly synthesized membrane proteins are incorporated into virions in the ER, but for most other enveloped viruses, membrane proteins must be transported to other compartments in the cell. These transport events are believed to involve the same vesicular transport machinery described for cellular proteins. Indeed many seminal studies of the exocytic pathway have exploited viral membrane proteins.

Formation of Enveloped Viruses

To assemble a replication-competent virus, the structural and genetic elements of a virion must be brought together at the same time and at the same cellular location. For enveloped viruses these events occur when the components are collected together on the cytoplasmic side of a specific cellular membrane system. The extent to which packaging of the viral nucleic acid occurs at the membrane varies for different viruses. For example, herpesvirus capsids assemble in the nucleus and are transported to the cytoplasm where, together with the tegument proteins, they bud through the membranes of intracellular organelles. With D-type retroviruses, e.g., Mason-Pfizer monkey virus (MPMV), capsids assemble in the cytosol and subsequently interact with membranes to form enveloped particles. But for many enveloped viruses the capsid assembly events are linked to budding and may well provide the driving force for membrane deformation. The membrane system used for assembly varies for different viruses and presumably confers some advantage for the biology of that virus. Thus, alphaviruses, such as SFV, assemble at the plasma membrane, flaviviruses at the ER, and coronaviruses at the Golgi apparatus. By contrast, HIV appears to use different membrane systems in different cell types, undergoing assembly at the plasma membrane of infected T cells and on late endosomes in macrophages. The process of enveloped virus budding and release has been likened to the formation of vesicles involved in intracellular transport, but with reverse topology, i.e., vesiculation occurs in an outward orientation rather than inward. This process is topologically similar to the formation of the small membrane vesicles that are frequently formed in endosomes and confer on these organelles the features that have led to them being termed multivesicular bodies (MVBs).

Assembly is driven to a large extent through protein-protein interactions between components of the viruses themselves, i.e., through self-assembly reactions. A role for cellular proteins in these events has not been well established. Studies with HIV and several other enveloped viruses, including Ebola and VSV, have identified a set of cellular proteins that are essential for the final steps in particle formation. Significantly, these proteins are also involved in the formation of MVBs. The Gag protein drives the formation of retroviral particles, and expression of Gag alone can generate virus-like particles. This assembly is dependent on a so-called late (L) domain in Gag and the presence of a specific four-amino-acid motif within this L domain. In HIV, for example, the L domain is contained within the Gag C-terminal peptide termed p6 and contains the motif PT/SAP. Other retroviruses also contain L domains, although they may vary slightly in amino acid sequence (e.g., PPxY). Significantly the L domain motifs can be moved around in the Gag sequence and exchanged between viruses, suggesting that they function in similar ways. These motifs mimic a sequence found in the cellular protein Hrs, an endosomal component involved in sorting ubiquitinated proteins and MVB formation. Through the PT/SAP-type motifs both Gag and Hrs recruit a cellular protein termed TSG101, a component of a protein complex termed ESCRT I (endosomal sorting complex required for transport I). This complex, together with three other functionally related complexes (ESCRT II and III, and the Hrs complex), has been implicated in the sorting of ubiquitinated receptors into the internal membranes of MVBs where they are subsequently degraded by lysosomal proteases. The viral Gag proteins have been able to mimic cellular mechanisms for recruiting ESCRT I at least to the membrane. In cells expressing Gag proteins with mutations in the L domain motif, particle assembly can be initiated but the near-complete particles fail to undergo scission from the membrane.

Thus ESCRT complexes provide an essential scission machinery that allows particles to be released from infected cells, and which is not carried by the virus itself. Alternate mechanisms for recruiting the ESCRT machinery appear to operate in the equine infectious anemia virus (EIAV). Whether other enveloped viruses, such as herpes or alphaviruses, that are not currently known to rely on these same cellular proteins require similar activities to complete their assembly remains unclear.

Roles for Raft Domains?

Treatments with cholesterol-depleting agents, such as methyl-β-cyclodextrin, have been found to interfere with the formation and release of a number of enveloped viruses, leading to the view that raft domains may facilitate virus assembly, perhaps by aiding the sorting of membrane proteins in the plasma membrane. The integrity of raft domains is crucial for the assembly of influenza viruses, which bud from the apical domains of epithelial cells that rely on raft-mediated sorting to establish their polarity, but the role of rafts in the assembly of other viruses remains less clear. The use of raft domains may well depend on the state of infection, i.e., when viral components are abundant, rafts may be less important but, by concentrating sparse components, rafts may aid assembly when viral components are scarce. Further work is required to fully appreciate the roles of lipid rafts in virus formation.

Assembly in the Endocytic Pathway

Enveloped viruses are usually thought to assemble at the plasma membrane or on organelles of the exocytic pathway, from where they are released by secretion. The endocytic pathway has not usually been considered as a site for virus assembly. If anything, these locations have been deemed detrimental, as newly formed viruses might be transferred to lysosomes where they would be degraded. For HIV, at least, virion assembly in infected macrophages can occur on late endosomes. The late endosome compartment in APCs is the site for antigen processing and loading of MHC II and has been termed the MHC II compartment (MIIC). As part of their function, MIICs must recycle peptide-loaded MHC II to the cell surface. MIICs are also known to form exosomes, small membrane vesicles that, like HIV, appear to rely on the ESCRT machinery for formation. Exosomes are released from APC and mediate as yet poorly understood roles in antigen presentation. Thus HIV particles formed in late endosomes have the potential to be released from infected cells. Significantly, this release may be initiated by interaction with T cells, raising the possibility that, under some circumstances, T cells might stimulate their own infection. Assembly of HCMV also appears to involve compartments of the endocytic pathway and in infected macrophages HCMV may use mechanisms similar to those described for HIV to facilitate cell-to-cell transmission.

Cell-to-Cell Transmission

The formation of virus particles is required to protect the viral genome during the cell-free stages of the virus life cycle. What actually happens during the transfer of virions from an infected cell to an uninfected one, in vivo, remains by and large a mystery. For a number of viruses there are indications that transfer occurs during close contact between the producer and target cell. This is exemplified by vaccinia virus (VV) where actin comets assembled by extracellular enveloped viruses (EEV) may propel newly assembled virions into adjacent uninfected cells (see below). The human T-cell leukemia virus (HTLV-1) can reorient the microtubule-based cytoskeleton in infected cells such that virus release is focused on a specific plasma membrane domain at the interface between the infected and uninfected cells. Similar studies have suggested that HIV may also be transferred at specialized junctions between infected and uninfected cells. The presence of actin, receptors, and the cellular proteins talin and LFA-1 to these regions of cell-cell interaction resembles the formation of the immunological synapse. Hence the terms virological synapse and infectious synapse have been coined. It is likely that other viruses, in particular, herpesviruses, may use similar mechanisms to enhance their transfer from cell to cell, and that these mechanisms may involve target cell-induced release of virus.

Movement of Viruses within Cells

The passive movement of particles >50 nm in diameter in the cytoplasm is restricted. During entry, viruses must be moved from the cell surface to the interior, where they must be targeted to compartments containing the machinery essential for their replication. Subsequently, assembled virions or viral components must be moved to sites where assembly and release occur. These translocation events involve interaction with cellular cytoskeletal systems. Cells contain three interrelated and dynamic cytoskeletal systems: (i) the microtubule system based on the GTPase tubulin, (ii) the actin-based microfilament system, and (iii) the intermediate filament system. The microtubule and microfilament systems have well-established roles in cell movement and intracellular transport, whereas the intermediate filament system provides cells with mechanical strength. Studies with drugs such as nocodazole (which depolymerizes microtubules) and the cytochalasins (which depolymerize actin filaments) have implicated the microtubule and microfilament systems in viral replication. These agents can have both positive and negative effects on virus production. However, the specific roles of the cytoskeletal elements have not always been clearly established. Moreover, it is often unclear whether drugs have a direct effect on virus translocation or indirect effects arising from, for example, changes in the organization of the cytoplasm. A number of studies using morphological, biochemical, and pharmacological approaches have started to reveal the molecular bases of viral translocation and the role of these events in viral replication.

Involvement of the Cytoskeleton in Virus Entry

The best-characterized example of cytoskeletal involvement in virus entry is that described for HSV. These enveloped viruses fuse at the surface of target cells, but the viral capsids must gain access to the nucleus for replication. Fluorescence experiments have shown that, following fusion, viral capsids are transported to the perinuclear region of the cell on microtubules from where they are recruited to NPCs. This minus end-directed translocation appears to involve dynein/dynactin and occurs at rates of 0.5 to 2.6 μm/s. While this microtubule-based translocation may not be essential for infection of, and/or replication in, tissue culture cells, it is likely to play a significant role in neurons, where initial fusion at nerve terminals in the periphery must then be followed by translocation of the viral capsid to the nucleus in the cell bodies in the spinal cord. Little is currently known of how HSV interacts with microtubules; however, Vp22, a major component of the tegument (a set of virally encoded proteins between the viral membrane and capsid of intact virions, see Figure 22.1), can bind tubulin. When virions are deleted for Vp22, the capsids cannot interact with microtubules and are severely compromised in their ability to undergo transport to the nucleus.

Similar translocation events have been described for adenoviruses and for two retroviruses, the human foamy virus and HIV. With HIV, roles for both the microtubule and actin-based cytoskeletons in the transfer of incoming reverse transcriptase complexes (RTCs, postfusion complexes containing the reverse-transcribed viral DNA, integrase, and other components of the viral core) to the nucleus have been reported. Ad2 movement on microtubules can also be bidirectional and is regulated through receptor ligation (see above). Ad2 can presumably engage either the minus end-directed dynein/dynactin motor, or an as yet uncharacterized plus end-directed motor. The purpose of the plus end-directed migration is unknown, but it may have a role in cell exit.

In these examples, viral components (namely, capsids or RTCs) are thought to interact with motor proteins and the appropriate cytoskeletal system. In other cases translocation of the virions within endocytic vesicles, prior to penetration, may involve the cytoskeleton. The interaction of endocytic vesicles with both the microtubule and microfilament systems is well documented. Reovirus and the canine parvovirus are transported, in endosomes, along microtubules toward the microtubule-organizing center (MTOC). Thus, viruses can capitalize on the normal translocation properties of the endocytic system to facilitate their delivery to appropriate sites in the cell.

Some viruses use the actin-based cytoskeleton during entry. Agents that disrupt actin microfilaments inhibit endocytosis of SV40 in caveolae. Perturbation of the actin cytoskeleton also inhibits HIV reverse transcription, suggesting that interaction with the cytoskeleton is required in some way to promote early postfusion events in HIV replication. By contrast, following entry to the cytoplasm, baculovirus capsids appear to propel themselves toward the nucleus by nucleating actin bundling and recruiting myosin motors. In cells treated with cytochalasin D or the myosin inhibitor butanedione monoxime, transport of capsids to the nucleus is inhibited.

Cytoskeletal systems may also impose a barrier to entry. Many cells contain a cortical cytoskeleton, composed of actin and associated proteins, that underlies and supports the plasma membrane. This web, which can be on the order of 100 nm thick in some cells, is likely to impose an additional barrier to incoming viral cores that have penetrated the plasma membrane, e.g., when SFV, which normally undergoes low pH-induced fusion in endocytic vesicles, is forced to fuse at the plasma membrane of Chinese hamster ovary cells (by incubation in acidic medium), it does not infect. It appears that the cortical cytoskeleton in these cells can prevent the viral core from gaining access to ribosomes that are required for uncoating and initiating infection (see above). Thus viruses that normally enter cells through the plasma membrane must have mechanisms to penetrate the cortex. These may involve interactions with microtubules or microfilaments to pull the particle through the web. Alternatively, viruses may be able to locally perturb the integrity of the cortex. It can be argued that viruses that rely on low pH to facilitate fusion do so because endocytic vesicles have a mechanism to bypass the cortical barrier that the viruses lack.

Involvement of the Cytoskeleton in Virus Egress

The cytoskeleton also plays roles in the formation and release of progeny virus. As with entry, the transport of vesicles along cytoskeletal elements, in particular, microtubules, can facilitate the delivery of viral components to specific compartments, such as the plasma membrane, where assembly and release can occur. The cortex can also present a barrier to virus release or may act as a scaffold on which viral components are brought together during assembly. The most dramatic example is perhaps that of VV.

VV generates several different forms of particle. Intracellular mature particles (IMVs) are enveloped by a cellular membrane derived from the intermediate compartment. A fraction of these particles undergo additional envelopment and acquire two more membranes from the *trans*-Golgi network (TGN) to form intracellular enveloped viruses (IEV). These IEVs can be released from cells by fusion of their outer membrane with the plasma membrane to generate EEVs. Transport of IMVs to the TGN and of IEVs to the cell surface involves microtubules. Little is known of how IMVs move, but translocation of IEVs involves a virally encoded protein, A36R. A36R is likely to recruit a plus end-directed microtubule motor, such as kinesin, to couple to and move along microtubules. After release the EEVs can remain bound to the outer surface of the plasma membrane and initiate actin comet formation to push the virions out from the cell at the tips of microvilli. Comet formation also depends on A36R. The switch from microtubules to microfilaments involves phosphorylation of two key tyrosine residues in A36R by Src family kinases located on the cytoplasmic side of the plasma membrane. When phosphorylated, A36R binds the adapter proteins Nck and Grb2 that recruit N-WASP, WIP, and ARP 2/3 to drive actin polymerization. The VV-induced comets are similar to those induced by *Listeria* and may have related functions in facilitating cell-cell transmission of virus.

Herpesviruses are likely to exploit the actin and microtubule-based cytoskeletons to move viral components to nerve terminals for assembly. The Vp22 protein discussed above as important for virus entry can interact with nonmuscle myosin 2 and release of virions from infected cells can be inhibited by the myosin inhibitor butanedione monoxime. The microtubule cytoskeleton has also been implicated in the assembly of D-type retroviruses. The capsids of MPMV are assembled in the perinuclear region of the cytoplasm close to the centrioles but become dispersed throughout the cytoplasm when microtubules are disrupted or dynein is inhibited. Subsequent release of these viruses appears to depend on trafficking of the viral envelope protein through the perinuclear recycling endosome compartment. HTLV-1 also appears to exploit the microtubule-based cytoskeleton to facilitate transfer of virus to uninfected cells via the "virological or infectious synapse" (see above).

Roles for the Intermediate Filament System

Although not involved in translocation, the intermediate filament system also has some roles in viral replication. For some viruses the assembly of new particles occurs in discrete regions of the cytoplasm termed "viral factories." For ASFV, the formation of these factories has been likened to the formation of aggresomes, aggregates of misfolded protein awaiting proteasome-mediated degradation. In cells overexpressing proteins (especially proteins containing mutations that prevent their correct folding), aggresomes are frequently seen in the perinuclear region of the cell around the MTOC. These aggresomes are surrounded by, and appear to be held within, a basket formed by collapse of the intermediate filament system. The factories formed by ASFV are also located in the perinuclear region of the cell around the MTOC and are surrounded by a vimentin-containing basket. Whether the intermediate filament system is essential for the formation and activities of these virus factories is at present unclear.

Viral Regulation of Cellular Protein Synthesis and Expression

Viruses can modulate the expression of host cell proteins through a plethora of different mechanisms, including alteration of transcriptional and translational activity. They can also influence protein stability posttranslationally. The time course of replication within infected cells varies dramatically for different viruses. Relatively simple agents, such as alphaviruses, replicate rapidly, producing large numbers of particles in a few hours before the cell kills itself (and the virus) by apoptosis. Other viruses exhibit longer courses of infection ranging from days to years (for viruses that establish latent infections). These viruses adopt a range of strategies to prevent cells undergoing apoptosis but must also find ways of defending themselves against the host's immune response. Complex viruses such as the herpes and poxviruses encode various membrane and secretory proteins that modulate the immune response, such as chemokine receptors, chemokines, chemokine-binding proteins, cytokines, etc., all of which aim to prevent immune effector cells sensing and destroying infected cells. An alternative strategy is to modulate the expression of the MHC molecules and, in particular, MHC I, whose function it is to flag to the immune system that a cell is infected. Herpesviruses have developed a number of strategies to inhibit MHC I antigen presentation. The EBV EBNA 1 protein inhibits the proteasome proteolytic activity that generates peptides for MHC I presentation. Both HSV ICP47 and the HCMV US6 proteins inhibit TAP, the transporter that carries peptides into the ER lumen for loading onto MHC I complexes. The adenovirus E3 protein can retain MHC I in the ER, and HCMV US2 and US11 direct newly synthesized MHC I for retrotranslocation and degradation by proteasomes.

Although genetically less complex, viruses such as HIV can also modulate antigen presentation by infected cells. The HIV Vpu protein induces proteasomal degradation of the receptor CD4, and makes MHC I less stable. The HIV Nef protein also has an influence on receptor expression by acting as a multifunctional adapter protein that couples various receptors to cellular trafficking machineries. For example, by interacting with CD4 and the AP-2 adapter complex of endocytic clathrin-coated pits, Nef mediates endocytosis of CD4 from the surface of infected cells. Subsequent association of Nef with COP proteins on endosomes prevents internalized CD4 recycling to the cell surface and causes it to be directed to lysosomes where it is degraded. Nef can also down-modulate MHC I and II. Down-modulation of MHC I occurs through a poorly understood interaction between Nef and the cellular PI3K regulated ARF6 and PACS-1 proteins, PACS-1 being responsible for targeting internalized MHC I to the TGN. Significantly, Nef can discriminate between the products of different MHC I alleles and only down-modulates HLA-A and -B molecules. HLA-C molecules lack a tyrosine residue in their short cytoplasmic domain that is crucial for Nef-induced down modulation. Thus these molecules remain at the cell surface to maintain recognition by natural killer cells. For human herpesvirus 8 (HHV8), the agent implicated in Karposi's sarcoma, the K3 and K5 proteins have features in common with a number of ubiquitin E3 ligases. These proteins down-modulate MHC I antigens through a ubiquitin-proteasome system that also involves late endocytic compartments. Thus, through a range of strategies and mechanisms viruses can modulate the cell surface expression of a variety of cellular proteins involved in immune recognition, signal transduction, and other events.

Signaling

As discussed above, signaling through viral receptors can influence virus entry and prime cells for viral replication. Modulation of cellular signaling pathways can also have a significant impact on replication and interaction with the host. Viruses have evolved many strategies to enhance their replication, nullify the effects of the innate and acquired immune responses, avoid induction of apoptosis, control the cell cycle in the host cell, or modulate the transcriptional and translational activities of the host cell. These events are controlled in different ways, from expression of specific viral gene products to interaction with cellular receptors or the transfer of signaling proteins between cells.

Viruses can transfer signaling proteins from cell to cell. HIV has been shown to incorporate ERK/MAPK and to deliver this protein to uninfected cells, where it influences infectivity early in the replication cycle. Other viruses may incorporate cellular membrane proteins into their own membrane and transfer these proteins to a host cell membrane during fusion, e.g., herpesviruses encode one or more heptahelical proteins with similarities to GPCRs. HCMV has four such proteins, two of which, UL33 and US27, are incorporated into the viral membrane. Although the ligands (if any) for these proteins have still to be identified and no signaling activity has yet been demonstrated, a third heptahelical protein (US28) is constitutively active in its signaling activity. This protein is also likely to be included in viral membranes and, when transferred to a new host cell, may be responsible for an observed increase in PI3K activity 15 to 30 min after entry into quiescent fibroblasts. The cellular kinases Akt and p70/S6K and the transcription factor NFκB are also activated in a PI3K-dependent manner.

Viruses can also activate a range of signaling pathways within the host cell. Many viruses encode proteins with transformation and oncogenic potential, the nonreceptor tyrosine kinase activities of oncogenic retroviruses and the GPCR ORF-74 of HHV8 being examples. P38/MAPK activation by rhinoviruses, herpesviruses, HIV, and adenoviruses can induce a range of effects, including cytokine production, cytoskeletal reorganization, and apoptosis. For adenoviruses, p38/MAPK activation can lead to the production of proinflammatory cytokines including TNF-α and IL-6 that might limit viral replication. Adenoviruses also use the p38/MAPK pathway to modulate transcription. For HIV, p38/MAPK activation can also regulate gene expression. Many viruses, including adenoviruses, herpesviruses (HCMV and HHV8), polyomaviruses, and the hepatitis B and C viruses, activate ERKs, which normally function downstream of various growth factor receptors and Ras to regulate cell growth and differentiation. HHV8 expresses the polymorphic kaposin, a membrane protein that can transform rat fibroblasts. Kaposin interacts with cytohesin-1, a guanine nucleotide exchange factor for the small GTPase Arf-1. Activated cytohesin-1 in turn activates ERK by an as yet unknown mechanism. SV40 activates ERK through the small tumor antigen that inhibits the protein phosphatase 2A, a negative ERK regulator.

Conclusions

Virus replication is tightly integrated into the properties of the host cell. Because of their ability to efficiently exploit specific cellular functions, viruses have been, and continue to be, very effective tools to study basic cellular functions. For example, attempts to understand the entry of alpha- and polyomaviruses have contributed significantly to our understanding of the endocytic mechanisms underlying both clathrin- and caveolin-mediated endocytosis. Indeed, virus systems of one sort or another underlie much of the current knowledge of molecular genetics and cell biology. Increased understanding of the interactions of viruses with cells will continue to enhance our understanding of basic cellular functions. Moreover, such studies will undoubtedly identify novel targets and strategies for combating viral infections or for using viruses to our own advantages.

Selected Readings

Greber, U. F. 2002. Signalling in viral entry. *Cell. Mol. Life Sci.* **59:**608–626.
> Recent review on signaling pathways activated by viruses during entry into cells.

Greber, U. F., and A. Fassati. 2003. Nuclear import of viral DNA genomes. *Traffic* **4:**136–143.
> Recent review on targeting of incoming viral genomes to the nucleus.

Heath, C. M., M. Windsor, and T. Wileman. 2001. Aggresomes resemble sites specialized for virus assembly. *J. Cell Biol.* **153:**449–455.
> Study of the organization of cytoplasmic replication sites for a large DNA virus.

Johnson, D. C., and M. T. Huber. 2002. Directed egress of animal viruses promotes cell-to-cell spread. *J. Virol.* **76:**1–8.
> Discussion of the mechanisms of cell-to-cell transfer of viruses.

Knipe, D. M., P. M. Howley, D. E. Griffin, R. A. Lamb, M. A. Martin, B. Roizman, and S. E. Straus (ed.). 2001. *Fields' Virology,* 4th ed. Lippincott Williams & Wilkins, Philadelphia, Pa.
> Standard virology text.

Kooyk, Y., B. Appelmelk, and T. B. Geijtenbeek. 2003. A fatal attraction: Mycobacterium tuberculosis and HIV-1 target DC-SIGN to escape immune surveillance. *Trends Mol. Med.* **9:**153–159.
> Review of the role of a C-type lectin, DC-SIGN, in viral and bacterial pathogenesis.

Marsh, M., and H. McMahon. 1999. The structural era of endocytosis. *Science* **285:**215–220.
> Review of interactions between proteins implicated in clathrin-mediated endocytosis.

McDonald, D., M. A. Vodicka, G. Lucero, T. M. Svitkina, G. G. Borisy, M. Emerman, and T. J. Hope. 2002. Visualization of the intracellular behavior of HIV in living cells. *J. Cell Biol.* **159:**441–452.
> Study of cellular mechanisms used by incoming HIV particles to aid transport to the nucleus. Implicates both actin and tubulin-based cytoskeletal systems.

Meier, O., K. Boucke, S. V. Hammer, S. Keller, R. P. Stidwill, S. Hemmi, and U. F. Greber. 2002. Adenovirus triggers macropinocytosis and endosomal leakage together with its clathrin-mediated uptake. *J. Cell Biol.* **158:**1119–11131.

Study of adenovirus entry via endocytosis. Indicates that infectious particles are internalized by clathrin-mediated endocytosis, but induce macropinocytosis. Moreover, this macropinocytic activity is required for penetration and viral replication.

Pelchen-Matthews, A., B. Kramer, and M. Marsh. 2003. Infectious HIV-1 assembles in late endosomes in primary macrophages. *J. Cell Biol.* **162:**443–455.

Demonstration that infectious HIV particles are assembled in late endosomes in infected macrophages.

Pornillos, O., J. E. Garrus, and W. I. Sundquist. 2002. Mechanisms of enveloped RNA virus budding. *Trends Cell Biol.* **12:**569–579.

Review of recent work on the assembly of enveloped viruses and the role of cellular ESCRT proteins in these events.

Rietdorf, J., A. Ploubidou, I. Reckmann, A. Holmstrom, F. Frischknecht, M. Zettl, T. Zimmermann, and M. Way. 2001. Kinesin-dependent movement on microtubules precedes actin-based motility of vaccinia virus. *Nat. Cell Biol.* **3:**992–1000.

Study of the role of microtubule-mediated translocation in vaccinia virus release.

Sodeik, B. 2000. Mechanisms of viral transport in the cytoplasm. *Trends Microbiol.* **8:**465–472.

Review of cellular transport systems used by viruses to facilitate entry, infection, and release.

23

Use of Simple Nonvertebrate Hosts To Model Mammalian Pathogenesis

COSTI D. SIFRI AND FREDERICK M. AUSUBEL

Microbial pathogenesis necessarily involves the interaction between two organisms. In many cases, it would be desirable to be able to genetically manipulate both a pathogen and its host to elucidate the molecular mechanisms underlying pathogen virulence and host defense. In general, however, mammals are not well suited for large-scale genetic studies because of the large numbers of animals that would be involved. As a consequence, for both practical and ethical reasons, no mammalian pathogenesis system has been subjected to systematic genetic analysis involving the generation and characterization of either avirulent pathogen mutants or host innate immune response mutants. In this chapter, we describe the unexpected finding that a variety of human bacterial and fungal pathogens can infect and cause disease in relatively simple nonmammalian model genetic hosts. In many instances, the interactions of pathogens with these alternative hosts mimic important features of mammalian pathogenesis.

Before describing these alternative pathogenesis models, it is instructive to examine what would be involved in carrying out comprehensive genetic analysis in a mammalian pathogenesis model. By genetic analysis, we mean the isolation and characterization of either pathogen mutants that exhibit an aberrant virulence phenotype or host mutants that exhibit an aberrant immune response phenotype. Identification and characterization of the mutated genes in these pathogen and host mutants help elucidate the molecular basis of pathogenesis and host defense.

To identify virulence-related genes in a pathogen, it is necessary to screen pathogen mutant libraries for clones that exhibit reduced virulence. Ideally, each putative mutant in the library would be tested in an individual host to determine whether virulence has been affected. Take the case of a bacterial pathogen with a genome about the same size as *Escherichia coli* strain K-12 (about 4.6 Mb) that encodes approximately 4,400 genes. A mutant library large enough to saturate such a genome (that is, large enough such that there is a 95% probability that each nonessential gene has been hit at least once) would have to contain approximately 20,000 independent transposon insertion strains. The reasoning is as follows: Although our hypothetical pathogen genome contains 4,400 genes, because bacteria are haploid, only insertions in nonessential genes can be recovered. It has been estimated that *E. coli* has about 400 essential genes. Assuming that transposon insertions occur at random, a statistical analysis shows that approximately five times as many mutant clones need to be screened as target genes to achieve ~95% probability of obtaining a hit in each nonessential gene. Moreover, each mutant clone would have to be tested for virulence in several animals (5 to 10) to achieve a minimal level of statistical significance. Thus, $5 \times 4,000$ (nonessential genes) $\times 5$ (or 10) = 100,000 to 200,000 animals, most likely mice, would have to be sacrificed to screen the genome of our hypothetical pathogen for virulence-related genes. This experiment is neither practical nor ethical.

What about carrying out host genetic analysis to identify components of the mammalian innate immune response? In recent years, many mouse mutants have become available and large-scale projects are under way to generate thousands of strains of mice, each of which contains a knockout mutation in a specific gene. In theory, this makes it possible to test each of these mutants for ones that exhibit an aberrant innate immune response. Mice appear to have at least 35,000 genes (although the actual number is still unknown and controversial) and, for each mutated gene, 5 to 10 mice would have to be tested to achieve statistical significance. Thus, to survey the entire mouse genome, as many as 350,000 mice would have to be sacrificed for each pathogenesis model used. The experimental situation is actually somewhat more complex than this. Because mice are diploids, in many cases it will be possible to test a heterozygous mutant when the homozygous knockout is lethal. Homozygous mutants sometimes exhibit a phenotype because of haploinsufficiency. Checking heterozygous mutants in those cases where the homozygous mutant is nonviable is important because some genes involved in innate immune responses may play essential roles in development or have organ-specific essential functions. Carrying out such genetic analysis on a genome-wide scale is a huge undertaking. Once again, this experimental strategy is neither feasible nor ethical.

Several innovative methods described in chapter 20 have been developed to overcome some of the practical and ethical limitations involved in the identification of bacterial virulence factors that would otherwise necessitate the use of large numbers of mammalian host animals. These methods greatly extend earlier microbe-centered approaches that involved screening for virulence factors based on altered biochemical phenotypes to those that take into account the critical role of the host in the disease process. Three such methods, in vivo expression technology (IVET), signature-tagged mutagenesis (STM), and differential fluorescence induction (DFI), allow the application of the powerful techniques of random mutagenesis and high-throughput screening for the identification of bacterial virulence factors in the context of actual host environments. IVET identifies genes that are specifically expressed in an animal host during pathogenesis, STM identifies genes in a relatively small pool that are required for replication in an animal host or cell line, and DFI utilizes green fluorescent protein and fluorescence-activated cell sorting to identify genes that are activated under specific conditions or in specific host cell types. In addition, transcriptional profiling analysis using DNA or oligonucleotide microarray analysis holds tremendous potential for identifying pathogen genes that are specifically activated or repressed during pathogenesis. Recently, the concept of STM has been extended to genome-wide analysis by making use of microarray technology. This latter technology, referred to as transposon site hybridization (TraSH), involves generating a large random library of transposon-generated mutants. By synthesizing an RNA probe using T7 RNA polymerase initiation sites at the ends of the transposon and probing a genome-wide DNA array, it is possible to determine which mutants fail to grow in a host animal or cell line.

IVET, STM, DFI, and TraSH allow pathogen genomes to be scanned for virulence-related genes without involving the use of large numbers of experimental animals. However, these approaches also have limitations. IVET and DFI analysis invariably identifies many genes that are up-regulated in the host environment but which are not directly involved in the pathogenic process. STM and TraSH will fail to identify pathogenesis-related mutants that can be complemented in *trans* by nonmutant cells in the mixed population of bacterial mutants used for the inoculum.

In this chapter, we describe an alternative method for identifying pathogen virulence factors that is based on the finding that many mammalian pathogens are also capable of causing disease in simple nonvertebrate hosts. Remarkably, many of the pathogen virulence factors that are involved in pathogenesis in mammalian hosts are also important for virulence in the nonvertebrate hosts, suggesting that the underlying mechanism of pathogenesis has been evolutionarily conserved. This makes it possible to utilize a variety of well-studied genetic hosts, including the fruit fly *Drosophila melanogaster*, the nematode *Caenorhabditis elegans*, and the plant *Arabidopsis thaliana*, as mimics of mammalian hosts to study various aspects of pathogenesis.

From the perspective of the pathogen, the advantage of using nonvertebrate "alternative" hosts as adjuncts to vertebrate animal models is that thousands of bacterial clones from a mutagenized library can be individually screened for avirulence using individual insects, nematodes, or plants as hosts. Assuming that overlapping sets of virulence factors are involved in pathogenesis in both vertebrate and nonvertebrate hosts, this genome-scanning procedure should identify the majority of pathogenicity-related genes. One advantage of this screening procedure is that the output of the screen, pathogenicity in a nonvertebrate host, may be a more direct measure of the process of pathogenicity in mammalian hosts than the more indirect IVET, STM, DFI, and TraSH methods. From the perspective of the hosts, the advantage of studying pathogenesis in model genetic organisms is that genetic analysis can be used to identify host genes involved in pathogen defense by screening for more susceptible or more resistant mutants.

Development of Alternative Host-Pathogen Models

Conservation of Bacterial Virulence Factors

As discussed above, models for human pathogens involving the infection of alternative nonmammalian hosts in which both the microbe and the host are genetically tractable would greatly facilitate our understanding of some of the universal mechanisms underlying host-pathogen interactions. Whether nonmammalian hosts can be used to model mammalian pathogenesis is directly related to the broader question of whether there are commonalities in the pathogenic mechanisms used by diverse bacteria irrespective of the hosts and whether evolutionarily diverse hosts utilize similar strategies to protect themselves from pathogen attack. The more commonalities there are, the more likely it is that pathogen virulence and host defense mechanisms have ancient evolutionary origins that can be studied in simple model hosts.

One line of evidence that suggests a high degree of similarity of virulence mechanisms is the conservation in both plant and animal pathogens of the type III secretion system (TTSS) that deliver proteinaceous virulence factors and toxins (effectors) directly into host cells (see chapter 15). TTSS has been studied most extensively in the human pathogens *Yersinia enterocolitica* and *Salmonella enterica* and in the plant pathogen *Pseudomonas syringae*. In the case of animal pathogens, the TTSS effectors target the host cytoskeleton, components of the host defense response, or key host signal transduction cascades. Less is known about the targets of the plant pathogen TTSS effectors, although a variety of recent evidence shows that the targets may be similar in both plant and animal cells. Is the conservation of TTSS systems evidence for evolutionary conservation of pathogenic mechanisms or a striking example of convergent evolution? Type III secretory systems appear to have evolved from ancestral mechanisms involved in flagellar biosynthesis, and it is certainly conceivable that type III secretory systems evolved independently in different bacterial species. In any case, even if TTSS is highly conserved, it does not necessarily mean that the effector proteins in different bacterial pathogens are also highly conserved. On the contrary, effector proteins may be highly specific to particular host species. Therefore, it is necessary to develop a model system or systems in which it is possible to directly test whether the same pathogen virulence factors are able to function in evolutionarily diverse hosts.

The first rigorous demonstration that virulence mechanisms are highly conserved in evolution involved the demonstration that several strains of the human opportunistic pathogen *Pseudomonas aeruginosa* can cause disease in both plants and animals. *P. aeruginosa* is a ubiquitous soil microorganism that is also an important human opportunistic pathogen, infecting a range of immunosuppressed individuals and cystic fibrosis patients. Although *P. aeruginosa* has not been described as an agricultural pathogen, it had been known for quite some time that some *P. aeruginosa* strains could also cause disease in plants, but the relevance of this observation to the evolution of pathogenic mechanisms had not been investigated. An important new discovery that was reported in 1995 was the finding that the same subset of *P. aeruginosa* virulence factors was required for full virulence in the plant *A. thaliana* and in a mouse pathogenesis model, suggesting that at the molecular level, plant and animal pathogenesis may be quite similar. In this study, it was reported that three well-known *Pseudomonas* virulence factors, ToxA, GacA, and PlcS, were required for maximum virulence in both a mouse burn model and an *Arabidopsis* leaf infection model.

The ability of *P. aeruginosa* to infect plants as well as animals was subsequently extended to show that the same clinical isolate of *P. aeruginosa* that can infect *Arabidopsis*, strain PA14, was also a pathogen of the nematode worm *C. elegans* as well as the insects *D. melanogaster* and *Galleria mellonella*. The plant, nematode, and insect pathogenesis models are described in detail below. Given these model hosts for *P. aeruginosa*, it was possible to design a more general test of whether *P. aeruginosa* virulence factors that function in pathogenesis in plants or nematodes are also involved in virulence in mammalian hosts. As depicted in Figure 23.1, after the generation of libraries of *P. aeruginosa* using random transposon mutagenesis, individual clones were screened for reduced virulence in plants and nematodes. The mutants identified in these screens were then tested in a mouse pathogenesis model. As summarized in Figure 23.2, most of the mutants identified by screening in plants and worms were also less virulent in mice,

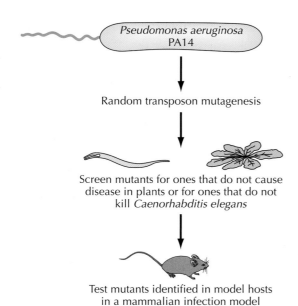

Figure 23.1 Identification of bacterial virulence factors by screening for avirulent mutants in model nonvertebrate hosts. A human bacterial pathogen, in this case *P. aeruginosa* strain PA14, which is also infectious in a nonvertebrate host (or hosts), is subjected to random transposon mutagenesis. Individual transposants are picked to 96- or 384-well microtiter plates and then tested individually in a *C. elegans* killing assay or in an *Arabidopsis* leaf infiltration assay. Mutants that are less pathogenic in *C. elegans* or *Arabidopsis* are characterized by determining the gene into which the transposon has inserted and verifying that the mutant phenotype corresponds to the transposon insertion. Finally, validated mutants are tested in a mouse pathogenesis model.

Figure 23.2 Isolation of *P. aeruginosa* mutants by direct screening in plants and nematodes and their phenotypes in plants, nematodes, and mice. Refer to Figure 23.1 for the overall experimental strategy. In the nematode assay, one gene was hit twice and one gene was identified in both the plant and nematode screens. Thus, the total number of genes identified is less than the number of mutants.

Host			
Assay	Growth in leaf	Nematode killing	Both screens
Number screened	2,500	5,500	8,000
Number of mutants isolated	9	14	23
Number of genes identified	9	13	21
	Number of genes identified, which when mutated resulted in decreased virulence on the hosts indicated on the left		
 (plant)	9	10	18
 (nematode)	5	13	17
 (mouse)	8	10	17

Table 23.1 *P. aeruginosa* PA14 virulence-related genes identified in nonvertebrate hosts

Category	No. of mutants	Genes
Regulators	4	*gacA, gacS, ptsP, lasR*
Membrane protein	1	*aefA*
Biosynthetic enzymes	2	*phzB, hrpM*
Modifying enzyme	1	*dsbA*
Multidrug transport	2	*mexA, mtrR*
Unknown proteins	11	?

demonstrating a high level of conservation of virulence factors for pathogenesis in diverse hosts. A list of some of the *P. aeruginosa* virulence-related genes identified in this screen is presented in Table 23.1. Importantly, about half of the mutants that were identified corresponded to previously uncharacterized genes. This experiment validated the use of these nonmammalian models for the efficient identification of previously unknown virulence factors important for mammalian pathogenesis. Because of the large number of mutants screened, it would not have been practical to do this experiment if it had been necessary to screen the mutants directly in mice.

Conservation of Host Innate Immune Factors

If pathogen virulence factors are highly conserved, are host defense mechanisms conserved as well? As described in chapter 18, a unique feature of the vertebrate immune system is acquired or adaptive immunity based on the clonal selection of T and B cells. The involvement of clonal selection presumably evolved after the diversification of vertebrates from invertebrates because it has not been found in nonvertebrates or plants. In contrast, the innate immune response, which is the first line of defense against invading microorganisms, appears to have more ancient origins. Indeed, recent progress in elucidating the molecular basis of innate immunity has been facilitated by the discovery that many features of the mammalian innate immune response are also present in *Drosophila*. As illustrated in Figure 23.3, comparing insects and mammals, clear similarities include the conservation of a family of so-called Toll-like receptors (TLRs) and the downstream signaling cascades leading to the activation of Rel-family transcription factors such as mammalian NF-κB. In both insects and mammals, TLRs are involved in the recognition of microbial pathogen-associated molecular patterns (PAMPs) that distinguish pathogens from host cells. Well-studied PAMPs include gram-negative lipopolysaccharide, gram-positive peptidoglycan, and eubacterial flagella. An interesting question relevant to this chapter is whether the conservation of innate immune functions also extends to nematodes and plants. Almost nothing is known about innate immunity in *C. elegans*, but as discussed in some detail below, recent work suggests that at least one feature of innate immune signaling, the involvement of a highly conserved mitogen-activated protein kinase (MAPK) cascade, has been conserved between mammals and nematodes (Figure 23.3). MAPK cascades are evolutionarily conserved protein modules that function in molecular signaling pathways in a variety of eukaryotes, including plants, yeasts, worms, flies, and mammals. MAPKs are protein kinases that activate regulatory proteins such as transcription factors by phosphorylating them. MAPKs are in turn phosphorylated by MAPK kinases (MAPKKs) and MAPKKs are phosphorylated by MAPKK kinases (MAPKKKs).

Plants also recognize PAMPs, including lipopolysaccharide (LPS) and bacterial flagellin. Recently, a signal transduction cascade that responds to bacterial flagellin has been identified in *Arabidopsis* that includes a transmembrane receptor kinase that bears some structural and functional similarities with TLRs and a downstream MAPK cascade (Figure 23.3). However, the components of this particular *Arabidopsis* MAPK cascade are not directly homologous to the MAPKs, MAPKKs, and MAPKKKs that function in mammalian innate immunity. Moreover, the so-called WRKY

Figure 23.3   Conservation of innate immunity signaling pathways in mammals, insects, nematodes, and plants. In mammals and insects, pathogen-associated molecular patterns (PAMPs), macromolecules that are synthesized by microbes but not host cells, are detected by a family of transmembrane leucine-rich repeat (LRR) Toll and Toll-like receptors (TLRs). The signal is then transduced via conserved adapter proteins including MyD88, leading to the activation of Rel/NF-κB transcription factors as well as a mitogen-activated protein kinase (MAPK) signaling cassette that includes the highly conserved p38 MAPK. Analogous pathways function in *Arabidopsis*, although the components of the *Arabidopsis* signaling pathways do not appear to be direct homologues of the proteins in the mammalian and insect pathways. For example, the *Arabidopsis* transmembrane LRR receptor kinase FLS2 is involved in the recognition of a 22-amino-acid component of the eubacterial flagellar protein. Mammals have a corresponding TLR that also responds to eubacterial flagellar protein. In comparison to mammals, insects, and plants, relatively little is known about the innate immune response in *C. elegans*. The *C. elegans* genome encodes a single Toll-like protein but does not encode transcription factors such as mammalian NF-κB or *Drosophila* DIF. *C. elegans* also has homologues of the mammalian TLR pathway components IRAK (interleukin-1 receptor-associated kinase; PIK-1 in worms) and TRAF6 (TNF receptor-associated factor; TRF-1 in worms). However, *C. elegans tol-1, pik-1,* or *trf-1* mutants do not exhibit an immunocompromised phenotype. Recent genetic analysis has led to the identification of a *C. elegans* p38 MAPK signaling cascade that is directly homologous to the mammalian MAPK cascade that plays a central role in mediating mammalian immunity. Moreover, a Toll interleukin-1 resistance (TIR) domain-containing protein, TIR-1, functions upstream of p38 in *C. elegans*, and it may couple a yet-to-be-identified PAMP receptor to the innate immune response.

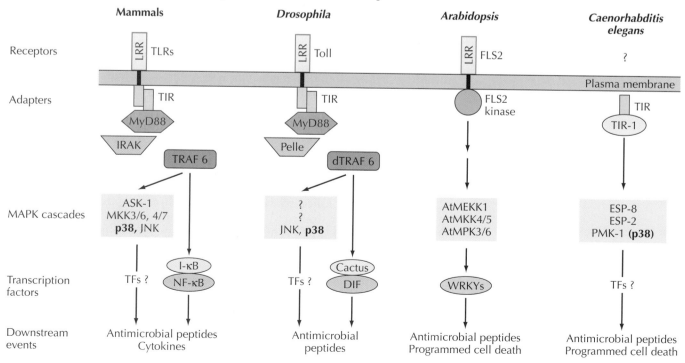

transcription factors that function downstream of the MAPK cascades in plants and the Rel-like transcription factor NF-κB that functions downstream in mammalian innate immunity, are not directly related. On the other hand, as illustrated in Figure 23.3, the overall organization of the immune signaling pathways in plants, insects, and mammals appears to be highly homologous.

One feature of the innate immune response that appears to be universally conserved in plants, invertebrate animals, and vertebrates is the production of a class of antimicrobial peptides called defensins in a pathogen-inducible manner. Defensins from all these species are folded similarly, suggesting that they may predate the evolutionary divergence of animals and plants. In insects and mammals, defensins are activated downstream of the TLR–NF-κB signaling pathway. In plants and nematodes, the pathways leading to defensin production have not been fully elucidated.

Caenorhabditis elegans as a Model Host

C. elegans is a soil-dwelling nonparasitic roundworm (nematode) that feeds on bacterial blooms found in decaying organic matter. Starting in the 1960s, Sydney Brenner (Nobel Laureate in Medicine or Physiology, 2002) advanced *C. elegans* as a model organism for the study of development and neurobiology. *C. elegans* was well suited for this purpose, since its transparent body allowed for direct microscopic observation of organogenesis and other biological processes. In addition, the nematode's small size (adults are 1 mm in length) and rapid generation time (3 days) allowed for the production of large populations of worms. In the laboratory, worms are easily maintained on agar plates or in liquid broth, where they feed on an auxotrophic *E. coli* strain, OP50. *C. elegans* has two dimorphic sexual forms—self-fertilizing hermaphrodites and males—which make isolation of mutant strains and genetic crosses straightforward. A wide assortment of genetic tools are available for the isolation, characterization, and mapping of mutant and transgenic *C. elegans*, and in 1998, the nematode became the first multicellular organism to have a completely sequenced genome.

In recent studies, a remarkably large number of human pathogens have been shown to kill *C. elegans* (Table 23.2). These studies initially focused on broad host range pathogens, such as *P. aeruginosa* and various *Burkholderia* species—opportunistic environmental pathogens that worms likely encounter in nature. Surprisingly, further investigations demonstrated that a number of narrow host range pathogens, including *S. enterica*, *Staphylococcus aureus*, and *Streptococcus pyogenes*, readily kill nematodes. In these investigations, pathogenesis was easily studied by replacing *E. coli*, the normal worm food source, with the bacterial pathogen of interest, and monitoring the health of the nematodes over the course of a few days (Figure 23.4). *C. elegans* is killed by two general mechanisms, depending on the pathogen species or strain and the growth conditions used. Toxin-mediated killing is characterized by rapid worm killing, usually within one day, and by the ability of the conditioned media to kill in the absence of live bacteria. Infection-associated killing usually occurs over the course of several days and thus far has been characterized by the accumulation and, in some cases, proliferation of bacteria within the worm digestive tract.

Strains of *P. aeruginosa* have been shown to kill worms by two distinct methods of intoxication. When *C. elegans* feeds on *P. aeruginosa* strain PA14 grown on high osmolarity rich media, worms die within a matter of hours.

Table 23.2   Model hosts and the human pathogens that infect them

Alternative host	Pathogen
Caenorhabditis elegans (nematode)	Gram-negative bacteria
	Burkholderia pseudomallei
	Burkholderia mallei
	Burkholderia cepacia
	Escherichia coli
	Pseudomonas aeruginosa
	Serratia marcescens
	Salmonella enterica
	Yersinia pseudotuberculosis
	Yersinia pestis
	Gram-positive bacteria
	Enterococcus faecalis
	Enterococcus faecium
	Staphylococcus aureus
	Streptococcus pneumoniae
	Streptococcus pyogenes
	Yeast
	Candida albicans
	Candida glabrata
	Cryptococcus neoformans
Drosophila melanogaster (fruit fly)	Gram-negative bacteria
	Escherichia coli
	Listeria monocytogenes
	Pseudomonas aeruginosa
	Serratia marcescens
	Gram-positive bacteria
	Enterococcus faecalis
	Mycobacterium
	Mycobacterium marinum
Galleria mellonella (greater wax moth larvae)	Gram-negative bacteria
	Pseudomonas aeruginosa
	Yeast
	Candida albicans
Arabidopsis thaliana (plant)	Gram-negative bacteria
	Pseudomonas aeruginosa
Dictyostelium discoideum (cellular slime mold)	Gram-negative bacteria
	Legionella pneumophila
	Pseudomonas aeruginosa
	Mycobacterium
	Mycobacterium marinum
	Yeast
	Cryptococcus neoformans
Saccharomyces cerevisiae (baker's yeast)	Gram-negative bacteria
	Pseudomonas aeruginosa
	Salmonella enterica
	Yersinia pseudotuberculosis
Danio rerio (zebrafish)	Gram-negative bacteria
	Salmonella enterica
	Gram-positive bacteria
	Streptococcus iniae
	Streptococcus pyogenes
	Mycobacterium
	Mycobacterium marinum

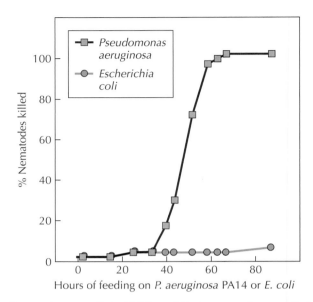

Figure 23.4   *P. aeruginosa*-mediated killing of *C. elegans* under so-called slow-killing conditions that involve an infection-like process. *C. elegans* strains die when they are transferred from a petri plate where they are feeding on their normal food source, *E. coli* strain OP50, to a petri plate containing a lawn of *P. aeruginosa*. The rate of killing depends on the strain of *P. aeruginosa* and on the medium on which the *P. aeruginosa* lawn was grown. When grown on minimal media and then fed to worms, *P. aeruginosa* accumulates in the *C. elegans* intestine and kills the worms by an infectious-like process over the course of a couple of days that requires live bacteria. In contrast, when grown in rich medium, *P. aeruginosa* appears to produce a variety of low-molecular-weight toxins, including phenazines, which kill the nematodes much more quickly. Live bacteria are not required for this latter type of toxin-mediated killing. Other human pathogens also kill *C. elegans*. The rate of killing varies from 24 h to several days, depending on the pathogen and the medium on which it is grown.

This killing, termed "fast killing," is due to the production of one or more diffusible, low-molecular-weight toxins of the pyocyanin-phenazine class. A different *P. aeruginosa* strain, PA01, kills nematodes within a few hours via a paralytic process, called "paralytic killing," by producing a large amount of hydrogen cyanide when grown on brain-heart infusion media. Other human pathogens, including *S. pyogenes* and *B. pseudomallei*, also produce low-molecular-weight toxins that kill *C. elegans*. *S. pyogenes* produces large quantities of hydrogen peroxide, but the *B. pseudomallei* toxin(s) remains to be elucidated.

In contrast to toxin-mediated killing, *P. aeruginosa* strain PA14 grown on a low osmolarity minimal medium kills *C. elegans* relatively slowly over the course of several days. This killing, called "slow killing," correlates with the accumulation of a large number of *P. aeruginosa* cells in the worm's intestine. Normally, when *C. elegans* is feeding on a nonpathogenic food source, very few if any bacterial cells accumulate in the intestine (Figure 23.5). In a process that is not completely understood, bacteria ingested through the mouth are mechanically disrupted in a specialized pharyngeal grinder organ before entering the digestive tract. Toxin-mediated "fast killing" and infection-mediated "slow killing," while both mediated by PA14, have been shown to be mechanistically distinct. Most PA14 mutants that are defective in toxin-mediated killing are still capable of infection-mediated killing and vice versa, yet a majority of both types of mutants have reduced virulence in a mouse burn model (Figure 23.2).

▶ *For Figure 23.5, see color insert.*

Interestingly, many other human pathogens that have been tested are capable of killing *C. elegans* by an infection-like process (Table 23.2). In the case of all of these pathogens, the worm intestinal tract accumulates bacteria, leading to intestinal distention (see Figure 23.5). Some bacteria, like *P. aeruginosa* and *S. aureus*, transiently colonize the alimentary tract and can be washed out of the intestinal tract if the worms are moved to a benign food source. Others, like *Salmonella* serovar Typhimurium and *Enterococcus faecalis*, persistently colonize and proliferate within the intestinal tract. Presumably, these latter bacteria specifically bind to intestinal tract cells, although the molecular characterization of this binding has yet to be identified for either organism. Although accumulation of bacteria in the *C. elegans* intestine is highly correlated with worm killing, accumulation in the intestinal tract is not sufficient for killing. *Enterococcus faecium* accumulates to high levels but does not kill. Thus far, no pathogen has been observed to invade intestinal cells. In all of the infection-killing models, worms become progressively more lethargic over the course of several days before dying. The mechanism of killing is likely specific to each pathogen, but some important common factors include secreted proteases (*S. aureus*, *E. faecalis*, and *P. aeruginosa*) and pore-forming toxins (*S. aureus* and *E. faecalis*). Virulence regulatory factors are also critically important for nematodical activity. These include two-component regulators (*gacA/gacS* of *P. aeruginosa* and *phoP/phoQ* of *Salmonella* serovar Typhimurium), quorum-sensing systems (*lasR* of *P. aeruginosa*, *agr* of *S. aureus*, and *fsr* of *E. faecalis*), and alternative sigma factors (*rpoN* of *P. aeruginosa*, *rpoS* of *Salmonella* serovar Typhimurium, and σ^B of *S. aureus*).

In an interesting nonlethal pathogenicity model, *Yersinia pestis* and *Y. pseudotuberculosis* have been shown to form dense biofilms around the *C. elegans* mouth. Biofilm formation depends on the hemin storage system, encoded by the *hmsHFRS* operon. While this dense biofilm does not directly kill the worm, it does block its digestive tract, thus inhibiting feeding and subsequent maturation. In an analogous fashion, *Y. pestis* blocks the digestive tract of fleas, the natural vector for *Y. pestis* transmission, in a *hmsHFRS*-dependent manner. This blockage leads to two actions that increase the transmission of the agent of plague. First, the flea starves, leading it to take more frequent blood meals. Second, the obstructed flea regurgitates thousands of bacteria into the skin of the host when it takes a meal. Investigation of the role of *Yersinia* biofilm formation and cuticle attachment in *C. elegans* may have broader implications in *Yersinia* interaction with fleas and other invertebrates.

Recently, the pathogenic yeasts *Cryptococcus neoformans* and *Candida* species have also been shown to kill *C. elegans*. Known yeast virulence determinants important for mammalian pathogenesis, such as virulence-associated signal transduction pathways, *C. neoformans* polysaccharide capsule, and *C. neoformans* melanin, are important for worm killing. Importantly, nonpathogenic yeast, such as *Cryptococcus laurentii* and *C. kuetzingii*, can be utilized by *C. elegans* as a food source. Nematode brood size and life span on these environmental fungi are equivalent to or better than those of nematodes raised on *E. coli* OP50, suggesting that yeast can serve as a nutritious food source for worms. Killing by *C. neoformans* and *Candida* appears to be primarily mediated by infection-like mechanisms, although *C. neoformans* polysaccharide capsule may be directly toxic to worms. Similar to the bacterial models, pathogenic yeasts such as *C. neoformans* accumulate in the *C. elegans* intestine, whereas nonpathogenic yeasts such as *C. laurentii* do not (Figure 23.5).

As described above and in Figure 23.3, in *Drosophila* and mammals, microbial infection elicits an evolutionarily conserved innate immune response, characterized by the recognition of pathogen-associated molecular patterns (PAMPS; e.g., LPS and peptidoglycan) by families of TLRs and the transcriptional activation of antimicrobial responses by Rel/NF-κB transcription factors. Surprisingly, even though *Drosophila* and *C. elegans* are thought to belong to sister phyla, it appears that a Toll-like signaling pathway does play a major role in *C. elegans* innate immunity. Bioinformatic analysis of the fully sequenced *C. elegans* genome identifies a single TLR and no Rel/NF-κB transcription factors. Moreover, mutation of the single *C. elegans* TLR does not alter the *C. elegans* response to bacterial pathogens.

Despite the apparent absence of a Toll–Rel/NF-κB signaling pathway, transcriptional profiling analysis has shown that *C. elegans* responds to bacterial pathogens by the activation of a variety of genes, homologues of which are known to be involved in antimicrobial responses in insects and mammals. In addition, presumptive immune-compromised *C. elegans* mutants have been isolated that are hypersusceptible to killing by a variety of bacterial and fungal pathogens. Two of these enhanced susceptibility to pathogen (*esp*) mutants were recently shown to correspond to mutations in genes encoding two components of a MAPK cascade. Importantly, as illustrated in Figure 23.3, the *C. elegans* MAPK cascade implicated in innate immune signaling is highly homologous to the mammalian p38 MAPK cascade known to play a key role in the response to LPS and other immune elicitors. Although relatively little is known about innate immunity in *C. elegans* compared with *Drosophila* and mammals, the work with the *esp* mutants demonstrates that *C. elegans* does indeed share components of an immune response with insects and mammals. Because *C. elegans* is such a versatile genetic organism, further genetic analysis in *C. elegans* is likely to lead to new insights into the evolution of innate immunity.

Drosophila melanogaster as a Model Host

The common fruit fly *D. melanogaster* is arguably the most extensively studied nonvertebrate model organism in modern biology. Its popularity dates back over a century, when it was adapted with great success by Thomas Morgan (Nobel Laureate in Medicine or Physiology, 1933) for fundamental genetic studies. Early scientific landmarks made with *Drosophila* include confirmation of the chromosomal theory of inheritance with the discovery of sex linkage and homologous recombination. Over the ensuing decades, its popularity grew as it proved to be a facile experimental organism given its small size (adults are 3 mm in length), ease of rearing in the laboratory, and short generation time. The natural habitat of *Drosophila* is spoiled fruit and other decaying organic matter. In this environment, the fruit fly encounters a wide variety of bacterial and fungal organisms that can act as potential pathogens. Although lacking in an adaptive B- and T-lymphocyte-based immune system, *Drosophila* possesses a potent inducible innate (i.e., germ line-encoded) immune system that responds with specificity to a broad spectrum of microbial pathogens. As described in the introduction to this chapter and in Figure 23.3, recent work has shown that the signaling pathways that activate the innate immune responses in fruit flies are remarkably similar to those of the mammalian innate immune system, suggesting that these systems are ancient in origin, having evolved prior to the divergence of invertebrates and vertebrates.

Most of the pathogenesis-related work that has been done with *Drosophila* has concerned host immunity. These experiments were primarily carried out with *E. coli*, which elicits a very strong immune response when injected into the *Drosophila* body cavity (septic injury model), and with the entomopathogenic fungus, *Beauveria bassiana*, which causes a natural infection in flies. Recently, however, investigators have also turned their attention to *Drosophila* for the study of bacterial virulence mechanisms, using the septic injury model originally developed for immunity studies. Fruit flies die approximately 24 h after being pricked in the dorsal thorax with a needle dipped in *P. aeruginosa*. In contrast, *E. coli* and *Stenotrophomonas maltophilia* are rapidly cleared from the thorax after inoculation and flies survive. The *P. aeruginosa* TTSS apparatus (see chapter 15) is required for normal fly killing, although the translocated bacterial effector proteins and the intracellular targets in *Drosophila* are not known. In addition, the *P. aeruginosa* chemotaxis-like gene cluster *pilGHIJKL-chpABCDE* is important but not required for fly killing. Strains with mutations in these gene clusters lack twitching motility; however, other twitching-motility mutants, including type IV pili regulatory (*pilR* and *fimS*) and structural genes (*pilE*, *pilW*, and *pilX*), are not impaired in fly killing, suggesting that twitching motility in and of itself is not required for the pathogenic process in insects. With a complete genome, a well-described innate immune system, and a facile assay, the fruit fly is an attractive host for a number of human pathogens.

Arabidopsis thaliana as a Model Host

In addition to *C. elegans* and *D. melanogaster*, a third model genetic host, the plant *A. thaliana*, is widely used for studies of bacterial and fungal pathogenesis. The study of plant pathology has a long history because of the relevance of plant pathogens to agricultural production. However, it was not until *Arabidopsis* was developed as a model laboratory system that the plant innate immune response to pathogen attack was subjected to systematic genetic analysis. *Arabidopsis* is a small plant in the mustard family that has a small stature (about 10 cm), fast generation time (3 to 4 weeks), copious production (thousands) of tiny (20 μg) seeds, mapped molecular markers, and a fully sequenced genome. It is also straightforward to generate transgenic *Arabidopsis* plants expressing a variety of pathogenesis-related genes. These genetic and genomic features of *Arabidopsis*, which are similar to those of *C. elegans* and *Drosophila*, facilitate the identification of defense-related mutants and the cloning of the corresponding genes. These features also facilitate the study of the role of pathogen virulence factors, especially protein effectors that are delivered directly to plant cells by the TTSS.

Because *Arabidopsis* has not been shown to be a host for specialized human pathogens, the relevance of *Arabidopsis* for this chapter is primarily its use as a host for *P. aeruginosa*. It is not likely that the ability of human clinical isolates of *P. aeruginosa* such as strain PA14 to infect plants is a laboratory artifact. *P. aeruginosa* infects and propagates within plant leaves and roots, similarly to well-studied plant pathogens such as *P. syringae*, and it is able to specifically penetrate plant cells via small holes that it forms in the cellulose cell walls (see Figure 23.6). When an *Arabidopsis* leaf is exposed to *P. aeruginosa* PA14, the bacterial cells attach to the leaf surface, congregate over the stomata (opening in the leaf surface where gas exchange occurs), invade through the stomata or wounds, colonize the intercellular spaces in

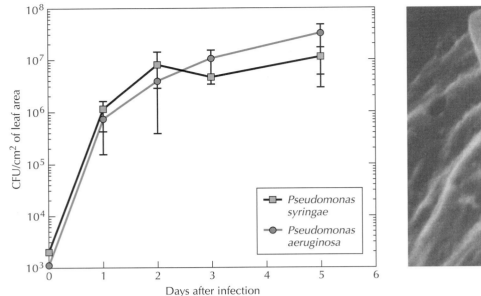

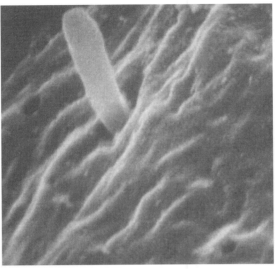

Figure 23.6 *P. aeruginosa* proliferates in plant leaves similarly to the well-characterized plant pathogen *Pseudomonas syringae,* and *P. aeruginosa* attaches perpendicularly to and perforates plant cell walls. As seen in the panel on the left, when infiltrated directly into an *Arabidopsis* leaf, *P. aeruginosa* proliferates in the intercellular spaces in the leaf similarly to the well-studied plant pathogen *P. syringae.* Once inside the leaf, *P. aeruginosa* attaches perpendicularly to plant cell walls and in some mesophyll cells gains entry via small holes that it makes in plant cell walls.

the leaf, and disrupt plant cell walls and membranes. PA14 also spreads throughout the leaf by traveling along the vascular structures in the leaf that are used to transport nutrients and water. The infection results in total maceration and rotting of the leaf and ultimately in systemic spread of PA14 to the rest of the plant. From these types of observations, there is no doubt that *P. aeruginosa* PA14 is a fully functional facultative pathogen of *Arabidopsis* that is capable of causing both local and systemic infection, which can result in the death of the infected plant.

P. aeruginosa PA14 will also infect and cause disease in a variety of other plants, including lettuce, which can be used very effectively to screen *P. aeruginosa* mutant libraries for less virulent mutants. Several *P. aeruginosa* clones can be efficiently and rapidly inoculated in a single lettuce leaf using tooth picks or disposable plastic pipette tips. Wild-type clones cause the formation of a brown disease lesion at the site of infection, whereas avirulent mutants fail to cause symptoms. Importantly, the ability of *P. aeruginosa* to cause disease in plants is not limited to PA14. Commonly studied *P. aeruginosa* strains, including PAO1 and PAK, also elicit disease in lettuce plants.

As summarized in Figure 23.3, the *Arabidopsis* innate immune response has been subjected to intense genetic analysis, primarily by the isolation and characterization of *Arabidopsis* mutants that are more resistant or more susceptible to pathogen attack. It is still an open question, however, whether elucidation of the molecular details of the plant defense response will yield new insights into the process of mammalian innate immunity.

Galleria mellonella (Wax Moth Caterpillar) as a Model Host

Drosophila is an ideal insect host for genetic analysis of the host innate immune response. However, because *Drosophila* is relatively small, it is difficult to inject bacteria directly into the body cavity. An attractive alternative insect host for experiments in which host genetic capability is not important is the greater wax moth caterpillar, *Galleria mellonella*. *G. mellonella* has been used as a model by insect physiologists for many years because it can be reared easily in captivity and because it is relatively large (250 mg), which enables the rapid injection of defined doses of bacteria by hand with a syringe (Figure 23.7), examination of the pathology of the infection, and the calculation of LD_{50}s. Moreover, there is an extensive literature on microbial pathogenesis in *G. mellonella*. Finally, *G. mellonella* is commercially available in both the United States and Europe at a low price from sports outlets because the larvae are commonly used as fishing bait. These attributes of *G. mellonella* prompted its development as a model system to study *P. aeruginosa* virulence.

Whereas *G. mellonella* larvae are impervious to infection by *P. aeruginosa* by dipping the larvae in a *P. aeruginosa* culture, as in *Drosophila*, even a single wild-type *P. aeruginosa* bacterium injected through the cuticle into the hemolymph can result in death. Following injection, *P. aeruginosa* replicates quickly in the larvae, reaching a final density of more than 10^{10} bacteria/g of larval tissue. Larval death occurs at a bacterial density of approximately 10^9 bacteria/g. Interestingly, when the LD_{50}s of a series of *P. aeruginosa* mutants that had been isolated previously using the *Arabidopsis* or *C. elegans* models were determined in *G. mellonella*, there was a significant correlation between an increased LD_{50} in *G. mellonella* larvae and reduced lethality in mice. This correlation, which is illustrated in Figure 23.8, indicates that the underlying mechanism of infection may be similar in mice and *G. mellonella* and that *G. mellonella* is an excellent model for the identification and study of bacterial virulence factors relevant to mammalian pathogenesis. In contrast, the severities of the effects of particular *P. aeruginosa* mutations do not correlate well in plants, nematodes, and mice. While it is well known that a

Figure 23.7 Wax moth caterpillars (*Galleria mellonella*) are large enough to be directly injected by hand with pathogenic bacteria. This allows very accurate determination of LD_{50}s. In the *P. aeruginosa* strain PA15, the LD_{50} varies from about 1 to 10 cells, depending on the particular experimental conditions.

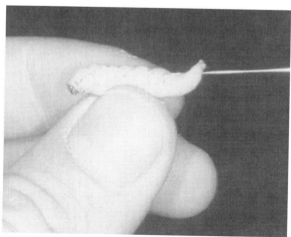

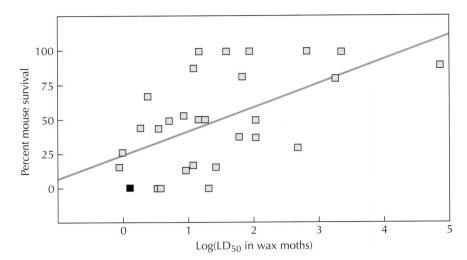

Figure 23.8 Relative importance of *P. aeruginosa* PA14 virulence factors in wax moths and mice. Each data point represents an individual *P. aeruginosa* PA14 mutant that was identified by screening in plants or nematodes (see Figures 23.1 and 23.2) and was subsequently tested for virulence in wax moth caterpillars and in mice. The black square represents wild-type PA14.

variety of mammalian pathogens can infect insects, including the bacterial pathogens *Proteus mirabilis, Proteus vulgaris,* and *Serratia marcescens* and the human fungal pathogens *Aspergillus fumigatus* and *Fusarium oxysporum,* the experiment shown in Figure 23.4 was the first study in which the virulence of otherwise isogenic microbial mutants was compared in mammalian and insect model systems.

A recent study comparing the role of *P. aeruginosa* TTSS effectors in plants, nematodes, insects, and mice illustrates the utility of the *G. mellonella* model. *P. aeruginosa* makes four known TTSS effectors, ExoU, ExoT, ExoY, and ExoS. In mammalian systems, these *P. aeruginosa* effector proteins have been shown to disrupt the actin cytoskeleton, interfere with cell matrix adherence, inhibit internalization by macrophages, and induce apoptosis. Most *P. aeruginosa* strains have the *exoS* or the *exoU* gene, but not both. The *P. aeruginosa* strain (PA14) that was used in the interspecies TTSS studies has the *exoU, exoT,* and *exoY* genes. Interestingly, no role for any of these genes could be demonstrated in pathogenesis in *C. elegans* or *Arabidopsis.* However, either *exoU* or *exoT,* but not both, was required for a significant level of pathogenicity in both *G. mellonella* and in a mammalian cytopathology assay.

Miscellaneous Hosts

A variety of other nonvertebrate organisms have been used as model hosts for human pathogens. For example, traditional cell culture-pathogen systems have been adapted with the use of free-living, genetically tractable unicellular organisms as model hosts. *D. discoideum* is a social amoeba, meaning that it has the remarkable capability for both unicellular and multicellular life forms. Its natural habitat is topsoil and decaying vegetation, where it feeds on bacteria and reproduces by binary fission. In the laboratory, *Dictyostelium* can be maintained on lawns of nonpathogenic bacteria, usually *Klebsiella aerogenes* or *E. coli,* and some anexic strains can grow on rich nutrient media. Importantly, the 34-Mb haploid genome of *Dic-*

tyostelium is being fully sequenced and can be manipulated by a wide array of genetic methodologies. These attributes have made *Dictyostelium* an inviting experimental model for developmental biologists interested in such fundamental processes as signal transduction, chemotaxis, phagocytosis, and vesicular trafficking. Recently, a number of investigators have turned their attention to using *Dictyostelium* as a model system for the study of pathogen-host interaction.

When cocultured with *Dictyostelium, Legionella pneumophila* is internalized and replicates within membrane-bound phagosomes closely associated with the amoebic rough endoplasmic reticulum. Vesicular trafficking is disrupted, preventing lysosome and endosome fusion with bacteria-laden phagosomes. These cellular processes mirror those observed after phagocytosis of *L. pneumophila* by human macrophages as well as by its natural host, freshwater amoebae (see chapter 10). Genetic analysis has identified a number of *Legionella* determinants important for survival and growth within macrophages. Genes of the *dot/icm* locus, which encode products of a bacterial secretion apparatus required for replication in macrophages, are also required for growth in *Dictyostelium*. Genetic screens for avirulent *L. pneumophila* may identify additional factors important for growth in *Dictyostelium* and possibly macrophages. From the host perspective, genetic screens for *Dictyostelium* mutants defective in supporting *L. pneumophila* growth may identify key amoebic targets for *Legionella* toxins, and homologues in macrophages may have corresponding functions.

Dictyostelium has also been used as a host for the extracellular pathogen, *P. aeruginosa*. In this system, replacement of the nonpathogenic bacterial food source with *P. aeruginosa* leads to rapid amoebic cell death. In contrast to the nonpathogenic bacteria used for nutriment, *P. aeruginosa* is not readily phagocytized by *Dictyostelium*. As discussed earlier, quorum-sensing and type III secretion systems are important for the pathogenic lifestyle of *P. aeruginosa* in a number of experimental models. In *Dictyostelium, P. aeruginosa* quorum-sensing and TTSS mutants are significantly impaired in their ability to prevent amoebic growth, although the mechanism of cell death remains an open question. With its ease of use and powerful genetic system, *Dictyostelium* may prove to be a valuable host model for a number of intracellular and extracellular human pathogens.

Another protist, the free-living solitary amoeba *Acanthamoeba castellanii*, can act as both an environmental and experimental host for a number of human pathogens, including *L. pneumophila, C. neoformans,* and *Mycobacterium avium,* among others. Like *Dictyostelium,* many of the cellular processes observed after phagocytosis of these pathogens by *Acanthemoeba* are similar to those described after macrophage phagocytosis. Preventing full development of *Acanthemoeba* as a model host organism, however, is the availability of only a limited repertoire of experimental genetic methods.

An alternative approach to investigating the function of microbial products that manipulate host cells is to express these bacterial toxins within a genetically tractable host system. The cellular targets of many bacterial proteins translocated by TTSSs are not known. This has led investigators to express TTSS-translocated toxins of *Yersinia* and *Salmonella* in *Saccharomyces cerevisiae.* Remarkably, the subcellular location of many of the bacterial proteins, fused to *Aqueoria victoria* green fluorescent protein (GFP), corresponded to those known locations in mammalian cells. Presumably, the structural motifs recognized by the introduced toxin are conserved between yeast and mammalian cells. Expression of bacterial products within yeast also allows eukaryotic cell death, a second marker of interaction between the bacterial proteins and host factors, to be directly accessed. It remains to be seen whether the protein-protein interactions identified for TTSS-translocated proteins of unknown function hold true in mammalian cells.

Moving in the opposite direction in terms of organismal complexity, fish are being used as simple vertebrate model hosts for a number of human pathogens. Fish offer a number of advantages as simple model host systems. One important difference with other simple hosts is that they have both adaptive and innate immune systems, including B and T lymphocytes, macrophages, granulocytes, and primitive circulatory systems. In addition, model fish commonly used in the laboratory, such as zebrafish (*Danio rerio*), are small, have rapid generation times, and are easy to experimentally manipulate.

Zebrafish have well-developed genetic systems and transparent bodies for much of their early development, allowing for the direct observation of organogenesis. Current efforts for zebrafish include ongoing genomic sequencing and expressed sequence tag (EST) projects. For these reasons, zebrafish have become a popular model for the study of a number of vertebrate developmental issues, including development of the immune system. Recently, *S. pyogenes* infection has been modeled in zebrafish. Fish injected with *S. pyogenes* into the peritoneum or dorsal muscle develop focal necrotic lesions with limited inflammatory cell infiltrates, which progress with direct extension to eventually cause fish death. *S. iniae,* a natural pathogen of fish that occasionally causes disease in humans, also causes death within a few days in wounded zebrafish, but infection is characterized by widespread systemic infection to nearly all organ tissues and a brisk pyogenic reaction. A mutant of the *S. pyogenes* regulatory gene *ropB* (also known as *rgg*) was shown to be important for fish killing. RopB is required for expression of streptococcal erythrogenic toxin B (SPE B), an extracellular cysteine protease implicated in tissue destruction and invasion in streptococcal necrotizing soft tissue infections. In a proof-of-principle experiment, a library of *S. pyogenes* mutants were screened in the model, and two mutants—one with a transposon insertion in peptidoglycan *N*-acetylglucosamine deacetylase and the other in a putative ATP-binding cassette (ABC) transporter—were identified as being attenuated.

Interactive Host-Pathogen Genetics

In addition to carrying out host genetic analysis, an advantage of using model genetic hosts for host-pathogen studies is the capability of utilizing an interactive genetic approach to uncover specific interactions between bacterial virulence determinants and corresponding host responses. Imagine a situation where a particular pathogen virulence factor targets a host defense function. A host mutant that specifically allowed the pathogen mutant to regain full virulence might correspond to the target of the virulence factor. A published example that illustrates this type of interactive genetics concerns *P. aeruginosa*-mediated killing of *C. elegans.* Pseudomonads secrete a variety of small toxic molecules, including redox-active phenazine derivatives. As described earlier in this chapter, *P. aeruginosa* kills *C. elegans* by at least two different mechanisms. One method, "fast killing," does not depend on live bacteria and involves low-molecular-weight toxins that kill relatively quickly. Because *P. aeruginosa* mutants that fail to secrete phenazines are not capable of carrying out rapid toxin-mediated killing, it was hypothesized that phenazines are involved in *C. elegans* killing. Phenazines are thought to be toxic to host cells because of the production of the reactive oxygen species generated when they react with molecular oxygen. Consistent with this model, *C. elegans* mutants that are resistant to oxidative stress, for example, the long-lived *age-1* mutant, are resistant to *P. aeruginosa* toxin-mediated killing, whereas *C. elegans* mutants such as *mev-1* and *rad-8*, which are more sensitive to oxidative stress, are more sensitive to toxin-mediated killing.

Interactive genetic analysis was also used to define the role of P-glycoproteins (PGP) in protecting *C. elegans* from the toxic effects of phenazines. PGPs are members of the ABC transporter family that play a key role in pumping toxic environmental molecules out of cells. *C. elegans* mutants containing a lesion in the *pgp-1* gene exhibit enhanced susceptibility to both wild-type *P. aeruginosa* strains and *P. aeruginosa* mutants that produce phenazines but which are normally less virulent in the *C. elegans* killing assay. On the other hand, *pgp-1* mutants do not exhibit enhanced susceptibility to *P. aeruginosa* phenazine mutants. One way to interpret these results is that PGP-1 specifically pumps phenazines out of *C. elegans* cells.

Conclusion

As described in this chapter and summarized in Table 23.3, simple nonvertebrate hosts can be used very effectively to model various aspects of mam-

Table 23.3   General characteristics of nonvertebrate host-pathogen model systems

Host	Route of infection	Duration of assay	Comments
Arabidopsis thaliana	Leaf infiltration	2–5 days	• Signs of infection (chlorosis, soft-rot) can be followed in real time • CFU/g of tissue can be obtained • Bacterial libraries can be screened initially in a variety of plants
Caenorhabditis elegans	Coculture	4–48 h (intoxication models) 2–10 days (infection models)	Intoxication models • Can be reproduced with toxin alone Infection models • Multifactorial pathogenesis • Signs of infection (colonization of the intestinal tract, biofilm formation on cuticle, lethargy) can be followed microscopically in real time • CFU/worm can be obtained
Drosophila melanogaster	Needle prick into dorsal thorax	1–7 days	• Well-studied innate immune system • CFU/fly can be obtained • Usually semiquantitative
Galleria mellonella (wax moth caterpillar)	Hemolymph injection with syringe	1–3 days	• Quantitative, LD$_{50}$s can be calculated • Experiments can be performed at 37°C • CFU/caterpillar can be obtained • No host genetics
Danio rerio (zebrafish)	Intraperitoneal or intramuscular injection	1–60 days	• Quantitative, LD$_{50}$s can be calculated • Adaptive and innate immune systems • Signs of infection (skin lesions, loss of scales) can be followed in real time • CFU/g of tissue (e.g., skin, heart, and kidney) can be obtained • Histopathology
Dictyostelium discoideum	Coculture	2–4 days	• Genetic model of phagocytosis

malian pathogenesis. Two major advantages of these alternative hosts in comparison with mammals have emerged. First, because large numbers of the simple hosts can be used, entire microbial genomes can be efficiently scanned for every gene that might play a role in virulence. Second, in the case of the model genetic hosts such as *Drosophila* and *C. elegans*, host genetic analysis can be used to dissect the molecular mechanisms underlying the host defense response. The use of alternative simple hosts for the study of human pathogens is now becoming relatively widespread, and it will be very interesting to see what new insights into the molecular basis of pathogenesis will be achieved by using this experimental approach.

Selected Readings

Aballay, A., P. Yorgey, and F. M. Ausubel. 2000. *Salmonella typhimurium* proliferates and establishes a persistent infection in the intestine of *Caenorhabditis elegans*. *Curr. Biol.* **10:**1539–1542.

Asai, T., G. Tena, J. Plotnikova, M. R. Willmann, W.-L. Chiu, L. Gomez-Gomez, T. Boller, F. M. Ausubel, and J. Sheen. 2002. MAP kinase signaling cascade in *Arabidopsis* innate immunity. *Nature* **415:**977–983.

> This work describes the identification and characterization of a MAP kinase cascade that induces the expression of *Arabidopsis* early-defense genes in response to bacterial flagellin.

Darby, C., C. L. Cosma, J. H. Thomas, and C. Manoil. 1999. Lethal paralysis of *Caenorhabditis elegans* by *Pseudomonas aeruginosa*. *Proc. Natl. Acad. Sci. USA* **96:**15202–15207.

Darby, C., J. W. Hsu, N. Ghori, and S. Falkow. 2002. *Caenorhabditis elegans*: plague bacteria biofilm blocks food intake. *Nature* **417:**243–244.

Fauvarque, M. O., E. Bergeret, J. Chabert, D. Dacheux, M. Satre, and I. Attree. 2002. Role and activation of type III secretion system genes in *Pseudomonas aeruginosa*-induced *Drosophila* killing. *Microb. Pathog.* **32:**287–295.

Garsin, D. A., C. D. Sifri, E. Mylonakis, X. Qin, K. V. Singh, B. E. Murray, S. B. Calderwood, and F. M. Ausubel. 2001. A simple host for identifying Gram-positive virulence factors. *Proc. Natl. Acad. Sci. USA* **98:**10892–10897.

Kim, D. H., R. Feinbaum, G. Alloing, F. E. Emerson, D. A. Garsin, H. Inoue, M. Tanaka-Hino, N. Hisamoto, K. Matsumoto, M.-W. Tan, and F. M. Ausubel. 2002. A conserved p38 MAP kinase signaling pathway in *Caenorhabditis elegans* innate immunity. *Science* **297:**623–626.

> In this paper, a screen of *C. elegans* mutants that were hypersusceptible to *Pseudomonas aeruginosa* infection led to the infection of a p38-like MAP kinase signaling pathway important in defense against gram-negative and gram-positive bacterial pathogens.

Lemaitre, B., J. Reichhard, and J. Hoffmann. 1997. *Drosophila* host defense: differential induction of antimicrobial peptide genes after infection by various classes of microorganisms. *Proc. Natl. Acad. Sci. USA* **94:**14614–14619.

Lesser, C. F., and S. I. Miller. 2001. Expression of microbial virulence proteins in *Saccharomyces cerevisiae* models mammalian infection. *EMBO J.* **20:**1840–1849.

Mahajan-Miklos, S., M.-W. Tan, L. G. Rahme, and F. M. Ausubel. 1999. Molecular mechanisms of bacterial virulence: *Caenorhabditis elegans* pathogenesis model. *Cell* **96:**47–56.

> This paper describes the *C. elegans-P. aeruginosa* fast-killing model and shows that the system could be used to identify virulence factors important for both *C. elegans* and mammalian pathogenesis. In addition, the work shows that the host and pathogen could be simultaneously examined genetically to determine the mechanisms of virulence of the *P. aeruginosa* phenazine toxins.

Mallo, G., C. Kurz, C. Couillault, N. Pujol, S. Granjeaud, Y. Kohara, and J. Ewbank. 2002. Inducible Antibacterial Defense System in *C. elegans*. *Curr. Biol.* **12:**1209–1214.

> In this paper, a cDNA microarray analysis of *C. elegans* demonstrated induction of several nematode lysozymes and lectins in response to *Serratia marcescens* infection. Previous re-

search had shown a similar pattern of expression under the control of the *C. elegans* TGF-β-related signaling molecule DBL-1. A *dbl-1* mutant was shown to be hypersusceptible to *S. marcescens* infection.

Mylonakis, E., F. M. Ausubel, J. R. Perfect, J. Heitman, and S. B. Calderwood. 2002. Killing of *Caenorhabditis elegans* by *Cryptococcus neoformans* as a model of yeast pathogenesis. *Proc. Natl. Acad. Sci. USA* **99:**15675–15680.

Neely, M. N., J. D. Pfeifer, and M. Caparon. 2002. Streptococcus-zebrafish model of bacterial pathogenesis. *Infect. Immun.* **70:**3904–3914.

Rahme, L., E. Stevens, S. Wolfort, J. Shao, R. Tompkins, and F. M. Ausubel. 1995. Common virulence factors for bacterial pathogenicity in plants and animals. *Science* **268:**1899–1902.

In this landmark paper, clinical *Pseudomonas aeruginosa* isolates were shown to cause soft-rot disease in *Arabidopsis thaliana* and severe disease in a murine infection model. Three *P. aeruginosa* pathogenicity-related genes (*toxA, plcS,* and *gacA*) were shown to be essential for infectivity in both plants and animals.

Solomon, J. M., A. Rupper, J. A. Cardelli, and R. R. Isberg. 2000. Intracellular growth of *Legionella pneumophila* in *Dictyostelium discoideum,* a system for genetic analysis of host-pathogen interaction. *Infect. Immun.* **68:**2939–2947.

Tan, M.-W., S. Mahajan-Miklos, and F. M. Ausubel. 1999. Killing of *C. elegans* by *P. aeruginosa* used to model mammalian bacterial pathogenesis. *Proc. Natl. Acad. Sci. USA* **96:**715–720.

Tzou, P., E. De Gegorio, and B. Lemaitre. 2002. How *Drosophila* combats microbial infection: a model to study innate immunity and host-pathogen interactions. *Curr. Opin. Microbiol.* **5:**102–110.

About the Contributors

Klaus Aktories is professor and director of the Institute of Pharmacology and Toxicology (Department I) at the University of Freiburg, Freiburg, Germany. He started his scientific career with studies on the regulation of adenylyl cyclase by G proteins, cholera toxin, and pertussis toxin. Since 1986, his main research topics have been bacterial protein toxins acting on actin and small GTPases. Major work focused on actin-ADP-ribosylating toxins, C3-like transferases which ADP-ribosylate Rho proteins, the family of large clostridial toxins, and toxins which deamidate Rho GTPases. He is interested in the structure-function analysis of the toxins and in their actions on signal transduction processes.

Frederick M. Ausubel is a Professor of Genetics at Harvard Medical School, Boston, Mass., and has been a member of the Department of Molecular Biology at Massachusetts General Hospital since its inception in 1982. His initial research centered on the molecular mechanisms of nitrogen fixation in plant symbionts and virulence in plant pathogens using the model plant *Arabidopsis thaliana*. Since 1995, his group has also focused on the study of human bacterial and fungal pathogenesis and host defense mechanisms using simple organisms, such as *Arabidopsis*, the nematode *Caenorhabditis elegans*, and the insects *Drosophila melanogaster* and *Galleria mellonella* (wax moth caterpillar), as alternative model hosts.

David N. Baldwin is at the Fred Hutchinson Cancer Research Center in Seattle, Wash. He received his Ph.D. in virology from the University of Washington, Seattle, Wash., for work on human foamy virus replication and subsequently conducted postdoctoral research at Stanford University School of Medicine on host cell transcriptional responses to infection by *Listeria monocytogenes* and at the Fred Hutchinson Cancer Research Center on genomics of *Helicobacter pylori*.

Kenneth Bell is a researcher at the Scottish Crop Research Institute, Invergowrie, Dundee, Scotland. He worked at the Sanger Institute Pathogen Sequencing Unit, Hinxton, Cambridge, United Kingdom, where he studied the genome sequences of *Erwinia carotovora* subsp. *atroseptica* and *Chlamydophila abortus*.

Stephen Bentley is a project manager in the Sanger Institute Pathogen Sequencing Unit, Cambridge, United Kingdom. His work covers many aspects of bacterial genome analysis, with particular emphasis on the evolution of chromosome structure.

Avri Ben-Ze'ev works in the Department of Molecular Cell Biology, Weizmann Institute of Science, Rehovot, Israel, studying different aspects of cell-cell interactions, especially the roles of β-catenin and plakoglobin in signaling and tumorigenesis. He received his Ph.D. at the Hebrew University of Jerusalem.

Alexander D. Bershadsky works in the Department of Molecular Cell Biology, Weizmann Institute of Science, Rehovot, Israel, studying the role of cytoskeleton and cell contractility in formation of focal adhesions and adherens junctions. He received his Ph.D. at the Cancer Research Center of the Russian Academy of Medical Sciences, Moscow.

Patrice Boquet is Professor of Bacteriology at the Nice University School of Medicine and works in an INSERM laboratory at the University of Nice Sophia Antipolis, France.

Michael Caparon is currently a Professor of Molecular Microbiology at the Washington University School of Medicine, St. Louis, Mo. Following undergraduate training at Michigan State University and graduate training at the University of Iowa, he began his work on the genetics and biology of the streptococci during his postdoctoral studies under the guidance of June Scott at Emory University. His work has focused on the development of streptococcal genetics and its application in understanding the interaction of *Streptococcus pyogenes* with host epithelial cells. His work has contributed to the identification of adhesins, the regulation of their expression, and their role in manipulating the signaling responses of epithelial cells.

Ana Cerdeño-Tarraga is a Senior Computer Biologist at the Wellcome Trust Sanger Institute Pathogen Sequencing Unit. Her work is focused mainly on pathogen genome annotation and analysis.

G. Singh Chhatwal is head of the Department of Microbial Pathogenesis and Vaccine Research at the National Research Centre for Biotechnology (GBF), Braunschweig, Germany. He also belongs to the Basic Sciences Faculty of the Technical University, Braunschweig. His research has focused mainly on the interaction of gram-positive bacteria with extracellular matrix and plasma proteins. His group was the first to identify SfbI, a fibronectin-binding protein, which is the major adhesin and invasin of group A streptococci. His current research is on the development of antiadhesive streptococcal vaccines. Before joining GBF, he was at the University of Giessen working on bacterial toxins and was involved in the identification of C3 botulinum toxin.

Su L. Chiang received her Ph.D. in microbiology and molecular genetics from Harvard Medical School, Boston, Mass. Her research has focused on

Vibrio cholerae pathogenesis and the development of transposon mutagenesis methods. She is the scientific assistant to the Chair of the Department of Microbiology and Molecular Genetics at Harvard Medical School.

Peter J. Christie received his Ph.D. in microbiology with Gary Dunny, Cornell University, Ithaca, N.Y., in 1987. He then trained as a postdoctoral fellow with Eugene Nester, University of Washington, Seattle, Wash., and Virginia Walbot, Stanford University, Stanford, Calif. In 1991, he joined the Microbiology and Molecular Genetics Department at The University of Texas Health Sciences Center, Houston, Tex. His laboratory characterizes the structure and function of the T-DNA transport system of *Agrobacterium tumefaciens.*

Pascale Cossart is the head of the Unité des Interactions Bactéries-Cellules, INSERM 604 in the Pasteur Institute, Paris, France. She is also a Howard Hughes International investigator and a member of the French Academy of Sciences. She has been working on the molecular basis of *Listeria monocytogenes* infection since 1986. Her studies range from genetics and genomics to cell biology and physiopathology. Major achievements of her group concern the analysis of the actin-based motility of *Listeria* and *Rickettsia,* the identification of the major invasion proteins of *Listeria* and of their mammalian receptors, the identification of an RNA thermosensor, the determination of the genome sequences of *L. monocytogenes* and *Listeria innocua,* and the discovery of how bacteria cross the intestinal and placental barriers.

Antonello Covacci was born in Tuscania, Italy, on 14 December 1957. During his medical studies, he was a junior fellow of the Sclavo Research Center (now Chiron Vaccines), Siena, Italy. He graduated summa cum laude from the University of Florence Medical School. His postdoctoral work was performed at the Hormone Research Institute, University of California, San Francisco. He is Executive Director of Research at . His laboratory is working on virulence mechanisms in *Helicobacter pylori* and *Streptococcus pneumoniae.*

Lisa Crossman is a Senior Computer Biologist at the Sanger Institute. She has previously worked on quorum sensing and aspects of the nitrogen cycle and is currently interested in the genomics of pathogenic and environmental bacteria.

Chantal de Chastellier is a Senior Researcher, employed by Institut National de la Santé et de la Recherche Médicale, Paris, France. She worked at the Institut Pasteur, Paris, France, from 1976 to 1989, and at the Necker Faculty of Medicine, Paris, France, in INSERM Unit 411, from 1989 to 2000. She joined Jean Pierre Gorvel's team at the Centre d'Immunologie de Marseille-Luminy, Marseille, France, in 2001. Here she is in charge of a program designed to elucidate the survival strategies of *Mycobacteria* within macrophages. She is also in charge of electron microscopy projects as an approach to the study of survival strategies of other pathogens (*Salmonella, Brucella, Chlamydia*). Her research activity first involved work on the ultrastructure of bacteria and then on endocytosis in the amoeba *Dictyostelium discoideum,* with special emphasis on the morphological characterization of organelles of the endocytic pathway, endomembrane trafficking, and exchange of endocytic contents. Since 1983, her work has specialized in endocytic and phagocytic processing in macrophages with the goal of characterizing the intraphagosomal survival strategies of pathogens.

Ennio De Gregorio completed his Ph.D. studies between 1996 and 2000 at EMBL, Gene Expression Program, Heidelberg, Germany, in the laboratory of M. W. Hentze. Here, he worked on the control of mRNA translation. From 2000 to 2003, he was a postdoctoral student at CNRS, Gif-sur-Yvette, France, in the laboratory of B. Lemaitre, where he studied the *Drosophila* innate immune response to infection. Since 2003, he has worked at Chiron Vaccines, Siena, Italy.

Michela Felberbaum-Corti has worked as a Ph.D. student in the Department of Biochemistry at the University of Geneva, Geneva, Switzerland, since 2000.

B. Brett Finlay is a Professor in the Biotechnology Laboratory and the Departments of Biochemistry and Molecular Biology and Microbiology and Immunology at the University of British Columbia. His research interests focus on host-pathogen interactions at the molecular level. By combining cell biology with microbiology, he has been at the forefront of the emerging field called cellular microbiology, making several fundamental discoveries in this field and publishing more than 200 papers. His laboratory studies several pathogenic bacteria, with *Salmonella* and pathogenic *Escherichia coli* interactions with host cells being the primary focus.

Raluca Flukiger-Gagescu received her Ph.D. from the University of Geneva, Geneva, Switzerland, in 2000. She then worked as associate editor for *Nature Reviews: Molecular Cell Biology* in London from 2000 to 2002. In 2003, she joined the technology transfer office of the University of Geneva (UNITEC), Geneva, Switzerland.

Åke Forsberg is research director at the Department of Medical Countermeasures at the Swedish Defence Research Agency, Umeå, Sweden. His work has focused on the regulation and function of virulence determinants in pathogenic bacteria. His current work involves characterization of virulence determinants of *Yersinia pestis*, *Pseudomonas aeruginosa*, and *Francisella tularensis*.

Matthew S. Francis obtained his Ph.D. at the Department of Microbiology and Immunology, University of Adelaide, Adelaide, Australia. Following a period of postdoctoral research with the *Yersinia* group at the Department of Molecular Biology, Umeå University, Umeå, Sweden, he established his own research group in Umeå. Using *Yersinia pseudotuberculosis* and *Pseudomonas aeruginosa* as model pathogens, he focuses his research on understanding the regulatory and structural mechanisms that define type III-mediated translocation of anti-host factors into target cells.

Benjamin Geiger works in the Department of Molecular Cell Biology, Weizmann Institute of Science, Rehovot, Israel, studying different aspects of cell-cell and cell-extracellular matrix interactions and their role in cell regulation and signaling. He received his Ph.D. at the Weizmann Institute of Science.

Jean Gruenberg received his Ph.D. from the University of Geneva, Geneva, Switzerland, in 1980. He was then a postdoctoral student at the University of California, Riverside, and at the EMBL from 1980 to 1987. After being a group leader at the EMBL from 1987 to 1993, he moved to the University of Geneva, where he has been a professor since 1994.

Matthew Holden is a Senior Computer Biologist in the Sanger Institute Pathogen Sequencing Unit. His work includes the annotation and analysis of microbial genomes, with particular interest in species evolution and gene transfer.

Ilona Idanpaan-Heikkila received medical training in Helsinki, Finland. She joined the laboratory of Elaine Tuomanen, then at the Rockefeller University. After a postdoctoral stay in the laboratory of Arturo Zychlinsky at the Skirball Institute, New York University Medical Center, she joined the vaccine division of SmithKline Beecham.

Frank Lafont has earned his Ph.D. and is a senior scientist at the Centre Médical Universitaire, University of Geneva. He has worked as a postdoctoral fellow at the EMBL, Heidelberg, Germany, on lipid raft-mediated sorting mechanisms in the laboratory of Kai Simons. He then established his own research project, focusing on the role of rafts in host-pathogen interactions, and joined the laboratory of F. G. van der Goot (Geneva, Switzerland) who is studying the raft-mediated mechanism of action of bacterial toxins.

Marc Lecuit has earned his M.D.-Ph.D. degree and is interested in developing the interface between clinical infectious diseases and cellular microbiology. He is working in the Department of Infectious Diseases at Necker Hospital and in the Pascale Cossart research unit (Unité des Interactions Bactéries-Cellules) at the Pasteur Institute. He is particularly interested in developing in vivo and ex vivo models for understanding the pathophysiology of human infectious diseases.

Stephen Lory is a Professor of Microbiology and Molecular Genetics at Harvard Medical School, Boston, Mass. His laboratory studies the molecular basis of pathogenesis of *Pseudomonas aeruginosa* infections in cystic fibrosis patients. He is also directing a drug-discovery effort in which functional genomics tools are used to identify new classes of antibiotics.

Mark Marsh graduated from University College London (UCL) and undertook postdoctoral work at the European Molecular Biology Laboratory (EMBL), Heidelberg, and Yale University Medical School, where he worked with Ari Helenius. He moved to the Institute of Cancer Research in London as a group leader and, in 1992, to the newly established Medical Research Council Laboratory for Molecular Cell Biology (MRC-LMCB) at UCL. He was appointed Professor of Molecular Cell Biology in 2000. His laboratory works on endocytosis and virus replication and is currently investigating the role of endocytic organelles in the assembly of human immunodeficiency virus and human cytomegalovirus.

Vega Masignani graduated with a degree in pharmaceutical chemistry and a Ph.D. in biotechnology from the University of Siena, Italy, with a thesis on the computational approach to the development of a novel *Neisseria meningitidis* protein-based vaccine. At present, she is staff scientist at the Cellular Microbiology and Bioinformatics Unit of Chiron Vaccines, conducting research in the field of computer analysis applied to microbial pathogenesis.

Frederick R. Maxfield is the Chair of the Department of Biochemistry, Weill Medical College, Cornell University, New York, N.Y. His research has focused on the use of quantitative fluorescence microscopy to study the properties of cells. Processes studied include endocytic membrane traffic, cell polarization and migration, cell signaling, cell adhesion, phagocytosis, and the cellular trafficking and distribution of lipids and cholesterol.

Sandra J. McCallum is with Labvelocity.com. She obtained her Ph.D. in biochemistry from Cornell University, Ithaca, N.Y., in the field of signal transduction by Rho family GTPases. Her postdoctoral research was conducted at Stanford University School of Medicine. Her work focused on the mechanism of actin-based motility by *Shigella flexneri*.

Timothy K. McDaniel is a Senior Scientist at Illumina, Inc., in San Diego, Calif., where he is developing biological applications on fiberoptic microarrays. Previously, he studied the genetics of *Helicobacter pylori* virulence as a postdoctoral fellow in the laboratory of Stanley Falkow at Stanford University. He received his Ph.D. in the laboratory of James Kaper at the University of Maryland, Baltimore, where he studied the genetics of enteropathogenic *Escherichia coli*.

Jeremy E. Moss received his M.D.-Ph.D. degree from the New York University Skirball Institute, New York, N.Y. He is completing his clinical training at Yale University. He received his B.A. degree from the Stanford University Program in Human Biology in 1993. During that time, he worked in the laboratory of Mark Holodniy analyzing mechanisms of human immunodeficiency virus drug resistance.

Julian Parkhill is a Senior Investigator at the Sanger Institute where he is responsible for the sequencing and analysis of bacterial genomes. His particular interest is in the evolutionary strategies and pathogenicity mechanisms of bacterial pathogens.

Dana Philpott is the leader of the Innate Immunity and Cell Signaling group at the Institut Pasteur.

Lynda M. Pierini is an assistant professor of microbiology and immunology in the Department of Surgery, Weill Medical College, Cornell University, New York, N.Y. Her research focuses on mammalian host-cell responses to bacterial invasion, including cell polarization, migration, and phagocytosis.

Javier Pizarro-Cerdà is a cell biologist and has been Chargé de Recherche at the Pasteur Institute in Pascale Cossart's laboratory since 2001. He received his M.Sc. from the University of Costa Rica working in Edgardo Moreno's laboratory and his Ph.D. from the University of Aix-Marseilles working with Jean-Pierre Gorvel. After studying the intracellular trafficking of *Brucella abortus*, he became interested in the early events of *Listeria monocytogenes* entry into target cells. Currently, he is studying *Listeria* entry into epithelial cells.

Mariagrazia Pizza obtained her Ph.D. in Chemistry and Pharmaceutical Technology at the University of Naples, Naples, Italy. Following stays at

EMBL, Heidelberg, Germany, and at the University of Naples, she joined the Laboratory of Molecular Biology at Sclavo Research Center, Siena, Italy. Her research interests have focused on the design and construction of nontoxic mutants of bacterial toxins (pertussis, cholera, and heat-labile toxins). Her main achievement has been the development of the first recombinant bacterial vaccine that contains a genetically detoxified form of pertussis toxin. She is Research Director at Chiron Vaccines.

Klaus T. Preissner is head of the Institute for Biochemistry, Medical Faculty, Justus-Liebig-University of Giessen, Giessen, Germany. The main focus of his research is dedicated to cellular adhesion mechanisms in the vascular system and in pathogen-host tissue interactions. His group identified novel mechanisms that are relevant for vascular remodeling processes related to angiogenesis and atherogenesis, inflammation, and humoral defense systems and characterized new initiation pathways in blood coagulation. Formerly, he joined the Department of Immunology, Scripps Clinic and Research Foundation, as a postdoctoral fellow and was group leader at the Clinical Research Unit for Blood Coagulation and Thrombosis as well as the Max-Planck-Institut, Kerckhoff-Klinik, Bad Nauheim, Germany.

Rino Rappuoli obtained his Ph.D. in Biological Sciences from the University of Siena, Italy. Following stays at the Rockefeller University and the Harvard Medical School, he returned to the Sclavo Research Center in Siena, first as head of the Laboratory of Bacterial Vaccines and subsequently as head of the Research and Development Division. His research interests include various aspects of bacterial pathogenesis, focusing on bacterial toxins and vaccines. His major achievements have been with the primary structures of diphtheria, pertussis, and *Helicobacter pylori* cytotoxins and the development of the first recombinant bacterial vaccine that contained a genetically detoxified form of pertussis toxin. Rino Rappuoli is Chief Scientific Officer and Vice President of Vaccines Research at Chiron Corporation.

Jennifer R. Robbins is in the Department of Biology, Xavier University, Cincinnati, Ohio, studying the effects of immune cell activation on calcium signaling. She received her Ph.D. in biochemistry at Stanford University School of Medicine, Stanford, Calif., studying the intercellular spread of *Listeria monocytogenes* by actin-based motility as well as generation of IcsA (VirG) polarity in *Shigella flexneri*.

David G. Russell is Professor and Chair of the Department of Microbiology and Immunology at the College of Veterinary Medicine, Cornell University. He trained at Imperial College, London, United Kingdom, and University of Kent, Canterbury, United Kingdom, before setting up an independent laboratory at the Max-Planck-Institut in Tuebingen, Germany. He has held faculty positions at NYU Medical Center, New York, N.Y., and at Washington University School of Medicine, St. Louis, MO. His research has always focused on the interface between the microbial pathogen and its host, and he has studied a range of eukaryote and prokaryote pathogens. His current research efforts are devoted entirely to the study of *Mycobacterium tuberculosis* and how the bacterium establishes, maintains, and transmits its infection.

Philippe J. Sansonetti earned his M.D. and is a Howard Hughes fellow, professor, and principal investigator at the Institut Pasteur, Paris, France.

He has been studying the mechanisms involved in the virulence of *Shigella,* the causative agent of bacillary dysentery, for the past 22 years. His laboratory has made key contributions in the identification and characterization of bacterial factors and cellular and tissue responses implied in the colonization of the colonic mucosa by *Shigella.*

Kurt Schesser is in the Department of Microbiology and Immunology at the University of Miami School of Medicine and leads a research group focused on host-microbe interactions. After receiving his Ph.D. in genetics at Oregon State University, he was a postdoctoral researcher in the group of Hans Wolf-Watz in Umeå, Sweden. Before accepting his current post, he led a research group in the Department of Cell and Molecular Biology, Lund University, Lund, Sweden.

Mohammed Sebaihia is involved in the annotation and analysis of bacterial genomes at the Sanger Institute.

Costi D. Sifri is an Instructor in Medicine at Harvard Medical School, Boston, Mass., and a Clinical Assistant in Medicine in the Division of Infectious Diseases at Massachusetts General Hospital. As a Howard Hughes Medical Institute Postdoctoral Fellow in the laboratories of Steven Calderwood and Frederick Ausubel at Harvard Medical School, he developed the use of *Caenorhabditis elegans* as a model host for a variety of gram-positive and other narrow host range pathogens. He is currently using these systems to study the genetic basis of *Staphylococcus* pathogenesis, *Candida* pathogenesis, and host innate immunity.

Frederick S. Southwick is Professor and Chief of the Division of Infectious Diseases, Department of Medicine, University of Florida College of Medicine, Gainesville, Fla. He is an infectious-diseases specialist with a long-standing interest in actin-based motility. His major research accomplishments include the discovery of CapG, a protein that caps the barbed ends of actin filaments in macrophages; demonstration of the association of defective neutrophil motility with impaired actin filament assembly in vivo; discovery of the central role of actin in the intracellular motility of *Listeria monocytogenes;* and identification of ABM-1 and ABM-2 oligoproline docking sites that regulate intracellular actin-based motility of *Listeria, Shigella,* and vaccinia virus.

Julie A. Theriot is in the Department of Biochemistry at Stanford University School of Medicine, Stanford, Calif., with a secondary appointment in the Department of Microbiology and Immunology. Her laboratory studies the actin-based motility of the bacterial pathogens *Listeria monocytogenes* and *Shigella flexneri.* She obtained her Ph.D. in cell biology at the University of California, San Francisco, and then completed her training as a Whitehead Fellow at the Whitehead Institute, Cambridge, Mass.

Nicholas Thomson is a project manager in the Sanger Institute Pathogen Sequencing Unit. His particular interest relates to gene acquisition and loss in the genomes of enteric bacteria.

Guy Tran Van Nhieu earned his Ph.D. and is an INSERM fellow at the Institut Pasteur. He joined the laboratory of Philippe J. Sansonetti in 1994, where he has been focusing on molecular and cellular aspects of *Shigella* host-cell invasion.

Emil R. Unanue is the Head of the Department of Pathology, Washington University School of Medicine, St. Louis, Mo. His work has centered on the analysis of antigen-presenting cells, particularly macrophages, in the immune response. His laboratory has shown the requirement for antigen processing for T-cell recognition. Their analysis of processing led to the findings that class II major histocompatibility complex molecules are peptide-binding molecules. His laboratory has also defined the role of the innate system in resistance to intracellular pathogens.

Raphael H. Valdivia is an Assistant Professor in the Department of Molecular Genetics and Microbiology at Duke University Medical Center, Durham, North Carolina. His laboratory studies the molecular basis of intracellular pathogenesis and the function of bacterial toxins by using model eukaryotic systems. Previously, he studied endosome dynamics in the yeast *Saccharomyces cerevisiae* in the laboratory of Randy Schekman at the University of California, Berkeley. He received his Ph.D. from Stanford University, where he studied *Salmonella* pathogenesis in the laboratory of Stanley Falkow.

Hans Wolf-Watz is professor at the Department of Molecular Biology, Umeå University, Umeå, Sweden. His work during the past 20 years has focused on the molecular mechanisms of bacterial virulence with an emphasis on the pathogenic *Yersinia* species. The major contributions to this field by his research team are the finding that *Yersinia* is an extracellular pathogen that blocks its own uptake into eukaryotic cells and the discovery of the type III-mediated translocation of virulence effector proteins into infected host cells.

Eli Zamir works in the Department of Molecular Cell Biology, Weizmann Institute of Science, Rehovot, Israel, studying the structure and dynamics of focal adhesions. He is a Ph.D. student in the laboratory of Benjamin Geiger.

Arturo Zychlinsky received his undergraduate education at the Escuela Nacional de Ciencias Biologicas, Mexico City, Mexico, and received his Ph.D. in immunology from the Rockefeller University, New York, N.Y. After a postdoctoral fellowship in the laboratory of Philippe Sansonetti at the Institut Pasteur, Paris, France, he joined the Skirball Institute and the Department of Microbiology at the New York University Medical Center, New York, N.Y. He currently is director at the Max Planck Institute for Infection Biology in Berlin, Germany.

Index